PRINCIPLES AND APPLICATIONS OF ELECTRICAL ENGINEERING

PRINCIPLES AND APPLICATIONS OF ELECTRICAL ENGINEERING

Giorgio Rizzoni
The Ohio State University

IRWIN

Homewood, IL 60430
Boston, MA 02116

Cover image: Frank Wing / The Image Bank

© RICHARD D. IRWIN, INC., 1993

Senior sponsoring editor: Tom Casson
Developmental editor: Kelley Butcher
Marketing manager: Robb Linsky
Project editor: Rita McMullen
Production manager: Ann Cassady
Designer: Heidi J. Baughman
Art manager: Kim Meriwether
Compositor: Technique Typesetting
Typeface: 10/12 Times Roman
Printer: R. R. Donnelley & Sons Company

Library of Congress Cataloging-in-Publication Data

Rizzoni, Giorgio.
   Principles and applications of electrical engineering / Giorgio
Rizzoni.
      p.   cm.
   Includes index.
   ISBN 0-256-07827-0 ISBN 0-256-12969-X (Int'l ed.)
   1. Electrical engineering.   I. Title.
TK146.R473   1993
621.3—dc20   92–13757

Printed in the United States of America
1 2 3 4 5 6 7 8 9 0 DOC 9 8 7 6 5 4 3 2

Ai miei genitori

About the Author

Giorgio Rizzoni is an assistant professor in the Department of Mechanical Engineering at The Ohio State University. He received his Ph.D. from the University of Michigan and then became a lecturer and research scientist in University of Michigan's EECS Department and assistant director of the Vehicular Electronics Laboratory. As a lecturer there, he taught the nonmajors course sequence and designed the related lab, for which he received Tau Beta Pi's College of Engineering Outstanding Teaching Award (1990).

In 1991, The National Science Foundation awarded Professor Rizzoni the Presidential Young Investigator Award, a research-based five-year grant. In 1992, Rizzoni received the Ralph R. Teetor Educational Award from the Society of Automotive Engineers (SAE) for his contribution to teaching, research, and student development. In addition to his teaching, Professor Rizzoni is active in ASME, IEEE, and SAE and has published a number of articles and papers.

Preface

The pervasive presence of electronic devices and instrumentation in all aspects of engineering design and analysis is one of the manifestations of the electronic revolution that has characterized the second half of the 20th century. Every aspect of engineering practice, and even of everyday life, has been affected in some way or another by electrical and electronic devices and instruments. Computers are perhaps the most obvious manifestations of this presence. However, many other areas of electrical engineering are also important to the practicing professional in a range of fields, from mechanical and industrial engineering to chemical, nuclear, and materials engineering, the aerospace and astronautical disciplines, and civil engineering. Engineers today must be able to communicate effectively within the interdisciplinary teams in which they work.

OBJECTIVES

The aim of *Principles and Applications of Electrical Engineering* is to provide the nonelectrical engineering student—typically in the third year of a four-year curriculum—with a foundation for understanding the basic principles of electrical and electronic systems. This book should help build a solid foundation for further study, either in the classroom or on the job.

It would be overly ambitious to attempt to summarize all important aspects of electrical engineering in a single book, without its taking on an undesired encyclopedic nature. Thus, this book differs in two respects in content and style of presentation from many of the leading textbooks that are currently available in the market. First, no attempt is made to provide an exhaustive treatment of electrical and electronic circuits, and of electrical machinery. Rather, the intent is to help the student in building a fundamental understanding of the basic principles, without clouding the main concepts with excessive detail. The reader interested in a more in-depth treatment of a particular subject will be able to refer to more advanced books, while the instructor who wishes to cover some particular area in greater depth will find that each chapter includes a number of homework problems designed for more advanced study. Second, a number of topics that are commonly found in some of the popular books that take a more encyclopedic approach to presenting the subject will not be found here. For example, topics such as control or communication systems are not included, although some discussion of relevant ideas is given when appropriate. The author believes that such subjects are best dealt with in a specially designed course (for example, many undergraduate engineering degree programs require a senior level course in control systems). Further, in attempting to include too many subjects, many authors have been forced to reduce the coverage of those essential topics that form the basis of electrical engineering. An honest effort has been made to avoid this shortcoming in the present book; a multitude of examples and exercises are provided in the early circuit analysis chapters to ensure a solid and complete understanding of the material that forms the basis of the entire discipline. This approach is based on the author's belief that learning the foundations of circuits and electronics and gaining a mastery of the fundamentals of AC and DC circuit analysis greatly simplifies the task of understanding the more advanced material.

The emphasis in *Principles and Applications of Electrical Engineering* is on teaching relevant electrical engineering concepts to engineers who will be users, not designers, of electrical, electromechanical, and electronic systems. Thus, the book provides a user's perspective of electrical engineering, focusing on those principles and applications that are likely to be of use to engineers of tomorrow. Electrical and electronic subsystems are an integral part of any complex plant or engineering system—from the self-contained power systems aboard a ship or an aircraft, to the control of industrial processes and power plants or the management of propulsion systems for automobiles and spacecrafts. In all of these applications, the common threads are *instrumentation,* which is required to perform measurement for process and plant performance evaluation and control, and the *devices,* which are used to implement the controls and which make up the measuring instruments.

Thus, this book will focus on the application of devices to specific measurements and control tasks and on the basic components of instrumentation and measurement systems that are likely to be encountered in practice.

ORGANIZATION AND CONTENT

The wide range of requirements that characterize four-year engineering programs in the United States presents a challenge to writing a useful and versatile book for the nonmajor audience. The following paragraphs provide a rationale for the organization and content of this book and suggest various sequences that may be adopted to match the requirements of varied curricula.

The book is structured in three parts: circuit analysis, electronics and computers, and electromechanics. It is difficult to decouple completely one subject from the others. Aside from the fact that circuit analysis forms the foundation of the entire book, the operation of modern electric machines cannot be separated from the solid-state electronic control circuits that drive them. Further, where possible, examples are given of the application of a specific circuit or device to measurements and instrumentation. Chapters 1 through 6 (and part of 10) cover the fundamentals; Chapters 7 through 12 cover electronics and computers; and Chapters 5 and 14 through 16 cover power systems and electric machines. The material in the first six chapters provides a basic understanding of DC and AC circuit analysis and of transient analysis and frequency response. The instructor who wishes to emphasize the role of operational amplifiers early will find that the first two sections of Chapter 10 can be covered early in conjunction with Chapter 3. Another alternative (employed by the author for several semesters) introduces the op-amp immediately after Chapter 4, to use it as a vehicle for presenting the concept of frequency response early on.

The electronics material in Part II can be covered in a variety of ways, to suit the requirements of different audiences. For example, a very light treatment in a one-quarter course might involve all or only parts of Chapters 8 and 9, while an entire course in electronics might include Chapters 7 through 13. A number of options are available to the instructor between these extremes. If time permits treatment of discrete electronic devices, Chapters 7, 8, and 9 provide substantial amounts of material; if the preferred emphasis is on IC electronics, Chapters 10 and 11 are self-contained and do not require a background in discrete electronics. If the instructor prefers to devote time to digital logic circuits and systems, parts of Chapter 11, and Chapters 12 and 13 should suit his or her needs. Instructors who teach a survey-oriented course will find that the chapters go into progressively more depth in the later sections, so that it is possible to cover only part of each chapter and achieve substantial coverage of different topics. It is conceivable, for example, to touch on part or all of the first 12 chapters in a one-semester course if desired.

Part III, dealing with electric machines, is geared toward providing the student with an understanding of the fundamental principles of operation of the principal electric machines, with some focus on the performance and selection criteria for different applications. In addition to an introduction to the basic properties of DC, synchronous, and induction motors, emphasis is also given to special-purpose machines, such as servos, stepper motors, brushless DC motors, and other devices commonly used in motion control systems, in the hope that the interested student may use this material as a stepping stone to more advanced courses.

The following chart illustrates possible coverage options. Each instructor will be able to tailor the book to the specific needs of his or her course.

	Suggested sequences
1-quarter course	Part I, Circuits, plus one chapter on electronics (Chapter 7 on Semiconductors and Diodes or Chapter 10 on Op-Amps, or Chapter 12 on Digital Logic Circuits).
1-semester course	Part I, Circuits, plus two introductory chapters on Electronics (Chapters 7 and 8, or Chapters 10 and 11, or Chapters 10 and 12).
2-quarter sequence	Part I, Circuits, plus chapters on energy systems (Chapters 5, 14, 15, and 16).
	or
	Part I, Circuits, plus Electronics, selected parts of Chapters 7 through 13.
2-semester sequence	Part I, Circuits, plus complete coverage of chapters on Energy Systems (Chapters 5, 14, 15, and 16).
	or
	Part I, Circuits, plus complete coverage of Electronics, Chapters 7 through 13.

PEDAGOGY

Each chapter has an introduction with chapter objectives, an abundance of illustrations (over 1,500 total), many completely worked examples (over 260 overall), drill exercises to reinforce concepts just learned, a chapter summary, and key terms lists. Also, there is a variety of homework problems at the end of each chapter. The problems are intended to assist the student in understanding and applying the basic concepts presented. More challenging problems are indicated with an asterisk (*). There are approximately 750 homework problems—an average of 50 per chapter. Answers to selected problems can be found at the back of the book.

In addition to an appendix on linear algebra, complex numbers, and the Laplace transform, there is an Engineer-in-Training (EIT) review appendix, designed to help students prepare for the Fundamentals of Engineering (FE) examination. Key topics found on the exam are identified and keyed to sections in the text that discuss these concepts. For each of the exam's nine subsections related to electrical engineering, there are two or more sample problems, representative of the type of problem the student will encounter on the FE exam. It is hoped that the nonmajor will be well prepared to succeed on the exam after studying from this book, with the appendix as a guide.

ACCURACY

Every example, drill exercise, and homework problem was double-checked for accuracy by two independent instructors (University of Arkansas and University of Michigan—Dearborn). In addition to being checked for errors, the manuscript was class tested at Ohio State University and University of Michigan—Ann Arbor. These schools used the manuscript in the classroom to test the readability, approach, and content; over 1,000 students used all or part of this book prior to publication. This rigorous process of checking the accuracy and approach of the book has greatly improved it.

SUPPLEMENTS

The Instructor's Manual, available to adopters from the publisher, contains complete solutions to the homework problems. Manufacturers' data sheets for important devices are included at the back of the Instructor's Manual; the instructor is permitted to reproduce these data sheets for classroom use. Transparency masters of important figures and example problems are also available.

ACKNOWLEDGMENTS

This book has been read critically by over 30 reviewers. Richard D. Irwin, Inc. and the author would like to thank the following reviewers for their contribution to *Principles and Applications of Electrical Engineering:*

    Joe Baker, University of Southern California, Los Angeles
    Kenneth Breeding, The Ohio State University
    Fred Brockhurst, Rose Hulman Institute of Technology
    Grigore Burdea, Rutgers University
    Don Calahan, University of Michigan
    Richard Cockrum, California State Polytechnic University
    Norman Cox, University of Missouri-Rolla
    Timothy Darling, University of California-Santa Barbara
    John Dollar, Kansas State University
    John Flemming, Texas A&M University

Edward Gardner, University of Colorado

Victor Gerez, Montana State University

James Holte, University of Minnesota

Mansell H. Hopkins, Virginia Polytechnic Institute and State University

John Hug, University of Arkansas

Charles Ih, University of Delaware

Sid James, Auburn University

Edward MacDonald, Drexel University

Dennis Malone, State University of New York-Buffalo

Dan McAuliff, Clemson University

Richard Miller, University of Central Florida

M. Paul Murray, Mississippi State University

Mesut Muslu, University of Wisconsin-Platteville

Robert Nethken, Louisiana State University

John O'Malley, University of Florida

Gerald Park, Michigan State University

William Potter, University of Washington

Ronald Scott, Northeastern University

W. David Stratton, Dartmouth College

Tony Valois, University of Minnesota

John Wait, University of Arizona

The author is indebted to these reviewers for their valuable insights and for the patient reviews that have greatly improved the accuracy and readability of this book. Special thanks also go to John Hug and Jim Daws for their hard work at uncovering errors and to Valdis Liepa and James G. Gottling for agreeing to use the manuscript in their classes.

The careful assistance provided by graduate students James C. Mejeur (University of Michigan) and Bahman Samimy and Jiande Wu (Ohio State University) in producing the solutions manual and by Ms Izena Goulding in typing the manuscript is gratefully acknowledged. The author would also like to recognize the continuing support of the project from three Department Chairs: Thomas B.A. Senior and Edward J. Davidson at the University of Michigan and Lawrence A. Kennedy at the Ohio State University.

Finally, the love, patience, and support provided by my wife, Kathryn, have made it possible to endure the long hours of work inevitably associated with such a challenging project.

Giorgio Rizzoni
The Ohio State University

Contents in Brief

Contents

PRINCIPLES AND APPLICATIONS
OF ELECTRICAL ENGINEERING

CHAPTER 1

Introduction to Electrical Engineering

The aim of this chapter is to introduce electrical engineering. The chapter is organized to provide the newcomer with a view of the different specialties making up electrical engineering and to place the intent and organization of the book into perspective. Perhaps the first question that surfaces in the mind of the student approaching the subject is, Why electrical engineering? Since this book is directed at a readership having a mix of engineering backgrounds other than electrical engineering, the question is well justified and deserves some discussion. The chapter defines the various branches of electrical engineering, showing some of the interactions among them, and illustrating by means of a simple example how electrical engineering is intimately connected to many other engineering disciplines. A brief historical perspective is also provided, to outline the growth and development of this relatively young engineering specialty. Next, the fundamental physical quantities and the system of units are defined, to set the stage for the chapters that follow. Finally, the organization of the book is discussed, to give the student, as well as the teacher, a sense of continuity in the development of the different subjects covered in Chapters 2 through 16.

1.1   ELECTRICAL ENGINEERING

The typical curriculum of an undergraduate electrical engineering student includes the subjects listed in Table 1.1. Although the distinction between some of these subjects is not always clear-cut, the table is sufficiently representative to serve our purposes. Figure 1.1 illustrates a possible interconnection between the disciplines of Table 1.1. The aim of this book is to introduce the non–electrical engineering student to those aspects of electrical engineering that are likely to be most relevant to his or her professional career. Virtually all of the topics of Table 1.1 will be touched on in the book,

Table 1.1   Electrical engineering disciplines

Circuit analysis
Electromagnetics
Solid-state electronics
Electric machines
Electric power systems
Digital logic circuits
Computer systems
Communication systems
Electro-optics
Instrumentation
Control systems

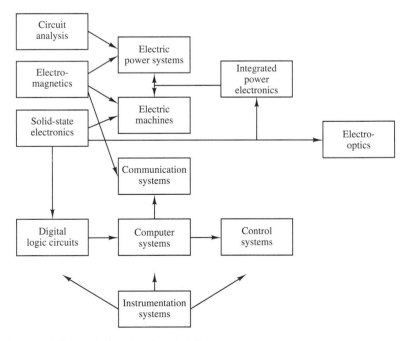

Figure 1.1 Electrical engineering disciplines

with varying degrees of emphasis. The following example illustrates the pervasive presence of electrical, electronic, and electromechanical devices and systems in a very common application: the automobile. As you read through the example, it will be instructive to refer to Figure 1.1 and Table 1.1.

Example   1.1   Electrical Systems in a Passenger Automobile

It may be enlightening to consider a familiar example to illustrate how the seemingly disparate specialties of electrical engineering actually interact to permit the operation of a very familiar engineering system: the automobile. Figure 1.2 presents a view of electrical engineering systems in a modern automobile. Even in older vehicles, the electrical system—in effect, an *electric circuit*—plays a very important part in the overall operation. An inductor coil generates a sufficiently high voltage to allow a spark to form across the spark plug gap, and to ignite the air and fuel mixture; the coil is supplied by a DC voltage provided by a lead-acid battery. In addition to providing the energy for the ignition circuits, the battery also supplies power to

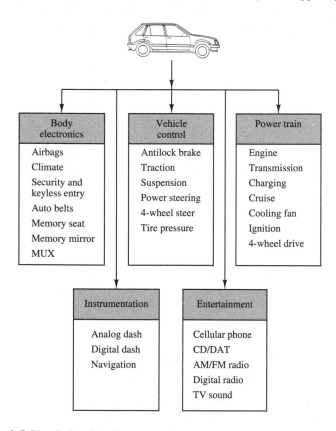

Figure 1.2 Electrical engineering systems in the automobile

many other electrical components, the most obvious of which are the lights, the windshield wipers, and the radio. Electric power is carried from the battery to all of these components by means of a wire harness, which constitutes a rather elaborate electrical circuit. In recent years, the conventional electrical ignition system has been supplanted by *electronic* ignition; that is, solid-state electronic devices called *transistors* have replaced the traditional breaker points. The advantage of transistorized ignition systems over the conventional mechanical ones is their greater reliability, ease of control, and life span (mechanical breaker points are subject to wear).

Other electrical engineering disciplines are fairly obvious in the automobile. The on-board radio receives electromagnetic waves by means of the antenna, and decodes the communication signals to reproduce sounds and speech of remote origin; other common *communication systems* that exploit *electromagnetics* are CB radios and the ever more common cellular phones. But this is not all! The battery is, in effect, a self-contained 12-VDC *electric power system,* providing the energy for all of the aforementioned functions. In order for the battery to have a useful lifetime, a charging system, composed of an alternator and of power electronic devices, is present in every automobile. The alternator is an *electric machine,* as are the motors that drive the power mirrors, power windows, power seats, and other convenience features found in luxury cars. Incidentally, the loudspeakers are also electric machines!

The list does not end here, though. In fact, some of the more interesting applications of electrical engineering to the automobile have not been discussed yet. Consider *computer systems.* You are certainly aware that in the last two decades, environmental concerns related to exhaust emissions from automobiles have led to the introduction of sophisticated engine emission *control systems.* The heart of such control systems is a type of computer called a *microprocessor.* The microprocessor receives signals from devices (called *sensors*) that measure relevant variables—such as the engine speed, the concentration of oxygen in the exhaust gases, the position of the throttle valve (i.e., the driver's demand for engine power), and the amount of air aspirated by the engine—and subsequently computes the optimal amount of fuel and the correct timing of the spark to result in the cleanest combustion possible under the circumstances. The measurement of the aforementioned variables falls under the heading of *instrumentation,* and the interconnection between the sensors and the microprocessor is usually made up of *digital circuits.* Finally, as the presence of computers on board becomes more pervasive—in areas such as anti-lock braking, electronically controlled suspensions, four-wheel steering systems, and electronic cruise control—communications among the various on-board computers will have to occur at faster and faster rates. Some day in the not-so-distant future, these communications may occur over a fiber optic network, and *electro-optics* will replace the conventional wire harness. It should be noted that electro-optics is already present in some of the more advanced displays that are part of an automotive instrumentation system.

1.2   BRIEF HISTORY OF ELECTRICAL ENGINEERING

The historical evolution of electrical engineering can be attributed, in part, to the work and discoveries of the people in the following list. You will find these scientists, mathematicians, and physicists referenced throughout the text.

William Gilbert (1540–1603), English physician, founder of magnetic science, published *De Magnete,* a treatise on magnetism, in 1600.

Charles A. Coulomb (1726–1806), French engineer and physicist, published the laws of electrostatics in seven memoirs to the French Academy of Science between 1785 and 1791. His name is associated with the units of charge.

James Watt (1736–1819), English inventor, developed the steam engine. His name is used to represent the units of power.

Alessandro Volta (17xx–18xx), Italian physicist, discovered the electric pile. The units of electric potential and the alternate name of this quantity (voltage) are named after him.

Hans Christian Oersted (1777–1851), Danish physicist, discovered the connection between electricity and magnetism in 1820. The units of magnetic field strength are after his name.

André Marie Ampère (1775–1836), French mathematician, chemist, and physicist, experimentally quantified the relationship between electric current and the magnetic field. His works were summarized in a treatise published in 1927. The units of electric current are named after him.

Georg Simon Ohm (1789–1854), German mathematician, investigated the relationship between voltage and current and quantified the phenomenon of resistance. His first results were published in 1827. His name is used to represent the units of resistance.

Michael Faraday (1791–1867), English experimenter, demonstrated electromagnetic induction in 1831. His electrical transformer and electromagnetic generator marked the beginning of the age of electric power. His name is associated with the units of capacitance.

Joseph Henry (1797–1878), American physicist, discovered self-induction around 1831, and his name has been designated to represent the units of inductance. He had also recognized the essential structure of the telegraph, which was later perfected by Samuel F. B. Morse.

Carl Friedrich Gauss (1777–1855), German mathematician, and **Wilhelm Eduard Weber** (1804–1891), German physicist, published a treatise in 1833 describing the measurement of the earth's magnetic field. The Gauss is a unit of magnetic field strength, while the Weber is a unit of magnetic flux.

James Clerk Maxwell (1831–1879), Scottish physicist, discovered the electromagnetic theory of light and the laws of electrodynamics. The modern theory of electromagnetics is entirely founded upon Maxwell's equations.

Ernst Werner Siemens (1816–1892) and **Wilhelm Siemens** (1821–1833), German inventors and engineers, contributed to the invention and development of electric machines, as well as to perfecting electrical science. The modern unit of conductance is named after them.

Heinrich Rudolph Hertz (1857–1894), German scientist and experimenter, discovered the nature of electromagnetic waves and published his findings in 1888. His name is associated with the units of frequency.

Nikola Tesla (1856–1943), Croatian inventor, emigrated to the United States in 1884, invented poliphase electric power systems and the induction motor. He pioneered modern AC electric power systems. His name is used to represent the units of magnetic flux density.

1.3   SYSTEM OF UNITS

This book employs the International System of Units (also called SI, from the French *Système International des Unités*). SI units are commonly adhered to by virtually all engineering professional societies. This section summarizes SI units and will serve as a useful reference in reading the book.

SI units are based on six fundamental quantities, listed in Table 1.2. All other units may be derived in terms of the fundamental units of Table 1.2. Since, in practice, one often needs to describe quantities that occur in large multiples or small fractions of a unit, standard prefixes are used to denote powers of 10 of SI (and derived) units. These prefixes are listed in Table 1.3. Note that, in general, engineering units are expressed in powers of 10 that are multiples of 3. For example, 10^{-4} s would be referred to as 100×10^{-6} s, or 100μs (or, less frequently, 0.1 ms).

Table 1.2   SI units

Quantity	Unit	Symbol
Length	Meter	m
Mass	Kilogram	kg
Time	Second	s
Electric current	Ampere	A
Temperature	Kelvin	K
Luminous intensity	Candela	cd

Table 1.3   Standard prefixes

Prefix	Symbol	Power
atto	a	10^{-18}
femto	f	10^{-15}
pico	p	10^{-12}
nano	n	10^{-9}
micro	μ	10^{-6}
milli	m	10^{-3}
centi	c	10^{-2}
deci	d	10^{-1}
deka	da	10
kilo	k	10^{3}
mega	M	10^{6}
giga	G	10^{9}
tera	T	10^{12}

1.4   ORGANIZATION OF THIS BOOK

This book is divided into three major sections: Circuits, Electronics and Computers, and Electromechanics. Circuit analysis forms the basis of the entire discipline of electrical engineering; thus, a substantial portion of the book is devoted to this subject. Since circuit analysis forms the core of the first course in electricity for non–electrical engineering majors, you will probably spend a significant amount of time working with Chapters 2 through 6. After completing these chapters, you will be able to analyze simple electric circuits, and will begin to appreciate the inner workings of many devices and systems you may have already encountered in other engineering classes and laboratories. However, circuit analysis does not end with Chapter 6! The material in the introductory chapters is absolutely necessary for understanding the fundamentals of electronic and electromechanical systems. This is true because simple electric circuits provide us with good models (i.e., approximations) of the behavior of more complex electronic and electromechanical devices. Thus, mastering the introductory materials will ensure that the more advanced topics of Chapters 7 through 16 will be more easily understood.

The second and third parts of the book are somewhat different in character from the circuit analysis chapters, in that it is not necessary to follow a prescribed order in covering various topics in electronics and electromechanics. Thus, some instructors will emphasize digital electronics and computer systems over analog electronic amplifiers, while some others may prefer to discuss electric motors and electric power systems. Since the interdependence of topics is relatively small past Chapter 6, you will find that this book will serve you as a useful reference for later classes as well. For example, you may be able to find valuable information to assist you in your senior design project.

To provide as much assistance as possible, this book contains a large number of examples and drill exercises; each important concept is reinforced by one or more examples, and a few drill exercises. Since the answers to many drill exercises are provided in the book, you should use these exercises as a check to verify complete understanding of the material. Using the drill exercises as an aid in reading and understanding the text will also help you in solving the homework problems. Many homework problems are simply extensions of the exercises and examples. Some are somewhat more advanced. As in many other subjects, repetition is essential to learning; this statement is particularly true in circuit analysis, where you need to quickly develop familiarity with many new ideas and build confidence in problem solving. In preparing this book, great effort has been spent in making this task as painless as possible.

HOMEWORK PROBLEMS

1.1 List five applications of electric motors in the common household.

1.2 By analogy with the discussion of electrical systems in the automobile, list examples of applications of the electrical engineering disciplines of Table 1.1 for each of the following engineering systems:

a. A ship.

b.   A commercial passenger aircraft.

c.   Your household.

1.3   Electric power systems provide energy in a variety of commercial and industrial settings. Make a list of systems and devices that receive electric power in:

a.   A large office building.

b.   A factory floor.

c.   A construction site.

PART

I

CIRCUITS

Fundamentals of Electric Circuits

T his chapter presents the fundamental laws of circuit analysis and serves as the foundation for the study of electrical circuits. The fundamental concepts developed in these first pages will be called upon throughout the book.

The chapter starts with definitions of charge, current, voltage, and power, and with the introduction of the basic laws of electrical circuit analysis: Kirchhoff's laws. Next, the basic circuit elements are introduced, first in their ideal form, then including the most important physical limitations. The elements discussed in the chapter include voltage and current sources, measuring instruments, and the ideal resistor. Once the basic circuit elements have been presented, the concept of an electrical circuit is introduced, and some simple circuits are analyzed using Kirchhoff's and Ohm's laws. The student should appreciate the fact that, although the material presented at this early stage is strictly introductory, it is already possible to discuss some useful applications of electric circuits to practical engineering problems. To this end, two examples are introduced which discuss simple resistive devices that can measure displacements and forces. The topics introduced in Chapter 2 form the foundations for the remainder of this book and should be mastered thoroughly. By the end of the chapter, you should have accomplished the following learning objectives:

- Application of Kirchhoff's and Ohm's laws to elementary resistive circuits.
- Power computation for a circuit element.
- Use of the passive sign convention in determining voltage and current directions.
- Solution of simple voltage and current divider circuits.
- Assigning node voltages and mesh currents in an electrical circuit.
- Writing the circuit equations for a linear resistive circuit by applying Kirchhoff's voltage law and Kirchhoff's current law.

2.1   CHARGE, CURRENT, AND KIRCHHOFF'S CURRENT LAW

Charles Coulomb (1736–1806).
Photo courtesy of French Embassy, Washington, D.C.

The earliest accounts of electricity date from about 2,500 years ago, when it was discovered that static charge on a piece of amber was capable of attracting very light objects, such as feathers. The word itself—*electricity*—originated about 600 B.C.; it comes from *elektron,* which was the ancient Greek word for amber. The true nature of electricity was not understood until much later, however. Following the work of Alessandro Volta[1] and his invention of the copper-zinc battery, it was determined that static electricity and the current that flows in metal wires connected to a battery are due to the same fundamental mechanism: the atomic structure of matter, consisting of a nucleus—neutrons and protons—surrounded by electrons. The fundamental electric quantity is **charge,** and the smallest amount of charge that exists is the charge carried by an electron, equal to

$$q_e = -1.602 \times 10^{-19} \text{ coulomb} \tag{2.1}$$

As you can see, the amount of charge associated with an electron is rather small. This, of course, has to do with the size of the unit we use to measure charge, the **coulomb (C),** named after Charles Coulomb.[2] However, the definition of the coulomb leads to an appropriate unit when we define electric current, since current consists of the flow of very large numbers of charge particles. The other charge-carrying particle in an atom, the proton, is assigned a positive sign, and the same magnitude. The charge of a proton is

$$q_p = +1.602 \times 10^{-19} \text{ coulomb} \tag{2.2}$$

Electrons and protons are often referred to as **elementary charges.**

[1] See brief biography on page 5.
[2] See brief biography on page 5.

Electric current is defined as the time rate of change of charge passing through a predetermined area. Typically, this area is the cross-sectional area of a metal wire; however, there are a number of cases we shall explore later in this book where the current-carrying material is not a conducting wire. Figure 2.1 depicts a macroscopic view of the flow of charge in a wire, where we imagine Δq units of charge flowing through the cross-sectional area A in Δt units of time. The resulting current, Δi, is then given by

Current $i = dq/dt$ is generated by the flow of charge through the cross-sectional area A in a conductor.

$$\Delta i = \frac{\Delta q}{\Delta t} \tag{2.3}$$

If we consider the effect of the enormous number of elementary charges actually flowing, we can write this relationship in differential form:

$$i = \frac{dq}{dt} \quad \frac{\text{C}}{\text{s}} \tag{2.4}$$

Figure 2.1 Current flow in an electric conductor

The units of current are called **amperes** (A), where 1 ampere = 1 coulomb/second. The name of the unit is a tribute to the French scientist André Marie Ampère.[3] The electric engineering convention states that the positive direction of current flow is that of positive charges. In metallic conductors, however, current is carried by negative charges; these charges are the free electrons in the conduction band, which are only weakly attracted to the atomic structure in metallic elements and are therefore easily displaced in the presence of electric fields.

Example   2.1

A cylindrical conductor is 1 m long and 2 mm in diameter and contains 10^{29} free carriers per cubic meter.

1. Find the total charge of the carriers in this wire.
2. If the wire is used in a circuit, find the current flowing in the wire if the average velocity of the carriers is 19.9×10^{-6} m/s.

Solution:

1. In order to compute the total charge contributed by the electrons, we first need to compute the volume of the conductor.

$$\text{Volume} = \text{Length} \times \text{Cross-sectional area}$$

$$= \pi r^2 L = \pi \left(\frac{2 \times 10^{-3}}{2} \right)^2 \tag{1}$$

Next we compute the charge by determining the total number of charge carriers

---

[3] See brief biography on page 5.

in the conductor as follows:

$$\text{Charge} = \text{Volume} \times \frac{\text{Charge}}{\text{Unit volume}}$$

$$Q = \pi \left(\frac{2 \times 10^{-3}}{2} \right)^2 (1)(-1.602 \times 10^{-19} \text{ C}) \left(10^{29} \frac{\text{carriers}}{\text{m}^3} \right)$$

$$= -50.33 \times 10^3 \text{ C}$$

2.   If the carriers move with an average velocity of 19.9×10^{-6} m/s, the magnitude of the total current flow in the wire can be computed by considering that current is the flow of charge per unit time:

$$\text{Current} = \text{Charge density per unit length (C/m)} \times \text{Carrier velocity (m/s)}$$

$$= \frac{50.33 \times 10^3}{1} \times 19.9 \times 10^{-6}$$

$$= 1 \text{ A}$$

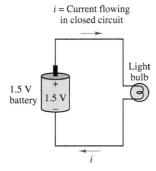

i = Current flowing in closed circuit

1.5 V battery

1.5 V

Light bulb

i

Figure 2.2 A simple electrical circuit

In order for current to flow there must exist a closed circuit. Figure 2.2 depicts a simple circuit, composed of a battery (e.g., a dry-cell or alkaline 1.5-V battery) and a light bulb.

Note that in the circuit of Figure 2.2, the current, i, flowing from the battery to the resistor is equal to the current flowing from the light bulb to the battery. In other words, no current (and therefore no charge) is "lost" around the closed circuit. This principle was observed by the German scientist G. R. Kirchhoff[4] and is now known as **Kirchhoff's current law (KCL)**. Kirchhoff's current law states that because charge cannot be created but must be conserved, *the sum of the currents at a node must equal zero* (in an electrical circuit, a **node** is the junction of two or more conductors). Formally:

$$\sum_{n=1}^{N} i_n = 0 \qquad \text{Kirchhoff's current law} \qquad (2.5)$$

The significance of Kirchhoff's current law is illustrated in Figure 2.3, where the simple circuit of Figure 2.2 has been augmented by the addition of two light bulbs (note how the two nodes that exist in this circuit have been emphasized by the shaded areas). In applying KCL, one usually defines currents entering the node as being negative and currents exiting a node as being positive. Thus, the resulting expression for the circuit of Figure 2.3 is:

$$-i + i_1 + i_2 + i_3 = 0$$

Kirchhoff's current law is one of the fundamental laws of circuit analysis, making it possible to express currents in a circuit in terms of each

---

[4] Gustav Robert Kirchhoff (1824–1887), a German scientist, who published the first systematic description of the laws of circuit analysis. His contribution—though not original in terms of its scientific content—forms the basis of all circuit analysis.

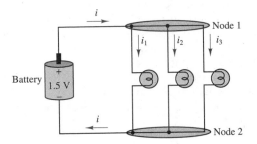

Illustration of KCL at node 1: $-i + i_1 + i_2 + i_3 = 0$

Figure 2.3 Illustration of Kirchhoff's current law

other; for example, one can express the current leaving a node in terms of all the other currents at the node. The ability to write such equations is a great aid in the systematic solution of large electric circuits. Much of the material presented in Chapter 3 will be an extension of this concept.

Example   2.2

An automotive battery provides an excellent intuitive illustration of Kirchhoff's current law. Assume for a moment that the only electrical circuits connected to the car battery are the headlights. How does the current provided by the battery divide between headlights?

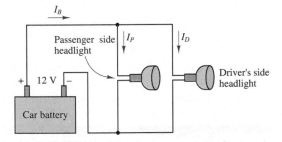

Figure 2.4 KCL in a practical setting

Solution:

The battery supplies power to the headlights by generating a current that flows through the filaments of the bulbs. Since the headlights are usually of the same type and both have the same power rating, it is reasonable to assume that the currents I_P and I_D must be the same in both magnitude and direction of current flow (incidentally, the power rating of a headlight is usually specified in watts). The battery, then, must supply the current to the headlights, and since the battery is connected so that 12 V is across both of the headlights (this is called a parallel connection), we reason that the total current out of the battery, I_B, is the sum of the two currents I_P and I_D, or:

$$I_B = I_P + I_D$$

If we wanted to consider a more realistic automotive circuit, we would have to account for the tail- and brake-light circuits, the turn signal lights, and a number of other electrical loads all connected in parallel across the battery. The electrical system in an automobile is an excellent example of how circuit analysis applies to real-life problems: the designer of the electrical system of an automobile would have to compute the individual current required by each load, evaluate the total capacity of the battery needed to furnish the current required by the loads under worst-case conditions, and select the wiring scheme and the wire gauges most appropriate for each load.

2.2 VOLTAGE AND KIRCHHOFF'S VOLTAGE LAW

Charge moving in an electric circuit gives rise to a current, as stated in the preceding section. Naturally, it must take some work, or energy, for the charge to move between two points in a circuit, say, from point a to point b. The total *work per unit charge* associated with the motion of charge between two points is called **voltage.** Thus, the units of voltage are those of energy per unit charge; they have been called **volts** in honor of Alessandro Volta:

$$1 \text{ volt} = \frac{1 \text{ joule}}{\text{coulomb}}$$

(2.6)

The voltage or **potential difference** between two points in a circuit indicates the energy required to move charge from one point to the other. As will be presently shown, the direction, or polarity, of the voltage is closely tied to whether energy is being dissipated or generated in the process. The seemingly abstract concept of work being done in moving charges can be directly applied to the analysis of electrical circuits; consider again the simple circuit consisting of a battery and a light bulb. The circuit is drawn again for convenience in Figure 2.5, and nodes are defined by the letters a and b. A series of carefully conducted experimental observations regarding the nature of voltages in an electric circuit led Kirchhoff to the formulation of the second of his laws, **Kirchhoff's voltage law, or KVL.** The principle underlying KVL is that no energy is lost or created in an electric circuit; in circuit terms, the sum of all voltages associated with sources must equal the sum of the load voltages, so that *the net voltage around a closed circuit is zero.* If this were not the case, we would need to find a physical explanation for the excess (or missing) energy not accounted for in the voltages around a circuit. Kirchhoff's voltage law may be stated in a form similar to that used for KCL:

$$\sum_{n=1}^{N} v_n = 0 \qquad \text{Kirchhoff's voltage law}$$

(2.7)

where the v_n are the individual voltages around the closed circuit. Making reference to Figure 2.5, we see that it must follow from KVL that the work generated by the battery is equal to the energy dissipated in the light bulb to

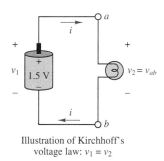

Illustration of Kirchhoff's voltage law: $v_1 = v_2$

Figure 2.5 Voltages around a circuit

sustain the current flow and to convert the electric energy to heat and light:

$$v_{ab} = v_{ba}$$

or

$$v_1 = v_2$$

One may think of the work done in moving charge from point a to point b and the work done moving it back from b to a as corresponding directly to the *voltages across individual circuit elements*. Let Q be the total charge that moves around the circuit per unit time, giving rise to the current i. Then the work done in moving Q from b to a (i.e., across the battery) is

$$W_{ba} = Q \times 1.5 \text{ V} \tag{2.8}$$

Similarly, work is done in moving Q from a to b, that is, across the light bulb. Note that the word *potential* is quite appropriate as a synonym of voltage, in that voltage represents the potential energy between two points in a circuit: if we remove the light bulb from its connections to the battery, there still exists a voltage across the (now disconnected) terminals b and a. This is illustrated in Figure 2.6.

Gustav Robert Kirchhoff (1824–1887). *Photo courtesy of Deutches Museum, Munich.*

The presence of a voltage, v_2, across the open terminals a and b indicates the potential energy that can enable the motion of charge once a closed circuit is established and allow current to flow.

Figure 2.6 Concept of voltage as potential difference

A moment's reflection upon the significance of voltage should suggest that it must be necessary to specify a sign for this quantity. Consider, again, the same dry-cell or alkaline battery, where, by virtue of an electrochemically induced separation of charge, a 1.5 V potential difference is generated. The potential generated by the battery may be used to move charge in a circuit. The rate at which charge is moved once a closed circuit is established (i.e., the current drawn by the circuit connected to the battery) depends now on the circuit element we choose to connect to the battery. Thus, while the voltage across the battery represents the potential for *providing energy* to a circuit, the voltage across the light bulb indicates the amount of work done in *dissipating energy*. In the first case, energy is generated; in the second, it is consumed (note that energy may also be stored, by suitable circuit elements yet to be introduced). This fundamental distinction requires attention in defining the sign (or polarity) of voltages.

We shall, in general, refer to elements that provide energy as **sources**, and to elements that dissipate energy as **loads**. Standard symbols for a generalized source-and-load circuit are shown in Figure 2.7. Formal definitions will be given in a later section.

A symbolic representation of the battery–light bulb circuit of Figure 2.5.

Figure 2.7 Sources and loads in an electrical circuit

Example   2.3

Many years ago, Christmas light manufacturers would manufacture strands of Christ-mas tree lights with the bulbs connected in a *series circuit,* as depicted in Figure 2.8. Since, by virtue of KVL, the net voltage around the loop is zero, the sum of the voltages across all of the light bulbs must equal the total voltage supplied by the 110 VAC source. If we assume that the bulbs are identical to each other, and if there were, say, 11 bulbs in a strand, then each bulb would have $\frac{1}{11}$ of the outlet's voltage across it, or 10 VAC.

Figure 2.8 Christmas tree lights in a series connection

Although the series connection is very simple to manufacture, it is not used too frequently, in practice. Why is it a poor configuration, from a practical standpoint? (Think what would happen if a light bulb burned out.)

2.3   IDEAL VOLTAGE AND CURRENT SOURCES

In the examples presented in the preceding sections, a battery was used as a source of energy, under the unspoken assumption that the voltage provided by the battery (e.g., 1.5 volts for a dry cell or alkaline battery, or 12 volts for an automotive lead-acid battery) is fixed. Under such an assumption, we implicitly treat the battery as an ideal source. In this section, we will formally define ideal sources. Intuitively, an ideal source is a source that can provide an arbitrary amount of energy. **Ideal sources** are divided into two types: voltage sources and current sources. Of these, you are probably more familiar with the first, since dry-cell, alkaline, and lead-acid batteries are all voltage sources (they are not ideal, of course). You might have to think harder to come up with a physical example that approximates the behavior of an ideal current source; however, reasonably good approximations of ideal current sources also exist. For instance, a voltage source connected in series with a circuit element that opposes a large resistance to the flow of current from the source provides a nearly constant—though small—current and therefore acts very nearly like an ideal current source.

Ideal Voltage Sources

An **ideal voltage source** is an electrical device that will generate a prescribed voltage at its terminals. The ability of an ideal voltage source to generate its output voltage is not affected by the current it must supply to the other circuit elements. Another way to phrase the same idea is as follows:

> An ideal voltage source provides a prescribed voltage across its terminals irrespective of the current flowing through it. The amount of current supplied by the source is determined by the circuit connected to it.

Figure 2.9 depicts various symbols for voltage sources that will be employed throughout this book. Note that the output voltage of an ideal source can be a function of time. In general, the following notation will be employed in the book, unless otherwise noted. A generic voltage source will be denoted by a lowercase v. If it is necessary to emphasize that the source produces a time-varying voltage, then the notation $v(t)$ will be employed. Finally, a constant (DC) voltage source will be denoted by the uppercase character V. Note that by convention the direction of positive current flow out of a voltage source is *out of the positive terminal*.

General symbol for ideal voltage source. $v_s(t)$ may be constant (DC source).

A special case: DC voltage source (ideal battery)

A special case: sinusoidal voltage source, $v_s(t) = V \cos \omega t$

Figure 2.9 Ideal voltage sources

The notion of an ideal voltage source is best appreciated within the context of the source-load representation of electrical circuits, which will frequently be referred to in the remainder of this book. Figure 2.10 depicts the connection of an energy source with a passive circuit (i.e., a circuit that can absorb and dissipate energy—for example, the headlights and light bulb of our earlier examples). Three different representations are shown to illustrate the conceptual, symbolic, and physical significance of this source-load idea.

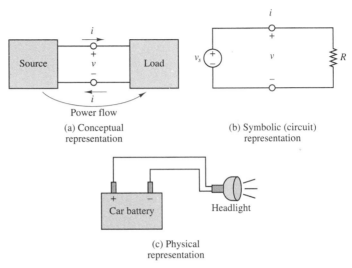

(a) Conceptual
representation

(b) Symbolic (circuit)
representation

(c) Physical
representation

Figure 2.10 Various representations of an electrical system

In the analysis of electrical circuits, we choose to represent the physical reality of Figure 2.10(c) by means of the approximation provided by ideal circuit elements, as depicted in Figure 2.10(b).

Ideal Current Sources

An **ideal current source** is a device that can generate a constant current independent of the circuit it is connected to. To do so, it must be able to generate an arbitrary voltage across its terminals. Figure 2.11 depicts the symbol used to represent ideal current sources. By analogy with the definition of the ideal voltage source stated in the previous section, we write:

Figure 2.11 Symbol for ideal current source

An ideal current source provides a prescribed current to any circuit connected to it. The voltage generated by the source is determined by the circuit connected to it.

The same uppercase and lowercase convention used for voltage sources will be employed in denoting current sources.

Dependent (Controlled) Sources

The sources described so far have the capability of generating a prescribed voltage or current independent of any other element within the circuit. Thus, they are termed *independent sources*. There exists another category of sources, however, whose output (current or voltage) is a function of some other voltage or current in a circuit. These are called **dependent (or controlled) sources.** A different symbol, in the shape of a diamond, is used to represent dependent sources and to distinguish them from independent sources. The symbols typically used to represent dependent sources are depicted in Figure 2.12; the table illustrates the relationship between the source voltage or current and the voltage or current it depends on—v_x or i_x, respectively—which can be any voltage or current in the circuit.

Source type	Relationship
Voltage controlled voltage source (VCVS)	$v_S = A\, v_x$
Current controlled voltage source (CCVS)	$v_S = A\, i_x$
Voltage controlled current source (VCCS)	$i_S = A\, v_x$
Current controlled current source (CCCS)	$i_S = A\, i_x$

Figure 2.12 Symbols for dependent sources

Dependent sources are very useful in describing certain types of electronic circuits. You will encounter dependent sources again in Chapters 8, 9 and 10, when electronic amplifiers are discussed.

2.4   ELECTRIC POWER AND SIGN CONVENTION

The definition of voltage as work per unit charge lends itself very conveniently to the introduction of power. Recall that power is defined as the work done per unit time. Thus, the power, P, either generated or dissipated by a circuit element can be represented by the following relationship:

$$\text{Power} = \frac{\text{Work}}{\text{Time}} = \frac{\text{Work}}{\text{Unit charge}}\frac{\text{Charge}}{\text{Time}} = \text{Voltage} \times \text{Current} \qquad (2.9)$$

Thus,

> The electrical power generated by an active element, or that dissipated or stored by a passive element, is equal to the product of the voltage across the element and the current flowing through it.

$$\boxed{P = VI} \qquad\qquad (2.10)$$

It is easy to verify that the units of voltage (joules/coulomb) times current (coulomb/second) are indeed those of power (joules/second, or watts).

It is important to realize that, just like voltage, power is a signed quantity, and that it is necessary to make a distinction between *positive* and *negative power*. This distinction can be understood with reference to Figure 2.13, in which a source and a load are shown side by side. The polarity of the voltage across the source and the direction of the current through it indicate that the voltage source *is doing work in moving charge from a lower potential to a higher potential.* On the other hand, the load is dissipating energy, because the direction of the current indicates that *charge is being displaced from a higher potential to a lower potential.* To avoid confusion with regard to the sign of power, the electrical engineering community uniformly adopts the **passive sign convention,** which simply states that *the power dissipated by a load is a positive quantity* (or, conversely, that the power generated by a source is a positive quantity). Another way of phrasing the same concept is to state that if current flows from a higher to a lower voltage (+ to −), the power dissipated will be a positive quantity.

Power dissipated = $v(-i) = (-v)i = -vi$ Power dissipated = vi
          Power generated = vi Power generated = $v(-i) = (-v)i = -vi$

Figure 2.13 The passive sign convention

It is important to note also that the actual numerical values of voltages and currents do not matter: once the proper reference directions have been established and the passive sign convention has been applied consistently, the answer will be correct regardless of the reference direction chosen. The following examples illustrate this point.

Example   2.4

$v_B = 12$ V $v_1 = 8$ V
$i = 0.1$ A $v_2 = 4$ V

Figure 2.14

This example will show that whichever reference direction is selected in a circuit, the same answer will be obtained, *provided that the passive sign convention is applied consistently.* The circuit of Figure 2.14 consists of one source (the battery) and two loads connected in series. We first arbitrarily select the clockwise direction for the current flowing in the series circuit. Once we have decided that the current flows in the clockwise direction, we can compute the power dissipated by each element. Since current flows from the negative to the positive terminal in the battery, the power dissipated by the battery will be:

$$P_B = (-v_B)(i) = (-12)(0.1) = -1.2 \text{ W}$$

That is, the battery *generates* 1.2 W. By the same convention, the two loads will dissipate power according to the relations.

$$P_1 = (v_1)(i) = (8)(0.1) = 0.8 \text{ W}$$

and

$$P_2 = (v_2)(i) = (4)(0.1) = 0.4 \text{ W}$$

Thus, energy is conserved, since the net power dissipation in the circuit is zero.

Now consider the same circuit but with the current direction arbitrarily given as counterclockwise, as shown in Figure 2.15. The voltages resulting from the assumed current direction are also shown in the figure. Note that, in keeping with the sign convention, the signs of the voltages are opposite to those of Figure 2.14. Thus, the battery voltage is now -12 V, and the voltages across the loads are -8 V and -4 V, respectively. Thus, according to the sign convention, the power dissipated by the battery is

$$P_B = (-v_B)(i) = -(-12)(-0.1) = -1.2 \text{ W}$$

while the power dissipated by the loads is given by

$$P_1 = (v_1)(i) = (-8)(-0.1) = 0.8 \text{ W}$$

and

$$P_2 = (v_2)(i) = (-4)(-0.1) = 0.4 \text{ W}$$

As in the previous case, the sign convention tells us that the battery is generating 1.2 W of power, which is dissipated by the two loads.

$v_B = -12$ V $v_1 = -8$ V
$i = -0.1$ A $v_2 = -4$ V

Figure 2.15

Example   2.5

This example illustrates how we can determine whether an unknown element is a source or a load simply by identifying voltage polarities and current directions. Assume that devices are available that allow us to measure voltages and currents in the unknown circuit of Figure 2.16(a). Using the diagram, determine which element is supplying power and which is dissipating it.

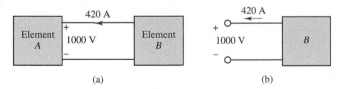

Figure 2.16

Solution:

Looking at element B, we recognize that this circuit element acts as a source, since current flows from a point of lower potential to one of higher potential, as shown in Figure 2.16(b). Therefore, we conclude that the power dissipated by element B is

$$P_B = (1,000)(-420) = -420 \text{ kW}$$

This means that element B is delivering 420 kW of power to element A.

DRILL EXERCISES

2.1 Compute the current flowing through each of the headlights of Example 2.2 if each headlight has a power rating of 50 W. How much power is the battery providing?

Answer: $I_P = I_D = 4.17$ A; 100 W

2.2 Determine which circuit element in the illustration is supplying power and which is dissipating power. Also determine the amount of power dissipated and supplied.

Answer: A, supplying 30.8 W; B, dissipating 30.8 W

2.3 If the battery in the accompanying diagram supplies a total of 10 mW to the three elements shown and $i_1 = 2$ mA and $i_2 = 1.5$ mA, what is the current i_3?

Answer: $i_3 = -1$ mA

If $i_1 = 1$ mA and $i_3 = 1.5$ mA, what is i_2?

Answer: $i_2 = 0$ mA

2.5   CIRCUIT ELEMENTS AND THEIR I-V CHARACTERISTICS

The relationship between current and voltage at the terminals of a circuit element defines the behavior of that element within the circuit. In this section we shall introduce a graphical means of representing the terminal characteristics of circuit elements. Figure 2.17 depicts the representation that will be employed throughout the chapter to denote a generalized circuit element:

the variable *i* represents the current flowing through the element, while *v* is the potential difference, or voltage, across the element.

Suppose now that a known voltage were imposed across a circuit element. The current that would flow as a consequence of this voltage, and the voltage itself, form a unique pair of values. If the voltage applied to the element were varied and the resulting current measured, it would be possible to construct a functional relationship between voltage and current known as the ***i-v* characteristic** (or **volt-ampere characteristic**). Such a relationship defines the circuit element, in the sense that if we impose any prescribed voltage (or current), the resulting current (or voltage) is directly obtainable from the *i-v* characteristic. A direct consequence is that the power dissipated (or generated) by the element may also be determined from the *i-v* curve.

Figure 2.17 Generalized representation of circuit elements

Figure 2.18 depicts an experiment for empirically determining the *i-v* characteristic of a tungsten filament light bulb. A variable voltage source is used to apply various voltages, and the current flowing through the element is measured for each applied voltage.

We could certainly express the *i-v* characteristic of a circuit element in functional form:

$$i = f(v) \qquad v = g(i) \tag{2.11}$$

In some circumstances, however, the graphical representation is more desirable, especially if there is no simple functional form relating voltage to current. The simplest form of the *i-v* characteristic for a circuit element is a straight line, that is,

$$i = kv \tag{2.12}$$

with *k* a constant. In the next section we shall see how this simple model of a circuit element is quite useful in practice and can be used to define the

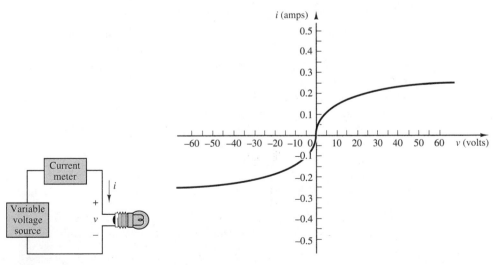

Figure 2.18 Volt-ampere characteristic of a tungsten light bulb

most common circuit elements: ideal voltage and current sources and the resistor.

We can also relate the graphical *i-v* representation of circuit elements to the power dissipated or generated by a circuit element. For example, the graphical representation of the light bulb *i-v* characteristic of Figure 2.18 illustrates that when a positive current flows through the bulb, the voltage is positive, and that, conversely, a negative current flow corresponds to a negative voltage. In both cases the power dissipated by the device is a positive quantity, as it should be, on the basis of the discussion of the preceding section, since the light bulb is a passive device. Note that the *i-v* characteristic appears in only two of the four possible quadrants in the *i-v* plane. In the other two quadrants, the product of voltage and current (i.e., power) is negative, and an *i-v* curve with a portion in either of these quadrants would therefore correspond to power generated. This is not possible for a passive load such as a light bulb; however, there are electronic devices that can operate, for example, in three of the four quadrants of the *i-v* characteristic and can therefore act as sources of energy for specific combinations of voltages and currents. An example of this dual behavior will be introduced in Chapter 7 where it is shown that the photodiode can act either in a passive mode (as a light sensor) or in an active mode (as a solar cell).

The *i-v* characteristics of ideal current and voltage sources can also be useful in visually representing their behavior. An ideal voltage source generates a prescribed voltage independent of the current drawn from the load; thus, its *i-v* characteristic is a straight vertical line with a voltage axis intercept corresponding to the source voltage. Similarly, the *i-v* characteristic of an ideal current source is a horizontal line with a current axis intercept corresponding to the source current. Figure 2.19 depicts this behavior.

i-v characteristic
of a 3 A current source

i-v characteristic
of a 6 V voltage source

Figure 2.19 *i-v* characteristics of ideal sources

2.6   RESISTANCE AND OHM'S LAW

When electric current flows through a metal wire or through other circuit elements, it encounters a certain amount of **resistance,** the magnitude of which depends on the electrical properties of the material. Resistance to the flow of current may be undesired—for example, in the case of lead wires and connection cable—or it may be exploited in an electrical circuit in a useful way. Nevertheless, practically all circuit elements exhibit some resistance; as a consequence, current flowing through an element will cause energy to be dissipated in the form of heat. An ideal **resistor** is a device that exhibits linear resistance properties according to **Ohm's law,** which states that

$$\boxed{V = IR \qquad \text{Ohm's law}} \tag{2.13}$$

that is, that the voltage across an element is directly proportional to the current flow through it. R is the value of the resistance in units of **ohms** (Ω), where

$$1\ \Omega = 1\ \text{V/A} \tag{2.14}$$

The resistance of a material depends on a property called **resistivity,** denoted by the symbol ρ; the inverse of resistivity is called **conductivity** and is denoted by the symbol σ. For a cylindrical resistance element (shown in Figure 2.20), the resistance is proportional to the length of the sample, l, and inversely proportional to its cross-sectional area, A, and conductivity, σ.

$$v = \frac{l}{\sigma A} i \qquad\qquad (2.15)$$

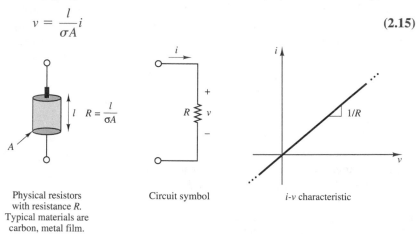

Physical resistors with resistance R. Typical materials are carbon, metal film.

Circuit symbol

i-v characteristic

Figure 2.20 The resistance element

It is often convenient to define the **conductance** of a circuit element as the inverse of its resistance. The symbol used to denote the conductance of an element is G, where

$$G = \frac{1}{R} \text{ siemens (S)} \qquad \text{where} \qquad 1 \text{ S} = 1 \text{ A/V} \qquad (2.16)$$

Thus, Ohm's law can be restated in terms of conductance as:

$$I = GV \qquad\qquad (2.17)$$

Ohm's law is an empirical relationship that finds widespread application in electrical engineering, because of its simplicity. It is, however, only an approximation of the physics of electrically conducting materials. Typically, the linear relationship between voltage and current in electrical conductors does not apply at very high voltages and currents. Further, not all electrically conducting materials exhibit linear behavior even for small voltages and currents. It is usually true, however, that for some range of voltages and currents, most elements display a linear i-v characteristic. Figure 2.21 illustrates how the linear resistance concept may apply to elements with nonlinear i-v characteristics, by graphically defining the linear portion of the i-v characteristic of two common electrical devices: the light bulb, which we have already encountered, and the semiconductor diode, which we shall study in greater detail in Chapter 7.

The typical construction and the circuit symbol of the resistor are shown in Figure 2.21. Resistors made of cylindrical sections of carbon (with

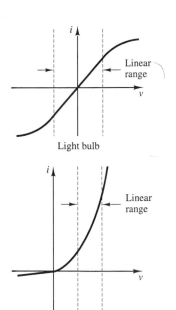

Light bulb

Exponential i-v characteristic (semiconductor diode)

Figure 2.21

resistivity $\rho = 3.5 \times 10^{-5}$ Ω-m) are very common and are commercially available in a wide range of values for several power ratings (as will be explained shortly). Another commonly employed construction technique for resistors employs metal film. A common power rating for resistors used in electronic circuits (e.g., in most consumer electronic appliances such as radios and television sets) is ¼ W. Table 2.1 lists the standard values for commonly used resistors and the color code associated with these values (i.e., the common combinations of the digits $b_1 b_2 b_3$ as defined in Figure 2.22). For example, if the first three color bands on a resistor show the colors red ($b_1 = 2$), violet ($b_2 = 7$), and yellow ($b_3 = 4$), the resistance value can be interpreted as follows:

$$R = 27 \times 10^4 = 270,000 \ \Omega = 270 \ \text{k}\Omega$$

In Table 2.1, the leftmost column represents the complete color code; columns to the right of it only show the third color, since this is the only one that changes. For example, a 10 Ω resistor has brown-black-*black* code, while a 100 Ω resistor has brown-black-*brown* code.

$b_4 \ b_3 \ b_2 \ b_1$

Color bands

black	0	blue	6
brown	1	violet	7
red	2	gray	8
orange	3	white	9
yellow	4	silver	10%
green	5	gold	5%

Resistor value = $(b_1 \ b_2) \times 10^{b_3}$; $b_4 = \%$ tolerance in actual value

Figure 2.22 Resistor color code

Table 2.1   Common resistor values (⅛-, ¼-, ½-, 1-, 2-W rating)

Ω	Code	Ω	Multiplier	kΩ	Multiplier	kΩ	Multiplier	kΩ	Multiplier
10	Brn-blk-blk	100	Brown	1.0	Red	10	Orange	100	Yellow
12	Brn-red-blk	120	Brown	1.2	Red	12	Orange	120	Yellow
15	Brn-grn-blk	150	Brown	1.5	Red	15	Orange	150	Yellow
18	Brn-gry-blk	180	Brown	1.8	Red	18	Orange	180	Yellow
22	Red-red-blk	220	Brown	2.2	Red	22	Orange	220	Yellow
27	Red-vlt-blk	270	Brown	2.7	Red	27	Orange	270	Yellow
33	Org-org-blk	330	Brown	3.3	Red	33	Orange	330	Yellow
39	Org-wht-blk	390	Brown	3.9	Red	39	Orange	390	Yellow
47	Ylw-vlt-blk	470	Brown	4.7	Red	47	Orange	470	Yellow
56	Grn-blu-blk	560	Brown	5.6	Red	56	Orange	560	Yellow
68	Blu-gry-blk	680	Brown	6.8	Red	68	Orange	680	Yellow
82	Gry-red-blk	820	Brown	8.2	Red	82	Orange	820	Yellow

In addition to the resistance in ohms, the maximum allowable power dissipation (or **power rating**) is typically specified for commercial resistors. Exceeding this power rating leads to overheating and can cause the resistor to literally start on fire. For a resistor R, the power dissipated is given by

$$P = VI = I^2 R = \frac{V^2}{R} \tag{2.18}$$

That is, the power dissipated by a resistor is proportional to the square of the current flowing through it, as well as the square of the voltage across it. The following example illustrates how one can make use of the power rating to determine whether a given resistor will be suitable for a certain application.

Example   2.6

A typical power rating (i.e., maximum power allowable for a device) for carbon resistors employed in low-power electronic circuits is ¼ W. What is the value of the smallest ¼-W resistor we can connect to a 1.5-V dry-cell battery?

Figure 2.23

Solution:

In order to compute the size of the resistor, we first calculate the power dissipated by the resistor as a function of its resistance:

$$P = VI = V\frac{V}{R} = \frac{V^2}{R} = \frac{1.5^2}{R}$$

Since the maximum allowable power dissipation is ¼ W, we can write

$$P_{max} = 0.25 = \frac{2.25}{R}$$

$$R = \frac{2.25}{0.25} = 9$$

Therefore R must be at least 9 Ω.

How would this result change if the voltage were doubled? We can easily calculate that the power dissipated is now given by $P = 3^2/R$, and therefore that R must be at least 36 Ω in order not to exceed its power rating. This result is shown in Figure 2.24. Note the effect of the square relationship: doubling the current increases R by a factor of 4.

Figure 2.24

Example   2.7   Resistive Displacement Transducer

A simple resistor can serve as an elementary displacement transducer. The principle illustrated in this example actually forms the basis of many commercial displacement transducers. The resistor in Figure 2.25 is a wire-wound resistor—that is, a resistor made of a tightly wound coil of thin wire—that is rated at 0.1 Ω per turn. The pointer attached to the moving cart makes contact with the resistor, thus closing the circuit at a distance x from the origin. Each turn occupies 0.5 mm, and the total

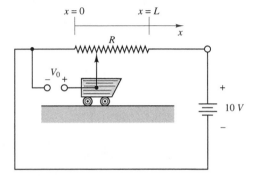

Figure 2.25 A simple resistive displacement transducer

length, L, of the resistor is 10 cm. Find the output voltage V_o as a function of the cart displacement, x.

Solution:

The total number of turns in the resistor is given by the following expression:

$$\text{Number of turns} = \frac{\text{Total distance}}{\text{Distance per turn}} = \frac{0.1 \text{ m}}{0.0005 \text{ m/turn}} = 200$$

which means that the total resistance from 0 to L is 20 Ω. Thus, we can determine the relationship between the position of the cart and the resistance between the origin and point x as follows:

$$R(x) = \frac{20}{10}x = 2x \ \Omega$$

where x is the distance traveled in centimeters. The output voltage can then be found from this expression if we consider Ohm's law:

$$I = \frac{10 \text{ V}}{20 \ \Omega} = 0.5 \text{ A}$$

and

$$V_o = IR(x) = (0.5)2x = x$$

Thus, the relative position of the moving cart, in cm, is numerically equal to the voltage V_o. This simple position-sensing concept finds common application in engineering.

Example 2.8   Resistance Strain Gauges

Another common application of the resistance concept to engineering measurements is the resistance **strain gauge.** Strain gauges are devices that are bonded to the surface of an object, and whose resistance varies as a function of the surface strain experienced by the object. Strain gauges may be used to perform measurements of strain, stress, force, torque, and pressure. Recall that the resistance of a cylindrical conductor of cross-sectional area A, length L, and conductivity σ is given by the expression

$$R = \frac{L}{\sigma A}$$

If the conductor is compressed or elongated as a consequence of an external force, its dimensions will change, and with them its resistance. In particular, if the conductor is stretched, its cross-sectional area will decrease and the resistance will increase. If the conductor is compressed, its resistance decreases, since the length, L, will decrease. The relationship between change in resistance and change in length is given by the gauge factor, G, defined by

$$G = \frac{\Delta R/R}{\Delta L/L}$$

and since the strain ε is defined as the fractional change in length of an object, by the formula

$$\varepsilon = \frac{\Delta L}{L}$$

the change in resistance due to an applied strain ε is given by the expression

$$\Delta R = R_0 G \varepsilon$$

where R_0 is the resistance of the strain gauge under no strain and is called the zero strain resistance. The value of G for resistance strain gauges made of metal foil is usually about 2.

Figure 2.26 depicts a typical foil strain gauge. The maximum strain that can be measured by a foil gauge is about 0.4 to 0.5 percent; that is, $\Delta L/L = 0.004\text{--}0.005$.

R_G Circuit symbol for the strain gauge

Metal-foil resistance strain gauge. The foil is formed by a photo-etching process and is less than 0.00002 in thick. Typical resistance values are 120, 350, and 1,000 Ω. The wide areas are bonding pads for electrical connections.

Figure 2.26 The resistance strain gauge

For a 120-Ω gauge, this corresponds to a change in resistance of the order of 0.96 to 1.2 Ω. Although this change in resistance is very small, it can be detected by means of suitable circuitry. Resistance strain gauges are usually connected in a circuit called the Wheatstone bridge, which we analyze later in this chapter.

Open and Short Circuits

Two convenient idealizations of the resistance element are provided by the limiting cases of Ohm's law as the resistance of a circuit element approaches zero or infinity. A circuit element with resistance approaching zero is called a **short circuit.** Intuitively, one would expect a short circuit to allow for unimpeded flow of current. In fact, metallic conductors (e.g., short wires of large diameter) approximate the behavior of a short circuit. Formally, a short circuit is defined as a circuit element across which the voltage is zero, regardless of the current flowing through it. Figure 2.27 depicts the circuit symbol for an ideal short circuit.

Physically, any wire or other metallic conductor will exhibit some resistance, though small. For practical purposes, however, many elements approximate a short circuit quite accurately under certain conditions. For example, a large diameter copper pipe is effectively a short circuit in the context of a residential electrical power supply, while in a low-power micro-electronic circuit (e.g., an FM radio) a short length of 24 gauge wire (refer to Table 2.2 for the resistance of 24 gauge wire) is a more than adequate short circuit.

The short circuit:
$R = 0$
$v = 0$ for any i

Figure 2.27 The short circuit

Table 2.2   Resistance of copper wire

AWG size	Number of strands	Diameter per strand (in)	Resistance per 1,000 ft (Ω)
24	Solid	0.0201	28.4
24	7	0.0080	28.4
22	Solid	0.0254	18.0
22	7	0.0100	19.0
20	Solid	0.0320	11.3
20	7	0.0126	11.9
18	Solid	0.0403	7.2
18	7	0.0159	7.5
16	Solid	0.0508	4.5
16	19	0.0113	4.7

A circuit element whose resistance approaches infinity is called an **open circuit.** Intuitively, one would expect no current to flow through an open circuit since it offers infinite resistance to any current. In an open circuit, we would expect to see zero current regardless of the externally applied voltage. Figure 2.28 illustrates this idea.

In practice, it is not too difficult to approximate an open circuit: any break in continuity in a conducting path amounts to an open circuit. The idealization of the open circuit, as defined in Figure 2.28, does not hold, however, for very high voltages. The insulating material between two insulated terminals will break down at a sufficiently high voltage. If the insulator is air, ionized particles in the neighborhood of the two conducting elements may lead to the phenomenon of arcing; in other words, a pulse of current may be generated that momentarily jumps a gap between conductors (thanks to this principle, we are able to ignite the air-fuel mixture in a spark-ignition internal combustion engine by means of spark plugs). The ideal open and short circuits are useful concepts and find extensive use in circuit analysis.

The open circuit:
$R \to \infty$
$i = 0$ for any v

Figure 2.28 The open circuit

Series Resistors and the Voltage Divider Rule

Although electrical circuits can take rather complicated forms, even the most involved circuits can be reduced to combinations of circuit elements *in parallel* and *in series*. Thus, it is important that you become acquainted with parallel and series circuits as early as possible, even before formally approaching the topic of network analysis. Parallel and series circuits have a direct relationship with Kirchhoff's laws. The objective of this section and the next is to illustrate two common circuits based on series and parallel combinations of resistors: the voltage and current dividers. These circuits form the basis of all network analysis; it is therefore important to master these topics as early as possible.

For an example of a series circuit, refer to the circuit of Figure 2.29, where a battery has been connected to resistors R_1, R_2, and R_3. The following definition applies:

Definition

Two or more circuit elements are said to be **in series** if the same current flows through each of the elements.

By applying KVL, you can verify that the sum of the voltages across the three resistors equals the voltage externally provided by the battery:

$$1.5 = v_1 + v_2 + v_3$$

The current i flows through each of the four series elements. Thus, by KVL,

$$1.5 = v_1 + v_2 + v_3$$

N series resistors are equivalent to a single resistor equal to the sum of the individual resistances.

Figure 2.29 Series circuits

and since, according to Ohm's law, the separate voltages can be expressed by the relations

$$v_1 = iR_1 \qquad v_2 = iR_2 \qquad v_3 = iR_3$$

we can therefore write

$$1.5 = i(R_1 + R_2 + R_3)$$

This simple result illustrates a very important principle: to the battery, the three series resistors appear as a single equivalent resistance of value R_{EQ}, where

$$R_{EQ} = R_1 + R_2 + R_3$$

The three resistors could thus be replaced by a single resistor of value R_{EQ} without changing the amount of current required of the battery. From this result we may extrapolate to the more general relationship defining the equivalent resistance of N series resistors:

$$R_{EQ} = \sum_{n=1}^{N} R_n \tag{2.19}$$

which is also illustrated in Figure 2.29. A concept very closely tied to series resistors is that of the **voltage divider.** This terminology originates from the observation that the source voltage in the circuit of Figure 2.29 divides among the three resistors according to KVL. If we now observe that the series current, i, is given by

$$i = \frac{1.5}{R_{EQ}} = \frac{1.5}{R_1 + R_2 + R_3}$$

we can write each of the voltages across the resistors as:

$$v_1 = iR_1 = \frac{R_1}{R_{EQ}}(1.5)$$

$$v_2 = iR_2 = \frac{R_2}{R_{EQ}}(1.5)$$

$$v_3 = iR_3 = \frac{R_3}{R_{EQ}}(1.5)$$

that is,

> the voltage across each resistor in a series circuit is directly proportional to the ratio of its resistance to the total series resistance of the circuit.

An instructive exercise consists of verifying the KVL is still satisfied, by adding the voltage drops around the circuit and equating their sum to the source voltage:

$$v_1 + v_2 + v_3 = \frac{R_1}{R_{EQ}}(1.5) + \frac{R_2}{R_{EQ}}(1.5) + \frac{R_3}{R_{EQ}}(1.5) = 1.5$$

since

$$R_{EQ} = R_1 + R_2 + R_3$$

Therefore, since KVL is satisfied, we are certain that the voltage divider rule is consistent with Kirchhoff's laws. By virtue of the voltage divider rule, then, we can always determine the proportion in which voltage drops are distributed around a circuit. This result will be useful in reducing complicated circuits to simpler forms. The general form of the voltage divider rule for a circuit with N series resistors and a voltage source is:

$$v_n = \frac{R_n}{R_1 + R_2 + \ldots + R_n + \ldots + R_N} v_S \qquad \text{Voltage divider} \qquad \textbf{(2.20)}$$

Example   2.9

This example will illustrate the notion of the voltage divider by means of a simple circuit. We seek to determine the voltage v_X in the circuit of Figure 2.30.

Solution:

If we select the reference direction indicated in Figure 2.30 by the direction of the current, we can assign voltage polarities around the circuit as shown. Then, application of KVL yields:

$$3.0 - v_{10} - v_6 - v_8 = 0$$

where v_{10} is the voltage across the 10-Ω resistor, according to the direction of the current indicated in the figure. Note that $v_X = v_8$. According to the voltage divider rule, then,

$$v_X = v_8 = \frac{8}{8 + 6 + 10}(3) = 1 \text{ V}$$

The technique employed in this example is representative of how the voltage divider rule is applied to series circuits. The key factor to be determined before applying this rule is whether the circuit is in fact a series circuit. We shall return to this point shortly.

Figure 2.30

Parallel Resistors and the Current Divider Rule

A concept analogous to that of the voltage divider may be developed by applying Kirchhoff's current law to a circuit containing only parallel resistances.

> **Definition**
>
> Two or more circuit elements are said to be **in parallel** if the same voltage appears across each of the elements.

Figure 2.31 illustrates the notion of parallel resistors connected to an ideal current source. Kirchhoff's current law requires that the sum of the currents

The voltage v appears across each parallel element; by KCL, $i_S = i_1 + i_S + i_3$

Figure 2.31 Parallel circuits

N resistors in parallel are equivalent to a single equivalent resistor with resistance equal to the inverse of the sum of the inverse resistances.

into, say, the top terminal of the circuit be zero:

$$i_s = i_1 + i_2 + i_3$$

But by virtue of Ohm's law we may express each current as follows:

$$i_1 = \frac{v}{R_1}$$

$$i_2 = \frac{v}{R_2}$$

$$i_3 = \frac{v}{R_3}$$

since, by definition, the same voltage, v, appears across each element. Kirchhoff's current law may then be restated as follows:

$$i_s = v\left(\frac{1}{R_1} + \frac{1}{R_2} + \frac{1}{R_3}\right)$$

Note that this equation can be also written in terms of a single equivalent resistance:

$$i_s = v\frac{1}{R_{EQ}}$$

where

$$\frac{1}{R_{EQ}} = \frac{1}{R_1} + \frac{1}{R_2} + \frac{1}{R_3}$$

As illustrated in Figure 2.31, one can generalize this result to an arbitrary number of resistors connected in parallel by stating that N resistors in parallel act as a single equivalent resistance, R_{EQ}, given by the expression

$$\frac{1}{R_{EQ}} = \frac{1}{R_1} + \frac{1}{R_2} + \cdots + \frac{1}{R_N} \tag{2.21}$$

or

$$R_{EQ} = \frac{1}{1/R_1 + 1/R_2 + \cdots + 1/R_N} \tag{2.22}$$

Very often in the remainder of this book we shall refer to the parallel combination of two or more resistors with the following notation:

$$R_1 \parallel R_2 \parallel \cdots$$

where the symbol $\parallel$ signifies "in parallel with."

From the results shown in equations 2.21 and 2.22, which were obtained directly from KCL, the **current divider** rule can be easily derived. Consider, again, the three-resistor circuit of Figure 2.31. From the expressions already derived from each of the currents, i_1, i_2, and i_3, we can write:

$$i_1 = \frac{v}{R_1}$$

$$i_2 = \frac{v}{R_2}$$

$$i_3 = \frac{v}{R_3}$$

and since $v = R_{EQ}i_s$, these currents may be expressed by:

$$i_1 = \frac{R_{EQ}}{R_1}i_s = \frac{1/R_1}{1/R_{EQ}}i_s = \frac{1/R_1}{1/R_1 + 1/R_2 + 1/R_3}i_s$$

$$i_2 = \frac{1/R_2}{1/R_1 + 1/R_2 + 1/R_3}i_s$$

$$i_3 = \frac{1/R_3}{1/R_1 + 1/R_2 + 1/R_3}i_s$$

One can easily see that the current in a parallel circuit divides in inverse proportion to the resistances of the individual parallel elements. The general expression for the current divider for a circuit with N parallel resistors is the following:

$$i_n = \frac{1/R_n}{1/R_1 + 1/R_2 + \cdots + 1/R_n + \cdots + 1/R_N}i_s \quad \text{Current divider} \quad \textbf{(2.23)}$$

Example 2.10 illustrates the application of the current divider rule.

Example   2.10

Find the current i_x in the circuit shown in Figure 2.32.

Solution:

The most direct solution is obtained by applying the current divider rule, where the 10-Ω resistance is associated with the desired current:

$$i_x = \frac{1/10}{1/10 + 1/2 + 1/20}(4) = (0.1538)4 = 0.6154 \text{ A}$$

An equivalent but less direct way of obtaining the same result would be to compute the equivalent resistance of the parallel circuit, determine the parallel voltage, v, and then compute the unknown current as

$$i_x = \frac{v}{10}$$

It is left as an exercise for you to demonstrate that this method is not as direct as the application of the current divider rule.

Figure 2.32

Much of the resistive network analysis that will be introduced in Chapter 3 is based on the simple principles of the voltage and current dividers introduced in this section. Unfortunately, practical circuits are rarely composed only of parallel or only of series elements. The following examples and drill exercises illustrate some simple and slightly more advanced circuits that combine parallel and series elements.

Example   2.11

Figure 2.33

This example illustrates the analysis of a circuit that contains a combination of parallel and series elements. The circuit of Figure 2.33 is neither a series nor a parallel circuit; thus, it is not possible to directly determine the voltage v or the current i by means of the voltage divider or current divider rule introduced earlier. How can v and i be determined?

Solution:

The circuit takes a much simpler appearance once it becomes evident that the same voltage appears across both R_2 and R_3 and, therefore, that these elements are in parallel. If these two resistors are replaced by a single equivalent resistor according to the procedures described in this section, the circuit of Figure 2.34 is obtained.

Figure 2.34

Note that now the equivalent circuit is a simple series circuit and the voltage divider rule can be applied to determine that:

$$v = \frac{R_2 \parallel R_3}{R_1 + R_2 \parallel R_3} v_S$$

while the current is found to be

$$i = \frac{v_S}{R_1 + R_2 \parallel R_3}$$

Example   2.12   The Wheatstone Bridge

The **Wheatstone bridge** is a resistive circuit that is frequently encountered in a variety of measurement circuits. The general form of the bridge circuit is shown in Figure 2.35(a), where R_1, R_2, and R_3 are known while R_x is an unknown resistance, to be determined. The circuit may also be redrawn as shown in Figure 2.35(b). The latter circuit will be used to demonstrate the use of the voltage divider rule in a mixed series-parallel circuit. The objective is to determine the unknown resistance, R_x.

1. Find the value of the voltage $v_{ab} = v_a - v_b$ in terms of the four resistances and the source voltage, v_S.

2. If $R_1 = R_2 = R_3 = 1\ \text{k}\Omega$, $v_S = 12$ V, and $v_{ab} = 12$ mV, what is the value of R_x?

(a)

Solution:

1. First, we observe that the circuit consists of the parallel combination of three subcircuits: the voltage source, the series combination of R_1 and R_2, and the series combination of R_3 and R_x. Since these three subcircuits are in parallel, the same voltage will appear across each of them, namely, the source voltage, v_S.

    Thus, the source voltage divides between each resistor pair, R_1-R_2 and R_3-R_x, according to the voltage divider rule: v_a is the fraction of the source voltage appearing across R_2, while v_b is the voltage appearing across R_x:

$$v_a = v_S \frac{R_2}{R_1 + R_2} \qquad \text{and} \qquad v_b = v_S \frac{R_x}{R_3 + R_x}$$

Finally, the voltage difference between points a and b is given by:

$$v_{ab} = v_a - v_b = v_S \left(\frac{R_2}{R_1 + R_2} - \frac{R_x}{R_3 + R_x} \right)$$

This result is very useful and quite general, and it finds application in numerous practical circuits.

(b)

Figure 2.35 Wheatstone bridge circuits

2. In order to solve for the unknown resistance, we substitute the numerical values in the preceding equation to obtain

$$0.012 = 12 \left(\frac{1{,}000}{2{,}000} - \frac{R_x}{1{,}000 + R_x} \right)$$

which may be solved for R_x to yield

$$R_x = 996\ \Omega$$

Example   2.13   The Strain Gauge Wheatstone Bridge

Strain gauges, which were introduced in Example 2.8, are frequently employed in the measurement of force. One of the simplest applications of strain gauges is in the measurement of the force applied to a cantilever beam, as illustrated in Figure 2.36. Four strain gauges are employed in this case, of which two are bonded to the upper surface of the beam at a distance L from the point where the external force, F, is applied and two are bonded on the lower surface, also at a distance L. Under the influence of the external force, the beam deforms and causes the upper gauges to extend and the lower gauges to compress. Thus, the resistance of the upper gauges will increase by an amount ΔR, and that of the lower gauges will decrease by an equal amount, assuming that the gauges are symmetrically placed. Let R_1 and R_4 be the upper gauges and R_2 and R_3 the lower gauges. Thus, under the influence of the external force, we have:

$$R_1 = R_4 = R_0 + \Delta R$$

$$R_2 = R_3 = R_0 - \Delta R$$

where R_0 is the zero strain resistance of the gauges (see Example 2.8). It can be shown from elementary statics that the relationship between the strain ε and a force F applied at a distance L for a cantilever beam is:

$$\varepsilon = \frac{6LF}{wh^2Y}$$

where h and w are as defined in Figure 2.36 and Y is the beam's modulus of elasticity.

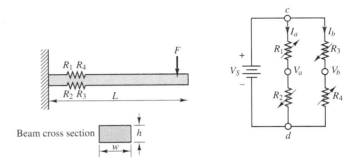

Figure 2.36 A force-measuring instrument

In the circuit of Figure 2.36, the currents i_a and i_b are given by

$$i_a = \frac{v_S}{R_1 + R_2} \quad \text{and} \quad i_b = \frac{v_S}{R_3 + R_4}$$

The bridge output voltage is defined by $v_o = v_b - v_a$ and may be found from the following expression:

$$
\begin{aligned}
v_o = i_b R_4 - i_a R_2 &= \frac{v_S R_4}{R_3 + R_4} - \frac{v_S R_2}{R_1 + R_2} \\
&= v_S \frac{R_0 + \Delta R}{R_0 + \Delta R + R_0 - \Delta R} - v_S \frac{R_0 - \Delta R}{R_0 + \Delta R + R_0 - \Delta R} \\
&= v_S \frac{\Delta R}{R_0} = v_S G \varepsilon
\end{aligned}
$$

where the expression for $\Delta R/R_0$ was obtained in Example 2.8. Thus, it is possible to obtain a relationship between the output voltage of the bridge circuit and the force, F, as follows:

$$v_o = v_S G \varepsilon = v_S G \frac{6LF}{wh^2Y} = \frac{6v_S GL}{wh^2Y} F = kF$$

where k is the calibration constant for this force transducer.

Strain gauge Wheatstone bridges are the workhorse of many instruments for the measurement of strain, stress, force, torque, pressure, and related quantities.

DRILL EXERCISES

2.4 Repeat Example 2.9 by reversing the reference direction of the current, to show that the same result is obtained.

2.5 The circuit in the accompanying illustration contains a battery, a resistor, and an unknown circuit element.

1. If the voltage $V_{battery}$ is 1.45 V and $i = 5$ mA, find the power supplied to or by the battery.
2. Repeat part 1 if $i = -2$ mA.

> Answer: $P_1 = 7.25 \times 10^{-3}$ W (supplied by)
> $P_2 = 2.9 \times 10^{-3}$ W (supplied to)

2.6 The battery in the accompanying circuit supplies power to the resistors R_1, R_2, and R_3. Use KCL to determine the current i_B, and find the power supplied by the battery if $V_{battery} = 3$ V.

> Answer: $i_B = 1.8$ mA
> $P_B = 5.4$ mW

2.7 Use the results of part 1 of Example 2.12 to find the condition for which the voltage $v_{ab} = v_a - v_b$ is equal to zero (this is called the balanced condition for the bridge). Does this result necessarily require that all four resistors be identical? Why?

> Answer: $R_1 R_x = R_2 R_3$

2.8 Verify that KCL is satisfied by the current divider rule and that the source current i_S divides in inverse proportion to the parallel resistors R_1, R_2, and R_3. (This should not be a surprise, since we would expect to see more current flow through the smaller resistance.)

2.9 Compute the full-scale (i.e., largest) output voltage for the force-measuring apparatus of Example 2.13. Assume that the strain gauge bridge is to measure forces ranging from 0 to 500 N, $L = 0.3$ m, $w = 0.05$ m, $h = 0.01$ m, $G = 2$, and the modulus of elasticity for the beam is 69×10^9 N/m^2 (aluminum). The source voltage is 12 V. What is the calibration constant of this force transducer?

> Answer: v_o (full scale) $= 62.6$ mV; $k = 0.125$ mV/N

2.10 Repeat the derivation of the current divider law by using conductance elements—that is, by replacing each resistance with its equivalent conductance, $G = 1/R$.

2.7   PRACTICAL VOLTAGE AND CURRENT SOURCES

The idealized models of voltage and current sources we discussed in Section 2.3 fail to take into consideration the finite-energy nature of practical voltage and current sources. The objective of this section is to extend the ideal models to models that are capable of describing the physical limitations of the voltage and current sources used in practice. Consider, for example, the model of an ideal voltage source shown in Figure 2.9. As the load resistance (R) decreases, the source is required to provide increasing amounts of current to maintain the voltage $v_S(t)$ across its terminals:

$$i(t) = \frac{v_S(t)}{R} \tag{2.24}$$

This circuit suggests that the ideal voltage source is required to provide an infinite amount of current to the load, in the limit as the load resistance approaches zero. Naturally, you can see that this is impossible; for example, think about the ratings of a conventional car battery: 12 V, 450 Ah (amphours). This implies that there is a limit (albeit a large one) to the amount of current a practical source can deliver to a load. Fortunately, it will not be necessary to delve too deeply into the physical nature of each type of source in order to describe the behavior of a practical voltage source: the limitations of practical sources can be approximated quite simply by exploiting the notion of the internal resistance of a source. Although the models described in this section are only approximations of the actual behavior of energy sources, they will provide good insight into the limitations of practical voltage and current sources. Figure 2.37 depicts a model for a practical voltage source; this is composed of an ideal voltage source, v_S, in series with a resistance, r_S. The resistance r_S in effect poses a limit to the maximum current the voltage source can provide:

$$i_{S\text{max}} = \frac{v_S}{r_S} \tag{2.25}$$

Typically, r_S is small. Note, however, that its presence affects the voltage across the load resistance: now this voltage is no longer equal to the source voltage. Since the current provided by the source is

$$i_S = \frac{v_S}{r_S + R_L} \tag{2.26}$$

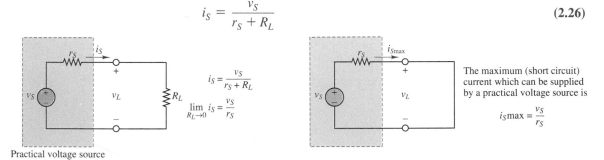

$$i_S = \frac{v_S}{r_S + R_L}$$

$$\lim_{R_L \to 0} i_S = \frac{v_S}{r_S}$$

Practical voltage source

The maximum (short circuit) current which can be supplied by a practical voltage source is

$$i_S\text{max} = \frac{v_S}{r_S}$$

Figure 2.37 Practical voltage source

the load voltage can be determined to be

$$v_L = i_S R_L$$
$$= v_S \frac{R_L}{r_S + R_L} \tag{2.27}$$

Thus, in the limit as the source internal resistance, r_S, approaches zero, the load voltage, v_L, becomes exactly equal to the source voltage. It should be apparent that a desirable feature of an ideal voltage source is a very small internal resistance, so that the current requirements of an arbitrary load may be satisfied. Often, the effective internal resistance of a voltage source is quoted in the technical specifications for the source, so that the user may take this parameter into account.

A similar modification of the ideal current source model is useful to describe the behavior of a practical current source. The circuit illustrated in Figure 2.38 depicts a simple representation of a practical current source, consisting of an ideal source in parallel with a resistor. Note that as the load resistance approaches infinity (i.e., an open circuit), the output voltage of the current source approaches its limit,

$$v_{S_{max}} = i_S r_S \tag{2.28}$$

A good current source should be able to approximate the behavior of an ideal current source. Therefore, a desirable characteristic for the internal resistance of a current source is that it be as large as possible.

A model for practical current sources consists of an ideal source in *parallel* with an internal resistance.

Maximum output voltage for practical current source with open-circuit load:

$$v_{S_{max}} = i_S \, r_S$$

Figure 2.38 Practical current source

2.8   MEASURING DEVICES

In this section, you should gain a basic understanding of the desirable properties of practical devices for the measurement of electrical parameters. The measurements most often of interest are those of current, voltage, power, and resistance. In analogy with the models we have just developed to describe the nonideal behavior of voltage and current sources, we shall similarly present circuit models for practical measuring instruments suitable for describing the nonideal properties of these devices.

The Ammeter

The **ammeter** is a device that, when connected in series with a circuit element, can measure the current flowing through the element. Figure 2.39 illustrates this idea. From Figure 2.39, two requirements are evident for obtaining a correct measurement of current:

1. The ammeter must be placed in series with the element whose current is to be measured (e.g., resistor R_2).

2. The ammeter should not restrict the flow of current (i.e., cause a voltage drop), or else it will not be measuring the true current flowing in the circuit. *An ideal ammeter has zero internal resistance.*

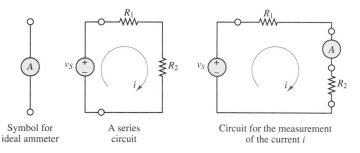

Symbol for
ideal ammeter

A series
circuit

Circuit for the measurement
of the current i

Figure 2.39 Measurement of current

The Voltmeter

The **voltmeter** is a device that can measure the voltage across a circuit element. Since voltage is the difference in potential between two points in a circuit, the voltmeter needs to be connected across the element whose voltage we wish to measure:

1. The voltmeter must be placed in parallel with the element whose voltage it is measuring.

2. The voltmeter should draw no current away from the element whose voltage it is measuring, or else it will not be measuring the true voltage across that element. Thus, *an ideal voltmeter has infinite internal resistance.*

Figure 2.40 illustrates these two points.

Once again, the definitions just stated for the ideal voltmeter and ammeter need to be augmented by considering the practical limitations of the devices. A practical ammeter will contribute some series resistance to the circuit in which it is measuring current; a practical voltmeter will not act

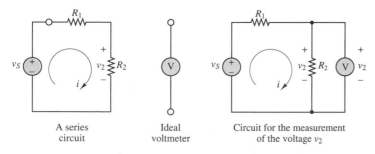

Figure 2.40 Measurement of voltage

as an ideal open circuit but will always draw some current from the measured circuit. The drill exercises verify that these practical restrictions do not necessarily pose a limit to the accuracy of the measurements obtainable with practical measuring devices, as long as the internal resistance of the measuring devices is known. Figure 2.41 depicts the circuit models for the practical ammeter and voltmeter.

All of the considerations that pertain to practical ammeters and voltmeters can be applied to the operation of a **wattmeter,** a measuring instrument that provides a measurement of the power dissipated by a circuit element, since the wattmeter is in effect made up of a combination of a voltmeter and an ammeter. Figure 2.42 depicts the typical connection of a wattmeter in the same series circuit used in the preceding paragraphs. In effect, the wattmeter measures the current flowing through the load and, simultaneously, the voltage across it and multiplies the two to provide a reading of the power dissipated by the load. The internal power consumption of a practical wattmeter is explored in the homework problems.

Practical
voltmeter

Practical
ammeter

Figure 2.41 Models for practical ammeter and voltmeter

Measurement of the power
dissipated in the resistor R_2:
$$P_2 = v_2 i$$

Internal wattmeter connections

Figure 2.42 Measurement of power

2.9   ELECTRICAL NETWORKS

In the previous sections we have outlined models for the basic circuit elements: sources, resistors, and measuring instruments. We have assembled all the tools and parts we need in order to define an **electrical network.** It is appropriate at this stage to formally define the elements of the electrical circuit; the definitions that follow are part of standard electrical engineering terminology.

Branch

A **branch** is any portion of a circuit with two terminals connected to it. A branch may consist of one or more circuit elements (Figure 2.43). In practice, any circuit element with two terminals connected to it is a branch.

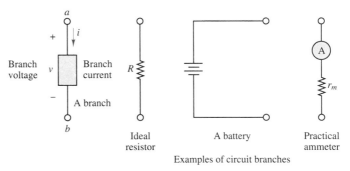

Figure 2.43 Definition of a branch

Node

A node is the junction of two or more branches (one often refers to the junction of only two branches as a *trivial node*). Figure 2.44 illustrates the concept. In effect, any connection that can be accomplished by soldering various terminals together is a node. It is very important to identify nodes properly in the analysis of electrical networks.

Figure 2.44 Definition of a node

Loop

A **loop** is any closed connection of branches that begins at one terminal of a branch and ends at the opposite terminal of the same branch. Various loop configurations are illustrated in Figure 2.45.

Examples:

Figure 2.45 Definition of a loop

Note how two different loops in the same circuit may include some of the same elements or branches.

Mesh

A **mesh** is a loop that does not contain other loops. Meshes are an important aid to certain analysis methods. In Figure 2.45, the circuit with loops 1, 2, and 3 consists of two meshes: loops 1 and 2 are meshes, but loop 3 is not a mesh, because it encircles both loops 1 and 2. The one-loop circuit of Figure 2.45 is also a one-mesh circuit. Figure 2.46 illustrates how meshes are simpler to visualize in complex networks than loops are.

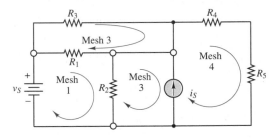

Figure 2.46 Definition of a mesh

How many loops can you identify in this four-mesh circuit? (Answer: 14)

Network Analysis

The analysis of an electrical network consists of determining each of the unknown branch currents and node voltages. It is therefore important to define all of the relevant variables as clearly as possible, and in systematic fashion. Once the known and unknown variables have been identified, a set of equations relating these variables is constructed, and these are solved by means of suitable techniques. The analysis of electrical circuits consists of writing the smallest set of equations sufficient to solve for all of the unknown variables. The procedures required to write these equations are the subject of Chapter 3 and are very well documented and codified in the form of simple rules. The analysis of electrical circuits is greatly simplified if some standard conventions are followed. The objective of this section is precisely to outline the preliminary procedures that will render the task of analyzing an electrical circuit manageable.

Circuit Variables

The first observation to be made is that the relevant variables in network analysis are the node voltages and the branch currents. This fact is really nothing more than a consequence of Ohm's law. Consider the branch depicted in Figure 2.47, consisting of a single resistor. Here, once a voltage v_R is defined across the resistor R, a current i_R will flow through the resistor, according to $v_R = i_R R$. But the voltage v_R, which causes the current to flow, is really the difference in electric potential between nodes a and b:

$$v_R = v_a - v_b \tag{2.29}$$

Figure 2.47 Variables in a network analysis problem

What meaning do we assign to the variables v_a and v_b? Was it not stated that voltage is a potential difference? Is it then legitimate to define the voltage at a single point (node) in a circuit? Whenever we reference the voltage at a node in a circuit, we imply an assumption that the voltage at that node is the potential difference between the node itself and a reference node called **ground,** which is located somewhere else in the circuit and which for convenience has been assigned a potential of zero volts. Thus, in Figure 2.47, the expression

$$v_R = v_a - v_b$$

really signifies that v_R is the difference between the voltage differences $v_a - v_c$ and $v_b - v_c$, where v_c is the (arbitrary) ground potential. Note that the equation $v_R = v_a - v_b$ would hold even if the reference node, c, were not assigned a potential of zero volts, since

$$v_R = v_a - v_b = (v_a - v_c) - (v_b - v_c) \tag{2.30}$$

What, then, is this ground or reference voltage?

Ground

The choice of the word *ground* is not arbitrary. This point can be illustrated by a simple analogy with the physics of fluid motion. Consider a tank of water, as shown in Figure 2.48, located at a certain height above the ground.

Figure 2.48 Analogy between electrical and earth ground

The potential energy due to gravity will cause water to flow out of the pipe at a certain flow rate. The pressure that forces water out of the pipe is directly related to the head, $(h_1 - h_2)$, in such a way that this pressure is zero when $h_2 = h_1$. Now the point h_3, corresponding to the ground level, is defined as having zero potential energy. It should be apparent that the pressure acting on the fluid in the pipe is really caused by the difference in potential energy, $(h_1 - h_3) - (h_2 - h_3)$. It can be seen, then, that it is not necessary to assign a precise energy level to the height h_3; in fact, it would be extremely cumbersome to do so, since the equations describing the flow of water would then be different, say, in Denver ($h_3 = 1,600$ m above sea level) from those that would apply in Miami ($h_3 = 0$ m above sea level). You see, then, that it is the relative difference in potential energy that matters in the water tank problem.

In analogous fashion, in every circuit a point can be defined that is recognized as "ground" and is assigned the electric potential of zero volts *for convenience*. Note that, unless they are purposely connected together, the grounds in two completely separate circuits are not necessarily at the same potential. This last statement may seem puzzling, but Example 2.14 should clarify the idea.

Example   2.14   Chassis and Earth Grounds

Automobiles and commercial vehicles are equipped with a lead-acid battery (usually 12 V) to provide electric power to a variety of accessories. The electrical circuits inside the vehicle are all tied to one common ground, which, for convenience, is none other than the metal body of the car (see Figure 2.49). The same notion extends to many electrically powered devices, where the case, enclosure, or chassis becomes a reference point for all electrical circuits. This type of ground connection takes the name of **chassis ground.**

Note that the car sits on an insulating support, that is, its tires. Therefore, the ground potential of the chassis does not necessarily have to be the same as the

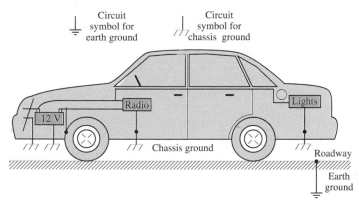

Figure 2.49 Chassis ground versus earth ground

ground potential of the earth! The latter quantity, in fact, takes a different name: it is commonly referred to as **earth ground.**

Often, chassis ground and earth ground are tied together. For example, a washing machine with internal circuitry tied to the metal enclosure (chassis) is also grounded to earth. This is a very good idea from the standpoint of safety. If the two grounds were not tied together, a potential difference might exist between earth and chassis ground, and the unaware user might risk electric shock by touching the washing machine while standing on a grounded surface, thus closing the loop between chassis and earth ground (see Figure 2.50). We shall be more precise with regard to grounding and safety issues when electric power is discussed, in Chapter 5. Note that in this book we shall not make a distinction between chassis and earth ground unless this is dictated by the requirements of a specific problem.

In a washing machine, the chassis ground is also tied to earth for safety reasons.

Figure 2.50 Grounding chassis to earth for safety

It is a useful exercise at this point to put the concepts illustrated in this chapter to practice by identifying the relevant variables in a few examples of electrical circuits. In the following examples, we shall illustrate how it is possible to define unknown voltages and currents in a circuit in terms of the source voltages and currents and the resistances in the circuit.

Example   2.15

Identify the branch and node voltages and the loop and mesh currents in the circuit of Figure 2.51.

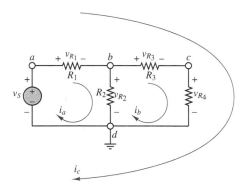

Figure 2.51

Solution:

The following node voltages may be identified:

Node voltages	Branch voltages
$v_a = v_S$ (source voltage)	$v_S = v_a - v_d = v_a$
$v_b = v_{R2}$	$v_{R1} = v_a - v_b$
$v_c = v_{R4}$	$v_{R2} = v_b - v_d = v_b$
$v_d = 0$ (ground)	$v_{R3} = v_b - v_c$
	$v_{R4} = v_c - v_d = v_c$

Currents i_a, i_b, and i_c are loop currents, but only i_a and i_b are mesh currents.

It should be clear at this stage that some method is needed to organize the wealth of information that can be generated simply by applying Ohm's law at each branch in a circuit. What would be desirable at this point is a means of reducing the number of equations needed to solve a circuit to the minimum necessary, that is, a method for obtaining N equations in N unknowns. The next chapter is devoted to the development of systematic circuit analysis methods that will greatly simplify the solution of electrical network problems.

DRILL EXERCISES

2.11   Write expressions for the voltage across each resistor in Example 2.15 in terms of the mesh currents.

> Answer: $v_{R1} = i_a R_1$; $v_{R2} = (i_a - i_b)R_2$; $v_{R3} = i_b R_3$; $v_{R4} = i_b R_4$

2.12   Apply KVL to the outer loop of Example 2.15.

> Answer: $v_S = v_{R1} + v_{R3} + v_{R4} = (i_c + i_a)R_1 + (i_c + i_b)R_3 + (i_c + i_b)R_4$

2.13   Write expressions for the current through each resistor in Example 2.15 in terms of the node voltages.

> Answer: $i_1 = \dfrac{v_a - v_b}{R_1}$; $i_2 = \dfrac{v_b - v_d}{R_2}$; $i_3 = \dfrac{v_b - v_c}{R_3}$; $i_4 = \dfrac{v_c - v_d}{R_4}$

CONCLUSION

The objective of this chapter was to introduce the background needed in the following chapters for the analysis of linear resistive networks. The fundamental laws of circuit analysis, Kirchhoff's current law, Kirchhoff's voltage law, and Ohm's law, were introduced, along with the basic circuit elements, and all were used to analyze the most basic circuits: voltage and current dividers. Measuring devices and a few other practical circuits employed in common engineering measurements were also introduced to provide a flavor of the applicability of these basic ideas to practical engineering problems. The remainder of the book draws on the concepts developed in this chapter. Mastery of the principles exposed in these first pages is therefore of fundamental importance.

KEY TERMS

HOMEWORK PROBLEMS

2.1 An automotive battery is rated at 120 A-h. This means that under certain test conditions it can output 1 A at 12 V for 120 h (under other test conditions, the battery may have other ratings).

a. How much total energy is stored in the battery?

b. If the headlights are left on overnight (8 h), how much energy will still be stored in the battery in the morning? (Assume a 150-W power rating for the headlights.)

2.2 A lightning bolt carries a current of 10,000 A and lasts for 100 μs.

a. How much charge is exchanged between the clouds and the ground?

b. If the charge differential between the clouds and the ground is caused by charge-deficient raindrops, determine the number of raindrops that must fall to create the charge separation that is neutralized by the lightning. Assume that each raindrop has a charge deficiency of 20 charges.

2.3 A car battery kept in storage in the basement needs recharging. If the voltage and the current provided by the charger during a charge cycle are as shown in Figure P2.1,

a. Find the total charge transferred to the battery.
b. Find the total energy transferred to the battery.

2.4 There are many battery-charging schemes used in battery chargers on the market. The charging scheme used in Figure P2.2 is an example of a constant-voltage charge with current limit. The charger voltage is such that the current into the battery does not exceed 100 mA, as shown in Figure P2.2. The charger's terminal voltage, however, does not exceed 9 V. The battery is charged for 6 hours. Find:

a. The total charge delivered to the battery.
b. The energy transferred to the battery during the charging cycle.

Hint: Recall that energy, w, is the integral of power, or $P = dw/dt$.

Figure P2.1

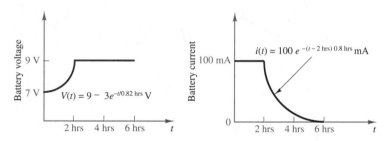

Figure P2.2

2.5 The battery-charging scheme shown in Figure P2.3 is an example of a constant-current charge cycle. The voltage at the battery's terminals is set by the charger so that the current into the battery is held constant at 40 mA. The battery is charged for 6 hours. Find

a. The total charge transferred to the battery.
b. The energy transferred to the battery during the 6-h charge cycle.

Hint: Recall that energy, w, is the integral of power, or $P = dw/dt$.

Figure P2.3

Figure P2.4

2.6   The battery-charging scheme shown in Figure P2.4 is called a *taper-current charge cycle*. The current starts off at its highest level and then decreases with time for the entire charge cycle as shown. Find:

a.   The total charge transferred to the battery.

b.   The energy transferred to the battery in the 12-h charge cycle.

Hint: Recall that energy, w, is the integral of power, or $P = dw/dt$.

2.7   The charge cycle shown in Figure P2.5 is an example of a *two-rate charge*. The current is held constant at 50 mA for 5 h. Then it is switched to 20 mA for the next 5 h. Find:

a.   The total charge transferred to the battery.

b.   The energy transferred to the battery.

Hint: Recall that energy, w, is the integral of power, or $P = dw/dt$.

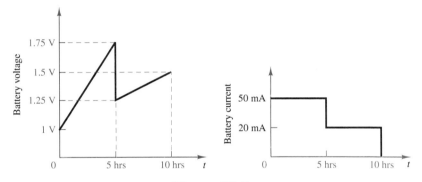

Figure P2.5

2.8   Apply KCL to find the current i in the circuit of Figure P2.6.

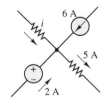

Figure P2.6   **Figure P2.7**

2.9   Apply KCL to find the current i in the circuit of Figure P2.7.

2.10   Apply KVL to find the voltage v in the circuit of Figure P2.8.

Figure P2.8

2.11   Find the power delivered by each source in the circuits of Figure P2.9.

2.12   In Figure 2.8 in the text, let $v_S = 110$ VAC and assume that the circuit wiring has a fixed resistance of 11 Ω. Plot the current i as a function of the resistance of a light bulb.

2.13   Find the power dissipated by the resistor in the circuit of Figure P2.10.

(a)

(b)

Figure P2.9

Figure P2.10   **Figure P2.11**

2.14   Determine which element in the circuit of Figure P2.11 is supplying power and which is dissipating power. Also determine the amount of power dissipated and supplied.

2.15   Determine which elements in the circuit of Figure P2.12 are supplying power and which are dissipating power. Also determine the amount of power dissipated and supplied.

Figure P2.12

2.16   Determine which elements in the circuit of Figure P2.13 are supplying power and which are dissipating power. Also determine the amount of power dissipated and supplied.

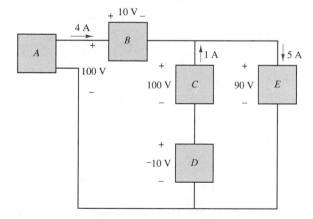

Figure P2.13

2.17 Determine which elements in the circuit of Figure P2.14 are supplying power and which are dissipating power. Also determine the amount of power dissipated and supplied.

Figure P2.14

Figure P2.15

2.18 Suppose one of the two headlights in Example 2.2 has been replaced with the wrong part and the 12-V battery is now connected to a 75-W and a 50-W headlight. What is the resistance of each headlight, and what is the total resistance seen by the battery?

2.19 What is the equivalent resistance seen by the battery of Example 2.2 if two 10-W taillights are added to the circuit?

2.20 Find the current i and the power delivered to the resistor in the circuit of Figure P2.15.

2.21 In the circuits of Figure P2.16, the directions of current and polarities of voltage have already been defined. Find the actual values of the indicated currents and voltages.

2.22 In the circuit of Figure P2.17, if $v_1 = v/8$ and the power delivered by the source is 8 mW, find R, v, v_1, and i.

Figure P2.17

Figure P2.18

Figure P2.16

2.23 Use Kirchhoff's current law to determine the current in each of the 30-Ω resistors in the circuit of Figure P2.18.

2.24 In the circuit of Figure P2.19, the direction of current and polarities of voltage have already been defined. Find the actual values of the indicated currents and voltages.

2.25 Use Kirchhoff's voltage law to determine the voltage across each of the resistors in the circuit of Figure P2.20.

Figure P2.19

Figure P2.20

Figure P2.21

2.26 Use Ohm's law and KVL to determine the voltage V_1 in the circuit of Figure P2.21.

2.27 Use Ohm's law and KCL to determine the current I_1 in the circuit of Figure P2.22.

Figure P2.22

2.28 The voltage divider network of Figure P2.23 is expected to provide 2.5 V at the output. The resistors, however, may not be exactly the same; that is, their tolerances are such that the resistances may not be exactly 10 kΩ.

a.   If the resistors have ± 10 percent tolerance, find the worst-case output voltages.
b.   Find these voltages for tolerances of ± 5 percent.

2.29 For the circuits of Figure P2.24, determine the resistor values (including the power rating) necessary to achieve the indicated voltages.
        Resistors are available in ⅛-, ¼-, ½-, and 1-W ratings.

Figure P2.23

Figure P2.24

Figure P2.25

2.30 For the circuit shown in Figure P2.25, find

a. The equivalent resistance seen by the source.

b. The current, i.

c. The power delivered by the source.

d. The voltages, v_1, v_2.

e. The minimum power rating required for R_1.

2.31 Find the equivalent resistance of the circuit of Figure P2.26 by combining resistors in series and in parallel.

Figure P2.26

Figure P2.27

2.32 Find the equivalent resistance seen by the source in Figure P2.27, and use the result to find i, i_1, and v.

2.33 Find the equivalent resistance seen by the source and the current i in the circuit of Figure P2.28.

Figure P2.28

Figure P2.29

2.34 In the circuit of Figure P2.29, the power absorbed by the 15-Ω resistor is 15 W. Find R.

2.35 In the circuit of Figure P2.30, find the equivalent resistance, where $R_1 = 5\ \Omega$, $R_2 = R_6 = 1\ k\Omega$, $R_3 = R_4 = 100\ \Omega$, and $R_5 = 9.1\ \Omega$.

Figure P2.30

2.36   In the circuit of Figure P2.31, find the equivalent resistance looking in at terminals a and b if terminals c and d are open and again if terminals c and d are shorted together. Also, find the equivalent resistance looking in at terminals c and d if terminals a and b are open and if terminals a and b are shorted together.

2.37   A *thermistor* is a device whose terminal resistance changes with the temperature of its surroundings. Its resistance is an exponential relationship:

$$R_{th}(T) = R_A e^{-\beta T}$$

where R_A is the terminal resistance at $T = 0°C$ and β is a material parameter.

a.   If $R_A = 100 \ \Omega$ and $\beta = 0.10/C°$, plot $R_{th}(T)$ versus T for $0 \le T \le 100°C$.
b.   The thermistor is placed in parallel with a resistor whose value is $100 \ \Omega$.
 i. Find an expression for the equivalent resistance.
 ii. Plot $R_{eq}(T)$ on the same plot you made in part (a).

2.38   A certain resistor has the following nonlinear characteristic:

$$R(x) = 100e^x$$

where x is a displacement (in cm). The nonlinear resistor is to be used to measure the voltage x in the circuit of Figure P2.32.

a.   If the total length of the resistor is 10 cm, find an expression for $v_{out}(x)$.
b.   If $v_{out} = 4$ V, what is the distance, x?

2.39   Find the currents i_1 and i_2 and the power delivered by the 2-A current source and the 10-V voltage source and the total power dissipated by the circuit of Figure P2.33. $R_1 = 32 \ \Omega$, $R_2 = R_3 = 6 \ \Omega$, and $R_4 = 50 \ \Omega$.

2.40   The circuit of Figure P2.34, which contains a dependent source, is a model for a transistor. Find the output voltage if $v_{in} = 0.5$ V. $R_S = 1,000 \ \Omega$, $R_b = 400 \ \Omega$, and $R_C = 200 \ \Omega$. $R_o = 980 \Omega$

correctly
done
in
test 1
review

Figure P2.31

Figure P2.32

Figure P2.33

Figure P2.34

2.41   Find the current i_o, in the circuit of Figure P2.35. Assume that $v_S = 0.2$ V, $R_S = 100 \ \Omega$, $R_1 = 10$ kΩ, $R_b = 500 \ \Omega$, $R_o = 5$ kΩ, and $R_D = 50 \ \Omega$.

Figure P2.35

2.42 Consider the circuit shown in Figure P2.36.

Figure P2.36

a. If $V_1 = 10.0$ V, $R_1 = 0.05$ Ω, and $R_L = 0.45$ Ω, find the load current I_L and the power dissipated by the load.

b. If we connect a second battery in parallel with battery 1 that has voltage $V_2 = 10$ V and $R_2 = 0.1$ Ω, will the load current I_L increase or decrease? Will the power dissipated by the load increase or decrease? By how much?

Figure P2.37

2.43 The circuit of Figure P2.37 is used to measure the internal impedance of a battery. The battery being tested is a zinc-carbon dry cell.

a. A fresh battery is being tested, and it is found that the voltage, V_{out}, is 1.64 V with the switch open and 1.63 V with the switch closed. Find the internal resistance of the battery.

b. The same battery is tested one year later, and V_{out} is found to be 1.6 V with the switch open but 0.17 V with the switch closed. Find the internal resistance of the battery.

2.44 Consider the practical ammeter, diagramed in Figure P2.38, consisting of an ideal ammeter in series with a 2-kΩ resistor. The meter sees a full-scale deflection when the current through it is 50 μA. If we wished to construct a multirange ammeter reading full-scale values of 1 mA, 10 mA, or 100 mA, depending on the setting of a rotary switch, what should R_1, R_2, and R_3 be?

Figure P2.38

2.45 A circuit that measures the internal resistance of a practical ammeter is shown in Figure P2.39, where $R_s = 10,000$ Ω, $V_s = 10$ V, and R_p is a variable resistor that can be adjusted at will.

Figure P2.39

a.   Assume that $r_a \ll 10,000 \; \Omega$. Estimate the current i.

b.   If the meter displays a current of 0.43 mA when $R_p = 7 \; \Omega$, find the internal resistance of the meter, r_a.

2.46   A practical voltmeter has an internal resistance r_m. What is the value of r_m if the meter reads 9.89 V when connected as shown in Figure P2.40?

2.47   Using the circuit of Figure P2.40, find the voltage that the meter reads if $V_s = 10$ V and R_s has the following values: $R_s = 0.1 r_m$, $0.3 r_m$, $0.5 r_m$, r_m, $3 r_m$, $5 r_m$, and $10 r_m$. How large (or small) should the internal resistance of the meter be relative to R_s?

2.48   A voltmeter is used to determine the voltage across a resistive element in the circuit of Figure P2.41. The instrument is modeled by an ideal voltmeter in parallel with a 97-kΩ resistor, as shown. The meter is placed to measure the voltage across R_3. Let $R_1 = 10$ kΩ, $R_S = 100$kΩ, $R_2 = 40$ kΩ, and $I_S = 90$ mA. Find the voltage across R_3 with and without the voltmeter in the circuit for the following values:

a. $R_3 = 100 \; \Omega$

b. $R_3 = 1 \; k\Omega$

c. $R_3 = 10 \; k\Omega$

d. $R_3 = 100 \; k\Omega$

Voltmeter

$R_S = 10 \; k\Omega$

$v_S = 10$ V

Figure P2.40

Figure P2.41

2.49 An ammeter is used as shown in Figure P2.42. The ammeter model consists of an ideal ammeter in series with a resistance. The ammeter model is placed in the branch as shown in the figure. Find the current through R_3 both with and without the ammeter in the circuit for the following values, assuming that $V_S = 10$ V, $R_S = 10$ Ω, $R_1 = 1$ kΩ, and $R_2 = 100$ Ω: (a) $R_3 = 1$ kΩ, (b) $R_3 = 100$ Ω, (c) $R_3 = 10$ Ω, (d) $R_3 = 1$ Ω.

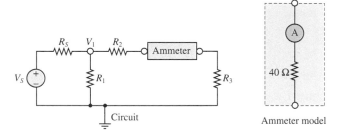

Figure P2.42

2.50 The d'Arsonval movement is a mechanism that moves on the basis of the current flowing through a meter. The movement's resistance is usually given by the voltage at which there is full-scale deflection and by the current at full-scale deflection. This movement can be used to make simple voltmeters when connected as shown in Figure P2.43. R_S is a resistance in series with the movement's internal resistance, R_m. R_S is chosen so that $V_{\text{voltmeter}}$ is equal to the desired full-scale reading when V_m is the movement's actual full-scale deflection voltage. In other words, V_m is a fraction of the external voltage, $V_{\text{voltmeter}}$. Find R_S so that the d'Arsonval movement deflects full scale when $V_{\text{voltmeter}}$ is 20 volts. V_m full scale = 0.01 V and I_m full scale = 0.5 mA.

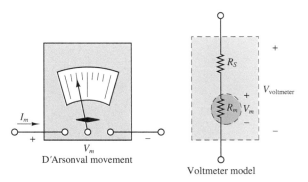

Figure P2.43

2.51 The d'Arsonval movement can be used to make a simple ammeter as shown in Figure P2.44. R_S is a resistance parallel to the movement's internal resistance, R_m. R_S is shown so that the current through R_m is equal to the full-scale deflection current while I_{ammeter} is the desired full-scale deflection current. Find R_S so that we have full-scale deflection at $I_{\text{ammeter}} = 100$ mA. V_m full scale $= 0.01$ V and I_m full scale $= 0.5$ mA.

2.52

a. With reference to Figure 2.42 in the text, find a general expression for the power internally dissipated by the wattmeter as a function of the resistors R_1 and R_2, assuming that the voltmeter may be treated as ideal but the ammeter has an internal resistance r_A.

b. Compute the percent error in the power measurement for this wattmeter if $R_1 = 50\ \Omega$, $R_2 = 10$ kΩ, and the internal ammeter resistance is 1.2 kΩ.

2.53   With reference to Figure 2.42 in the text,

a. Find a general expression for the power internally dissipated by the wattmeter as a function of the resistors R_1 and R_2, assuming that the voltmeter has an internal resistance r_V and the ammeter has an internal resistance r_A.

b. Compute the percent error in the power measurement for this wattmeter if $R_1 = 50\ \Omega$, $R_2 = 10$ kΩ, $r_A = 1.2$ kΩ, and $r_V = 1$ MΩ.

2.54   For the wattmeter of Figure 2.42, using the results of problem 2.53, generate a graph of the percent error in the power measurement as a function of R_2, assuming that R_2 can vary between 1 and 100 kΩ. Can you draw a general conclusion with regard to the accuracy of power measurements?

Figure P2.44

CHAPTER

3

Resistive Network Analysis

T his chapter will illustrate the fundamental techniques for the analysis of resistive circuits. The methods introduced are based on the circuit laws presented in Chapter 2: Kirchhoff's and Ohm's laws. The main thrust of the chapter is to introduce and illustrate various methods of circuit analysis that will be applied throughout the book.

The first topic is the analysis of resistive circuits by the methods of mesh currents and node voltages; these are fundamental techniques, which should be mastered as early as possible. The second topic is a brief introduction to the principle of superposition. Section 3.5 introduces another fundamental concept in the analysis of electrical circuits: the reduction of an arbitrary circuit to equivalent circuit forms (Thévenin and Norton equivalent circuits). In this section it will be shown that it is generally possible to reduce all linear circuits to one of two equivalent forms, and that any linear circuit analysis problem can be reduced to a simple voltage or current divider problem. The Thevenin and Norton equivalent representation of electrical circuits naturally leads to the description of electrical circuits in terms of sources and loads. This notion, in turn, leads to the analysis of the transfer of power between a source and a load, and of the phenomenon of source loading. Finally, some simple graphical and numerical techniques are introduced for the analysis of nonlinear circuit elements.

Upon completing this chapter, you should have developed confidence in your ability to compute numerical solutions for a wide range of resistive circuits. Good familiarity with the techniques illustrated in this chapter will greatly simplify the study of AC circuits in Chapter 4. The objective of the chapter is to develop a solid understanding of the following topics:

- Node voltage and mesh current analysis.
- The principle of superposition.
- Thévenin and Norton equivalent circuits.
- Numerical and graphical (load-line) analysis of nonlinear circuit elements.

3.1   THE NODE VOLTAGE METHOD

Chapter 2 introduced the essential elements of network analysis, paving the way for a systematic treatment of the analysis methods that will be introduced in this chapter. You are by now familiar with the application of the three fundamental laws of network analysis: KCL, KVL, and Ohm's law; these will be employed to develop a variety of solution methods that can be applied to linear resistive circuits. The material presented in the following sections presumes good understanding of Chapter 2. You can resolve many of the doubts and questions that may occasionally arise by reviewing the material presented in the preceding chapter.

Node voltage analysis is the most general method for the analysis of electrical circuits. In this section, its application to linear resistive circuits will be illustrated. The **node voltage method** is based on defining the voltage at each node as an independent variable. One of the nodes is selected as a **reference node** (usually—but not necessarily—ground), and each of the other node voltages is referenced to this node. Once each node voltage is defined, Ohm's law may be applied between any two adjacent nodes in order to determine the current flowing in each branch. In the node voltage method, *each branch current is expressed in terms of one or more node voltages;* thus, currents do not explicitly enter into the equations. Figure 3.1 illustrates how one defines branch currents in this method. You may recall a similar description given in Chapter 2.

Once each branch current is defined in terms of the node voltages, Kirchhoff's current law is applied at each node. The particular form of KCL employed in nodal analysis equates the sum of the currents into the node to the sum of the currents leaving the node:

$$\sum i_{\text{in}} = \sum i_{\text{out}} \tag{3.1}$$

In the node voltage method, we assign the node voltages v_a and v_b; the branch current flowing from a to b is then expressed in terms of these node voltages.

$$i = \frac{v_a - v_b}{R}$$

Figure 3.1 Branch current formulation in nodal analysis

By KCL: $i_1 = i_2 + i_3$. In the node voltage method, we express KCL by $\frac{v_a - v_b}{R_1} = \frac{v_b - v_c}{R_2} + \frac{v_b - v_d}{R_3}$.

Figure 3.2 Use of KCL in nodal analysis

Figure 3.2 illustrates this procedure.

The systematic application of this method to a circuit with n nodes would lead to writing n linear equations. However, one of the node voltages is the reference voltage and is therefore already known, since it is usually assumed to be zero (recall that the choice of reference voltage is dictated mostly by convenience, as explained in Chapter 2). Thus, we can write $n - 1$ *independent linear equations* in the $n - 1$ independent variables (the node voltages). Nodal analysis provides the minimum number of equations required to solve the circuit, since any branch voltage or current may be determined from knowledge of the node voltages. At this stage, you might wish to review Example 2.15, to verify that, indeed, knowledge of the node voltages is sufficient to solve for any other current or voltage in the circuit.

The nodal analysis method may also be defined as a sequence of steps, as outlined in the following box:

Node Voltage Analysis Method

1. Select a reference node (usually ground). All other node voltages will be referenced to this node.
2. Define the remaining $n - 1$ node voltages as the independent variables.
3. Apply KCL at each of the $n - 1$ nodes, expressing each current in terms of the adjacent node voltages.
4. Solve the linear system of $n - 1$ equations in $n - 1$ unknowns.

Following the procedure outlined in the box guarantees that the correct solution to a given circuit will be found, provided that the nodes are properly identified and KCL is applied consistently. As an illustration of the method, consider the circuit shown in Figure 3.3. The circuit is shown in two different forms to illustrate equivalent graphical representations of the same circuit.

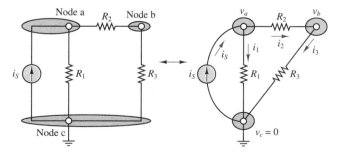

Figure 3.3 Illustration of nodal analysis

The circuit on the right leaves no question where the nodes are. The direction of current flow is selected arbitrarily (assuming that i_S is a positive current). Application of KCL at node a yields:

$$i_S = i_1 + i_2 \tag{3.2}$$

whereas, at node b,

$$i_2 = i_3 \tag{3.3}$$

It is instructive to verify (at least the first time the method is applied) that it is not necessary to apply KCL at the reference node. The equation obtained at node c,

$$i_1 + i_3 = i_S \tag{3.4}$$

is not independent of equations 3.2 and 3.3; in fact, it may be obtained by adding the equations obtained at nodes a and b (verify this, as an exercise). This observation confirms the statement made earlier:

> in a circuit containing n nodes, we can write at most $n - 1$ independent equations.

Now, in applying the node voltage method, the currents i_1, i_2, and i_3 are expressed as functions of v_a, v_b, and v_c, the independent variables. Ohm's law requires that i_1, for example, be given by

$$i_1 = \frac{v_a - v_c}{R_1} \tag{3.5}$$

since it is the potential difference, $v_a - v_c$, across R_1 that causes the current i_1 to flow from node a to node c. Similarly,

$$i_2 = \frac{v_a - v_b}{R_2}$$
$$i_3 = \frac{v_b - v_c}{R_3} \tag{3.6}$$

Substituting the expressions for the three currents in the nodal equations (equations 3.2 and 3.3), we obtain the following relationships:

$$i_S = \frac{v_a}{R_1} + \frac{v_a - v_b}{R_2} \tag{3.7}$$

$$\frac{v_a - v_b}{R_2} = \frac{v_b}{R_3} \tag{3.8}$$

Equations 3.7 and 3.8 may be obtained directly from the circuit, with a little practice. Note that these equations may be solved for v_a and v_b, assuming that i_S, R_1, R_2, and R_3 are known. The same equations may be reformulated as follows:

$$\left(\frac{1}{R_1} + \frac{1}{R_2}\right) v_a + \left(-\frac{1}{R_2}\right) v_b = i_S$$

$$\left(-\frac{1}{R_2}\right) v_a + \left(\frac{1}{R_2} + \frac{1}{R_3}\right) v_b = 0$$

(3.9)

The following examples further illustrate the application of the method.

Example   3.1

Solve for all unknown currents and voltages in the circuit of Figure 3.4, using nodal analysis.

Solution:

The circuit of Figure 3.4 is shown again in Figure 3.5, with a graphical indication of how KCL will be applied to determine the nodal equations. Note that we have selected to ground the lower part of the circuit, resulting in a reference voltage of zero at that node.

Applying KCL at node 1 yields:

$$0.01 = \frac{v_1 - 0}{1,000} + \frac{v_1 - v_2}{10,000} + \frac{v_1 - v_2}{2,000}$$

whereas at node 2 we have

$$\frac{v_1 - v_2}{10,000} + \frac{v_1 - v_2}{2,000} = \frac{v_2}{2,000} + 0.05$$

Note how each current has been expressed directly in terms of the node voltages. Since the direction of the currents has been (arbitrarily) assigned in Figure 3.5, writing the nodal equations becomes no more than a simple application of KCL and Ohm's law.

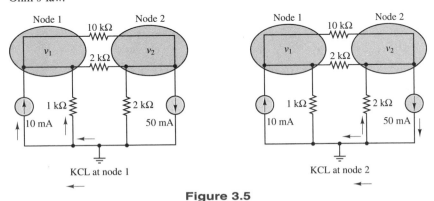

Figure 3.5

The next task is to write the equations in a more systematic fashion. From node 1, we obtain

$$\left(\frac{1}{1,000} + \frac{1}{10,000} + \frac{1}{2,000}\right) v_1 + \left(-\frac{1}{10,000} - \frac{1}{2,000}\right) v_2 = 0.01$$

Figure 3.4

while at node 2 we have

$$\left(-\frac{1}{10,000} - \frac{1}{2,000}\right) v_1 + \left(\frac{1}{10,000} + \frac{1}{2,000} + \frac{1}{2,000}\right) v_2 = -0.05$$

With some manipulation, the equations finally lead to the following form:

$$1.6v_1 - 0.6v_2 = 10$$
$$-0.6v_1 + 1.1v_2 = -50$$

These equations may be solved simultaneously to obtain

$$v_1 = -13.57 \text{ V}$$
$$v_2 = -52.86 \text{ V}$$

Knowing the node voltages, we can determine each of the branch currents and voltages in the circuit. For example, the current through the 10-kΩ resistor is given by:

$$i_{10 \text{ k}\Omega} = \frac{v_1 - v_2}{10,000} = 3.93 \text{ mA}$$

indicating that the initial (arbitrary) choice of direction for this current was opposite to the actual direction of current flow. As another example, consider the current through the 1-kΩ resistor:

$$i_{1 \text{ k}\Omega} = \frac{v_1}{1,000} = -13.57 \text{ mA}$$

In this case, the current is positive, indicating that current actually flows from node 1 to ground, as it should, since the voltage at node 1 is positive with respect to ground. You may continue the branch-by-branch analysis started in this example, to verify that the solution obtained in the example is indeed correct.

Example   3.2

Write the node voltage equations for the circuit shown in Figure 3.6. Assume that $R_1 = 1 \text{ k}\Omega$, $R_2 = 500 \ \Omega$, $R_3 = 2.2 \text{ k}\Omega$, $R_4 = 4.7 \text{ k}\Omega$, $i_a = 1 \text{ mA}$, and $i_b = 2 \text{ mA}$.

Solution:

To write the node equations, we start by selecting the reference node (step 1). Figure 3.7 illustrates that two nodes remain after the selection of the reference node. Let us label these a and b and define voltages v_a and v_b (step 2).

Figure 3.6

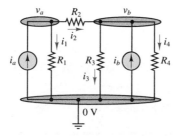

Figure 3.7

Next, we apply KCL at each of the nodes, a and b (step 3):

$$i_a = \frac{v_a}{R_1} + \frac{v_a - v_b}{R_2} \qquad \text{(node } a\text{)}$$

$$\frac{v_a - v_b}{R_2} + i_b = \frac{v_b}{R_3} + \frac{v_b}{R_4} \qquad \text{(node } b\text{)}$$

and rewrite the equations to obtain a linear system:

$$\left(\frac{1}{R_1} + \frac{1}{R_2} \right) v_a + \left(-\frac{1}{R_2} \right) v_b = i_a$$

$$\left(-\frac{1}{R_2} \right) v_a + \left(\frac{1}{R_2} + \frac{1}{R_3} + \frac{1}{R_4} \right) v_b = i_b$$

Substituting the numerical values in these equations, we get

$$3 \times 10^{-3} v_a - 2 \times 10^{-3} v_b = 1 \times 10^{-3}$$

$$-2 \times 10^{-3} v_a + 2.67 \times 10^{-3} v_b = 2 \times 10^{-3}$$

or

$$3 v_a - 2 v_b = 1$$

$$-2 v_a + 2.67 v_b = 2$$

The solution $v_a = 1.667$ V, $v_b = 2$ V may then be obtained by solving the system of equations.

Example   3.3

The system of equations generated in Example 3.2 may also be solved by using linear algebra methods, by recognizing that the system of equations can be written as:

$$\begin{bmatrix} 3 & -2 \\ -2 & 2.67 \end{bmatrix} \begin{bmatrix} v_a \\ v_b \end{bmatrix} = \begin{bmatrix} 1 \\ 2 \end{bmatrix}$$

By using Cramer's rule (see Appendix A), the solution for the two unknown variables, v_a and v_b, can be written from equation 3.12:

$$v_a = \frac{\begin{vmatrix} 1 & -2 \\ 2 & 2.67 \end{vmatrix}}{\begin{vmatrix} 3 & -2 \\ -2 & 2.67 \end{vmatrix}} = \frac{(1)(2.67) - (-2)(2)}{(3)(2.67) - (-2)(-2)} = \frac{6.67}{4} = 1.667$$

$$v_b = \frac{\begin{vmatrix} 3 & 1 \\ -2 & 2 \end{vmatrix}}{\begin{vmatrix} 3 & -2 \\ -2 & 2.67 \end{vmatrix}} = \frac{(3)(2) - (-2)(1)}{(3)(2.67) - (-2)(-2)} = \frac{8}{4} = 2$$

The result is the same as in Example 3.2.

Nodal Analysis with Voltage Sources

It would appear from the examples just shown that the node voltage method is very easily applied when current sources are present in a circuit. This is, in fact, the case, since current sources are directly accounted for by KCL. Some confusion occasionally arises, however, when voltage sources are present in a circuit analyzed by the node voltage method. In fact, the presence of voltage sources actually simplifies the calculations. To further illustrate this point, consider the circuit of Figure 3.8. Note immediately that one of the node voltages is known already! The voltage at node a is forced to be equal to

Figure 3.8 Nodal analysis with voltage sources

that of the voltage source; that is, $v_a = v_S$. Thus, only two nodal equations will be needed, at nodes b and c:

$$\frac{v_S - v_b}{R_1} = \frac{v_b}{R_2} + \frac{v_b - v_c}{R_3} \qquad \text{(node } b\text{)}$$

$$\frac{v_b - v_c}{R_3} + i_S = \frac{v_c}{R_4} \qquad \text{(node } c\text{)}$$

(3.10)

Rewriting these equations, we obtain:

$$\left(\frac{1}{R_1} + \frac{1}{R_2} + \frac{1}{R_3}\right) v_b + \left(-\frac{1}{R_3}\right) v_c = \frac{v_S}{R_1}$$

$$\left(-\frac{1}{R_3}\right) v_b + \left(\frac{1}{R_3} + \frac{1}{R_4}\right) v_c = i_S$$

(3.11)

Note how the term v_S/R_1 on the right-hand side of the first equation is really a current, as is dimensionally required by the nature of the node equations.

Example   3.4

For the circuit of Figure 3.9, find the voltages at node a and node b.

Solution:

Applying KCL at node a yields:

$$-0.002 = \frac{v_a - 0}{1,000} + \frac{v_a - v_b}{2,000}$$

whereas at node b, we have

$$\frac{v_a - v_b}{2{,}000} = \frac{v_b - 3}{3{,}000}$$

Reformulating the last two equations, we derive the following system:

$$1.5v_a - 0.5v_b = -2$$
$$-0.5v_a + 0.833v_b = 1$$

Solving the last set of the equations, we obtain the following values:

$$v_a = -1.167 \text{ V} \qquad \text{and} \qquad v_b = 0.5 \text{ V}$$

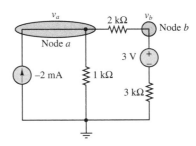

Figure 3.9

DRILL EXERCISES

3.1   Find the current i_L in the circuit shown, using the node voltage method.

Answer: 0.2857 A

3.2   Find the voltage v_x by the node voltage method for the circuit shown.

Answer: −18 V

3.3   Show that the answer to Example 3.2 is correct by applying KVL around one or more meshes.

3.2   THE MESH CURRENT METHOD

The second method of circuit analysis discussed in this chapter, which is in many respects analogous to the method of node voltages, employs **mesh currents** as the independent variables. The idea is to write the appropriate number of independent equations, using mesh currents as the independent variables. This method is not as broadly applicable as nodal analysis, because there exists a class of circuits, called nonplanar circuits, that is not amenable to analysis by mesh currents (these circuits may be solved by nodal analysis). Analysis by mesh currents consists of defining the currents around the individual meshes as the independent variables. Subsequent application of Kirchhoff's voltage law around each mesh provides the desired system of equations.

In the mesh current method, we observe that a current flowing through a resistor in a specified direction defines the polarity of the voltage across the resistor, as illustrated in Figure 3.10, and that the sum of the voltages around a closed circuit must equal zero, by KVL. Once a convention is established regarding the direction of current flow around a mesh, simple application of KVL provides the desired equation. Figure 3.11 illustrates this point.

The number of equations one obtains by this technique is equal to the number of meshes in the circuit. All branch currents and voltages may subsequently be obtained from the mesh currents, as will presently be shown. Since meshes are easily identified in a circuit, this method provides a very efficient and systematic procedure for the analysis of electrical circuits. The following box outlines the procedure used in applying the mesh current method to a linear circuit.

The current i, defined as flowing from left to right, establishes the polarity of the voltage across R.

Figure 3.10 Basic principle of mesh analysis

Once the direction of current flow has been selected, KVL requires that $v_1 = v_2 + v_3$.

A mesh

Figure 3.11 Use of KVL in mesh analysis

Mesh Current Analysis Method

1. Define each mesh current consistently. We shall always define mesh currents clockwise, for convenience.
2. Apply KVL around each mesh, expressing each voltage in terms of one or more mesh currents.
3. Solve the resulting linear system of equations with mesh currents as the independent variables.

In mesh analysis, it is important to be consistent in choosing the direction of current flow. To avoid confusion in writing the circuit equations, mesh currents will be defined exclusively clockwise when we are using this method. To illustrate the mesh current method, consider the simple two-mesh circuit shown in Figure 3.12. This circuit will be used to generate two equations in the two unknowns, the mesh currents i_1 and i_2. It is instructive to first consider each mesh by itself. Beginning with mesh 1, note that the

Figure 3.12 A two-mesh circuit

Mesh 1: KVL requires that
$v_S = v_1 + v_2$, where $v_1 = i_1R_1$,
$v_2 = (i_1 - i_2)R_2$.

Figure 3.13 Assignment
of currents and voltages around
mesh 1

voltages around the mesh have been assigned in Figure 3.13 according to the direction of the mesh current, i_1. Recall that as long as signs are assigned consistently, an arbitrary direction may be assumed for any current in a circuit; if the resulting numerical answer for the current is negative, then the chosen reference direction is opposite to the direction of actual current flow. Thus, one need not be concerned about the actual direction of current flow in mesh analysis, once the directions of the mesh currents have been assigned. The correct solution will result, eventually.

According to the sign convention, then, the voltages v_1 and v_2 are defined as shown in Figure 3.13. Now, it is important to observe that while mesh current i_1 is equal to the current flowing through resistor R_1 (and is therefore also the branch current through R_1), it is not equal to the current through R_2. The branch current through R_2 is the difference between the two mesh currents, $i_1 - i_2$. Thus, since the polarity of the voltage v_2 has already been assigned, according to the convention discussed in the previous paragraph, it follows that the voltage v_2 is given by:

$$v_2 = (i_1 - i_2)R_2 \tag{3.12}$$

Finally, the complete expression for mesh 1 is:

$$v_S = i_1R_1 + (i_1 - i_2)R_2 \tag{3.13}$$

Mesh 2: KVL requires that

$$0 = v_2 + v_3 + v_4,$$

where

$$v_2 = (i_2 - i_1)R_2,$$

$$v_3 = i_2R_3,$$

$$v_4 = i_2R_4$$

Figure 3.14 Assignment
of currents and voltages around
mesh 2

The same line of reasoning applies to the second mesh. Figure 3.14 depicts the voltage assignment around the second mesh, following the clockwise direction of mesh current i_2. The mesh current i_2 is also the branch current through resistors R_3 and R_4; however, the current through the resistor that is shared by the two meshes, R_2, is now equal to $(i_2 - i_1)$, and the voltage across this resistor is

$$v_2 = (i_2 - i_1)R_2 \tag{3.14}$$

and the complete expression for mesh 2 is

$$0 = (i_2 - i_1)R_2 + i_2R_3 + i_2R_4 \tag{3.15}$$

Why is the expression for v_2 obtained in equation 3.14 different from equation 3.12? The reason for this apparent discrepancy is that the voltage assignment for each mesh was dictated by the (clockwise) mesh current. Thus, since the mesh currents flow through R_2 in opposing directions, the voltage assignments for v_2 in the two meshes will also be opposite. This is perhaps a potential source of confusion in applying the mesh current method; you should be very careful to carry out the assignment of the voltages around each mesh separately.

Combining the equations for the two meshes, we obtain the following system of equations:

$$\begin{aligned}(R_1 + R_2)i_1 - R_2i_2 &= v_S \\ -R_2i_1 + (R_2 + R_3 + R_4)i_2 &= 0\end{aligned} \tag{3.16}$$

These equations may be solved simultaneously to obtain the desired solution, namely, the mesh currents, i_1 and i_2. You should verify that knowledge of the mesh currents permits determination of all the other voltages and currents in the circuit. The following examples further illustrate some of the details of this method.

Example   3.5

In this example, we analyze a simple two-mesh circuit, shown in Figure 3.15, by means of the mesh current method.

Solution:

The circuit of Figure 3.15 will yield two equations in two unknowns, i_1 and i_2. It is instructive to consider each mesh separately in writing the mesh equations; to this end, Figure 3.16 depicts the appropriate voltage assignments around the two meshes, based on the assumed directions of the mesh currents. From Figure 3.16, we write the mesh equations:

$$5i_1 + 10(i_1 - i_2) = 10 - 9$$
$$5i_2 + 5i_2 + 10(i_2 - i_1) = 9 - 1$$

Rearranging the linear system of the equation, we obtain

$$15i_1 - 10i_2 = 1$$
$$-10i_1 + 20i_2 = 8$$

which can be solved to obtain i_1 and i_2:

$$i_1 = 0.5 \text{ A} \qquad \text{and} \qquad i_2 = 0.65 \text{ A}$$

Figure 3.15

Analysis of mesh 1   Analysis of mesh 2

Figure 3.16

Example   3.6

Write the mesh equations for the circuit of Figure 3.17.

Solution:

Starting from mesh 1, we apply KVL using the direction of mesh current i_1 as our reference for assignment voltages:

$$12 = 3(i_1 - i_3) + 8(i_1 - i_2)$$

For mesh 2, we obtain:

$$6 = 6(i_2 - i_3) + 8(i_2 - i_1)$$

and, finally, for mesh 3:

$$0 = 3(i_3 - i_1) + 6(i_3 - i_2) + 4i_3$$

These equations can be rearranged in standard form to obtain

$$(3 + 8)i_1 - 8i_2 - 3i_3 = 12$$
$$-8i_1 + (6 + 8)i_2 - 6i_3 = 6$$
$$-3i_1 - 6i_2 + (3 + 6 + 4)i_3 = 0$$

You may verify that KCL holds at any one of the nodes, as a test to check that the answer is indeed correct.

Figure 3.17

A comparison of this result with the analogous result obtained by the node voltage method reveals that we are using Ohm's law in conjunction with KVL (in contrast with the use of KCL in the node voltage method) to determine the minimum set of equations required to solve the circuit.

Mesh Analysis with Current Sources

Mesh analysis is particularly effective when applied to circuits containing voltage sources exclusively; however, it may be applied to mixed circuits, containing both voltage and current sources, if care is taken in identifying the proper current in each mesh. The method is illustrated by solving the circuit shown in Figure 3.18. The first observation in analyzing the circuit is that the presence of the current source requires that the following relationship hold true:

$$i_1 - i_2 = 2 \text{ A} \tag{3.17}$$

If the unknown voltage across the current source is labeled v_X, application of KVL around mesh 1 yields:

$$10 = 5i_1 + v_X \tag{3.18}$$

Figure 3.18 Mesh analysis with current sources

while KVL around mesh 2 dictates that

$$v_X = 2i_2 + 4i_2 \qquad\qquad (3.19)$$

Substituting equation 3.19 in equation 3.18, and using equation 3.17, we can then obtain the system of equations

$$5i_1 + 6i_2 = 10$$
$$-i_1 + i_2 = -2 \qquad\qquad (3.20)$$

which we can solve to obtain

$$i_1 = 2 \text{ A}$$
$$i_2 = 0 \text{ A} \qquad\qquad (3.21)$$

Note also that the voltage across the current source may be found by using either equation 3.18 or equation 3.19; for example, using equation 3.19,

$$v_X = 6i_2 = 0 \text{ V} \qquad\qquad (3.22)$$

The following example further illustrates the solution of this type of circuit.

Example   3.7

Solve the circuit of Figure 3.19, using mesh analysis.

Figure 3.19

Solution:

Starting from mesh 1, we can immediately see that the current source forces

$$i_1 = 0.5 \text{ A}$$

There is no need to write any further equations around mesh 1, since the value of mesh current i_1 is already known. In this respect, the current source has in fact simplified the problem! Applying KVL around mesh 2, we obtain the following equation:

$$6(i_2 - i_3) + 8(i_2 - i_1) = 6$$

while around mesh 3 we can write

$$3(i_3 - i_1) + 6(i_3 - i_2) + 4i_3 = 0$$

Rearranging the equations and substituting the known value of i_1, we obtain a system of two equations in two unknowns:

$$14i_2 - 6i_3 = 10$$
$$-6i_2 + 13i_3 = 1.5$$

which can be solved to obtain

$$i_2 = 0.95 \text{ A} \qquad i_3 = 0.55 \text{ A}$$

As usual, you should verify that the solution is correct by applying KCL.

DRILL EXERCISES

3.4 Find the unknown voltage, v_X, by mesh current analysis in the circuit of Figure 3.20.

Answer: 5 V

3.5 Find the unknown current I_X using mesh current methods in the circuit of Figure 3.21.

Figure 3.21

Answer: 2 A

3.6 Show that the answer given in Example 3.6 is correct, by applying KCL at each node.

Figure 3.20

3.3 NODAL AND MESH ANALYSIS WITH CONTROLLED SOURCES

The methods just described also apply, with relatively minor modifications, in the presence of dependent (controlled) sources. Solution methods that allow for the presence of controlled sources will be particularly useful in the study of *transistor amplifiers* in Chapter 8. Recall from the discussion in Section 2.3 that a dependent source is a source that generates a voltage or current that depends on the value of another voltage or current in the circuit. When a dependent source is present in a circuit to be analyzed by node or mesh analysis, one can initially treat it as an ideal source and write the node or mesh equations accordingly. In addition to the equation obtained in this fashion, there will also be an equation relating the dependent source to one of the circuit voltages or currents. This **constraint equation** can then be substituted in the set of equations obtained by the techniques of nodal and mesh analysis, and the equations can subsequently be solved for the unknowns.

It is important to remark that once the constraint equation has been substituted in the initial system of equations, the number of unknowns remains unchanged. Consider, for example, the circuit of Figure 3.22, which

Figure 3.22 Circuit with dependent source

is a simplified model of a bipolar transistor amplifier (transistors will be introduced in Chapter 8). In the circuit of Figure 3.22, two nodes are easily recognized, and therefore nodal analysis is chosen as the preferred method. Applying KCL at node 1, we obtain the following equation:

$$i_S = v_1 \left(\frac{1}{R_S} + \frac{1}{R_b} \right) \tag{3.23}$$

KCL applied at the second node yields:

$$-75i_b = \frac{v_2}{R_C} \tag{3.24}$$

Next, it should be observed that the current i_b can be determined by means of a simple current divider:

$$i_b = i_S \frac{1/R_b}{1/R_b + 1/R_S} = i_S \frac{R_S}{R_b + R_S} \tag{3.25}$$

and therefore a system of two equations is obtained:

$$
\begin{aligned}
i_S &= v_1 \left(\frac{1}{R_S} + \frac{1}{R_b} \right) \\
-75i_S \frac{R_S}{R_b + R_S} &= \frac{v_2}{R_C}
\end{aligned}
\tag{3.26}
$$

which can be used to solve for v_1 and v_2. Note that, in this particular case, the two equations are independent of each other. The following example illustrates a case in which the resulting equations are not independent.

Figure 3.23

Example   3.8

Solve for the unknown mesh currents in the circuit of Figure 3.23. The dependent source v_X depends on the mesh current i_2 according to the relation $v_X = 0.5i_2$.

Solution:

According to the solution technique proposed in the opening of this section, we first write the conventional mesh equations, treating v_X as an ideal independent source:

$$v_X - 6 = 5i_1 \qquad \text{(mesh 1)}$$
$$6 = (2 + 4)i_2 \qquad \text{(mesh 2)}$$

We then recognize that $v_X = 0.5i_2$ and substitute this quantity in the first equation:

$$0.5i_2 - 6 = 5i_1 \qquad \text{(mesh 1)}$$

From the equation for mesh 2, we obtain

$$i_2 = 1 \text{ A}$$

and can the solve for i_1 from the equation for mesh 1:

$$i_1 = \frac{-5.5}{5} = -1.1 \text{ A}$$

We may verify that the solution is correct by applying KVL around the first mesh:

$$v_x - 6 = 5i_1$$
$$0.5i_2 - 6 = 5i_1$$
$$0.5(1) - 6 = 5(-1.1)$$
$$-5.5 = -5.5$$

showing that the solution is consistent.

Remarks on Node Voltage and Mesh Current Methods

The techniques presented in this section and the two preceding sections find use more generally than just in the analysis of resistive circuits. These methods should be viewed as general techniques for the analysis of any linear circuit; they provide systematic and effective means of obtaining the minimum number of equations necessary to solve a network problem. Since these methods are based on the fundamental laws of circuit analysis, KVL and KCL, they also apply to any electrical circuit, even circuits containing nonlinear circuit elements, such as those to be introduced later in this chapter.

You should master both methods as early as possible. Proficiency in these circuit analysis techniques will greatly simplify the learning process for more advanced concepts.

DRILL EXERCISES

3.7 The current source i_X is related to the voltage v_X in Figure 3.24 by the relation

$$i_X = \frac{v_X}{3}$$

Find the voltage across the 8-Ω resistor by nodal analysis.

Answer: 12 V

Figure 3.24   Figure 3.25

3.8 Find the unknown current i_x in Figure 3.25 using the mesh current method. The dependent voltage source is related to the current i_{12} through the 12-Ω resistor by $v_X = 2i_{12}$.

Answer: 1.39 A

3.4   THE PRINCIPLE OF SUPERPOSITION

This brief section discusses a concept that is frequently called upon in the analysis of linear circuits. Rather than a precise analysis technique, like the mesh current and node voltage methods, the principle of superposition is a conceptual aid that can be very useful in visualizing the behavior of a circuit containing multiple energy sources. The principle of superposition applies exclusively to linear circuits, and may be stated as follows:

In a linear circuit containing N sources, each branch voltage and current is the sum of N voltages and currents each of which may be computed by setting all but one source equal to zero and solving the circuit containing that single source.

An elementary illustration of the concept may easily be obtained by simply considering a circuit with two sources connected in series, as shown in Figure 3.26.

The net current through R is the sum of the individual source currents: $i = i_{B1} + i_{B2}$.

Figure 3.26 The principle of superposition

The circuit of Figure 3.26 is more formally analyzed as follows. The current, i, flowing in the circuit on the left-hand side of Figure 3.27 may be expressed as:

$$i = \frac{v_{B1} + v_{B2}}{R} = \frac{v_{B1}}{R} + \frac{v_{B2}}{R} = i_{B1} + i_{B2} \qquad (3.27)$$

Figure 3.26 also depicts the circuit as being equivalent to the combined effects of two circuits, each containing a single source. In each of the two subcircuits, a short circuit has been substituted for the missing battery. This should appear as a sensible procedure, since a short circuit—by definition—will always "see" zero voltage across itself, and therefore this procedure is equivalent to "zeroing" the output of one of the voltage sources.

If, on the other hand, one wished to cancel the effects of a current source, it would stand to reason that an open circuit could be substituted for the current source, since an open circuit is by definition a circuit element through which no current can flow (and which will therefore generate zero current). These basic principles are used frequently in the analysis of circuits, and are summarized in Figure 3.27.

1. In order to set a voltage source equal to zero, we replace it with a short circuit.

A circuit   The same circuit with $v_S = 0$

2. In order to set a current source equal to zero, we replace it with an open circuit.

A circuit   The same circuit with $i_S = 0$

Figure 3.27 Zeroing voltage and current sources

The principle of superposition can easily be applied to circuits containing multiple sources and is sometimes an effective solution technique. More often, however, other methods result in a more efficient solution. Example 3.9 further illustrates the use of superposition to analyze a simple network. The drill exercises at the end of the section illustrate the fact that superposition is often a cumbersome solution method.

Example   3.9

Use the circuits of Figure 3.27 and the principle of superposition to determine the current through resistor R_2. Assume that $v_S = 6$ V, $i_S = 0.5$ A, $R_1 = 50$ Ω, and $R_2 = 100$ Ω.

Solution:

The current through resistor R_2 in the circuit of Figure 3.27 is due to the superposition of the effects of the voltage and current sources. Let the notation i_{i2} and i_{v2} denote the currents through resistor R_2 due to the current and voltage sources, respectively. Then

$$i_{i2} = 0.5 \times \frac{R_1}{R_1 + R_2} = (0.5)\frac{1}{3} = \frac{1}{6} \text{ A}$$

by the current divider rule, and

$$i_{v2} = 6 \times \frac{R_2}{R_1 + R_2} \times \frac{1}{R_2} = 6 \times \frac{100}{150} \times \frac{1}{100} = \frac{1}{25} \text{ A}$$

by the voltage divider rule and Ohm's law. Finally, by the principle of superposition, the current through resistor R_2, i_2, is given by

$$i_2 = i_{i2} + i_{v2} = \frac{1}{6} + \frac{1}{25} = 0.207 \text{ A}$$

You should verify that this solution is correct by using nodal or mesh analysis.

DRILL EXERCISES

3.9   Find the voltages v_a and v_b for the circuits of Example 3.4 by superposition.

3.10   Repeat Drill Exercise 3.2, using superposition. This exercise illustrates that superposition is not necessarily a computationally efficient solution method.

3.11   Solve Example 3.5, using superposition.

3.12   Solve Example 3.7, using superposition.

3.5   ONE-PORT NETWORKS AND EQUIVALENT CIRCUITS

You may recall that in the discussion of ideal sources in Chapter 2, the flow of energy from a source to a load was described in a very general form, by showing the connection of two "black boxes" labeled source and load (see Figure 2.10). In the same figure, two other descriptions were shown: a symbolic one, depicting an ideal voltage source and an ideal resistor; and a physical representation, in which the load was represented by a headlight and the source by an automotive battery. Whatever the form chosen for source-load representation, each block—source or load—may be viewed as a two-terminal device, described by an *i-v* characteristic. This general circuit representation is shown in Figure 3.28. This representation is called **one-port network** and is particularly useful for introducing the notion of equivalent circuits. Note that the network of Figure 3.28 is completely described by its *i-v* characteristic; this point is best illustrated by an example.

Figure 3.28 One-port network

Example   3.10

Consider the circuit shown in Figure 3.29. If we wished to determine, say, the current demand placed on the voltage source by the resistive network, we would compute the value of the current provided by the source, *i*, according to the following expression:

$$i = \frac{v_S}{1/(1/R_1 + 1/R_2 + 1/R_3)}$$

Figure 3.29 Illustration of equivalent-circuit concept

It should be apparent that *insofar as the source is concerned,* it is irrelevant whether the load consists of three resistors in parallel or of a single equivalent resistance given by the expression

$$R_{eq} = \frac{1}{1/R_1 + 1/R_2 + 1/R_3}$$

In either case, we may represent the load by means of a single resistor of value R_{eq}, as shown in Figure 3.30.

Similarly, if the voltage source consists of a 6-V battery, *insofar as the load is concerned* it is irrelevant whether the source circuit consists of a single 6-V battery or of four 1.5-V batteries connected in series.

Load circuit

Equivalent load circuit

Figure 3.30 Equivalent load resistance concept

For the remainder of this chapter, we shall focus on developing techniques for computing equivalent representations of linear networks. Such representations will be useful in deriving some simple—yet general—results for linear circuits, as well as analyzing simple nonlinear circuits.

Thévenin and Norton Equivalent Circuits

This section discusses one of the most important topics in the analysis of electrical circuits: the concept of an **equivalent circuit.** It will be shown that it is always possible to view even a very complicated circuit in terms of much simpler *equivalent* source and load circuits, and that the transformations leading to equivalent circuits are easily managed, with a little practice. In studying node voltage and mesh current analysis, you may have observed that there is a certain correspondence (called **duality**) between current sources and voltage sources, on the one hand, and parallel and series circuits, on the other. This duality appears again very clearly in the analysis of equivalent circuits: it will shortly be shown that equivalent circuits fall into one of two classes, involving either voltage or current sources and (respectively) either series or parallel resistors, reflecting this same principle of duality. The discussion of equivalent circuits begins with the statement of two very important theorems, summarized in Figures 3.31 and 3.32.

Figure 3.31 Illustration of Thévenin theorem

Figure 3.32 Illustration of Norton theorem

The Thévenin Theorem

As far as a load is concerned, any network composed of ideal voltage and current sources, and of linear resistors, may be represented by an equivalent circuit consisting of an ideal voltage source, v_T, in series with an equivalent resistance, R_T.

The Norton Theorem

As far as a load is concerned, any network composed of ideal voltage and current sources, and of linear resistors, may be represented by an equivalent circuit consisting of an ideal current source, i_N, in parallel with an equivalent resistance, R_N.

The first obvious question to arise is, how are these equivalent source voltages, currents, and resistances computed? The next few sections illustrate the computation of these equivalent circuit parameters, mostly through examples. A substantial number of drill exercises are also provided, with the following caution: the only way to master the computation of Thévenin and Norton equivalent circuits is by patient repetition.

Determination of Norton or Thévenin Equivalent Resistance

Complete circuit

Circuit with load removed for computation of R_T. The voltage source is replaced by a short circuit.

Figure 3.33 Computation of Thévenin resistance

The first step in computing a Thévenin or Norton equivalent circuit consists of finding the equivalent resistance presented by the circuit at its terminals. This is done by setting all sources in the circuit equal to zero and computing the effective resistance between terminals. The voltage and current sources present in the circuit are set to zero by the same technique used with the principle of superposition: voltage sources are replaced by short circuits, current sources by open circuits. To illustrate the procedure, consider the simple circuit of Figure 3.33; the objective is to compute the equivalent resistance the load R_L "sees" at port a-b.

In order to compute the equivalent resistance, we remove the load resistance from the circuit and replace the voltage source, v_S, by a short circuit. At this point—seen from the load terminals—the circuit appears as shown in Figure 3.34. You can see that R_1 and R_2 are in parallel, since

Figure 3.34 Equivalent resistance seen by the load

they are connected between the same two nodes. If the total resistance between terminals a and b is denoted by R_T, its value can be determined as follows:

$$R_T = R_3 + R_1 \parallel R_2 \tag{3.28}$$

An alternative way of viewing R_T is depicted in Figure 3.35, where a hypothetical 1-A current source has been connected to the terminals a and b. The voltage v_X appearing across the a-b pair will then be numerically equal to R_T (only because $i_S = 1$ A!). With the 1-A source current flowing in the circuit, it should be apparent that the source current encounters R_3 as a resistor in series with the parallel combination of R_1 and R_2, prior to completing the loop.

   Summarizing the procedure, we can produce a set of simple rules as an aid in the computation of the Thévenin (or Norton) equivalent resistance for a linear resistive circuit:

Computation of Equivalent Resistance of a One-Port Network

1. Remove the load.
2. Zero all voltage and current sources.
3. Compute the total resistance between load terminals, *with the load removed*. This resistance is equivalent to that which would be encountered by a current source connected to the circuit in place of the load.

What is the total resistance the current i_S will encounter in flowing around the circuit?

$$R_T = R_1 \| R_2 + R_3$$

Figure 3.35 An alternative method for determining the Thévenin resistance

We note immediately that this procedure yields a result that is independent of the load. This is a very desirable feature, since once the equivalent resistance has been identified for a source circuit, the equivalent circuit remains unchanged if we connect a different load. The following examples further illustrate the procedure.

Example   3.11

Find the Thévenin equivalent resistance seen by the load resistance, R_L, in the circuit of Figure 3.36.

Figure 3.36

Solution:

After replacing the current source with an open circuit, the circuit appears as shown in Figure 3.37. Looking into terminals a and b, we then give the equivalent resistance by

$$R_T = ((20 \| 20) + 10) \| 20 + 10$$

or

$$R_T = 20 \ \Omega$$

10 Ω   10 Ω   a

20 Ω   20 Ω   20 Ω

b

Figure 3.37

Example   3.12

For the circuit shown in Figure 3.38, find the Thévenin equivalent resistance seen by the load resistance, R_L.

Figure 3.38

Solution:

Figure 3.39

First we replace the voltage source with a short circuit and the current source with an open circuit. The circuit now has the appearance of Figure 3.39. Looking into terminals a and b, we then give the equivalent resistance by

$$R_T = ((2 \,\|\, 2) + 1) \,\|\, 2$$

or

$$R_T = 1 \ \Omega$$

As a final note, it should be remarked that the Thévenin and Norton equivalent resistances are one and the same quantity:

$$R_T = R_N \tag{3.29}$$

Therefore, the preceding discussion holds whether we wish to compute a Norton or a Thévenin equivalent circuit. From here on we shall use the notation R_T exclusively, for both Thévenin and Norton equivalents. Drill Exercise 3.13 will give you an opportunity to explain why the two equivalent resistances are one and the same.

DRILL EXERCISES

3.13 Apply the methods described in this section to show that $R_T = R_N$ in the circuits of Figure 3.40.

Figure 3.40

3.14 Find the Thévenin resistance of the circuit of Figure 3.41 seen by the load resistor, R_L.

Figure 3.41

Answer: $R_T = 2.5$ k Ω

3.15 Find the Thévenin resistance seen by the load resistor, R_L, in the circuit of Figure 3.42.

Figure 3.42

Answer: $R_T = 7$ Ω

3.16 For the circuit of Figure 3.43, find the Thévenin resistance seen by the load resistor, R_L.

Figure 3.43

Answer: $R_T = 4.0$ kΩ

3.17   For the circuit of Figure 3.44, find the Thévenin resistance seen by the load resistor, R_L.

Figure 3.44

Answer: $R_T = 7.06 \ \Omega$

Computing the Thévenin Voltage

This section describes the computation of the Thévenin equivalent voltage, v_T, for an arbitrary linear resistive circuit. The Thévenin equivalent voltage is defined as follows:

> The equivalent (Thévenin) source voltage is equal to the **open circuit voltage** present at the load terminals with the load removed.

Figure 3.45 Equivalence of open-circuit and Thévenin voltage

This states that in order to compute v_T, it is sufficient to remove the load and to compute the open-circuit voltage at the one-port terminals. Figure 3.45 illustrates that the open-circuit voltage, v_{OC}, and the Thévenin voltage, v_T, must be the same if the Thévenin theorem is to hold. This is true because in the circuit consisting of v_T and R_T, the voltage v_{OC} must equal v_T, since no current flows through R_T and therefore the voltage across R_T is zero. Kirchhoff's voltage law confirms that

$$v_T = R_T(0) + v_{OC} = v_{OC} \tag{3.30}$$

The actual computation of the open-circuit voltage is best illustrated by examples; there is no substitute for practice in becoming familiar with these computations. To summarize the main points in the computation of open-circuit voltages, consider the circuit of Figure 3.33, shown again in Figure 3.46 for convenience. Recall that the equivalent resistance of this circuit was given by $R_T = R_3 + R_1 \| R_2$. To compute v_{OC}, we disconnect the load, as shown in Figure 3.47, and immediately observe that no current flows through R_3, since there is no closed circuit connection at that branch. Therefore, v_{OC} must be equal to the voltage across R_2, as illustrated in Figure 3.48. Since the only closed circuit is the mesh consisting of v_S, R_1, and R_2,

Figure 3.46

the answer we are seeking may be obtained by means of a simple voltage divider:

$$v_{OC} = v_{R2} = v_S \frac{R_2}{R_1 + R_2}$$

It is instructive to review the basic concepts outlined in the example by considering the original circuit and its Thévenin equivalent side by side, as shown in Figure 3.49. The two circuits of Figure 3.49 are equivalent in the sense that the current drawn by the load, i_L, is the same in both circuits, that current being given by:

$$i_L = v_S \cdot \frac{R_2}{R_1 + R_2} \cdot \frac{1}{(R_3 + R_1 \| R_2) + R_L} = \frac{v_T}{R_T + R_L} \tag{3.31}$$

Figure 3.47

Figure 3.48

A circuit   Its Thévenin equilalent

Figure 3.49 A circuit and its Thévenin equivalent

The computation of Thévenin equivalent circuits is further illustrated in the following examples.

Example   3.13

Compute the open-circuit voltage for the circuit of Figure 3.44.

Solution:

With the load open-circuited, the circuit becomes a two-mesh circuit, as shown in Figure 3.50. The open-circuit voltage, v_{OC}, is equal to the voltage across the 20-Ω resistor and is therefore given by the expression

$$v_{OC} = 20i_2$$

Writing the mesh equations for the circuit, we obtain

$$(10 + 1)i_1 - 10i_2 = 12$$
$$-10i_1 + 40i_2 = 0$$

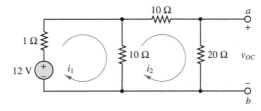

Figure 3.50

or, in matrix form,

$$\begin{bmatrix} 11 & -10 \\ -10 & 40 \end{bmatrix} \begin{bmatrix} i_1 \\ i_2 \end{bmatrix} = \begin{bmatrix} 12 \\ 0 \end{bmatrix}$$

The system of equations can be solved for i_2 (for example, using Cramer's rule—see Appendix A) to obtain

$$i_2 = 0.3529 \text{ A}$$

or

$$v_{OC} = 20i_2 = 7.06 \text{ V}$$

Example   3.14

In Figure 3.51, replace the network to the left of terminals a and b by its Thévenin equivalent circuit, and use the result to find the load current, i.

Figure 3.51

Solution:

First, we replace the voltage source with a short circuit and the current source by an open circuit. The circuit appears as shown in Figure 3.52. Looking into terminals a and b, we then give the equivalent resistance by

$$R_T = 12 \,\|\, 4 = 3 \ \Omega$$

Figure 3.52

For the open-circuit voltage, we can immediately observe that v_{OC} must be equal to the voltage across the 12-Ω resistor. Mesh analysis can be applied to find v_{OC} as shown in Figure 3.53. Applying KVL to meshes 1 and 2, we obtain

$$4i_1 - 24 + v_1 = 0$$
$$12i_2 - v_1 = 0$$

Figure 3.53

Also, we observe that the current source constrains the mesh currents to satisfy the relationship

$$i_2 - i_1 = 3$$

Rearranging the three equations, we can formulate a system of two equations in two unknowns:

$$4i_1 + 12i_2 = 24$$
$$i_2 - i_1 = 3$$

Solving the two equations, we obtain the following values for the currents i_1 and i_2:

$$i_1 = -\frac{3}{4} \text{ A}$$
$$i_2 = \frac{9}{4} \text{ A}$$

Observing that the open-circuit voltage is actually equal to the voltage across the 12-Ω resistor, we find the Thévenin equivalent voltage to be

$$v_T = v_{OC} = 12i_2 = 27 \text{ V}$$

Finally, to determine the load current, i, we connect the load to the Thévenin equivalent circuit, as shown in Figure 3.54, and compute

$$i = \frac{v_{OC}}{R_T + 6} = \frac{27}{3 + 6} = 3 \text{ A}$$

Thévenin equivalent

Figure 3.54

DRILL EXERCISES

3.18 With reference to Figure 3.46, find the load current, i_L, by mesh analysis, if $v_S = 10$ V, $R_1 = R_3 = 50$ Ω, $R_2 = 100$ Ω, $R_L = 150$ Ω.

> Answer: $i_L = 0.02857$ A

3.19 Find the Thévenin equivalent circuit seen by the load resistor, R_L, for the circuit of Figure 3.55.

Figure 3.55

> Answer: $R_T = 30$ Ω; $v_{OC} = v_T = 5$ V

3.20 Find the Thévenin equivalent circuit for the circuit of Figure 3.56.

Figure 3.56

Answer: $R_T = 10\ \Omega$; $v_{OC} = v_T = 0.704$ V

Computing the Norton Current

The computation of the Norton equivalent current is very similar in concept to that of the Thévenin voltage. The following definition will serve as a starting point:

Definition

The Norton equivalent current is equal to the **short-circuit current** that would flow were the load replaced by a short circuit.

An explanation for the definition of the Norton current is easily found by considering, again, an arbitrary one-port network, as shown in Figure 3.57, where the one-port network is shown together with its Norton equivalent circuit.

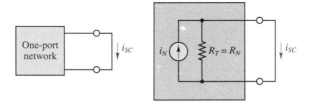

Figure 3.57 Illustration of Norton equivalent circuit

It should be clear that the current, i_{SC}, flowing through the short circuit replacing the load is exactly the Norton current, i_N, since all of the source current in the circuit of Figure 3.57 must flow through the short circuit. Consider the circuit of Figure 3.58, shown with a short circuit in place of the load resistance. Any of the techniques presented in this chapter could be employed to determine the current i_{SC}. In this particular case, mesh analysis is a convenient tool, once it is recognized that the short-circuit current is a mesh current. Let i_1 and $i_2 = i_{SC}$ be the mesh currents in the circuit of

Figure 3.58 Computation of Norton current

Figure 3.58. Then, the following mesh equations can be derived and solved for the short-circuit current:

$$(R_1 + R_2)i_1 - R_2 i_{SC} = v_S$$
$$-R_2 i_1 + (R_2 + R_3)i_{SC} = 0$$

An alternative formulation would employ nodal analysis to derive the equation

$$\frac{v_S - v}{R_1} = \frac{v}{R_2} + \frac{v}{R_3}$$

leading to

$$v = v_S \frac{R_2 R_3}{R_1 R_3 + R_2 R_3 + R_1 R_2}$$

Recognizing that $i_{SC} = v/R_3$, we can determine the Norton current to be:

$$i_N = \frac{v}{R_3} = \frac{v_S R_2}{R_1 R_3 + R_2 R_3 + R_1 R_2}$$

Thus, conceptually, the computation of the Norton current simply requires identifying the appropriate short-circuit current. The following example further illustrates this idea.

Example   3.15

Find the Norton equivalent circuit to the left of terminals a and b for the network of Figure 3.59.

Figure 3.59

Solution:

First, we replace the voltage source with a short circuit, and the current source by an open circuit. The circuit now has the appearance of Figure 3.60 and permits the calculation of the Thévenin resistance:

$$R_T = 6 \parallel 3 + 2 = 2 + 2 = 4$$

To find the short-circuit, or Norton, current we can short-circuit terminals a and b and use mesh analysis to calculate i_N, as suggested in Figure 3.61. Using mesh analysis,

Figure 3.60   **Figure 3.61**

we may obtain the following equations:

$$i_1 = 2 \text{ A}$$

$$9i_2 - 3i_{SC} = 18$$

$$5i_{SC} - 3i_2 = 0$$

We may then solve these to compute

$$i_1 = 2 \text{ A}, \qquad i_2 = 2.5 \text{ A}, \qquad \text{and} \qquad i_{SC} = 1.5 \text{ A}$$

The complete equivalent circuit is shown in Figure 3.62.

Figure 3.62 Norton equivalent circuit

Source Transformations

This section illustrates **source transformations,** a procedure that may be very useful in the computation of equivalent circuits, permitting, in some circumstances, replacement of current sources with voltage sources, and vice versa. The Norton and Thévenin theorems state that any one-port network can be represented by a voltage source in series with a resistance, or by a current source in parallel with a resistance, and that either of these representations is equivalent to the original circuit, as illustrated in Figure 3.63.

Thévenin equivalent   Norton equivalent

Figure 3.63 Equivalence of Thévenin and Norton representations

An extension of this result is that any circuit in Thévenin equivalent form may be replaced by a circuit in Norton equivalent form, provided that we use the following relationship:

$$v_T = R_T i_N \tag{3.32}$$

Thus, the subcircuit to the left of the dashed line in Figure 3.64 may be replaced by its Norton equivalent, as shown in the figure. Then, the com-

Figure 3.64 Effect of source transformation

putation of i_{SC} becomes very straightforward, since the three resistors are in parallel with the current source and therefore a simple current divider may be used to compute the short-circuit current. Observe that the short-circuit current is the current flowing through R_3; therefore,

$$i_{SC} = i_N = \frac{1/R_3}{1/R_1 + 1/R_2 + 1/R_3} \frac{v_S}{R_1} = \frac{v_S R_2}{R_1 R_3 + R_2 R_3 + R_1 R_2} \tag{3.33}$$

which is the identical result obtained for the same circuit in the preceding section, as you may easily verify. This source transformation method can be very useful, if employed correctly. Figure 3.65 shows how one can recognize subcircuits amenable to such source transformations. Example 3.16 is a numerical example illustrating the procedure.

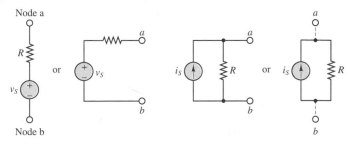

Thévenin subcircuits   Norton subcircuits

Figure 3.65 Subcircuits amenable to source transformation

Example   3.16

This example will illustrate the connection between Thévenin and Norton source transformations and the computation of Thévenin equivalents. The aim is to determine the Norton equivalent for the circuit shown in Figure 3.66.

Figure 3.66

Solution:

With the 50-V and 5-V sources replaced by short circuits and the 0.5-A source replaced by an open circuit, we can find the Thévenin resistance to be

$$R_T = 160 + 200 \,\|\, 100 \,\|\, 100 = 200 \; \Omega$$

To take advantage of the source transformation technique, it is useful to sketch the circuit again, emphasizing the location of the nodes, as shown in Figure 3.67. Note

Figure 3.67

that we have purposely separated the points a' and b' from the points a'' and b'' in the circuit, *even though these are the same nodes*. Now, the branch containing the 50-V source in series with the 100-Ω resistor clearly appears between nodes a'' and b'' and can therefore be replaced by an equivalent Norton circuit, with equivalent current source of strength $50/100$, in parallel with a 100-Ω resistor. Similarly, the branch containing the 5-V source in series with the 200-Ω resistor can be manipulated to obtain the circuit shown in Figure 3.68.

Observe how all of the elements to the left of the 160-Ω resistor are now in parallel! An additional manipulation, consisting of placing all of the sources next to each other in the diagram, illustrates the simple nature of the original circuit in this new light (Figure 3.69), revealing that if we simply sum (algebraically) the three current sources and perform a conversion from current to voltage source, we can finally obtain the Norton equivalent circuit of Figure 3.70.

Figure 3.68

Figure 3.69

Figure 3.70

Experimental Determination of Thévenin and Norton Equivalents

The idea of equivalent circuits as a means of representing complex and sometimes unknown networks is useful not only analytically, but in practical engineering applications as well. It is very useful to have a measure, for example, of the equivalent internal resistance of an instrument, so as to have an idea of its power requirements and limitations. Fortunately, Thévenin and Norton equivalent circuits can also be evaluated experimentally by means of very simple techniques. The basic idea is that the Thévenin voltage is an open-circuit voltage and that the Norton current is a short-circuit current. It should therefore be possible to conduct appropriate measurements to determine these quantities. Once v_T and i_N are known, we can determine the Thévenin resistance of the circuit being analyzed according to the relationship

$$R_T = \frac{v_T}{i_N} \tag{3.34}$$

How are v_T and i_N measured, then?

Figure 3.71 illustrates the measurement of the open-circuit voltage and short-circuit current for an arbitrary network connected to any load and also illustrates that the procedure requires some special attention, because of the nonideal nature of any practical measuring instrument. The figure clearly illustrates that in the presence of finite meter resistance, r_m, one must take this quantity into account in the computation of the short-circuit current and open-circuit voltage; v_{OC} and i_{SC} appear between quotation marks in the figure specifically to illustrate that the measured "open-circuit voltage" and "short-circuit current" are in fact affected by the internal resistance of the measuring instrument.

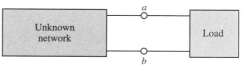

An unknown network connected to a load

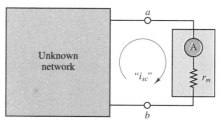

Network connected for measurement of short-circuit current

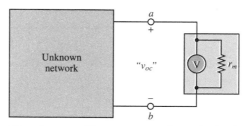

Network connected for measurement of open-circuit voltage

Figure 3.71 Measurement of open-circuit voltage and short-circuit current

You should verify that the following expressions for the true short-circuit current and open-circuit voltage apply (see the material on nonideal measuring instruments in Section 2.8):

$$i_N = \text{``}i_{SC}\text{''} \left(1 + \frac{r_m}{R_T} \right)$$
$$v_T = \text{``}v_{OC}\text{''} \left(1 + \frac{R_T}{r_m} \right)$$

$$(3.35)$$

where i_N is the ideal Norton current, v_T the Thévenin voltage, and R_T the true Thévenin resistance. If you recall the earlier discussion of the properties of ideal ammeters and voltmeters, you will recall that for an ideal ammeter, r_m should approach zero, while in an ideal voltmeter, the internal resistance should approach an open circuit (infinity); thus, the two expressions just given permit the determination of the true Thévenin and Norton equivalent sources from an (imperfect) measurement of the open-circuit voltage and short-circuit current, providing that the internal meter resistance, r_m, is known. Note also that, in practice, the internal resistance of voltmeters is sufficiently high to be considered infinite relative to the equivalent resistance of most practical circuits; on the other hand, it is impossible to construct an ammeter that does not have a finite, nonzero internal resistance. If the internal ammeter resistance is known, however, a reasonably accurate measurement of short-circuit current may be obtained. The following example illustrates the point.

Example   3.17

Assume that the measurements of Figure 3.71 are performed with an ideal voltmeter and an ammeter with internal resistance equal to 15 Ω. If the readings obtained are "i_{SC}"= 3.75 mA and "v_{OC}"= 6.5 V, what is the Thévenin resistance of the unknown network?

Solution:

Using the circuit of Figure 3.71 and replacing the unknown source network with its Thévenin equivalent, as shown in Figure 3.72, it should be apparent that

Network connected for measurement of   Network connected for measurement of
short-circuit current (practical ammeter)   open-circuit voltage (ideal voltmeter)

Figure 3.72

$$\frac{v_{OC}}{i_{SC}} = R_T + r_m$$

Thus,

$$R_T = \frac{"v_{OC}"}{"i_{SC}"} - r_m = 1{,}733.33 - 15 = 1{,}718.33 \ \Omega$$

One last comment is in order concerning the practical measurement of the internal resistance of a network. In most cases, it is not advisable to actually short-circuit a network by inserting a series ammeter as shown in Figure 3.71; permanent damage to the circuit or to the ammeter may be a consequence. For example, imagine that you wanted to estimate the internal resistance of an automotive battery; connecting a laboratory ammeter between the battery terminals would surely result in immediate loss of the instrument. Most ammeters are not designed to withstand currents of such magnitude. Thus, the experimenter should pay attention to the capabilities of the ammeters and voltmeters used in measurements of this type, as well as to the (approximate) power ratings of any sources present. However, there are established techniques especially designed to measure large currents.

3.6   MAXIMUM POWER TRANSFER

The reduction of any linear resistive circuit to its Thévenin or Norton equivalent form is a very convenient conceptualization, as far as the computation of load-related quantities is concerned. One such computation is that of the power absorbed by the load. The Thevenin and Norton models imply that some of the power generated by the source will necessarily be dissipated by the internal circuits within the source. Given this unavoidable power loss, a logical question to ask is, How much power can be transferred to the load from the source under the most ideal conditions? Or, alternatively, What is the value of the load resistance that will absorb maximum power from the source? The answer to these questions is contained in the **maximum power transfer theorem,** which is the subject of the present section.

The model employed in the discussion of power transfer is illustrated in Figure 3.73, where a practical source is represented by means of its Thévenin equivalent circuit. The maximum power transfer problem is easily

Given v_T and R_T, what value of R_L will allow for maximum power transfer?

Figure 3.73 Power transfer between source and load

formulated if we consider that the power absorbed by the load, P_L, is given by the expression

$$P_L = i_L^2 R_L \tag{3.36}$$

and that the load current is given by the familiar expression

$$i_L = \frac{v_T}{R_L + R_T} \tag{3.37}$$

Combining the two expressions, we can compute the load power as

$$P_L = \frac{v_T^2}{(R_L + R_T)^2} R_L \tag{3.38}$$

To find the value of R_L that maximizes the expression for P_L (assuming that V_T and R_T are fixed), the simple maximization problem

$$\frac{dP_L}{dR_L} = 0 \tag{3.39}$$

must be solved. Computing the derivative, we obtain the following expression:

$$\frac{dP_L}{dR_L} = \frac{v_T^2(R_L + R_T)^2 - 2v_T^2 R_L(R_L + R_T)}{(R_L + R_T)^4} \tag{3.40}$$

which leads to the expression

$$(R_L + R_T)^2 - 2R_L(R_L + R_T) = 0 \tag{3.41}$$

It is easy to verify that the solution of this equation is:

$$R_L = R_T \tag{3.42}$$

Thus, in order to transfer maximum power to a load, the equivalent source and load resistances must be **matched,** that is, equal to each other.

Example   3.18

A greatly simplified model of a stereo amplifier and a speaker is shown in Figure 3.74. In order to transfer maximum power to the speaker, we should select $R_L = R_T$, as illustrated in the preceding discussion. Suppose, instead, that, not knowing that the internal resistance of the amplifier is $R_T = 8\ \Omega$, we have connected a mismatched speaker, with $R_L = 16\ \Omega$. We would like to determine how much more power we could be delivering to the speaker if its resistance were matched to that of the amplifier.

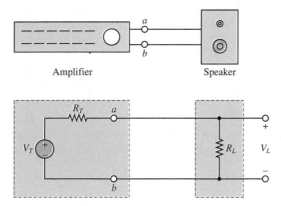

Figure 3.74 A simplified model of an audio system

Solution:

With the 16-Ω speaker, the power delivered by the amplifier is computed as follows:

$$v_L = \frac{R_L}{R_L + R_T}v_T = \frac{2}{3}v_T$$

$$P_L = \frac{v_L^2}{R_L} = \frac{4}{9R_L}v_T^2 = (0.0278)v_T^2$$

If the speaker resistance were matched to the amplifier resistance, that is, equal to 8 Ω, the power delivered to the speaker under matched load conditions—which we will call P_L'— would be greater, according to the following calculation. Since the load resistance is matched to the source, we have

$$v_L' = \frac{v_T}{2}$$

and therefore

$$P_L' = \frac{v_L'^2}{R_L} = \frac{1}{4R_L}v_T^2 = (0.03125)v_T^2$$

with an increase in power equal to $\frac{0.03125-0.0278}{0.0278} \times 100 = 12.5$ percent. In reality, an audio amplifier is not well represented by a single resistance, but requires a somewhat more complex circuit model. The audiophile can find more interesting tidbits concerning loudspeakers and audio circuits in Chapters 4, 6, and 14.

Figure 3.75 Source loading effects

A problem related to power transfer is that of **source loading.** This phenomenon, which is illustrated in Figure 3.75, may be explained as follows: when a practical voltage source is connected to a load, the current that flows from the source to the load will cause a voltage drop across the internal source resistance, v_{int}; as a consequence, the voltage actually seen by the load will be somewhat lower than the *open-circuit voltage* of the source. As stated earlier, the open-circuit voltage is equal to the Thévenin voltage. The extent of the internal voltage drop within the source depends on the amount of current drawn by the load. With reference to Figure 3.75, this internal drop is equal to iR_T, and therefore the load voltage will be:

$$v_L = v_T - iR_T \tag{3.43}$$

It should be apparent that it is desirable to have as small an internal resistance as possible in a practical voltage source.

In the case of a current source, the internal resistance will draw some current away from the load because of the presence of the internal source resistance; this current is denoted by i_{int} in Figure 3.75. Thus, the load will receive only part of the *short-circuit current* available from the source (the Norton current):

$$i_L = i_N - \frac{v}{R_T} \tag{3.44}$$

It is therefore desirable to have a very large internal resistance in a practical current source. You may wish to refer back to the discussion of practical sources to verify that the earlier interpretation of practical sources can be expanded in light of the more recent discussion of equivalent circuits.

DRILL EXERCISES

3.21   A practical voltage source has an internal resistance of 1.2 Ω and generates a 30-V output under open-circuit conditions. What is the smallest load resistance we can connect to the source if we do not wish the load voltage to drop by more than 2 percent with respect to the source open-circuit voltage?

> Answer: 58.8 Ω

3.22 A practical current source has an internal resistance of 12 kΩ and generates a 200-mA output under short-circuit conditions. What percent drop in load current will be experienced (with respect to the short-circuit condition) if a 200 Ω load is connected to the current source?

> Answer: 1.64%

3.7   NONLINEAR CIRCUIT ELEMENTS

Until now the focus of this chapter has been on linear circuits, containing ideal voltage and current sources, and on linear resistors. In effect, one reason for the simplicity of some of the techniques illustrated in the earlier part of this chapter is the ability to utilize Ohm's law as a simple, linear description of the i-v characteristic of an ideal resistor. In many practical instances, however, the engineer is faced with elements exhibiting a nonlinear i-v characteristic. This section explores two methods for analyzing nonlinear circuit elements.

Description of Nonlinear Elements

There are a number of useful cases in which a simple functional relationship exists between voltage and current in a nonlinear circuit element. For example, Figure 3.76 depicts an element with an exponential i-v characteristic, described by the following equations:

$$i = I_0 e^{\alpha v} \qquad v > 0$$
$$i = -I_0 \qquad v \le 0 \tag{3.45}$$

Figure 3.76 A nonlinear element: the "exponential resistor"

There exists, in fact, a circuit element (the semiconductor diode) that very nearly satisfies this simple relationship. The difficulty in the i-v relationship of equation 3.45 is that it is not possible, in general, to obtain a closed-form analytical solution, even for a very simple circuit.

    With the knowledge of equivalent circuits you have just acquired, one approach to analyzing a circuit containing a nonlinear element might be to treat the nonlinear element as a load, and to compute the Thévenin equivalent of the remaining circuit, as shown in Figure 3.77. Applying KVL, the following equation may then be obtained:

$$v_T = R_T i_X + v_X \tag{3.46}$$

To obtain the second equation needed to solve for both the unknown voltage, v_X, and the unknown current, i_X, it is necessary to resort to the i-v description

Nonlinear element as a load. We wish to solve for v_X and i_X.

Figure 3.77 Representation of nonlinear element in a linear circuit

of the nonlinear element, namely, equation 3.45. If, for the moment, only positive voltages are considered, the circuit is completely described by the following system:

$$i_X = I_0 e^{\alpha v_X} \qquad v > 0$$
$$v_T = R_T i_X + v_X \tag{3.47}$$

The two parts of equation 3.47 represent a system of two equations in two unknowns; however, one of these equations is nonlinear. If we solve for the load voltage and current, for example, by substituting the expression for i_X in the linear equation, we obtain the following expression:

$$v_T = R_T I_0 e^{\alpha v_X} + v_X \tag{3.48}$$

or

$$v_X = v_T - R_T I_0 e^{\alpha v_X} \tag{3.49}$$

Equations 3.48 and 3.49 do not have a closed-form solution; that is, they are *transcendental equations* (why?). How can v_X be found? One possibility is to generate a solution numerically, by guessing an initial value (e.g., $v_X = 0$) and iterating until a sufficiently precise solution is found. This solution is explored further in the homework problems. Another method is based on a graphical analysis of the circuit and is described in the following section.

Graphical (Load-Line) Analysis of Nonlinear Circuits

The nonlinear system of equations of the previous section may be analyzed in a different light, by considering the graphical representation of equation 3.46, which may also be written as follows:

$$i_X = -\frac{1}{R_T} v_X + \frac{v_T}{R_T} \tag{3.50}$$

We notice first that equation 3.50 describes the behavior of any load, linear or nonlinear, since we have made no assumptions regarding the nature of the load voltage and current. Second, it is the equation of a line in the i_X-v_X plane, with slope $-1/R_T$ and i_X intercept V_T/R_T. This equation is referred to as the **load-line equation;** its graphical interpretation is very useful and is shown in Figure 3.78.

The load-line equation is but one of two i-v characteristics we have available, the other being the nonlinear-device characteristic of equation 3.45. The intersection of the two curves yields the solution of our nonlinear system of equations. This result is depicted in Figure 3.79.

Finally, another important point should be emphasized: the linear network reduction methods introduced in the preceding sections can always be

Figure 3.78 Load line

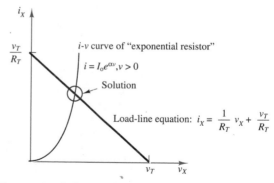

Figure 3.79 Graphical solution of equations 3.48

employed to reduce any circuit containing a single nonlinear element to the Thévenin equivalent form, as illustrated in Figure 3.80. The key is to identify the nonlinear element and to treat it as a load. Thus, the equivalent-circuit solution methods developed earlier can be very useful in simplifying problems in which a nonlinear load is present. Example 3.19 illustrates this point.

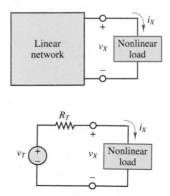

Figure 3.80 Transformation of nonlinear circuit to Thévenin equivalent

Example   3.19

The Thévenin equivalent model of a generator has been found experimentally to be $v_T = 15$ V, $R_T = 30$ Ω, using the methods described in Section 3.5. The generator is connected to a nonlinear load whose i-v characteristic has also been measured and is available in graphical form, as shown in Figure 3.81. Find the power dissipated by the nonlinear load.

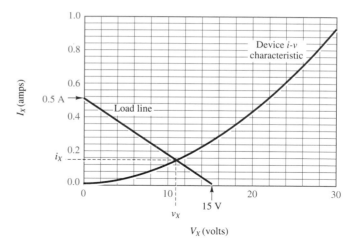

Figure 3.81

Solution:

We can model the circuit as shown in Figure 3.80. The objective is to determine the voltage v_X and the current i_X using graphical methods. The load-line equation for the circuit is given by the expression

$$i_X = -\frac{1}{R_T}v_X + \frac{v_T}{R_T}$$

or

$$i_X = -\frac{1}{30}v_X + \frac{15}{30}$$

This equation represents a line in the i_X-v_X plane, with i_X intercept at 0.5 A and v_X intercept at 15 V. In order to determine the operating point of the circuit, we superimpose the load line on the device i-v characteristic and determine the solution by finding the intersection of the load line with the device curve. Inspection of the graph reveals that the intersection point is given approximately by

$$i_X = 0.14 \text{ A} \qquad v_X = 11 \text{ V}$$

and therefore the power dissipated by the nonlinear load is

$$P_X = 0.14 \times 11 = 1.54 \text{ W}$$

It is important to observe that the result obtained in this example is, in essence, a description of experimental procedures, indicating that the analytical concepts developed in this chapter also apply to practical measurements.

CONCLUSION

The objective of this chapter was to provide a practical introduction to the analysis of linear resistive circuits. The emphasis on examples is important at this stage, since we believe that familiarity with the basic circuit analysis techniques will greatly ease the task of learning more advanced ideas in circuits and electronics. In particular, your goal at this point should be to have mastered four analysis methods, summarized as follows:

1. *Node voltage and mesh current analysis.* These methods are analogous in concept; the choice of a preferred method depends on the specific circuit. They are generally applicable to the circuits we will analyze in this book and are amenable to solution by matrix methods.
2. *The principle of superposition.* This is primarily a conceptual aid that may simplify the solution of circuits containing multiple sources. It is usually not an efficient method.
3. *Thévenin and Norton equivalents.* The notion of equivalent circuits is at the heart of circuit analysis. Complete mastery of the reduction of linear resistive circuits to either equivalent form is a must.
4. *Numerical and graphical analysis.* These methods apply in the case of nonlinear circuit elements. The load-line analysis method is intuitively appealing and will be employed again in this book to analyze electronic devices.

The material covered in this chapter will be essential to the development of more advanced techniques throughout the remainder of the book.

KEY TERMS

Node Voltage Method   65
Reference Node   65
Mesh Current   73
Constraint Equation   79
One-Port Network   84

Equivalent Circuit   85
Duality   85
Open-Circuit Voltage   90
Short-Circuit Current   94
Source Transformation   96

Maximum Power Transfer Theorem   102
Matched Resistances   103
Source Loading   104
Load-Line Equation   106

HOMEWORK PROBLEMS

3.1 Use node voltage analysis to find the voltages V_1 and V_2 for the circuit of Figure P3.1.

Figure P3.1   Figure P3.2

3.2 Using node voltage analysis, find the voltages V_1 and V_2 for the circuit of Figure P3.2.

3.3 Find the currents i_1 and i_2 for the circuit of Figure 3.18, using node voltage analysis.

3.4 Find the currents i_1, i_2, and i_3 for the circuit of Figure 3.19, using node voltage analysis.

3.5 Find the voltage v_1 and v_2 for the circuit of Figure 3.16, using node voltage analysis.

3.6 Find the unknown current i_X for the circuit of Figure 3.21, using node voltage analysis.

3.7 Find the unknown voltage v_X for the circuit of Figure 3.20, using node voltage analysis.

3.8 The circuit shown in Figure P3.3 is a Wheatstone bridge circuit. Use node voltage analysis to determine V_a and V_b, and thus determine $V_a - V_b$.

3.9 Find the power delivered to the load resistor, R_L, for the circuit of Figure P3.4, using node voltage analysis, given that $R_I = R_1 = 2\ \Omega$, $R_V = R_2 = R_L = 4\ \Omega$, $V_S = 4$ V, and $I_S = 0.5$ A.

Figure P3.3

Figure P3.4

Figure P3.5

*3.10

a. For the circuit of Figure P3.5, write the node equations necessary to find voltages V_1, V_2, and V_3. Note that $G = 1/R =$ conductance. From the results, note the interesting form that the matrices $[G]$ and $[I]$ have taken in the equation $[G][V] = [I]$, where

$$[G] = \begin{bmatrix} g_{11} & g_{12} & \cdots & g_{1n} \\ g_{21} & g_{22} & & \\ \vdots & & \ddots & \\ g_{n1} & & & g_{nn} \end{bmatrix} \quad \text{and} \quad [I] = \begin{bmatrix} I_1 \\ I_2 \\ \vdots \\ I_n \end{bmatrix}$$

b. Write the matrix form of the node voltage equations again, using the following formulas:

$g_{ii} = \Sigma$ conductances connected to node i

$g_{ij} = -\Sigma$ (conductance *shared* by nodes i and j)

$I_i = \Sigma$ all *source* currents into node i

3.11 Using mesh current analysis, find the currents I_1 and I_2 and the voltage across the 20-Ω resistor for the circuit of Figure P3.6.

Figure P3.6

3.12 Using mesh current analysis, find the currents I_1 and I_2 and the voltage across the 10-Ω resistor in the circuit of Figure P3.7.

3.13 Find the voltage across the 1-kΩ resistor in Figure 3.9, using mesh current analysis (assume node a is above ground).

3.14 Find the voltage across the 10-Ω resistor of Drill Exercise 3.2, using mesh current analysis (assume the voltage drops from left to right).

3.15 Using mesh current analysis, find the currents I_1, I_2, and I_3 and the voltage across the 40-Ω resistor in the circuit of Figure P3.8 (assume polarity according to I_2).

Figure P3.7

Figure P3.8

3.16 Find the voltage across the 10-kΩ resistor in Figure 3.4, using mesh current analysis (assume the voltage drops from left to right).

3.17 Find the voltage across resistor R_2 in Figure 3.6, using mesh current analysis (assume the voltage drops from left to right).

3.18 Using mesh current analysis, solve for all unknown currents and voltages in Figure 3.36, given that $R_L = 10\Omega$.

3.19 Using mesh current analysis, solve for all unknown currents and voltages in Figure 3.38, given that $R_L = 2\Omega$.

3.20 Using mesh current analysis, solve for all unknown currents and voltages of the circuit given in Drill Exercise 3.14. Assume that $R_L = 5$ kΩ.

Figure P3.9

3.21 Using mesh current analysis, find the voltage across the 1-Ω resistor in Figure 3.44, given that $R_L = 10\ \Omega$.

3.22 Find the unknown current I_x and the unknown voltage V_x, using mesh current analysis, for the circuit of Figure P3.9. The dependent voltage source is related to the currents I_x and I_2 by $V_x = 5(I_2 - I_x)$.

3.23 For the Wheatstone bridge circuit of Figure P3.3, use mesh current methods to determine V_a and V_b and thus to determine $V_a - V_b$.

*3.24

a.   For the circuit of Figure P3.10, write the mesh equations in matrix form. Notice the form of the $[R]$ and $[V]$ matrices in $[R][I] = [V]$, where

$$[R] = \begin{bmatrix} r_{11} & r_{12} & \cdots & r_{1n} \\ r_{21} & r_{22} & & \\ \vdots & & \ddots & \\ r_{n1} & & & r_{nn} \end{bmatrix} \quad \text{and} \quad [V] = \begin{bmatrix} V_1 \\ V_2 \\ \vdots \\ V_n \end{bmatrix}$$

b.   Write the matrix form of the mesh equations again by using the following formulas:

$$r_{ii} = \Sigma \text{ resistances around loop } i$$
$$r_{ij} = -\Sigma \text{ resistances } shared \text{ by loops } i \text{ and } j$$
$$V_i = \Sigma \text{ sum of source voltages around loop } i$$

Figure P3.10

Figure P3.11

3.25 Find the Thévenin equivalent of the circuit of Figure P3.11.

3.26 Find the Norton equivalent of the circuit of Figure P3.12.

3.27 Find the Thévenin equivalent voltage for the circuit of Drill Exercise 3.14.

3.28 Find the Norton equivalent current for the circuit of Drill Exercise 3.14.

3.29 Find the Thévenin equivalent voltage for the circuit of Drill Exercise 3.15.

3.30 Find the Norton equivalent current for the circuit of Drill Exercise 3.15.

3.31 Find the Thévenin equivalent voltage for the circuit of Drill Exercise 3.16.

3.32 Find the Norton equivalent current for the circuit of Drill Exercise 3.16.

3.33 Find the Thévenin equivalent voltage for the circuit of Drill Exercise 3.17.

3.34 Find the Norton equivalent current for the circuit of Drill Exercise 3.17.

Figure P3.12

3.35 Find the Thévenin equivalent circuit that the load sees for the circuit of Figure P3.13.

Figure P3.13

[Figure P3.14 diagram]

Figure P3.14

3.36 Find the Thévenin equivalent resistance seen by the load resistor, R_L, in the circuit of Figure P3.14.

3.37 Find the Thévenin equivalent of the circuit of Figure P3.15.

Figure P3.15

3.38 Find the Thévenin equivalent of the circuit of Figure P3.16, where $R_1 = 10 \, \Omega$, $R_2 = 20 \, \Omega$, $R_g = 0.1 \, \Omega$, and $R_p = 1 \, \Omega$.

Figure P3.16

3.39 Consider the circuit of Figure P3.17.

a. Find the power delivered to the load resistor, R_L.
b. Find the power dissipated by the source.
c. Find the Thévenin equivalent circuit of the source.
d. Find the power dissipated by the Thévenin resistance, R_T.
e. Is the power dissipated by the source as found in part b the same as the power dissipated in R_T? Can we say that the Thévenin equivalent circuit is adequate to describe the internal power consumption of the source?

Figure P3.17

3.40 The Wheatstone bridge circuit shown in Figure P3.18 is used in a number of practical applications. One traditional use is in determining the value of an unknown resistor, R_x.

a.   Find the value of the voltage $V_{ab} = V_a - V_b$ in terms of R, R_x and V_S.
b.   If $R = 1\ k\Omega$, $V_S = 12$ V, and $V_{ab} = 12$ mV, what is the value of R_x?

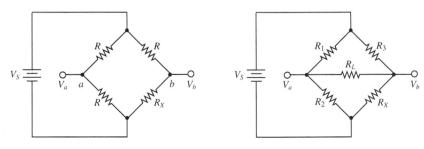

Figure P3.18   Figure P3.19

3.41   It is sometimes useful to compute a Thévenin equivalent circuit for a Wheatstone bridge. For the circuit of Figure P3.19,

a.   Find the Thévenin equivalent circuit seen by the load resistor, R_L.
b.   If $V_S = 12$ V, $R_1 = R_2 = R_3 = 1\ k\Omega$, and R_X is the resistance found in part b of the previous problem, use the Thévenin equivalent to compute the power dissipated by R_L, if $R_L = 500\ \Omega$.
c.   Find the power dissipated by the Thévenin equivalent resistance, R_T, with R_L included in the circuit.
d.   Find the power dissipated by the bridge without the load resistor in the circuit.

3.42   A resistive "pad," shown in Figure P3.20, is often used to make sources and loads that have different resistances appear to each other to be of compatible magnitude. If R_S, the internal source resistance, is 600 Ω, $R_1 = 600\ \Omega$, $R_2 = 1.2\ \Omega$, and $R_3 = 1\ k\Omega$,

a.   Find the equivalent circuits seen by the source and load, respectively. (Note that $R_2 \ll R_3 + R_L$.)
b.   Also find the Thévenin voltage that the load sees, as a function of V_S.

Figure P3.20

*3.43 Find the Thévenin resistance seen by the load, R_L, as well as the open-circuit voltage, v_{OC}, for the circuit of Figure P3.21.

Figure P3.21

*3.44   As discussed in Appendix A, a system of equations can be expressed in a matrix form such as

$$[A][x] = [u]$$

and solved by taking the inverse of A and premultiplying through the equation:

$$[A]^{-1}[A][x] = [A]^{-1}[u]$$
$$[x] = [A]^{-1}[u]$$

Thus, we can see that since we have a system of equations when we write node voltage and mesh current equations, we can also write

$$[G][v_x] = [i]$$
$$[R][i_x] = [v]$$

Use nodal analysis to find the matrices required to solve the circuit shown in Figure P3.21. Solve the circuit using computer aids to calculate the matrix inverses. [*Hint:* You may find source transformations useful.]

*3.45   For the circuit of Figure P3.22, use mesh current analysis to find the matrices required to solve the circuit, and solve for the unknown currents. [*Hint:* You may find source transformations useful.]

Figure P3.22

*3.46   Solve Example 3.6 using matrix techniques.

Figure P3.23

3.47 For the circuit of Figure P3.23,

a. Find an expression for the power dissipated by the load as a function of R_L.

b. Plot the power dissipated by the load versus the load resistance. Determine the value of R_L corresponding to maximum load power.

3.48 The circuit of Figure P3.24 has a "square law resistor" as a load, with i-v characteristic given by

$$I^2 = \frac{V}{R'}$$

If $V_T = 12$ V, $R_T = 20$ Ω, and $R' = 10$ Ω, find the output voltage, V_2, analytically *and* graphically.

Figure P3.24

3.49 A nonlinear element is placed in the circuit of Figure P3.25. The nonlinear element has the i-v characteristics shown in Figure P3.26. Use load-line analysis to determine the operating point of the element and thus V_{out}.

Figure P3.25

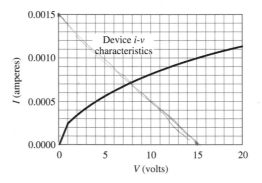

Figure P3.26

3.50   A nonlinear element is placed in the circuit of Figure P3.27. The nonlinear element has the i-v characteristics shown in Figure P3.28. Use load-line analysis to determine the operating point of the element and thus V_{out}.

Figure P3.27

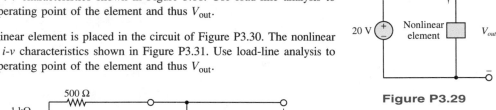

Device i-v characteristics

Figure P3.28

3.51   A nonlinear element is placed in the circuit of Figure P3.29. The nonlinear element has the i-v characteristics shown in Figure 3.81. Use load-line analysis to determine the operating point of the element and thus V_{out}.

3.52   A nonlinear element is placed in the circuit of Figure P3.30. The nonlinear element has the i-v characteristics shown in Figure P3.31. Use load-line analysis to determine the operating point of the element and thus V_{out}.

Figure P3.29

Figure P3.30

Device i-v characteristics

Figure P3.31

3.53 A thermistor is a device whose resistance and therefore whose $i\text{-}v$ characteristics change with temperature. Shown in Figure P3.32 are a family of $i\text{-}v$ characteristic curves for a certain thermistor. The thermistor is connected to the circuit of Figure P3.33.

Figure P3.32

Figure P3.33

a.   If the voltage V_{out} is 3 V, what is the temperature?
b.   If the temperature is $-10°$ C, what is the output voltage?
c.   Plot V_{out} versus T for the range given in the plot.

*3.54 The voltage-current relationship of a semiconductor diode may be approximated by the expression

$$i_D = I_{SAT} \left(\exp \left\{ \frac{v_D}{kT/q} \right\} - 1 \right)$$

where

$$I_{SAT} = 10^{-12} \text{ A}$$

$$\frac{kT}{q} = 0.0259 \text{ V} \qquad \text{at room temperature}$$

and V_D is the voltage across the diode.

a.   Given the circuit of Figure P3.34 use graphical analysis to find the diode current i_0 and the diode voltage v_D if $R_T = 22 \ \Omega$ and $V_T = 12$ V.
b.   Write a FORTRAN or BASIC program that will find the diode voltage v_D and the current i_0, using the flowchart in Figure P3.34.

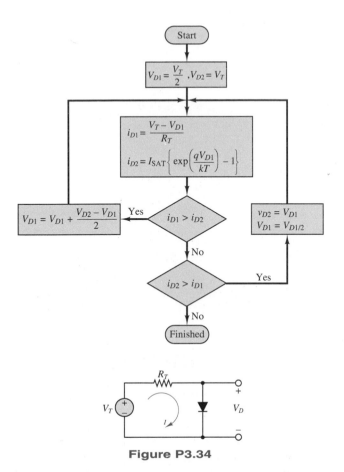

Figure P3.34

3.55 We have seen that some devices do not have a linear current-voltage characteristic for all i and v—that is, R is not constant for all values of current and voltage.
For many devices, however, we can estimate the characteristics by piecewise linear
approximation. For a portion of the characteristic curve around an operating point,
the slope of the curve is relatively constant. The inverse of this slope at the operating
point is defined as "incremental resistance," R_{inc}:

$$R_{inc} = \left. \frac{dV}{dI} \right|_{[V_0, I_0]} \approx \left. \frac{\Delta V}{\Delta I} \right|_{[V_0, I_0]}$$

where $[V_0, I_0]$ is the operating point of the circuit.

a. For the circuit of Figure P3.35, find the operating point of the element that has
    the characteristic curve shown.

b. Find the incremental resistance of the nonlinear element at the operating point
    of part a.

c. If V_T were increased to 20 V, find the new operating point and the new incre-
    mental resistance.

Figure P3.35

CHAPTER

4

AC Network Analysis

In this chapter we introduce energy-storage elements, dynamic circuits, and the analysis of circuits excited by sinusoidal voltages and currents. Sinusoidal (or AC) signals constitute the most important class of signals in the analysis of electrical circuits. The simplest reason is that virtually all of the electric power used in households and industries comes in the form of sinusoidal voltages and currents.

The chapter is arranged as follows. First, energy-storage elements are introduced, and time-dependent signal sources and the concepts of average and root-mean-square (rms) values are discussed. Next, we analyze the circuit equations that arise when time-dependent signal sources excite circuits containing energy-storage elements; in the course of this discussion, it will become apparent that differential equations are needed to describe the dynamic behavior of these circuits. The remainder of the chapter is devoted to the development of circuit analysis techniques that greatly simplify the solution of dynamic circuits for the special case of sinusoidal signal excitation; the more general analysis of these circuits will be completed in Chapter 6.

120

By the end of the chapter, you should have mastered a number of concepts that will be used routinely in the remainder of the book; these are summarized as follows:

- Definition of the *i-v* relationship for inductors and capacitors.
- Computation of rms values for periodic waveforms.
- Representation of sinusoidal signals by complex phasors.
- Impedance of common circuit elements.
- AC circuit analysis by Kirchhoff's laws and equivalent-circuit methods.

4.1   ENERGY-STORAGE (DYNAMIC) CIRCUIT ELEMENTS

The ideal resistor was introduced through Ohm's law in Chapter 2 as a useful idealization of many practical electrical devices. However, in addition to resistance to the flow of electric current, which is purely a dissipative (i.e., an energy-loss) phenomenon, electric devices may also exhibit energy-storage properties, much in the same way a spring or a flywheel can store mechanical energy. Two distinct mechanisms for energy storage exist in electric circuits: **capacitance** and **inductance,** both of which lead to the storage of energy in an electromagnetic field. For the purpose of this discussion, it will not be necessary to enter into a detailed electromagnetic analysis of these devices. Rather, two ideal circuit elements will be introduced to represent the ideal properties of capacitive and inductive energy storage: the **ideal capacitor** and the **ideal inductor.** It should be stated clearly that ideal capacitors and inductors do not exist, strictly speaking; however, just like the ideal resistor, these "ideal" elements are very useful for understanding the behavior of physical circuits. In practice, any component of an electric circuit will exhibit some resistance, some inductance, and some capacitance—that is, some energy dissipation and some energy storage.

The situation depicted in this introduction is completely analogous to the description commonly given of mechanical systems, which involves frictional, elastic, and inertial components (namely, dampers, springs, and masses). The analogy with mechanical systems will be developed in parallel with the description of the new circuit elements.

The Ideal Capacitor

A physical capacitor is a device that can store energy in the form of a charge separation when appropriately polarized by an electric field (i.e., a voltage). The simplest capacitor configuration consists of two parallel conducting plates of cross-sectional area *A*, separated by air (or another

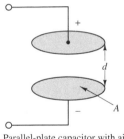

Parallel-plate capacitor with air
gap d (air is the dielectric)

$C = \dfrac{\varepsilon A}{d}$

ε = permittivity of air

$= 8.854 \times 10^{-12} = \dfrac{F}{m}$

Circuit
symbol

Figure 4.1 Structure of
parallel-plate capacitor

dielectric[1] material, such as mica or Teflon). Figure 4.1 depicts a typical configuration and the circuit symbol for a capacitor.

The presence of an insulating material between the conducting plates does not allow for the flow of DC current; thus, *a capacitor acts as an open circuit in the presence of DC currents.* However, if the voltage present at the capacitor terminals changes as a function of time, so will the charge that has accumulated at the two capacitor plates, since the degree of polarization is a function of the applied electric field, which is time-varying. In a capacitor, the charge separation caused by the polarization of the dielectric is proportional to the external voltage, that is, to the applied electric field:

$$Q = CV \tag{4.1}$$

where the parameter C is called the *capacitance* of the element and is a measure of the ability of the device to accumulate, or store, charge. The unit of capacitance is the coulomb/volt and is called the **farad (F)**. The farad is an unpractically large unit; therefore it is common to use microfarads $(1\ \mu F = 10^{-6}\ F)$ or picofarads $(1\ pF = 10^{-12} F)$. From equation 4.1 it becomes apparent that if the external voltage applied to the capacitor plates changes in time, so will the charge that is internally stored by the capacitor:

$$q(t) = Cv(t) \tag{4.2}$$

Thus, although no current can flow through a capacitor if the voltage across it is constant, a time-varying voltage will cause charge to vary in time.

The change with time in the stored charge is analogous to a current. You can easily see this by recalling the definition of current given in Chapter 2, where it was stated that

$$i(t) = \frac{dq(t)}{dt} \tag{4.3}$$

that is, that electric current corresponds to the time rate of change of charge. Differentiating equation 4.2, one can obtain a relationship between the current and voltage in a capacitor:

$$\boxed{i(t) = C\,\frac{dv(t)}{dt}} \tag{4.4}$$

Equation 4.4 is the defining circuit law for a capacitor. It may be of interest to examine the analogy between equation 4.4 and the equation describing a force acting on a mass. Let $u(t)$ represent the velocity of the mass and $f(t)$ represent the force acting on it. Then Newton's law dictates that

$$f(t) = M\,\frac{du(t)}{dt} \tag{4.5}$$

---

[1] A dielectric material is a material that is not an electrical conductor but contains a large number of electric dipoles, which become polarized in the presence of an electric field.

This principle is illustrated in Figure 4.2.

If the differential equation that defines the *i-v* relationship for a capacitor is integrated, one can obtain the following relationship for the voltage across a capacitor:

$$v_C(t) = \frac{1}{C} \int_{-\infty}^{t} i_C \, dt \qquad (4.6)$$

Equation 4.6 indicates that the capacitor voltage depends on the past current through the capacitor, up until the present time, t. Of course, one does not usually have precise information regarding the flow of capacitor current for all past time, and so it is useful to define the initial voltage (or *initial condition*) for the capacitor according to the following, where t_0 is an arbitrary initial time:

$$V_0 = v_C(t = t_0) = \frac{1}{C} \int_{-\infty}^{t_0} i_C \, dt \qquad (4.7)$$

The capacitor voltage is now given by the expression

$$v_C(t) = \frac{1}{C} \int_{t_0}^{t} i_C \, dt + V_0 \qquad t \geq t_0 \qquad (4.8)$$

The significance of the initial voltage, V_0, is simply that at time t_0 some charge is stored in the capacitor, giving rise to a voltage, $v_C(t_0)$, according to the relationship $Q = CV$. Knowledge of this initial condition is sufficient to account for the entire past history of the capacitor current.

From the standpoint of circuit analysis, it is important to point out that capacitors connected in series and parallel can be combined to yield a single equivalent capacitance. The rule of thumb, which is illustrated in Figure 4.3, is the following:

The defining equation for the capacitance circuit element is analogous to the equation of motion of a mass acted upon by a force.

$$i_c(t) = C \frac{dv_c(t)}{dt}$$

$$f = M \frac{du}{dt}$$

Figure 4.2 Defining equation for the ideal capacitor, and analogy with force-mass system

Capacitors in parallel add. Capacitors in series combine according to the same rules used for resistors connected in parallel.

Capacitances in series combine like resistors in parallel

Capacitances in parallel add

Figure 4.3 Combining capacitors in a circuit

Example   4.1

Calculate the current through a capacitor from knowledge of its terminal voltage. Assume that a 0.1 μF capacitor is charged by an exponential voltage, $v(t)$, shown in Figure 4.4. If $v(t) = 5\left(1 - e^{-t/10^{-6}}\right)$, with $t \geq 0$, what is the current flowing through the capacitor for $t \geq 0$?

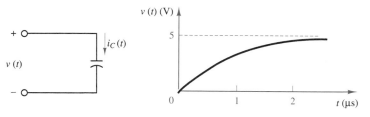

Figure 4.4

Solution:

Using the defining relationship for the capacitor, we may obtain the current by differentiating the voltage:

$$i_C = C\,\frac{dv(t)}{dt}$$

$$= \left(10^{-7}\right)\left(\frac{5}{10^{-6}}\right)e^{-t/10^{-6}}$$

$$= 0.5e^{-t/10^{-6}}\ \text{A}$$

Figure 4.5

A sketch of the capacitor current is shown in Figure 4.5. Note how this current jumps instantaneously to 0.5 A as the voltage begins to slowly rise exponentially: the ability of capacitor currents to change suddenly is an important property of capacitors.

It is also worth noting that, as the voltage approaches the constant value 5 V, the capacitor reaches its maximum charge storage capability for that voltage (since $Q = CV$) and no more current flows through the capacitor. The total charge stored is $Q = 0.5 \times 10^{-6}$ C. This is a fairly small amount of charge, but it can produce a substantial amount of current for a brief period of time. For example, the fully charged capacitor could provide 100 mA of current for a period of time equal to 5 μs:

$$\Delta I = \frac{\Delta Q}{\Delta t} = \frac{0.5 \times 10^{-6}}{5 \times 10^{-6}} = 0.1\ \text{A}$$

There are many useful applications of this energy-storage property of capacitors in practical circuits.

Example   4.2

To illustrate the effect of initial conditions in a capacitor, we provide a simple example in which a 1,000-μF capacitor has been charged to 2 V. At $t = t_0 = 0$, a current of 10 mA is injected into the capacitor; the current is held constant for 1 second. Sketch the capacitor voltage and current waveforms.

Solution:

Integrating the capacitor current from $t = 0$ to $t = 1$ is very simple, since the current is constant in this interval. The general expression for the capacitor voltage is:

$$v_C(t) = \frac{1}{C} \int_{t_0}^{t} i_C \, dt + V_0 \qquad t \geq t_0$$

and therefore,

$$v_C(t) = \frac{1}{C} \int_{0}^{t} I \, dt + 2$$

and

$$v_C(t) = \frac{I}{C} t + 2 \qquad 1 \geq t \geq 0$$

$$= 10t + 2$$

Once the current stops, at $t = 1$ s, the capacitor voltage cannot develop any further but remains now at the maximum value it reached at $t = 1$ s: $v_C(t = 1) = 12$ V. Note that this final value of the capacitor voltage after the current source has stopped charging the capacitor depends on two factors: (1) the initial value of the capacitor voltage and, (2) the history of the capacitor current. Figure 4.6 depicts the two waveforms.

Figure 4.6

Physical capacitors are rarely constructed of two parallel plates separated by air, because this configuration yields very low values of capacitance, unless one is willing to tolerate very large plate areas. In order to increase the capacitance (i.e., the ability to store energy), physical capacitors are often made of tightly rolled sheets of metal film, with a dielectric (paper or Mylar) sandwiched in between. Table 4.1 illustrates typical values, materials, maximum voltage ratings, and useful frequency ranges for various types of capacitors. The voltage rating is particularly important, because any insulator will break down if a sufficiently high voltage is applied across it.

Table 4.1   Capacitors

Material	Capacitance range	Maximum voltage (V)	Frequency range (Hz)
Mica	1 pF to 0.1 μF	100–600	10^3–10^{10}
Ceramic	10 pF to 1 μF	50–1,000	10^3–10^{10}
Mylar	0.001 μF to 10 μF	50–500	10^2–10^8
Paper	1,000 pF to 50 μF	100–105	10^2–10^8
Electrolytic	0.1 μF to 0.2 F	3–600	10–10^4

Energy Storage in Capacitors

You may recall that the capacitor was described as an energy-storage element earlier in this section. An expression for the energy stored in the capacitor, $W_C(t)$, may be derived easily if we recall that energy is the integral of power, and that the instantaneous power in a circuit element is equal to the product of voltage and current:

$$
\begin{aligned}
W_C(t) &= \int P_C(t)\, dt \\
&= \int v_C(t) i_C(t)\, dt \\
&= \int v_C(t) C \frac{dv_C(t)}{dt}\, dt
\end{aligned}
\tag{4.9}
$$

$$\boxed{W_C(t) = \tfrac{1}{2} C v_C^2(t) \qquad \text{Energy stored in a capacitor}}$$

Example 4.3 illustrates the calculation of the energy stored in a capacitor. It will also be observed that the expression for the stored energy in a capacitor is analogous to the expression for the kinetic energy of the mass of Figure 4.2, where

$$W_M(t) = \tfrac{1}{2} M u^2(t) \tag{4.10}$$

It should be apparent that these energy equations suggest an analogy in Figure 4.2 between the velocity of the mass and the voltage across the capacitor.

Example 4.3

Consider a 10-μF capacitor that has been charged by a 12-V battery. The charge stored in the capacitor is given by the expression

$$Q = CV = 10^{-5} \times 12$$
$$= 120\mu C$$

The energy stored in the capacitor is

$$W_C = \tfrac{1}{2}Cv_C^2 = 720 \times 10^{-6} \text{ J} = 720 \ \mu\text{J}$$

Example   4.4   Capacitive Displacement Transducer and Microphone

As shown in Figure 4.1, the capacitance of a parallel-plate capacitor is given by the expression

$$C = \frac{\varepsilon A}{d}$$

where ε is the **permittivity** of the dielectric material, A the area of each of the plates, and d their separation. The permittivity of air is $\varepsilon_0 = 8.854 \times 10^{-12}$F/m, so that two parallel plates of area 1 m^2, separated by a distance of 1 mm, would give rise to a capacitance of $8.854 \times 10^{-3} \mu$F, a very small value for a very large plate area. This relative inefficiency makes parallel-plate capacitors impractical for use in electronic circuits. On the other hand, parallel-plate capacitors find application as *motion transducers,* that is, as devices that can measure the motion or displacement of an object. In a capacitive motion transducer, the air gap between the plates is designed to be variable, typically by fixing one plate and connecting the other to an object in motion. Using the capacitance value just derived for a parallel-plate capacitor, one can obtain the expression

$$C = \frac{8.854A}{x}$$

where C is the capacitance in pF, A is the area of the plates in mm^2, and x is the (variable) distance in mm. It is important to observe that the change in capacitance caused by the displacement of one of the plates is nonlinear, since the capacitance varies as the inverse of the displacement. For small displacements, however, the capacitance varies approximately in a linear fashion.

The *sensitivity, S*, of this motion transducer is defined as the slope of the change in capacitance per change in displacement, x, according to the relation

$$S = \frac{dC}{dx} = -\frac{8.854A}{x^2} \frac{\text{pF}}{\text{mm}}$$

Thus, the sensitivity increases for small displacements. This behavior can be verified by plotting the capacitance as a function of x and noting that as x approaches zero, the slope of the nonlinear $C(x)$ curve becomes steeper (thus the greater sensitivity). Figure 4.7 depicts this behavior for a transducer with area equal to 10 mm^2.

This simple capacitive displacement transducer actually finds use in the popular *capacitive (or condenser) microphone,* in which the sound pressure waves act to displace one of the capacitor plates. The change in capacitance can then be converted into a change in voltage or current by means of a suitable circuit. An extension of this concept that permits measurement of differential pressures is shown in simplified form in Figure 4.8. In the figure, a three-terminal variable capacitor is shown to be made up of two fixed surfaces (typically, spherical depressions ground into glass disks and coated with a conducting material) and of a deflecting plate (typically made of steel) sandwiched between the glass disks. Pressure inlet orifices are provided, so

Figure 4.7 Response of a capacitive displacement transducer

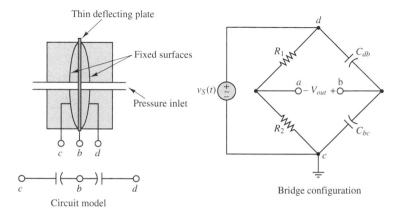

Figure 4.8 Capacitive microphone, and related bridge circuit

that the deflecting plate can come into contact with the fluid whose pressure it is measuring. When the pressure on both sides of the deflecting plate is the same, the capacitance between terminals b and d, C_{bd}, will be equal to that between terminals b and c, C_{bc}. If any pressure differential exists, the two capacitances will change, with an increase on the side where the deflecting plate has come closer to the fixed surface and a corresponding decrease on the other side.

This behavior is ideally suited for the application of a bridge circuit, similar to the Wheatstone bridge circuit studied in Chapters 2 and 3, and also shown in Figure 4.8. In the bridge circuit, the output voltage, v_{out}, is precisely balanced when the differential pressure across the transducer is zero, but it will deviate from zero whenever the two capacitances are not identical because of a pressure differential across the transducer. We shall analyze the bridge circuit in a later example.

The Ideal Inductor

The ideal inductor is an element that has the ability to store energy in a magnetic field. Inductors are typically made by winding a coil of wire around a **core,** which can be an insulator or a ferromagnetic material, as shown in Figure 4.9. When a current flows through the coil, a magnetic field is established, as you may recall from early physics experiments with electromagnets. In an ideal inductor, the resistance of the wire is zero, so that a constant current through the inductor will flow freely without causing a voltage drop. In other words, *the ideal inductor acts as a short circuit in the presence of DC currents*. If a time-varying voltage is established across the inductor, a corresponding current will result, according to the following relationship:

$$v_L(t) = L\frac{di_L}{dt} \tag{4.11}$$

where L is called the *inductance* of the coil and is measured in **henrys (H),** where

$$1 \text{ H} = 1 \text{ V-s/A} \tag{4.12}$$

Henrys are reasonable units for practical inductors; millihenrys (mH) and microhenrys (μH) are also used.

It is instructive to compare equation 4.11, which defines the behavior of an ideal inductor, with the expression relating capacitor current and voltage:

$$i_C(t) = C\frac{dv_C}{dt} \tag{4.13}$$

We note that the roles of voltage and current are reversed in the two elements, but that both are described by a differential equation of the same form. This *duality* between inductors and capacitors can be exploited to derive the same basic results for the inductor that we already have for the capacitor simply by replacing the capacitance parameter, C, with the inductance, L, and voltage with current (and vice versa) in the equations we derived for the capacitor. Thus, the inductor current is found by integrating the voltage across the inductor:

$$i_L(t) = \frac{1}{L}\int_{-\infty}^{t} v_L\,dt \tag{4.14}$$

If the current flowing through the inductor at time $t = t_0$ is known to be I_0, with

$$I_0 = i_L(t = t_0) = \frac{1}{L}\int_{-\infty}^{t_0} v_L\,dt \tag{4.15}$$

then the inductor current can be found according to the equation

$$i_L(t) = \frac{1}{L}\int_{t_0}^{t} v_L\,dt + I_0 \qquad t \geq t_0 \tag{4.16}$$

It is instructive to compare the behavior of the inductor with that of a linear spring subject to a force. Figure 4.10 illustrates the analogy between the circuit equation for the ideal inductor and the equation of motion of a linear spring with coefficient K, subject to an external force, f. Note the

Figure 4.9 Iron-core inductor

The defining equation for the inductance circuit element is analogous to the equation of motion of a spring acted upon by a force.

Figure 4.10 Defining equation for the ideal inductor and analogy with force-spring system

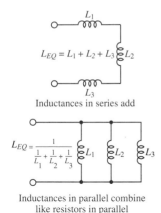

$$L_{EQ} = L_1 + L_2 + L_3$$

Inductances in series add

$$L_{EQ} = \cfrac{1}{\frac{1}{L_1} + \frac{1}{L_2} + \frac{1}{L_3}}$$

Inductances in parallel combine like resistors in parallel

Figure 4.11 Combining inductors in a circuit

analogy between the force acting on the spring (equation 4.17) and the current through the inductor in equation 4.14:

$$f(t) = K \int u \, dt \qquad\qquad (4.17)$$

This analogy will be developed further when the energy-storage properties of the ideal inductor are discussed. Series and parallel combinations of inductors behave like resistors, as illustrated in Figure 4.11, and stated as follows:

> Inductors in series add. Inductors in parallel combine according to the same rules used for resistors connected in parallel.

Example   4.5

An inductor with $L = 10$ mH is subject to the time-varying current $i(t)$ shown in Figure 4.12. Find the voltage across the inductor.

Solution:

The voltage may be found by differentiating the current in a piecewise fashion:

$$v_L(t) = \begin{cases} L\dfrac{di_L(t)}{dt} = 0 \text{ V} & t < 1 \text{ ms} \\[2mm] (0.01)\dfrac{0.1}{4 \times 10^{-3}} = 0.25 \text{ V} & 1 \le t \le 5 \text{ ms} \\[2mm] 0 & 5 < t \le 9 \text{ ms} \\[2mm] -(0.01)\dfrac{0.1}{4 \times 10^3} = -0.25 \text{ V} & 9 \le t \le 13 \text{ ms} \\[2mm] 0 & t \ge 13 \text{ ms} \end{cases}$$

Graphically, we may represent the inductor voltage as shown in Figure 4.13.

$i(t)$

$+$

$L \ v_L(t)$

$-$

$i(t)$ (A)

0.1

0 1   5   9   13 t (ms)

Figure 4.12

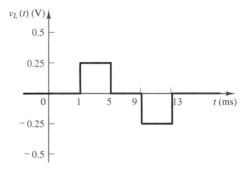

Figure 4.13

Example   4.6

Assume that the inductor of Example 4.5 is connected to a voltage source that provides a constant voltage equal to -10 mV for a period of 1 s, starting at $t = 0$. If no current is flowing through the inductor prior to $t = 0$, sketch the inductor current as a function of time.

Solution:

Integrating the inductor voltage from $t = 0$ to $t = 1$ yields:

$$i_L(t) = \frac{1}{L} \int_0^t V \, dt + 0 \qquad t \geq 0$$

and

$$i_L(t) = \frac{V}{L} t \qquad 1 \geq t \geq 0$$

$$= -t$$

Thus, the inductor current will decrease linearly to a value of -1 A, as shown in Figure 4.14.

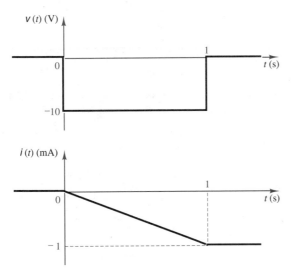

Figure 4.14

Energy Storage in Inductors

The magnetic energy stored in an ideal inductor may be found from a power calculation by following the same procedure employed for the ideal capacitor.

The instantaneous power in the inductor is given by

$$P_L(t) = i_L(t)v_L(t)$$

$$= i_L(t)L\frac{di_L(t)}{dt} \qquad (4.18)$$

$$= \frac{d}{dt}\left[\frac{1}{2}Li_L^2(t)\right]$$

Integrating the power, we obtain the total energy stored in the inductor, as shown in equation 4.19:

$$W_L(t) = \int P_L(t)\,dt = \int \frac{d}{dt}\left[\frac{1}{2}Li_L^2(t)\right]\,dt \qquad (4.19)$$

$$\boxed{W_L(t) = \tfrac{1}{2}Li_L^2(t) \qquad \text{Energy stored in an inductor}}$$

Note, once again, the duality with the expression for the energy stored in a capacitor, in equation 4.9. Further, the expression for the energy stored in the inductor may be compared with the energy stored in the spring of Figure 4.10, given the following equation:

$$W_K = \tfrac{1}{2}Kf^2 \qquad (4.20)$$

Example   4.7

In this example we compute the energy stored in the inductor of Example 4.5. First, we compute the current for each part of the current waveform:

$$i_L(t) = \begin{cases} 0\text{ A} & t < 1\text{ ms} \\ \left(\frac{0.1}{0.004}\right)\left(t - 10^{-3}\right) & 1 \le t < 5\text{ ms} \\ 0.1 & 5 \le t < 9\text{ ms} \\ \left(-\frac{0.1}{0.004}\right)\left(t - 13 \times 10^{-3}\right) & 9 \le t < 13\text{ ms} \\ 0 & t \ge 13\text{ ms} \end{cases}$$

Then we compute the energy according to equation 4.19:

$$W_L(t) = \begin{cases} 0 & t < 1\text{ ms} \\ 3.125(t - 10^{-3})^2 & 1 \le t < 5\text{ ms} \\ 0.00005 & 5 \le t < 9\text{ ms} \\ -3.125(t - 13 \times 10^{-3})^2 & 9 \le t < 13\text{ ms} \\ 0 & t \ge 13\text{ ms} \end{cases}$$

Note that energy is stored in the inductor only when a current is flowing through it. No energy is stored when $i_L(t) = 0$. It is also useful to observe that energy is stored in the inductor when a DC (constant) current flows through it, although there is no voltage induced across the inductor, since $v_L = L\frac{di_L}{dt} = 0$ for a constant current.

Analogy between Electrical and Mechanical Elements

To complete the analogy between electrical and mechanical systems developed in this section (see Figures 4.2 and 4.10), we again call attention to the role of the resistor as a power-dissipating element, and to its mechanical counterpart, the **damper.** Figure 4.15 illustrates the fundamental relationship

The defining equation for the resistance circuit element is analogous to the equation of motion of a damper acted upon by a force.

Figure 4.15 Analogy between electrical and mechanical elements

between voltage and current in a resistor, on the one hand, and force and velocity in a damper, on the other. Note that the basic form of the equations is the same. The analogies between mechanical and electrical systems are summarized in Table 4.2.

Table 4.2   Analogy between electrical and mechanical variables

Mechanical system	Electrical system
Force, f(N)	Current, i (A)
Velocity, u (m/s)	Voltage, v (V)
Damping, B (N-s/m)	Conductance, $\frac{1}{R}$ (S)
Compliance, $\frac{1}{K}$ (m/N)	Inductance, L (H)
Mass, M (kg)	Capacitance, C (F)

DRILL EXERCISES

4.1   The current waveform shown in Figure 4.16 flows through a 50-mH inductor. Plot the inductor voltage, $v_L(t)$.

Figure 4.16   **Figure 4.17**

4.2   The voltage wave form of Figure 4.17 appears across a 1,000-μF capacitor. Plot the capacitor current, $i_C(t)$.

4.3　Calculate the energy stored in the inductor at $t = 3$ ms by the waveform of Drill Exercise 4.1.

4.4　Perform the calculation of Drill Exercise 4.3 for the capacitor if $v_C(0) = 0$ V.

4.5　Compute and plot the inductor energy and power for the case of Drill Exercise 4.1.

4.2　TIME-DEPENDENT SIGNAL SOURCES

In Chapter 2, the general concept of an ideal energy source was introduced. In the present chapter, it will be useful to specifically consider sources that generate time-varying voltages and currents and, in particular, sinusoidal sources. Figure 4.18 illustrates the convention that will be employed to denote time-dependent signal sources.

Generalized time-dependent sources　　　Sinusoidal source

Figure 4.18 Time-dependent signal sources

　　One of the most important classes of time-dependent signals is that of **periodic signals.** These signals appear frequently in practical applications and are a useful approximation of many physical phenomena. A periodic signal $x(t)$ is a signal that satisfies the following equation:

$$x(t) = x(t + nT) \qquad n = 1, 2, 3, \dots \qquad (4.21)$$

where T is the **period** of $x(t)$. Figure 4.19 illustrates a number of the periodic waveforms that are typically encountered in the study of electrical circuits. Waveforms such as the sine, triangle, square, pulse, and sawtooth waves are provided in the form of voltages (or, less frequently, currents) by commercially available **signal** (or **waveform**) **generators.** Such instruments allow for selection of the waveform peak amplitude, and of its period.

　　As stated in the introduction, sinusoidal waveforms constitute by far the most important class of time-dependent signals. Figure 4.20 depicts the relevant parameters of a sinusoidal waveform. A generalized sinusoid is defined as follows:

$$x(t) = A \cos(\omega t + \phi) \qquad (4.22)$$

where A is the **amplitude,** ω the **radian frequency,** and ϕ the **phase.** Figure 4.20 summarizes the definitions of A, ω, and ϕ for the waveforms

$$x_1(t) = A \cos(\omega t) \qquad \text{and} \qquad x_2(t) = A \cos(\omega t + \phi)$$

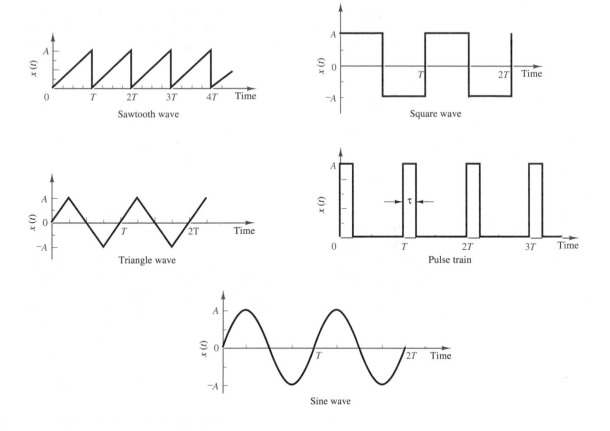

Figure 4.19 Periodic signal waveforms

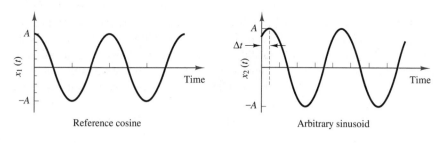

Figure 4.20 Sinusoidal waveforms

where

$$f = \text{natural frequency} = \frac{1}{T} \text{ (cycles/s, or Hz)}$$

$$\omega = \text{radian frequency} = 2\pi f \text{ (radians/s)} \qquad \textbf{(4.23)}$$

$$\phi = 2\pi \frac{\Delta T}{T} \text{ (radians)} = 360\frac{\Delta T}{T} \text{ (degrees)}$$

The phase shift, ϕ, permits the representation of an arbitrary sinusoidal signal. Thus, the choice of the reference cosine function to represent sinusoidal signals—arbitrary as it may appear at first—does not restrict the ability to represent all sinusoids. For example, one can represent a sine wave in terms of a cosine wave simply by introducing a phase shift of $\pi/2$ radians:

$$A \sin \omega t = A \cos \left(\omega t + \frac{\pi}{2} \right) \tag{4.24}$$

It is important to note that, although one usually employs the variable ω (in units of radians per second) to denote sinusoidal frequency, it is common to refer to natural frequency, f, in units of cycles per second, or **Hertz (Hz)**. The reader with some training in music theory knows that a sinusoid represents what in music is called a *pure tone;* an A-440, for example, is a tone at a frequency of 440 Hz. It is important to be aware of the factor of 2π that differentiates radian frequency (in units of rad/s) from natural frequency (in units of Hz). The distinction between the two units of frequency—which are otherwise completely equivalent—is whether one chooses to define frequency in terms of revolutions around a trigonometric circle (in which case the resulting units are rad/s), or to interpret frequency as a repetition rate (cycles/second), in which case the units are Hz. The relationship between the two is the following:

$$\boxed{\omega = 2\pi f} \tag{4.25}$$

Why Sinusoids?

You should by now have developed a healthy curiosity about why so much attention is being devoted to sinusoidal signals. Perhaps the simplest explanation is that the electric power used for industrial and household applications worldwide is generated and delivered in the form of either 50- or 60-Hz voltages and currents. Chapter 5 will provide more detail regarding the analysis of electric power circuits. The more ambitious reader may explore the box "Fourier Analysis" to obtain a more comprehensive explanation of the importance of sinusoidal signals. It should be remarked that the methods developed in this section and the subsequent sections apply to many engineering systems, not just to electrical circuits, and will be encountered again in the study of dynamic-system modeling and control systems.

Average and RMS Values

Now that a number of different signal waveforms have been defined, it is appropriate to define suitable measurements for quantifying the strength of a time-varying electrical signal. The most common types of measurements are the **average** (or **DC**) **value** of a signal waveform—which corresponds to just measuring the mean voltage or current over a period of time—and the **root-mean-square** (or **rms**) **value,** which takes into account the fluctuations of the signal about its average value. Formally, the operation of computing the average value of a signal corresponds to integrating the signal waveform over some (presumably, suitably chosen) period of time. We define the time-averaged value of a signal $x(t)$ as

$$\langle x(t) \rangle = \frac{1}{T} \int_0^T x(t)\, dt \tag{4.26}$$

where T is the period of integration. Figure 4.21 illustrates how this process does, in fact, correspond to computing the average amplitude of $x(t)$ over a period of T seconds.

Figure 4.21 Averaging a signal waveform

Example   4.8

In this example, we illustrate the computation of the time average of a periodic signal by integration over one period. We seek to find the average value of the signal

$$x(t) = 10\cos(100t)$$

Solution:

Since the signal is periodic, with period

$$T = \frac{2\pi}{\omega} = \frac{2\pi}{100}$$

we need only integrate over one period of the waveform to obtain the average value:

$$\langle x(t) \rangle = \frac{100}{2\pi} \int_0^{2\pi/100} 10\cos(100t)\, dt = \frac{10}{2\pi}(\sin 2\pi - \sin 0) = 0$$

This result states an important conclusion: that the average value of the reference cosine is identically zero, independent of its amplitude and frequency.

The result of Example 4.8 can be generalized to state that

$$\langle A\cos(\omega t + \phi) \rangle = 0 \tag{4.27}$$

a result that might be perplexing at first: if any sinusoidal voltage or current has zero average value, is its average power equal to zero? Clearly, the answer must be no. Otherwise, it would be impossible to illuminate households and

Fourier Analysis

In this brief introduction to **Fourier theory,** we shall explain in an intuitive manner how it is possible to represent many signals by means of the superposition of various sinusoidal signals of different amplitude, phase, and frequency. Any periodic finite-energy signals may be expressed by means of an infinite sum of sinusoids, as illustrated in the following paragraphs.

Consider a periodic waveform, $x(t)$. Its Fourier series representation is defined below by the infinite summation of sinusoids at the frequencies $n\omega_0$ (integer multiples of the **fundamental frequency,** ω_0), with amplitudes A_n and phases ϕ_n.

$$x(t) = x(t + T_0) \qquad T_0 = \text{period} \qquad \textbf{(4.28)}$$

$$x(t) = \sum_{n=0}^{\infty} A_n \cos\left(\frac{2\pi nt}{T_0} + \phi_n\right) \qquad \textbf{(4.29)}$$

One could also write the term $2\pi n/T_0$ as $n\omega_0$, where

$$\omega_0 = \frac{2\pi}{T_0} = 2\pi f_0 \qquad \textbf{(4.30)}$$

is the fundamental (radian) frequency and the frequencies $2\omega_0$, $3\omega_0$, $4\omega_0$, and so on, are called its **harmonics.**

The notion that a signal may be represented by sinusoidal components is particularly useful, and not only in the study of electrical circuits—in the sense that we need only understand the response of a circuit to an arbitrary sinusoidal excitation in order to be able to infer the circuit's response to more complex signals. In fact, the frequently employed *sinusoidal frequency response* discussed in this chapter and in Chapter 6 is a function that enables us to explain how a circuit would respond to a signal made up of a superposition of sinusoidal components at various frequencies. These sinusoidal components form the **spectrum** of the signal, that is, its frequency composition; the amplitude and phase of each of the sinusoids contribute to the overall "character" of the signal, in the same sense as the timbre of a musical instrument is made up of the different harmonics that are generated when a note is played (the timbre is what differentiates, for example, a viola from a cello or a violin). An example of the amplitude spectrum of a "square-wave" signal is shown in Figure 4.22. In order to further illustrate how the superposition of sinusoids can give rise to a signal that at first might appear substantially different from a sinusoid, the evolution of a sine wave into a square wave is displayed in Figure 4.23, as more Fourier components are added. The first picture represents the fundamental component,

Square wave

Fourier spectrum of square wave

Figure 4.22 Amplitude spectrum of square wave

Fourier Analysis (continued)

that is, the sinusoid that has the same frequency as the square wave. Then one harmonic at a time is added, up to the fifth nonzero component (the ninth frequency component; see Figure 4.22), illustrating how, little by little, the rounded peaks

of the sinusoid transform into the flat top of the square wave!

Although this book will not deal with the mathematical aspects of Fourier series, it is important to recognize that this analysis tool provides excellent motivation for the study of sinusoidal signals.

Jean Baptiste Joseph Fourier (1768–1830), French mathematician and physicist who formulated the Fourier series. *Photo courtesy of Deutches Museum, Munich.*

Three frequency components

Fundamental frequency

Four frequency components

Two frequency components

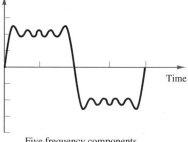

Five frequency components

Figure 4.23 Evolution of a square wave from its Fourier components streets and power industrial machinery with 60-Hz sinusoidal current! There must be another way, then, of quantifying the strength of an AC signal.

Very conveniently, a useful measure of the voltage of an AC waveform is the root-mean-square, or rms, value of the signal, $x(t)$, defined as follows:

$$x_{rms} = \sqrt{\frac{1}{T} \int_0^T x^2(t)\, dt} \tag{4.31}$$

Note immediately that if $x(t)$ is a voltage, the resulting x_{rms} will also have units of volts. If you analyze equation 4.31, you can see that, in effect, the rms value consists of the square *root* of the average (or *mean*) of the *square* of the signal. Thus, the notation *rms* indicates exactly the operations performed on $x(t)$ in order to obtain its rms value.

Example   4.9

In this example we illustrate the procedure one follows in computing the rms value of a signal, by computing the rms value of the sinusoidal current

$$i(t) = I \cos(\omega t)$$

Applying the definition of rms value, we compute

$$i_{rms} = \sqrt{\frac{\omega}{2\pi} \int_0^{2\pi/\omega} I^2 \cos^2 \omega t\, dt}$$

$$= \sqrt{\frac{\omega}{2\pi} \int_0^{2\pi/\omega} I^2 \left(\frac{1}{2} + \frac{1}{2}\cos^2 \omega t\right) dt}$$

$$= \sqrt{\frac{I^2}{2} + \frac{\omega}{2\pi} \int_0^{2\pi/\omega} \frac{I^2}{2}\cos 2\omega t\, dt}$$

At this point, we recognize that the integral under the square root sign is equal to zero, since we are averaging the signal $\cos 2\omega t$ over two periods. Therefore, we conclude that the rms value of the sinusoidal current is

$$i_{rms} = \frac{I}{\sqrt{2}} = 0.707\, I$$

where I is the *peak value* of $i(t)$. This result is clearly independent of the frequency and phase of the waveform.

The preceding example illustrates how the rms value of a sinusoid is proportional to its peak amplitude. The factor of $0.707 = 1/\sqrt{2}$ is a useful number to remember, since it applies to any sinusoidal signal. It is not, however, generally applicable to signal waveforms other than sinusoids, as the Drill Exercises will illustrate.

DRILL EXERCISES

4.6 Express the voltage $v(t) = 155.6 \sin(377t + 60°)$ in cosine form. You should note that the radian frequency $\omega = 377$ will recur very often, since $377 = 2\pi 60$; that is, 377 is the radian equivalent of the natural frequency of 60 cycles/second, which is the frequency of the electric power generated in North America.

4.7 Compute the average value of the sawtooth waveform shown in Figure 4.24.

Figure 4.24

Answer: $\langle v(t) \rangle = 2.5$ V

4.8 Compute the average value of the shifted triangle wave shown in Figure 4.25.

Figure 4.25

Answer: $\langle v(t) \rangle = 1.5$ V

4.9 Find the rms value of the sawtooth wave of Drill Exercise 4.7.

Answer: 2.89 V

4.10 Find the rms value of the half cosine wave shown in Figure 4.26.

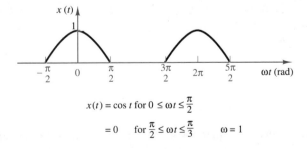

$$x(t) = \cos t \text{ for } 0 \le \omega t \le \frac{\pi}{2}$$

$$= 0 \quad \text{for } \frac{\pi}{2} \le \omega t \le \frac{\pi}{3} \qquad \omega = 1$$

Figure 4.26

Answer: 0.5 V

4.3   SOLUTION OF CIRCUITS CONTAINING DYNAMIC ELEMENTS

The first two sections of this chapter introduced energy-storage elements and time-dependent signal sources. The next logical task is to analyze the behavior of circuits containing such elements. The major difference between the analysis of the resistive circuits studied in Chapters 2 and 3 and the circuits we will explore in the remainder of this chapter is that now the equations that result from applying Kirchhoff's laws are differential equations, as opposed to the algebraic equations obtained in solving resistive circuits. Consider, for example, the circuit of Figure 4.27, which consists of the series connection of a voltage source, a resistor, and a capacitor. Applying KVL around the loop, we may obtain the following equation:

A circuit containing energy-storage elements is described by a differential equation. The differential equation describing the series RC circuit shown is
$$\frac{di_C}{dt} + \frac{1}{RC}i_C = \frac{dv_S}{dt}$$

Figure 4.27 Circuit containing energy-storage element

$$v_S(t) = v_R(t) + v_C(t) \tag{4.32}$$

Observing that $i_R = i_C$, equation 4.32 may be combined with the defining equation for the capacitor (equation 4.6) to obtain

$$v_S(t) = Ri_C(t) + \frac{1}{C}\int_{-\infty}^{t} i_C \, dt \tag{4.33}$$

Equation 4.33 is an integral equation, which may be converted to the more familiar form of a differential equation by differentiating both sides of the equation, and recalling that

$$\frac{d}{dt}\left(\int_{-\infty}^{t} i_C(t) \, dt\right) = i_C(t) \tag{4.34}$$

to obtain the following differential equation:

$$\frac{di_C}{dt} + \frac{1}{RC}i_C = \frac{1}{R}\frac{dv_S}{dt} \tag{4.35}$$

where the argument (t) has been dropped for ease of notation.

Observe that in equation 4.35, the independent variable is the series current flowing in the circuit, and that this is not the only equation that describes the series RC circuit. If, instead of applying KVL, for example, we had applied KCL at the node connecting the resistor to the capacitor, we would have obtained the following relationship:

$$i_R = \frac{v_S - v_C}{R} = i_C = C\frac{dv_C}{dt} \tag{4.36}$$

or

$$\frac{dv_C}{dt} + \frac{1}{RC}v_C = \frac{1}{RC}v_S \tag{4.37}$$

Note the similarity between equations 4.35 and 4.37. The left-hand side of both equations is identical, except for the independent variable, while the right-hand side takes a slightly different form. The solution of either equation is sufficient, however, to determine all voltages and currents in the circuit.

The following example illustrates the derivation of the differential equation
for another simple circuit containing an energy-storage element.

Example   4.10

In this example we illustrate the derivation of the differential equation governing the
behavior of the circuit shown in Figure 4.28. Assume that $R_1 = 10 \ \Omega$, $R_2 = 5 \ \Omega$,
and $L = 0.4$ H; $v_S(t)$ is an arbitrary function of time.

Solution:

One possible approach to deriving the differential equation for the circuit is to apply
KCL to the node at the top right corner of the circuit, obtaining the equation

$$\frac{v_S - v_L}{R_1} = i_L + \frac{v_L}{R_2}$$

Figure 4.28

(since the voltage across R_2 is the same as the inductor voltage). Next, the i-v
relationship for the inductor (equation 4.11) may be used to eliminate the variable
v_L from the equation:

$$\frac{v_S}{R_1} - \frac{L}{R_1}\frac{di_L}{dt} = i_L + \frac{L}{R_2}\frac{di_L}{dt}$$

to obtain the final form of the equation:

$$\frac{di_L}{dt} + \frac{R_1 R_2}{L(R_1 + R_2)}i_L = \frac{R_2}{L(R_1 + R_2)}v_S$$

or

$$\frac{di_L}{dt} + 8.33 i_L = 0.833 v_S$$

The preceding discussion and accompanying example have illustrated
how application of the circuit laws to circuits containing energy-storage elements leads to the formulation of ordinary differential equations. The order
of the equation depends on the number of energy-storage elements present
in the circuit; for the sake of simplicity, only first-order circuits have been
considered in this discussion. You may recall from an earlier course in differential equations that the solution of ordinary differential equations depends
in part on the forcing function (in this case, the voltage sources in the circuits
of Figures 4.27 and 4.28); thus, the complete solution of equation 4.37, for
example, can be obtained only if the function $v_S(t)$ is known. In practical
electrical engineering circuits, two types of time-dependent source signals are
commonly encountered: sinusoidal sources and switched DC sources. The
former type is of great practical interest to all engineers, since any machine
or tool that is powered by the local electric utility employs 60-Hz sinusoidal
voltages and currents; the latter type of signal is encountered whenever a DC
electric source (e.g., a battery) is switched on or off. The remainder of the

chapter will be exclusively concerned with the analysis of circuits excited by sinusoidal sources. These circuits are commonly referred to as **AC circuits.** It will also be shown that when the excitation source (or forcing function, to use the terminology of differential equations) is a sinusoidal signal, it is not necessary to formulate and solve the differential equation describing the circuit, thanks to a special notation that permits the user to formulate the circuit equations using techniques similar to those studied in Chapter 3. The general solution of dynamic circuits will be explored in Chapter 6.

Solution of Circuits Excited by Sinusoidal Sources

Consider again the circuit of Figure 4.27, where now the external source produces a sinusoidal voltage, described by the expression

$$v_S(t) = V \cos (\omega t) \tag{4.38}$$

Substituting the expression $V \cos (\omega t)$ in place of the source voltage, $v_S(t)$, in the differential equation obtained earlier (equation 4.37), we obtain the following differential equation:

$$\frac{d}{dt}v_C + \frac{1}{RC}v_C = \frac{1}{RC}V \cos \omega t \tag{4.39}$$

Since the forcing function is a sinusoid, the solution may also be assumed to be of the same form. The most general expression for $v_C(t)$ is then the following:

$$v_C(t) = A \sin \omega t + B \cos \omega t \tag{4.40}$$

which is equivalent to

$$v_C(t) = C \cos (\omega t + \phi) \tag{4.41}$$

Substituting equation 4.40 in the differential equation for $v_C(t)$ and solving for the coefficients A and B yields the expression

$$A\omega \cos \omega t - B\omega \sin \omega t + \frac{1}{RC}(A \sin \omega t + B \cos \omega t) = \frac{1}{RC}V \cos \omega t \tag{4.42}$$

and if the coefficients of like terms are grouped, the following equation is obtained:

$$\left(\frac{A}{RC} - B\omega \right) \sin \omega t + \left(A\omega + \frac{B}{RC} - \frac{V}{RC} \right) \cos \omega t = 0 \tag{4.43}$$

The coefficients of $\sin \omega t$ and $\cos \omega t$ must both be identically zero in order for equation 4.43 to hold. Thus,

$$\frac{A}{RC} - B\omega = 0$$

and $\tag{4.44}$

$$A\omega + \frac{B}{RC} - \frac{V}{RC} = 0$$

The unknown coefficients, A and B, may now be determined by solving equation 4.44, to obtain:

$$A = \frac{V\omega RC}{1 + \omega^2(RC)^2}$$

$$B = \frac{V}{1 + \omega^2(RC)^2}$$

(4.45)

Thus, the solution for $v_C(t)$ may be written as follows:

$$v_C(t) = \frac{V\omega RC}{1 + \omega^2(RC)^2} \sin \omega t + \frac{V}{1 + \omega^2(RC)^2} \cos \omega t$$

(4.46)

This equation is plotted in Figure 4.29.

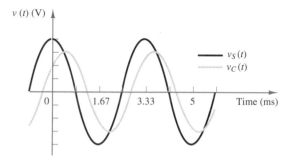

Figure 4.29 Waveforms for the AC circuit of Figure 4.27

The solution method outlined in the previous paragraphs can become quite complicated for circuits containing a large number of elements; in particular, one may need to solve higher-order differential equations if more than one energy-storage element is present in the circuit. A simpler and preferred method for the solution of AC circuits will be presented in the next section. This brief section has provided a simple, but complete, illustration of the key elements of AC circuit analysis. These can be summarized in the following statement:

> In a sinusoidally excited linear circuit, all branch voltages and currents are *sinusoids* at the *same frequency* as the excitation signal. The amplitudes of these voltages and currents are a *scaled* version of the excitation *amplitude,* and the voltages and currents may be *shifted in phase* with respect to the excitation signal.

These observations indicate that three parameters uniquely define a sinusoid: *frequency, amplitude,* and *phase.* But if this is the case, is it necessary to carry the "excess luggage," that is, the sinusoidal functions? Might it be possible to simply keep track of the three parameters just mentioned? Fortunately, the answers to these two questions are no and yes, respectively. The next section will describe the use of a notation that, with the aid of complex algebra, eliminates the need for the sinusoidal functions of time, and for the formulation and solution of differential equations, permitting the use of simpler algebraic methods.

DRILL EXERCISES

4.11 Show that the solution to either equation 4.35 or equation 4.37 is sufficient to compute all of the currents and voltages in the circuit of Figure 4.27.

4.12 Show that the equality

$$A \sin \omega t + B \cos \omega t = C \cos (\omega t + \phi)$$

holds if

$$A = -C \sin \phi$$
$$B = C \cos \phi$$

or, conversely, if

$$C = \sqrt{A^2 + B^2}$$
$$\phi = \tan^{-1} \left(\frac{-A}{B} \right)$$

4.13 Use the result of Drill Exercise 4.12 to compute C and ϕ as functions of V, ω, R, and C in equation 4.46.

Answer: $C = \dfrac{V}{\sqrt{1+(\omega RC)^2}}$

$\phi = \tan^{-1}(-\omega RC)$

4.14 Sketch $v_S(t)$ and $v_C(t)$ of the circuit of Figure 4.27 if $R = 1$ kΩ, $C = 0.47\mu$F, and $v_S(t) = 10\cos(\omega t)$, with $\omega = 600\pi$ rad/s. Compute the exact amplitude and time delay of $v_C(t)$.

Answer: The curves are displayed in Figure 4.29.

4.4 PHASORS AND IMPEDANCE

In this section, we introduce an efficient notation to make it possible to represent sinusoidal signals as *complex numbers,* and to eliminate the need for solving differential equations. The student who needs a brief review of complex algebra will find a reasonably complete treatment in Appendix A, including solved examples and drill exercises. For the remainder of the chapter, it will be assumed that you are familiar with both the rectangular and the polar forms of complex number coordinates, with the conversion between these two forms, and with the basic operations of addition, subtraction, multiplication, and division of complex numbers.

Euler's Identity

Named after the Swiss mathematician Leonhard Euler[2] (the last name is pronounced "Oiler"), Euler's identity forms the basis of phasor notation. Simply stated, the identity defines the **complex exponential**[3] $e^{j\theta}$ as a point in the complex plane, which may be represented by real and imaginary components:

$$e^{j\theta} = \cos\theta + j\sin\theta \tag{4.47}$$

Figure 4.30 illustrates how the complex exponential may be visualized as a point (or vector, if referenced to the origin) in the complex plane. Note immediately that the magnitude of $e^{j\theta}$ is equal to 1:

$$\left| e^{j\theta} \right| = 1 \tag{4.48}$$

since

$$\left| \cos\theta + j\sin\theta \right| = \cos^2\theta + \sin^2\theta = 1 \tag{4.49}$$

and note also that writing Euler's identity corresponds to equating the polar

Leonhard Euler (1707–1783).
Photo courtesy of Deutches Museum, Munich.

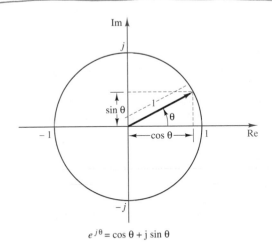

$$e^{j\theta} = \cos\theta + j\sin\theta$$

Figure 4.30 Euler's identity

---

[2] Swiss mathematician whose work forms the basis of phasor notation.
[3] Note that in electrical engineering notation the symbol j, rather than i, is used to denote the square root of (-1); this is required because the symbol i is used to denote current.

form of a complex number to its rectangular form. For example, consider a vector of length A making an angle θ with the real axis. Equation 4.50 illustrates the relationship between the rectangular and polar forms.

$$Ae^{j\theta} = A\cos\theta + jA\sin\theta \tag{4.50}$$

In effect, Euler's identity is simply a trigonometric relationship in the complex plane.

Phasors

To see how complex numbers can be used to represent sinusoidal signals, rewrite the expression for a generalized sinusoid in light of Euler's equation:

$$A\cos(\omega t + \phi) = \text{Re}\left[Ae^{j(\omega t + \phi)}\right] \tag{4.51}$$

This equality is easily verified by expanding the right-hand side, as follows:

$$\text{Re}[Ae^{j(\omega t + \phi)}]$$
$$= \text{Re}[A\cos(\omega t + \phi) + jA\sin(\omega t + \phi)]$$
$$= A\cos(\omega t + \phi)$$

We see, then, that *it is possible to express a generalized sinusoid as the real part of a complex vector* whose **argument, or angle,** is given by $(\omega t + \phi)$ and whose length, or **magnitude,** is equal to the peak amplitude of the sinusoid. The **complex phasor** corresponding to the sinusoidal signal $A\cos(\omega t + \phi)$ is therefore defined to be the complex number $Ae^{j\phi}$:

$$Ae^{j\phi} = \text{complex phasor notation for } A\cos(\omega t + \phi) \tag{4.52}$$

It is important to explicitly point out that this is a *definition*. Phasor notation arises from equation 4.51; however, this expression is simplified (for convenience, as will be promptly shown) by removing the "real part of" operator (Re) and factoring out and deleting the term $e^{j\omega t}$. Equation 4.53 illustrates the simplification.

$$A\cos(\omega t + \phi) = \text{Re}[Ae^{j(\omega t + \phi)}] = \text{Re}[Ae^{j\phi}e^{j\omega t}] \tag{4.53}$$

The reason for this simplification is simply mathematical convenience, as will become apparent in the examples; you will have to remember that the $e^{j\omega t}$ term that was removed from the complex form of the sinusoid is really still present, indicating the specific frequency of the sinusoidal signal, ω. With these caveats, you should now be prepared to use the newly found phasor to analyze AC circuits. The following comments summarize the important points developed thus far in the section.

1. Any sinusoidal signal may be mathematically represented in one of two ways: a **time-domain form,**

$$v(t) = A\cos(\omega t + \phi)$$

and a **frequency-domain** (or **phasor**) **form,**

$$\mathbf{V}(\omega) = Ae^{j\phi}$$

Note the "ω" in the notation $\mathbf{V}(\omega)$, indicating the $e^{j\omega t}$ dependence of the phasor. In the remainder of this chapter, bold uppercase quantities will be employed to indicate phasor voltages or currents.

2. A phasor is a complex number, expressed in polar form, consisting of a *magnitude* equal to the peak amplitude of the sinusoidal signal and a *phase angle* equal to the phase shift of the sinusoidal signal *referenced to a cosine signal.*

3. When using phasor notation, it is important to make a note of the specific frequency, ω, of the sinusoidal signal, since this is not explicitly apparent in the phasor expression.

Example   4.11

Two sinusoidal sources of equal frequency, but of different phase, can be added using phasor notation. Let $v_1(t)$ and $v_2(t)$ be the source voltages defined by

$$v_1(t) = 15 \cos\left(377t + \frac{\pi}{4}\right)$$

and

$$v_2(t) = 15 \cos\left(377t + \frac{\pi}{12}\right)$$

If the two sources are connected in series, as shown in Figure 4.31, what is the resultant voltage, $v_S(t)$?

Figure 4.31

Solution:

Write $v_1(t)$ and $v_2(t)$ in phasor form:

$$\mathbf{V}_1(\omega) = 15e^{j45°}$$
$$\mathbf{V}_2(\omega) = 15e^{j15°}$$

To compute the complex addition of $\mathbf{V}_1$ and $\mathbf{V}_2$, the phasors must first be converted to rectangular form:

$$\mathbf{V}_1 = 15e^{j15°} = 14.49 + j3.88$$
$$\mathbf{V}_2 = 15e^{j45°} = 10.61 + j10.61$$

so that

$$V_S = V_1 + V_2 = 25.10 + j14.49 = 28.98e^{j30°}$$

Having computed the phasor sum, we may convert V_S to time-domain form:

$$v_S(t) = 29.98 \cos(377t + 30°)$$

Note that the same result could have been obtained by adding the two sinusoids in the time domain, utilizing trigonometric identities. Expanding $v_1(t)$ and $v_2(t)$, we obtain

$$v_1(t) = 15 \cos(\omega t + 45°) = 15 \cos(45°) \cos(\omega t) - 15 \sin(45°) \sin(\omega t)$$
$$v_2(t) = 15 \cos(\omega t + 15°) = 15 \cos(15°) \cos(\omega t) - 15 \sin(15°) \sin(\omega t)$$

Then, combining like terms to get

$$v_1(t) + v_2(t) = 15(\cos 15° + \cos 45°) \cos \omega t - 15(\sin 15° + \sin 45°) \sin \omega t$$

we obtain

$$v_1(t) + v_2(t) = 15(1.673 \cos \omega t - 0.966 \sin \omega t)$$
$$= 15\sqrt{(1.673)^2 + (0.966)^2} \times \cos\left(\omega t + \tan^{-1}\left(\frac{0.966}{1.673}\right)\right)$$
$$= 15(1.932 \cos(\omega t + 30°))$$
$$= 29 \cos(\omega t + 30°)$$

which is, of course, the same answer we obtained using phasors. Which of the two solution methods required less labor?

It should be apparent by now that phasor notation can be a very efficient technique to solve AC circuit problems. The following sections will continue developing this new method to build your confidence in using it.

Superposition of AC Signals

Example 4.11 explored the combined effect of two sinusoidal sources of different phase and amplitude, but of the same frequency. It is important to realize that the simple answer obtained there does not apply to the superposition of two (or more) sinusoidal sources that *are not at the same frequency.* In this subsection, the case of two sinusoidal sources oscillating at different frequencies will be used to illustrate how phasor analysis can deal with this more general case.

The circuit shown in Figure 4.32 depicts a source excited by two current sources connected in parallel, where

$$i_1(t) = A_1 \cos(\omega_1 t)$$
$$i_2(t) = A_2 \cos(\omega_2 t)$$
(4.54)

Figure 4.32 Superposition of AC currents

The load current is equal to the sum of the two source currents; that is,

$$i_L(t) = i_1(t) + i_2(t)$$
(4.55)

or, in phasor form,

$$\mathbf{I}_L = \mathbf{I}_1 + \mathbf{I}_2 \qquad (4.56)$$

At this point, you might be tempted to write $\mathbf{I}_1$ and $\mathbf{I}_2$ in a more explicit phasor form as

$$\mathbf{I}_1 = A_1 e^{j0}$$
$$\mathbf{I}_2 = A_2 e^{j0} \qquad (4.57)$$

and to add the two phasors using the familiar techniques of complex algebra. However, this approach *would be incorrect*. Whenever a sinusoidal signal is expressed in phasor notation, the term $e^{j\omega t}$ is intrinsically present, where ω is the actual radian frequency of the signal. In our example, the two frequencies are not the same, as can be verified by writing the phasor currents in the form of equation 4.51:

$$\mathbf{I}_1 = \mathrm{Re}\left[A_1 e^{j0} e^{j\omega_1 t}\right]$$
$$\mathbf{I}_2 = \mathrm{Re}\left[A_2 e^{j0} e^{j\omega_2 t}\right] \qquad (4.58)$$

Since phasor notation does not *explicitly* include the $e^{j\omega t}$ factor, this can lead to serious errors if you are not careful! The two phasors of equation 4.57 cannot be added, but must be kept separate; thus, the only unambiguous expression for the load current in this case is equation 4.55. In order to complete the analysis of any circuit with multiple sinusoidal sources at different frequencies using phasors, it is necessary to solve the circuit separately for each signal and then add the individual answers obtained for the different excitation sources. Example 4.12 illustrates the response of a circuit with two separate AC excitations using AC superposition.

Example   4.12

The circuit of Figure 4.33 is excited by two sources, both sinusoidal. The current source has a frequency of 100 Hz and can be expressed as

$$i_S(t) = 0.5 \cos(2\pi 100 t)$$

or

$$\mathbf{I}_S = 0.5\angle 0°$$

The voltage source has a frequency of 1,000 Hz and can be expressed as

$$v_S(t) = 20 \cos(2\pi 1{,}000 t)$$

or

$$\mathbf{V}_S = 20\angle 0°$$

Find the voltages $v_{R_2}(t)$ and $v_{R_1}(t)$.

$R_1 = 150\ \Omega,\ R_2 = 50\ \Omega$

Figure 4.33

Solution:

Since we have sources of different frequency, we are forced to find a separate solution for each source. First, we consider the effect of the current source. To do so, we replace the voltage source with a short circuit, as shown in Figure 4.34. It should be clear that we have reduced the circuit to a current divider, and that the voltages indicated in the figure are given by

Figure 4.34

$$\mathbf{V}_{R_1}(\mathbf{I}_S) = \mathbf{I}_{R_1}R_1$$

$$\mathbf{V}_{R_1}(\mathbf{I}_S) = \mathbf{I}_{R_2}R_2$$

$$\mathbf{V}_{R_1}(\mathbf{I}_S) = \mathbf{I}_S\frac{R_2}{R_2 + R_1}R_1 = 0.5\angle0°\left(\frac{50}{50 + 150}\right)(150)$$

$$= 18.75\angle0°$$

$$\mathbf{V}_{R_2}(\mathbf{I}_S) = \mathbf{I}_S\frac{R_1}{R_2 + R_1}R_2 = 0.5\angle0°\left(\frac{150}{50 + 150}\right)(50)$$

$$= 18.75\angle0°$$

Next, we consider the effect of the voltage source. Replacing the current source, with an open circuit, we obtain the voltage divider shown in Figure 4.35, from which we obtain the expressions

Figure 4.35

$$\mathbf{V}_{R_1}(\mathbf{V}_S) = \mathbf{V}_S\frac{R_1}{R_2 + R_1} = 20\angle0°\left(\frac{150}{50 + 150}\right)$$

$$= 15\angle0°$$

$$\mathbf{V}_{R_2}(\mathbf{V}_S) = -\mathbf{V}_S\frac{R_2}{R_2 + R_1} = -20\angle0°\left(\frac{50}{50 + 150}\right)$$

$$= -5\angle0° = 5\angle180°$$

Adding the contribution from each of the sources and converting to time-domain representation yields

$$\mathbf{V}_{R_1} = \mathbf{V}_{R_1}(\mathbf{I}_S) + \mathbf{V}_{R_1}(\mathbf{V}_S)$$

$$v_{R_1}(t) = 18.75\cos(2\pi100t) + 15\cos(2\pi1{,}000t)$$

and

$$\mathbf{V}_{R_2}(t) = \mathbf{V}_{R_2}(\mathbf{I}_S) + \mathbf{V}_{R_2}(\mathbf{V}_S)$$

$$v_{R_2}(t) = 18.75\cos(2\pi100t) + 5\cos(2\pi1{,}000t + 180°)$$

It is important to observe that it is not possible to simplify these expressions any further, since the two sinusoidal components are at different frequencies.

Impedance

We now analyze the *i-v* relationship of the three ideal circuit elements in light of the new phasor notation. The result will be a new formulation in which resistors, capacitors, and inductors will be described in the same notation. A direct consequence of this result will be that the circuit theorems of Chapter

3 will be extended to AC circuits. In the context of AC circuits, any one of the three ideal circuit elements defined so far will be described by a parameter called **impedance**, which may be viewed as a *complex resistance*. The impedance concept is equivalent to stating that capacitors and inductors act as *frequency-dependent resistors*, that is, as resistors whose resistance is a function of the frequency of the sinusoidal excitation. Figure 4.36 depicts the same circuit represented in conventional form (left-hand side) and in phasor-impedance form (right-hand side); the latter representation explicitly shows phasor voltages and currents and treats the circuit element as a generalized "impedance." It will presently be shown that each of the three ideal circuit elements may be represented by one such impedance element.

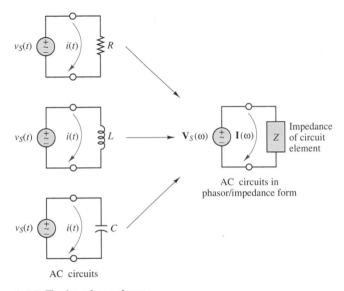

Figure 4.36 The impedance element

Let the source voltage in the circuit of Figure 4.36 be defined by

$$v_S(t) = A \cos \omega t \quad \text{or} \quad \mathbf{V}_S(\omega) = A e^{j0°} \tag{4.59}$$

without loss of generality. Then the current $i(t)$ is defined by the *i-v* relationship for each circuit element. Let us examine the frequency-dependent properties of the resistor, inductor, and capacitor, one at a time.

The Resistor

Ohm's law dictates the well-known relationship $v = iR$. In the case of sinusoidal sources, then, the current flowing through the resistor of Figure 4.36 may be expressed as

$$i(t) = \frac{v_S(t)}{R} = \frac{A}{R} \cos(\omega t) \tag{4.60}$$

Converting the voltage $v_S(t)$ and the current $i(t)$ to phasor notation, we obtain the following expressions:

$$\mathbf{V}_S(\omega) = Ae^{j0°}$$

$$\mathbf{I}(\omega) = \frac{A}{R}e^{j0°} \tag{4.61}$$

Finally, the *impedance* of the resistor is defined as the ratio of the phasor voltage across the resistor to the phasor current flowing through it, and the symbol Z_R is used to denote it:

$$\boxed{Z_R(\omega) = \frac{\mathbf{V}_S(\omega)}{\mathbf{I}(\omega)} = R \qquad \text{Impedance of a resistor}} \tag{4.62}$$

Equation 4.62 corresponds to Ohm's law in phasor form, and the result should be intuitively appealing: Ohm's law applies to a resistor independently of the particular form of the voltages and currents (whether AC or DC, for instance). The ratio of phasor voltage to phasor current has a very simple form in the case of the resistor. In general, however, the impedance of an element is a complex function of frequency, as it must be, since it is the ratio of two phasor quantities, which are frequency-dependent. This property will become apparent when the impedances of the inductor and capacitor are defined.

The Inductor

Recall the defining relationships for the ideal inductor (equations 4.11 and 4.14), repeated here for convenience:

$$v_L(t) = L\frac{di_L(t)}{dt}$$

$$i_L(t) = \frac{1}{L}\int v_L\,dt \tag{4.63}$$

Let $v_L(t) = v_S(t)$ and $i_L(t) = i(t)$ in the circuit of Figure 4.36. Then the following expression may be derived for the inductor current:

$$i_L(t) = i(t) = \frac{1}{L}\int v_S(t)\,dt$$

$$i_L(t) = \frac{1}{L}\int A\cos\omega t\,dt \tag{4.64}$$

$$= \frac{A}{\omega L}\sin\omega t$$

Note how a dependence on the radian frequency of the source is clearly present in the expression for the inductor current. Further, the inductor current is shifted in phase (by 90°) with respect to the voltage. This fact can be seen by writing the inductor voltage and current in time-domain form:

$$v_S(t) = v_L(t) = A\cos\omega t$$

$$i(t) = i_L(t) = \frac{A}{\omega L}\cos(\omega t - 90°) \tag{4.65}$$

It is evident that the current is not just a scaled version of the source voltage, as it was for the resistor. Its magnitude depends on the frequency, ω, and it is shifted (delayed) in phase by 90°. Using phasor notation, equations 4.65 become

$$\mathbf{V}_S(\omega) = Ae^{j0°}$$
$$\mathbf{I}(\omega) = \frac{A}{\omega L}e^{-j90°} \tag{4.66}$$

Thus, the impedance of the inductor is defined as follows:

$$\boxed{Z_L(\omega) = \frac{\mathbf{V}_S(\omega)}{\mathbf{I}(\omega)} = \omega Le^{j90°} = j\omega L \qquad \text{Impedance of an inductor}} \tag{4.67}$$

Note that the inductor now appears to behave like a *complex frequency-dependent resistor*, and that the magnitude of this complex resistor, ωL, is proportional to the signal frequency, ω. Thus, an inductor will "impede" current flow in proportion to the sinusoidal frequency of the source signal. This means that at low signal frequencies, an inductor acts somewhat like a short circuit, while at high frequencies it tends to behave more as an open circuit. Another important point is that *the impedance of an inductor is always positive,* since both L and ω are positive numbers. You should verify that the units of impedance for an inductor are also ohms.

The Capacitor

An analogous procedure may be followed to derive the equivalent result for a capacitor. Beginning with the defining relationships for the ideal capacitor,

$$i_C(t) = C\frac{dv_C(t)}{dt}$$
$$v_C(t) = \frac{1}{C}\int i_C(t)\,dt \tag{4.68}$$

with $i_C = i$ and $v_C = v_S$ in Figure 4.36, the capacitor current may be expressed as:

$$
\begin{aligned}
i_C(t) &= C\frac{dv_C(t)}{dt} \\
&= C\frac{d}{dt}(A\cos\omega t) \\
&= -C(A\omega\sin\omega t) \\
&= \omega CA\cos(\omega t + 90°)
\end{aligned} \tag{4.69}
$$

so that, in phasor form,

$$\mathbf{V}_S(\omega) = Ae^{j0°}$$
$$\mathbf{I}(\omega) = \omega CAe^{j90°} \tag{4.70}$$

The impedance of the ideal capacitor, $Z_C(\omega)$, is therefore defined as follows:

$$Z_C(\omega) = \frac{\mathbf{V}_S(\omega)}{\mathbf{I}(\omega)} = \frac{1}{\omega C}e^{-j90°} = \frac{-j}{\omega C} = \frac{1}{j\omega C} \qquad \text{Impedance of a capacitor} \qquad (4.71)$$

where we have used the fact that $1/j = e^{-j90°} = -j$. Thus, the impedance of a capacitor is also a frequency-dependent complex quantity, with the impedance of the capacitor varying as an inverse function of frequency. Thus, a capacitor acts like a short circuit at high frequencies, whereas it behaves more like an open circuit at low frequencies. Another important point is that _the impedance of a capacitor is always negative,_ since both C and ω are positive numbers. You should verify that the units of impedance for a capacitor are ohms. Figure 4.37 depicts $Z_C(\omega)$ in the complex plane, alongside $Z_R(\omega)$ and $Z_L(\omega)$.

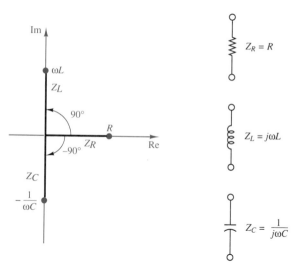

Figure 4.37 Impedances of R, L, and C in the complex plane

The impedance parameter defined in this section is extremely useful in solving AC circuit analysis problems, because it will make it possible to take advantage of most of the network theorems developed for DC circuits by replacing resistances with complex-valued impedances. The examples that follow illustrate how branches containing series and parallel elements may be reduced to a single equivalent impedance, much in the same way resistive circuits were reduced to equivalent forms. It is important to emphasize that although the impedance of simple circuit elements is either purely real (for resistors) or purely imaginary (for capacitors and inductors), the general definition of impedance for an arbitrary circuit must allow for the possibility of having both a real and an imaginary part, since practical circuits are made up of more or less complex interconnections of different circuit elements. In its most general form, the impedance of a circuit element is defined as the sum of a real part and an imaginary part:

$$Z(\omega) = R(\omega) + jX(\omega) \qquad (4.72)$$

where R is called the **AC resistance** and X is called the **reactance.** The frequency dependence of R and X has been indicated explicitly, since it is possible for a circuit to have a frequency-dependent resistance. The examples illustrate how a complex impedance containing both real and imaginary parts arises in a circuit.

Example   4.13

Find the equivalent impedance of the branch shown in Figure 4.38 at the frequency $\omega = 377$ rad/s.

$R_1 = 50\ \Omega$

$C_1 = 470\ \mu F$

Figure 4.38

Solution:

To obtain the equivalent impedance, we combine the individual impedances of the two elements in parallel:

$$Z_1 = R_1 \parallel \frac{1}{j\omega C_1} = \frac{R_1 \frac{1}{j\omega C_1}}{R_1 + \frac{1}{j\omega C_1}} = \frac{R_1}{1 + j\omega C_1 R_1}$$

Thus,

$$Z_1(\omega = 377) = \frac{50}{1 + j8.86} = \frac{50(1 - j8.86)}{(1 + j8.86)(1 - j8.86)}$$

$$= \frac{50 - j443}{1 + 78.5} = 0.629 - j5.57$$

How would the answer change if the frequency doubled?

Note that the answer indicates that the impedance corresponding to the parallel RC circuit is equivalent to the series connection of a resistance (0.629 Ω) with a reactance ($-j5.57$ Ω). What circuit element corresponds to this reactance? Since the reactance is negative, the corresponding circuit element must be a capacitor; this is true because the impedance of a capacitor is always expressed by

$$Z_C = \frac{-j}{\omega C}$$

Thus, we have

$$\frac{1}{\omega C} = 5.57$$

and since $\omega = 377$, the capacitance corresponding to the given reactance is

$$C = \frac{1}{5.57\omega} = 476 \ \mu F$$

Example   4.14

Figure 4.39

Find the equivalent impedance of the circuit shown in Figure 4.39 at the frequency $\omega = 10{,}000$ rad/s.

Solution:

Let us first consider the parallel R_2-C combination, to obtain the equivalent parallel impedance $Z_\| = Z_{R_2} \| Z_C$:

$$Z_{R_2} \| Z_C = \frac{Z_{R_2}Z_C}{Z_{R_2} + Z_C} = Z_\|$$

$$Z_\| = \frac{50(-j10)}{50 - j10}$$

$$= \frac{500e^{-j90°}}{51e^{-j11.31°}} = 9.8e^{-j78.69°}$$

$$= 1.92 - j9.61$$

Thus, $Z_\|$ may be considered the combination of two elements, a 1.92-Ω resistor and a capacitor with reactance equal to

$$X_\| = \frac{1}{\omega C} = 9.61$$

which, at the frequency of 10,000 radians per second, corresponds to a capacitance

$$C = 10.41 \mu F$$

It is very instructive to stop for a second to consider the fact that the parallel combination of a resistor and a capacitor can be converted to a completely equivalent series combination *at a given frequency* and that the resulting equivalent capacitance and resistance differ from the original values. (In this particular case, the capacitance dominates, because of the high frequency. What would the equivalent resistance and capacitance be at a frequency of 10 rad/s?)

The resulting circuit is shown in Figure 4.40, with the associated complex impedance values. If we combine the real and imaginary parts of the series elements of the branch, we have the total equivalent impedance Z_{EQ}, given by

$$Z_{EQ} = (100 + 1.92) + j(100 - 9.61)$$

$$= 101.91 + j90.39$$

100 Ω

$j\,100$ Ω

1.92 Ω

$-j9.61$ Ω

Figure 4.40

This impedance is the total equivalent impedance of the circuit branch. Note that this impedance is inductive (why?).

Example   4.15

The inductor, which we have so far modeled as an ideal circuit element, cannot always be idealized in practice, because of the finite resistance of the wire used in the coil. A more accurate model for an inductor is shown in Figure 4.41. The inductor model consists of an ideal inductor and two resistances that represent the resistance of the wire used in the windings of the inductor. R_1 and R_2 are usually considered to be equal, with each resistance equal to half the total winding resistance. Practical inductors are often constructed in the form of a *toroid* (doughnut), as shown in Figure 4.42.

Figure 4.41

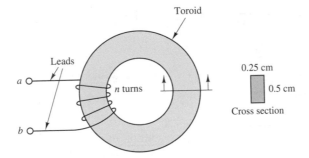

Figure 4.42 A practical inductor

1. If the toroid is wound with 250 turns of 30 gauge wire, what is the total resistance of the windings, including lead length of 10 cm per lead? Find R_1 and R_2? (30 gauge wire has a resistance of 105 Ω/1,000 ft.)
2. If the inductance is 0.098 H and we assume that in order for the equivalent circuit to be "mostly inductive," the reactance of the inductor must be at least 10 times larger than the resistance of the lead wires, over what frequency range will the equivalent circuit be considered "mostly inductive"?

Solution:

1. For the 250-turn coil, we can compute the length of wire used in the windings as l_w and the lead length as l_c, with

$$l_w = 250 \times [2(0.25 \text{ cm}) + 2(0.5 \text{ cm})] = 375 \text{ cm}$$

$$l_c = 2(10) \text{ cm} = 20 \text{ cm}$$

Total length $= 395$ cm

$= 155.5$ in.

$= 12.96$ ft

Thus the total resistance is given by:

$$\text{Total resistance} = \frac{150 \ \Omega}{1,000 \ \text{ft}} \times 12.96 \ \text{ft}$$

$$R_{\text{total}} = 1.944 \ \Omega$$

$$R_1 = R_2 = \frac{R_{\text{total}}}{2} = 0.972 \ \Omega$$

2. To determine the desired frequency range, we require that the reactance of the inductor be greater than or equal to 10 times the resistance of the lead wires:

$$\omega L \geq 10(1.944)$$

$$\omega \geq \frac{10(1.944)}{0.098}$$

which in turn suggests that the desired frequency range is

$$\omega \geq 198.4 \ \text{rad/s}$$

or

$$f \geq 31.6 \ \text{Hz}$$

Admittance

In Chapter 3, it was shown that the solution of certain circuit analysis problems was handled more easily in terms of conductances than resistances. In AC circuit analysis, an analogous quantity may be defined, the reciprocal of complex impedance. Just as the conductance, G, of a resistive element was defined as the inverse of the resistance, the **admittance** of a branch is defined as follows:

$$Y = \frac{1}{Z} \ \text{S} \tag{4.73}$$

Note immediately that whenever Z is purely real—that is, when $Z = R + j0$—the admittance Y is identical to the conductance G. In general, however, Y is the complex number

$$Y = G + jB \tag{4.74}$$

where G is called the **AC conductance** and B is called the **susceptance;** the latter plays a role analogous to that of reactance in the definition of impedance. Clearly, G and B are related to R and X. However, this relationship is not as simple as an inverse. Let $Z = R + jX$ be an arbitrary impedance. Then, the corresponding admittance is

$$Y = \frac{1}{Z} = \frac{1}{R + jX} \tag{4.75}$$

In order to express Y in the form $Y = G + jB$, we multiply numerator and denominator by $R - jX$:

$$Y = \frac{1}{R + jX} \frac{R - jX}{R - jX} = \frac{R - jX}{R^2 + X^2}$$

$$= \frac{R}{R^2 + X^2} - j\frac{X}{R^2 + X^2}$$

(4.76)

and conclude that

$$G = \frac{R}{R^2 + X^2}$$

(4.77)

$$B = \frac{-X}{R^2 + X^2}$$

Notice in particular that *G is not the reciprocal of R in the general case!*

The following example illustrates the determination of Y for some common circuits.

Example   4.16

This example illustrates how one can compute the admittance of a circuit. We consider the impedance seen at the input terminals of a one-port circuit and compute the admittance as the inverse of the impedance. The example is carried out for the two circuits of Figure 4.43.

Solution:

In Figure 4.43(a), the impedance Z_{ab} is

$$Z_{ab} = R + j\omega L$$

and therefore the admittance of the circuit is

$$Y_{ab} = \frac{1}{R + j\omega L} = \frac{R - j\omega L}{R^2 + (\omega L)^2}$$

In Figure 4.43(b), the impedance Z_{ab} is

$$Z_{ab} = R \parallel \frac{1}{j\omega C} = \frac{R}{1 + j\omega RC}$$

and so the admittance in the circuit is

$$Y_{ab} = \frac{1 + j\omega RC}{R} = \frac{1}{R} + j\omega C$$

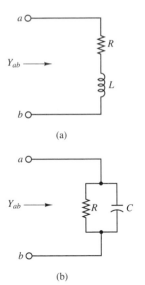

(a)

(b)

Figure 4.43

If the source frequency were 1,000 Hz and the resistors had values of $R_a = 50\ \Omega$ and $R_b = 100\ \Omega$, with $L = 16$ mH and $C = 3\ \mu$F, we could compute the admittances in Figure 4.43(a) and (b) as follows:

$$Y_{ab(a)} = \frac{1}{50 + j100.5} = 0.891 \times 10^{-3} \angle -63.6° = 0.4 \times 10^{-3} - j0.4 \times 10^{-3}$$

$$Y_{ab(b)} = \frac{1}{100} + j0.01885 = 0.0213 \angle 62° = 1 \times 10^{-3} + j19 \times 10^{-3}$$

$$= \frac{1}{100} + j \times 1,000 \times 3 \times 10^{-6} = 0.01 + j0.003 = 0.0104 \angle 16.7°$$

Example   4.17   Capacitive Displacement Transducer

In Example 4.4, the idea of a capacitive displacement transducer was introduced when we considered a parallel-plate capacitor composed of a fixed plate and a movable plate. The capacitance of this variable capacitor was shown to be a *nonlinear* function of the position of the movable plate, x (see Figure 4.7). In this example, we show that under certain conditions the impedance of the capacitor varies as a *linear* function of displacement—that is, the movable-plate capacitor can serve as a linear transducer.

Recall the expression derived in Example 4.4:

$$C = \frac{8.854A}{x}$$

where C is the capacitance in pF, A is the area of the plates in mm^2, and x is the (variable) distance in mm. If the capacitor is placed in an AC circuit, its impedance will be determined by the expression

$$Z_C = \frac{1}{j\omega C}$$

so that

$$Z_C = \frac{x}{j\omega 8.854A}$$

Figure 4.44 Bridge circuit for capacitive displacement transducer

Thus, at a fixed frequency ω, the impedance of the capacitor will vary linearly with displacement. This property may be exploited in the bridge circuit of Example 4.4, where a differential pressure transducer was shown as being made of two movable plate capacitors, such that if the capacitance of one increased as a consequence of a pressure differential across the transducer, the capacitance of the other had to decrease by a corresponding amount (at least for small displacements). The circuit is shown again in Figure 4.44, where two resistors have been connected in the bridge along with the variable capacitors (denoted by $C(x)$). The bridge is excited by a sinusoidal source.

Using phasor notation, we can express the output voltage as follows:

$$\mathbf{V}_{out}(j\omega) = \mathbf{V}_S(j\omega) \left(\frac{Z_{C_{bc}}(x)}{Z_{C_{db}(x)} + Z_{C_{bc}(x)}} - \frac{R_2}{R_1 + R_2} \right)$$

If the nominal capacitance of each movable-plate capacitor with the diaphragm in the center position is given by

$$C = \frac{\varepsilon A}{d}$$

where d is the nominal (undisplaced) separation between the diaphragm and the fixed surfaces of the capacitors (in mm), the capacitors will see a change in capacitance given by

$$C_{db} = \frac{\varepsilon A}{d - x} \quad \text{and} \quad C_{bc} = \frac{\varepsilon A}{d + x}$$

when a pressure differential exists across the transducer, so that the impedances of the variable capacitors change according to the displacement:

$$Z_{C_{db}} = \frac{d - x}{j\omega 8.854A} \quad \text{and} \quad Z_{C_{bc}} = \frac{d + x}{j\omega 8.854A}$$

and we obtain the following expression for the phasor output voltage:

$$\mathbf{V}_{out}(j\omega) = \mathbf{V}_S(j\omega)\left(\frac{\frac{d+x}{j\omega 8.854A}}{\frac{d-x}{j\omega 8.854A} + \frac{d+x}{j\omega 8.854A}} - \frac{R_2}{R_1 + R_2}\right)$$

$$= \mathbf{V}_S(j\omega)\left(\frac{1}{2} + \frac{x}{2d} - \frac{R_1}{R_1 + R_2}\right)$$

$$= \mathbf{V}_S(j\omega)\frac{x}{2d}$$

if we choose $R_1 = R_2$. Thus, the output voltage will vary as a scaled version of the input voltage in proportion to the displacement. A typical $v_{out}(t)$ is displayed in Figure 4.45 for a 0.05-mm "triangular" diaphragm displacement, with $d = 0.5$ mm and $\mathbf{V}_S$ a 25-Hz sinusoid with 1 V amplitude.

Displacement input

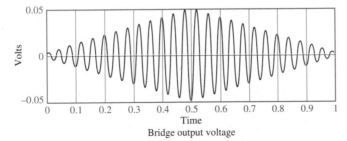

Bridge output voltage

Figure 4.45 Displacement input and bridge output voltage for capacitive displacement transducer

---

DRILL EXERCISES

4.15   Add the sinusoidal voltages $v_1(t) = A\cos(\omega t + \phi)$ and $v_2(t) = B\cos(\omega t + \theta)$ using phasor notation, and then convert back to time-domain form, for:

a. $A = 1.5$, $\phi = 10°$; $B = 3.2$, $\theta = 25°$.
b. $A = 50$, $\phi = -60°$; $B = 24$, $\theta = 15°$.

> Answer: (a) $v_1 + v_2 = 4.67\cos(\omega t + 20.2°)$;
>   (b) $v_1 + v_2 = 60.8\cos(\omega t - 37.6°)$

4.16   Add the sinusoidal currents $i_1(t) = A\cos(\omega t + \phi)$ and $i_2(t) = B\cos(\omega t + \theta)$ for:

a. $A = 0.09,\ \phi = 72°;\ B = 0.12,\ \theta = 20°.$

b. $A = 0.82,\ \phi = -30°;\ B = 0.5,\ \theta = -36°.$

> Answer: (a) $i_1 + i_2 = 0.19\cos(\omega t + 42.0°)$;
>   (b) $i_1 + i_2 = 1.32\cos(\omega t - 32.3°)$

4.17   Compute the equivalent impedance of the circuit of Example 4.14 for $\omega = 1{,}000$ and $100{,}000$ rad/s.

> Answer: $Z(1{,}000) = 140 - j10;\ Z(100{,}000) = 100 + j999$

4.18   Compute the equivalent admittance of the circuit of Example 4.14.

> Answer: $Y_{EQ} = 5.492 \times 10^{-3} - j4.871 \times 10^{-3}$

4.19   Calculate the equivalent series capacitance of the circuit of Example 4.14 at the frequency $\omega = 10$ rad/s.

> Answer: $X_{\parallel} = 0.25;\ C = 0.4$ F

4.5   AC CIRCUIT ANALYSIS METHODS

This section will illustrate how the use of phasors and impedance facilitates the solution of AC circuits by making it possible to use the same solution methods developed in Chapter 3 for DC circuits. The AC circuit analysis problem of interest in this section consists of determining the unknown voltages (or currents) in a circuit containing linear passive circuit elements (R, L, C) and excited by a sinusoidal source. Figure 4.46 depicts one such circuit, represented in both conventional time-domain form and phasor-impedance form.

The first step in the analysis of an AC circuit is to note the frequency of the sinusoidal excitation. Next, all sources are converted to phasor form,

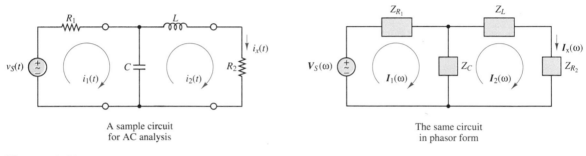

A sample circuit
for AC analysis

The same circuit
in phasor form

Figure 4.46 An AC circuit

and each circuit element to impedance form. This is illustrated in the phasor circuit of Figure 4.46. At this point, if the excitation frequency, ω, is known numerically, it will be possible to express each impedance in terms of a known amplitude and phase, and a numerical answer to the problem will be found. It does often happen, however, that one is interested in a more general circuit solution, valid for an arbitrary excitation frequency. In this latter case, the solution becomes a function of ω. This point will be developed further in Chapter 6, where the concept of sinusoidal frequency response is discussed.

With the problem formulated in phasor notation, the resulting solution will be in phasor form and will need to be converted to time-domain form. In effect, the use of phasor notation is but an intermediate step that greatly facilitates the computation of the final answer. In summary, here is the procedure that will be followed to solve an AC circuit analysis problem. Example 4.18 illustrates the various aspects of this method.

1. Identify the sinusoidal source(s) and note the excitation frequency.
2. Convert the source(s) to phasor form.
3. Represent each circuit element by its impedance.
4. Solve the resulting phasor circuit, using appropriate network analysis tools.
5. Convert the (phasor-form) answer to its time-domain equivalent, using equation 4.53.

Example   4.18

In this example, the method of circuit analysis described in this section is applied to solve for the unknown source current $i_S(t)$ in the circuit shown in Figure 4.47.

Figure 4.47

Solution:

From the time-domain form of the circuit, we deduce that

$$\omega = 100 \text{ rad/s}$$

$$V_S(\omega) = 10e^{j0} \text{ V}$$

$$Z_C(\omega) = \frac{1}{j\omega C} = \frac{-j}{100 \times 100 \times 10^{-6}} = -j100$$

The corresponding phasor circuit is shown in Figure 4.48.

Figure 4.48

In order to solve the circuit by means of the techniques illustrated in the preceding sections, we recognize that the impedances Z_3 and Z_2 are in parallel; thus,

$$Z_2 \parallel Z_3 = \frac{Z_2 Z_3}{Z_2 + Z_3} = \frac{-j2 \times 10^4}{200 - j100}$$

$$= \frac{2 \times 10^4 e^{-j90°}}{22,361 e^{-j26.57°}}$$

$$= 40 - j80$$

The resulting equivalent circuit is shown in Figure 4.49, and the total series impedance seen by the source is given by the expression

$$Z_S = Z_1 + Z_2 \parallel Z_3 = 90 - j80$$

Finally, the phasor current $\mathbf{I}_S$ is given by the ratio of the source voltage to the equivalent impedance:

$$\mathbf{I}_S = \frac{10e^{j0°}}{90 - j80} = \frac{10e^{j0°}}{120e^{-j41.6°}} = 0.083e^{+j41.6°}$$

In order to complete the problem, we need to convert the phasor solution to time-domain form:

$$\mathbf{I}_S(\omega = 100) = 0.083e^{+j41.6°} \Rightarrow i_S(t) = 0.083\cos(100t + 41.6°)$$

How would the answer differ if the excitation frequency were changed to $\omega = 200$?

Figure 4.49

Example   4.19

This example illustrates how to obtain the response of the circuit of Example 4.18 to a sinusoidal source of arbitrary amplitude A, phase ϕ, and frequency ω. Let the source excitation be $v_S(t) = A\cos(\omega t + \phi)$. The phasor form of the source will then be $\mathbf{V}_S(\omega) = Ae^{j\phi}$.

Solution:

Using the same expression as in Example 4.18 to find the equivalent parallel impedance, we note that now the impedance of the capacitor is a function of frequency, since no value has been specified for ω:

$$Z_3 = \frac{1}{j\omega C} = \frac{10^4}{j\omega}$$

Using the results of the preceding example, we find the equivalent parallel impedance of Z_2 and Z_3 to be

$$Z_2 \parallel Z_3 = \frac{200 \times 10^4/j\omega}{200 + 10^4/j\omega} = \frac{2 \times 10^6}{10^4 + 200 \times j\omega}$$

The total series impedance may then be computed as follows:

$$Z_1 + Z_2 \parallel Z_3 = 50 + \frac{2 \times 10^6}{10^4 + 200 \times j\omega} = \frac{2.5 \times 10^6 + j\omega \times 10^4}{10^4 + 200 \times j\omega}$$

Thus, the phasor current $\mathbf{I}_S$ is given by

$$\mathbf{I}_S(\omega) = \frac{\mathbf{V}_S(\omega)}{Z_1 + Z_2 \parallel Z_3} = Ae^{j\phi}\frac{10^4 + 200 \times j\omega}{2.5 \times 10^6 + j\omega \times 10^4}$$

The numerical expression for $\mathbf{I}_S(\omega)$ can now be obtained for any excitation amplitude A, frequency ω, and phase ϕ.

By now it should be apparent that the laws of network analysis introduced in Chapter 3 are also applicable to phasor voltages and currents. This fact suggests that it may be possible to extend the node and mesh analysis methods developed earlier to circuits containing phasor sources and impedances, although the resulting simultaneous complex equations are difficult to solve without the aid of a computer, even for relatively simple circuits. On the other hand, it is very useful to extend the concept of equivalent circuits to the AC case, and to define complex Thévenin (or Norton) equivalent impedances. The fundamental difference between resistive and AC equivalent circuits is that the AC Thévenin (or Norton) equivalent circuits will be frequency-dependent and complex-valued. In general, then, one may think of the resistive circuit analysis of Chapter 3 as a special case of AC analysis, where all impedances are real.

AC Equivalent Circuits

In Chapter 3, we demonstrated that it was convenient to compute equivalent circuits, especially in solving for load-related variables. Figure 4.50 depicts the two representations analogous to those developed in Chapter 3. Figure 4.50(a) shows an *equivalent load,* as viewed by the source, while Figure 4.50(b) shows an *equivalent source* circuit, from the perspective of the load.

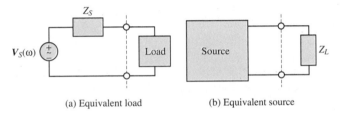

(a) Equivalent load   (b) Equivalent source

Figure 4.50 AC equivalent circuits

Impedances in series add:

Impedances in parallel behave like resistors in parallel:

Admittances in parallel add:

Admittances in series behave like conductances in series:

Figure 4.51 Rules for impedance and admittance reduction

In the case of linear resistive circuits, the equivalent load circuit can always be expressed by a single equivalent resistor, while the equivalent source circuit may take the form of a Norton or a Thévenin equivalent. This section extends these concepts to AC circuits and demonstrates that the notion of equivalent circuits applies to phasor sources and impedances as well. The techniques described in this section are all analogous to those used for resistive circuits, with resistances replaced by impedances, and arbitrary sources replaced by phasor sources. The principal difference between resistive and AC equivalent circuits will be that the latter are frequency-dependent. Figure 4.51 summarizes the fundamental principles used in computing an AC equivalent circuit. Note the definite analogy between impedance and resistance elements, and between conductance and admittance elements.

The computation of an equivalent impedance is carried out in the same way as that of equivalent resistance in the case of resistive circuits:

1. Short-circuit all voltage sources, and open-circuit all current sources.
2. Compute the equivalent impedance between load terminals, with the load disconnected.

In order to compute the Thévenin or Norton equivalent form, we recognize that the Thévenin equivalent voltage source is the open-circuit voltage at the load terminals and the Norton equivalent current source is the short-circuit current (the current with the load replaced by a short circuit). Figure 4.52 illustrates these points by outlining the steps in the computation of an equivalent circuit. The remainder of the section will consist of examples aimed at clarifying some of the finer points in the calculation of such equivalent circuits. Note how the initial circuit reduction proceeds exactly as in the case

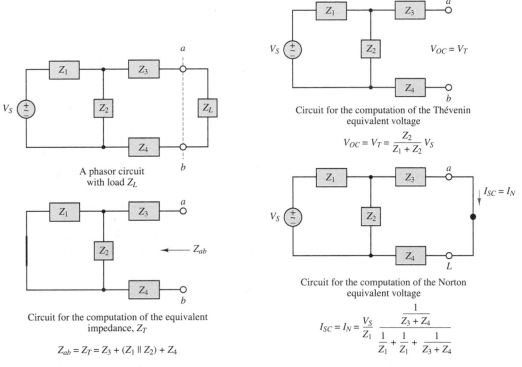

A phasor circuit
with load Z_L

Circuit for the computation of the Thévenin
equivalent voltage

$$V_{OC} = V_T = \frac{Z_2}{Z_1 + Z_2} V_S$$

Circuit for the computation of the equivalent
impedance, Z_T

$$Z_{ab} = Z_T = Z_3 + (Z_1 \parallel Z_2) + Z_4$$

Circuit for the computation of the Norton
equivalent voltage

$$I_{SC} = I_N = \frac{V_S}{Z_1} \frac{\dfrac{1}{Z_3 + Z_4}}{\dfrac{1}{Z_1} + \dfrac{1}{Z_1} + \dfrac{1}{Z_3 + Z_4}}$$

Figure 4.52 Reduction of AC circuit to equivalent form

of a resistive circuit; the details of the complex algebra required in the
calculations are explored in the examples.

Example   4.20

In this example, we illustrate the computation of the Thévenin equivalent of a simple
AC circuit, shown in Figure 4.53.

Solution:

Note that, in solving problems of this nature, one may choose between two algebraic
techniques, both leading to the same final answer. One uses the method of computing
the ratio of two complex numbers in rectangular form; the other uses the polar form
of the complex numbers. It should be remarked that, in general, the rectangular form
is more convenient for adding or subtracting complex numbers, while the polar form
is easier to use in computing products and ratios. However, these operations are
frequently mixed together in the solution of AC circuit problems, and the preferred
method depends on the user.

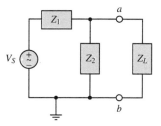

$V_S = 110\angle 0°$ $Z_1 = 5\ \Omega$ $Z_2 = j20\ \Omega$

Figure 4.53

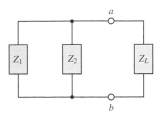

Figure 4.54

To compute the Thévenin impedance, Z_T, we replace each voltage source with a short circuit and each current source with an open circuit, obtaining the circuit of Figure 4.54. The Thévenin impedance is then computed much in the same way as Thévenin resistances were computed in Chapter 3. After removing the load impedance, Z_L, we see that the remaining impedances appear in parallel with respect to the load terminals:

$$Z_T = Z_1 \parallel Z_2$$

Since $Z_1 = 50\ \Omega$ and $Z_2 = j20\ \Omega$, we can write

$$Z_T = \frac{(5)(j20)}{5 + j20}$$

To obtain a physically meaningful expression for the impedance Z_T, we need to reduce this expression to one of the form $R_T + jX_T$; that is, we need to reduce the denominator of the expression to a real number. One method that will accomplish this consists of multiplying the numerator and the denominator by the complex conjugate of the denominator. The effect of this operation is to convert the denominator into a pure real quantity, reducing the ratio of two complex numbers to a single complex number divided by a real number:

$$Z_T = \frac{(5)(j20)}{(5 + j20)}\frac{(5 - j20)}{(5 - j20)} = \frac{2{,}000 + j500}{425} = 4.71 + j1.176$$

An alternative method of accomplishing the same objective consists of converting both numerator and denominator to polar form (i.e., magnitude and phase) and computing the ratio of the complex numbers in polar form. The latter is a simple task, since the ratio of two complex numbers in polar form is a complex number with magnitude equal to the ratio of the magnitudes and phase equal to the difference between the phases:

$$Z_T = \frac{(5)(j20)}{(5 + j20)} = \frac{100e^{j90°}}{20.615e^{j76°}} = 4.85e^{j14°} = 4.71 + j1.176$$

The two techniques are equivalent, although occasional savings in computation may be experienced by using one rather than the other.

Now, recall that the Thévenin voltage, V_T, was simply the open-circuit voltage between terminals a and b. If we were to replace the impedances Z_1 and Z_2 with resistors R_1 and R_2, then, using the methods of Chapter 3, we could compute the Thévenin voltage as

$$\mathbf{V}_T = \frac{R_2}{R_1 + R_2}\mathbf{V}_S$$

The Thévenin phasor voltage for a complex circuit is found in a similar fashion, by computing the (complex) voltage divider ratio

$$\mathbf{V}_T = \frac{Z_2}{Z_1 + Z_2}\mathbf{V}_S$$

or

$$\mathbf{V}_T = \frac{j20}{5 + j20}110e^{j0°}$$

Note that the voltage source has a zero phase angle, so that the source is expressed by $\mathbf{V}_S = 110e^{j0°}$.

The computation of $\mathbf{V}_T$ can also be carried out in two different ways:

$$\mathbf{V}_T = \frac{(j2{,}220)\,(5-j20)}{(5+j20)\,(5-j20)} = \frac{44{,}000+j11{,}000}{425}$$

$$= 103.5 + j25.88 = 106.7e^{j14°}$$

or

$$\mathbf{V}_T = \frac{(j2{,}220)}{(5+j20)} = \frac{2{,}200e^{j90°}}{20.165e^{j76°}} = 106.7e^{j14°}$$

The complete Thévenin equivalent circuit can finally be drawn as shown in Figure 4.55.

Figure 4.55

Example   4.21

This example illustrates the computation of a Norton equivalent. The objective is to find the Norton equivalent circuit seen by the load impedance, Z_L, for the circuit shown in Figure 4.56, given the following source and impedance values:

$$i_S(t) = 50\cos(\omega t + 45°) \qquad Z_1 = 400\ \Omega \qquad Z_2 = j10\ \Omega \qquad Z_3 = -j20\ \Omega$$

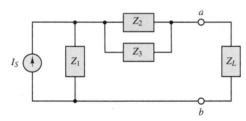

Figure 4.56

Solution:

Begin by determining the Thévenin (or Norton) impedance, Z_T. Replacing the current source with an open circuit yields the circuit of Figure 4.57, from which it should be apparent that the Thévenin impedance is given by

$$Z_T = Z_1 + Z_2 \,\|\, Z_3$$

or

$$Z_T = 400 + \frac{(j10)(-j20)}{(j10-j20)} = 400 + \frac{200}{-j10} = 400 + j20$$

The Norton current source is simply the short-circuit current between the load terminals (replace Z_L with a short circuit). In the circuit of Figure 4.56, this current is determined by means of the current divider between impedance Z_1 and the parallel combination of impedances Z_3 and Z_2, with $\mathbf{I}_{SC}$ the current flowing through $Z_2 \,\|\, Z_3$:

$$\mathbf{I}_{SC} = \mathbf{I}_N = \frac{Z_1}{Z_1 + Z_2 \,\|\, Z_3}\mathbf{I}_S$$

$$= \frac{400}{400 + j20}\,50e^{j45°} = 49.94e^{j42.14°}$$

The final appearance of the Norton equivalent representation is shown in Figure 4.58.

Figure 4.57

Figure 4.58

DRILL EXERCISES

4.20   Compute the magnitude of the current $\mathbf{I}_S(\omega)$ of Example 4.19 if $A = 1$ and $\phi = 0$, for $\omega = 10$, 10^2, 10^3, 10^4, and 10^5 rad/s. Can you explain these results intuitively? [*Hint:* Evaluate the impedance of the capacitor relative to that of the two resistors at each frequency.]

> Answer: $|\mathbf{I}_S| = 0.0041$ A; 0.0083 A; 0.0194 A; 0.02 A; 0.02 A

4.21   Find the voltage across the capacitor in Example 4.18.

> Answer: $7.43e^{-j21.8°}$ V

4.22   Determine the Norton current in Example 4.20.

> Answer: $22e^{j0°}$ A

4.23   Determine the Thévenin voltage in Example 4.21.

> Answer: $20{,}000e^{j45°}$ V

CONCLUSION

In this chapter we have introduced concepts and tools useful in the analysis of AC circuits. The importance of AC circuit analysis cannot be overemphasized, for a number of reasons. First, circuits made up of resistors, inductors, and capacitors constitute reasonable models for more complex devices, such as transformers, electric motors, and electronic amplifiers. Second, sinusoidal signals are ever present in the analysis of many physical systems, not just circuits. The skills developed in Chapter 4 will be called upon in the remainder of the book. In particular, they form the basis of Chapters 5 and 6.

- In addition to elements that dissipate electric power, there are also electric energy-storage elements. The ideal inductor and capacitor are ideal elements that represent the energy-storage properties of electric circuits.
- Since the *i-v* relationship for the ideal capacitor and the ideal inductor consists of a differential equation, application of the fundamental circuit laws in the presence of such dynamic circuit elements leads to the formulation of differential equations.
- For the very special case of sinusoidal sources, the differential equations describing circuits containing dynamic elements can be converted into algebraic equations and solved using techniques similar to those employed in Chapter 3 for resistive circuits.

- Sinusoidal voltages and currents can be represented by means of complex phasors, which explicitly indicate the amplitude and phase of the sinusoidal signal and implicitly denote the sinusoidal frequency dependence.
- Circuit elements can be represented in terms of their impedance, which may be conceptualized as a frequency-dependent resistance. The rules of circuit analysis developed in Chapters 2 and 3 can then be employed to analyze AC circuits by using impedance elements as complex resistors. Thus, the only difference between the analysis of AC and resistive circuits lies in the use of complex algebra instead of real algebra.

KEY TERMS

Capacitance   *121*

Inductance   *121*

Ideal Capacitor   *121*

Ideal Inductor   *121*

Dielectric   *122*

Farad (F)   *122*

Permittivity   *127*

Inductive Core   *129*

Henry (H)   *129*

Damper   *133*

Periodic Signal   *134*

Period   *134*

Signal (or Waveform) Generator   *134*

Amplitude   *134*

Radian Frequency   *134*

Phase   *134*

Hertz (Hz)   *136*

Average (or DC) Value   *136*

Root-Mean-Square (or RMS) Value   *136*

Fourier Theory   *138*

Fundamental Frequency   *138*

Harmonics   *138*

Spectrum   *138*

AC Circuits   *144*

Complex Exponential   *147*

Argument   *148*

Magnitude (of a Complex Vector)   *148*

Complex Phasor   *148*

(Phase) Angle   *148*

Time-Domain Form   *148*

Frequency-Domain (Phasor) Form   *149*

Impedance   *153*

AC Resistance   *157*

Reactance   *157*

Admittance   *160*

AC Conductance   *160*

Susceptance   *160*

HOMEWORK PROBLEMS

4.1   Find the rms value of the waveform shown in Figure P4.1.

4.2   Find the rms voltage of the waveform of Figure P4.2.

Figure P4.1

Figure P4.2

4.3   Find the rms value of the waveform shown in Figure P4.3.

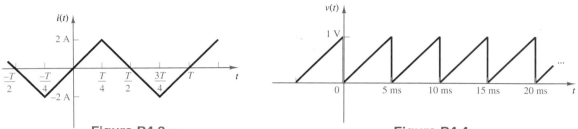

Figure P4.3   **Figure P4.4**

4.4 Find the rms value of the waveform of Figure P4.4.

4.5 Find the rms value of the waveform of Figure P4.5.

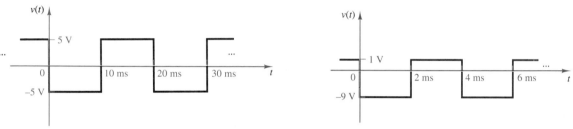

Figure P4.5   **Figure P4.6**

4.6 Find the rms value of the waveform of Figure P4.6.

4.7 Find the rms value of the waveform of Figure P4.7.

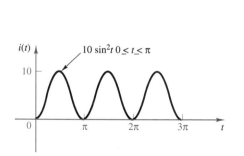

Figure P4.7   **Figure P4.8**

4.8 Find the rms value of $x(t)$ if $x(t)$ is a sinusoid that is offset by a DC value;

$$x(t) = 2 \sin(\omega t) + 2.5$$

4.9 For the waveform of Figure P4.8:

a.   Find the rms voltage.

b.   If θ_1 is $\pi/2$, what is the rms current of this waveform?

4.10 Find the rms value of the waveform of Figure P4.9 if:

a. $\theta = 30°$.
b. $\theta = 60°$.
c. $\theta = 90°$.

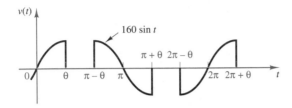

Figure P4.9

4.11 Find the average power dissipated by the resistor of the circuit of Figure P4.10 if

a. $v(t) = 10$ V.
b. $v(t) = 10\sqrt{2}\cos 200\pi t$ V.

Note that the DC voltage of part a is the same as the rms voltage of the source in part b.

Figure P4.10

4.12 A capacitor carries a charge of 20 mC, and the voltage across it is 60 V. Find the capacitance and stored energy.

4.13 Use the defining law for a capacitor to find the current $i_C(t)$ corresponding to the voltage shown in Figure P4.11. Sketch your results.

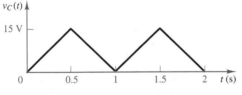

Figure P4.11

4.14 If a sine wave of current $i_C(t) = I_m \sin \omega t$ passes through a capacitor $C = 1$ F, draw the voltage waveform.

4.15 If a cosine wave of current $i_L(t) = I_m \cos \omega t$ flows through an inductor $L = 1$ H, draw the voltage waveform.

4.16 A spring-loaded scale mechanism is being used to measure the weight of an object placed on the platform shown in Figure P4.12. The position of the plunger is directly proportional to the force exerted on the platform by the weight of the object. The variable capacitor has a flat plate at its bottom that is electrically isolated from its case. The face of the variable capacitor's plunger is also flat and electrically isolated from the variable capacitor.

Find the capacitance of the parallel plates as a function of the distance x shown in Figure P4.12 if the dielectric material is air.

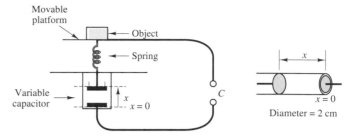

Figure P4.12

4.17 The voltage across a 20-μF capacitor takes the following values. Calculate the sinusoidal expression of current through the capacitor in each case.

a. $v_C(t) = 30 \sin 50t$ V

b. $v_C(t) = 10 \cos 30t$ V

c. $v_C(t) = -60 \sin 10t$ V

d. $v_C(t) = 4 \cos (100t + \pi/2)$ V

4.18 The current through a 10-mH inductor takes the values shown. Calculate the sinusoidal expression of voltage across the inductor in each case.

a. $i_L(t) = 10 \sin 50t$ mA

b. $i_L(t) = 4 \cos 200t$ mA

c. $i_L(t) = 20 \sin (10t + \pi/4)$ mA

d. $i_L(t) = -6 \sin 40t$ mA

4.19 Use the defining law for an inductor to find the current $i_L(t)$ as shown in Figure P4.13. Sketch your results.

Figure P4.13

4.20 The voltage $v(t)$ shown in Figure P4.14 is applied to a 10-mH inductor. Find the current through the inductor.

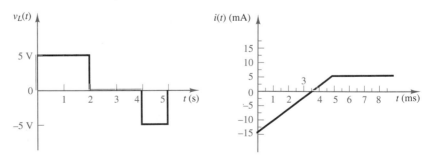

Figure P4.14   **Figure P4.15**

4.21 The current waveform shown in Figure P4.15 flows through a 2-H inductor. Plot the inductor voltage $v_L(t)$, assuming $i(t) = 0$ for t.

4.22   An exponential current $i(t) = 4e^{-t/10}$ A is applied to a 10-mH inductor at $t = 0$. Find the resulting voltage across the inductor, and draw the waveform (assume $i_L(0) = 0$ A).

4.23   The voltage waveform shown in Figure P4.16 appears across a 100-mH inductor and a 500-μF capacitor. Plot the capacitor and inductor currents, $i_C(t)$ and $i_L(t)$ assuming $i_L(0) = 0$ A.

4.24   Calculate the energy stored in the capacitor at $t = 3$ ms by the waveform of Drill Exercise 4.1.

4.25   Calculate and plot the energy stored in the inductor for the case of Drill Exercise 4.1.

4.26   Calculate the energy stored in the capacitor shown in Problem 4.13.

4.27   Calculate the energy stored in the inductor shown in Problem 4.19.

4.28   For the circuit shown in Figure P4.17, find v_{out} by writing the differential equation. Assume that $v_{out}(t)$ has the form $v_{out} = A \sin \omega t + B \cos \omega t$, and, using trigonometric identities, find $v_{out}(t)$ in the form $v_{out}(t) = D \cos (\omega t + \phi)$.

4.29   Write the differential equation for the circuit shown in Figure P4.18, using the capacitor voltage as the independent variable.

Figure P4.16

Figure P4.17

Figure P4.18

Figure P4.19

4.30   Find the output voltage, $v_{out}(t)$, for the circuit shown in Figure P4.19, by assuming that the solution has the form $v_{out}(t) = D \cos (\omega t + \phi)$.

4.31   Refer to the circuit of Figure P4.17. If $R = 40$ kΩ, $C = 0.05$ μF, $V_1 = 20$ V, and $\omega = 377$ rad/s, find $v_{out}(t)$ using phasor techniques.

4.32   Find the output voltage, $v_{out}(t)$, for the circuit shown in Figure P4.20, by assuming that the solution has the form $v_{out}(t) = D \cos (\omega t + \phi)$.

4.33

a.   Find the current flowing in the series RLC circuit shown in Figure P4.21, by writing the differential equation and assuming that the solution has the form $i(t) = A \sin \omega t + B \cos \omega t$.

b.   What is the inductor voltage, $v_L(t)$?

c.   What is the resistor voltage, $v_R(t)$?

Figure P4.20

Figure P4.21

4.34

a.   For the parallel RLC circuit shown in Figure P4.22, find the output voltage, $v_{out}(t)$, by writing the differential equation and assuming that $v_{out}(t)$ has the form $v_{out}(t) = A \sin \omega t + B \cos \omega t$.

b.   What is the current $i_R(t)$?

c.   What is the current $i_L(t)$?

Figure P4.22

4.35 Write and solve the differential equation for $v_{\text{out}}(t)$ for the circuit shown in Figure P4.17 if $R = 10\ \Omega$, $C = 20\mu\text{F}$, and $v_{\text{in}}(t) = 10\cos(2{,}500t)$.

4.36 For the circuit shown in Figure P4.23, find the differential equation necessary to determine $v_C(t)$.

Figure P4.23

4.37 Express the following current in phasor notation:

$$i(t) = 0.5\cos(2\pi 10t + 45°)\ \text{A}$$

Sketch the waveform along with the waveform $\cos(2\pi 10t)$ on the same axis as precisely as you can.

4.38 Find the phasor form of the following functions:

a. $v(t) = 155\cos(377t - 25°)\ \text{V}$
b. $v(t) = 5\sin(1{,}000t - 40°)\ \text{V}$
c. $i(t) = 10\cos(10t + 63°) + 15\cos(10t - 42°)\ \text{A}$
d. $i(t) = 460\cos(500\pi t - 25°) - 220\sin(500\pi t + 15°)\ \text{A}$

4.39 Convert the following complex numbers to polar form:

a. $4 + j4$
b. $-3 + j4$
c. $j + 2 - j4 - 3$

4.40 Convert the following to polar form by first multiplying rectangular coordinates and then converting to polar form:

a. $(50 + j10)(4 + j8)$
b. $(j2 - 2)(4 + j5)(2 + j7)$

4.41 Complete the following exercises in complex arithmetic.

a. Find the complex conjugate of $(4 + j4)$, $(2 - j8)$, $(-5 + j2)$.
b. Convert the following to polar form by multiplying the numerator and denominator by the complex conjugate of the denominator and then performing the conversion to polar coordinates: $\frac{1+j7}{4+j4}$, $\frac{j4}{2-j8}$, $\frac{1}{-5+j2}$.
c. Repeat part b but this time convert to polar coordinates before performing the division.

4.42 Given the two voltages

$$v_1(t) = 10\cos(\omega t + 30°) \quad \text{and} \quad v_2(t) = 20\cos(\omega t + 60°)$$

Find $v(t) = v_1(t) + v_2(t)$ using

a.   Trigonometric identities.

b.   Phasors.

4.43   For the circuit shown in Figure P4.24, find the frequency that causes the equivalent impedance to appear to be purely resistive.

Figure P4.24

4.44

a.   Find the equivalent impedance Z_L shown in Figure P4.25, as shown by the source, if the frequency of the source is 377 rad/s.

b.   If we wanted the source to see the load as completely resistive, what value of capacitance should we place between terminals a and b as shown in Figure P4.26? [*Hint:* Find an expression for the equivalent impedance Z_L, and then find C so that $\phi = 0$.]

c.   What is the actual impedance that the source sees with the capacitor included in the circuit?

Figure P4.25

Figure P4.26

4.45   The capacitor model we have used so far has been an ideal circuit element. A more accurate model for a capacitor is shown in Figure P4.27. The ideal capacitor, C, has a large resistance, R_C, in parallel with it. R_C models the leakage current through the capacitor. R_1 and R_2 represent the line resistances, and L_1 and L_2 model the line inductances.

a.   If $C = 1\ \mu F$, $R_C = 100\ M\Omega$, $R_1 = R_2 = 1\ \mu\Omega$, and $L_1 = L_2 = 0.1\ \mu H$, find the equivalent impedance seen by terminals a and b as a function of frequency, ω.

b.   Find the range of frequencies for which Z_{ab} is capacitive (i.e., $X_{ab} > 10R_{ab}$).

[*Hint:* Assume that R_C is much greater than $1/\omega C$ so that we can replace R_C by an infinite resistance in part b.]

4.46   Find the equivalent impedance as seen by terminals a and b for the circuit shown in Figure P4.28. That is, find the series equivalent circuit for a resistor in parallel with an inductor (or find R_2 and L_2 in terms of R_1, L_1, and ω).

Figure P4.27

Figure P4.28

4.47   Solve Problem 4.28 using phasor techniques.

4.48   Solve Problem 4.29 using phasor techniques.

4.49   Solve Problem 4.30 using phasor techniques.

4.50   Solve Problem 4.33 using phasor techniques.

4.51   For the circuit in Example 4.12, R_2 is replaced by a capacitor $C = 1\mu F$. Find the voltages $v_{R1}(t)$ and $v_C(t)$, assuming the polarities of Figure 4.33.

4.52   Find the sinusoidal steady-state outputs for each of the circuits shown in Figure P4.29.

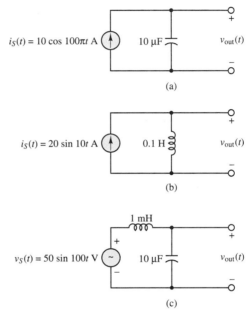

(a)

(b)

(c)

Figure P4.29

4.53   For the circuit in Example 4.12, R_2 is replaced by an inductor $L = 0.1$ mH. Find the voltages $v_{R1}(t)$ and $v_L(t)$, assuming the polarities of Figure 4.33.

4.54   Determine the voltage across the inductor in the circuit shown in Figure P4.30.

4.55   Find the current through the resistor in the circuit shown in Figure P4.31.

Figure P4.30

Figure P4.31

Figure P4.32

4.56   Find the voltage v_{CL} shown in Figure P4.32.

4.57 Find $v_{\text{out}}(t)$ for the circuit shown in Figure P4.33.

4.58 Determine the Thévenin equivalent circuit shown in Figure P4.34 as seen by the load if

a. $v_S(t) = 10 \cos (1,000t)$.

b. $v_S(t) = 10 \cos (1,000,000t)$.

Figure P4.33

$$R_S = R_L = 500 \ \Omega$$
$$L = 10 \text{ mH}$$
$$C = 0.1 \ \mu\text{F}$$

Figure P4.34

4.59 Find the Thévenin equivalent of the circuit shown in Figure P4.35 as seen by the load resistor.

Figure P4.35

4.60 Solve for $i(t)$ in the circuit of Figure P4.36, using phasor techniques if $v_S(t) = 2 \cos (2t)$, $R_1 = 4 \ \Omega$, $R_2 = 4 \ \Omega$, $L = 2$ H, and $C = \frac{1}{4}$ F.

*4.61 Using mesh current analysis, determine the currents $i_1(t)$ and $i_2(t)$ in the circuit shown in Figure P4.37.

Figure P4.36

Figure P4.37

*4.62 Using node voltage methods, determine the voltages $v_1(t)$ and $v_2(t)$ in the circuit shown in Figure P4.38.

$$R_1 = 10\ \Omega$$
$$R_2 = 40\ \Omega$$
$$C = 500\ \mu\text{F}$$
$$L = 0.2\ \text{H}$$

Figure P4.38

*4.63   The circuit shown in Figure P4.39 is a Wheatstone bridge that will allow you to determine the reactance of an inductor or a capacitor. The circuit is adjusted by changing R_1 and R_2 until v_{ab} is zero.

a.   Assuming that the circuit is balanced, that is, that $v_{ab} = 0$, determine X_4 in terms of the circuit elements.

b.   If $C_3 = 4.7\,\mu\text{F}$, $L_3 = 0.098$ H, $R_1 = 100\ \Omega$, $R_2 = 1\ \Omega$, $v_S(t) = 24\sin(2{,}000t)$, and $v_{ab} = 0$, what is the reactance of the unknown circuit element? Is it a capacitor or an inductor? What is its value?

c.   What frequency should be avoided by the source in this circuit, and why?

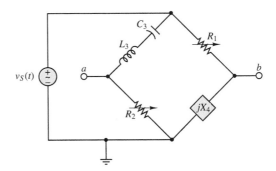

Figure P4.39

4.64   Compute the Thévenin impedance seen by resistor R_2 in Problem 4.60.

4.65   Compute the Thévenin impedance seen by resistor R_2 in Problem 4.61.

4.66   Compute the Thévenin voltage seen by the inductance, L, in Problem 4.62.

4.67   Compute the Thévenin voltage seen by resistor R_2 in Problem 4.60.

4.68   Compute the Thévenin voltage seen by resistor R_2 in Problem 4.61.

4.69   Find the Norton equivalent circuit seen by resistor R_2 in Problem 4.60.

4.70   Find the Norton equivalent circuit seen by resistor R_2 in Problem 4.61.

AC Power

T he aim of this chapter is to introduce the student to simple AC power calculations, and to the generation and distribution of electric power. The chapter builds on the material developed in Chapter 4—namely, phasors and complex impedance—and paves the way for the material on electric machines in Chapters 14, 15, and 16.

The chapter starts with the definition of AC average and complex power and illustrates the computation of the power absorbed by a complex load; special attention is paid to the calculation of the power factor, and to power factor correction. The next subject is a brief discussion of ideal transformers and of maximum power transfer. This is followed by an introduction to three-phase power. The chapter ends with a discussion of electric power generation and distribution.

Upon completing this chapter, you should have mastered the following basic concepts:

- Calculation of real and reactive power for a complex load.
- Operation of ideal transformers.
- Impedance matching and maximum power transfer.

• Basic notions of residential circuit wiring, including grounding and safety.

• Configuration of electric power distributon networks.

5.1   POWER IN AC CIRCUITS

The objective of this section is to introduce the notion of AC power. As already mentioned in Chapter 4, 50- or 60-Hz AC power constitutes the most common form of electric power; in this section, the phasor notation developed in Chapter 4 will be employed to analyze the power absorbed by both resistive and complex loads.

Instantaneous and Average Power

From Chapter 4, you already know that when a linear electric circuit is excited by a sinusoidal source, all voltages and currents in the circuit are also sinusoids of the same frequency as that of the excitation source. Figure 5.1

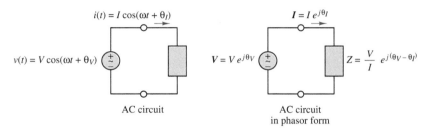

Figure 5.1 Circuit for illustration of AC power

depicts the general form of a linear AC circuit. The most general expressions for the voltage and current delivered to an arbitrary load are as follows:

$$v(t) = V \cos (\omega t + \theta_V)$$
$$i(t) = I \cos (\omega t + \theta_I)$$

(5.1)

where V and I are the peak amplitudes of the sinusoidal voltage and current, respectively, and θ_V and θ_I are their phase angles. Two such waveforms are plotted in Figure 5.2, with unit amplitude and with phase angles $\theta_V = \pi/6$ and $\theta_I = \pi/3$.

Since the **instantaneous power** dissipated by a circuit element is given by the product of the instantaneous voltage and current, it is possible to obtain a general expression for the power dissipated by an AC circuit element:

$$p(t) = v(t)i(t)$$
$$= VI \cos (\omega t + \theta_V) \cos (\omega t + \theta_I)$$

(5.2)

Equation 5.2 can be further simplified with the aid of trigonometric identities

Figure 5.2 Current and voltage waveforms for illustration of AC power

to yield

$$p(t) = \frac{VI}{2}\cos(\theta) + \frac{VI}{2}\cos(2\omega t + \theta_V + \theta_I)$$

$$= \frac{VI}{2}\cos(\theta) + \frac{VI}{2}\cos(2\omega t + \theta_V + \theta_I + \theta_I - \theta_I) \qquad (5.3)$$

$$= \frac{VI}{2}\cos(\theta) + \frac{VI}{2}\cos(2\omega t + \theta + 2\theta_I)$$

where $\theta = \theta_V - \theta_I$ is the difference in phase between voltage and current. Note further that if one chooses to reference all phase angles to θ_I, then it may be assumed without loss of generality that $\theta_I = 0$. Thus, Equation 5.3 becomes

$$p(t) = \frac{VI}{2}\cos(\theta) + \frac{VI}{2}\cos(2\omega t + \theta) \qquad (5.4)$$

Equation 5.4 illustrates how the instantaneous power dissipated by an AC circuit element is equal to the sum of an average component, $\frac{VI}{2}\cos(\theta)$, plus a sinusoidal component, $\frac{VI}{2}\cos(2\omega t + \theta)$, oscillating at a frequency double that of the original source frequency.

The instantaneous and average power are plotted in Figure 5.3 for the signals of Figure 5.2. The **average power** corresponding to the voltage and current signals of equation 5.1 can be obtained by integrating the instantaneous power over one cycle of the sinusoidal signal. Let $T = 2\pi/\omega$ represent one cycle of the sinusoidal signals. Then the average power, P_{av}, is given by the integral of the instantaneous power, $p(t)$, over one cycle:

$$P_{av} = \frac{1}{T}\int_0^T p(t)\,dt$$

$$= \frac{1}{T}\int_0^T \frac{VI}{2}\cos(\theta)\,dt + \frac{1}{T}\int_0^T \frac{VI}{2}\cos(2\omega t + \theta_V + \theta_I)\,dt \quad (5.5)$$

$$P_{av} = \frac{VI}{2}\cos(\theta) \qquad \text{Average power}$$

since the second integral is equal to zero.

Figure 5.3 Instantaneous and average power dissipation corresponding to the signals plotted in Figure 5.2

As shown in Figure 5.1, the same analysis carried out in equations 5.1 to 5.4 can also be repeated using phasor analysis. In phasor notation, the current and voltage of equation 5.1 are given by

$$\mathbf{V}(\omega) = Ve^{j\theta_V}$$
$$\mathbf{I}(\omega) = Ie^{j\theta_I} \tag{5.6}$$

Note further that the impedance of the circuit element shown in Figure 5.1 is defined by the phasor voltage and current of equation 5.6 to be

$$Z(\omega) = \frac{V}{I}e^{j(\theta_V - \theta_I)} = |Z|\, e^{j\theta_Z} \tag{5.7}$$

and therefore that the phase angle of the impedance is

$$\theta_Z = \theta_V - \theta_I = \theta \tag{5.8}$$

The expression for the average power obtained in equation 5.4 can therefore also be represented using phasor notation, as follows:

$$P_{\text{av}} = \frac{1}{2}\frac{V^2}{|Z|}\cos\theta = \frac{1}{2}I^2\,|Z|\cos\theta \tag{5.9}$$

AC Power Notation

It has already been noted that AC power systems operate at a fixed frequency; in North America, this frequency is 60 cycles per second (Hz), corresponding to a radian frequency

$$\omega = 2\pi \cdot 60 = 377 \text{ rad/s} \tag{5.10}$$

In Europe and most other parts of the world, AC power is generated at a frequency of 50 Hz (this is the reason why some appliances will not operate under one of the two systems). It will therefore be understood that for the remainder of this chapter the radian frequency, ω, is fixed at 377 rad/s. With knowledge of the radian frequency of all voltages and currents, it will always be possible to compute the exact magnitude and phase of any impedance in a circuit.

A second point concerning notation is related to the factor $\frac{1}{2}$ in equation 5.9. It is customary in AC power analysis to employ the rms value of the AC voltages and currents in the circuit (see Section 4.2). Use of the rms value eliminates the factor $\frac{1}{2}$ in power expressions and leads to considerable simplification. Thus, the following expressions will be used in this chapter:

$$V_{rms} = \frac{V}{\sqrt{2}} = \tilde{V} \tag{5.11}$$

$$I_{rms} = \frac{I}{\sqrt{2}} = \tilde{I} \tag{5.12}$$

$$P_{av} = \frac{1}{2}\frac{V^2}{|Z|}\cos\theta = \frac{\tilde{V}^2}{|Z|}\cos\theta$$
$$= \frac{1}{2}I^2|Z|\cos\theta = \tilde{I}^2|Z|\cos\theta = \tilde{V}\tilde{I}\cos\theta \tag{5.13}$$

Figure 5.4 illustrates the so-called **impedance triangle,** which provides a convenient graphical interpretation of impedance as a vector in the complex plane. From the figure, it is simple to verify that

$$R = |Z|\cos\theta \tag{5.14}$$

$$X = |Z|\sin\theta \tag{5.15}$$

Finally, phasor voltages and currents will be denoted by means of the rms amplitude:

$$\mathbf{V} = V_{rms}e^{j\theta_V} = \tilde{V}e^{j\theta_V} = \tilde{V}\angle\theta_V \tag{5.16}$$

and

$$\mathbf{I} = I_{rms}e^{j\theta_I} = \tilde{I}e^{j\theta_I} = \tilde{I}\angle\theta_I \tag{5.17}$$

Figure 5.4 Impedance triangle

In other words, throughout the remainder of this chapter the symbols $\tilde{V}$ and $\tilde{I}$ will denote the rms value of a voltage or a current. Also note the use of the symbol $\angle$ to represent the complex exponential. Thus, the sinusoidal waveform corresponding to the phasor current $\mathbf{I} = \tilde{I}\angle\theta_I$ corresponds to the time-domain waveform

$$i(t) = \sqrt{2}\tilde{I}\cos(\omega t + \theta_I) \tag{5.18}$$

and the sinusoidal form of the phasor voltage $\mathbf{V} = \tilde{V}\angle\theta_V$ is

$$v(t) = \sqrt{2}\tilde{V}\cos(\omega t + \theta_V) \tag{5.19}$$

Example   5.1

This example illustrates the computation of average power for a typical load consisting of an inductor in series with a resistor. This type of load is actually quite common in practice, being representative of the circuit behavior of many electric motors. The circuit is shown in Figure 5.5.

The circuit can be analyzed by phasor notation to compute the current $i(t)$ that flows through the series RL load. The radian frequency of the source voltage is $\omega = 377$ rad/s; thus, the load impedance is:

14.14 sin (ωt)
($\omega = 377$ rad/s)

$i(t)$

$\dfrac{4}{377}$ H $\dfrac{4}{377}$ Ω

Figure 5.5

$$Z = 4 + j3 = \sqrt{16 + 9}e^{j\tan^{-1}(3/4)}$$
$$= 5e^{j36.9°}$$

The phasor source voltage may be expressed in rms form as

$$\mathbf{V} = 10e^{-j90°}$$

since

$$14.14\sin\omega t = 10\sqrt{2}\cos(\omega t - 90°)$$

Thus the phasor current must be

$$\mathbf{I} = \frac{\mathbf{V}}{Z} = \frac{10e^{-j90°}}{5e^{j36.9°}} = 2e^{-j126.9°}$$

which corresponds to the time-domain current

$$i(t) = 2\sqrt{2}\cos(\omega t - 126.9°)$$

The two functions

$$v(t) = 10\sqrt{2}\cos(\omega t - 90°)$$

and

$$i(t) = 2\sqrt{2}\cos(\omega t - 126.9°)$$

and the instantaneous power

$$p(t) = 2\sqrt{2}\cos(\omega t - 126.9°)10\sqrt{2}\cos(\omega t - 90°)$$

are sketched in Figure 5.6.

Voltage and current waveforms for Example 5.1

Instantaneous and average power for Example 5.1

Figure 5.6

Example   5.2

This example illustrates the computation of average power for a parallel RC load. The circuit is shown in Figure 5.7. Note that in the circuit of Figure 5.7, the AC source is modeled as having an internal resistance; this resistance affects the magnitude and phase of the load voltage and causes a power loss.

Figure 5.7

Solution:

The first step in the computation of average power consists of the calculation of the load impedance:

$$Z_L = R \, \| \, \frac{1}{j\omega C} = \frac{R}{1 + j\omega RC} = \frac{16}{1 + j0.6032} = 11.7315 - j7.0764 = 13.7e^{-j31.1°}$$

Thus, $|Z_L| = 13.7$ and $\theta = -31.1°$.

The second step consists of determining the load voltage. Application of the voltage divider rule yields the expression

$$V_L = 110e^{j0°} \frac{Z_L}{2 + Z_L} = 110e^{j0°} \frac{11.7315 - j7.0764}{13.7315 - j7.0764}$$

$$= (110e^{j0°})(0.8869e^{-j3.83°}) = 97.6e^{-j3.83°}$$

Given the impedance and the load voltage, it is then possible to directly apply equation 5.8 to compute the average power:

$$P_{av} = \frac{\tilde{V}_L^2}{|Z_L|} \cos\theta = \frac{(97.6)^2}{13.6} \cos(-31.1°) = 549.9 \text{ W}$$

Alternatively, the same solution could be obtained by first determining the load current and using the expression

$$P_{av} = \tilde{I}^2 |Z| \cos\theta$$

Example   5.3

Find the power dissipated by the complex load impedance seen by the 60-Hz AC source in the circuit shown in Figure 5.8. Assume that the rms source voltage is $V = 110e^{j0°}$.

Solution:

The circuit is shown in complex form in Figure 5.8. The equivalent impedance seen by the source is equal to the parallel combination of the two branches:

$$jX_C = \frac{1}{j\omega C} = -j\frac{1}{377 \times 470 \times 10^{-6}} = -j5.64$$

$$jX_L = j\omega L = j377 \times 0.05 = j18.85$$

$$Z_{eq} = (10 + j18.85) \| (-j5.64)$$

$$= \frac{(10 + j18.85)(-j5.64)}{10 + j18.85 - j5.64}$$

$$= \frac{106.31 - j56.4}{10 + j13.21} = \frac{120.35\angle-27.95°}{16.568\angle52.87°}$$

$$= 7.26\angle-80.8° = 1.16 - j7.17$$

Thus, the equivalent impedance seen by the source consists of an AC resistance of 1.16 Ω and a (capacitive) reactance of −7.17 Ω, shown in Figure 5.9.
Finally, the average power dissipated by the load is

$$P_{av} = \frac{V^2}{|Z|} \cos(\theta) = \frac{110^2}{7.26} \cos(-80.6°) = 265.7 \text{ W}$$

An AC circuit

Its complex form

Figure 5.8

Figure 5.9

Power Factor

The phase angle of the load impedance plays a very important role in the absorption of power by a load impedance. As illustrated in equation 5.13 and in the preceding examples, the average power dissipated by an AC load is dependent on the cosine of the angle of the impedance. To recognize the importance of this factor in AC power computations, the term $\cos(\theta)$ is referred to as the **power factor (pf).** Note that the power factor is equal to 0 for a purely inductive or capacitive load and equal to 1 for a purely resistive load; in every other case,

$$0 < pf < 1 \tag{5.20}$$

Two equivalent expressions for the power factor are given in the following:

$$pf = \cos(\theta) = \frac{P_{av}}{\tilde{V}\tilde{I}} \qquad \text{Power factor} \tag{5.21}$$

where $\tilde{V}$ and $\tilde{I}$ are the rms values of the load voltage and current.

DRILL EXERCISES

5.1   Show that the equalities in equation 5.9 hold when phasor notation is used.

5.2   Consider the circuit shown in Figure 5.10. Find the load impedance of the circuit, and compute the average power dissipated by the load.

Answer: $Z = 4.8e^{-j33.5°}\,\Omega$; $P_{av} = 2,102.1$ W

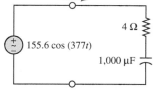

Figure 5.10

5.3   Use the expression $P_{av} = \tilde{I}^2|Z|\cos\theta$ to compute the average power dissipated by the load of Example 5.2.

5.4   Compute the power dissipated by the internal source resistance in Example 5.2.

Answer: 101.46 W

5.5   Compute the power factor for an inductive load with $L = 100$ mH and $R = 0.4\,\Omega$. Assume $\omega = 377$ rad/s.

Answer: $pf = \cos 89.36° = 0.0105$

5.2   COMPLEX POWER

The expression for the instantaneous power given in equation 5.4 may be further expanded to provide further insight into AC power. Using trigonometric identities, we obtain the following expressions:

$$
\begin{aligned}
p(t) &= \frac{\tilde{V}^2}{|Z|}[\cos\theta + \cos\theta\cos(2\omega t) + \sin\theta\sin(2\omega t)] \\
&= \tilde{I}^2|Z|[\cos\theta + \cos\theta\cos(2\omega t) + \sin\theta\sin(2\omega t)] \\
&= \tilde{I}^2|Z|\cos\theta(1 + \cos 2\omega t) + \tilde{I}^2|Z|\sin\theta\sin(2\omega t)
\end{aligned}
\tag{5.22}
$$

Recalling the geometric interpretation of the impedance Z of Figure 5.4, you may recognize that

$$
|Z|\cos\theta = R
$$

and $\hspace{9cm}$ (5.23)

$$
|Z|\sin\theta = X
$$

are the resistive and reactive components of the load impedance, respectively. On the basis of this fact, it becomes possible to write the instantaneous power as:

$$
\begin{aligned}
p(t) &= \tilde{I}^2 R(1 + \cos 2\omega t) + \tilde{I}^2 X\sin 2\omega t \\
&= \tilde{I}^2 R + \tilde{I}^2 R\cos 2\omega t + \tilde{I}^2 X\sin 2\omega t
\end{aligned}
\tag{5.24}
$$

The physical interpretation of this expression for the instantaneous power should be intuitively appealing at this point. As equation 5.23 suggests, the instantaneous power dissipated by a complex load consists of the following three components:

1. An average component, which is constant; this is called the *average power* and is denoted by the symbol P_{av}:

    $$
    P_{av} = \tilde{I}^2 R
    \tag{5.25}
    $$

    where $R = \text{Re}\,[Z]$.

2. A time-varying (sinusoidal) component with zero average value that is contributed by the power fluctuations in the resistive component of the load and is denoted by $p_R(t)$:

    $$
    \begin{aligned}
    p_R(t) &= \tilde{I}^2 R\cos 2\omega t \\
    &= P_{av}\cos 2\omega t
    \end{aligned}
    \tag{5.26}
    $$

3. A time-varying (sinusoidal) component with zero average value, due to the power fluctuation in the reactive component of the load and denoted by $p_X(t)$:

    $$
    \begin{aligned}
    p_X(t) &= \tilde{I}^2 X\sin 2\omega t \\
    &= Q\sin 2\omega t
    \end{aligned}
    \tag{5.27}
    $$

    where $X = \text{Im}\,[Z]$ and Q is called the **reactive power**. *Note that since reactive elements can only store energy and not dissipate it, there is no net average power absorbed by X.*

Since P_{av} corresponds to the power absorbed by the load resistance, it is also called the **real power,** measured in units of watts (W). On the other hand, Q takes the name of *reactive power,* since it is associated with the load reactance. Table 5.1 shows the general methods of calculating P and Q.

Table 5.1   Real and reactive power

Real power, P_{av}	Reactive power, Q
$\tilde{V}\tilde{I}\cos(\theta)$	$\tilde{V}\tilde{I}\sin(\theta)$
$\tilde{I}^2 R$	$\tilde{I}^2 X$

The units of Q are **volt-amperes reactive,** or **VAR.** Note that Q represents an exchange of energy between the source and the reactive part of the load; thus, no net power is gained or lost in the process, since the average reactive power is zero. In general, it is desirable to minimize the reactive power in a load. Example 5.5 will explain the reason for this statement.

The computation of AC power is greatly simplified by defining a fictitious but very useful quantity called the **complex power,** S:

$$\boxed{S = \mathbf{VI}^*}$$ (5.28)

where the asterisk denotes the complex conjugate (see Appendix A). You may easily verify that this definition leads to the convenient expression

$$S = \tilde{V}\tilde{I}\cos\theta + j\tilde{V}\tilde{I}\sin\theta = \tilde{I}^2 R + j\tilde{I}^2 X = \tilde{I}^2 Z$$

or (5.29)

$$S = P_{av} + jQ$$

The complex power S may be interpreted graphically as a vector in the complex plane, as shown in Figure 5.11.

The magnitude of S, $|S|$, is measured in units of **volt-amperes (VA)** and is called **apparent power,** because this is the quantity one would compute by measuring the rms load voltage and currents without regard for the phase angle of the load. Note that the right triangle of Figure 5.11 is similar to the right triangle of Figure 5.4, since θ is the load impedance angle. The complex power may also be expressed by the product of the square of the rms current through the load and the complex load impedance:

$$S = \tilde{I}^2 Z$$

or (5.30)

$$\tilde{I}^2 R + j\tilde{I}^2 X = \tilde{I}^2 Z$$

or, equivalently, by the ratio of the rms voltage across the load to the complex conjugate of the load impedance:

$$S = \frac{\tilde{V}^2}{Z^*}$$ (5.31)

$|S| = \sqrt{P_{av}^2 + Q^2} = \tilde{V}\cdot\tilde{I}$

$P_{av} = \tilde{V}\tilde{I}\cos\theta$

$Q = \tilde{V}\tilde{I}\sin\theta$

Figure 5.11 The complex power triangle

The power triangle and complex power greatly simplify load power calculations, as illustrated in the following examples.

Example   5.4

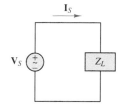

Figure 5.12

Calculate the complex power for the circuit shown in Figure 5.12, given the following voltage and current waveforms:

$$v(t) = 100 \cos (\omega t + 15°) \qquad i(t) = 2 \cos (\omega t - 15°)$$

Solution:

First, we convert the voltage and current into rms phasor form:

$$\mathbf{V} = \frac{100}{\sqrt{2}} \angle 15° \qquad \mathbf{I} = \frac{2}{\sqrt{2}} \angle -15°$$

Then the real and reactive load power are computed using equation 5.13. The average power is found to be

$$P_{av} = \tilde{V}\tilde{I} \cos \theta \qquad \left(\theta = \theta_V - \theta_I \right)$$

$$= \frac{100}{\sqrt{2}} \times \frac{2}{\sqrt{2}} \times \cos (15° - (-15°)) = 86.6 \text{ W}$$

while the reactive power is

$$Q = \frac{100}{\sqrt{2}} \times \frac{2}{\sqrt{2}} \times \sin (15° - (-15°)) = 50 \text{ VAR}$$

Next, the same quantities are computed using equation 5.28:

$$S = \mathbf{VI}^* = P_{av} + jQ$$

$$= \left(\frac{100}{\sqrt{2}} e^{j15°} \right) \left(\frac{2}{\sqrt{2}} e^{-j15°} \right)^* = 100e^{j30°} = 86.6 + j50$$

or

$$P_{av} = 86.6 \text{ W} \qquad Q = 50 \text{ VAR}$$

Note that the computation using complex power yields both real and reactive power in a single calculation.

Example   5.5

The objective of this example is to illustrate the effect of a reactive load element on the power dissipated by a load. We seek to find the real and reactive instantaneous power dissipated by the load in the circuit of Figure 5.13 (all voltages and currents are rms and have frequency 60 Hz).

Figure 5.13

Solution:

We start by expressing the load current as the ratio of the source voltage to the total series impedance of the circuit:

$$Z = R_S + R_L + X_L$$

$$= 2 + 5 + \frac{1}{j377 \times 2,000 \times 10^{-6}}$$

$$= 7.13e^{-j10.76°}$$

$$\mathbf{V}_S = 110e^{j0°}$$

$$\mathbf{I} = \frac{\mathbf{V}_S}{Z} = \frac{110e^{j0°}}{7.13e^{-j10.76°}}$$

$$= 15.44e^{j10.76°}$$

The power factor of the load is

$$\cos \theta = 0.967$$

The AC load resistance is equal to

$$R = 5 \ \Omega$$

and the load reactance is

$$X = -1.33 \ \Omega$$

while the rms load current is

$$\tilde{I} = 15.44 \ \text{A}$$

Thus, we find the real power absorbed by the resistive component of the load:

$$P_{\text{av}} = \tilde{I}^2 R = 1,192 \ \text{W}$$

and the reactive power:

$$Q = \tilde{I}^2 X = -316 \ \text{VAR}$$

Although the reactive power does not contribute to any average power dissipation in the load, it may have an adverse effect on power consumption, because it increases the overall rms current flowing in the circuit. Recall from Example 5.2 that the presence of any source resistance (typically, the resistance of the line wires in AC power circuits) will cause a loss of power; the power loss due to this line resistance is unrecoverable and constitutes a net loss for the electric company, since the user never receives this power. The following example illustrates quantitatively the effect of such **line losses** in an AC circuit.

Example   5.6

Figure 5.14

This example illustrates how, by reducing the reactive power at the load, it is possible to improve the **power transfer** from a source to a load. Consider the AC source shown in Figure 5.14, modeled by an ideal source with internal resistance (including the wires connecting the source to the load) of 4 Ω. The load consists of a parallel RL combination, with $R = 10$ Ω and $X = 6$ Ω.

1. Compute the load impedance seen by the source.
2. Find the real and reactive power of the load.
3. Determine the percent real power transfer from the source to the load with and without the inductor included in the circuit.

Solution:

1. The equivalent load impedance is given by

$$Z = 10 \,\|\, j6 = \frac{(j6)(10)}{j6 + 10} = \frac{60\angle90°}{11.66\angle31°}$$

$$= 5.145\angle59° \ \Omega$$

2. To find the real and reactive power, we compute the load voltage. By the voltage divider rule, we obtain:

$$\mathbf{V} = \frac{Z}{Z+4}\mathbf{V}_S = \frac{5.145\angle59°}{2.647 + j4.406 + 4}\mathbf{V}_S$$

$$= \frac{5.145\angle59°}{6.647 + j4.4118}\mathbf{V}_S = \frac{5.145\angle59°}{7.978\angle33.57°}\mathbf{V}_S$$

$$= 70.94\angle25.46°$$

The real power absorbed by the load is found to be

$$P = \frac{\tilde{V}^2}{R} = \frac{(70.94)^2}{10} = 503.2 \text{ W}$$

and the reactive power is

$$Q = \frac{\tilde{V}^2}{X} = \frac{(70.9)^2}{6} = 838.7 \text{ VAR}$$

3. Without the inductor in the circuit, as shown in Figure 5.15, we would compute the following results:

Figure 5.15

$$P = \tilde{I}^2 10 = \left(\frac{\tilde{V}_S}{10 + 4}\right)^2 10 = \left(\frac{110}{14}\right)^2 (10) = 617.3 \text{ W}$$

and

$$Q = 0$$

The total power generated by the source is the product of the source current and the source voltage (since these are in phase when the inductor is not in the circuit):

$$P_S = \tilde{V}\tilde{I} = 110 \left(\frac{110}{14}\right) \text{ W}$$

$$P_S = 864.3 \text{ W}$$

so that the percent power transfer in the absence of the inductive component of the load is

$$100\% \frac{P_L}{P_S} = 71.4\%$$

With the inductor in the circuit, we need to consider the phasor form of the source voltage and load current:

$$\mathbf{V}_S = 110\angle 0°$$

$$\mathbf{I}_S = \frac{\mathbf{V}_S}{Z + 4} = 13.79\angle -33.57°$$

Hence

$$P_S = 110 \times 13.79 \times \cos(-33.57°)$$

$$= 1{,}263.7 \text{ W}$$

Thus the percent power transfer is now

$$\% \text{ power transfer} = 100\% \frac{P_L}{P_S} = 100\% \frac{503.2}{1{,}263.7} = 39.82\%$$

Observe that, although the inductor does not dissipate any real power, its presence greatly reduces the power that can be absorbed by the load and increases the total power generated by the source.

Power Factor, Revisited

The power factor, defined earlier as the cosine of the angle of the load impedance, plays a very important role in AC power. A power factor close to unity signifies an efficient transfer of energy from the AC source to the load, while a small power factor corresponds to inefficient use of energy, as illustrated in Example 5.6. It should be apparent that if a load requires a fixed amount of real power, P, the source will be providing the smallest amount of current when the power factor is greatest, that is when $\cos \theta = 1$. If the power factor is less than unity, some additional current will be drawn by the source, lowering the efficiency of power transfer from the source to the load. However, it will be shown shortly that it is possible to correct the power factor of a load by adding an appropriate reactive component to the load itself.

Since the reactive power, Q, is related to the reactive part of the load, its sign depends on whether the load reactance is inductive or capacitive. This leads to the following important statement:

> If the load has an inductive reactance, then θ is positive and the current *lags* (or *follows*) the voltage. Thus, when θ and Q are positive, the corresponding power factor is termed *lagging*. Conversely, a capacitive load will have a negative Q, and hence a negative θ. This corresponds to a *leading* power factor, meaning that the load current *leads* the load voltage.

Table 5.2 illustrates the concept and summarizes all of the important points so far. In the table, the phasor voltage **V** has a zero phase and the current phasor is referenced to the phase of **V**.

Table 5.2   Important facts related to complex power

	Resistive load	Capacitive load	Inductive load
Ohm's law	$\mathbf{V}_L = Z_L \mathbf{I}_L$	$\mathbf{V}_L = Z_L \mathbf{I}_L$	$\mathbf{V}_L = Z_L \mathbf{I}_L$
Complex impedance	$Z_L = R_L$	$Z_L = R_L - jX_L$	$Z_L = R_L + jX_L$
Phase angle	$\theta_V - \theta_I = \theta = 0$	$\theta_V - \theta_I = \theta < 0$	$\theta_V - \theta_I = \theta > 0$
Complex plane sketch			
Explanation	The current is in phase with the voltage.	The current "leads" the voltage.	The current "lags" the voltage.
Power factor	Unity	Leading, < 1	Lagging, < 1
Reactive power	0	Negative	Positive

The following examples illustrate the computation of complex power for a simple circuit.

Example   5.7

Find the reactive and real power dissipated by the circuit of Figure 5.16, and draw the associated power triangle.

Solution:

The phasor current **I** flowing in the series circuit is found as follows:

$$\mathbf{I} = \frac{60\angle 0°}{Z_{eq}} = \frac{60\angle 0°}{3 + j9 - j5} = \frac{60\angle 0°}{5\angle 53.13°} = 12\angle -53.13° \text{ A}$$

and the complex power S is given by the expression

$$S = \tilde{I}^2 Z = (12^2)(3 + j4) = (144)(3 + j4) = 432 + j576$$
$$= P + jQ$$

If we observe that the total reactive power must be the sum of the reactive power in the inductor and in the capacitor, we can write

$$Q = Q_C + Q_L$$

and compute each quantity, as follows:

$$Q_C = (144)(-5) = -720 \text{ VAR}$$
$$Q_L = (144)(9) = 1,296 \text{ VAR}$$
$$Q = Q_L + Q_C = 1,296 - 720 = 576 \text{ VAR}$$

Next, we draw the power triangle (Figure 5.17), showing how the complex power, S, is a vector in the complex plane equal to the vector sum of the three components, corresponding to the three elements.

60 VAC

Figure 5.16

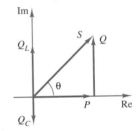

Note: $S = P_R + jQ_C + jQ_L$

Figure 5.17

The distinction between leading and lagging power factors made in Table 5.1 is important, because it corresponds to opposite signs of the reactive power: Q is positive if the load is inductive ($\theta > 0$) and the power factor is lagging; Q is negative if the load is capacitive and the power factor is leading ($\theta < 0$). It is therefore possible to improve the power factor of a load according to a procedure called **power factor correction**—that is, by placing a suitable reactance in parallel with the load so that the reactive power component generated by the additional reactance is of opposite sign to the original load reactive power. Most often the need is to improve the power factor of an inductive load, because many common industrial loads consist of electric motors, which are predominantly inductive loads. This improvement may be accomplished by placing a capacitance in parallel with the load. The following example illustrates a typical power factor correction for an industrial load.

Figure 5.18

Example   5.8

For the circuit shown in Figure 5.18:

1.   Calculate the complex power for the load.
2.   Correct the power factor by adding a suitable reactance in parallel with the load.

Solution:

1.   The circuit of Figure 5.18 is an inductive load. The total impedance is

$$Z = R + jX_L = 50 + j86.7 \ \Omega = 100\angle 60°$$

The power factor is then

$$pf = \cos \theta = \cos 60° = 0.5 \text{ (lagging)}$$

The current drawn from the source by the load is

$$\mathbf{I}_S = \frac{\mathbf{V}_S}{Z} = \frac{117\angle 0°}{100\angle 60°} = 1.17\angle -60°$$

and the average power is found to be

$$P = \tilde{V}_S \tilde{I}_S \cos \theta = 117 \times 1.17 \cos 60° = 68.4 \text{ W}$$

while the reactive power is

$$Q_L = \tilde{V}_S \tilde{I}_S \sin \theta = 117 \times 1.17 \sin 60° = 119 \text{ VAR}$$

The power triangle is shown in Figure 5.19.

2.   The unity power factor for the circuit can be obtained by simply reducing the power factor angle θ to 0°. This can be accomplished by adding a capacitor to the circuit that requires -119 VAR of reactive power. The capacitive power and the inductive power will then cancel each other in the power triangle, resulting in a unity power factor, as shown in Figure 5.20.

The value of capacitive reactance, X_C, required to cancel the reactive power due to the inductance is found most easily by observing that the total reactive power in the circuit must be the sum of the reactive power due to the capacitance and that due to the inductance. Observing that the capacitor sees the

Figure 5.19

Figure 5.20 Power factor correction

same voltage as the *RL* load, because of the parallel connection, we can write

$$X_C = \frac{\tilde{V}_S^2}{Q_C} = \frac{117^2}{119} = 115 \ \Omega$$

From the expression for the reactance, it is then possible to compute the value of the capacitor that will cancel the reactive power due to the inductor:

$$C = \frac{1}{\omega X_C} = \frac{1}{377 \times 115} = 23.1 \ \mu F$$

The reactive component of power needed by the inductance is now balanced by the capacitance, and all the power delivered by the source is real power. The power factor is 1.

Example   5.9

With reference to Example 5.8, consider the series capacitor connection shown in Figure 5.21, which is proposed as an alternative to the parallel connection to correct the power factor. Is this a feasible alternative to the solution of Example 5.8?

Figure 5.21

Solution:

The series load impedance, Z', may be computed as follows:

$$Z' = R + jX_L - jX_C = 50 + j86.7 - j86.7 = 50$$

while the series current in the circuit is

$$I_S = \frac{V_S}{Z'} = \frac{117\angle 0^\circ}{50} = 2.34 \ A$$

Note the significant increase in the current drawn from the source! The addition of the series capacitor is not an acceptable method for power factor correction, because the circuit current through the original *RL* load has increased by a factor of 2, from 1.17 A to 2.34 A. It is important to observe that the voltage across the *RL* load has also increased to:

$$I_S(R + j\omega L) = 2.34\angle 30^\circ \times 100\angle 60^\circ = 234\angle 90^\circ \ V$$

which is also approximately double the load voltage prior to the addition of the series capacitor. It should be apparent that the power factor correction of an inductive load must be accomplished by connecting a capacitance in parallel with the original load, or else the current drawn by the source will be increased excessively.

External connections

Wattmeter coils (inside)

Figure 5.22

Source   Load

$v_S(t) = 156 \cos(377t)$
$R_1 = 10\ \Omega$
$R_2 = 5\ \Omega$
$L = 20\ \text{mH}$

Figure 5.24

Example   5.10

The instrument used to measure power is called a *wattmeter*. The external part of a wattmeter consists of four connections and a metering mechanism that displays the amount of real power dissipated by a circuit. The external and internal appearance of a wattmeter is depicted in Figure 5.22. Inside the wattmeter are two coils: a current-sensing coil, and a voltage-sensing coil. In this example, we assume for simplicity that the impedance of the current-sensing coil, C_I, is zero and the impedance of the voltage-sensing coil, C_V, is infinite. In practice, this will not necessarily be true; some correction mechanism will be required to account for the impedance of the sensing coils.

A wattmeter should be connected as shown in Figure 5.23, to provide both current and voltage measurements. We see that the current-sensing coil is placed in series with the load and the voltage-sensing coil is placed in parallel with the load. In this manner, the wattmeter is seeing the current through the voltage across the

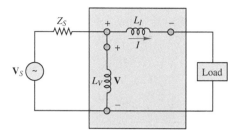

Figure 5.23

load. Remember that the power dissipated by a circuit element is related to these two quantities. The wattmeter, then, is constructed to provide a readout of the product of the rms values of the load current and the voltage, which is the real power absorbed by the load: $P = \text{Re}(S) = \text{Re}(\mathbf{VI}^*)$.

1. For the circuit shown in Figure 5.24, show the connections of the wattmeter, and find the power dissipated by the load.

2. Show the connections that will determine the power dissipated by R_2. What should the meter read?

Solution:

1. To measure the power dissipated by the load, we must have the current through and the voltage across the entire load circuit. This means that the wattmeter must be connected as shown in Figure 5.25. The wattmeter should read:

$$P = \text{Re}(\mathbf{V}_S\mathbf{I}^*)$$

$$= \text{Re}\left\{ \left(\frac{156}{\sqrt{2}}\angle 0^\circ\right) \left(\frac{\frac{156}{\sqrt{2}}\angle 0^\circ}{R_1 + R_2 + j\omega L}\right)^* \right\}$$

$$= \text{Re}\left\{ 110\angle 0^\circ \left(\frac{110\angle 0^\circ}{15 + j7.54}\right)^* \right\}$$

$$\text{Re}\left\{110\angle 0°\left(\frac{110\angle 0°}{16.79\angle 26.69°}\right)^{*}\right\} = \text{Re}\left(\frac{110^{2}}{16.79\angle -26.69°}\right)$$

$$= \text{Re}\,(720.67\angle 26.69°)$$

$$= 643.88 \text{ W}$$

2. To measure the power dissipated by R_2 alone, we must measure the current through R_2 and the voltage across R_2 *alone*. The connection is shown in Figure 5.26. The meter will read

$$P = \tilde{I}^{2}R_{2}$$

$$= \left(\frac{110}{(15^{2} + 7.54^{2})^{1/2}}\right)^{2} \times 5$$

$$= \left(\frac{110^{2}}{15^{2} + 7.54^{2}}\right) \times 5 = 215 \text{ W}$$

Figure 5.25

Figure 5.26

The measurement and correction of the power factor for the load are an extremely important aspect of any engineering application in industry that requires the use of substantial quantities of electric power. In particular, industrial plants, construction sites, heavy machinery, and other heavy users of electric power must be aware of the power factor their loads present to the electric utility company. As was already observed, a low power factor results in greater current draw from the electric utility, and in greater line losses. Thus, computations related to the power factor of complex loads are of great practical utility to any practicing engineer. To provide you with deeper insight into calculations related to power factor, a few more advanced examples are given in the remainder of the section.

Example   5.11

A capacitor is being used to correct the power factor to unity. The circuit is shown in Figure 5.27. The capacitor value is varied, and measurements of the total current are taken. Explain how it is possible to "zero in" on the capacitance value necessary to bring the power factor to unity just by monitoring the current $\mathbf{I}_S$.

Figure 5.27

Solution:

The current through the load is

$$\mathbf{I}_L = \frac{\tilde{V}_S\angle 0°}{R + j\omega L} = \frac{\tilde{V}_S}{R^2 + \omega^2 L^2}(R - j\omega L)$$

$$= \frac{\tilde{V}_S R}{R^2 + \omega^2 L^2} - j\frac{\tilde{V}_S\omega L}{R^2 + \omega^2 L^2}$$

The current through the capacitor is

$$\mathbf{I}_C = \frac{\tilde{V}_S\angle 0°}{1/j\omega C} = j\tilde{V}_S\omega C$$

The source current to be measured is

$$\mathbf{I}_S = \mathbf{I}_L + \mathbf{I}_C = \frac{\tilde{V}_S R}{R^2 + \omega^2 L^2} + j\left(\tilde{V}_S\omega C - \frac{\tilde{V}_S\omega L}{R^2 + \omega^2 L^2}\right)$$

The magnitude of the source current is

$$\tilde{I}_S = \sqrt{\left(\frac{\tilde{V}_S R}{R^2 + \omega^2 L^2}\right)^2 + \left(\tilde{V}_S\omega C - \frac{\tilde{V}_S\omega L}{R^2 + \omega^2 L^2}\right)^2}$$

We know that when the load is a pure resistance, the current and voltage are in phase, the power factor is 1, and all the power delivered by the source is dissipated by the load as real power. This corresponds to equating the imaginary part of the expression for the source current to zero, or, equivalently, to the following expression:

$$\frac{\tilde{V}_S\omega L}{R^2 + \omega^2 L^2} = \tilde{V}_S\omega C$$

in the expression for $\tilde{I}_S$. Thus, the magnitude of the source current is actually a minimum when the power factor is unity! It is therefore possible to "tune" a load to a unity *pf* by observing the readout of the ammeter while changing the value of capacitor and selecting the capacitor value that corresponds to the lowest source current value.

Example 5.12

A 100-kW load with *pf* = 0.7 (lagging) is connected to a 480-V (rms) source. A capacitor is connected in parallel with the load to improve the power factor, as shown in Figure 5.28.

1. Determine the reactive power when no capacitor is connected to the load.
2. Determine the required capacitance to bring the power factor to unity.

Solution:

1. The magnitude of the complex power, *S*, may be found by computing the ratio of the rated power to the power factor as shown in the power triangle:

Figure 5.28

$$|S| = \frac{P}{\cos \theta} = \frac{P}{pf} = \frac{10^5}{0.7}$$

$$= 142{,}857 \text{ VA}$$

Since the power factor is lagging, the reactive power is positive, and we can compute the reactive power according to the relation

$$Q = \tilde{V}\tilde{I} \sin \theta = |S| \sin (\arccos (0.7))$$

$$= 102 \text{ kVAR}$$

2. There are two approaches to solving this part of the problem. One is to determine the value of a capacitor whose reactance will cancel that of the rest of the circuit. A second method, usually more tedious in calculations, is to find a capacitor whose value will make the load seem purely resistive. It is then simply a matter of finding the capacitance that causes the impedance (or, equivalently, the admittance) of the load to be purely real; this corresponds to equating the imaginary part to zero.

Using the first method to solve this problem, we know that placing a capacitor in parallel with the load will not change the current in the load, nor will it change the voltage across the load. This makes it clear that the reactive power in the load will be 102 kVAR with or without the power factor correction capacitor. To cancel this reactive power, we require that

$$Q_C = -Q_L = -102 \text{ kVAR}$$

$$Q_C = -\frac{\tilde{V}^2}{X_C}$$

$$X_C = -\frac{1}{\omega C}$$

$$C = -\frac{Q_C}{\omega \tilde{V}^2} = \frac{102 \times 10^3}{377 \times 480^2} = 1{,}174 \ \mu\text{F}$$

To determine the capacitance required to obtain a power factor of unity using the second method, we first compute the current flowing in the circuit without the capacitor:

$$\mathbf{I}_S^* = \frac{S}{\mathbf{V}} = \frac{100{,}000 + j102{,}000}{480\angle 0^\circ} = \frac{142{,}840\angle 45.6^\circ}{480\angle 0^\circ}$$

$$\mathbf{I}_S = 297.6\angle -45.6^\circ \text{ A}$$

From knowledge of the source voltage and current, we can then infer the value of the load impedance:

$$Z_L = \frac{\mathbf{V}}{\mathbf{I}_S} = \frac{480\angle 0^\circ}{297.6\angle -45.6^\circ}$$

$$= 1.613\angle + 45.6^\circ$$

$$= 1.13 + j1.15$$

With a capacitor in parallel with the load, the new load impedance is Z_L':

$$Z_L' = Z_L \| -jX_C$$

$$= \frac{-jX_C Z_L}{Z_L - jX_C}$$

$$= \frac{1.613 X_C \angle -44.43°}{1.13 + j(1.15 - X_C)}$$

$$= \frac{(1.613 X_C \angle -44.43°)[1.13 - j(1.15 - X_C)]}{1.13^2 + (1.15 - X_C)^2}$$

To bring the power factor to unity, we require the imaginary part of Z_L' to be zero. Neglecting the denominator in the expression for Z_L', the requirement $X_L' = 0$ is satisfied by the following expression:

$$-1.613 X_C[(0.7)(1.13) + (1.15 - X_C)(0.714)] = 0$$

or

$$0.7(1.13) + 0.714(1.15 - X_C) = 0$$

Solving for the capacitive reactance, X_C, we obtain:

$$X_C = 2.26 = \frac{1}{\omega C}$$

$$C = \frac{1}{2.26(377)} = 1.174 \times 10^{-3} \text{ F}$$

$$= 1,174 \ \mu\text{F}$$

Thus, a 1,174 μF capacitor placed in parallel with the load will reduce the reactive power to zero. Following the power factor correction, the current supplied by the source is given by:

$$I_S' = \frac{V}{Z_L'}$$

where Z_L' can be calculated to be 5.77 Ω by substituting the value $X_C = 2.26$ in the earlier expression. Thus,

$$I_S' = \frac{480 \angle 0°}{5.77} = 83.2 \text{ A}$$

Example   5.13

A second load, with $P = 50$ kW, $pf = 0.95$ (leading), is also connected to the source in Example 5.12, in parallel with the first load. The overall circuit is shown in Figure 5.29.

Figure 5.29

1. Determine the size of the capacitance needed to obtain an overall power factor of unity.
2. Draw the phasor diagram of the two load currents $\mathbf{I}_1$ and $\mathbf{I}_2$, the capacitor current $\mathbf{I}_C$, and the source current $\mathbf{I}_S$.

Solution:

1. The circuit of Example 5.12 is now modified as shown in Figure 5.29. We can find the currents $\mathbf{I}_1$ and $\mathbf{I}_2$ by the following calculations:

$$\mathbf{I} = \frac{P}{pf_1 \cdot \mathbf{V}} \angle \cos^{-1}(pf)$$

$$\mathbf{I}_1 = \frac{10^5}{(0.7)480\angle0°} \angle -\cos^{-1}(0.7) = 297.6\angle-45.6° \text{ A}$$

$$\mathbf{I}_2 = \frac{5 \times 10^4}{(0.95)480\angle0°} \angle \cos^{-1}(0.95) = 109.65\angle18.2° \text{ A}$$

so that the current drawn by the combined loads prior to connecting the power factor correction capacitor is

$$\mathbf{I}_S = \mathbf{I}_L = \mathbf{I}_2 + \mathbf{I}_1 = (104.17 + j34.23 + 208.3 - j212.5) \text{ A}$$

$$= 312.5 - j178.3 = 359.76\angle-29.7° \text{ A}$$

To have unity power factor, we must cancel the reactive power of the loads. We already know the reactive power of the first load from the previous example:

$$Q_1 = 102{,}000 \text{ VAR}$$

For the second load,

$$|S_2| = \frac{P_2}{pf_2} = \frac{50 \text{ kW}}{0.95}$$

$$= 52{,}632 \text{ VA}$$

Since the power factor is leading, the reactive power is negative; thus,

$$Q_2 = |S_2| \sin(-\cos^{-1}(0.95))$$

$$= -16{,}434 \text{ VAR}$$

and the total reactive power of the combined loads is

$$Q_1 + Q_2 = 85{,}566 \text{ VAR}$$

To find the capacitance value required to cancel this reactive power, we apply the same method as in the previous example:

$$Q_C = -\frac{\tilde{V}^2}{X_C}$$

$$X_C = -\frac{1}{\omega C}$$

$$C = -\frac{Q_C}{\omega \tilde{V}^2} = \frac{85,566}{377 \times 480^2} = 985 \ \mu\text{F}$$

2. We need find only the capacitor current, since we know the two load currents already:

$$\mathbf{I}_C = \frac{\mathbf{V}_S}{\mathbf{Z}_C} = \frac{480\angle 0°}{2.69\angle - 90°} = 178.3\angle 90°\ \text{A}$$

Then the new source current, $\mathbf{I}'_S$, is the sum of the capacitor and load currents:

$$\mathbf{I}'_S = \mathbf{I}_C + \mathbf{I}_L = \mathbf{I}_C + \mathbf{I}_1 + \mathbf{I}_2$$

$$= j178.3 + 312.5 - j178.3$$

$$= 312.5\ \text{A}$$

Note that the current drawn from the source after power factor correction, $\mathbf{I}'_S$, is lower than the current required of the source prior to insertion of the capacitor, $\mathbf{I}_S$. The phasor diagram corresponding to these calculations is shown in Figure 5.30.

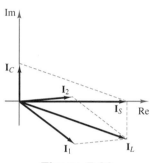

Figure 5.30

DRILL EXERCISES

5.6 Compute the power factor for the load of Example 5.6 with and without the inductance in the circuit.

5.7 Show that one can also express the instantaneous power for an arbitrary complex load $Z = |Z|\angle\theta$ as

$$p(t) = \tilde{I}^2|Z|\cos\theta + \tilde{I}^2|Z|\cos(2\omega t + \theta)$$

5.8 Determine the power factor for the load in the circuit of Figure 5.31, and state whether it is leading or lagging for the following conditions:

a. $v_S(t) = 540\cos(\omega t + 15°)$ V

    $i(t) = 2\cos(\omega t + 47°)$ A

b. $v_S(t) = 155\cos(\omega t - 15°)$ V

    $i(t) = 2\cos(\omega t - 22°)$ A

Figure 5.31

Answer: (a) 0.848, lead; (b) 0.9925, lag

5.9 Determine whether the load is capacitive or inductive for the circuit of Figure 5.32 if

a. $pf = 0.87$ (leading)

b. $pf = 0.42$ (leading)

c. $v_S(t) = 42 \cos (\omega t)$

   $i_S(t) = 4.2 \sin (\omega t)$

d. $v_S(t) = 10.4 \cos (\omega t - 12°)$

   $i_S(t) = 0.4 \cos (\omega t - 12°)$

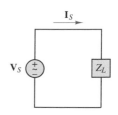

Figure 5.32

> Answer: (a) capacitive; (b) capacitive; (c) inductive; (d) neither (resistive)

5.10 Prove that the power factor is indeed 1 after the addition of the parallel capacitor in Example 5.8.

5.11 Compute the magnitude of the current drawn from the source after the power factor correction in the circuit of Example 5.8.

5.3   TRANSFORMERS

AC circuits are very commonly connected to each other by means of **transformers.** A transformer is a device that couples two AC circuits magnetically rather than through any direct conductive connection and permits a "transformation" of the voltage and current between one circuit and the other (for example, by matching a high-voltage, low-current AC output to a circuit requiring a low-voltage, high-current source). Transformers play a major role in electric power engineering and are a necessary part of the electric power distribution network. The objective of this section is to introduce the ideal transformer and the concepts of impedance reflection and impedance matching. The physical operations of practical transformers, and more advanced models, will be discussed in Chapter 14.

The Ideal Transformer

The ideal transformer consists of two coils that are coupled to each other by some magnetic medium. There is no electrical connection between the coils. The coil on the input side is termed the **primary,** and that on the output side the **secondary.** The primary coil is wound so that it has n_1 turns, while the secondary has n_2 turns. We define the **turns ratio** N as

$$N = \frac{n_2}{n_1} \tag{5.32}$$

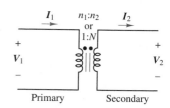

Figure 5.33 Ideal transformer

Figure 5.33 illustrates the convention by which voltages and currents are usually assigned at a transformer. The dots in Figure 5.33 are related to the polarity of the coil voltage: coil terminals marked with a dot have the same polarity.

Since an ideal inductor acts as a short circuit in the presence of DC currents, transformers do not perform any useful function when the primary voltage is DC. However, when a time-varying current flows in the primary winding, a corresponding time-varying voltage is generated in the secondary because of the magnetic coupling between the two coils. This behavior is due to Faraday's law, as will be explained in Chapter 13. The relationship between primary and secondary current in an ideal transformer is very simply stated as follows:

$$
\begin{aligned}
\mathbf{V}_2 &= N\mathbf{V}_1 \\
\mathbf{I}_2 &= \frac{\mathbf{I}_1}{N}
\end{aligned}
\tag{5.33}
$$

An ideal transformer multiplies a sinusoidal input voltage by a factor of N and divides a sinusoidal input current by a factor of N.

If N is greater than 1, the output voltage is greater than the input voltage and the transformer is called a **step-up transformer.** If N is less than 1, then the transformer is called a **step-down transformer,** since $\mathbf{V}_2$ is now smaller than $\mathbf{V}_1$. An ideal transformer can be used in either direction (i.e., either of its coils may be viewed as the input side or primary). Finally, a transformer with $N = 1$ is called an **isolation transformer** and may perform a very useful function if one needs to electrically isolate two circuits from each other; note that any DC currents at the primary will not appear at the secondary coil. An important property of ideal transformers is conservation of power; one can easily verify that an ideal transformer conserves power, since

$$
S_1 = \mathbf{I}_1^* \mathbf{V}_1 = N\mathbf{I}_2^* \frac{\mathbf{V}_2}{N} = \mathbf{I}_2^* \mathbf{V}_2 = S_2
\tag{5.34}
$$

That is, the power on the primary side power equals that on the secondary.

In many practical circuits, the secondary is tapped at two different points, giving rise to two separate output circuits, as shown in Figure 5.34.

Figure 5.34 Center-tapped transformer

The most common instance is the **center-tapped transformer,** which splits the secondary voltage into two equal voltages of half the original amplitude. The most common occurrence of this type of transformer is found at the entry of a power line into a household, where the 240-VAC line is split into two 120-VAC lines (this may help explain why both 240-VAC and 120-VAC power are present in your house).

Impedance Reflection and Power Transfer

As stated in the preceding paragraphs, transformers are commonly used to couple one AC circuit to another. A very common and rather general situation is that depicted in Figure 5.35, where an AC source, represented by its Thévenin equivalent, is connected to an equivalent load impedance by means of a transformer.

Figure 5.35 Operation of an ideal transformer

It should be apparent that expressing the circuit in phasor form does not alter the basic properties of the ideal transformer, as illustrated in the following equation:

$$\mathbf{V}_1 = \frac{\mathbf{V}_2}{N} \qquad \mathbf{I}_1 = N\mathbf{I}_2$$
$$\mathbf{V}_2 = N\mathbf{V}_1 \qquad \mathbf{I}_2 = \frac{\mathbf{I}_1}{N}$$

(5.35)

These expressions are very useful in determining the equivalent impedance seen by the source and by the load, on opposite sides of the transformer. At the primary connection, the equivalent impedance seen by the source must equal the ratio of $\mathbf{V}_1$ to $\mathbf{I}_1$:

$$Z' = \frac{\mathbf{V}_1}{\mathbf{I}_1}$$

(5.36)

which can be written as:

$$Z' = \frac{\mathbf{V}_2/N}{N\mathbf{I}_2} = \frac{1}{N^2}\frac{\mathbf{V}_2}{\mathbf{I}_2}$$

(5.37)

But the ratio $\mathbf{V}_2/\mathbf{I}_2$ is by definition the load impedance, Z_L. Thus,

$$Z' = \frac{1}{N^2}Z_L$$

(5.38)

That is, the AC source "sees" the load impedance reduced by a factor of $1/N^2$.

The load impedance also sees an equivalent source. The open-circuit voltage is given by the expression

$$\mathbf{V}_{OC} = N\mathbf{V}_1 = N\mathbf{V}_S \tag{5.39}$$

since there is no voltage drop across the source impedance in the circuit of Figure 5.35. The short-circuit current is given by the expression

$$\mathbf{I}_{SC} = \frac{\mathbf{V}_S}{Z_S}\frac{1}{N} \tag{5.40}$$

and the load sees a Thévenin impedance equal to

$$Z'' = \frac{\mathbf{V}_{OC}}{\mathbf{I}_{SC}} = \frac{N\mathbf{V}_S}{\frac{\mathbf{V}_S}{Z_S}\frac{1}{N}} = N^2 Z_S \tag{5.41}$$

Thus the load sees the source impedance multiplied by a factor of N^2. Figure 5.36 illustrates this **impedance reflection** across a transformer. It is

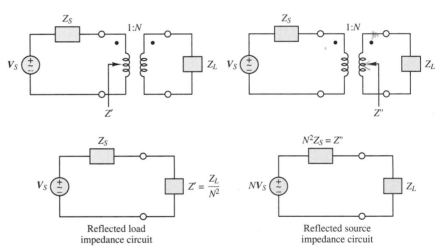

Figure 5.36 Impedance reflection across a transformer

very important to note that an ideal transformer changes the magnitude of the load impedance seen by the source by a factor of $1/N^2$. This property naturally leads to the discussion of power transfer, which we consider next.

Recall that in DC circuits, given a fixed equivalent source, maximum power is transferred to a resistive load when the latter is equal to the internal resistance of the source; achieving an analogous maximum power transfer condition in an AC circuit is referred to as **impedance matching**. Consider the general form of an AC circuit, shown in Figure 5.37, and assume that the source impedance, Z_S, is given by

Figure 5.37 The maximum power transfer problem in AC circuits

$$Z_S = R_S + jX_S \tag{5.42}$$

The problem of interest is often that of selecting the load resistance and reactance that will maximize the real (average) power absorbed by the load. Note that the requirement is to maximize the real power absorbed by the load. Thus, the problem can be restated by expressing the real load power in terms of the impedance of the source and load. The real power absorbed by the load is

$$P_L = \tilde{V}_L \tilde{I}_L \cos \theta = \text{Re}\,(\mathbf{V}_L \mathbf{I}_L^*) \tag{5.43}$$

where

$$\mathbf{V}_L = \frac{Z_L}{Z_S + Z_L} \mathbf{V}_S \tag{5.44}$$

and

$$\mathbf{I}_L^* = \left(\frac{\mathbf{V}_S}{Z_S + Z_L}\right)^* = \frac{\mathbf{V}_S^*}{(Z_S + Z_L)^*} \tag{5.45}$$

Thus, the complex load power is given by

$$S_L = \mathbf{V}_L \mathbf{I}_L^* = \frac{Z_L \mathbf{V}_S}{Z_S + Z_L} \times \frac{\mathbf{V}_S^*}{(Z_S + Z_L)^*} = \frac{\mathbf{V}_S^2}{|Z_S + Z_L|^2} Z_L \tag{5.46}$$

and the average power by

$$\begin{aligned}
P_L &= \text{Re}\,(\mathbf{V}_L \mathbf{I}_L^*) = \text{Re}\left(\frac{\mathbf{V}_S^2}{|Z_S + Z_L|^2}\right) \text{Re}\,(Z_L) \\
&= \frac{\tilde{V}_S^2}{(R_S + R_L)^2 + (X_S + X_L)^2} \,\text{Re}\,(Z_L) \tag{5.47} \\
&= \frac{\tilde{V}_S^2 R_L}{(R_S + R_L)^2 + (X_S + X_L)^2}
\end{aligned}$$

The expression for P_L is maximized by selecting appropriate values of R_L and X_L; it can be shown that the average power is greatest when $R_L = R_S$ and $X_L = -X_S$, that is, when the load impedance is equal to the complex conjugate of the source impedance, as shown in the following equation:

$$\begin{aligned}
Z_L &= Z_S^* \\
R_L &= R_S \qquad X_L = -X_S
\end{aligned} \tag{5.48}$$

> When the load impedance is equal to the complex conjugate of the source impedance, the load and source impedances are matched and maximum power is transferred to the load.

In many cases, it may not be possible to select a matched load impedance, because of physical limitations in the selection of appropriate

transformer use
when you cannot get
max power out of a
circuit

components. In these situations, it is possible to use the impedance reflection properties of a transformer to maximize the transfer of AC power to the load. The circuit of Figure 5.38 illustrates how the reflected load impedance, as seen by the source, is equal to Z_L/N^2, so that maximum power transfer occurs when

$$\frac{Z_L}{N^2} = Z_S^*$$

$$R_L = N^2 R_S$$

$$X_L = -N^2 X_S$$

(5.49)

Source   Transformer   Load   Equivalent circuit referred
                                        to transformer primary

Figure 5.38 Maximum power transfer in an AC circuit with a transformer

---

Example   5.14

This example illustrates the transfer of maximum AC power with the aid of a transformer. A 240-VAC, 60-Hz source with the internal impedance shown in Figure 5.39 is connected to a load that has a resistive component of 400 Ω. Find the transformer turns ratio N and load reactance that allow for maximum power transfer.

Figure 5.39

Solution:

For maximum power transfer, we require that

$$R_L = N^2 R_S$$

$$N^2 = \frac{R_L}{R_S} = \frac{400}{10} = 40$$

$$N = \sqrt{40} = 6.325$$

and we must also have

$$X_L = -N^2 X_S$$

with

$$X_S = (\omega)(0.1 \text{ H}) = 37.7 \ \Omega$$

Thus,

$$X_L = -40 \times 37.7$$
$$= -1,508$$

This means that the load reactive element should be a capacitor whose value is

$$C_L = \frac{-1}{X_L \omega} = 1.76 \ \mu\text{F}$$

DRILL EXERCISES

5.12   If the transformer shown in Figure 5.40 is ideal, find the turns ratio, N, that will ensure maximum power transfer to the load. Assume that $Z_S = 1,800 \ \Omega$ and $Z_L = 8 \ \Omega$.

> Answer: $N = 0.0667$

5.13   If the circuit of Drill Exercise 5.12 has $Z_L = (2 + j10) \ \Omega$ and the turns ratio of the transformer is $N = 5.4$, what should Z_S be in order to have maximum power transfer?

> Answer: $Z_S = 0.0686 - j0.3429 \ \Omega$

Figure 5.40

5.4   THREE-PHASE POWER

The material presented so far in this chapter has dealt exclusively with **single-phase AC power,** that is, with single sinusoidal sources. In fact, most of the AC power used today is generated and distributed as **three-phase power,** by means of an arrangement in which three sinusoidal voltages are generated out of phase with each other. The primary reason is efficiency: the weight of the conductors and other components in a three-phase system is much lower than in a single-phase system delivering the same amount of power. Further, while the power produced by a single-phase system has a pulsating nature (recall the results of Section 5.1), a three-phase system can deliver a steady, constant supply of power. For example, later in this section it will be shown that a three-phase generator producing three **balanced voltages**—that is, voltages of equal amplitude and frequency displaced in phase by 120°—has the property of delivering constant instantaneous power.

To begin the discussion of three-phase power, consider a three-phase source connected in the **wye** (or **Y**) **configuration,** as shown in Figure 5.41. Each of the three voltages is 120°out of phase with the others, so that, using

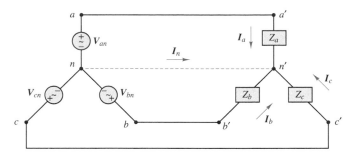

Figure 5.41 Balanced three-phase AC circuit

phasor notation, we may write:

$$\begin{aligned}
\mathbf{V}_{an} &= \tilde{V}_{an}\angle 0° \\
\mathbf{V}_{bn} &= \tilde{V}_{bn}\angle -120° \\
\mathbf{V}_{cn} &= \tilde{V}_{cn}\angle -240° = \tilde{V}_{cn}\angle 120°
\end{aligned} \tag{5.50}$$

where the quantities $\tilde{V}_{an}$, $\tilde{V}_{bn}$, and $\tilde{V}_{cn}$ are rms values and are equal to each other. To simplify the notation, it will be assumed from here on that

$$\tilde{V}_{an} = \tilde{V}_{bn} = \tilde{V}_{cn} = \tilde{V} \tag{5.51}$$

Chapter 15 will discuss how three-phase AC electric generators may be constructed to provide such balanced voltages. In the circuit of Figure 5.41, the resistive loads are also wye-connected and balanced (i.e., equal). The three AC sources are all connected together at a node called the *neutral node,* denoted by *n.* The voltages $\mathbf{V}_{an}$, $\mathbf{V}_{bn}$, and $\mathbf{V}_{cn}$ are called the **phase voltages** and form a balanced set in the sense that

$$\mathbf{V}_{an} + \mathbf{V}_{bn} + \mathbf{V}_{cn} = 0 \tag{5.52}$$

This last statement is easily verified by sketching the phasor diagram. The sequence of phasor voltages shown in Figure 5.42 is usually referred to as the **positive (or *abc*) sequence.**

Consider now the "lines" connecting each source to the load and observe that it is possible to also define **line voltages** (also called *line-to-line voltages*) by considering the voltages between the lines aa' and bb', aa' and cc', and bb' and cc'. Since the line voltage, say, between aa' and bb' is given by

$$\mathbf{V}_{ab} = \mathbf{V}_{an} + \mathbf{V}_{nb} = \mathbf{V}_{an} - \mathbf{V}_{bn} \tag{5.53}$$

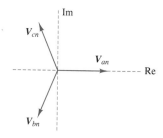

Figure 5.42 Positive, or *abc,* sequence for balanced three-phase voltages

the line voltages may be computed relative to the phase voltages as follows:

$$\begin{aligned}
\mathbf{V}_{ab} &= \tilde{V}\angle 0° - \tilde{V}\angle -120° = \sqrt{3}\tilde{V}\angle 30° \\
\mathbf{V}_{bc} &= \tilde{V}\angle -120° - \tilde{V}\angle 120° = \sqrt{3}\tilde{V}\angle -90° \\
\mathbf{V}_{ca} &= \tilde{V}\angle 120° - \tilde{V}\angle 0° = \sqrt{3}\tilde{V}\angle 150°
\end{aligned} \tag{5.54}$$

It can be seen, then, that the magnitude of the line voltages is equal to $\sqrt{3}$

times the magnitude of the phase voltages. It is instructive, at least once, to point out that the circuit of Figure 5.41 can be redrawn to have the appearance of the circuit of Figure 5.43.

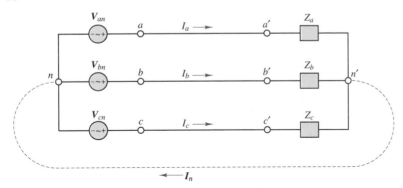

Figure 5.43 Balanced three-phase AC circuit (redrawn)

One of the important features of a balanced three-phase system is that it does not require a fourth wire (the neutral connection), since the current $\mathbf{I}_n$ is identically zero (for balanced load $Z_a = Z_b = Z_c = Z$). This can be shown by applying KCL at the neutral node n:

$$
\begin{aligned}
\mathbf{I}_n &= (\mathbf{I}_a + \mathbf{I}_b + \mathbf{I}_c) \\
&= \frac{1}{Z}(\mathbf{V}_{an} + \mathbf{V}_{bn} + \mathbf{V}_{cn}) \\
&= 0
\end{aligned}
\tag{5.55}
$$

Another, more important characteristic of a balanced three-phase power system may be illustrated by simplifying the circuits of Figures 5.41 and 5.43 by replacing the balanced load impedances with three equal resistances, R. With this simplified configuration, one can show that the total power delivered to the balanced load by the three-phase generator is constant. This is an extremely important result, for a very practical reason: delivering power in a smooth fashion (as opposed to the pulsating nature of single-phase power) reduces the wear and stress on the generating equipment. Although we have not yet discussed the nature of the machines used to generate power, a useful analogy here is that of a single-cylinder engine versus a perfectly balanced V-8 engine. To show that the total power delivered by the three sources to a balanced resistive load is constant, consider the instantaneous power delivered by each source:

$$
\begin{aligned}
p_a(t) &= \frac{\tilde{V}^2}{R}(1 + \cos 2\omega t) \\
p_b(t) &= \frac{\tilde{V}^2}{R}[1 + \cos(2\omega t - 120°)] \\
p_c(t) &= \frac{\tilde{V}^2}{R}[1 + \cos(2\omega t + 120°)]
\end{aligned}
\tag{5.56}
$$

A delta-connected three-phase generator with line voltages V_{ab}, V_{bc}, V_{ca}

Figure 5.44 Delta-connected generators

The total instantaneous load power is then given by the sum of the three contributions:

$$p(t) = p_a(t) + p_b(t) + p_c(t)$$

$$= \frac{3\tilde{V}^2}{R} + \frac{\tilde{V}^2}{R}[\cos 2\omega t + \cos (2\omega t - 120°)$$

$$+ \cos (2\omega t + 120°)] \tag{5.57}$$

$$= \frac{3\tilde{V}^2}{R} = \text{ a constant!}$$

You may wish to verify that the sum of the trigonometric terms inside the brackets is identically zero.

It is also possible to connect the three AC sources in a three-phase system in a so-called **delta** (or **Δ**) **connection,** although in practice this configuration is rarely used. Figure 5.44 depicts a set of three delta-connected generators.

Example 5.15

This example illustrates the effect of line impedance in a balanced wye-wye system, as well as a solution method that involves computations for only a single phase. Compute the power delivered to the load shown by the three-phase generator in the circuit shown in Figure 5.45, where the phase voltage is 480 V. Assume

$$Z_y = 2 + j4 = 4.47\angle 63.43°\ \Omega$$

$$\mathbf{V}_{an} = 480\angle 0°\qquad \mathbf{V}_{bn} = 480\angle -120°\qquad \mathbf{V}_{cn} = 480\angle 120°$$

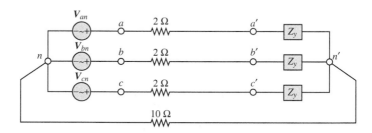

Figure 5.45

Solution:

In approaching the solution of this problem, we note that, since the circuit is perfectly balanced, the current through the 10-Ω resistor is zero, and therefore the voltage across it is zero. Further, by virtue of the symmetry of the circuit, it is possible to solve the problem by considering just one phase, as shown in Figure 5.46.

Note that when we redraw the a phase of the circuit, we connect n' and n together. This is done to show that n and n' are at the same voltage for a balanced wye-connected load. The result is a simple series circuit representing one phase of the three-phase circuit.

Figure 5.46 One phase of the three-phase circuit

The average power in phase a is given by

$$P_a = \tilde{I}^2 R_L$$

where the rms current $\tilde{I}$ is

$$|\mathbf{I}| = \left| \frac{\mathbf{V}_a}{Z_y + 2} \right| = \left| \frac{480\angle 0°}{2 + 2 + j4} \right| = \left| \frac{480\angle 0°}{5.66\angle 45°} \right| = |84.85\angle -45°|$$

so that

$$P_a = (84.85)^2 \times 2 = 14{,}400 \text{ W}$$

The calculations are similar for the b and c phases, so that

$$
\begin{aligned}
P_{\text{Total}} &= P_a + P_b + P_c \\
&= 3P_a \\
&= 43.2 \text{ kW}
\end{aligned}
$$

Balanced Wye Loads

In the previous section we performed some power computations for a purely resistive balanced wye load. We shall now generalize those results for an arbitrary balanced complex load. Consider again the circuit of Figure 5.41, where now the balanced load consists of the three complex impedances

$$Z_a = Z_b = Z_c = Z_y = |Z_y|\angle\theta \tag{5.58}$$

From the diagram of Figure 5.41, it can be verified that each impedance sees the corresponding phase voltage across itself; thus, since the currents $\mathbf{I}_a$, $\mathbf{I}_b$, and $\mathbf{I}_c$ have the same rms value, $\tilde{I}$, the phase angles of the currents will differ by $\pm 120°$. It is therefore possible to compute the power for each phase by considering the phase voltage (equal to the load voltage) for each impedance, and the associated line current. Let us denote the complex power for each phase by S:

$$S = \mathbf{V} \cdot \mathbf{I}^* \tag{5.59}$$

so that

$$
\begin{aligned}
S &= P + jQ \\
&= \tilde{V}\tilde{I}\cos\theta + j\tilde{V}\tilde{I}\sin\theta
\end{aligned}
\tag{5.60}
$$

where $\tilde{V}$ and $\tilde{I}$ denote, once again, the rms values of each phase voltage and line current. Consequently, the total real power delivered to the balance wye load is $3P$, and the total reactive power is $3Q$. Thus, the total complex power, S_T, is given by

$$
\begin{aligned}
S_T &= P_T + jQ_T = 3P + j3Q \\
&= \sqrt{(3P)^2 + (3Q)^2}\angle\theta
\end{aligned}
\tag{5.61}
$$

and the apparent power is

$$|S_T| = 3\sqrt{(VI)^2\cos^2\theta + (VI)^2\sin^2\theta}$$
$$= 3VI$$

and the total real and reactive power may be expressed in terms of the apparent power:

$$P_T = |S_T|\cos\theta$$
$$Q_T = |S_T|\sin\theta \tag{5.62}$$

Balanced Delta Loads

In addition to a wye connection, it is also possible to connect a balanced load in the delta configuration. A wye-connected generator and a delta-connected load are shown in Figure 5.47.

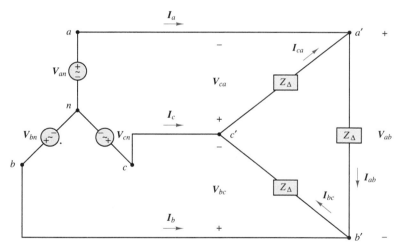

Figure 5.47 Balanced wye generators with balanced delta load

It should be noted immediately that now the corresponding line voltage (not phase voltage) appears across each impedance. For example, the voltage across $Z_{c'a'}$ is $\mathbf{V}_{ca}$. Thus, the three load currents are given by the following expressions:

$$\mathbf{I}_{ab} = \frac{\mathbf{V}_{ab}}{Z_\Delta} = \frac{\sqrt{3}V\angle 30°}{|Z_\Delta|\angle\theta}$$

$$\mathbf{I}_{bc} = \frac{\mathbf{V}_{bc}}{Z_\Delta} = \frac{\sqrt{3}V\angle -90°}{|Z_\Delta|\angle\theta} \tag{5.63}$$

$$\mathbf{I}_{ca} = \frac{\mathbf{V}_{ca}}{Z_\Delta} = \frac{\sqrt{3}V\angle 150°}{|Z_\Delta|\angle\theta}$$

To understand the relationship between delta-connected and wye-connected loads, it is reasonable to ask the question, For what value of Z_Δ would a delta-connected load draw the same amount of current as a wye-connected load with impedance Z_y for a given source voltage? This is equivalent to asking what value of Z_Δ would make the line currents the same in both circuits (compare Figure 5.43 with Figure 5.47).

The line current drawn, say, in phase a by a wye-connected load is

$$(\mathbf{I}_{an})_y = \frac{\mathbf{V}_{an}}{Z} = \frac{\tilde{V}}{|Z_y|}\angle -\theta \tag{5.64}$$

while that drawn by the delta-connected load is

$$
\begin{aligned}
(\mathbf{I}_a)_\Delta &= \mathbf{I}_{ab} - \mathbf{I}_{ca} \\
&= \frac{\mathbf{V}_{ab}}{Z_\Delta} - \frac{\mathbf{V}_{ca}}{Z_\Delta} \\
&= \frac{1}{Z_\Delta}(\mathbf{V}_{an} - \mathbf{V}_{bn} - \mathbf{V}_{cn} + \mathbf{V}_{an}) \\
&= \frac{1}{Z_\Delta}(2\mathbf{V}_{an} - \mathbf{V}_{bn} - \mathbf{V}_{cn}) \\
&= \frac{3\mathbf{V}_{an}}{Z_\Delta} = \frac{3\mathbf{V}}{|Z_\Delta|}\angle -\theta
\end{aligned}
\tag{5.65}
$$

One can readily verify that the two currents $(\mathbf{I}_a)_\Delta$ and $(\mathbf{I}_a)_y$ will be equal if the magnitude of the delta-connected impedance is 3 times larger than Z_y:

$$Z_\Delta = 3Z_y \tag{5.66}$$

This result also implies that a delta load will necessarily draw 3 times as much current (and therefore absorb 3 times as much power) as a wye load with the same branch impedance.

Example 5.16

The generator of Example 5.15 is now required to deliver power to an additional delta load, connected in parallel with the wye load of Example 5.15. If

$$Z_\Delta = 5 - j2$$

what is the total power absorbed by the two parallel loads?

Solution:

The circuit of Example 5.15 is sketched in Figure 5.48 with the additional delta load. Assume that

$$Z_y = 2 + j4 \ \Omega$$

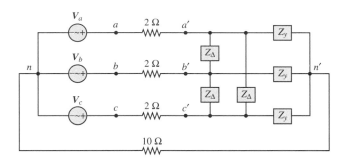

Figure 5.48 AC circuit with delta and wye loads

and

$$Z_\Delta = 5 - j2 \ \Omega$$

One method of solving this problem is to convert the balanced delta load to an equivalent wye load, using the method outlined in the preceding paragraphs. Figure 5.49 illustrates how this conversion can take place.

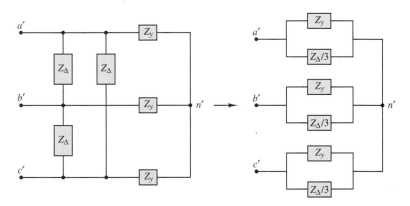

Figure 5.49 Conversion of delta load to equivalent wye load

Now we need to perform the calculations for only one phase, recognizing that the wye load appears in parallel with the delta-to-wye transformed load, as shown in Figure 5.50. If we compute the effective impedance at each phase, we have

$$Z_y \parallel \frac{Z_\Delta}{3} = (2 + j4) \parallel \left(\frac{5 - j2}{3} \right)$$

$$= \frac{(2 + j4)\left(\frac{5}{3} - j\frac{2}{3} \right)}{\left(2 + j\frac{5}{3} \right) + j\left(4 - \frac{2}{3} \right)} = \frac{10 + 8 - j4 + j20}{11 + j10}$$

$$= \frac{18 + j16}{11 + j10} = \frac{24.08\angle 41.6^\circ}{14.87\angle 42.3^\circ} = 1.62\angle -0.64^\circ$$

Figure 5.50 Per-phase circuit

and therefore a current

$$\mathbf{I}_a = \frac{480\angle0°}{2 + 1.62\angle-0.64°} = \frac{480\angle0°}{2 + 1.62 - j0.018}$$

$$= \frac{480\angle0°}{3.62\angle-0.29°} = 132.6\angle0.29°$$

yielding the single-phase power

$$P_a = \tilde{I}_a^2 R_L = (132.6)^2 \times 1.62 = 28,482 \text{ W}$$

The total power will then be 3 times P_a, or

$$P_{\text{Total}} = 52.44 \text{ kW}$$

DRILL EXERCISES

5.14   Find the power lost in the lines in the circuit of Example 5.15.

> Answer: $P_{\text{Loss}} = 43.2$ kW

5.15   Draw the phasor diagram and power triangle for a single phase and compute the power delivered to the balanced load of Example 5.15 is the lines have zero resistance and $Z_L = 1 + j3$ Ω.

> Answer: $\mathbf{V}_a = 480\angle0°$ V; $\mathbf{I}_a = 151.8\angle-71.6°$ A;
> $S_a = 23,040$ W $+ j69,120$ VAR

5.16   Show that the voltage across each branch of the balanced wye load is equal to the corresponding phase voltage (e.g., the voltage across Z_a is $\mathbf{V}_a$).

5.17   Prove that the sum of the instantaneous powers absorbed by the three branches in a balanced wye-connected load is constant and equal to $3\tilde{V}\tilde{I}\cos\theta$.

5.18   Derive an expression for the rms line current of a delta load in terms of the rms line current for a wye load with the same branch impedances (i.e., $Z_y = Z_\Delta$) and same source voltage.

5.19   The equivalent wye load of Example 5.16 is connected in a delta configuration. Compute the line currents.

Answer: $\mathbf{I}_a = 243\angle 108°$ A; $\mathbf{I}_b = 200\angle 124°$ A; $\mathbf{I}_c = 197\angle 19°$ A

5.5   RESIDENTIAL WIRING; GROUNDING AND SAFETY

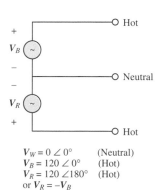

$V_W = 0 \angle 0°$ (Neutral)
$V_B = 120 \angle 0°$ (Hot)
$V_R = 120 \angle 180°$ (Hot)
or $V_R = -V_B$

Figure 5.51 Line voltage convention for residential circuits

Common residential electric power service consists of a three-wire AC system supplied by the local power company. The three wires originate from a utility pole and consist of a neutral wire, which is connected to earth ground, and two "hot" wires. Each of the hot lines supplies 120 V rms to the residential circuits; the two lines are 180° out of phase, for reasons that will become apparent during the course of this discussion. The phasor line voltages, shown in Figure 5.51, are usually referred to by means of a subscript convention derived from the color of the insulation on the different wires: *W* for white (neutral), *B* for black (hot), and *R* for red (hot). This convention is adhered to uniformly.

The voltages across the hot lines are given by:

$$\mathbf{V}_B - \mathbf{V}_R = \mathbf{V}_{BR} = \mathbf{V}_B - (-\mathbf{V}_B) = 2\mathbf{V}_B = 240\angle 0° \qquad (5.67)$$

Thus, the voltage between the hot wires is actually 240 V rms. Appliances such as electric stoves, air conditioners, and heaters are powered by the 240-V rms arrangement. On the other hand, lighting and all of the electric outlets in the house used for small appliances are powered by a single 120-V rms line.

The use of 240-V rms service for appliances that require a substantial amount of power to operate is dictated by power transfer considerations. Consider the two circuits shown in Figure 5.52. In delivering the necessary power to a load, a lower line loss will be incurred with the 240-V rms wiring, since the power loss in the lines (the I^2R **loss,** as it is commonly referred to) is directly related to the current required by the load. In an effort to minimize line losses, the size of the wires is increased for the lower voltage case. This typically reduces the wire resistance by a factor of two. In the circuit on the left, assuming $R_S/2 = 0.01$ Ω, the current required by the 10-kW load is approximately 83.3 A, while in the circuit on

Figure 5.52 Line losses in 120-VAC and 240-VAC circuits

the right, with $R_S = 0.02 \ \Omega$, it is approximately half as much (41.7 A). [You should be able to verify that the approximate I^2R losses are 69.4 W in the circuit on the left, and 34.7 W in the circuit on the right.] Limiting the I^2R losses is important from the viewpoint of efficiency, besides reducing the amount of heat generated in the wiring for safety considerations. Figure 5.53 shows some typical wiring configurations for a home. Note that several circuits are wired and fused separately.

Figure 5.53 A typical residential wiring arrangement

Today, most homes have three wire connections to their outlets. The outlets appear as sketched in Figure 5.54. Then why are both the ground and neutral connections needed in an outlet? The answer to this question is *safety:* the ground connection is used to connect the chassis of the appliance to earth ground. Without this provision, the appliance chassis could be at any

Figure 5.54 A three-wire outlet

potential with respect to ground, possibly even at the hot wire's potential if a segment of the hot wire were to lose some insulation and come in contact with the inside of the chassis! Poorly grounded appliances can thus be a significant hazard. Figure 5.55 illustrates schematically how, even though the

Figure 5.55 Unintended connection

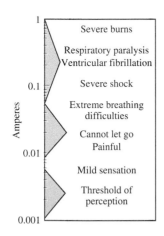

Figure 5.56 Physiological effects of electric currents

chassis is intended to be insulated from the electric circuit, an unintended connection (represented by the dashed line) may occur, for example, because of corrosion or a loose mechanical connection. A path to ground might be provided by the body of a person touching the chassis with a hand. In the figure, such an undesired ground loop current is indicated by I_G. In this case, the ground current I_G would flow directly through the body to ground and could be harmful.

In some cases the danger posed by such undesired ground loops can be great, leading to death by electric shock. Figure 5.56 describes the effects of electric currents on an average male when the point of contact is dry skin. Particularly hazardous conditions are liable to occur whenever the natural resistance to current flow provided by the skin breaks down, as would happen in the presence of water. The **ground fault circuit interrupter,** labeled **GFCI** in Figure 5.53, is a special safety circuit used primarily with outdoor circuits and in bathrooms, where the risk of death by electric shock is greatest. Its application is best described by an example.

Consider the case of an outdoor pool surrounded by a metal fence, which uses an existing light pole for a post, as shown in Figure 5.57. The

Figure 5.57 Outdoor pool

light pole and the metal fence can be considered as forming a chassis. If the fence were not properly grounded all the way around the pool and if the light fixture were poorly insulated from the pole, a path to ground could easily be created by an unaware swimmer reaching, say, for the metal gate. A GFCI provides protection from potentially lethal ground loops, such as this one, by sensing both the hot-wire (B) and the neutral (W) currents. If the difference between the hot-wire current, I_B, and the neutral current, I_W, is more than a few milliamperes, then the GFCI disconnects the circuit nearly instantaneously. Any significant difference between the hot and neutral (return-path) currents means that a second path to ground has been created (by the unfortunate swimmer, in this example) and a potentially dangerous condition has arisen. Figure 5.58 illustrates the idea. GFCI are typically resettable circuit breakers, so that one does not need to replace a fuse every time the GFCI circuit is enabled.

Figure 5.58 Use of a GFCI in a potentially hazardous setting

DRILL EXERCISE

5.20 Use the circuit of Figure 5.52 to show that the I^2R losses will be higher for a 120-V service appliance than a 240-V service appliance if both have the same power usage rating.

5.6   GENERATION AND DISTRIBUTION OF AC POWER

We now conclude the discussion of power systems with a brief description of the various elements of a power system. Electric power originates from a variety of sources; in Chapter 15, electric generators will be introduced as a means of producing electric power from a variety of energy-conversion processes. In general, electric power may be obtained from hydroelectric, thermoelectric, geothermal, wind, solar, and nuclear sources. The choice of a given source is typically dictated by the power requirement for the given application, and by economic and environmental factors. In this section, the structure of an AC power network, from the power-generating station to the residential circuits discussed in the previous section, is briefly outlined.

A typical generator will produce electric power at 18 kV, as shown in the diagram of Figure 5.59. To minimize losses along the conductors,

Figure 5.59 Structure of an AC power distribution network

the output of the generators is processed through a step-up transformer to achieve line voltages of hundreds of kilovolts (345 kV, in Figure 5.59). Without this transformation, the majority of the power generated would be lost in the **transmission lines** that carry the electric current from the power station.

The local electric company operates a power-generating plant that is capable of supplying several hundred megavolt-amperes (MVA) on a three-phase basis. For this reason, the power company uses a three-phase step-up transformer at the generation plant to increase the line voltage to around 345 kV. One can immediately see that at the rated power of the generator (in MVA) there will be a significant reduction of current beyond the step-up transformer.

Beyond the generation plant, an electric power network distributes energy to several **substations.** This network is usually referred to as the **power grid.** At the substations, the voltage is stepped down to a lower level (10 to 150 kV, typically). Some very large loads (for example, an industrial plant) may be served directly from the power grid, although most loads are supplied by individual substations in the power grid. At the local substations (one of which you may have seen in your own neighborhood), the voltage is stepped down further by a three-phase step-down transformer to 4,800 V. These substations distribute the energy to residential and industrial customers. To further reduce the line voltage to levels that are safe for residential use, step-down transformers are mounted on utility poles. These drop the voltage to the 120/240-V three-wire single-phase residential service discussed in the previous section. Industrial and commercial customers receive 460- and/or 208-V three-phase service.

CONCLUSION

This chapter introduced the essential elements leading to the analysis of AC power systems. Single-phase AC power, ideal transformers, and three-phase power were discussed. A brief review of residential circuit wiring and safety, and a description of an electric distribution network, were also given to underscore the importance of these concepts in electric power.

- The power dissipated by a load in an AC circuit consists of the sum of an average and a fluctuating component. In practice, the average power is the quantity of interest.
- AC power can best be analyzed with the aid of complex notation. Complex power is defined as the product of the phasor load voltage and the complex conjugate of the phasor load current. Complex power consists of the sum of a real component (the average, or real, power) and an imaginary component (reactive power). Real power corresponds to the electric power for which a user is billed by a utility company; reactive power corresponds to energy storage and cannot be directly used.

- Although reactive power is of no practical use, it does cause an undesirable increase in the current that must be generated by the electric company, resulting in additional line losses. Thus, it is customary to try to reduce reactive power. A measure of the presence of reactive power at a load is the power factor, equal to the cosine of the angle of the load impedance. By adding a suitable reactance to the load, it is possible to attain power factors close to ideal (unity). This procedure is called *power factor correction.*

- Electric power is most commonly generated in three-phase form, for reasons of efficiency. Three-phase power entails the generation of three 120° out-of-phase AC voltages of equal amplitude, so that the instantaneous power is actually constant. Three-phase sources and loads can be configured in either wye or delta configurations; of these, the wye form is more common. The calculation of currents, voltages, and power in three-phase circuits is greatly simplified if one uses per-phase calculations.

KEY TERMS

HOMEWORK PROBLEMS

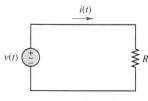

Figure P5.1

5.1   Find the peak instantaneous power, P, and the average power, P_{av}, if $v(t) = 80 \sin 377t$ and $R = 2$ kΩ for the circuit shown in Figure P5.1.

5.2   The heating element in a soldering iron has a resistance of 391 Ω. Find the average power dissipated in the soldering iron if it is connected to a voltage source of 117 V rms .

5.3   The heating element in an electric heater has a resistance of 10 Ω. Find the power dissipated in the heater when it is connected to a voltage source of 240 V rms.

5.4 Find the peak instantaneous power and the average power dissipated in a 200-Ω resistor connected across a voltage source $v(t) = 10 \sin 10^5 t$ V.

5.5 Find the average power supplied to a 5-Ω resistor by each of the periodic voltage waveforms shown in Figure P5.2.

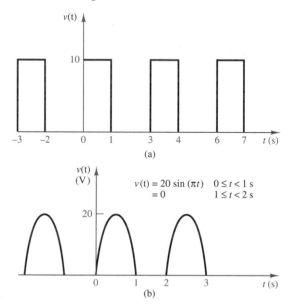

(a)

$v(t) = 20 \sin (\pi t)$ $0 \le t < 1$ s
$= 0$ $1 \le t < 2$ s

(b)

Figure P5.2

5.6 A current source $i(t)$ is connected to a 100-Ω resistor. Find the average power delivered to the resistor, given that $i(t)$ is:

a. $4 \cos 100t$ A
b. $4 \cos (100t - 50°)$ A
c. $4 \cos 100t - 3 \cos (100t - 50°)$ A
d. $4 \cos 100t - 3$ A

5.7 Find the rms value of each of the following periodic currents:

a. $\cos 377t + \cos 377t$
b. $\cos 2t + \sin 2t$
c. $\cos 377t + 1$
d. $\cos 2t + \cos (2t + 135°)$
e. $\cos 2t + \cos 3t$

5.8 A current of 10 A flows when a single-phase circuit is placed across a 220-V rms source. The current lags the voltage by 60°. Find the power drawn by the current, and the power factor.

5.9 A single-phase circuit is placed across a 120-V rms, 60-Hz source, with an ammeter, a voltmeter, and a wattmeter connected. The instruments indicate 12 A, 120 V, and 800 W, respectively. Find

a. The power factor.
b. The phase angle.
c. The impedance.
d. The resistance.

5.10 The nameplate on a single-phase induction machine reads 2 horsepower (output), 110 V rms, 60 Hz, and 24 A rms. Find the power factor of the machine if the efficiency at rated output is 80 percent. [*Note:* 1 horsepower = 0.746 kW.]

5.11 For the following numerical values, determine the average power, P, the reactive power, Q, and the complex power, S, of the circuit shown in Figure P5.3. Note: phasor quantities are rms.

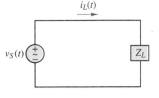

Figure P5.3

a. $v_S(t) = 650 \cos (377t)$ V

$i_L(t) = 20 \cos (377t - 10°)$ A

b. $\mathbf{V}_S = 460\angle 0°$ V

$\mathbf{I}_L = 14.14\angle -45°$ A

c. $\mathbf{V}_S = 100\angle 0°$ V

$\mathbf{I}_L = 8.6\angle -86°$ A

d. $\mathbf{V}_S = 208\angle -30°$ V

$\mathbf{I}_L = 2.3\angle -63°$ A

5.12 For the circuit of Figure P5.3, determine the power factor for the load and state whether it is leading or lagging for the following conditions:

a. $v_S(t) = 540 \cos (\omega t + 15°)$ V

$i_L(t) = 20 \cos (\omega t + 47°)$ A

b. $v_S(t) = 155 \cos (\omega t - 15°)$ V

$i_L(t) = 20 \cos (\omega t - 22°)$ A

c. $v_S(t) = 208 \cos (\omega t)$ V

$i_L(t) = 1.7 \sin (\omega t + 175°)$ A

d. $Z_L = (48 + j16)$ Ω

5.13 For the circuit of Figure P5.3, determine whether the load is capacitive or inductive for the circuit shown if

a. $pf = 0.87$ (leading)
b. $pf = 0.42$ (leading)
c. $v_S(t) = 42 \cos (\omega t)$

$i_L(t) = 4.2 \sin (\omega t)$

d. $v_S(t) = 10.4 \cos (\omega t - 12°)$

$i_L(t) = 0.4 \cos (\omega t - 12°)$

5.14 Given the waveform of the voltage source and the circuit shown in Figure P5.4,

a. Find the rms voltage, $\tilde{V}_S$, of $v_S(t)$.
b. If the voltage, $v_S(t)$, is applied to a 10-Ω resistor, what is the average power supplied to the resistor?

Figure P5.4

5.15 In the circuit shown in Figure P5.5, two loads are connected in parallel across a 208-V rms source. Load 1 absorbs 2 kW and 1.4 kVAR. Load 2 absorbs 4 kVA at 0.7 pf (leading). Find the impedance that is equivalent to the two parallel loads.

Figure P5.5

5.16 For the circuit of Figure P5.3, a source of strength $\mathbf{V}_S = 120$ V rms delivers $\mathbf{I}_S = 1.94\angle40°$ A rms to a load.

a. Determine the complex power delivered to the load.
b. Find the power dissipated by the load.
c. What is the equivalent impedance of the load?

5.17 The circuit shown in Figure P5.6(a) contains two resistors. The circuit shown in Figure P5.6(b) has an impedance of value $= -j4$ Ω replacing R_1.

$R_1 = 4\,\Omega$ $R_2 = 4\,\Omega$ $V_S = 40 \angle 0°$ volts

Figure P5.6(a)

Figure P5.6(b)

a. Find $\mathbf{V}_{out}$ for Figure P5.6(a).
b. What is the power dissipated by R_2 in this circuit?
c. Find $\mathbf{V}_{out}$ for Figure P5.6(b).
d. What is the power dissipated by R_2 in this circuit?
e. In which circuit does R_2 dissipate more power?

$V_S = 55\angle0°$

$25\ \Omega$

I

$22\ \mu F$

Figure P5.7

5.18   Find the instantaneous real and reactive power supplied by the source in the circuit shown in Figure P5.7. $\omega = 337$ rad/s.

5.19   The circuit shown in Figure P5.8 is to be used on two different sources, each with the same amplitude but at different frequencies.

a.   Find the instantaneous real and reactive power if $v_S(t) = 120\cos 377t$ (i.e., the frequency is 60 Hz).

b.   Find the instantaneous real and reactive power if $v_S(t) = 120\cos 314t$ (i.e., the frequency is 50 Hz).

L

C

$V_S(t)$

R

$C = 265$ mF $L = 25.55$ mH
$R = 10\ \Omega$

Figure P5.8

a

R_1

L_1

b

a

R_2 L_2

b

Figure P5.9

5.20   Use complex power considerations to determine the parallel-circuit equivalent of a series RL circuit; that is, find R_2 and L_2 in terms of R_1, L_1, and ω by computing the complex power for each circuit (for arbitrary voltage or current) and equating the two. The circuits are shown in Figure P5.9.

5.21   A load impedance, $Z_L = 10 + j3\ \Omega$, is connected to a source with line resistance equal to $1\ \Omega$, as shown in Figure P5.10. Calculate the following values:

a.   The average power delivered to the load.

b.   The average power absorbed by the line.

c.   The apparent power supplied by the generator.

d.   The power factor of the load.

e.   The power factor of line plus load.

I_S
Line
$R = 1\ \Omega$

$V_S = 230\angle0°$ Z_L Load

Figure P5.10

5.22   The load Z_L in the circuit of Figure P5.10 consists of a 25-Ω resistor in parallel with a 100-mF capacitor. Calculate

a.   The apparent power delivered to the load.

b.   The apparent power supplied by the source.

c.   The power factor of the load.

I_S

$R = 20\ \Omega$

$V_S = 50$ V

$-jX_C = -j34.6\ \Omega$

Figure P5.11

5.23   Calculate the apparent power, real power, and reactive power for the circuit shown in Figure P5.11. Draw the power triangle.

5.24   A single-phase motor draws 220 W at a power factor of 80 percent (lagging) when connected across a 200-V, 60-Hz source. A capacitor is connected in parallel with the load to give a unity power factor, as shown in Figure P5.12. Find the required capacitance.

5.25 Suppose that the electricity in your home has gone out and the power company will not be able to have you hooked up again for several days. The freezer in the basement contains several hundred dollars' worth of food that you cannot afford to let spoil. You have also been experiencing very hot, humid weather and would like to keep one room air-conditioned with a window air conditioner, as well as run the refrigerator in your kitchen. When the appliances are on, they draw the following currents (all values are rms):

Air conditioner: 9.6 A @120V $pf = 0.90$ (lagging)

Freezer: 4.2 A @120V $pf = 0.87$ (lagging)

Refrigerator: 3.5 A @120V $pf = 0.80$ (lagging)

In the worst-case scenario, how much power must an emergency generator supply?

*5.26 The circuit shown in Figure P5.13 contains a parallel R-L-C circuit. Two tests are performed on the circuit, the first at 1,000 Hz and the second at 100 Hz. The following data are recorded from the two tests:

1,000-Hz test: $v_{in}(t) = 100 \sin (6{,}283t)$ V

$i_{in}(t) = 44.51 \sin (6{,}283t + 77°)$ A

$P_{in} = 500$ W

100-Hz test: $v_{in}(t) = 50 \sin (628t)$ V

$i_{in}(t) = 75.6 \sin (628t - 85.3°)$ A

$P_{in} = 155$ W

Determine the value of the circuit elements.

*5.27 Two motors are connected in parallel as shown in Figure P5.14(a). Both motors are rated for 6.8 A and 120 V. This means that each of these motors draws 6.8 A when the motor is supplying rated power to a mechanical system. The plots in Figure P5.14(b) show the power factor of the identical motors as a function of the mechanical load placed on the motor, and the current into each motor as a function of the same load.

a.   How much average power is lost if both motors are coasting (0 percent load)?

b.   If the motors are 80 percent efficient at full load—that is, if 80 percent of the electrical power is converted to mechanical power—how much power in watts is delivered to the mechanical systems under full load?

Figure P5.12

Figure P5.13

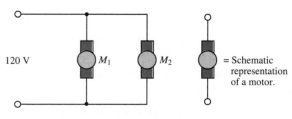

Figure P5.14(a)

c.   If motor 1 is operating at 75 percent of full load capability and motor 2 is operating at 95 percent of full load, find the power factor for the system.

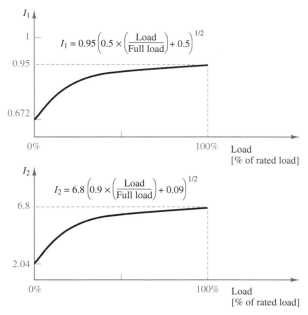

Figure P5.14(b)

5.28   The motor inside a blender can be modeled as a resistance in series with an inductance, as shown in Figure P5.15.

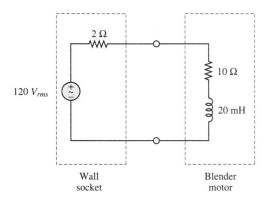

Figure P5.15

a.   What is the average power, P_{av}, dissipated in the load?

b.   What is the motor's power factor?

c.   What value of capacitor when placed in parallel with the motor will change the power factor to 0.9 (lagging)?

5.29   A resistance and a capacitor are connected in series across a 220-V rms, 60-Hz source. The resistance is adjusted so that an ammeter in the circuit indicates

8 A rms and a wattmeter indicates 250 W. Find the resistance and the capacitance of the circuit.

5.30   If the circuits shown in Figure P5.16 are to be at unity power factor, find C_P and C_S.

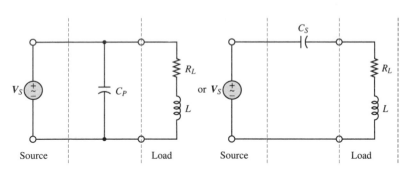

$$R_L = 5\ \Omega,\ X_L = 5\ \Omega,\ v_S(t) = 100 \sin(377t)$$

Figure P5.16

5.31   For the circuit shown in Figure P5.17,

a.   Find the Thévenin equivalent circuit for the source.

b.   Find the power dissipated by the load resistor.

c.   What value of load impedance would permit maximum power transfer?

5.32   A center-tap transformer has the schematic representation shown in Figure P5.18. The primary-side voltage is stepped down to a secondary-side voltage, V_{sec}, by a ratio of $n : 1$. On the secondary side, $V_{sec1} = V_{sec2} = \frac{1}{2}V_{sec}$.

a.   If $V_{prim} = 120\angle 32°$ V and $n = 9$, find V_{sec}, V_{sec1}, and V_{sec2}.

b.   What must n be if $V_{prim} = 208\angle 10°$ V and we desire $|V_{sec2}|$ to be 8.7 V?

5.33   The transformer shown in Figure P5.19 has several sets of windings on the secondary side. The windings have the following turns ratios:

$$a : N = 1/15$$
$$b : N = 1/4$$
$$c : N = 1/12$$
$$d : N = 1/18$$

If $V_{prim} = 120$ V, find and draw the connections that will allow you to construct the following voltage sources:

a. $24.67\angle 0°$ V

b. $36.67\angle 0°$ V

c. $18\angle 0°$ V

d. $54.67\angle 180°$ V

5.34   An ideal transformer is rated to deliver 400 kVA at 460 V to a customer as shown in Figure P5.20.

Figure P5.17

Figure P5.18

Figure P5.19

a. How much current can the transformer supply to the customer?
b. If the customer's load is purely resistive (i.e., if $pf = 1$), what is the maximum power that the customer can receive?
c. If the customer's power factor is 0.8 (lagging), what is the maximum usable power the customer can receive?
d. What is that maximum power if the pf is 0.7 (lagging)?
e. If the customer requires 300 kW to operate, what is the minimum power factor with the given size transformer?

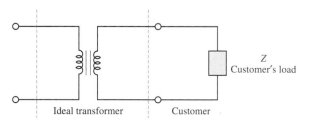

Figure P5.20

5.35 For the ideal transformer shown in Figure P5.21, find $v_o(t)$ if $v_S(t)$ is 294 cos 377t.

Figure P5.21

$R_S = 1800 \, \Omega$ $R_L = 8 \, \Omega$

Figure P5.22

5.36 If the transformer shown in Figure P5.22 is ideal, find the turns ratio $N = \frac{1}{n}$ that will provide maximum power transfer to the load.

5.37 The wire that connects an antenna on your roof to the TV set in your den is a 300-Ω wire, as shown in Figure P5.23(a). This means that the impedance seen

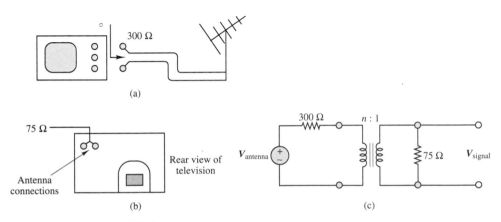

Figure P5.23

by the connections on your set is 300 Ω. Your TV, however, has a 75-Ω impedance connection, as shown in Figure P5.23(b). To achieve maximum power transfer from the antenna to the television set, you place an ideal transformer between the antenna and the TV as shown in Figure P5.23(c). What is the turns ratio, $N = \frac{1}{n}$, needed to obtain maximum power transfer?

5.38 The circuit in Figure P5.24 shows the use of ideal transformers for impedance matching. You have a limited choice of turns ratios among available transformers. Suppose you can find transformers with turns ratios of 2:1, 7:2, 120:1, 3:2, and 6:1. If Z_L is $475\angle-25°$ Ω and Z_{ab} must be $267\angle-25°$, find the combination of transformers that will provide this impedance. (You may assume that polarities are easily reversed on these transformers.)

Figure P5.24

Figure P5.25

5.39 If a transformer is modeled as shown in Figure P5.25, find the actual voltage on the secondary side of the transformer.

*5.40 If we knew that the transformer shown in Figure P5.26 was to deliver 50 A at 110 V rms with a certain resistive load, what rms phasor voltage source, V_S, would provide this voltage and current?

Figure P5.26

*5.41 A method for determining the equivalent circuit of a transformer consists of two tests: the open-circuit test, and the short-circuit test. The open-circuit test, shown in Figure P5.27(a), is usually done by applying rated voltage to the primary side of the transformer while leaving the secondary side open. The current into the primary side is measured, as is the power dissipated.

The short-circuit test, shown in Figure P5.27(b), is performed by increasing the primary voltage until rated current is going into the transformer while the secondary side is short-circuited. The current into the transformer, the applied voltage, and the power dissipated are measured.

The equivalent circuit of a transformer is shown in Figure P5.27(c), where r_w and L_w represent the winding resistance and inductance, respectively, and r_c and L_c

Figure P5.27

represent the losses in the core of the transformer and the inductance of the core.
The ideal transformer is also included in the model.

With the open-circuit test, we may assume that $\mathbf{I}_P = \mathbf{I}_S = 0$. Then all of the
current that is measured is directed through the parallel combination of r_c and L_c. We
also assume that $|r_c \, \| \, j\omega L_c|$ is much greater than $r_w + j\omega L_w$. Using these assumptions
and the open-circuit test data, we can find the resistance r_c and inductance L_c.

In the short-circuit test, we assume that $\mathbf{V}_{\text{secondary}}$ is zero, so that the voltage
on the primary side of the ideal transformer is also zero, causing no current flow
through the $r_c - L_c$ parallel combination. Using this assumption with the short-circuit
test data, we are able to find the resistance r_w and inductance L_w.

Using the following test data, find the equivalent circuit of the transformer:

Open-circuit test: $V = 241$ V

$I = 0.95$ A

$P = 32$ W

Short-circuit test: $V = 5$ V

$I = 5.25$ A

$P = 26$ W

Both tests were made at $\omega = 377$ rad/s.

*5.42   Using the methods of Problem 5.41 and the following data, find the equivalent circuit of the transformer tested:

Open-circuit test: $V_P = 4{,}600$ V

$I_{OC} = 0.7$ A

$P = 200$ W

Short-circuit test: $P = 50$ W

$$V_P = 5.2 \text{ V}$$

The transformer is a 460-kVA transformer, and the tests are performed at 60 Hz.

5.43   The magnitude of the phase voltage of a balanced three-phase wye system is 100 V. Express each phase and line voltage in both polar and rectangular coordinates.

5.44   The phase currents in a four-wire wye-connected load are as follows:

$$\mathbf{I}_{an} = 10\angle 0°, \mathbf{I}_{bn} = 12\angle 150°, \mathbf{I}_{cn} = 8\angle 165°$$

Determine the current in the neutral wire.

5.45   For the circuit shown in Figure P5.28, we see that each voltage source has a phase difference of 120° in relation to the others.

a.   Find $\mathbf{V}_{RW}$, $\mathbf{V}_{WB}$, and $\mathbf{V}_{BR}$, where $\mathbf{V}_{RW} = \mathbf{V}_R - \mathbf{V}_W$, $\mathbf{V}_{WB} = \mathbf{V}_W - \mathbf{V}_B$, and $\mathbf{V}_{BR} = \mathbf{V}_B - \mathbf{V}_R$.

b.   Repeat part a, using the calculations

$$\mathbf{V}_{RW} = \mathbf{V}_R\sqrt{3}\angle -30°$$

$$\mathbf{V}_{WB} = \mathbf{V}_W\sqrt{3}\angle -30°$$

$$\mathbf{V}_{BR} = \mathbf{V}_B\sqrt{3}\angle -30°$$

c.   Compare the results of part a with the results of part b.

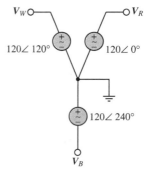

Figure P5.28

5.46   For the three-phase circuit shown in Figure P5.29, find the currents $\mathbf{I}_W$, $\mathbf{I}_B$, $\mathbf{I}_R$, and $\mathbf{I}_N$.

Figure P5.29   **Figure P5.30**

5.47   For the circuit shown in Figure P5.30, find the currents $\mathbf{I}_R$, $\mathbf{I}_W$, $\mathbf{I}_B$, and $\mathbf{I}_N$.

***5.48**   If we model each winding of a three-phase motor like the circuit shown in Figure P5.31(a) and connect the windings as shown in Figure P5.31(b), we have the three-phase circuit shown in Figure P5.31(c). The motor can be constructed so that $R_1 = R_2 = R_3$ and $L_1 = L_2 = L_3$, as is the usual case. If we connect the motor as shown in Figure P5.31(c), find the currents $\mathbf{I}_R$, $\mathbf{I}_W$, $\mathbf{I}_B$, and $\mathbf{I}_N$, assuming that the resistances are 40 Ω each and each inductance is 5 mH. The frequency of each of the sources is 60 Hz.

Figure P5.31

5.49 With reference to the motor of Problem 5.48,

a. How much power (in watts) is delivered to the motor?
b. What is the motor's power factor?
c. Why is it common in industrial practice *not* to connect the ground lead to motors of this type?

5.50 A three-phase balanced load draws 40 A per terminal when connected to a 440-V three-phase system. The power factor of the load is 0.8, lagging. Determine the power drawn by the load.

5.51 Find the apparent power and the real power delivered to the load in the Y-Δ circuit shown in Figure P5.32. What is the power factor?

Figure P5.32

5.52 The electric power company is concerned with the loading of its transformers. Since it is responsible to a large number of customers, it must be certain that
it can supply the demands of *all* customers. The power company's transformers will
deliver rated kVA to the secondary load. However, if the demand were to increase to
a point where greater than rated current were required, the secondary voltage would
have to drop below rated value. Also, the current would increase, and with it the
I^2R losses (due to winding resistance), possibly causing the transformer to overheat.
Unreasonable current demand could be caused, for example, by excessively low
power factors at the load.

The customer, on the other hand, is not greatly concerned with an inefficient
power factor, provided that sufficient power reaches the load. To make the customer
more aware of power factor considerations, the power company may install a penalty
on the customer's bill. A typical penalty–power factor chart is shown in Table 5.3.
Power factors below 0.7 are not permitted. A 25 percent penalty will be applied to
any billing after two consecutive months in which the customer's power factor has
remained below 0.7.

Table 5.3

Power factor	Penalty
0.850 and higher	None
0.8 to 0.849	1%
0.75 to 0.799	2%
0.7 to 0.749	3%

Courtesy of Detroit Edison.

The Y-Y circuit shown in Figure P5.33 is representative of a three-phase motor
load.

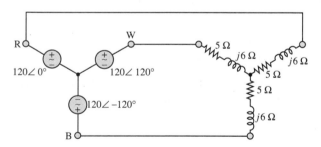

Figure P5.33

a. Find the total power supplied to the motor.
b. Find the power converted to mechanical energy if the motor is 80 percent effi-
    cient.
c. Find the power factor.
d. Does the company risk facing a power factor penalty on its next bill if all the
    motors in the factory are similar to this one?

*5.53 To correct the power factor problems of the motor in Problem 5.52, the
company has decided to install capacitors as shown in Figure P5.34.

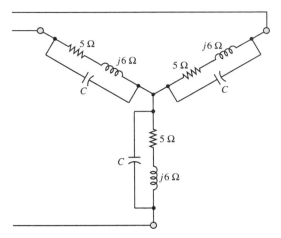

Figure P5.34

a. What capacitance must be installed to achieve a unity power factor if the line frequency is 60 Hz?

b. Repeat part a if the power factor is to be 0.85 (lagging).

*5.54   For the wye circuit shown in Figure P5.35(a),

a. Find the total complex power.

b. Find the real power dissipated.

For the delta circuit shown in Figure P5.35(b),

c. Find the total complex power.

d. Find the real power dissipated.

[*Note:* A common method of starting a motor with reduced voltage is called the *star-delta,* or *wye-delta,* method, where the motor is first connected in a wye configuration and then switched to the delta after starting.]

Figure P5.35

*5.55   The voltage supplied to an industrial plant can be viewed as a single three-phase supply with line voltage 208 V. This supply does not supply power to just one piece of machinery. It must supply power to several different motors that are neither all identical nor connected in the same manner and are switched on and off. The plant we are considering has three types of three-phase motors. A Δ-connected motor (load) has the impedances shown in Figure P5.36(a) when running under normal conditions.

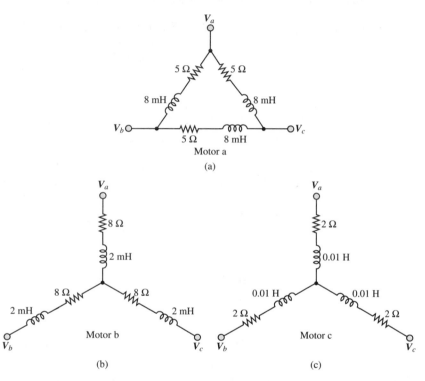

Figure P5.36

A wye-connected load has the impedances shown in Figure P5.36(b) when running under normal conditions. A second wye-connected load has the impedances shown in Figure P5.36(c) when running under normal conditions. The plant contains:

    8 motors of type a
   15 motors of type b
   30 motors of type c

During a typical shift of 8 hours, the plant runs

6 motors of type a	8 h
2 motors of type a	4 h
1 motor of type b	8 h
14 motors of type b	2 h
30 motors of type c	6 h

a. Find the average power delivered to each of the three motor types, that is, the average power delivered to a single type a, b, and c motor.

b. Find the reactive power delivered to each of the three motor types.

c. Which of the motors has the best power factor (closest to unity)?

d. From the data given for a typical shift and with the motors running under normal conditions, find the energy in kilowatt-hours consumed by the plant.

*5.56 The circuit shown in Figure P5.37 is a Y-Δ-Y connected three-phase circuit. The primaries of the transformers are wye-connected, the secondaries are delta-connected, and the load is wye-connected. Find the currents I_{RP}, I_{WP}, I_{BP}, I_A, I_B, and I_C.

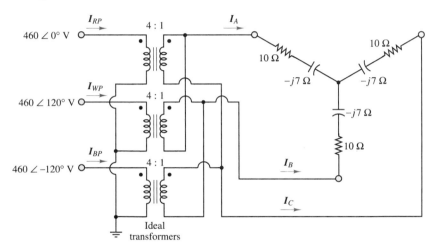

Figure P5.37

*5.57 A three-phase motor is modeled by the wye-connected circuit shown in Figure P5.38. At $t = t_1$, a line fuse is blown (modeled by the switch). Find the

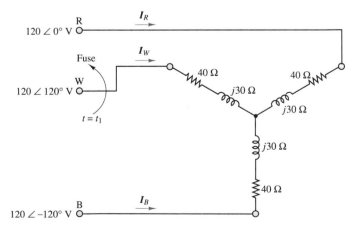

Figure P5.38

line currents I_R, I_W, and I_B and the power dissipated by the motor for the following conditions:

a. $t \ll t_1$

b. $t \gg t_1$

5.58   For the circuit shown in Figure P5.39, find the currents I_A, I_B, I_C, and I_N, and the real power dissipated by the load.

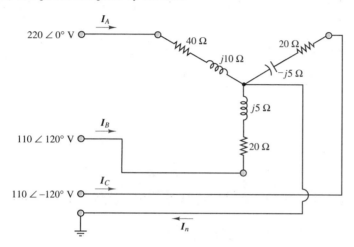

Figure P5.39

CHAPTER

6

Frequency Response and Transient Analysis

hapter 4 introduced the notions of energy-storage elements and dynamic circuit equations and developed appropriate tools (complex algebra and phasors) for the solution of AC circuits. In Chapter 5, AC analysis techniques were applied to the study of electric power. The aim of the present chapter is twofold: first, to exploit AC circuit analysis methods to study the frequency response of electric circuits; and second, to continue the discussion of dynamic circuit equations for the purpose of analyzing the transient response of electrical circuits.

It is common, in engineering problems, to encounter phenomena that are frequency-dependent. For example, structures vibrate at a characteristic frequency when excited by wind forces (some high-rise buildings experience perceptible oscillation!). The propeller on a ship excites the shaft at a vibration frequency related to the engine's speed of rotation and to the number of blades on the propeller. An internal combustion engine is excited periodically by the combustion events in the individual cylinder, at a frequency determined by the firing of the cylinders. Wind blowing across a pipe excites a resonant vibration that is perceived as sound (wind instruments operate on this principle). Electrical circuits are no different from other dynamic systems in this respect, and a large body of knowledge has been developed for understanding the frequency response of electrical circuits, mostly based on the ideas behind phasors and impedance. These ideas, and the concept of filtering, will be explored in this chapter.

In addition to analyzing the frequency response of electrical circuits, it is often necessary to characterize the response of a circuit or system to abrupt changes in input, such as, for example, the change that occurs when power to a circuit is initially turned on. Sections 6.3 through 6.5 are devoted to the analysis of the transient response of simple first- and second-order linear circuits.

The ideas developed in this chapter will also be applied, by analogy, to the analysis of other physical systems (e.g., hydraulic, thermal, and mechanical systems), to illustrate the generality of the concepts. By the end of the chapter, you should be able to:

- Compute the frequency response function for an arbitrary circuit.
- Use knowledge of the frequency response to determine the output of a circuit.
- Compute expressions for the transient and steady-state response of first- and second-order circuits.
- Recognize the analogy between electrical circuits and other dynamic systems.

6.1   SINUSOIDAL FREQUENCY RESPONSE

The **sinusoidal frequency response** (or, simply, **frequency response**) of a circuit provides a measure of how the circuit responds to sinusoidal inputs of arbitrary frequency. In other words, given the input signal amplitude, phase, and frequency, knowledge of the frequency response of a circuit permits the computation of the output signal. Suppose, for example, that you wanted to determine how the load voltage or current varied in response to different excitation signal frequencies in the circuit of Figure 6.1. An analogy could

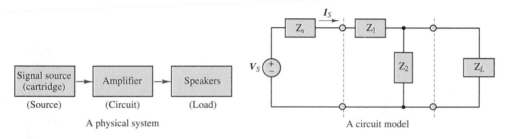

A physical system   A circuit model

Figure 6.1 A circuit model

be made, for example, with how a speaker (the load) responds to the audio signal generated by a phonograph cartridge (the source) when an amplifier (the circuit) is placed between the two.[1] In the circuit of Figure 6.1, the signal source circuitry is represented by its Thévenin equivalent. Recall that the impedance Z_S presented by the source to the remainder of the circuit is a function of the frequency of the source signal (Section 4.4). For the purpose of illustration, the amplifier circuit is represented by the idealized connection of two impedances, Z_1 and Z_2, and the load is represented by an additional impedance, Z_L. What, then, is the frequency response of this circuit? The following is a fairly general definition:

> The frequency response of a circuit is a measure of the variation of a load-related voltage or current as a function of the amplitude, phase, and frequency of the excitation signal.

According to this definition, frequency response could be defined in a variety of ways. For example, we might be interested in determining how the load voltage varies as a function of the source voltage. Then, analysis of the circuit of Figure 6.1 might proceed as follows.

To express the frequency response of a circuit in terms of variation in output voltage as a function of source voltage, we use the general formula

$$H_V(\omega) = \frac{\mathbf{V}_L(\omega)}{\mathbf{V}_S(\omega)} \tag{6.1}$$

One method that allows for representation of the load voltage as a function of the source voltage (this is, in effect, what the frequency response of a circuit implies) is to describe the source and attached circuit by means of the Thevenin equivalent circuit. (This is not the only useful technique; the node voltage or mesh current equations for the circuit could also be employed.) Figure 6.2 depicts the original circuit of Figure 6.1 with the load removed, ready for the computation of the Thévenin equivalent.

Next, an expression for the load voltage, $\mathbf{V}_L$, may be found by connecting the load to the Thévenin equivalent source circuit and by computing

---

[1] In reality, the circuitry in a hi-fi stereo system is far more complex than the circuits that will be discussed in this chapter and in the homework problems. However, from the standpoint of intuition and everyday experience, the audio analogy provides a useful example; it allows you to build a quick feeling for the idea of frequency response. Practically everyone has an intuitive idea of bass, mid range, and treble as coarsely defined frequency regions in the audio spectrum. The material presented in the next few sections should give you a more rigorous understanding of these concepts.

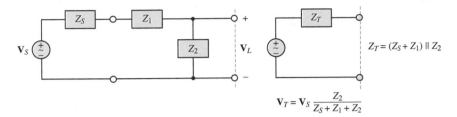

Figure 6.2 Thévenin equivalent source circuit

the result of a simple voltage divider, as illustrated in Figure 6.3 and the
following equation:

$$
\begin{aligned}
\mathbf{V}_L &= \frac{Z_L}{Z_L + Z_T}\mathbf{V}_T \\
&= \frac{Z_L}{Z_L + \frac{(Z_S+Z_1)Z_2}{Z_S+Z_1+Z_2}} \cdot \frac{Z_2}{Z_S + Z_1 + Z_2}\mathbf{V}_S \\
&= \frac{Z_L Z_2}{Z_L(Z_S + Z_1 + Z_2) + (Z_S + Z_1)Z_2}\mathbf{V}_S
\end{aligned}
$$

(6.2)

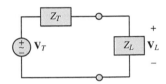

Figure 6.3 Complete
equivalent circuit

Thus, the frequency response of the circuit, as defined in equation 6.1, is
given by the expression

$$
\frac{\mathbf{V}_L}{\mathbf{V}_S}(\omega) = H_V(\omega) = \frac{Z_L Z_2}{Z_L(Z_S + Z_1 + Z_2) + (Z_S + Z_1)Z_2}
$$

(6.3)

The expression for $H_V(\omega)$ could be evaluated for any given $\mathbf{V}_S(\omega)$ (i.e., for
any given source signal amplitude, phase, and frequency) to determine what
the resultant load voltage would be. Note that $H_V(\omega)$ is a complex quantity
(dimensionless, because it is the ratio of two voltages), and that it therefore
follows that

$\mathbf{V}_L(\omega)$ is a phase-shifted and amplitude-scaled version of $\mathbf{V}_S(\omega)$:

$$
\mathbf{V}_L(\omega) = H_V(\omega) \cdot \mathbf{V}_S(\omega)
$$

(6.4)

$$
V_L e^{j\phi_L} = |H_V| e^{j\phi_H} \cdot V_S e^{j\phi_S}
$$

(6.5)

or

$$
V_L e^{j\phi_L} = |H_V| V_S e^{j(\phi_H + \phi_S)}
$$

(6.6)

where

$$
V_L = |H_V| \cdot V_S
$$

and

$$\phi_L = \phi_H + \phi_S \qquad (6.7)$$

Thus, the effect of inserting a linear circuit between a source and a load is best understood by considering that, at any given frequency, ω, the load voltage is a sinusoid at the same frequency as the source voltage, with amplitude given by $V_L = |H_V| \cdot V_S$ and phase equal to $\phi_L = \phi_H + \phi_S$, where $|H_V|$ is the magnitude of the frequency response and ϕ_H its phase angle. Both $|H_V|$ and ϕ_H are functions of frequency.

Example   6.1

Figure 6.4

Compute the frequency response $H_V(\omega)$ for the circuit of Figure 6.4.

Solution:

The impedances corresponding to the circuit elements of Figure 6.4 are $Z_1 = 1,000$, $Z_2 = 10^{-5}\omega$, $Z_L = 10,000$, and $Z_S = 0$. To represent the load voltage as a function of the source voltage, we need to describe the source and attached circuit by means of the Thévenin equivalent circuit. Figure 6.5 shows the equivalent circuit with the load removed, ready for the computation of the Thévenin circuit.

Figure 6.5

Next, we connect the load to the Thévenin equivalent and find an expression for $\mathbf{V}_L$ by computing the result of a voltage divider:

$$\mathbf{V}_L = \frac{Z_L}{Z_L + Z_T}\mathbf{V}_T$$

$$= \frac{Z_L}{Z_L + \frac{Z_1 Z_2}{Z_1 + Z_2}} \cdot \frac{Z_2}{Z_1 + Z_2}\mathbf{V}_S$$

$$= \frac{Z_L Z_2}{Z_L(Z_1 + Z_2) + Z_1 Z_2}\mathbf{V}_S$$

Thus, the frequency response of the circuit, as defined before, is given by the expression

$$\frac{\mathbf{V}_L}{\mathbf{V}_S}(\omega) = H_V(\omega) = \frac{Z_L Z_2}{Z_L(Z_1 + Z_2) + Z_1 Z_2}$$

Replacing Z_1, Z_2, and Z_L by their numerical values, we can obtain $H_V(\omega)$ as

$$H_V(\omega) = \frac{100}{j\omega + 110}$$

$H_V(\omega)$ can also be represented in polar form as

$$H_V(\omega) = \frac{100}{\sqrt{\omega^2 + 110^2}\,e^{j\,\arctan(\omega/110)}}$$

or

$$= \frac{100}{\sqrt{\omega^2 + 110^2}}e^{-j\,\arctan(\omega/110)} = \frac{100}{\sqrt{\omega^2 + 110^2}}\angle - \arctan(\omega/110)$$

The importance and usefulness of the frequency response concept lies in its ability to summarize the response of a circuit in a single function of frequency, $H(\omega)$, which can predict the load voltage or current at any frequency, given the input. Note that the frequency response of a circuit can be defined in four different ways:

$$
\begin{array}{ll}
H_V(\omega) = \dfrac{\mathbf{V}_L(\omega)}{\mathbf{V}_S(\omega)} & H_I(\omega) = \dfrac{\mathbf{I}_L(\omega)}{\mathbf{I}_S(\omega)} \\[2ex]
H_Z(\omega) = \dfrac{\mathbf{V}_L(\omega)}{\mathbf{I}_S(\omega)} & H_Y(\omega) = \dfrac{\mathbf{I}_L(\omega)}{\mathbf{V}_S(\omega)}
\end{array}
\tag{6.8}
$$

If $H_V(\omega)$ and $H_I(\omega)$ are known, one can directly derive the other two expressions:

$$H_Z(\omega) = \frac{\mathbf{V}_L(\omega)}{\mathbf{I}_S(\omega)} = Z_L(\omega)\frac{\mathbf{I}_L(\omega)}{\mathbf{I}_S(\omega)} = Z_L(\omega)H_I(\omega) \tag{6.9}$$

$$H_Y(\omega) = \frac{\mathbf{I}_L(\omega)}{\mathbf{V}_S(\omega)} = \frac{1}{Z_L(\omega)}\frac{\mathbf{V}_L(\omega)}{\mathbf{V}_S(\omega)} = \frac{1}{Z_L(\omega)}H_V(\omega) \tag{6.10}$$

With these definitions in hand, it is now possible to introduce one of the central concepts of electrical circuit analysis: **filters.** The concept of filtering an electrical signal will be discussed in the next section.

Example   6.2

Compute the frequency response $H_Z(\omega)$ for the circuit of Figure 6.6.

Solution:

The frequency response $H_Z(\omega) = \mathbf{V}_S(\omega)/\mathbf{I}_L(\omega)$ may be found in a relatively simple manner by observing that (1) the load current is related to the source current by a current divider and (2) the load voltage is equal to the product of the 4-kΩ resistor and the load current. Formally,

$$\mathbf{I}_L(\omega) = \frac{1,000}{1,000 + 4,000 + j\omega \times 2 \times 10^{-3}}\mathbf{I}_S(\omega)$$

and

$$\mathbf{V}_L(\omega) = 4,000\mathbf{I}_L(\omega) = \frac{4 \times 10^6}{5 \times 10^3 + j\omega \times 2 \times 10^{-3}} = \frac{0.8 \times 10^3}{1 + j\omega \times 0.4 \times 10^{-6}}$$

Figure 6.6

Finally,

$$H_Z(\omega) = \frac{\mathbf{V}_S(\omega)}{\mathbf{I}_L(\omega)} = \frac{1,000}{1 + j\omega \times 0.5 \times 10^{-6}}$$

6.2   FILTERS

There are a host of practical, everyday applications that involve filters of one kind or another. Just to mention two, filtration systems are used to eliminate impurities from drinking water, and sunglasses are used to filter out eye-damaging ultraviolet radiation and to reduce the intensity of sunlight reaching the eyes. An analogous concept applies to electrical circuits: it is possible to *attenuate* (i.e., reduce in amplitude) or altogether eliminate signals of unwanted frequencies, such as those that may be caused by electrical noise or other forms of interference. This section will be devoted to the analysis of electrical filters.

Low-Pass Filters

RC low-pass filter. The circuit preserves lower frequencies while attenuating the frequencies above the cutoff frequency, $\omega_0 = 1/RC$. The voltages $\mathbf{V}_i$ and $\mathbf{V}_o$ are the filter input and output voltages, respectively.

Figure 6.7 A simple *RC* filter

Figure 6.7 depicts a simple ***RC* filter** and denotes its input and output voltages by $\mathbf{V}_i$ and $\mathbf{V}_o$. The frequency response for the filter may be obtained by considering the function

$$H(\omega) = \frac{\mathbf{V}_o}{\mathbf{V}_i}(\omega) \tag{6.11}$$

and noting that the output voltage may be expressed as a function of the input voltage by means of a voltage divider, as follows:

$$\mathbf{V}_o(\omega) = \mathbf{V}_i(\omega)\frac{1/j\omega C}{R + 1/j\omega C} = \mathbf{V}_i(\omega)\frac{1}{1 + j\omega RC} \tag{6.12}$$

Thus, the frequency response of the *RC* filter is

$$\frac{\mathbf{V}_o}{\mathbf{V}_i}(\omega) = \frac{1}{1 + j\omega CR} \tag{6.13}$$

An immediate observation upon studying this frequency response is that if the signal frequency, ω, is zero, the response of the filter is equal to 1. That is, the filter is passing all of the input. Why? To answer this question, we note that at $\omega = 0$, the impedance of the capacitor $1/j\omega C$, becomes infinite. Thus, the capacitor acts as an open circuit, and the output voltage equals the input:

$$\mathbf{V}_o(\omega = 0) = \mathbf{V}_i(\omega = 0) \tag{6.14}$$

Since a signal at sinusoidal frequency equal to zero is a DC signal, this filter circuit does not in any way affect DC voltages and currents. As the signal frequency increases, the magnitude of the frequency response decreases, since the denominator increases with ω. More precisely, equations 6.15 to 6.18 describe the magnitude and phase of the frequency response of the RC filter:

$$H(\omega) = \frac{\mathbf{V}_o}{\mathbf{V}_i}(\omega) = \frac{1}{1 + j + \omega CR}$$

$$= \frac{1}{\sqrt{1 + (\omega CR)^2}} \frac{e^{j0°}}{e^{j \arctan(\omega CR/1)}} \tag{6.15}$$

$$= \frac{1}{\sqrt{1 + (\omega CR)^2}} \cdot e^{-j \arctan(\omega CR)}$$

or

$$H(\omega) = |H(\omega)| \, e^{j\phi_H(\omega)} \tag{6.16}$$

with

$$|H(\omega)| = \frac{1}{\sqrt{1 + (\omega CR)^2}} = \frac{1}{\sqrt{1 + (\omega/\omega_0)^2}} \tag{6.17}$$

and

$$\phi_H(\omega) = -\arctan(\omega CR) = -\arctan\left(\frac{\omega}{\omega_0}\right) \tag{6.18}$$

with

$$\omega_0 = \frac{1}{RC} \tag{6.19}$$

The simplest way to envision the effect of the filter is to think of the phasor voltage $\mathbf{V}_i = V_i e^{j\phi_i}$ scaled by a factor of $|H|$ and shifted by a phase angle ϕ_H by the filter *at each frequency,* so that the resultant output is given by $\mathbf{V}_o = \mathbf{V}_o e^{j\phi_o}$, with

$$V_o = |H| \cdot V_i$$
$$\phi_o = \phi_H + \phi_i \tag{6.20}$$

and where $|H|$ and ϕ_H are functions of frequency. The frequency ω_0 is called the **cutoff frequency** of the filter and, as will presently be shown, gives an indication of the filtering characteristics of the circuit.

It is customary to represent $H(\omega)$ in two separate plots, representing $|H|$ and ϕ_H as functions of ω. These are shown in Figure 6.8 in normalized form—that is, with $|H|$ and ϕ_H plotted versus ω/ω_0, corresponding to a cutoff frequency $\omega_0 = 1$ rad/s. Note that, in the plot, the frequency axis has been scaled logarithmically. This is a common practice in electrical engineering, because it allows viewing a very broad range of frequencies on the same plot without excessively compressing the low-frequency end of the plot. The frequency response plots of Figure 6.8 are commonly employed to describe the frequency response of a circuit, since they can provide a clear idea at a glance of the effect of a filter on an excitation signal. For example, the RC filter of Figure 6.7 has the property of "passing" signals at low frequencies ($\omega \ll 1/RC$) and of filtering out signals at high frequencies ($\omega \gg 1/RC$). This type of filter is called a **low-pass filter.** The cutoff frequency $\omega = 1/RC$ has a special significance in that it represents—approximately—the point where the filter begins to filter out the higher-frequency signals. The value of $H(\omega)$ at the cutoff frequency is $1/\sqrt{2} = 0.707$. Note how the cutoff frequency depends exclusively on the values of R and C. Therefore, one can adjust the filter response as desired simply by selecting appropriate values for C and R, and therefore choose the desired filtering characteristics.

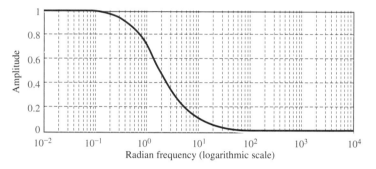

Magnitude response of R-C low-pass filter

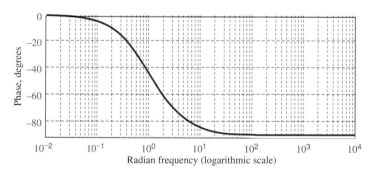

Phase response of R-C low-pass filter

Figure 6.8 Magnitude and phase response plots for RC filter

Example   6.3

To develop a better understanding of frequency response, it is instructive to compute the effects of the RC filter circuit of Figure 6.7 on a sinusoidal input signal at frequencies above and below the cutoff frequency. Assume that $R = 1\ k\Omega$ and $C = 0.47\ \mu F$. Compute the response of the filter at the frequencies 60 Hz and 10,000 Hz. Also compute the output voltage if the input voltage (at each of these two frequencies) has a peak amplitude of 5 V and zero phase angle.

Solution:

Observe, first, that the cutoff frequency of the filter is

$$\omega_0 = \frac{1}{RC} = 2{,}127.7\ \text{rad/s}$$

or

$$f_0 = \frac{\omega_0}{2\pi} = 338.6\ \text{Hz}$$

Evaluating the amplitude at the first frequency, $\omega = 2\pi \times 60$, we obtain

$$\left| \frac{V_o}{V_i}(\omega = 2\pi \times 60) \right| = \frac{1}{\sqrt{1 + \left(\frac{\omega}{\omega_0} \right)^2}} = \frac{1}{1 + \left(\frac{377}{2{,}127.7} \right)^2} = 0.9847$$

The phase angle may also be computed at the same frequency:

$$\phi(\omega = 2\pi \times 60) = -\arctan\left(\frac{377}{2{,}127.7} \right) = -10.05°$$

This result corresponds to stating that if the input voltage to the filter were a sinusoid with zero phase and, say, a 5-V amplitude, the output voltage would consist of a sinusoid at the same frequency with amplitude 0.9847×5 volts and phase equal to $-10.05°$:

$$v_i(t) = 5\cos(377t)$$

$$v_o(t) = 4.9233\cos(377t - 10.05°)$$

At the frequency $\omega = 20{,}000\pi$ rad/s (10,000 Hz), the magnitude and phase response are computed to be

$$\left| \frac{V_o}{V_i}(\omega = 2\pi \times 10{,}000) \right| = \frac{1}{\sqrt{1 + \left(\frac{20{,}000\pi}{2{,}127.7} \right)^2}} = 0.03384$$

and

$$\phi(\omega = 2\pi \times 10{,}000) = -\arctan\left(\frac{20{,}000\pi}{2{,}127.7} \right) = -88°$$

so that the input and output voltages would be, respectively,

$$v_i(t) = 5\cos(20{,}000\pi t)$$

$$v_o(t) = 0.169\cos(20{,}000\pi t - 88°)$$

The magnitude and phase responses of the filter are plotted in Figure 6.9.

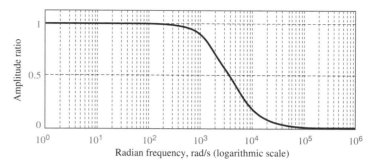

Magnitude response of R-C filter of Example 6.3

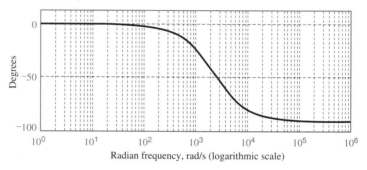

Phase response of R-C filter of Example 6.3

Figure 6.9 Response of *RC* filter of Example 6.3

Example   6.4

This example illustrates quantitatively the computation of the frequency response of the circuit of Figure 6.10. In this example, we assume that $R_S = 50 \ \Omega$, $R_1 = 200 \ \Omega$, $R_L = 500 \ \Omega$, and $C = 10 \ \mu F$.

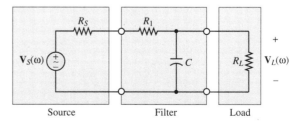

Figure 6.10 *RC* filter inserted in a circuit

Solution:

The circuit of Figure 6.10 represents a more realistic circuit than the simple *RC* filter discussed earlier in this section. It includes the internal resistance of the source (assumed to be 50 Ω) and a separate load resistance. Although the circuit of Figure

6.10 may seem more complex than that of Figure 6.7 at first, it will be shown that the two circuits may be reduced to the same form with judicious use of equivalent-circuit concepts. Observe, first, that the circuit may be represented directly in Thevenin equivalent form by combining resistors R_1 and R_S and representing the parallel combination of the capacitor and the load resistor as a single impedance, as shown in Figure 6.11. The equivalent circuit is described by the following circuit elements:

$$R' = R_S + R_1$$

$$Z' = R_L \parallel \frac{1}{j\omega C} = \frac{R_L}{1 + j\omega C R_L}$$

Now, the frequency response of the entire circuit—including the source, filter, and load—may be computed as follows:

$$\frac{\mathbf{V}_L}{\mathbf{V}_S}(\omega) = \frac{Z'}{R' + Z'} = \frac{R_L/(1 + j\omega C R_L)}{R_S + R_1 + R_L/(1 + j\omega C R_L)}$$

$$= \frac{R_L}{R_L + R_S + R_1 + j\omega C R_L(R_S + R_1)}$$

$$= \frac{R_L/(R_L + R')}{1 + j\omega C(R_L \parallel R')}$$

Compare this last frequency response expression with equation 6.13:

$$H(\omega) = \frac{R_L/(R_L + R')}{1 + j\omega C(R_L \parallel R')} = \frac{k}{1 + j\omega C R_{eq}} = \frac{0.667}{1 + j(\omega/600)}$$

versus

$$\frac{\mathbf{V}_o}{\mathbf{V}_i}(\omega) = \frac{1}{1 + j\omega C R} \qquad (6.13)$$

Note the similarity between the two expressions: the denominator is virtually the same if we recognize that the parallel combination of R' and R_L (indicated as R_{eq} corresponds to the resistor R in equation 6.13, and the constant k replaces the 1 in the numerator. Further, the cutoff frequency takes the same form as in equation 6.19:

$$\omega_0 = \frac{1}{C(R_L \parallel R')} = \frac{1}{C R_{eq}}$$

Figure 6.11 Equivalent-circuit representation of Figure 6.10

Example 6.5 Wheatstone Bridge Filter

The Wheatstone bridge circuit of Examples 2.12 and 2.13 is used in a number of instrumentation applications, including the measurement of force (see Example 2.13, describing the strain gauge bridge). Figure 6.12 depicts the appearance of the bridge circuit. When undesired noise and interference are present in the measurement, it is often appropriate to use a low-pass filter to reduce the effect of the noise. The capacitor that is connected to the output terminals of the bridge in Figure 6.12 constitutes an effective and simple low-pass filter, in conjunction with the bridge resistance. Assume that the average resistance of each leg of the bridge is 350 Ω (a standard value for strain gauges) and that we desire to measure a sinusoidal force at a frequency of 30 Hz. From prior measurements, it has been determined that a filter

Filter capacitor

$V_{out} = V_a - V_b$

Wheatstone bridge
equivalent circuit

Figure 6.12 Wheatstone
bridge with equivalent circuit
and simple capacitive filter

with a cutoff frequency of 300 Hz is sufficient to reduce the effects of noise. Choose a capacitor that matches this filtering requirement.

Solution:

By evaluating the Thévenin equivalent circuit for the Wheatstone bridge, calculating the desired value for the filter capacitor becomes relatively simple, as illustrated at the bottom of Figure 6.12. The Thévenin resistance for the bridge circuit may be computed by short-circuiting the two voltage sources and removing the capacitor placed across the load terminals:

$$R_T = R_1 \parallel R_2 + R_3 \parallel R_4 = 350 \parallel 350 + 350 \parallel 350 = 350 \ \Omega$$

Since the required cutoff frequency is 300 Hz, the capacitor value can be computed from the expression

$$\omega_0 = \frac{1}{R_T C} = 2\pi \times 300$$

or

$$C = \frac{1}{R_T \omega_0} = \frac{1}{350 \times 2\pi \times 300} = 1.51 \ \mu F$$

The frequency response of the bridge circuit is of the same form as equation 6.13:

$$\frac{V_{out}}{V_T}(\omega) = \frac{1}{1 + j\omega C R_T}$$

It can be evaluated at the frequency of 30 Hz to verify that the attenuation and phase shift at the desired signal frequency are minimal:

$$\frac{V_{out}}{V_T}(\omega = 2\pi \times 30) = \frac{1}{1 + j2\pi \times 30 \times 1.51 \times 10^{-6} \times 350}$$
$$= 0.9951\angle{-5.7°}$$

Figure 6.13 depicts the appearance of a 30-Hz sinusoidal signal before and after the addition of the capacitor to the circuit.

Figure 6.13 Unfiltered and filtered bridge output

Much more complex low-pass filters than the simple *RC* combinations shown so far can be designed by using appropriate combinations of various circuit elements. The synthesis of such advanced filter networks is beyond the scope of this book; however, we shall discuss the practical implementation of some commonly used filters in Chapters 10 and 11, in connection with the discussion of the operational amplifier. The next two sections extend the basic ideas introduced in the preceding pages to high- and band-pass filters—that is, to filters that emphasize the higher frequencies or a band of frequencies, respectively.

High-Pass Filters

Just as you can construct a simple filter that preserves low frequencies and attenuates higher frequencies, you can easily construct a **high-pass filter** that passes mainly those frequencies *above a certain cutoff frequency*. The analysis of a simple high-pass filter can be conducted by analogy with the preceding discussion of the low-pass filter. Consider the circuit shown in Figure 6.14. The frequency response for the high-pass filter, $H(\omega) = \frac{V_o}{V_i}(\omega)$, may be obtained by noting that

$$V_o(\omega) = V_i(\omega)\frac{R}{R + 1/j\omega C} = V_i(\omega)\frac{j\omega CR}{1 + j\omega CR} \qquad (6.21)$$

Thus, the frequency response of the filter is:

$$\frac{V_o}{V_i}(\omega) = \frac{j\omega CR}{1 + j\omega CR} \qquad (6.22)$$

which can be expressed in magnitude-and-phase form by

$$H(\omega) = \frac{V_o}{V_i}(\omega) = \frac{j\omega CR}{1 + j\omega CR} = \frac{\omega CRe^{j90°}}{\sqrt{1 + (\omega CR)^2}e^{j\arctan(\omega CR/1)}}$$

$$= \frac{1}{\sqrt{1 + (\omega CR)^2}} \cdot e^{j(90° - \arctan(\omega CR))} \qquad (6.23)$$

or

$$H(\omega) = |H|\, e^{j\phi_H}$$

with

$$H(\omega) = \frac{\omega CR}{\sqrt{1 + (\omega CR)^2}} \qquad (6.24)$$

$$\phi_H(\omega) = 90° - \arctan(\omega CR)$$

You can verify by inspection that the amplitude response of the high-pass filter will be zero at $\omega = 0$ and will asymptotically approach 1 as ω approaches infinity, while the phase shift is 90° at $\omega = 0$ and tends to zero for increasing ω. Amplitude and phase response curves for the high-pass filter are shown in Figure 6.15. These plots have been normalized to have the filter cutoff frequency $\omega_0 = 1$ rad/s. Note that, once again, it is possible to define a cutoff frequency at $\omega_0 = 1/RC$ in the same way as was done for the low-pass filter.

RC high-pass filter. The circuit preserves higher frequencies while attenuating the frequencies below the cutoff frequency, $\omega_0 - 1/RC$.

Figure 6.14 High-pass filter

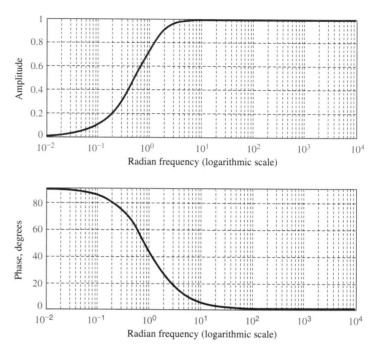

Figure 6.15 Frequency response of a high-pass filter

Example   6.6

This example will analyze the frequency response of a high-pass RC circuit. Assume that a 0.199-μF capacitor and a 200-Ω resistor are employed in constructing a high-pass filter similar to that of Figure 6.14. The cutoff frequency of the filter is

$$\omega_0 = \frac{1}{RC} = 2\pi \times 4{,}000 \text{ rad/s}$$

Thus, the frequency response of equation 6.22 becomes

$$\frac{\mathbf{V}_o}{\mathbf{V}_i}(\omega) = \frac{j\omega/(2\pi \times 4{,}000)}{1 + j\omega/(2\pi \times 4{,}000)}$$

The magnitude and phase of this frequency response are given by

$$\left|\frac{\mathbf{V}_o}{\mathbf{V}_i}(\omega)\right| = \frac{\omega/\omega_0}{\sqrt{1 + (\omega/\omega_0)^2}}$$

and

$$\phi(\omega) = 90° - \arctan\left(\frac{\omega}{\omega_0}\right)$$

and are plotted in Figure 6.16. We can compute the frequency response of the filter at the frequencies $\omega = 2\pi \times 100$ rad/s and $\omega = 2\pi \times 10{,}000$ rad/s to evaluate its

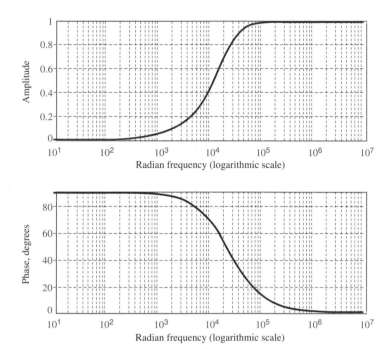

Figure 6.16 Response of high-pass filter of Example 6.6

performance:

$$\left| \frac{\mathbf{V}_o}{\mathbf{V}_i}(\omega = 2\pi \times 100) \right| = \frac{100/4{,}000}{\sqrt{1 + (100/4{,}000)^2}} = 0.025$$

$$\phi(\omega = 2\pi \times 100) = 90\,° - \arctan\left(\frac{100}{4{,}000}\right) = 88.6°$$

and

$$\left| \frac{\mathbf{V}_o}{\mathbf{V}_i}(\omega = 2\pi \times 10{,}000) \right| = \frac{10{,}000/4{,}000}{\sqrt{1 + (10{,}000/4{,}000)^2}} = 0.928$$

$$\phi(\omega = 2\pi \times 10{,}000) = 90° - \arctan\left(\frac{10{,}000}{4{,}000}\right) = 21.8°$$

It should be apparent that, while at the higher frequency most of the amplitude of the input signal is preserved in the output, at the much lower frequency of 100 Hz less than 3 percent of the input signal amplitude passes through the filter. The high-pass filter analyzed in this example would "pass" only the *treble range* of the audio spectrum.

Band-Pass Filters

Building on the principles developed in the preceding sections, we can also construct a circuit that acts as a **band-pass filter,** passing mainly those

RLC band-pass filter. The
circuit preserves frequencies
within a band.

Figure 6.17 RLC band-
pass filter

frequencies *within a certain frequency range*. The analysis of a simple *second-order* band-pass filter (i.e., a filter with two energy-storage elements) can be conducted by analogy with the preceding discussions of the low-pass and high-pass filters. Consider the circuit shown in Figure 6.17, and the related frequency response function for the filter $H(\omega) = \frac{V_o}{V_i}(\omega)$. Noting that

$$
\begin{aligned}
\mathbf{V}_o(\omega) &= \mathbf{V}_i(\omega) \cdot \frac{R}{R + 1/j\omega C + j\omega L} \\
&= \mathbf{V}_i(\omega) \frac{j\omega CR}{1 + j\omega CR + (j\omega)^2 LC}
\end{aligned}
\tag{6.25}
$$

we may write the frequency response of the filter as

$$
\frac{\mathbf{V}_o}{\mathbf{V}_i}(\omega) = \frac{j\omega CR}{1 + j\omega CR + (j\omega)^2 LC}
\tag{6.26}
$$

Equation 6.26 can often be factored into the following form:

$$
\frac{\mathbf{V}_o}{\mathbf{V}_i}(\omega) = \frac{jA\omega}{(j\omega/\omega_1 + 1)(j\omega/\omega_2 + 1)}
\tag{6.27}
$$

where ω_1 and ω_2 are the two frequencies that determine the **pass-band** (or **bandwidth**) of the filter—that is, frequency range over which the filter "passes" the input signal—and A is a constant that results from the factoring. An immediate observation we can make is that if the signal frequency, ω, is zero, the response of the filter is equal to zero, since at $\omega = 0$ the impedance of the capacitor, $1/j\omega C$, becomes infinite. Thus, the capacitor acts as an open circuit, and the output voltage equals zero. Further, we note that the filter output in response to an input signal at sinusoidal frequency approaching infinity is again equal to zero. This result can be verified by considering that as ω approaches infinity, the impedance of the inductor becomes infinite, that is, an open circuit. Thus, the filter cannot pass signals at very high frequencies. In an intermediate band of frequencies, the band-pass filter circuit will provide a variable attenuation of the input signal, dependent on the frequency of the excitation. This may be verified by taking a closer look at equation 6.27:

$$
\begin{aligned}
H(\omega) = \frac{\mathbf{V}_o}{\mathbf{V}_i}(\omega) &= \frac{jA\omega}{(j\omega/\omega_1 + 1)(j\omega/\omega_2 + 1)} \\
&= \frac{A\omega e^{j90°}}{\sqrt{1 + \left(\frac{\omega}{\omega_1}\right)^2}\sqrt{1 + \left(\frac{\omega}{\omega_2}\right)^2} e^{j\arctan(\omega/\omega_1)}e^{j\arctan(\omega/\omega_2)}} \\
&= \frac{A\omega}{\sqrt{\left[1 + \left(\frac{\omega}{\omega_1}\right)^2\right]\left[1 + \left(\frac{\omega}{\omega_2}\right)^2\right]}} \cdot e^{j[90° - \arctan(\omega/\omega_1) - \arctan(\omega/\omega_2)]}
\end{aligned}
\tag{6.28}
$$

Equation 6.28 is of the form $H(\omega) = |H| e^{j\phi_H}$, with

$$|H(\omega)| = \frac{A\omega}{\sqrt{\left[1 + \left(\frac{\omega}{\omega_1}\right)^2\right]\left[1 + \left(\frac{\omega}{\omega_2}\right)^2\right]}}$$

(6.29)

and

$$\phi_H(\omega) = 90° - \arctan\left(\frac{\omega}{\omega_1}\right) - \arctan\left(\frac{\omega}{\omega_2}\right)$$

The magnitude and phase plots for the frequency response of the band-pass filter of Figure 6.17 are shown in Figure 6.18. These plots have been normalized to have the filter pass-band centered at the frequency $\omega = 1$ rad/s.

The frequency response plots of Figure 6.18 suggest that, in some sense, the band-pass filter acts as a combination of a high-pass and a low-pass filter. As illustrated in the previous cases, it should be evident that one can adjust the filter response as desired simply by selecting appropriate values for L, C, and R.

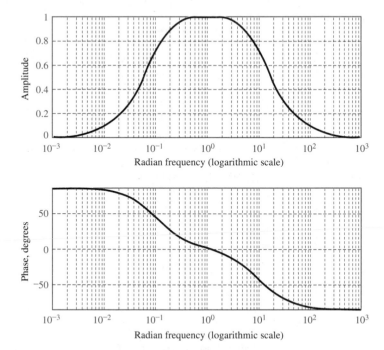

Figure 6.18 Frequency response of *RLC* band-pass filter

Example   6.7

This example illustrates quantitatively the computation of the frequency response of a band-pass RLC circuit. The frequency response of the band-pass filter is essentially determined by the cutoff frequencies, ω_1 and ω_2. Thus, we can write the frequency response of the band-pass filter as

$$\frac{\mathbf{V}_o}{\mathbf{V}_i}(\omega) = \frac{j\omega CR}{1 + j\omega CR + (j\omega)^2 LC}$$

The general form of this response is

$$H(\omega) = \frac{\mathbf{V}_o}{\mathbf{V}_i}(\omega) = \frac{\omega CRe^{j90°}}{\sqrt{(1 - \omega^2 LC)^2 + (\omega CR)^2}\,e^{j\,\arctan[\omega CR/(1-\omega^2 LC)]}}$$

with

$$|H(\omega)| = \frac{\omega CR}{\sqrt{(1 - \omega^2 LC)^2 + (\omega CR)^2}}$$

$$\phi(\omega) = 90° - \arctan\left(\frac{\omega CR}{1 - \omega^2 LC}\right)$$

It is very instructive to consider the effect of varying the series resistance on the width of the pass-band (or bandwidth) of the filter. Figures 6.19 and 6.20 depict the band-pass filter magnitude and phase response for the following two sets of

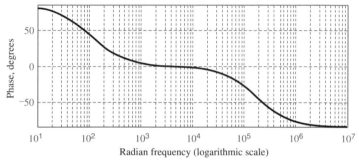

Figure 6.19 Frequency response of broad-band band-pass filter of Example 6.7

component values: $R = 1$ kΩ, $L = 5$ mH, $C = 10$ μF, and $R = 10$ Ω, $L = 5$ mH, $C = 10$ μF. For the larger value of R (1 kΩ), the filter response allows a relatively broad band of frequencies to be passed; filters with a response similar to that shown in Figure 6.19 are thus usually referred to as **broad-band filters.** For the smaller R (10 Ω), only a very narrow band of frequencies is allowed through the filter. For this reason, filters such as the one shown in Figure 6.20 are referred to as **narrow-band filters.**

It should be apparent that, while at the higher and lower frequencies most of the amplitude of the input signal is filtered from the output, at the mid-band frequency (4,500 rad/s) most of the input signal amplitude passes through the filter. The first band-pass filter analyzed in this example would "pass" the *mid-band range* of the audio spectrum, while the second would pass only a very narrow band of frequencies around the **center frequency** of 4,500 rad/s. Such narrow-band filters find application in **tuning circuits,** such as those employed in conventional AM radios (although at frequencies much higher than that of the present example). In a tuning circuit, a narrow band filter is used to tune in a frequency associated with the **carrier** of a radio station (for example, for a station found at a setting of "AM 820," the carrier wave transmitted by the radio station is at a frequency of 820 kHz). By using a variable capacitor, it is possible to tune in a range of carrier frequencies and therefore select the preferred station. Other circuits are then used to decode the actual speech or music signal modulated on the carrier wave; some of these will be discussed in Chapters 7 and 9.

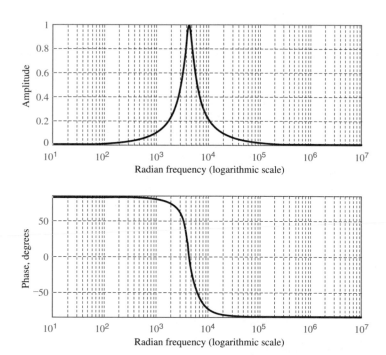

Figure 6.20 Frequency response of narrow-band band-pass filter of Example 6.7

Example   6.8   AC Line Interference Filter

One very useful application of narrow-band filters is in rejecting interference due to the AC line power. An undesired 60-Hz signal originating in the AC line power can cause serious interference in delicate instruments by disturbing the desired signals. In medical instruments such as the electrocardiograph, 60-Hz notch filters are often provided to reduce the effect of this interference upon delicate measurements. Figure 6.21 depicts a circuit representing the situation just described by means of two series voltage sources: one representing the desired signal, the other representing the undesired 60-Hz interference. This example will illustrate the design of a 60-Hz notch filter capable of reducing the interference caused by the AC line power. Assume that $R_S = 50\ \Omega$ and $R_L = 150\ \Omega$ in the circuit of Figure 6.21.

Figure 6.21 60-Hz notch filter

Solution:

The **notch filter** shown in Figure 6.21 utilizes the parallel combination of an inductor and a capacitor to attain the desired band-rejection effect. To understand how this takes place, consider the combined impedance of the parallel LC circuit:

$$Z_\| = Z_L \parallel Z_C = \frac{j\omega L/j\omega C}{j\omega L + 1/j\omega C} = \frac{j\omega L}{1 - \omega^2 LC}$$

and note that when $\omega^2 LC = 1$, or $\omega = 1/\sqrt{LC}$, the impedance of the circuit is infinite! The frequency

$$\omega_R = \frac{1}{\sqrt{LC}}$$

is called the **resonant frequency** of the circuit. If the capacitor and the inductor were selected to have a resonant frequency of 60 Hz, then, the series circuit would show an infinite impedance to 60-Hz currents, while passing currents at other frequencies. Of course, we shall see that, because some finite resistance always exists in the circuit, some 60-Hz current will always pass through the notch filter. However, notch filters can be quite selective in rejecting one frequency component without excessively disturbing the desired signal components.

To analyze the complete circuit, we compute the frequency response of the circuit, $H_V(\omega) = \frac{V_o}{V_i}(\omega)$, where $V_o = V_L$ and $V_i = V_{60\text{ Hz}} + V_S$. $H_V(\omega)$ can be determined by applying the voltage divider rule, to obtain

$$H_V(\omega) = \frac{V_o}{V_i}(\omega) = \frac{R_L}{R_L + R_S + Z_\|} = \frac{150}{200 + j\omega L/(1 - \omega^2 LC)}$$

To obtain the desired rejection of the 60-Hz signal, we need to select appropriate component values to satisfy the relationship

$$\omega_R = \frac{1}{\sqrt{LC}} = 2\pi \times 60$$

Selecting $L = 100$ mH leads to the following value for C:

$$C = \frac{1}{L\omega_R^2} = 70.36\ \mu F$$

The frequency response of the complete circuit is plotted in Figure 6.22. Note that

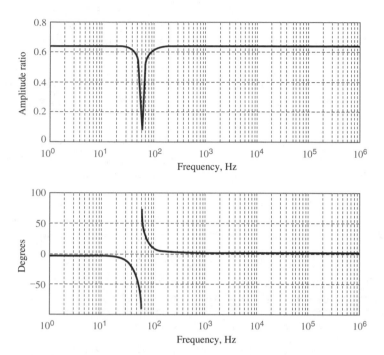

Figure 6.22 Frequency response of 60-Hz notch filter

the magnitude response takes a sharp dip around 60 Hz, and that the phase angle is equal to zero, except for a small region around 60 Hz.

Example   6.9   Displacement Transducer

This example illustrates the application of the frequency response idea to a practical displacement transducer. The frequency response of a *seismic displacement transducer* is analyzed, and it is shown that there is an analogy between the equations describing the mechanical transducer and those that describe a second-order electrical circuit.

The configuration of the transducer is shown in Figure 6.23. The transducer is housed in a case rigidly affixed to the surface of a body whose motion is to be measured. Thus, the case will experience the same displacement as the body, x_i.

Figure 6.23 Seismic displacement transducer

Inside the case, a small mass, M, rests on a spring characterized by stiffness K, placed in parallel with a damper, B. The wiper arm of a potentiometer is connected to the floating mass, M; the potentiometer is attached to the transducer case, so that the voltage V_o is proportional to the *relative displacement of the mass with respect to the case*, x_o.

The equation of motion for the mass-spring-damper system may be obtained by summing all the forces acting on the mass M:

$$Kx_o + B\frac{dx_o}{dt} = M\frac{d^2x_M}{dt^2} = M\left(\frac{d^2x_i}{dt^2} - \frac{d^2x_o}{dt^2}\right)$$

where we have noted that the motion of the mass is equal to the difference between the motion of the case and the motion of the mass relative to the case itself; that is,

$$x_M = x_i - x_o$$

If we assume that the motion of the mass is sinusoidal, we may use phasor analysis to obtain the frequency response of the transducer by defining the phasor quantities

$$\mathbf{X}_i(\omega) = |X_i|\,e^{j\phi_i} \quad \text{and} \quad \mathbf{X}_o(\omega) = |X_o|\,e^{j\phi_o}$$

The assumption of a sinusoidal motion may be justified in light of the discussion of Fourier analysis in Chapter 4. If we then recall (from Chapter 4) that taking the derivative of a phasor corresponds to multiplying the phasor by $j\omega$, we can rewrite the second-order differential equation as follows:

$$M(j\omega)^2\mathbf{X}_o(\omega) + B(j\omega)\mathbf{X}_o(\omega) + K\mathbf{X}_o(\omega) = M(j\omega)^2\mathbf{X}_i(\omega)$$

$$(-\omega^2 M + j\omega B + K)\mathbf{X}_o(\omega) = -\omega^2 M\mathbf{X}_i(\omega)$$

and we can write an expression for the frequency response:

$$\frac{\mathbf{X}_o(\omega)}{\mathbf{X}_i(\omega)} = H(\omega) = \frac{-\omega^2 M}{-\omega^2 M + j\omega B + K}$$

The frequency response of the transducer is plotted in Figure 6.24 for the component values $M = 0.005$ kg and $K = 1,000$ N/m and for three values of B:

> $B = 10$ N/ms   (dotted line)
>
> $B = 2$ N/ms   (solid line)

and

> $B = 1$ N/ms   (dashed line)

The transducer clearly displays a high-pass response, indicating that for a sufficiently high input signal frequency, the measured displacement (proportional to the voltage V_o) is equal to the input displacement, x_i, which is the desired quantity. Note how sensitive the frequency response of the transducer is to changes in damping: as B changes from 2 to 1, a sharp **resonant peak** appears around the frequency $\omega = 316$ rad/s (approximately 50 Hz). As B increases to a value of 10, the amplitude response curve shifts to the right. Thus, this transducer, with the preferred damping given by $B = 2$, would be capable of correctly measuring displacements at frequencies above a minimum value, about 1,000 rad/s (or 159 Hz). The choice of $B = 2$ as the preferred design may be explained by observing that, ideally, we would like to

$B = 1$ —— $B = 2$ —— $B = 10$ ·······

Figure 6.24 Frequency response of seismic transducer

obtain a flat magnitude frequency response curve at all frequencies. The magnitude response that most closely approximates the ideal case in Figure 6.24 corresponds to $B = 2$. This concept is commonly applied to a variety of vibration measurements.

We now illustrate how a second-order electrical circuit will exhibit the same type of response as the seismic transducer. Consider the circuit shown in Figure 6.25. The frequency response for the circuit may be obtained by using the principles developed in the preceding sections:

$$\frac{\mathbf{V}_o}{\mathbf{V}_i}(\omega) = \frac{j\omega L}{R + 1/j\omega C + j\omega L} = \frac{(j\omega L)(j\omega C)}{j\omega CR + 1 + (j\omega L)(j\omega C)}$$

$$= \frac{-\omega^2 L}{-\omega^2 L + j\omega R + 1/C}$$

Figure 6.25 Electrical circuit analog of the seismic transducer

Comparing this expression with the frequency response of the seismic transducer,

$$\frac{\mathbf{X}_o(\omega)}{\mathbf{X}_i(\omega)} = H(\omega) = \frac{-\omega^2 M}{-\omega^2 M + j\omega B + K}$$

we find that there is a definite resemblance between the two. In fact, it is possible to draw an analogy between input and output motions and input and output voltages. Note also that the mass, M, plays a role analogous to that of the inductance, L. The damper, B, acts in analogy with the resistor, R; and the spring, K, is analogous to the inverse of the capacitance, C. This analogy between the mechanical system and the electrical circuit derives simply from the fact that the equations describing the two systems have the same form. Engineers often use such analogies to construct electrical *models,* or *analogs,* of physical systems. For example, to study the behavior of a large mechanical system, it might be easier and less costly to start by modeling the mechanical system with an inexpensive electrical circuit and testing the model, rather than the full-scale mechanical system.

Decibel (dB) Plots

Frequency response plots are often displayed in the form of logarithmic plots, where the horizontal axis represents frequency on a logarithmic scale (to base 10) and the vertical axis represents the amplitude of the frequency response, in units of **decibels (dB).** In a dB plot, the ratio $|V_{out}/V_{in}|$ is given in units of decibels (dB), where

$$\left|\frac{V_{out}}{V_{in}}\right|_{dB} = 20 \log_{10} \left|\frac{V_{out}}{V_{in}}\right| \tag{6.30}$$

and this is plotted as a function of frequency on a $\log_{10}$ scale. Note that the use of decibels implies that one is measuring a *ratio.* Decibel plots are usually displayed on semilogarithmic paper, with decibels on the linear axis and frequency on the logarithmic axis.

Let us examine the appearance of dB plots for typical low-pass and high-pass filter circuits. From Figure 6.26, we can see that both plots have a

dB plot of low-pass filter

dB plot of high-pass filter

Figure 6.26 dB magnitude plots of low- and high-pass filters very simple appearance: either the low-frequency part of the plot (for a low-pass filter) or the high-frequency part (for the high-pass filter) is well approximated by a flat line, indicating that for some range of frequencies, the filter has a constant amplitude response, equal to 1. Further, the filter cutoff frequency, ω_0, appears quite clearly as the approximate frequency where the filter response starts to fall. The response of the circuit decreases (or increases) with a constant slope with respect to ω (on a logarithmic scale). For the high-pass and low-pass filters described earlier, this slope is equal to ± 20 dB/decade ($-$ for the low-pass filter, $+$ for the high-pass), where a **decade** is a range of frequencies f_1 to f_2 such as that

$$\frac{f_2}{f_1} = 10 \qquad\qquad\qquad (6.31)$$

What kind of decrease in gain is -20 dB/decade? The expression

$$\left. |H(\omega)| \right|_{dB} = -20 \text{ dB} \qquad\qquad\qquad (6.32)$$

means that

$$-20 = 20 \log_{10} |H(\omega)|$$

or   (6.33)

$$|H(\omega)| = 0.1$$

That is, the gain decreases by a factor of 10 for every increase in frequency of a factor of 10. You see how natural these units are. Further, if ω_0 is known, a plot of $|H(\omega)|_{dB}$ versus ω (on a logarithmic scale) may be readily sketched using the asymptotic approximations of two straight lines, one of slope zero and the other with slope equal to -20 dB/decade, and with intersection at ω_0. The homework problems and exercises provide a good number of practical examples of this technique.

DRILL EXERCISES

6.1   Derive an expression for $H_I(\omega) = \frac{I_L}{I_S}(\omega)$ for the circuit of Figure 6.1.

> Answer: $H_I(\omega) = \dfrac{Z_2}{Z_L + Z_2}$

6.2   Use the method of node voltages to derive $H_Y(\omega)$ for the circuit of Figure 6.1.

6.3   Use the method of mesh currents to derive $H_V(\omega)$ for the circuit of Figure 6.1.

6.4   Connect the filter of Example 6.3 to a 1-V sinusoidal source with internal resistance of 50 Ω to form a circuit similar to that of Figure 6.10. Determine the circuit cutoff frequency, ω_0.

> Answer: $\omega_0 = 2{,}026.3$ rad/s

(a)

6.5 Determine the cutoff frequency for each of the four "prototype" filters shown in Figure 6.27. Which are high-pass and which are low-pass?

> Answer: (a) $\omega_0 = \frac{1}{RC}$ (low); (b) $\omega_0 = \frac{R}{L}$ (high);
> (c) $\omega_0 = \frac{1}{RC}$ (high); (d) $\omega_0 = \frac{R}{L}$ (low)

(b)

6.6 Show that it is possible to obtain a high-pass filter response simply by substituting an inductor for the capacitor in the circuit of Figure 6.7. Derive the frequency response for the circuit.

> Answer: $H(\omega) = \dfrac{\omega L/R}{\sqrt{1 + (\omega L/R)^2}}$ $\phi(\omega) = 90\,° + \arctan \dfrac{-\omega L}{R}$

(c)

6.7 Determine the cutoff frequency for the high-pass RC filter shown in Figure 6.28. [*Hint:* Find the frequency response in the form $j\omega a/(1 + j\omega b)$ first where a and b are constants related to R_1, R_2, and C_1, and then solve numerically.] Sketch the amplitude and frequency responses.

> Answer: $\omega_0 = 35.46$ rad/s

(d)

Figure 6.27

6.8 A simple RC low-pass filter is constructed using a 10-μF capacitor and a 2.2-kΩ resistor. Over what range of frequencies will the output of the filter be within 1 percent of the input signal amplitude (i.e., when will $V_L \geq 0.99V_S$)?

> Answer: $0 \leq \omega \leq 6.48$ rad/s

6.9 Compute the frequency at which the phase shift introduced by the circuit of Example 6.3 is equal to $-10°$.

> Answer: $\omega = 375.17$ rad/s

$R_1 = 100\ \Omega$
$C = 47\ \mu F$
$R_2 = 500\ \Omega$

Figure 6.28

6.10 Compute the frequency at which the output of the circuit of Example 6.3 is attenuated by 10 percent (i.e., $V_L = 0.9V_S$).

> Answer: $\omega = 1{,}030.49$ rad/s

6.11 Compute the frequency at which the output of the circuit of Example 6.6 is attenuated by 10 percent (i.e., $V_L = 0.9V_S$).

> Answer: $\omega = 51{,}878$ rad/s

6.12 Compute the frequency at which the phase shift introduced by the circuit of Example 6.6 is equal to $20°$.

> Answer: $\omega = 69{,}032$ rad/s

6.13 Compute the frequencies ω_1 and ω_2 for the band-pass filter of Example 6.7 (with $R = 1$ kΩ) by equating the magnitude of the band-pass filter frequency response to $1/\sqrt{2}$ (this will result in a quadratic equation in ω, which can be solved for the two frequencies).

Answer: $\omega_1 = 99.95$ rad/s; $\omega_2 = 200.1$ k rad/s

6.3   TRANSIENT ANALYSIS

In analyzing the frequency response of AC circuits earlier in this chapter, we made the assumption that the particular form of the voltages and currents in the circuit was sinusoidal. For this very useful special case it was found that it was not necessary to describe the circuit by means of differential equations. Instead, by employing complex algebra and the phasor representation for sinusoids, the analysis of such circuits could be completed using the methods and techniques of Chapter 4. But how should one analyze circuits in which voltages and currents are not sinusoidal? In the case of periodic signals, it may be possible to resort to the idea of a Fourier series, which was briefly introduced in Chapter 4: a periodic signal can be expressed as the sum of a (possibly very large) number of sinusoids at various amplitudes, frequencies, and phases:

$$v(t) = \sum_{n=1}^{N} a_n \cos(\omega_n t + \phi_n) \tag{6.34}$$

In this case, sinusoidal analysis is still applicable, and the newly acquired understanding of the frequency response of a circuit may be used to obtain a complete solution by the superposition of the individual responses at each frequency.

There are many signals, however, for which the steady-state sinusoidal representation is not adequate. In particular, the Fourier method of analysis does not apply to **transient signals,** that is, voltages and currents that vary as a function of time as a consequence of a sudden change in the input. Figure 6.29 illustrates the appearance of the voltage across some hypothetical load when a DC and an AC source, respectively, are abruptly switched on at time $t = 2$ s. The waveforms in Figure 6.29 can be subdivided into three regions: a *steady-state region* for $0 \leq t \leq 0.2$ s; a *transient region,* for 0.2 s $\leq t \leq 2$ s (approximately); and a new steady-state region for $t > 2$ s, where the waveform reaches a new steady-state DC or sinusoidal condition. The objective of *transient analysis* is to describe the behavior of a voltage or current during the transition that takes place between two different steady-state conditions.

You already know how to analyze circuits in a sinusoidal steady-state by means of phasors. The material presented in the remainder of this chapter will provide the tools necessary to describe the *transient response* of circuits

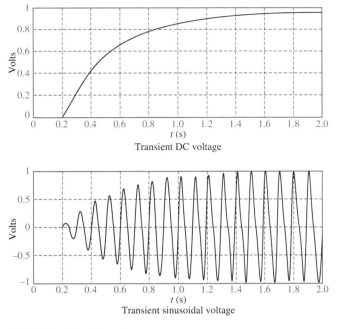

Transient DC voltage

Transient sinusoidal voltage

Figure 6.29 Examples of transient response

containing resistors, inductors, and capacitors. A general example of the type of circuit that will be discussed in this section is shown in Figure 6.30. The switch indicates that we turn the battery power on at time $t = 0$. Transient behavior may be expected whenever a source of electrical energy is switched on or off, whether it be AC or DC. A typical example of the transient response to a switched DC voltage would be what occurs when the ignition circuits in an automobile are turned on, so that a 12-V battery is suddenly connected to a large number of electrical circuits. The degree of complexity in transient analysis depends on the number of energy-storage elements in the circuit; the analysis can become quite involved for high-order circuits. In this chapter, we shall analyze only first- and second-order circuits—that is, circuits containing one or two energy-storage elements, respectively. In electrical engineering practice, we would typically resort to computer-aided analysis for higher-order circuits.

A convenient starting point in approaching the transient response of electrical circuits is to consider the general model shown in Figure 6.31, where the circuits in the box consist of a combination of resistors connected to a *single energy-storage element,* either an inductor or a capacitor. Regardless of how many resistors the circuit contains, it is a **first-order circuit.** In general, the response of a first-order circuit to a switched DC source will appear in one of the two forms shown in Figure 6.32, which represent, in order, a **decaying exponential** and a **rising exponential** waveform. In the

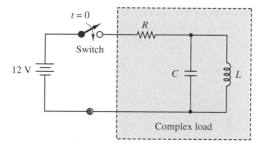

Figure 6.30 Circuit with switched DC excitation

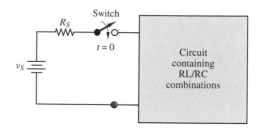

Figure 6.31 A general model of the transient analysis problem

Figure 6.32 Decaying and rising exponential responses

next sections, we will systematically analyze these responses by recognizing that they are exponential in nature and can be computed very easily once we have the proper form of the differential equation describing the circuit.

6.4   TRANSIENT RESPONSE OF FIRST-ORDER CIRCUITS

Application of the basic circuit laws—Kirchhoff's laws, Ohm's law, and the defining relationships for the ideal inductor and capacitor—to circuits containing energy-storage elements naturally leads to the formulation of differential equations, as was explained in Section 4.3. In Section 4.4, it was found that whenever the voltages and currents in such a circuit are sinusoidal, it is possible to circumvent solving the resulting circuit differential equation by employing phasor notation. In general, however, *it is not possible to employ phasor analysis when the circuit currents and voltages are not steady-state sinusoids.* Thus, in the transient analysis of electrical circuits, it

Figure 6.33 A simple *RL* circuit

will be necessary to formulate a solution to the differential circuit equations. In the following sections, we shall analyze the solution of such equations for first- and second-order circuits.

Consider the simple series *RL* circuit of Figure 6.33. Application of KVL yields the expression

$$v_S(t) = v_R(t) + v_L(t) \tag{6.35}$$

Since the defining relationships for the ideal resistor and ideal inductor are $v_R(t) = Ri(t)$ and $v_L(t) = L\, di(t)/dt$, respectively, the following differential equation results from application of KVL:

$$\frac{di(t)}{dt} + \frac{R}{L}i(t) = \frac{1}{L}v_S(t) \tag{6.36}$$

This equation governs the relationship between the source voltage and the series current, and therefore it also indirectly explains the relationship between $v_S(t)$ and the resistor and inductor voltage. Similar equations can also be obtained for other simple *RL* and *RC* circuits (see Drill Exercises 6.14 to 6.16).

Natural Response of First-Order Circuits

RC circuit: $\dfrac{dv_C}{dt} + \dfrac{1}{RC}v_C = \dfrac{1}{RC}v_S$

RL circuit: $\dfrac{di_L}{dt} + \dfrac{R}{L}i_L = \dfrac{1}{L}v_S$

Figure 6.34 Differential equations of first-order circuits

Figure 6.34 compares the *RL* circuit of Figure 6.33 with the general form of the series *RC* circuit, showing the corresponding differential equation. From Figure 6.34, it is clear that equation 6.37 is in the general form of the equation for any first-order circuit:

$$\frac{dx(t)}{dt} + sx(t) = f(t) \tag{6.37}$$

where f is the **forcing function** and $x(t)$ represents either $v_C(t)$ or $i_L(t)$. The constant s is the inverse of the parameter τ, called the **time constant** of the system: $s = 1/\tau$.

To gain some insight into the solution of this equation, consider first the **natural solution,** or **natural response,** of the equation,[2] which is obtained by setting the forcing function equal to zero. This solution, in effect, describes the response of the circuit in the absence of a source and is therefore characteristic of all *RL* and *RC* circuits, regardless of the nature of the excitation. Thus, we are interested in the solution of the equation

$$\frac{dx_N(t)}{dt} + sx_N(t) = 0 \tag{6.38}$$

or

$$\frac{dx_N(t)}{dt} = -sx_N(t) \tag{6.39}$$

---

[2] Mathematicians usually refer to the unforced solution as the **homogeneous solution.**

where the subscript N has been chosen to denote the natural solution. One can easily verify by substitution that the general form of the solution of the homogeneous equation for a first-order circuit must be exponential in nature, that is, that

$$x_N(t) = Ke^{-st} \qquad (6.40)$$

To evaluate the constant K, we need to know the **initial condition.** The initial condition is related to the energy stored in the capacitor or inductor, as will be further explained shortly. Knowing the value of the capacitor voltage or inductor current at $t = 0$ allows for the computation of the constant K, as follows:

$$x_N(t = 0) = Ke^{-0} = K = x_0 \qquad (6.41)$$

Thus, the natural solution, which depends on the initial condition of the circuit at $t = 0$, is given by the expression

$$x_N(t) = x_0 e^{-st} \qquad (6.42)$$

where, once again, $x_N(t)$ represents either the capacitor voltage or the inductor current and x_0 is the initial condition (i.e., the value of the capacitor voltage or inductor current at $t = 0$). The next section reviews some fundamental ideas related to energy storage to provide some insight into the significance of the natural solution.

Energy Storage in Capacitors and Inductors

Before delving into the complete solution of the differential equation describing the response of first-order circuits, it will be helpful to review some basic results pertaining to the response of energy-storage elements to DC sources. This knowledge will later greatly simplify the complete solution of the differential equation describing a circuit. Consider, first, a capacitor, which accumulates charge according to the relationship $Q = CV$. The charge accumulated in the capacitor leads to the storage of energy according to the following equation:

$$W_C = \frac{1}{2}Cv_C^2(t) \qquad (6.43)$$

To understand the role of stored energy, consider, as an illustration, the simple circuit of Figure 6.35, where a capacitor is shown to have been connected to a battery, V_B, for a long time. The capacitor voltage is therefore equal to the battery voltage: $v_C(t) = V_B$. The charge stored in the capacitor (and the corresponding energy) can be directly determined using equation 6.43. Suppose, next, that at $t = 0$ the capacitor is disconnected from the battery and connected to a resistor, as shown by the action of the switches in Figure 6.35. The resulting circuit would be governed by the RC differential equation described earlier (see Figure 6.34, with $v_S = 0$),

Figure 6.35 Decay through a resistor of energy stored in a capacitor

subject to the initial condition $v_C(t = 0) = V_B$. Thus, according to the results of the preceding section, the capacitor voltage would decay exponentially according to the following equation:

$$v_C(t) = V_B e^{-t/RC} \tag{6.44}$$

Physically, this exponential decay signifies that the energy stored in the capacitor at $t = 0$ is dissipated by the resistor at a rate determined by the time constant of the circuit, $\tau = RC$. Intuitively, the existence of a closed circuit path allows for the flow of a current, thus draining the capacitor of its charge. All of the energy initially stored in the capacitor is eventually dissipated by the resistor.

A very analogous reasoning process explains the behavior of an inductor. Recall that an inductor stores energy according to the expression

$$W_L = \frac{1}{2} L i_L^2(t) \tag{6.45}$$

Thus, in an inductor, energy storage is associated with the flow of a current (note the dual relationship between i_L and v_C). Consider the circuit of Figure 6.36, which is similar to that of Figure 6.35 except that the battery has been replaced with a current source, and the capacitor with an inductor. For $t < 0$, the source current, I_B, flows through the inductor, and energy is thus stored; at $t = 0$, the inductor current is equal to I_B. At this point, the current source is disconnected by means of the right-hand switch and a resistor is simultaneously connected to the inductor, to form a closed circuit. The inductor current will now continue to flow through the resistor, which dissipates the energy stored in the inductor. By the reasoning in the preceding discussion, the inductor current will decay exponentially:

$$i_L(t) = I_B e^{-tR/L} \tag{6.46}$$

That is, the inductor current will decay exponentially from its initial condition, with a time constant $\tau = L/R$. Example 6.10 further illustrates the significance of the time constant in a first-order circuit.

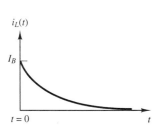

Figure 6.36 Decay through a resistor of energy stored in an inductor

Example   6.10

This example focuses on the concept of a time constant and aims to build an intuitive understanding of this very important electrical engineering concept. A certain radioactive sample has a half-life of 35 s. The half-life is defined as the time $(t_1 - t_0)$ it takes for the radioactivity to decay to one half of its initial value (its value at $t = t_0$). The radioactivity of this element is modeled by the inductor current, $i_L(t)$, in an RL circuit. We assume that the current $i_L(t)$ is directly proportional to the radioactivity of the element.

1. Find the time constant of radioactivity decay.
2. If $R = 0.1 \ \Omega$ in the circuit model, what should L be?
3. If $i(0) = I_o$, find the current $i_L(t)$ at the following times: (a) $t = \tau$; (b) $t = 2 \tau$; (c) $t = 3 \tau$; (d) $t = 4 \tau$; (e) $t = 5 \tau$

Solution:

Let the symbol γ denote the radioactivity of the sample.

1. The general form of the exponential radioactivity decay is $\gamma(t) = \gamma(0)e^{-t/\tau}$. Therefore, knowing the half-life of the radioactive sample, we can write

$$\tfrac{1}{2}\gamma(t_0) = \gamma(t_0)e^{-(t_1-t_0)/\tau}$$

and we can compute the time constant, τ, according to the equation

$$\frac{t_1 - t_0}{\tau} = \log_e\left(\frac{1}{2}\right)$$

Since $(t_1 - t_0)$ corresponds, by definition, to the half-life, which is 35 s, we can write

$$\frac{1}{\tau} = \frac{\log_e\left(\frac{1}{2}\right)}{-35}$$

or $\tau = t_{1/2}\log_e(2)$, where $t_{1/2} = $ half decay time. Thus,

$$\tau = 50.5 \text{ s}$$

2. In the circuit analogy, $\tau = L/R$, so that $L = \tau R = 5.05$ H.

t	$i_L(t)$
τ	$0.368I_0$
2τ	$0.135I_0$
3τ	$0.0498I_0$
4τ	$0.0183I_0$
5τ	$0.006I_0$

Forced and Complete Response of First-Order Circuits

In the preceding section, the natural response of a first-order circuit was found by setting the forcing function equal to zero and considering the energy initially stored in the circuit as the driving force. The **forced response,** $x_F(t)$, of the **inhomogeneous equation**

$$\frac{dx_F(t)}{dt} + \frac{1}{\tau}x_F(t) = f(t) \tag{6.47}$$

is defined as the response to a particular forcing function $f(t)$, *without regard for the initial conditions.*[3] Thus, the forced response depends exclusively on the nature of the forcing function. The distinction between natural and forced response is particularly useful because it clarifies the nature of the transient response of a first-order circuit: the voltages and currents in the circuit are

---

[3] Mathematicians call this solution the **particular solution.**

due to the superposition of two effects, the presence of *stored energy* (which can either decay or further accumulate) and the action of *external sources* (forcing functions). The natural response considers only the former, while the forced response describes the latter. The sum of these two responses forms the **complete response** of the circuit:

$$x(t) = x_N(t) + x_F(t) \tag{6.48}$$

The forced response depends, in general, on the form of the forcing function, $f(t)$. For the purpose of the present discussion, it will be assumed that $f(t)$ is a DC source switched on at $t = 0$, that is, that

$$f(t) = F \qquad t \geq 0 \tag{6.49}$$

In this case, the differential equation describing the circuit may be written as follows:

$$\frac{dx_F}{dt} = -\frac{x_F}{\tau} + F \qquad t \geq 0 \tag{6.50}$$

For the case of a DC forcing function, the form of the forced solution is also a constant. Substituting $x_F(t) = X_F = $ constant in the inhomogeneous differential equation, we obtain

$$0 = -\frac{x_F}{\tau} + F$$

or $\qquad\qquad\qquad\qquad\qquad\qquad\qquad\qquad\qquad\qquad\qquad\qquad$ (6.51)

$$x_F = \tau F$$

Thus, the complete solution of the original differential equation subject to initial condition $x(t = 0) = x_0$ and to a DC forcing function F for $t \geq 0$ is

$$x(t) = x_N(t) + x_F(t)$$

or $\qquad\qquad\qquad\qquad\qquad\qquad\qquad\qquad\qquad\qquad\qquad\qquad$ (6.52)

$$x(t) = Ke^{-t/\tau} + \tau F$$

where the constant K can be determined from the initial condition $x(t = 0) = x_0$:

$$x_0 = K + \tau F$$
$$K = x_0 - \tau F \tag{6.53}$$

Electrical engineers often classify this response as the sum of a **transient response** and a **steady-state response,** rather than a sum of a natural response and a forced response. The transient response is the response of the circuit following the switching action *before the exponential decay terms have died out;* that is, the transient response is the sum of the natural and forced responses during the transient readjustment period we have just described. The steady-state response is the response of the circuit *after all of the exponential terms have died out.* Equation 6.52 could therefore be rewritten as

$$x(t) = x_T(t) + x_{SS} \tag{6.54}$$

where

$$x_T(t) = (x_0 - \tau F)e^{-t/\tau}$$

and in the case of a DC excitation, F,   (6.55)

$$x_{SS}(t) = \tau F = X_\infty$$

Note that the transient response is not equal to the natural response, but it includes part of the forced response. The representation in equations 6.55 is particularly convenient, because it allows for solution of the differential equation that results from describing the circuit *by inspection*. The key to solving first-order circuits subject to DC transients by inspection is in considering *two separate circuits:* the circuit prior to the switching action, to determine the initial conditions, x_0; and the circuit following the switching action, to determine the time constant of the circuit, τ, and the steady-state (final) condition, X_∞. Having determined these three values, you can write the solution directly in the form of equation 6.54, and you can then evaluate it using the initial condition to determine the constant K.

To summarize, the transient behavior of a circuit can be characterized in three stages. Prior to the switching action, the circuit is in a steady-state condition (the initial condition, determined by x_0). For a period of time following the switching action, the circuit sees a transient readjustment, which is the sum of the effects of the natural response and of the forced response. Finally, after a suitably long time (which depends on the time constant of the system), the natural response decays to zero (i.e., the term $e^{-t/\tau} \to 0$ as $t \to \infty$) and the new steady-state condition of the circuit is equal to the forced response: as $t \to \infty$, $x(t) \to x_F(t)$. You may recall that this is exactly the sequence of events described in the introductory paragraphs of Section 6.3. Analysis of the circuit differential equation has formalized our understanding of the transient behavior of a circuit.

Continuity of Capacitor Voltages and Inductor Currents

As has already been stated, the primary variables employed in the analysis of circuits containing energy-storage elements are *capacitor voltages* and *inductor currents*. This choice stems from the fact that the energy-storage process in capacitors and inductors is closely related to these respective variables. The amount of charge stored in a capacitor is directly related to the voltage present across the capacitor, while the energy stored in an inductor is related to the current flowing through it. A fundamental property of inductor currents and capacitor voltages makes it easy to identify the initial condition and final value for the differential equation describing a circuit: *capacitor voltages and inductor currents cannot change instantaneously.* An instantaneous change in either of these variables would require an infinite amount of power. Since power equals energy per unit time, it follows that a truly instantaneous change in energy (i.e., a finite change in energy in zero time) would require infinite power.

Another approach to illustrating the same principle is as follows. Consider the defining equation for the capacitor:

$$i_C(t) = C\frac{dv_C(t)}{dt} \tag{6.56}$$

and assume that the capacitor voltage, $v_C(t)$, can change instantaneously, say, from 0 to V volts, as shown in Figure 6.37. The value of dv_C/dt at $t = 0$ is simply the slope of the voltage, $v_C(t)$, at $t = 0$. Since the slope is infinite at that point, because of the instantaneous transition, it would require an infinite amount of current for the voltage across a capacitor to change instantaneously. But this is equivalent to requiring an infinite amount of power,

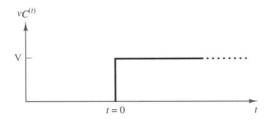

Figure 6.37 Abrupt change in capacitor voltage

since power is the product of voltage and current. A similar argument holds if we assume a "step" change in inductor current from, say, 0 to I amperes: an infinite voltage would be required to cause an instantaneous change in inductor current. This simple fact is extremely useful in determining the response of a circuit. Its immediate consequence is that *the value of an inductor current or a capacitor voltage just prior to the closing (or opening) of a switch is equal to the value just after the switch has been closed (or opened).* Formally,

$$v_C(0^+) = v_C(0^-) \tag{6.57}$$

$$i_L(0^+) = i_L(0^-) \tag{6.58}$$

where the notation 0^+ signifies "just after $t = 0$" and 0^- means "just before $t = 0$."

Example   6.11

The continuity of inductor currents can be used to determine the initial condition and the final value of the inductor current in a circuit. Consider the circuit of Figure 6.38, where the switch opens at $t = 0$, disconnecting the current source. Find the initial condition and final value of the inductor current. Assume that $I_S = 10$ mA.

Solution:

At $t = 0^-$, the current source has been connected to the circuit for a very long time. Thus, the inductor acts as a short circuit, and all of the source current, I_S, flows

Figure 6.38

through the inductor. Thus,

$$i_L(0^-) = I_S$$

Further, the voltage across the resistor must be zero, since

1. $v_L(t) = L\frac{di_L}{dt} = 0$, or
2. The current through R is zero.

At $t = 0$, the switch opens, and we can immediately state that

$$i_L(0^+) = i_L(0^-) = I_S$$

because of the principle of continuity of the inductor current. The new circuit, for $t \geq 0$, is shown in Figure 6.39, where the presence of a current through the inductor at $t = 0$ represents the initial condition for the circuit, indicating that a current was flowing through the inductor just before the switch was opened. Note that, since the inductor current cannot change instantaneously, its direction is as indicated in Figure 6.39.

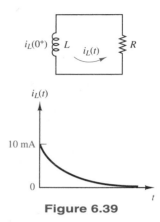

Figure 6.39

Complete Solution of First-Order Circuits

In this section, we illustrate the application of the principles put forth in the preceding sections by presenting a number of examples. The first example summarizes the complete solution of a simple *RC* circuit.

Example   6.12

This example will illustrate the complete solution of a first-order circuit. In Figure 6.40, the capacitor is assumed to have been previously charged, so that $v_C(t = 0) = 5$ V. Find an expression for the capacitor voltage. Assume that $R = 1$ kΩ and $C = 470$ μF, and let the battery voltage be $V_B = 12$ V.

Solution:

To compute all of the voltages and currents in the circuit, we need to solve for the capacitor voltage; all other circuit variables of interest can be derived from $v_C(t)$. For $t \leq 0$, the capacitor voltage is

$$v_C(t) = 5 \text{ V} \qquad t \leq 0$$

At $t = 0$, the switch is closed, resulting in the following differential equation:

$$\frac{dv_C}{dt} + \frac{1}{RC}v_C = \frac{1}{RC}V_B$$

where we recognize the following:

$$x = v_C \qquad \tau = RC \qquad F = \frac{1}{RC}V_B$$

The natural response of the circuit is of the form

$$x_N = v_{CN} = Ke^{-t/\tau} = Ke^{-t/RC}$$

Figure 6.40

while the forced response is

$$x_F = v_{CF} = \tau F = V_B$$

Thus, the complete response of the circuit is

$$x(t) = v_C(t) = Ke^{-t/RC} + V_B = Ke^{-t/RC} + 12$$

At this point, we must apply the initial condition, $v_C(0) = 5$, to determine the value of the constant K:

$$v_C(0) = K + V_B$$

or

$$K = v_C(0) - V_B = 5 - 12 = -7$$

We can then write the complete response as

$$v_C(t) = [v_C(0) - V_B]e^{-t/RC} + V_B$$
$$= -7e^{-t/0.47} + 12$$

The different components of the response are shown graphically in Figure 6.41.

Response of RC circuit to DC transient

Figure 6.41

To summarize, the following steps have led to the solution:

1. Write the differential equation for the circuit for $t \geq 0$.
2. Determine the initial condition.
3. Compute the time constant of the circuit for $t \geq 0$.
4. Write the complete solution as the sum of natural and forced responses.
5. Apply the initial condition *to the complete solution* to determine the constant K.

Example   6.13

The transient solution for a first-order circuit can be easily determined by following the three simple steps outlined in the previous section. These steps consist of: (1) determining the initial condition; (2) determining the final value; and (3) writing the solution according to equations 6.54 and 6.55. The last step requires that we recognize the appropriate time constant. Assume that a capacitor has been charged to $v_C = 3$ V by means of a separate circuit (not shown). The capacitor is then connected to a resistive circuit at $t = 0$, as shown in Figure 6.42. Find $v_C(t)$ for $t \geq 0$.

Figure 6.42

Solution:

We know immediately that $v_C(0^-) = v_C(0^+) = 3$ V. To determine the time constant of the circuit and the steady-state value of $v_C(t)$, it is helpful to compute the Thévenin equivalent for the resistive circuit for $t \geq 0$. This is shown in Figure 6.43, where:

$$V_T = \frac{100}{150} \times 12 \text{ V} = 8 \text{ V}$$

$$R_T = 50 \parallel 100 \ \Omega = 3.33 \ \Omega$$

The differential equation for the RC circuit for $t \geq 0$ can now be written as

$$\frac{dv_C}{dt} + \frac{1}{R_T C} v_C = \frac{V_T}{R_T C}$$

Recognizing that the time constant is equal to the product of the Thévenin equivalent resistance and the capacitance, we find the time constant to be

$$\tau = 0.333 \times 10^{-3} \text{ s}$$

Further, we can also see that at steady state (i.e., as $t \to \infty$), the capacitor voltage will be equal to the Thévenin voltage, since the capacitor acts as an open circuit with respect to DC currents:

$$v_C(\infty) = V_{C\infty} = V_T = 8 \text{ V}$$

Then, by direct comparison with the general form of the solution (equation 6.55), we write

$$v_C(t) = (3 - 8)e^{-t/0.333 \times 10^{-3}} + 8$$

$$= 8 - 5e^{-t/0.333 \times 10^{-3}}$$

Figure 6.43

The results of Example 6.13 indicate that even when faced with more involved circuits, we can obtain solutions of first-order circuits subject to DC transients by inspection,[4] provided that we carefully identify the Thévenin

---

[4] We have examined only DC transients for conceptual clarity. The case of switched sinusoids is treated in a few advanced homework problems for this chapter.

equivalent circuit seen by the energy-storage element. Figure 6.44 illustrates how every *RL* or *RC* circuit can effectively be reduced to a simple series connection of the energy-storage element to a Thévenin equivalent circuit (or reduced to a parallel connection to a Norton equivalent, by duality). Thus, once you have mastered the solution of the simple circuits presented so far, the extension of these ideas to more complex circuits simply requires identifying the Thévenin equivalent circuit seen by the energy-storage element *before and after* the switching action. It is important to emphasize that, in general, the two equivalent circuits seen by the energy-storage element are not the same, since the switching may disconnect or add several circuit elements. Therefore, it must be remarked, again, that the equivalent circuit prior to switching should be used to compute the initial condition, while the equivalent circuit after switching is useful in computing the steady-state value and the time constant of the circuit.

Figure 6.44 Equivalent-circuit representation of first-order circuits

To illustrate the procedure, we next reduce an *RC* circuit to a form consisting of a Thévenin equivalent circuit with a capacitor appearing as the load. The objective is to find the capacitor voltage for $t \geq 0$ for the circuit of Figure 6.45. Before the switch closes—that is, at $t < 0$—the capacitor sees the series combination of V_2 and R_2 only; since this circuit is already in Thévenin form, we recognize that at steady state $v_C(0^-) = V_2$. For $t \geq 0$, we resketch the circuit as shown in Figure 6.46. Then we compute the Thévenin equivalent, using a source transformation technique (from Chapter

Figure 6.45 A more involved *RC* circuit

Figure 6.46 The circuit of Figure 6.45 for $t \geq 0$

3), as shown in Figure 6.47, where

$$V_T = \left(\frac{V_1}{R_1} + \frac{V_2}{R_2} \right) \cdot R_T \tag{6.59}$$

and

$$R_T = R_1 \parallel R_2 \parallel R_3 \tag{6.60}$$

Figure 6.47 Reduction of the circuit of Figure 6.46 to Thévenin equivalent form

Finally, we connect the capacitor to the equivalent circuit and recognize that, as $t \to \infty$, the capacitor voltage is equal to the Thévenin voltage $V_{C\infty} = V_T$, obtaining the circuit of Figure 6.48.

To determine the time constant, we write the differential equation

$$\frac{dv_C}{dt} + \frac{1}{R_T C} v_C = V_T \tag{6.61}$$

which suggests (as should already be apparent) that $\tau = R_T C = (R_1 \parallel R_2 \parallel R_3)C$. The complete solution is given by the expression

$$v_C(t) = (V_2 - V_T)e^{-t/R_T C} + V_T \tag{6.62}$$

The procedure just shown may be followed for RL circuits as well as RC circuits to obtain a standard circuit that involves only a single equivalent source and its associated resistance.

Figure 6.48 The circuit of Figure 6.45 in equivalent form for $t \geq 0$

Example   6.14

For a numerical example that shows how the methodology of Example 6.12 also applies to circuits containing inductors, determine the inductor current for $t > 0$ if the switch opens at $t = 0$ in the circuit of Figure 6.49. Assume that $L = 10$ mH.

Figure 6.49

Solution:

Since the inductor acts as a short circuit for $t < 0$ (assuming that the switch has been closed for a long time), the equivalent circuit for $t < 0$ is the one shown in Figure 6.50, where the thick line indicates that the inductor has been replaced with a short circuit. To understand this simplification, recall that an inductor acts as a short circuit in the presence of steady-state DC currents and therefore no current can reach the circuit to the right of the inductor at $t < 0$.

Figure 6.50

On the basis of the circuit in Figure 6.50, we can compute the initial condition:

$$i_L(0^-) = \frac{5}{2} = 2.5 \text{ A}$$

At $t = 0$, the switch is opened, and we can sketch the equivalent circuit for $t \geq 0$, shown in Figure 6.51. Note that the direction of the inductor current cannot change,

Figure 6.51

according to the continuity principle, and has been indicated accordingly. Because no source is present at $t \geq 0$, the inductor current must decay to zero; thus,

$$I_{L\infty} = 0$$

To write the differential equation, we apply KVL to the equivalent circuit for $t > 0$. The circuit is redrawn in Figure 6.52 to show the polarity of the voltages:

$$v_L(t) = -v_R(t)$$

Figure 6.52

(Convince yourself of the minus sign!) Since the resistor voltage is

$$v_R(t) = R_{eq}i_L(t)$$

and since

$$v_L(t) = L\frac{di_L}{dt}$$

we obtain the equation

$$L\frac{di_L}{dt} + R_{eq}i_L(t) = 0$$

or

$$\frac{di_L}{dt} + \frac{R_{eq}}{L}i_L(t) = 0 \qquad t \geq 0$$

so that the time constant is found to be $\tau = L/R_{eq} = 10^{-3}$ s. The complete solution is

$$i_L(t) = (i_L(0) - i_{LSS})e^{-t/\tau} + i_{LSS}$$
$$= 2.5e^{-1,000t}$$

Example 6.15   Pulse Response

A problem of great practical importance is the transmission of voltage *pulses* along
cables. Short voltage pulses are used to represent the two-level binary signals that
are characteristic of digital computers; it is often necessary to transmit such voltage
pulses over a long distance through **coaxial cables,** which are characterized by a
finite resistance and by a certain capacitance, usually expressed in units of pF/m.
A simplified model of a long coaxial cable is shown in Figure 6.53; it has the
appearance of a low-pass filter. If a 100-m cable has a capacitance of 40 pF/m and
a series resistance of 0.2 Ω/m, what will the output pulse look like after traveling
the length of the cable? Assume the input pulse has a duration of 0.1 μs and has an
amplitude of 5 V. The load resistance is 150 Ω.

Figure 6.53 Pulse transmission in a coaxial cable

Solution:

The Thévenin equivalent circuit seen by the capacitor will vary, depending on whether
the pulse is "on" or "off." In the former case, the equivalent resistance consists of
the parallel combination of R_1 and R_L; in the latter case, since the switch is open, the
capacitor is only connected to R_L. Thus, the effect of the pulse is to charge C through
the parallel combination of R_1 and R_L during the "on" time ($0 \leq t < 0.1$ μs); the
capacitor will then discharge through R_L during the "off" time. This behavior is
depicted by the circuit model of Figure 6.53, in which the pulse signal is represented
by a 5 V battery in series with a switch.

The charging time constant of the coaxial cable equivalent circuit when the
switch is closed is therefore given by

$$\tau_{on} = (R_1 \parallel R_L)C = 17.65 \times 4000 \times 10^{-12} = 7.06 \times 10^{-8} \text{ s}$$

and the transient response of the cable during the "on" time is:

$$v_L(t) = 4.41(1 - e^{-t/\tau}) = 4.41(1 - e^{-1.42 \times 10^7 t}) \qquad 0 \leq t \leq 0.1 \ \mu\text{s}$$

where $V_T = \frac{R_L}{R_1 + R_L} \times 5 = 4.41$ V. At $t = 0.1$ μs we calculate the load voltage to be:

$$v_L(0.1 \; \mu s) = 4.41(1 - e^{-1.42}) = 3.35 \; V$$

For $t \geq 0.1$ μs, the output voltage will naturally decay to zero, starting from the initial condition, $v_L(0.1 \; \mu s)$, with a time constant τ_{off} equal to:

$$\tau_{off} = R_L C = 150 \times 4000 \times 10^{-12} = 0.6 \; \mu s$$

The load voltage will therefore decay according to the following expression:

$$v_L(t) = 3.35 \left(e^{-1.67 \times 10^6 (t - 0.1 \times 10^{-6})} \right) \qquad t > 0.1 \; \mu s$$

The appearance of the response is shown in Figure 6.54. It should be apparent that as the cable becomes longer, both R_1 and C increase, and therefore the output voltage will respond more slowly to the input pulse; according to the simple model of a long coaxial cable given in this example, there is a limit to the maximum distance over which a voltage pulse can be transmitted by cable.

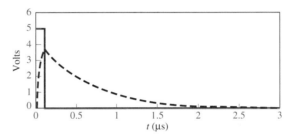

Figure 6.54 Pulse response of 100-m-long coaxial cable

The response of first-order circuits to switching transients is analogous to the behavior of other dynamical systems that are described by first-order differential equations. The following examples illustrate how the analysis methods developed so far in this chapter could also be employed to describe the dynamics of other systems. These examples further reinforce the notion that electrical circuit analysis methods are applicable to much more than just electrical circuits.

Example   6.16   Fluid Flow

Consider a tank filled with water, and with input flow rate equal to F_i. Let the output flow rate be equal to F_o. If we assume that the output flow rate is linearly related to the fluid level (or *head*), h, we can write $F_o = h/R$, where R is related to the diameter of the pipe. Figure 6.55 depicts the simple hydraulic system.

In effect, the *fluid resistance* parameter R is analogous to the resistance parameter in an electrical circuit, since the diameter of the pipe limits the flow out of the tank much as a resistor limits the flow of an electrical current. In other words, an analogy could be constructed in which the pipe acts as a resistor and the fluid flow acts as current. Similarly, the head, h, corresponds to the electric potential in a circuit: it

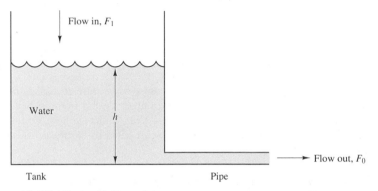

Figure 6.55 A hydraulic first-order system

enables the fluid to flow, much as the potential difference between two points in a circuit enables a current flow.

But if fluid flow behaves in a manner analogous to current (which is defined as the flow of charge), we may think of the fluid as being analogous to charge. Then, in effect, the water tank acts as a capacitor; that is, it stores "charge" (fluid). We can already realize that there is a clear analogy between variables in the fluid flow problem and the circuit problem. This analogy is formally illustrated in Table 6.1.

Table 6.1   Analogy between electrical and hydraulic circuits

Type of variable	Hydraulic variable	Corresponding electrical variable
Flow	Output flow rate, F_o	Current, i
Potential	Fluid level, h	Voltage, v
Resistance	Fluid resistance, R	Electrical resistance, R
Energy storage	Storage of fluid, A	Capacitance, C
Energy storage	Inertance, L	Inductance, L

To unravel the analogy between the fluid storage parameter, A, and the capacitance, C, let A be the cross-sectional area of the tank. Since the volume of the tank is equal to the product of the head and the cross-sectional area, we obtain the equation

$$\text{Volume} = Ah$$

which is analogous to the equation $Q = CV$ in a capacitor. If we write the mass balance equation, by equating the rate of change of fluid volume to the net difference between input and output flow, we have

$$\frac{d\,\text{volume}}{dt} = \frac{d(Ah)}{dt} = F_i - F_o = F_i - \frac{h}{R}$$

or

$$A\frac{dh}{dt} + \frac{h}{R} = F_i$$

where F_i (the input flow rate) is the forcing function. Does this equation look familiar?

Example   6.17   Heat Flow

A simplified interpretation of a heat flow problem gives rise to another interesting analogy. Consider a house as a heat-storage element: the heat capacity of the house is analogous to capacitance, in the sense that the house retains part of the heat generated by the heating system. However, a temperature difference, ΔT, between the inside and outside of the house causes heat to flow outward from the house. The insulation present in the walls, the roof, and other areas, limits the flow of heat. From this simple description, we may infer the analogies between electrical and thermal systems listed in Table 6.2.

Table 6.2   Analogy between thermal and electrical systems

Thermal	Electrical
Heat	Charge
Heat flow	Current
Temperature difference	Voltage
Ambient temperature	Ground
Heat capacity	Capacitance
Thermal resistance	Resistance
—	Inductance[a]

[a] No direct physical analogy.

One interpretation of the heat flow analogy is obtained by revisiting Newton's law of cooling (though, admittedly, in a greatly simplified form), which states that the heat flow, q, is proportional to the rate of change of temperature with respect to distance. We may approximate the derivative, to obtain a linear relationship between heat flow and change in temperature:

$$q = K\frac{dT}{dx} \approx K\frac{\Delta T}{\Delta x} = \frac{K}{\Delta x}\Delta T$$

Thus, $K/\Delta x$ plays the role of conductance in our thermal analogy, and Newton's law of cooling (although in a simplified form) turns out to be a thermal version of Ohm's law![5]

Figure 6.56

DRILL EXERCISES

6.14   Write the differential equation for the circuit shown in Figure 6.56.

Answer: $\frac{dv_C}{dt} + \frac{1}{RC}v_C = \frac{1}{RC}v_S$

[5] In reality, heat flow is more properly described by partial differential equations, since it is a distributed phenomenon. However, the approximation used here is adequate to illustrate the analog.

6.15   Write the differential equation for the circuit shown in Figure 6.57.

Answer: $\frac{dv_C}{dt} + \frac{1}{RC}v_C = \frac{1}{C}i_S$

6.16   Write the differential equation for the circuit shown in Figure 6.58.

Answer: $\frac{di_L}{dt} + \frac{R}{L}i_L = \frac{R}{L}i_S$

6.17   It is instructive to repeat the analysis of Example 6.11 for a capacitive circuit. For the circuit shown in Figure 6.59, compute the quantities $v_C(0^-)$ and $i_R(0^+)$, and sketch the response of the circuit, that is, $v_C(t)$, if the switch opens at $t = 0$.

Answer: $v_C(0^-) = 8$ V and $i_R(0^+) = 2$ A

6.18   The circuit of Figure 6.60 has a switch that can be used to connect and disconnect a battery. The switch has been open for a very long time. At $t = 0$, the switch closes, and then at $t = 50$ ms, the switch opens again. Assume that $R_1 = R_2 = 1{,}000$ Ω, $R_3 = 500$ Ω, and $C = 25\ \mu$F.

a.   Determine the capacitor voltage as a function of time.
b.   Plot the capacitor voltage from $t = 0$ to $t = 100$ ms.

Answer: $v_C = 7.5 - 7.5e^{-t/0.025}$ V, $0 \le t < 0.05$ s;
$v_C = 6.485e^{-(t-0.05)/0.0375}$ V, $t \ge 0.05$ s

6.19   If the 10-mA current source is switched on at $t = 0$ in the circuit of Figure 6.61, how long will it take for the capacitor to charge to 90 percent of its final voltage?

Figure 6.61

Answer: $t_{90\%} = 12.5\ \mu$s

6.20   Find the time constant for the circuit shown in Figure 6.62.

Answer: 545 μs

Figure 6.57

Figure 6.58

Figure 6.59

Figure 6.60

Figure 6.62

6.21 Compute the maximum amplitude of the output pulse in Example 6.15 if the length of the coaxial cable is 300 m.

Answer: 0.649 V

6.22 Repeat the calculations of Example 6.15 if the load resistance is 1,000 Ω. What is the effect of this change?

6.5   TRANSIENT RESPONSE OF SECOND-ORDER CIRCUITS

In many practical applications, understanding the behavior of first- and second-order systems is often all that is needed to describe the response of a physical system to external excitation. In this section, we discuss the solution of the second-order differential equations that characterize second-order circuits.

Deriving the Differential Equations for Second-Order Circuits

A simple way of introducing second-order circuits consists of replacing the box labeled "energy-storage element" in Figure 6.30 with a combination of two energy-storage elements, as shown in Figure 6.63. Note that two different

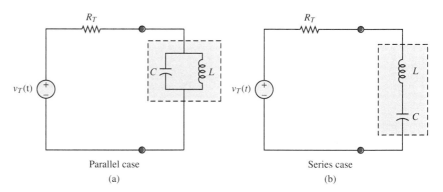

Parallel case   Series case
(a)   (b)

Figure 6.63 Second-order circuit, series case

cases are considered, depending on whether the energy-storage elements are connected in series or in parallel.

Consider the parallel case first, which has been redrawn in Figure 6.64 for clarity. Practice and experience will eventually suggest the best method for writing the circuit equations. At this point, the most sensible procedure consists of applying the basic circuit laws to the circuit of Figure 6.64. Start with KVL around the left-hand loop:

Figure 6.64 Parallel case

$$v_T(t) = R_T i_S(t) + v_C(t) \tag{6.63}$$

Then apply KCL to the top node, to obtain

$$i_S(t) = i_C(t) + i_L(t) \tag{6.64}$$

Further, KVL applied to the right-hand loop yields

$$v_C(t) = v_L(t) \tag{6.65}$$

It should be apparent that we have all the equations we need (in fact, more). Using the defining relationships for capacitor and inductor, we can express equation 6.64 as

$$i_S(t) = \frac{v_T(t) - v_C(t)}{R_T} = C\frac{dv_C}{dt} + i_L(t) \tag{6.66}$$

and equation 6.65 becomes

$$v_C(t) = L\frac{di_L}{dt} \tag{6.67}$$

Substituting equation 6.67 in equation 6.66, we can obtain a differential circuit equation in terms of the variable $i_L(t)$:

$$\frac{1}{R_T}v_T(t) - \frac{L}{R_T}\frac{di_L}{dt} = LC\frac{d^2 i_L}{dt^2} + i_L(t) \tag{6.68}$$

or

$$\frac{d^2 i_L}{dt^2} + \frac{1}{R_T C}\frac{di_L}{dt} + \frac{1}{LC}i_L = \frac{v_T(t)}{R_T LC} \tag{6.69}$$

The solution to this differential equation (which depends, as in the case of first-order circuits, on the initial conditions and on the forcing function) completely determines the behavior of the circuit. By now, two questions should have appeared in your mind:

1. Why is the differential equation expressed in terms of $i_L(t)$? (Why not $v_C(t)$?)
2. Why did we not use equation 6.63 in deriving equation 6.69?

In response to the first question, it is instructive to note that, knowing $i_L(t)$, we can certainly derive any one of the voltages and currents in the circuit. For example,

$$v_C(t) = v_L(t) = L\frac{di_L}{dt} \tag{6.70}$$

$$i_C(t) = C\frac{dv_C}{dt} = LC\frac{d^2 i_L}{dt^2} \tag{6.71}$$

To answer the second question, note that equation 6.69 is not the only form the differential circuit equation can take. By using equation 6.63 in conjunction with equation 6.64, one could obtain the following equation:

$$v_T(t) = R_T[i_C(t) + i_L(t)] + v_C(t) \tag{6.72}$$

Upon differentiating both sides of the equation and appropriately substituting from equation 6.66, the following second-order differential equation in v_C would be obtained:

$$\frac{d^2v_C}{dt^2} + \frac{1}{R_TC}\frac{dv_C}{dt} + \frac{1}{LC}v_C = \frac{1}{R_TC}\frac{dv_T(t)}{dt} \tag{6.73}$$

Note that the left-hand side of the equation is identical to equation 6.69, except that v_C has been substituted for i_L. The right-hand side, however, differs substantially from equation 6.69, because the forcing function is the derivative of the equivalent voltage.

Since all of the desired circuit variables may be obtained either as a function of i_L or as a function of v_C, the choice of the preferred differential equation depends on the specific circuit application and we conclude that there is no unique method to arrive at the final equation. As a case in point, consider the two circuits depicted in Figure 6.65. If the objective of the analysis were to determine the output voltage, v_{out}, then for the circuit in Figure 6.65(a), one would choose to write the differential equation in v_C, since $v_C = v_{out}$. In the case of Figure 6.65(b), however, the inductor current would be a better choice, since $v_{out} = R_Ti_{out}$.

(a)   (b)

Figure 6.65 Two second-order circuits

Natural Response of Second-Order Circuits

From the previous discussion, we can derive a general form for the governing equation of a second-order circuit:

$$\frac{d^2x(t)}{dt^2} + a\frac{dx(t)}{dt} + bx(t) = f(t) \tag{6.74}$$

It is now appropriate to derive a general form for the solution. The same classification used for first-order circuits is also valid for second-order circuits. Therefore, the complete solution of the second-order equation is the sum of the natural and forced responses:

$$x(t) = \underset{\text{Natural response}}{x_N(t)} + \underset{\text{Forced response}}{x_F(t)} \tag{6.75}$$

where the natural response is the solution of the homogeneous equation without regard for the forcing function (i.e., with $f(t) = 0$) and the forced response is the solution of the forced equation with no consideration of the effects of the initial conditions. Once the general form of the complete response is found, the unknown constants are evaluated subject to the initial conditions, and the solution can then be divided into transient and steady-state parts, with

$$x(t) = \underset{\text{Transient part}}{x_T(t)} + \underset{\text{Steady-state part}}{x_{SS}(t)} \tag{6.76}$$

The aim of this section is to determine the natural response, which satisfies the homogeneous equation:

$$\frac{d^2 x_N(t)}{dt^2} + b\frac{dx_N(t)}{dt} + cx_N(t) = 0 \tag{6.77}$$

Just as in the case of first-order circuits, $x_N(t)$ takes on an exponential form:

$$x_N(t) = Ke^{st} \tag{6.78}$$

where (s) is an arbitrary constant to be determined as part of the solution. This is easily verifiable by direct substitution in the differential equation:

$$s^2 Ke^{st} + bsKe^{st} + cKe^{st} = 0 \tag{6.79}$$

and since it is possible to divide both sides by Ke^{st}, the natural response of the differential equation is, in effect, determined by the solution of the quadratic equation

$$s^2 + bs + c = 0 \tag{6.80}$$

This polynomial in the variable s is called the **characteristic polynomial** of the differential equation. Thus, the natural response, $x_N(t)$, is of the form

$$x_N(t) = K_1 e^{s_1 t} + K_2 e^{s_2 t} \tag{6.81}$$

where the exponents s_1 and s_2 are found by applying the quadratic formula to the characteristic polynomial:

$$s_{1,2} = -\frac{b}{2} \pm \frac{1}{2}\sqrt{b^2 - 4c} \tag{6.82}$$

The exponential solution in terms of the exponents $s_{1,2}$ can take different forms depending on whether the roots of the quadratic equation are real or complex. As an example, consider the parallel circuit of Figure 6.64, and the governing differential equation, 6.69. The natural response for $i_L(t)$ in this case is the solution of the following equation:

$$\frac{d^2 i_{LT}}{dt^2} + \frac{1}{RC}\frac{di_{LT}}{dt} + \frac{1}{LC}i_{LT} = 0 \tag{6.83}$$

where $R = R_T$ in Figure 6.64. The solution of equation 6.83 is determined by solving the quadratic equation

$$s^2 + \frac{1}{RC}s + \frac{1}{LC} = 0 \tag{6.84}$$

The roots are

$$s_{1,2} = -\frac{1}{2RC} \pm \frac{1}{2}\sqrt{\left(\frac{1}{RC}\right)^2 - \frac{4}{LC}} \tag{6.85}$$

where

$$s_1 = -\frac{1}{2RC} + \frac{1}{2}\sqrt{\left(\frac{1}{RC}\right)^2 - \frac{4}{LC}} \tag{6.86a}$$

$$s_2 = -\frac{1}{2RC} - \frac{1}{2}\sqrt{\left(\frac{1}{RC}\right)^2 - \frac{4}{LC}} \tag{6.86b}$$

The key to interpreting this solution is to analyze the term under the square root sign; we can readily identify three cases:

- Case I:

$$\left(\frac{1}{RC}\right)^2 > \frac{4}{LC} \tag{6.87}$$

s_1 and s_2 are real and distinct roots: $s_1 = \alpha_1$ and $s_2 = \alpha_2$.

- Case II:

$$\left(\frac{1}{RC}\right)^2 = \frac{4}{LC} \tag{6.88}$$

s_1 and s_2 are real, repeated roots: $s_1 = s_2 = \alpha$.

- Case III:

$$\left(\frac{1}{RC}\right)^2 < \frac{4}{LC} \tag{6.89}$$

s_1, s_2 are complex conjugate roots: $s_1 = s_2^* = \alpha \pm j\beta$.

For each of these three cases, as we shall see, the solution of the differential equation takes a different form. The remainder of this section will explore the three different cases that can arise.

Example   6.18

Consider the circuit of Figure 6.66. Find the natural response for $i_L(t)$.

Figure 6.66

Solution:

Looking at the circuit, we can immediately see that

$$R_T = 8 \text{ k}\Omega \parallel 8 \text{ k}\Omega = 4 \text{ k}\Omega$$

The natural response for $i_L(t)$, in this case, is the solution to the homogeneous equation

$$\frac{d^2 i_L(t)}{dt^2} + \frac{1}{R_T C}\frac{d i_L(t)}{dt} + \frac{1}{LC}i_L(t) = 0$$

This equation can be determined by solving the quadratic equation

$$s^2 + \frac{1}{R_T C}s + \frac{1}{LC} = 0$$

Thus, for the given values, we obtain the following quadratic equation:

$$s^2 + \frac{1}{0.04}s + \frac{1}{0.00001} = 0$$

which has roots

$$s_{1,2} = 12.5 \pm \frac{1}{2}\sqrt{\left(\frac{1}{0.04}\right)^2 - \frac{4}{10^{-5}}}$$

with

$$s_1 = 12.5 + j315.98$$
$$s_2 = 12.5 - j315.98$$

Finally, the natural response for $i_L(t)$ is of the form

$$i_L(t) = K_1 e^{s_1 t} + K_2 e^{s_2 t}$$

or

$$i_L(t) = K_1 e^{(12.5+j315.98)t} + K_2 e^{(12.5-j315.98)t}$$

Here, K_1 and K_2 are constants that can be found from given initial conditions.

Although the previous example dealt with a specific circuit, one can extend the result by stating that the natural response of any second-order system can be described by one of the following three expressions:

- Case I. Real, distinct roots: $s_1 = \alpha_1$, $s_2 = \alpha_2$.

$$x_N(t) = K_1 e^{\alpha_1 t} + K_2 e^{\alpha_2 t} \tag{6.90}$$

- Case II. Real, repeated roots: $s_1 = s_2 = \alpha$.

$$x_N(t) = K_1 e^{\alpha t} + K_2 t e^{\alpha t} \tag{6.91}$$

- Case III. Complex conjugate roots: $s_1 = \alpha + j\beta$, $s_2 = \alpha - j\beta$.

$$x_N(t) = K_1 e^{(\alpha+j\beta)t} + K_2 e^{(\alpha-j\beta)t} \tag{6.92}$$

The solution of the homogeneous second-order differential equation will now be discussed for each of the three cases.

Overdamped Solution

The case of real and distinct roots yields the so-called **overdamped solution,** which consists of a sum of real exponentials. An overdamped system naturally decays to zero in the absence of a forcing function, according to the expression

$$x_N(t) = K_1 e^{-\alpha_1 t} + k_2 e^{-\alpha_2 t} \tag{6.93}$$

Note that α_1 and α_2 are the reciprocals of two time constants:

$$\tau_1 = \frac{1}{\alpha_1} \qquad \tau_2 = \frac{1}{\alpha_2} \tag{6.94}$$

so that the behavior of an overdamped system may be portrayed, for example, as in Figure 6.67 ($K_1 = K_2 = 1$, $\alpha_1 = 5$, and $\alpha_2 = 2$ in the figure).

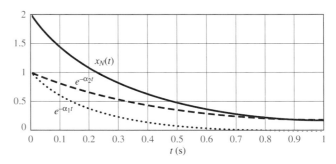

Figure 6.67 Response of overdamped second-order circuit

Critically Damped Solution

When the roots are real and repeated, the natural solution is said to be **critically damped,** and is of the form

$$x_N(t) = K_1 e^{-\alpha t} + K_2 t e^{-\alpha t} \tag{6.95}$$

The first term, $K_1 e^{-\alpha t}$, is the familiar exponential decay term. The term $K_2 t e^{-\alpha t}$, on the other hand, has a behavior that differs from a decaying exponential: for small t, the function t grows faster than $e^{-\alpha t}$ decays, so that the function initially increases, reaches a maximum at $t = 1/\alpha$, and finally decays to zero. Figure 6.68 depicts the critically damped solution for $K_1 = K_2 = 1$, $\alpha = 5$.

Underdamped Solution

A slightly more involved form of the natural response of a second-order circuit occurs when the roots of the characteristic polynomial form a complex conjugate pair, that is, $s_1 = s_2^*$. In this case, the solution is said to be **underdamped.** The solution for $x_N(t)$, then, is of the form

$$x_N(t) = K_1 e^{s_1 t} + K_2 e^{s_2 t} \tag{6.96}$$

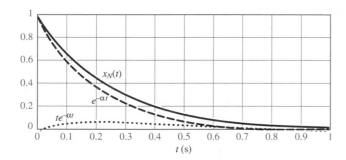

Figure 6.68 Response of critically damped second-order circuit

or

$$x_N(t) = K_1 e^{\alpha t} e^{j\beta t} + K_2 e^{\alpha t} e^{-j\beta t} \tag{6.97}$$

where $s_1 = \alpha + j\beta$ and $s_2 = \alpha - j\beta$. What is the significance of the complex exponential in the case of underdamped natural response? Recall Euler's identity, which was introduced in Chapter 4:

$$e^{j\theta} = \cos\theta + j\sin\theta \tag{6.98}$$

If we assume for the moment that $K_1 = K_2 = K$, then the natural response takes the form

$$\begin{aligned} x_N(t) &= K e^{\alpha t}(e^{j\beta t} + e^{-j\beta t}) \\ &= K e^{\alpha t}(2\cos\beta t) \end{aligned} \tag{6.99}$$

Thus, in the case of complex roots, the natural response of a second-order circuit can have oscillatory behavior! The function $2KAe^{\alpha t}\cos\beta t$ is a **damped sinusoid**; it is depicted in Figure 6.69 for $\alpha = -5$, $\beta = 50$, and $A = 0.5$. Note that, in general, K_1 and K_2 will not be equal; nonetheless, the underdamped response will still display damped sinusoidal oscillations.

Figure 6.69 Response of underdamped second-order circuit

Forced and Complete Response of Second-Order Circuits

Once we obtain the natural response using the techniques described in the preceding section, we may find the forced response using the same method employed for first-order circuits. Once again, we shall limit our analysis to a switched DC forcing function, for the sake of simplicity (the form of the forced response when the forcing function is a switched sinusoid is explored in the homework problems). The form of the forced differential equation is

$$\frac{d^2x(t)}{dt^2} + a\frac{dx(t)}{dt} + bx(t) = F \tag{6.100}$$

where F is a constant. Therefore we assume a solution of the form $x_F(t) = X_F = $ constant, and we substitute in the forced equation to find that

$$X_F = \frac{F}{b} \tag{6.101}$$

Finally, in order to compute the complete solution, we sum the natural and forced responses, to obtain

$$x(t) = x_N(t) + x_F(t) = K_1 e^{s_1 t} + K_2 e^{s_2 t} + \frac{F}{b} \tag{6.102}$$

For a second-order differential equation, we need two initial conditions to solve for the constants K_1 and K_2. These are the values of $x(t)$ at $t = 0$ and of the derivative of $x(t)$, dx/dt, at $t = 0$. To complete the solution, we therefore need to solve the two equations

$$x(t = 0) = x_0 = K_1 + K_2 + \frac{F}{b} \tag{6.103}$$

and

$$\frac{dx}{dt}(t = 0) = \dot{x}_0 = s_1 K_1 + s_2 K_2 \tag{6.104}$$

To summarize, we must follow these steps to obtain the complete solution of a second-order circuit excited by a switched DC source:

1. Write the differential equation for the circuit.
2. Find the roots of the characteristic polynomial.
3. Find the forced response.
4. Write the complete solution as the sum of natural and forced responses.
5. Apply the initial conditions *to the complete solution* to determine the constants K_1 and K_2.

Although these steps are straightforward, the successful application of this technique will require some practice, especially the determination of the initial conditions and the computation of the constants. There is no substitute for practice in gaining familiarity with these techniques! The following examples should be of help in illustrating the methods just described.

Example   6.19

Determine the complete response of the circuit shown in Figure 6.70. Assume that some energy is stored in the capacitor, so that $v_C(0^-) = 5$ V. No energy could be stored in the inductor for $t < 0$ since no current flows in the circuit until the switch is closed. The inductor current, $i(t)$, and the capacitor voltage, $v_C(t)$, are the outputs to be determined.

$R = 5000\ \Omega$ $L = 1$ H $C = 1\ \mu$F $V_S = 25$ V

Figure 6.70

Solution:

We begin by applying KVL around the circuit for $t \geq 0$:

$$v_C(t) + v_R(t) + v_L(t) = V_S$$

or

$$\frac{1}{C} \int i\, dt + Ri + L\frac{di}{dt} = V_S$$

Differentiating once yields the second-order differential equation

$$\frac{d^2 i}{dt^2} + \frac{R}{L}\frac{di}{dt} + \frac{1}{LC} i = \frac{dV_S}{dt} = 0$$

from which we determine the characteristic polynomial

$$s^2 + \frac{R}{L}s + \frac{1}{LC} = 0$$

the solution to which is given by

$$s_{1,2} = -\frac{R}{2L} \pm \frac{\sqrt{\left(\frac{R}{L}\right)^2 - \frac{4}{LC}}}{2} = -2{,}500 \pm \frac{\sqrt{5{,}000^2 - 4 \times 10^6}}{2}$$

or,

$$s_1 = -208.7, \quad s_2 = -4{,}791.3$$

which are real, distinct roots. This indicates that we have an overdamped circuit with a natural response

$$i_N(t) = K_1 e^{-208.7t} + K_2 e^{-4{,}791.3t}$$

The forced response is found to be zero, since $F = dV_S/dt = 0$ for $t > 0$. This result should not be surprising, since the capacitor will act as an open circuit at steady state and therefore no current can flow through the inductor due to the forcing function alone (however, transient current will flow). Thus, the complete solution is

$$i(t) = K_1 e^{-208.7t} + K_2 e^{-4,791.3t}$$

To evaluate the coefficients K_1 and K_2, we consider the initial conditions, $i(0^+)$ and $\frac{di}{dt}(0^+)$. The current in the inductor cannot change instantaneously, so that

$$i(0^-) = i(0^+) = 0 \text{ A}$$

and

$$i(0^-) = K_1 e^0 + K_2 e^0 = 0$$
$$0 = K_1 + K_2$$

or

$$K_1 = -K_2$$

To find $\frac{di}{dt}(0^+)$, we observe that $\frac{di}{dt}(0^+) = \frac{1}{L}v_L(0^+)$, from the defining relationship for the inductor, $v_L = L\, di/dt$. Now, the voltage across the inductor can change instantaneously, unlike the inductor current, and so we cannot state that $v_L(0^-) = v_L(0^+)$. To find $v_L(0^+)$, however, we can apply KVL once again for $t = 0^+$:

$$V_S = v_C(0^+) + v_R(0^+) + v_L(0^+)$$

But

$$v_R(0^+) = i(0^+)R = 0$$

so that

$$v_L(0^+) = 25 - v_C(0^+) = 20$$

and we conclude that

$$\frac{di}{dt}(0^+) = \frac{v_L(0^+)}{L} = 20$$

Applying this second initial condition to the solution of the differential equation at $t = 0^+$, we obtain

$$\frac{di}{dt}(0^+) = K_1(-208.7) + K_2(-4,791.3) = 20$$

Solving the two equations in K_1 and K_2 thus obtained, we find that

$$-4,791.3K_2 - 208.7K_2 = 20$$
$$K_2 = -\frac{20}{5,000} = -0.004$$

and

$$K_1 = -K_2 = 0.004$$

Thus, the complete solution is

$$i(t) = 0.004e^{-4791.3t} + 0.004e^{-208.7t}$$

Finally, to compute $v_C(t)$, we could integrate $i(t)$, but it is simpler to apply KVL once again, to obtain

$$v_C(t) = V_S - v_R(t) - v_L(t)$$
$$= 25 - Ri(t) - L\frac{di}{dt}$$

where

$$L\frac{di}{dt} = -19.2e^{-4791.3t} + 0.835e^{-208.7t}$$

and

$$Ri(t) = 20e^{-4791.3t} - 20e^{-208.7t}$$

Thus,

$$v_C(t) = 25 - 20e^{-4791.3t} + 20e^{-208.7t} + 19.2e^{-4791.3t} - 0.835e^{-208.7t}$$

$$= 25 - 0.8e^{-4791.3t} + 19.165e^{-208.7t}$$

Example 6.20

This example illustrates the solution method for a critically damped circuit. Assuming that the initial inductor current and capacitor voltage are both zero, determine $v(t)$ for $t \geq 0$ in the circuit of Figure 6.71

$$L = 2\,\text{H} \quad C = 2\,\mu\text{F} \quad R = 500\,\Omega \quad I_S = 5\,\text{A}$$

Figure 6.71

Solution:

Applying KCL for $t \geq 0$ results in the following equation:

$$i_L(t) + i_R(t) + i_C(t) = I_S$$

or

$$-\frac{1}{L}\int v(t)\,dt - \frac{v(t)}{R} - C\frac{dv(t)}{dt} = -I_S$$

Differentiating once gives us a second-order differential equation:

$$-C\frac{d^2v}{dt} - \frac{1}{R}\frac{dv}{dt} - \frac{1}{L}v = 0$$

or

$$\frac{d^2v}{dt} + \frac{1}{RC}\frac{dv}{dt} + \frac{1}{LC}v = 0$$

which results in the characteristic polynomial

$$s^2 = \frac{1}{RC}s + \frac{1}{LC} = 0$$

The solution to the quadratic equation is

$$s_{1,2} = -\frac{1}{2RC} \pm \frac{\sqrt{\left(\frac{1}{RC}\right)^2 - \frac{4}{LC}}}{2} = -500 \pm \frac{\sqrt{(10^3)^2 - 10^6}}{2} = -500 \pm 0$$

or

$$s_1 = s_2 = -500$$

Thus, we see that we have real, repeated roots. The general solution to the differential equation of this type was given by equation 6.95:

$$x(t) = K_1 e^{st} + K_2 t e^{st}$$
$$= K_1 e^{-500t} + K_2 t e^{-500t}$$

The coefficients K_1 and K_2 are determined on the basis of the initial conditions. Since the voltage across the capacitor cannot change instantaneously, we know that

$$v(0^-) = v(0^+) = 0$$

and

$$v(0) = 0 = K_1 e^0 + K_2(0)e^0$$

or

$$K_1 = 0$$

The capacitor current, however, can change instantaneously; thus, we cannot assume that $i_C(0^+) = i_C(0^-)$. By applying KCL, we can see that at $t = 0^+$,

$$i_C(0^+) + i_R(0^+) + i_L(0^+) = I_S$$

But

$$i_R(0^+) = \frac{v(0^+)}{R} = \frac{0}{R} = 0$$

so that

$$i_C(0) = -I_S$$

Since $i_C(t)$ is given by

$$i_C(t) = C\frac{dv_C}{dt}$$

we can write

$$i_C(t) = C[K_1(-500)e^{-500t} + K_2 e^{-500t} + K_2(-500)te^{-500t}]$$

$$i_C(0) = C[K_1(-500)e^0 + K_2e^0 + K_2(-500)(0)e^0] = I_S$$
$$5 = C[K_1(-500) + K_2]$$

and since we have found K_1 to be zero, we can solve for K_2 as follows:

$$K_2 = \frac{5}{C} = 2.5 \times 10^6$$

Finally, the complete solution for $v(t)$ is

$$v(t) = 2.5 \times 10^6 te^{-500t}$$

Example   6.21

The circuit shown in Figure 6.72 is identical to the band-pass filter of Example 6.7. This example illustrates the transient analysis of a second-order band-pass filter when subjected to a switched DC input. We shall consider the case where $R = 10\ \Omega$, $L = 5$ mH, and $C = 10\ \mu F$, which corresponds to an underdamped second-order circuit. Assume zero initial conditions for both energy-storage elements—that is, $v_C(0^-) = i_L(0^-) = 0$.

Figure 6.72

Solution:

To determine the load voltage, $v_L(t)$, at $t \geq 0$, we observe that the desired output is $v_L(t) = Ri_L(t)$. Thus, the preferred circuit variable in this case is the inductor current. Applying KVL for $t \geq 0$, we write

$$12 = L\frac{di_L}{dt} + v_C + Ri_L$$

Observing that

$$v_C = \frac{1}{C}\int i_L\, dt$$

we obtain the equation for $i_L(t)$:

$$12 = L\frac{di_L}{dt} + \frac{1}{C}\int i_L\, dt + Ri_L$$

which can be differentiated to obtain the second-order differential equation

$$\frac{d^2 i_L}{dt^2} + \frac{R}{L}\frac{di_L}{dt} + \frac{1}{LC}i_L = 0$$

Thus, the characteristic polynomial for the equation is given by

$$s^2 + \frac{R}{L}s + \frac{1}{LC} = 0$$

or

$$s^2 + 2{,}000s + 2 \times 10^7 = 0$$

The solution of the quadratic equation reveals that the characteristic polynomial is that of an underdamped circuit whose roots are

$$s_{1,2} = \alpha \pm j\beta = -1{,}000 \pm j4{,}359$$

Thus, the general form of the natural response is

$$i_L(t) = K_1 e^{\alpha + j\beta t} + K_2 e^{\alpha - j\beta t}$$

Since the steady-state response must be equal to zero (no current can flow through the series circuits as $t \to \infty$), the complete response is equal to the natural response. To find the constants K_1 and K_2, we must use the initial conditions for the inductor current, $i_L(0)$ and $\frac{di_L}{dt}(0)$. The first condition is easily identified:

$$i_L(0) = K_1 + K_2 = 0$$

or

$$K_1 = -K_2$$

To identify the second condition, we note that the inductor voltage is $v_L = L\,di_L/dt$. Thus, at $t = 0^+$, we can apply KVL to obtain

$$12 = L\frac{di_L}{dt}(0^+) + v_C(0^+) + Ri_L(0^+)$$

and since $v_C(0^-) = i_L(0^-) = 0$, we obtain $\frac{di_L}{dt}(0^+) = \frac{12}{L} = 2{,}400$. Now we can apply the second initial condition:

$$\frac{di_L}{dt}(0^+) = s_1 K_1 + s_2 K_2$$

$$2{,}400 = K_1[(-1{,}000 + j4{,}359) - (-1{,}000 - j4{,}359)]$$

or

$$K_1 = \frac{2{,}400}{j8{,}718} = -j0.2753$$

$$K_2 = -K_1 = j0.2753$$

The complete response is then given by

$$i_L(t) = -j0.2753e^{(-1{,}000 + j4{,}359)t} + j0.2753e^{(-1{,}000 - j4{,}359)t}$$
$$= 0.2753e^{-1{,}000t}(-je^{j4{,}359t} + je^{-j4{,}359t})$$
$$= (0.2753e^{-1{,}000t})2\sin(4{,}359t)$$

The output voltage, $v_L(t) = Ri_L(t)$, is plotted in Figure 6.73.

Figure 6.73 Transient response of band-pass filter

Example   6.22

Throughout the preceding examples, we have studied methods for computing the transient response of a second-order circuit. It is also possible, however, to determine the parameters of a circuit by studying its transient response to a known input. Suppose, for example, we wished to determine the (unknown) values of an inductor and a capacitor, by placing these elements in the circuit shown in Figure 6.74. The source voltage is $V_S = 10$ V, and at $t = 0$, the switch closes. The voltage across the known resistor ($R = 100$ Ω) is measured and recorded; the results are displayed in Figure 6.75. Is it possible to determine the values of C and L from the measured response?

Figure 6.74 A second-order circuit with unknown L and C

Figure 6.75 Response of second-order circuit with unknown parameters

Solution:

The differential equation for the series RLC circuit is:

$$\frac{d^2 i}{dt} + \frac{R}{L}\frac{di}{dt} + \frac{1}{LC}i = 0$$

with the associated characteristic polynomial

$$s^2 + \frac{R}{L}s + \frac{1}{LC} = 0$$

The roots of this quadratic equation are

$$s_{1,2} = -\frac{R}{2L} \pm \frac{\sqrt{\left(\frac{R}{L}\right)^2 - \frac{4}{LC}}}{2}$$

From the appearance of the response measured across the 100-Ω resistor, it is clear that we are faced with an underdamped circuit. Thus, the response is described by an equation of the form

$$v(t) = K_1 e^{s_1 t} + K_2 e^{s_2 t}$$

where $s_1 = \alpha + j\beta$, $s_2 = \alpha - j\beta$, and

$$\alpha = -\frac{R}{2L}$$

$$\beta = \frac{\sqrt{(R/L)^2 - 4/LC}}{2}$$

This expression can also be written as

$$v(t) = K_1 e^{\alpha t} \sin \beta t + K_2 e^{\alpha t} \cos \beta t$$
$$= e^{\alpha t} A \cos (\beta t + \phi)$$

where $e^{\alpha t}$ is the envelope of the response, β the frequency of the damped oscillations, and ϕ a phase angle resulting from combining the sine and cosine terms. From the measured resistor voltage, we see that the envelope can be characterized using the local maxima of the voltages at $t_0 = 0.2$ ms and $t_1 = 1.2$ ms:

$$v_R(t_0) = 1.24 \text{ V}$$
$$v_R(t) = 0.455 \text{ V}$$

Since an envelope can be drawn as shown in Figure 6.76, we can write the equation of the envelope using two points:

Figure 6.76

$$v_{en}(t) = A e^{\alpha(t - t_0)}$$

where A is the amplitude at the first local maximum. If we write the equation for point b, we have

$$v_{en}(t) = A e^{\alpha(t_1 - t_0)}$$

or

$$0.455 = 1.24 e^{\alpha(1.2 - 0.2) \times 10^{-3}}$$

Now we solve for α:

$$10^{-3} \times \alpha = \log_e \left(\frac{0.455}{1.25} \right)$$
$$\alpha = \frac{1}{0.001} \log_e \left(\frac{0.455}{1.24} \right)$$
$$= -1,014$$

Since

$$\alpha = -\frac{R}{2L} = -1,014$$

and

$$R = 100 \ \Omega$$

we obtain

$$L = \frac{R}{2(1,014)} = \frac{100}{2(1,014)}$$

or

$$= 0.05 \text{ H}$$

Now we consider the frequency of the damped oscillations, β. Since we have already found the local maxima, we will use these to determine the period of the sinusoid (we could also use the zero crossings, however). The period of the sinusoid is estimated to be

$$T = t_1 - t_0 = 1 \text{ ms}$$

and

$$f = \frac{1}{T} = 1{,}000 \text{ Hz}$$

Converting to rad/s, we have

$$\omega = 2\pi f$$

$$\omega = \beta = 6{,}283 \text{ rad/s}$$

Now, since

$$\beta = \frac{\sqrt{(R/L)^2 - 4/LC}}{2}$$

and we know R, L, and β, we can solve for C:

$$(2\beta)^2 = \left(\frac{R}{L}\right)^2 - \frac{4}{LC}$$

$$\frac{4}{LC} = \left(\frac{R}{L}\right)^2 - (2\beta)^2$$

$$C = \frac{4}{L}\left[\frac{1}{(R/L)^2 - (2\beta)^2}\right] = \frac{4}{0.05}\left[\frac{1}{(100/0.05)^2 - (2 \times 6{,}283)^2}\right]$$

$$= 0.494 \ \mu\text{F}$$

Thus, the measured transient response of a circuit permits the determination of unknown circuit elements.

DRILL EXERCISES

6.23 Derive the differential equation for the series circuit of Figure 6.63(b). Show that one can write the equation either as

$$\frac{d^2 v_C}{dt^2} + \frac{R_T}{L}\frac{dv_C}{dt} + \frac{1}{LC}v_C = \frac{1}{LC}v_T(t)$$

or as

$$\frac{d^2 i_L}{dt^2} + \frac{R_T}{L}\frac{di_L}{dt} + \frac{1}{LC}i_L = \frac{1}{L}\frac{dv_T(t)}{dt}$$

6.24 Determine the roots of the characteristic equation of the series RLC circuit of Figure 6.63(b) with $R = 100 \ \Omega$, $C = 10 \ \mu\text{F}$, and $L = 1 \ \text{H}$.

> Answer: $-50 \pm j312.25$

6.25 For the series RLC circuit of Figure 6.63(b), with $L = 1 \ \text{H}$, and $C = 10 \ \mu\text{F}$, find the ranges of values of R for which the circuit response is overdamped and underdamped, respectively.

> Answer: Overdamped: $R > 632.46 \ \Omega$; underdamped: $R < 632.46 \ \Omega$

6.26 For the seismic transducer of Example 6.9, show that the three cases given by $B = 10$, $B = 2$, and $B = 1$ correspond to an overdamped, critically damped, and underdamped second-order system, respectively.

6.27 For what value of R is the band-pass filter of Figure 6.17 critically damped?

> Answer: $2\sqrt{L/C}$

CONCLUSION

- In many practical applications it is important to analyze the *frequency response* of a circuit, that is, the response of the circuit to sinusoidal signals of different frequencies. This can be accomplished quite effectively by means of the phasor analysis methods developed in Chapter 4, where the radian frequency, ω, is now a variable. The frequency response of a circuit is then defined as the ratio of an output phasor quantity (voltage or current) to an input phasor quantity (voltage or current), as a function of frequency.

- One of the primary applications of frequency analysis is in the study of electrical *filters,* that is, circuits that can selectively attenuate signals in certain frequency regions. Filters can be designed, using standard resistors, inductors, and capacitors, to have one of four types of characteristics: low-pass, high-pass, band-pass, and band-reject. Such filters find widespread application in many practical engineering applications that involve signal conditioning. Filters will be studied in more depth in Chapters 10 and 11.

- Although the analysis of electrical circuits by means of phasors, that is, steady-state sinusoidal voltages and currents, is quite useful in many applications, there are situations where these methods are not appropriate. In particular, when a circuit (or other system) is subject to an abrupt change in input voltage or current, different analysis methods must be employed to determine the *transient response* of the circuit. In this chapter, we have studied the analysis methods that are required to determine the transient response of first and second order circuits (that is, circuits containing one or two energy storage elements, respectively). These methods involve identifying the differential equation that describes the circuit during the transient period and recognizing important parameters, such as the time constant of a first order circuit and the damping ratio and natural frequency of a second order circuit.

KEY TERMS

HOMEWORK PROBLEMS

6.1

a.   Determine the frequency response $\mathbf{V}_{out}(\omega)/\mathbf{V}_{in}(\omega)$ for the circuit of Figure P6.1.

b.   Plot the magnitude and phase of the circuit for frequencies between 1 and 100 rad/s on graph paper, with a linear scale for frequency.

c.   Repeat part b, using semilog paper. (Place the frequency on the logarithmic axis.)

d.   Plot the magnitude response on semilog paper with magnitude in dB.

Figure P6.1

6.2   Repeat Problem 6.1 for $C = 10 \ \mu F$ and $R = 10 \ k\Omega$.

6.3   Repeat Problem 6.1 for the circuit of Figure P6.2.

6.4   Repeat Problem 6.1 for the circuit of Figure P6.3.

Figure P6.2

Figure P6.3

Figure P6.4

6.5   Repeat Problem 6.1 for the circuit of Figure P6.4.

6.6   Repeat Problem 6.5, again using Figure P6.4, for $C = 10 \ \mu F$ and $R = 10 \ k\Omega$.

Figure P6.5

6.7

a. Determine the frequency response, $\mathbf{V}_{out}(\omega)/\mathbf{V}_{in}(\omega)$, for the circuit of Figure P6.5.

b. Plot the magnitude and phase of the circuit for frequencies between 100 krad/s and 10 Mrad/s on graph paper, with a linear scale for frequency.

c. Repeat part b, using semilog paper. (Place frequency on the logarithmic axis.)

d. Plot the magnitude response on semilog paper with magnitude in dB.

6.8 Repeat Problem 6.7 for the circuit of Figure P6.6.

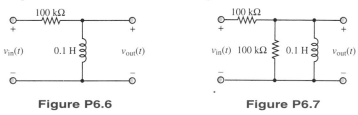

Figure P6.6   **Figure P6.7**

6.9 Repeat Problem 6.7 for the circuit of Figure P6.7.

6.10 Determine the frequency response $\mathbf{I}_L(\omega)/\mathbf{V}_S(\omega)$ for the circuit of Figure P6.8.

Figure P6.8   **Figure P6.9**

6.11 Compute and plot the frequency response (magnitude and phase) $\mathbf{V}_3(\omega)/\mathbf{V}_S(\omega)$ for the circuit of Figure P6.9.

6.12 Determine the component of the output voltage, $v_C(t)$, in the circuit of Figure P6.10 due to

$R_1 = R_2 = 1,000\ \Omega\ C = 1\ \mu F$
$V_{S1}(t) = 10\cos(500t)\ V$
$V_{S2}(t) = 10\cos(5000t)\ V$

Figure P6.10

a. $v_{S_1}(t)$

b. $v_{S_2}(t)$

c. Define the voltage attenuation as $|\mathbf{V}_C(\omega)|/|\mathbf{V}_S(\omega)|$. Find the attenuation of *each* of the sources for this circuit. Also compute the attenuation in dB.

6.13

a. Determine the frequency response, $\mathbf{V}_{out}(\omega)/\mathbf{V}_{in}(\omega)$, for the circuit of Figure 6.17 (in the text), where $R_1 = 1,000\ \Omega$, $C = 1\ \mu F$, and $L = 1\ H$.

b. What is the center frequency of this band-pass filter circuit?

c. Plot the frequency response (magnitude and phase) of this circuit on semilog paper in units of dB.

d. Define the *3-dB bandwidth* of the filter as the frequency span between points of 3-dB attenuation. Determine the 3-dB bandwidth of the filter analytically and from the plot of part c.

6.14 Repeat Problem 6.13 for $R = 10$ kΩ, $C = 100$ μF, and $L = 0.1$ H.

6.15 Repeat Problem 6.13 for $R = 500$ Ω, $C = 20$ μF, and $L = 1$ H.

6.16 Repeat Problem 6.13 for $R = 5$ Ω, $C = 20$ μF, and $L = 1$ H.

6.17

a. What is the equivalent impedance, Z_{ab}, of the filter of Figure P6.11?

b. At what frequency does the magnitude of the impedance go to infinity?

$L = 10$ μH $C = 0.1$ μF

Figure P6.11

6.18 Repeat Problem 6.17 for $C = 10$ μF and $L = 0.1$ mH.

6.19 What is the frequency response, $\mathbf{V}_{out}(\omega)/\mathbf{V}_S(\omega)$, for the circuit of Figure P6.12? Sketch the frequency response of the circuit (magnitude and phase) if $R_S = R_L = 5,000$ Ω, $L = 10$ μH, and $C = 0.1$ μF.

Figure P6.12

Figure P6.13

6.20 A phonograph driving an audio amplifier might be modeled as shown in the circuit of Figure P6.13. A "bypass" capacitor is placed between terminals 1 and 2 to reduce pickup from a nearby broadcast station, represented in the drawing as the interference voltage, $\mathbf{V}_{int}$.

a. If $\mathbf{V}_{L0}$ is the voltage at the amplifier with the switch open and $\mathbf{V}_L$ is the same voltage with the switch closed, find $\mathbf{V}_L/\mathbf{V}_{L0}$ in terms of ω if $R_S = R_L = 10^4$ Ω and $C = 10^{-9}$ F.

b. If the broadcast station's frequency is 10^7 rad/s, find the magnitude of $\mathbf{V}_L/\mathbf{V}_{L0}$.

c. Find $|\mathbf{V}_L/\mathbf{V}_{L0}|$ if the broadcast station's frequency is 10^5 rad/s.

*6.21 Many stereo speakers are two-way speaker systems; that is, they have a woofer for low-frequency sounds and a tweeter for high-frequency sounds. To get the proper separation of frequencies going to the woofer and to the tweeter, crossover circuitry is used. A crossover circuit is effectively a band-pass, high-pass, or low-pass filter. The system model is shown in Figure P6.14.

a. If $L = 2$ mH, $C = 125$ μF, and $R_S = 4$ Ω, find the load impedance as a function of frequency. At what frequency is maximum power transfer obtained?

b. Plot the magnitude and phase of the currents through the woofer and tweeter as a function of frequency.

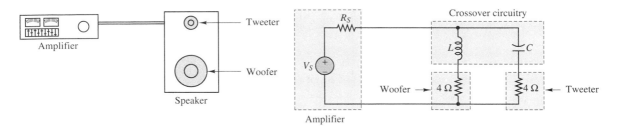

Figure P6.14

6.22 The same *LC* filter section of Problem 6.19 is used in the circuit of Figure P6.15.

$R_S = R_L = 500 \ \Omega \ ; L = 10 \ \text{mH} \ ; C = 0.1 \ \mu\text{F}$

Figure P6.15

a.   Compute the frequency response of this circuit, $\mathbf{V}_{out}(\omega)/\mathbf{V}_S(\omega)$.
b.   Plot the frequency response of the circuit.

6.23 It is very common to see interference caused by the power lines, at a frequency of 60 Hz. This problem outlines the design of a notch filter, shown in Figure P6.16, to reject a band of frequencies around 60 Hz.

Figure P6.16

a.   Write the impedance function for the filter of Figure P6.16 (the resistor r_L represents the internal resistance of a practical inductor).
b.   For what value of C will the center frequency of the filter equal 60 Hz if $L = 100$ mH and $R_L = 5 \ \Omega$?
c.   Would the "sharpness," or selectivity, of the filter increase or decrease if R_L were to increase?
d.   Assume that the filter is used to eliminate the 60-Hz noise from a signal generator with output frequency of 1 KHz. Evaluate the frequency response $\frac{\mathbf{V}_L}{\mathbf{V}_{in}} \ (\omega)$ at both frequencies if:

$$v_g(t) = \sin{(2\pi 1{,}000t)} \text{ V} \qquad r_g = 50 \ \Omega$$
$$v_n(t) = 3\sin{(2\pi 60t)} \qquad R_L = 300 \ \Omega$$

and if L and C are as in part b.

e. Plot the magnitude frequency response $\left|\frac{V_L}{V_{in}}(\omega)\right|$ in dB versus log (ω), and indicate the value of $\left.\left|\frac{V_L}{V_{in}}\right|\right|_{dB}$ at the frequencies 60 Hz and 1,000 Hz on your plot.

*6.24 The circuit of Figure P6.17 is representative of an amplifier-speaker connection. The crossover circuit (filter) is a low-pass filter that is connected to a woofer. The filter's topography is known as a π network.

a. Find the frequency response $\mathbf{V}_o(\omega)/\mathbf{V}_S(\omega)$.

b. If $C_1 = C_2 = C$, $R_S = R_L = 600 \ \Omega$, and $1/\sqrt{LC} = R/L = 1/RC = 2{,}000\pi$, plot $|\mathbf{V}_o(\omega)/\mathbf{V}_S(\omega)|$ in dB versus frequency (logarithmic scale) in the range 100 Hz $\leq f \leq$ 10,000 Hz.

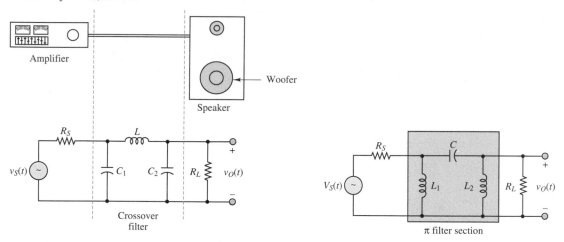

| Figure P6.17 | Figure P6.18 |

*6.25 The π filter of the circuit of Figure P6.18 is a high-pass filter that may be used to pass signals to the tweeter portion of a speaker.

a. Find the frequency response $\mathbf{V}_o(\omega)/\mathbf{V}_S(\omega)$.

b. If $L_1 = L_2 = L$, $R_S = R_L = 600 \ \Omega$, and $1/\sqrt{LC} = R/L = 1/RC = 2{,}000\pi$, plot $|\mathbf{V}_o(\omega)/\mathbf{V}_S(\omega)|$ in dB versus frequency (logarithmic scale) in the range 100 Hz $\leq f \leq$ 10,000 Hz.

*6.26 The circuit of Figure P6.19 is representative of an amplifier-speaker connection (see the left side of Figure P6.19). The crossover circuit (filter) is a high-pass filter that is connected to a tweeter. The filter's topography is known as a T network.

a. Find the frequency response $\mathbf{V}_o(\omega)/\mathbf{V}_S(\omega)$.

b. If $C_1 = C_2 = C$, $R_L = R_S = 600 \ \Omega$, and $1/\sqrt{LC} = R/L = 1/RC = 2{,}000\pi$, plot $|\mathbf{V}_o(\omega)/\mathbf{V}_S(\omega)|$ in dB versus frequency (logarithmic scale) in the range 100 Hz $\leq f \leq$ 10,000 Hz.

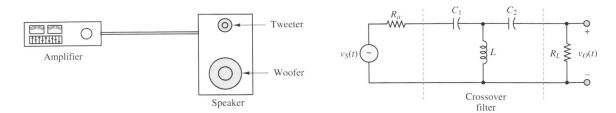

Figure P6.19

*6.27 The T filter of the circuit of Figure P6.20 is a low-pass filter that may be used to pass signals to the woofer portion of a speaker.

a. Find the frequency response $\mathbf{V}_o(\omega)/\mathbf{V}_S(\omega)$.

b. If $L_1 = L_2 = L$, $R_S = R_L = 600\ \Omega$, $1/\sqrt{LC} = R/L = 1/RC = 2{,}000\pi$, plot $|\mathbf{V}_o(\omega)/\mathbf{V}_S(\omega)|$ in dB versus frequency (logarithmic scale) in the range 100 Hz $\leq f \leq$ 10,000 Hz.

Figure P6.20

Figure P6.21

6.28 For the circuit shown in Figure P6.21, assume switch S_1 is always open and switch S_2 closes at $t = 0$.

a. Find the inductor current at $t = 0^+$.

b. Find the time constant, τ, for $t \geq 0$.

c. Find an expression for $i_L(t)$ and sketch the function.

d. Find $i_L(t)$ for each of the following values of t: 0, τ, 2τ, 5τ, 10τ.

6.29 For the circuit shown in Figure P6.21, assume that switch S_1 is always open; switch S_2 has been closed for a long time, and opens at $t = 0$.

a. Find the inductor current at $t = 0^+$.

b. Find the time constant, τ, for $t \geq 0$.

c. Find an expression for $i_L(t)$ and sketch the function.

d. Find $i_L(t)$ for each of the following values of t: 0, τ, 2τ, 5τ, 10τ.

6.30 For the circuit shown in Figure P6.21, assume switch S_1 is always open; switch S_2 has been closed for a long time and opens at $t = 0$. At $t = t_1 = 3\tau$, switch S_2 closes again.

a.   Find the inductor current at $t = 0^+$.
b.   Find the expression for $i_L(t)$ for $t > 0$ and sketch the function.

6.31   Assume both switches S_1 and S_2 in Figure P6.21 close at $t = 0$.

a.   Find the inductor current at $t = 0^+$.
b.   Find the time constant, τ, for $t \geq 0$.
c.   Find an expression for $i_L(t)$ and sketch the function.
d.   Find $i_L(t)$ for each of the following values of t: 0, τ, 2τ, 5τ, 10τ.

6.32   Assume both switches S_1 and S_2 in Figure P6.21 have been closed for a long time, and that switch S_1 opens at $t = 0^+$.

a.   Find the inductor current at $t = 0^+$.
b.   Find an expression for $i_L(t)$ and sketch the function.
c.   Find $i_L(t)$ for each of the following values of t: 0, τ, 2τ, 5τ, 10τ.

6.33   For the circuit shown in Figure P6.22, assume that switch S_1 is always open and that switch S_2 closes at $t = 0$.

Figure P6.22

a.   Find the capacitor voltage, $v_C(t)$, at $t = 0^+$.
b.   Find the time constant, τ, for $t \geq 0$.
c.   Find an expression for $v_C(t)$ and sketch the function.
d.   Find $v_C(t)$ for each of the following values of t: 0, τ, 2τ, 5τ, 10τ.

6.34   For the circuit shown in Figure P6.22, assume that switch S_1 is always open; switch S_2 has been closed for a long time, and opens at $t = 0$.

a.   Find the capacitor voltage, $v_C(t)$, at $t = 0^+$.
b.   Find the time constant, τ, for $t \geq 0$.
c.   Find an expression for $v_C(t)$ and sketch the function.
d.   Find $v_C(t)$ for each of the following values of t: 0, τ, 2τ, 5τ, 10τ.

6.35   For the circuit of Figure P6.22, assume that switch S_2 is always open, and that switch S_1 has been closed for a long time and opens at $t = 0$. At $t = t_1 = 3\tau$, switch S_1 closes again.

a.   Find the capacitor voltage, $v_C(t)$, at $t = 0^+$.

b.   Find the expression for $v_C(t)$ for $t > 0$ and sketch the function.

6.36   Assume both switches S_1 and S_2 in Figure P6.22 close at $t = 0$.

a.   Find the capacitor voltage, $v_C(t)$, at $t = 0^+$.

b.   Find the time constant, τ, for $t \geq 0$.

c.   Find an expression for $v_C(t)$ and sketch the function.

d.   Find $v_C(t)$ for each of the following values of t: 0, τ, 2τ, 5τ, 10τ.

6.37   Assume both switches S_1 and S_2 in Figure P6.22 have been closed for a long time and switch S_2 opens at $t = 0^+$.

a.   Find the capacitor voltage, $v_C(t)$, at $t = 0^+$.

b.   Find an expression for $v_C(t)$ and sketch the function.

c.   Find $v_C(t)$ for each of the following values of t: 0, τ, 2τ, 5τ, 10τ.

Figure P6.23

C_S = heat capacity of burner
R_S = heat loss of burner
    = 1.5 Ω

Figure P6.24

6.38   For the circuit of Figure P6.23, determine the time constants τ and τ' before and after the switch opens, respectively. $R_S = 4$ kΩ, $R_1 = 2$ kΩ, $R_2 = R_3 = 6$ kΩ, and $C = 1$ μF.

6.39   Use an analogy between electrical circuits and thermal conduction to analyze the behavior of a pot heating on an electric stove. We can model the heating element as shown in the circuit of Figure P6.24. Find the "heat capacity" of the burner, C_S, if the burner reaches 90 percent of the desired temperature in 10 seconds.

6.40   With a pot placed on the burner of Problem 6.39, we can model the system as shown in the circuit of Figure P6.25. Some loss between the burner and pot is modeled by the series resistance R_L. The pot is modeled by a heat-storage element, C_p, and a loss element, R_p.

Figure P6.25

a.   Find the final temperature of the water in the pot—that is, find $v(t)$ as $t \to \infty$—if $I_S = 75$ A, $C_p = 80$ F, $R_L = 0.8$ Ω, $R_p = 2.5$ Ω, and the burner is the same as in the previous problem.

b.   How long will it take for the water to reach 80 percent of its final temperature? [*Hint:* Neglect C_S, since $C_S \ll C_p$.]

*6.41   The circuit of Figure P6.26 is used as a variable delay in a burglar alarm. The alarm is a siren with internal resistance of 1 kΩ. The alarm will not sound until the current i_L exceeds 100 μA. Find the range of the variable resistor, R, for which the delay is between 1 and 2 seconds. This problem will require a graphical or numerical solution. Assume the capacitor is initially uncharged.

Figure P6.26

*6.42 Find the voltage across C_1 in the circuit of Figure P6.27 for $t > 0$. $C_1 = 5\ \mu F$ and $C_2 = 10\ \mu F$. Assume the capacitors are initially uncharged.

Figure P6.27   **Figure P6.28**

6.43 For the circuit of Figure P6.28, find the initial current through the inductor, the final current through the inductor, and the expression for $i_L(t)$ for $t \geq 0$.

6.44 At $t = 0$, the switch in the circuit of Figure P6.29 opens. At $t = 10$ s, the switch closes.

a. What is the time constant for $9 < t < 10$ s?

b. What is the time constant for $t > 10$ s?

6.45 The circuit of Figure P6.30 includes a model of a voltage-controlled switch. When the voltage across the capacitor reaches 7 V, the switch is closed. When the capacitor voltage reaches 0.5 V, the switch opens. Assume that the capacitor voltage is initially $V_C = 0.5$ V and that the switch has just opened.

a. Sketch the capacitor voltage versus time, showing explicitly the periods when the switch is open and when the switch is closed.

b. What is the period of the voltage waveform across the 10-Ω resistor?

6.46 The circuit of Figure P6.31 models the charging circuit of an electronic flash. When you turn on the flash, you must wait until a light turns on to tell you that the capacitor is completely charged.

a. If the light that tells you that the flash is ready turns on when $V_C = 0.99(7.5$ V), how long will you have to wait to take your picture?

b. If the shutter button stays closed for $\frac{1}{30}$ s, how much energy is delivered to the flashbulb (R_2)? Assume that the capacitor has completely charged.

c. If you don't wait till the flash is fully charged and you take the picture 3 seconds after the flash is turned on, how much energy is delivered to R_2?

Figure P6.29

Figure P6.30

$C = 1{,}500\ \mu F,\ R_1 = 1{,}000\ \Omega,\ R_2 = 1\ \Omega$

Figure P6.31

Figure P6.32

6.47 The circuit of Figure P6.32 has a switch that connects and disconnects a battery to and from the rest of the circuit. At $t < 0$, the switch has been open for a very long time. At $t = 0$, the switch closes, and then at $t = 50$ ms, the switch opens again.

a. Determine the capacitor voltage as a function of time (don't forget that there are three periods of time that must be considered here).

b. Plot the capacitor voltage from $t = 0$ to $t = 100$ ms.

6.48 The ideal current source in the circuit of Figure P6.33 switches between several current levels, as shown in the graph. Determine and sketch the voltage across the inductor, $v_L(t)$ for t between 0 and 2 seconds. Note that there are six different time segments to consider for this problem. When you consider each time interval, you must, in effect, switch your origin. That is, if we assume that $t_0 = 0$ for the origin for one interval and that the next interval begins at t_1, you might write the exponential term in the solution as $e^{-\frac{(t-t_1)}{\tau}}$ for the second interval. You may assume that the current source has been at zero for a very long time before $t = 0$.

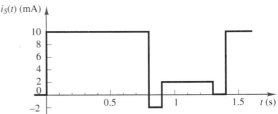

$R = 500\ \Omega,\quad L = 50\ H$

Figure P6.33

$L_1 = 1\ H\ L_2 = 5\ H$

Figure P6.34

6.49 At $t = 0$, the switch in the circuit of Figure P6.34 closes. Assume that $i_L(0) = 0$ A. For $t \geq 0$,

a. Find $i_L(t)$.

b. Find $v_{L_1}(t)$.

6.50 Repeat Drill Exercise 6.23 for the circuit of Figure 6.63(a) with $R_T = 1\ k\Omega$, $C = 10\ \mu F$, $L = 1\ H$, and $v_T = 15$ V.

6.51 Repeat Drill Exercise 6.23 for the circuit of Figure 6.63(a) with $R_T = 10\ k\Omega$, $C = 100\ \mu F$, $L = 10$ mH, and $v_T = 12$ V.

6.52 Find i for $t > 0$ in the circuit of Figure P6.35 if $i(0) = 4$ A and $v(0) = 6$ V.

6.53 Find v for $t > 0$ in the circuit of Figure P6.36 if the circuit is in steady state at $t = 0^-$.

Figure P6.35

Figure P6.36   **Figure P6.37**

6.54 Find i for $t > 0$ in the circuit of Figure P6.37 if the circuit is in steady state at $t = 0^-$.

6.55 Find i for $t > 0$ in the circuit of Figure P6.38 if the circuit is in steady state at $t = 0^-$.

Figure P6.38

6.56 Find v for $t > 0$ in the circuit of Figure P6.39 if the circuit is in steady state at $t = 0^-$.

Figure P6.39

6.57 The circuit of Figure P6.40 is in steady state at $t = 0^-$. Find v for $t > 0$ if L is (a) 2.4 H, (b) 3 H, and (c) 4 H.

6.58 Find v for $t > 0$ in the circuit of Figure P6.41 if the circuit is in steady state at $t = 0^-$.

Figure P6.40

Figure P6.41

6.59   An unknown element is placed in the circuit of Figure P6.42(a). At $t = 0$, the switch is closed. If $i(t)$ has the waveforms shown in Figures P6.42(b), state what the element, and give its value for each of the cases in the figure.

Figure P6.42(a)

Figure P6.42(b)

6.60   A simple RC circuit must be constructed with a time constant $\tau = RC_{eq} = 36.5$ ms. The circuit should have the form shown in Figure P6.43. Assume that capacitors are available in the following sizes: 10 μF, 0.5 μF, and 5 μF. Show the values in the RC circuit that will provide this time constant.

Figure P6.43

PART

II

ELECTRONICS

CHAPTER

7

Semiconductors and Diodes

This chapter introduces semiconductor-based electronic devices, and in so doing, it provides a transition between the fundamentals of electrical circuit analysis and the study of electronic circuits. Although the theme of this chapter seems somewhat different from the circuit analysis of the first six chapters, it is important to recognize that the analysis of electrical circuits is still at the core of the material. For example, we will explain the operation of diodes using load-line analysis (explored in Chapter 3), in conjunction with linear circuit models containing resistors, capacitors, and voltage and current sources. In fact, the primary emphasis in this chapter and Chapters 8 and 9 will be the use of linear circuit models for understanding and analyzing the behavior of more complex nonlinear electronic devices; we will show how it is possible to construct models of devices having nonlinear i-v characteristics by means of linear circuits. The alternative to the modeling approach would be to conduct an in-depth study of the physics of each class of device: diodes, bipolar transistors, field-effect devices, and other types of semiconductors. Such an approach is neither practical nor fruitful from the viewpoint of this book, since it would entail lengthy explanations and require a significant background in semiconductor physics. Thus, the approach here will be to first provide a qualitative understanding of the physics of each family of devices, and then to quickly turn to describing the devices in terms of their *i-v* characteristics and simple circuit models, to illustrate their analysis and applications.

The chapter starts with a discussion of semiconductors and of the *pn* junction and the semiconductor diode. The second part of the chapter is devoted to a study of diode circuit models, and numerous practical applications. By the end of Chapter 7, you should have accomplished the following objectives:

- A qualitative understanding of electrical conduction in semiconductor materials.
- The ability to explain the *i-v* characteristic of a semiconductor diode (or of a *pn* junction).
- The ability to use the ideal, offset, and piecewise linear diode models in simple circuits.
- The ability to analyze diode rectifier, peak limiter, peak detector, and regulator circuits and the behavior of LEDs and photocells.

7.1   ELECTRICAL CONDUCTION IN SEMICONDUCTOR DEVICES

This section briefly introduces the mechanism of conduction in a class of materials called **semiconductors.** Semiconductors are typically materials consisting of group IV elements and having electrical properties falling somewhere between those of conducting and of insulating materials. As an example, consider the conductivity of three common materials. Copper, a good conductor, has a conductivity of 0.59×10^6 S/cm; glass, a common insulator, may range between 10^{-16} and 10^{-13} S/cm; while silicon, a semiconductor, has a conductivity that varies from 10^{-8} to 10^{-1} S/cm. You see, then, that the name *semiconductor* is an appropriate one.

A conducting material is characterized by a large number of conduction-band electrons, which have a very weak bond with the basic structure of the material. Thus, an electric field easily imparts energy to the outer electrons in a conductor and enables the flow of electric current. In a semiconductor, on the other hand, one needs to consider the lattice structure of the material, which in this case is characterized by **covalent bonding.** Figure 7.1 depicts the lattice arrangement for silicon (Si), one of the more common semiconductors. At sufficiently high temperatures, thermal energy causes the atoms in the lattice to vibrate; when sufficient kinetic energy is present, some of the valence electrons break their bonds with the lattice structure and become available as conduction electrons. These **free electrons** enable current flow in the semiconductor. It should be noted that in a conductor valence electrons have a very loose bond with the nucleus and are therefore available for conduction to a much greater extent than valence electrons in

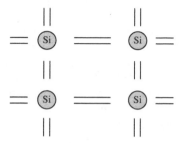

Figure 7.1 Lattice structure of silicon, with four valence electrons

a semiconductor. One important aspect of this type of conduction is that the number of charge carriers depends on the amount of thermal energy present in the structure. Thus, many semiconductor properties are a function of temperature.

The free valence electrons are not the only mechanism of conduction in a semiconductor, however. Whenever a free electron leaves the lattice structure, it creates a corresponding positive charge within the lattice. Figure 7.2 depicts the situation in which a covalent bond is missing because of the departure of a free electron from the structure. The vacancy caused by the departure of a free electron is called a **hole.** Note that wherever a hole is present, we have, in effect, a positive charge. The positive charges also contribute to the conduction process, in the sense that if a valence-band electron "jumps" to fill a neighboring hole, thereby neutralizing a positive charge, it correspondingly creates a new hole at a different location. Thus, the effect is equivalent of that of a positive charge moving to the right, in the sketch of Figure 7.2. This phenomenon becomes relevant when an external electric field is applied to the material. It is important to point out here that the **mobility**—that is, the ease with which charge carriers move across the lattice—differs greatly for the two types of carriers. Free electrons can move

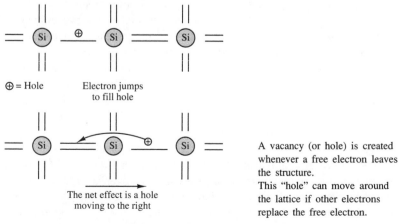

⊕ = Hole   Electron jumps
                 to fill hole

The net effect is a hole
moving to the right

A vacancy (or hole) is created whenever a free electron leaves the structure.
This "hole" can move around the lattice if other electrons replace the free electron.

Figure 7.2 Free electrons and "holes" in the lattice structure far more easily around the lattice than holes. To appreciate this, consider the fact that a free electron has already broken the covalent bond, whereas for a hole to travel through the structure, an electron must overcome the covalent bond each time the hole jumps to a new position.

According to this relatively simplified view of semiconductor materials, we can envision a semiconductor as having two types of charge carriers—holes and free electrons—which travel in opposite directions when the semiconductor is subject to an external electric field, giving rise to a net flow of current in the direction of the electric field. Figure 7.3 illustrates the concept.

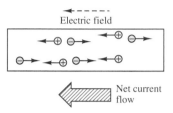

An external electric field forces holes to migrate to the left and free electrons to the right. The net current flow is to the left.

Figure 7.3 Current flow in a semiconductor

An additional phenomenon, called **recombination,** reduces the number of charge carriers in a semiconductor. Occasionally, a free electron traveling in the immediate neighborhood of a hole will recombine with the hole, to form a covalent bond. Whenever this phenomenon takes place, two charge carriers are lost. However, in spite of recombination, the net balance is such that a number of free electrons always exist at a given temperature. These electrons are therefore available for conduction. The number of free electrons available for a given material is called the **intrinsic concentration,** n_i. For example, at room temperature, silicon has

$$n_i = 1.5 \times 10^{16} \text{electrons/m}^3 \tag{7.1}$$

Note that there must be an equivalent number of holes present as well.

Semiconductor technology rarely employs pure, or intrinsic, semiconductors. To control the number of charge carriers in a semiconductor, the process of **doping** is usually employed. Doping consists of adding impurities to the crystalline structure of the semiconductor. The amount of these impurities is controlled, and the impurities can be one of two types. If the dopant is an element from the fifth column of the periodic table (e.g., arsenic), the end result is that wherever an impurity is present, an additional free electron is available for conduction. Figure 7.4 illustrates the concept. The elements providing the impurities are called **donors** in the case of group V elements, since they "donate" an additional free electron to the lattice structure. An equivalent situation arises when group III elements (e.g., indium) are used to dope silicon. In this case, however, an additional hole is created by the doping element, which is called an **acceptor,** since it accepts a free electron from the structure and generates a hole in doing so.

An additional free electron is created when Si is "doped" with a group V element.

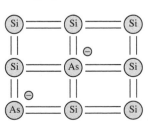

Figure 7.4 Doped semiconductor

Semiconductors doped with donor elements conduct current predominantly by means of free electrons and are therefore called **_n_-type semiconductors.** When an acceptor element is used as the dopant, holes constitute the most common carrier, and the resulting semiconductor is said to be a **_p_-type semiconductor.** Doping usually takes place at such levels that the concentration of carriers due to the dopant is significantly greater than the intrinsic concentration of the original semiconductor. If n is the total number of free electrons and p that of holes, then in an n-type doped semiconductor, we have

$$n \gg n_i \qquad\qquad\qquad\qquad\qquad (7.2)$$

and

$$p \ll p_i \qquad\qquad\qquad\qquad\qquad (7.3)$$

Thus, free electrons are the **majority carriers** in an n-type material, while holes are the **minority carriers.** In a p-type material, the majority and minority carriers are reversed.

Doping is a standard practice for a number of reasons. Among these are the ability to control the concentration of charge carriers, and the increase in the conductivity of the material that results from doping.

7.2   THE _pn_ JUNCTION AND THE SEMICONDUCTOR DIODE

A simple section of semiconductor material does not in and of itself possess properties that make it useful for the construction of electronic circuits. However, when a section of p-type material and a section of n-type material are brought in contact to form a **_pn_ junction,** a number of interesting properties arise. The pn junction forms the basis of the **semiconductor diode,** a widely used circuit element.

Figure 7.5 depicts an idealized pn junction, where on the p side, we see a dominance of positive charge carriers, or holes, and on the n side, the free electrons dominate. Now, in the neighborhood of the junction, in a small section called the **depletion region,** the mobile charge carriers (holes

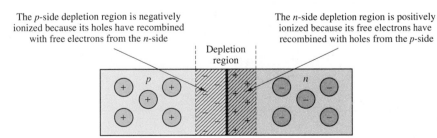

Figure 7.5 A pn junction

and free electrons) come into contact with each other and recombine, thus leaving virtually no charge carriers at the junction. What is left in the depletion region, in the absence of the charge carriers, is the lattice structure of the n-type material on the right, and of the p-type material on the left. But the n-type material, deprived of the free electrons, which have recombined with holes in the neighborhood of the junction, is now positively ionized. Similarly, the p-type material at the junction is negatively ionized, because holes have been lost to recombination. The net effect is that, while most of the material (p- or n-type) is charge-neutral because the lattice structure and the charge carriers neutralize each other (on average), the depletion region sees a separation of charge, giving rise to an electric field pointing from the n side to the p side. The charge separation therefore causes a **contact potential** to exist at the junction. This potential is typically on the order of a few tenths of a volt and depends on the material (about 0.6 to 0.7 V for silicon). The contact potential is also called the *offset voltage, V_γ*.

In effect, then, if one were to connect the two terminals of the pn junction to each other, to form a closed circuit, two currents would be present. First, a small current, called **reverse saturation current, I_0,** exists because of the presence of the contact potential and the associated electric field. In addition, it also happens that holes and free electrons with sufficient thermal energy can cross the junction. This current across the junction flows opposite to the reverse saturation current and is called **diffusion current, I_d.** Of course, if a hole from the p side enters the n side, it is quite likely that it will quickly recombine with one of the n-type carriers on the n side. One way to explain diffusion current is to visualize the diffusion of a gas in a room: gas molecules naturally tend to diffuse from a region of higher concentration to one of lower concentration. Similarly, the p-type material—for example—has a much greater concentration of holes than the n-type material. Thus, some holes will tend to diffuse into the n-type material across the junction, although only those that have sufficient (thermal) energy to do so will succeed. Figure 7.6 illustrates this process.

The phenomena of drift and diffusion help explain how a pn junction behaves when it is connected to an external energy source. Consider the diagrams of Figure 7.7, where a battery has been connected to a pn junction

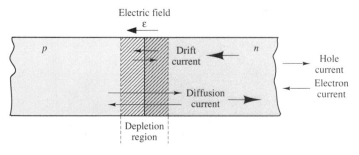

Figure 7.6 Drift and diffusion currents in a pn junction

in the **reverse-biased** direction (Figure 7.7(a)), and in the **forward-biased** direction (Figure 7.7(b)). We assume that some suitable form of contact between the battery wires and the semiconductor material can be established (this is called an **ohmic contact**). The effect of a reverse bias is to increase the contact potential at the junction. Now, the majority carriers trying to diffuse across the junction need to overcome a greater barrier (a larger potential) and a wider depletion region. Thus, the diffusion current becomes negligible. The only current that flows under reverse bias is the very small reverse saturation current, so that the diode current, i_D (defined in the figure), is

$$i_D = -I_0 \qquad (7.4)$$

When the *pn* junction is forward-biased, the contact potential across the junction is lowered (note that V_B acts in opposition to the contact potential). Now, the diffusion of majority carriers is aided by the external voltage source; in fact, the diffusion current increases as a function of the applied voltage, according to the expression

$$I_d = I_0 e^{q v_D / kT} \qquad (7.5)$$

where V_D is the voltage across the *pn* junction, $k = 1.381 \times 10^{-23} J/K$ is Boltzmann's constant, q the charge of one electron, and T the temperature of the material in kelvins (K). Thus, the net diode current under forward bias is given by the expression

$$\boxed{i_D = I_d - I_0 = I_0(e^{q v_D / kT} - 1)} \qquad \text{Diode equation} \qquad (7.6)$$

which is known as the **diode equation.** Figure 7.8 depicts the diode *i-v* characteristic described by the diode equation for a fairly typical silicon diode for positive diode voltages. Since the reverse saturation current, I_0 is typically very small (10^{-9} to 10^{-15} A), the expression

$$i_D = I_0 e^{q v_D / kT} \qquad (7.7)$$

is a good approximation if the diode voltage, v_D, is greater than a few tenths of a volt.

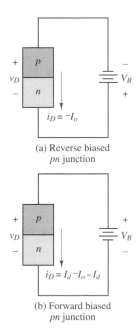

(a) Reverse biased
pn junction

(b) Forward biased
pn junction

Figure 7.7 Forward- and reversed-biased *pn* junctions

Figure 7.8 Semiconductor diode *i-v* characteristic

The arrow in the circuit symbol for the diode indicates the direction of current flow when the diode is forward-biased.

Circuit symbol

Ohmic contacts

p-n junction

Figure 7.9 Semiconductor diode circuit symbol

The ability of the *pn* junction to essentially conduct current in only one direction—that is, to conduct only when the junction is forward-biased—makes it valuable in circuit applications. A device having a single *pn* junction and ohmic contacts at its terminals, as described in the preceding paragraphs, is called a *semiconductor diode,* or simply *diode.* As will be shown later in this chapter, it finds use in many practical circuits. The circuit symbol for the diode is shown in Figure 7.9, alongside with a sketch of the *pn* junction.

Figure 7.10 summarizes the behavior of the semiconductor diode in terms of its *i-v* characteristic; it will become apparent later that this *i-v* characteristic plays an important role in constructing circuit models for the diode. Note that a third region appears in the diode *i-v* curve that has not been discussed yet. The **reverse-breakdown** region to the far left of the curve represents the behavior of the diode when a sufficiently high reverse bias is applied. Under such a large reverse bias (greater in magnitude than the voltage V_Z), the diode conducts current again, this time *in the reverse direction.* To explain the mechanism of reverse conduction, one needs to visualize the phenomenon of *avalanche breakdown.* When a very large negative bias is applied to the *pn* junction, sufficient energy is imparted to charge carriers that reverse current can flow, well beyond the normal reverse saturation current. In addition, because of the large electric field, electrons are energized to such levels that if they collide with other charge carriers at a lower energy level, some of their energy is transferred to the carriers with lower energy, and these can now contribute to the reverse conduction process, as well. This process is called *impact ionization.* Now, these new carriers may also have enough energy to energize other low-energy electrons by impact ionization, so that once a sufficiently high reverse bias is provided, this process of conduction takes place very much like an avalanche: a single electron can ionize several others.

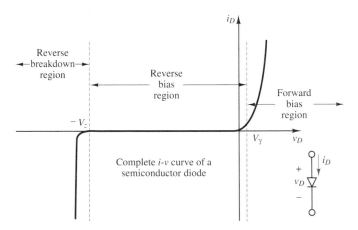

Figure 7.10 *i-v* characteristic of semiconductor diode

The phenomenon of **Zener breakdown** is related to avalanche breakdown. It is usually achieved by means of heavily doped regions in the neighborhood of the metal-semiconductor junction (the ohmic contact). The high density of charge carriers provides the means for a substantial reverse breakdown current to be sustained, at a nearly constant reverse bias, the **Zener voltage,** V_Z. This phenomenon is very useful in applications where one would like to hold some load voltage constant—for example, in **voltage regulators,** which are discussed in a later section.

To summarize the behavior of the semiconductor diode, it is useful to refer to the sketch of Figure 7.10, observing that when the voltage across the diode, v_D, is greater than the offset voltage, V_γ, the diode is said to be forward-biased and acts nearly as a short circuit, readily conducting current. When v_D is between V_γ and the Zener breakdown voltage, $-V_Z$, the diode acts very much like an open circuit, conducting a small reverse current, I_0, of the order of only nanoamperes (nA). Finally, if the voltage v_D is more negative than the Zener voltage, $-V_Z$, the diode conducts again, this time in the reverse direction.

7.3 CIRCUIT MODELS FOR THE SEMICONDUCTOR DIODE

From the viewpoint of a *user* of electronic circuits (as opposed to a *designer*), it is often sufficient to characterize a device in terms of its *i-v* characteristic, using either load-line analysis or appropriate circuit models to determine the operating currents and voltages. This section will show how it is possible to use the *i-v* characteristics of the semiconductor diode to construct simple yet useful *circuit models*. Depending on the desired level of detail, it is possible to construct *large-signal models* of the diode, which describe the gross behavior of the device in the presence of relatively large voltages and currents, or *small-signal models,* which are capable of describing the behavior of the diode in finer detail and, in particular, the response of the diode to small changes in the average diode voltage and current. From the user's standpoint, these circuit models greatly simplify the analysis of diode circuits and make it possible to effectively analyze relatively "difficult" circuits simply by using the familiar circuit analysis tools of Chapter 3. The first two major divisions of this section will describe different diode models and the assumptions under which they are obtained, to provide the knowledge you will need to select and use the appropriate model for a given application.

Large-Signal Diode Models

Ideal Diode Model

Our first large-signal model treats the diode as a simple on-off device (much like a check valve in hydraulic circuits). Figure 7.11 illustrates how, on a

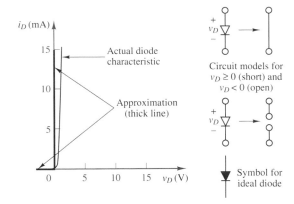

Figure 7.11 Large-signal on-off diode model

large scale, the *i-v* characteristic of a typical diode may be approximated by an open circuit when $v_D < 0$ and by a short circuit when $v_D \geq 0$ (recall the *i-v* curves of the ideal short and open circuits presented in Chapter 2). The analysis of a circuit containing a diode may be greatly simplified by using the short-circuit–open-circuit model. From here on, this diode model will be known as the **ideal diode model.** In spite of its simplicity, the ideal diode model (indicated by the symbol shown in Figure 7.11) can be very useful in analyzing diode circuits. Consider the circuit shown in Figure 7.12, which contains a 1.5-V battery, an ideal diode, and a 1-kΩ resistor. A technique will now be developed to determine whether the diode is conducting or not, with the aid of the ideal diode model.

Assume first that the diode is conducting (or, equivalently, that $v_D \geq 0$). This enables us to substitute a short circuit in place of the diode, as shown in Figure 7.13. Since the diode is now represented by a short circuit, $v_D = 0$; this is consistent with the initial assumption (i.e., diode "on"), since the diode is assumed to conduct for $v_D \geq 0$ and since $v_D = 0$ does not contradict the assumption. The series current in the circuit (and through the diode) is $i_D = 1.5/1,000 = 1.5$ mA. To summarize, the assumption that the diode is on in the circuit of Figure 7.13 allows us to assume a positive (clockwise) current in the circuit. Since the direction of the current and the diode voltage are consistent with the assumption that the diode is on ($v_D \geq 0$, $i_D > 0$), it must be concluded that the diode is conducting.

Suppose, now, that the diode had been assumed to be off. In this case, the diode would be represented by an open circuit, as shown in Figure 7.14. Applying KVL to the circuit of Figure 7.14 reveals that the voltage v_D must equal the battery voltage, or $v_D = 1.5$ V, since the diode is assumed to be an open circuit and no current flows through the circuit:

$$1.5 = v_D + 1,000i_D = v_D \tag{7.8}$$

But the result $v_D = 1.5$ V is contrary to the initial assumption (i.e., $v_D < 0$). Thus, assuming that the diode is an open circuit leads to an inconsistent

Figure 7.12 Circuit containing ideal diode

Figure 7.13 Circuit of Figure 7.12, assuming that the ideal diode conducts

Figure 7.14 Circuit of Figure 7.12, assuming that the ideal diode does not conduct

answer. Clearly, the assumption must be incorrect, and therefore the diode must be conducting.

This method may seem somewhat trivial, because of the simplicity of the circuit used in the illustration, but it can be very useful in more involved circuits, where it is not quite so obvious whether a diode is seeing a positive or a negative bias. The method is particularly effective in these cases, since one can make an educated guess whether the diode is on or off and solve the resulting circuit to verify the correctness of the initial assumption. Some solved examples are perhaps the best way to illustrate the concept.

Example   7.1

The method just illustrated in this section will now be applied to a slightly more complicated circuit, shown in Figure 7.15. Determine whether the ideal diode in the circuit is conducting.

Figure 7.15

Solution:

To determine the condition of the ideal diode, let us initially assume that it does not conduct, and let us replace the diode with an open circuit, as shown in Figure 7.16.

Figure 7.16

Since no current flows in the mesh on the right-hand side of the circuit of Figure 7.16, we may compute the voltage across the (shunt) 10-Ω resistor to be 8 V, by the voltage divider rule. Then, applying KVL around the right-hand mesh, we obtain:

$$8 = v_D + 11$$

or

$$v_D = -3$$

This result is consistent with the initial assumption (diode not conducting, or $v_D <$ 0). Therefore, the diode does not conduct.

To verify the result just obtained, we may reverse the original assumption and presume that the diode is conducting. Figure 7.17 depicts the appearance of the circuit

Figure 7.17

with the diode replaced by a short circuit. If this assumption were correct, we should be able to verify that the mesh current i_2 (equal to the diode current) is positive. With a short circuit in place of the diode, this is a simple two-mesh problem:

$$(10 + 5)i_1 + (-10)i_2 = 12$$
$$(-10)i_1 + (10 + 10)i_2 = -11$$

which can be solved for i_2 by using Cramer's rule (see Appendix A):

$$i_2 = \frac{\det \begin{bmatrix} 15 & 12 \\ -10 & -11 \end{bmatrix}}{\det \begin{bmatrix} 15 & -10 \\ -10 & 20 \end{bmatrix}} = -0.225 \text{ A}$$

or by substituting one equation in the other, as preferred. In any case, the resulting current is $i_2 = -0.225$ A. This result corresponds to stating that a reverse current is flowing through the diode, which is not possible. Thus, the assumption that the diode is on was incorrect, and the diode cannot conduct.

Example 7.2

Determine whether the ideal diode in the circuit of Figure 7.18 is conducting.

Figure 7.18

Solution:

Assume that the diode is off, and replace it with an open circuit, as shown in Figure 7.19. Then compute the current flowing in the single loop that remains:

$$i = \frac{(12 - 11)}{(5 + 4)} = \frac{1}{9} \text{ A}$$

Thus, the potential at the "node" between resistors is $12 - \frac{5}{9} = 11 + \frac{4}{9}$ V. This corresponds to a strong reverse bias for the diode, since the negative terminal of the diode is well above ground. Thus, the diode cannot conduct, and the assumption is correct.

Figure 7.19

One of the important applications of the semiconductor diode is **rectification** of AC signals, that is, the ability to convert an AC signal with zero average (DC) value to a signal with a nonzero DC value. The application of the semiconductor diode as a rectifier is very useful in obtaining DC voltage supplies from the readily available AC line voltage. Here, we illustrate the basic principle of rectification, using an ideal diode—for simplicity, and also because the large-signal' model is appropriate when the diode is used in applications involving large AC voltage and current levels.

Consider the circuit of Figure 7.20, where an AC source, $v_i = 155.56 \cdot \sin \omega t$, is connected to a load by means of a series ideal diode. From the analysis of Example 7.1, it should be apparent that the diode will conduct only during the positive half-cycle of the sinusoidal voltage—that is, that the condition $v_D \geq 0$ will be satisfied only when the AC source voltage is positive—and that it will act as an open circuit during the negative half-cycle of the sinusoid ($v_D < 0$). Thus, the appearance of the load voltage will be as shown in Figure 7.21, with the negative portion of the sinusoidal waveform cut off. The rectified waveform clearly has a nonzero DC (average) voltage,

Figure 7.20

Figure 7.21 Ideal diode rectifier input and output voltages

whereas the average input waveform voltage was zero. When the diode is conducting, or $v_D \geq 0$, the unknowns v_L and i_D can be found by using the following equations:

$$i_D = \frac{v_i}{R_L} \qquad \text{when} \qquad v_i > 0 \tag{7.9}$$

and

$$v_L = i_D R_L \tag{7.10}$$

The load voltage, v_L, and the input voltage, v_i are sketched in Figure 7.21. From the previous equation, it is obvious that the current waveform has the same shape as the load voltage. The average value of the load voltage is obtained by integrating the load voltage over one period and dividing by the period, 2π:

$$v_{\text{load,DC}} = \frac{1}{2\pi} \int_0^\pi 155.56 \sin \omega t \, dt = \frac{155.56}{\pi} = 49.52 \text{ V} \tag{7.11}$$

The circuit of Figure 7.20 is called a **half-wave rectifier,** since it preserves only half of the waveform. This is not usually a very efficient way of rectifying an AC signal, since half the energy in the AC signal is not recovered. It will be shown in a later section that it is possible to recover also the negative half of the AC waveform by means of a **full-wave rectifier.**

Offset Diode Model

While the ideal diode model is useful in approximating the large-scale characteristics of a physical diode, it does not account for the presence of an offset voltage, which is an unavoidable component in semiconductor diodes (recall the discussion of the contact potential in Section 7.2). The **offset diode model** consists of an ideal diode in series with a battery of strength equal to the offset voltage (we shall use the value $V_\gamma = 0.6$ V for silicon diodes, unless otherwise indicated). The effect of the battery is to shift the characteristic of the ideal diode to the right on the voltage axis, as shown in Figure 7.22. This model is a better approximation of the large-signal behavior of a semiconductor diode than the ideal diode model.

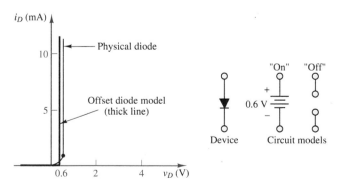

Figure 7.22 Offset diode model

According to the offset diode model, the diode of Figure 7.22 acts as an open circuit for $v_D < 0.6$ V, and it behaves like a 0.6-V battery for $v_D \geq 0.6$ V. The equations describing the offset diode model are as follows:

$$v_D \geq 0.6 \text{ V} \qquad \text{Diode} \rightarrow \text{0.6-V battery}$$
$$v_D < 0.6 \text{ V} \qquad \text{Diode} \rightarrow \text{Open circuit}$$

(7.12)

The diode offset model may be represented by an ideal diode in series with a 0.6-V ideal battery, as shown in Figure 7.23. Use of the offset diode model is best described by means of examples.

Figure 7.23 Offset diode as an extension of ideal diode model

Example   7.3

In this example, we illustrate the effect of the offset voltage on a diode half-wave rectifier using the offset diode model. Consider the circuit of Figure 7.24, where a sinusoidal source of peak amplitude 3 V is exciting a series-connected diode-resistor circuit. Next to the circuit we show its equivalent representation when the diode is replaced by the equivalent circuit of the offset diode model. Note that, in the equivalent circuit, we need only determine whether the voltage v_D, across the ideal diode, is positive or negative to compute a value for i. In other words, once the circuit model has replaced the diode in the circuit, the problem resembles an ideal diode circuit problem, except for the presence of an additional voltage source, representing the diode offset voltage.

Actual circuit   Circuit with offset diode model

Figure 7.24

Assume that the diode is initially off. Then the circuit will appear as shown in Figure 7.25(a). In this case, since no current flows around the circuit, the voltage drop across the diode must be

$$v_D = v_S - 0.6 \text{ V}$$

and for our answer to be consistent with the assumption that the diode is off, the voltage v_D must be negative:

$$v_D < 0 \qquad \text{or} \qquad v_S < 0.6 \text{ V}$$

Thus, whenever the source voltage is less than the offset voltage (0.6 V), the diode does not conduct, and no current flows in the circuit. If, on the other hand, the source voltage is greater than 0.6 V, the diode conducts, and the circuit of Figure 7.25(b) applies. In this case, the current flowing in the circuit is

$$i = \frac{v_S - 0.6}{R}$$

(a) Diode off

(b) Diode on

Figure 7.25

The load voltage, v_R, must then be given by

$$v_R = iR = v_S - 0.6$$

The quantities v_S and v_R are sketched in Figure 7.26.

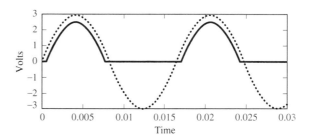

Figure 7.26 Source voltage and rectified voltage for the circuit of Figure 7.24

A glance at Figure 7.26 reveals that, in addition to rectifying the sinusoidal waveform, the offset diode model predicts a downward shift of the load waveform, caused by the offset voltage. This shift is particularly visible in this case, because the excitation signal amplitude is of the same order of magnitude as the offset voltage. If the source voltage were, say, of the order of hundreds of volts, the 0.6-V shift would hardly be noticeable.

Example   7.4

Use the offset diode model to determine the value of v_1 for which the diode D_1 will first conduct in the circuit of Figure 7.27.

Figure 7.27

Solution:

The circuit of Figure 7.27 is shown in Figure 7.28 with the diode replaced by the circuit model. It is safe to assume that for v_1 zero or negative, the diode is off. Thus, a reasonable approach to determine when the diode will turn on is to assume that the diode is off and to write the corresponding circuit equations. Then, it will be possible to solve for the values of v_1 that will cause v_{D_1} to become greater than zero. Note that assuming the diode to be initially off simplifies the problem, since no current flows in the circuit. Applying KVL to each of the loops with this assumption, we obtain

$$v_1 = v_{D_1} + 0.6 + 2$$

and

$$v_0 = 2$$

According to the first equation, then,

$$v_{D_1} = v_1 - 2.6$$

and the condition for the diode to conduct is

$$v_1 > 2.6 \text{ V}$$

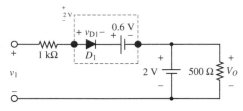

Figure 7.28

Small-Signal Diode Models

As one examines the diode *i-v* characteristic more closely, it becomes apparent that the short-circuit approximation is not adequate to represent the *small-signal behavior* of the diode. The term *small-signal behavior* usually signifies the response of the diode to small time-varying signals that may be superimposed on the average diode current and voltage. Figure 7.8 was a close-up view of a silicon diode *i-v* curve. From this figure, it should be apparent that the short-circuit approximation is not very accurate when a diode's behavior is viewed on an expanded scale. To a first-order approximation, however, the *i-v* characteristic resembles that of a resistor (i.e., is linear) for voltages greater than the offset voltage. Thus, it may be reasonable to model the diode as a resistor (instead of a short circuit) *once it is conducting,* to account for the slope of its *i-v* curve. In the following discussion, the method of load-line analysis (which was introduced in Chapter 3) will be exploited to determine the **small-signal resistance** of a diode.

Consider the circuit of Figure 7.29, which represents the Thévenin equivalent circuit of an arbitrary linear resistive circuit connected to a diode. The equations describing the operation of the circuit are

$$v_T = i_D R_T + v_D \tag{7.13}$$

which arises from application of KVL; and the diode equation,

$$i_D = I_0(e^{qv_D/KT} - 1) \tag{7.14}$$

Figure 7.29 Diode circuit for illustration of load-line analysis

Although we thus have two equations in two unknowns, these cannot be solved analytically, since one of the equations contains v_D in exponential form. As discussed in Chapter 3, two methods exist for the solution of *transcendental equations* of this type: graphical and numerical. In the present case, only the graphical solution shall be considered. The graphical solution is best understood if we associate a curve in the i_D-v_D plane with each of the two preceding equations. The diode equation gives rise to the familiar curve of Figure 7.8. The *load-line equation,* obtained by KVL, is the equation of a line with slope $-1/R$ and ordinate intercept given by V_T/R_T.

$$i_D = -\frac{1}{R_T}v_D + \frac{1}{R_T}V_T \qquad \text{Load line equation} \qquad (7.15)$$

The superposition of these two curves gives rise to the plot of Figure 7.30, where the solution to the two equations is graphically found to be the pair of values (I_Q, V_Q). The intersection of the two curves is called the **quiescent (operating) point,** or **Q point.** The voltage $v_D = V_Q$ and the current $i_D = I_Q$ are the actual diode voltage and current when the diode is connected as in the circuit of Figure 7.29. Note that this method is also useful for circuits containing a larger number of elements, provided that we can represent these circuits by their Thévenin equivalents, with the diode appearing as the load.

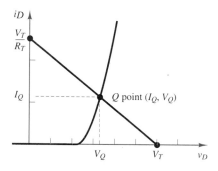

Figure 7.30 Graphical solution of equations 7.13 and 7.14

Example   7.5

In this example, load-line analysis is employed to find the diode current and voltage in order to compute the total power output of the battery in the circuit of Figure 7.31. The diode used is an MV122, which has the small-signal *i-v* characteristics shown in Figure 7.32.

Figure 7.31

Figure 7.32 MV122 diode *i-v* curve

Solution:

First, the Thévenin equivalent circuit as seen by the diode is computed:

$$R_T = 20 + 20 + (10 \parallel 50) = 48.33 \ \Omega$$

and

$$V_T = \frac{10}{60}(12) = 2 \ V$$

The Thévenin equivalent circuit is now shown in Figure 7.33.

Figure 7.33

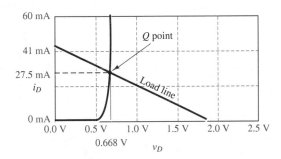

Figure 7.34 Superposition of load line and diode *i-v* curve

Next, the load line is plotted, superimposed on the *i-v* characteristic curve; the combined curves are shown in Figure 7.34. The line intercepts the v_D axis at 2.0 V and the i_D axis at 41 mA. We see from the sketch that the load line intersects the diode curve at approximately 0.67 V and 27.5 mA. The voltage across the 10-Ω resistor is

$$V_{10 \ \Omega} = 40 I_Q + V_Q = 1.77 \ V$$

Thus, the current through the 10-Ω resistor is 0.177 A. The total current out of the source is therefore computed to be

$$I_B = 0.177 \ A + 0.0275 \ A = 0.2046 \ A$$

and therefore the total power supplied to the circuit by the battery is

$$P_B = 12 \times 0.2046 = 2.455 \ W$$

Piecewise Linear Diode Model

The graphical solution of diode circuits can be somewhat tedious, and its accuracy is limited by the resolution of the graph; it does, however, provide insight into the **piecewise linear diode model.** In the piecewise linear model, the diode is treated as an open circuit in its off state, and as a linear resistor in series with V_γ in the on state. Figure 7.35 illustrates the graphical appearance of this model. Note that the straight line that approximates the "on" part of the diode characteristic is tangent to the Q point. Thus, in the neighborhood of the Q point, the diode does act as a linear small-signal resistance, with slope given by $\frac{1}{r_D}$, where

$$\frac{1}{r_D} = \left. \frac{\partial i_D}{\partial v_D} \right|_{(I_Q, V_Q)} \tag{7.16}$$

That is, it acts as a linear resistance whose *i-v* characteristic is the tangent to the diode curve at the operating point. The tangent is extended to meet the voltage axis, thus defining the intersection as the diode offset voltage. Thus, rather than represent the diode by a short circuit in its forward-biased state, we treat it as a linear resistor, with resistance r_D. The piecewise linear model offers the convenience of a linear representation once the state of the diode is established, and of a more accurate model than either the ideal or the offset diode model. This model is very useful in illustrating the performance of diodes in real-world applications.

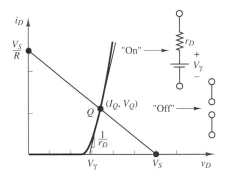

Figure 7.35 Piecewise linear diode model

Example　7.6

In this example, we compute the incremental (small-signal) resistance of a diode at a given operating point. Assume that a certain diode has $I_0 = 10^{-14}$ A and $kT/q = \alpha = 0.025$ V at room temperature. Use the approximate exponential diode equation to compute the incremental resistance of the diode at the operating point defined by $I_Q = 50$ mA.

Solution:

The approximate exponential diode equation is given by

$$i_D = I_0 e^{v_D/\alpha}$$

while the incremental resistance can be found from the expression

$$\frac{1}{r_D} = \left. \frac{\partial i_D}{\partial v_D} \right|_{(I_Q, V_Q)}$$

Differentiating the expression for i_D with respect to v_D, we can obtain

$$\frac{1}{r_D} = \frac{I_0}{\alpha}e^{v_D/\alpha}$$

To determine the operating voltage of the diode, we may solve the diode equation with $i_D = 50$ mA:

$$\log_e\left(\frac{0.05}{10^{-14}}\right) = \frac{v_D}{0.025}$$

which can be evaluated to obtain

$$v_D = 0.731 \text{ V}$$

so that

$$\frac{1}{r_D} = \frac{10^{-14}}{0.025}e^{0.731/0.025} = 2.00$$

or

$$r_D = 0.5 \ \Omega$$

Example   7.7

In this example, we use the piecewise linear model of Figure 7.35 to analyze a slightly more realistic rectifier circuit, shown in Figure 7.36. Determine the rectified load voltage, v_L, if the source voltage is $v_S(t) = 10\cos(\omega t)$.

Solution:

Figure 7.37 depicts the rectifier circuit with the diode replaced by its piecewise linear model. Note that the circuit of Figure 7.37 contains an ideal diode with linear resistors

Figure 7.36

Figure 7.37

and voltage sources, and we can therefore determine the state of the diode using the same method explored earlier for the ideal and offset diode models. In particular, if KVL is applied to the circuit, the following expression is obtained:

$$v_S = v_1 + v_2 + v_D + 0.6 + v_L$$

or

$$v_D = v_S - v_1 - v_2 - 0.6 - v_L$$

To determine the state of the diode, we reason as follows. Since, for negative voltages (i.e., during the negative half-cycle of the sine wave), the diode is off, it will be sufficient to determine the point at which the diode first turns on. Assume, therefore, that the diode is off (corresponding to the negative half-cycle of the source voltage); then, no current flows in the series circuit, and the voltages v_1, v_2, and v_L are all zero. Thus, when the diode is not conducting, the following expression holds:

$$v_D = v_S - 0.6$$

and the condition for the ideal diode to conduct is

$$v_D \geq 0 \quad \text{or} \quad v_S \geq 0.6 \text{ V}$$

Thus, the diode will conduct when $v_S = 0.6$ V. Once the diode conducts, the expression for the load voltage can be easily obtained by the voltage divider rule, considering that the ideal diode behaves like a short circuit. On the other hand, the load voltage is zero for $v_S < 0.6$ V, since the diode does not conduct and no current reaches the load. The complete expression for the load voltage is therefore given by:

$$v_L = \frac{10}{10 + 1 + 0.5} (v_S - 0.6) = 8.7 \cos(\omega t) - 0.52 \quad v_S \geq 0.6 \text{ V}$$

$$v_L = 0 \quad v_S < 0.6 \text{ V}$$

The source and load voltages are plotted in Figure 7.38. Note that, because of the combined effect of the offset voltage and the voltage divider, the peak voltage of the rectified waveform is only 8.18 V.

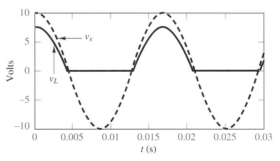

Source voltage and rectified load voltage

Figure 7.38

DRILL EXERCISES

7.1 Repeat Example 7.2, assuming that the diode is conducting, and show that this assumption leads to an inconsistent result.

7.2 Compute the DC value of the rectified waveform for the circuit of Figure 7.20 for $v_i = 52 \cos \omega t$ V.

Answer: 16.55 V

7.3   Use load-line analysis to determine the operating point (Q point) of the diode in the circuit of Figure 7.39. The diode has the characteristic curve shown in Figure 7.32.

> Answer: $V_Q = 0.65$ V; $I_Q = 37$ mA

Figure 7.39

7.4   Compute the incremental resistance of the diode of Example 7.6 if the current through the diode is 250 mA.

> Answer: 0.1 Ω

7.5   Consider a half-wave rectifier similar to that of Figure 7.20, with $v_S(t) = 18 \cos(t)$, and a 4-Ω load resistor. Sketch the output waveform if the piecewise linear diode model is used to describe the operation of the diode, with $r_D = 1$ Ω. What is the peak value of the rectifier waveform?

> Answer: 13.92 V

7.6   Determine which of the diodes in the circuit of Figure 7.40 conducts. Each diode has an offset voltage of 0.6 V.

> Answer: Both diodes conduct.

Figure 7.40

7.7   Determine which of the diodes in Figure 7.41 conduct for the following voltages (in V): (a) $v_1 = 0$, $v_2 = 0$; (b) $v_1 = 5$, $v_2 = 5$; (c) $v_1 = 0$, $v_2 = 5$; (d) $v_1 = 5$, $v_2 = 0$. Treat the diodes as ideal.

Figure 7.41

> Answer: (a) Neither conducts. (b) Both conduct.
>   (c) Only D_2 conducts. (d) Only D_1 conducts.

7.4   PRACTICAL DIODE CIRCUITS

This section illustrates some of the applications of diodes to practical engineering circuits. The nonlinear behavior of diodes, especially the rectification property, makes these devices valuable in a number of applications. In this section, more advanced rectifier circuits (the full-wave rectifier and the bridge rectifier) will be explored, as well as *limiter* and *peak detector* circuits. These circuits will be analyzed by making use of the circuit models developed in the preceding sections; as stated earlier, these models are more than adequate to develop an understanding of the operation of diode circuits.

In addition to the operation of diodes as rectifiers and limiters, there is another useful class of applications that takes advantage of the reverse-breakdown characteristic of the semiconductor diode discussed in the opening section. The phenomenon of Zener breakdown is exploited in a class of devices called **Zener diodes,** which enjoy the property of a sharp reverse-bias breakdown with relatively constant breakdown voltage. These devices are used as voltage regulators, that is, to provide a nearly constant output (DC) voltage from a voltage source whose output might ordinarily fluctuate substantially (for example, a rectified sinusoid).

The Full-Wave Rectifier

The half-wave rectifier discussed earlier is one simple method of converting AC energy to DC energy. The need for converting one form of electrical energy into the other arises frequently in practice. The most readily available form of electric power is AC (the standard 110- or 220-V rms AC line power), but one frequently needs a DC power supply, for applications ranging from the control of certain types of electric motors to the operation of electronic circuits such as those that will be discussed in Chapters 8 through 13.

The half-wave rectifier, however, is not a very efficient AC-DC conversion circuit, because it fails to utilize half the energy available in the AC waveform, by not conducting current during the negative half-cycle of the AC waveform. The full-wave rectifier shown in Figure 7.42 offers a substantial improvement in efficiency over the half-wave rectifier. The first section of the full-wave rectifier circuit includes an AC source and a center-tapped transformer (see Chapter 5) with $1 : 2N$ turns ratio. The purpose of the transformer is to obtain the desired voltage amplitude prior to rectification. Thus, if the peak amplitude of the AC source voltage is v_S, the amplitude of the voltage across each half of the output side of the transformer will be Nv_S; this scheme permits scaling the source voltage up or down (depending on whether N is greater or less than 1), according to the specific requirements of the application. In addition to scaling the source voltage, the transformer also isolates the rectifier circuit from the AC source voltage, since there is no direct electrical connection between the input and output of a transformer.

In the analysis of the full-wave rectifier, the diodes will be treated as ideal, since in most cases the source voltage is the AC line voltage (110 V rms, 60 HZ) and therefore the offset voltage is negligible in comparison. The key to the operation of the full-wave rectifier is to note that during the positive half-cycle of v_S, the top diode is forward-biased while the bottom diode is reverse-biased; therefore, the load current during the positive

Figure 7.42 Full-wave rectifier

half-cycle is

$$i_L = i_1 = \frac{Nv_S}{R_L} \qquad v_S \geq 0 \qquad\qquad (7.17)$$

while during the negative half-cycle, the bottom diode conducts and the top diode is off, and the load current is given by

$$i_L = i_2 = \frac{Nv_S}{R_L} \qquad v_S < 0 \qquad\qquad (7.18)$$

Note that the direction of i_L is always positive, because of the manner of connecting the diodes (when the top diode is off, i_2 is forced to flow from $+$ to $-$ across R_L).

The source voltage, the load voltage, and the currents i_1 and i_2 are shown in Figure 7.43 for a load resistance $R_L = 1\ \Omega$ and $N = 1$. The full-wave rectifier results in a twofold improvement in efficiency over the half-wave rectifier introduced earlier.

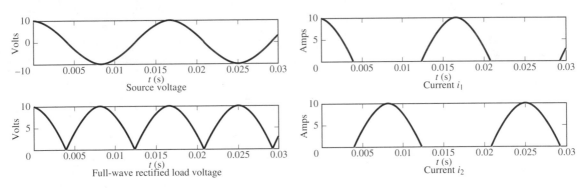

Figure 7.43 Full-wave rectifier current and voltage waveforms ($R_L = 1\ \Omega$)

The Bridge Rectifier

Another rectifier circuit commonly available "off the shelf" as a single *integrated circuit package*[1] is the *bridge rectifier,* which employs four diodes in a bridge configuration, similar to the Wheatstone bridge already explored in Chapter 3. Figure 7.44 depicts the bridge rectifier, along with the associated integrated circuit (IC) package.

The analysis of the bridge rectifier is simple to understand by visualizing the operation of the rectifier for the two half-cycles of the AC waveform separately. The key is that, as illustrated in Figure 7.45, diodes D_1 and D_3 conduct during the positive half-cycle, while diodes D_2 and D_4 conduct during the negative half-cycle. Because of the structure of the bridge, the

Bridge rectifier

Corresponding IC package

Figure 7.44 Full-wave bridge rectifier

---

[1] An integrated circuit is a collection of electronic devices interconnected on a single silicon chip.

During the positive half-cycle of $v_S(t)$, D_1 and D_3 are forward-biased and $i_L = v_S(t)/R_L$ (ideal diodes). During the negative half-cycle of $v_S(t)$, D_2 and D_4 are forward-biased and $i_L = v_S(t)/R_L$ (ideal diodes).

Figure 7.45 Diodes conduct

flow of current through the load resistor is in the same direction (from c to d) during both halves of the cycle; hence, the full-wave rectification of the waveform. The original and rectified waveforms are shown in Figure 7.46(a) for the case of ideal diodes and a 30-V peak AC source. Figure 7.46(b) depicts the rectified waveform if we assume diodes with a 0.6-V offset voltage. Note that the waveform of Figure 7.46(b) is not a pure rectified sinusoid any longer: the effect of the offset voltage is to shift the waveform downward by twice the offset voltage. This is most easily understood by considering that the load seen by the source during either half-cycle consists of two diodes in series with the load resistor.

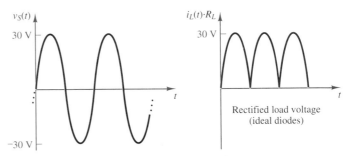

Figure 7.46(a) Bridge rectifier output with ideal diodes

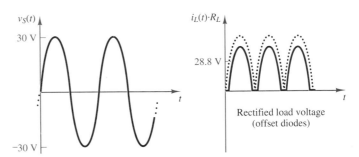

Figure 7.46(b) Bridge rectifier output accounting for diode offset voltage

Although the conventional and bridge full-wave rectifier circuits effectively convert AC signals that have zero average, or DC, value to a signal with a nonzero average voltage, either rectifier's output is still an oscillating waveform. Rather than provide a smooth, constant voltage, the full-wave rectifier generates a sequence of sinusoidal pulses at a frequency double that of the original AC signal. The *ripple*—that is, the fluctuation about the mean voltage that is characteristic of these rectifier circuits—is undesirable if one desires a true DC supply. A simple yet effective means of eliminating most of the ripple (i.e., AC component) associated with the output of a rectifier is to take advantage of the energy-storage properties of capacitors to filter out the ripple component of the load voltage. A low-pass filter that preserves the DC component of the rectified voltage while filtering out components at frequencies at or above twice the AC signal frequency would be an appropriate choice to remove the ripple component from the rectified voltage. In most practical applications of rectifier circuits, the signal waveform to be rectified is the 60-Hz, 110-V rms line voltage. The ripple frequency is, therefore, $f_{ripple} = 120$ Hz, or $\omega_{ripple} = 2\pi \cdot 120$ rad/s. A low-pass filter is required for which

$$\omega_0 \ll \omega_{ripple} \tag{7.19}$$

For example, the filter could be characterized by

$$\omega_0 = 2\pi \cdot 2 \text{ rad/s}$$

A simple low-pass filter circuit similar to those studied in Chapter 6 that accomplishes this task is shown in Figure 7.47.

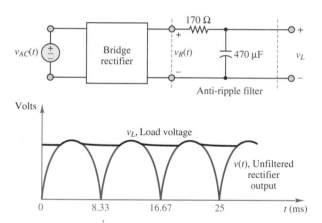

Figure 7.47 Bridge rectifier with filter circuit

DC Power Supplies and Voltage Regulation

The principal application of rectifier circuits is in the conversion of AC to DC power. A circuit that accomplishes this conversion is usually called a **DC power supply.** In power supply applications, transformers are employed to obtain an AC voltage that is reasonably close to the desired DC supply voltage. DC power supplies are very useful in practice: many familiar electrical and electronic appliances (e.g., radios, personal computers, TVs) require DC power to operate. For most applications, it is desirable that the DC supply be as steady and ripple-free as possible. To ensure that the DC voltage generated by a DC supply is constant, the DC supply is made up of voltage regulators, that is, devices that can hold a DC load voltage relatively constant in spite of possible fluctuations in the DC supply. This section will describe the fundamentals of voltage regulators.

A typical DC power supply is made up of the components shown in Figure 7.48. In the figure, a transformer is shown connecting the AC source to the rectifier circuit to permit scaling of the AC voltage to the desired level. For example, one might wish to step the 110-V rms line voltage down to 24 V rms by means of a transformer prior to rectification and filtering, to eventually obtain a 12-VDC regulated supply (*regulated* here means that the output voltage is a DC voltage that is constant and independent of load and supply variations). Following the step-down transformer are a bridge rectifier, a filter capacitor, a voltage regulator, and, finally, the load.

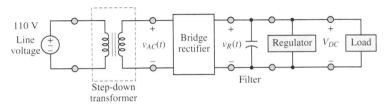

Figure 7.48 DC power supply

The most common device employed in voltage regulation schemes is the Zener diode. Zener diodes function on the basis of the reverse portion of the *i-v* characteristic of the diode discussed in Section 7.2. Figure 7.10 in Section 7.2 illustrates the general characteristic of a Zener diode, with forward offset voltage V_γ and reverse Zener voltage V_Z. Note how steep the *i-v* characteristic is at the Zener breakdown voltage, indicating that in the Zener breakdown region the diode can hold a very nearly constant voltage for a large range of currents. This property makes it possible to use the Zener diode as a voltage regulator.

The operation of the Zener diode may be analyzed by considering three modes of operation:

Figure 7.49 Zener diode model for forward bias

1. For $v_D \geq V_\gamma$, the device acts as a conventional forward-biased diode (Figure 7.49).

2. For $V_Z < v_D < V_\gamma$, the diode is reverse-biased but Zener breakdown has not taken place yet. Thus, it acts as an open circuit.

3. For $v_D \leq V_Z$, Zener breakdown occurs and the device holds a nearly constant voltage, $-V_Z$ (Figure 7.50).

The combined effect of forward and reverse bias may be lumped into a single model with the aid of ideal diodes, as shown in Figure 7.51.

To illustrate the operation of a Zener diode as a voltage regulator, consider the circuit of Figure 7.52, where the unregulated DC source, V_S, is regulated to the value of the Zener voltage V_Z. Note how the diode must be connected "upside down" to obtain a positive regulated voltage. Note also that if v_S is greater than V_Z, it follows that the Zener diode is in its reverse-breakdown mode. Thus, one need not worry whether the diode is conducting

Figure 7.50 Zener diode model for reverse bias

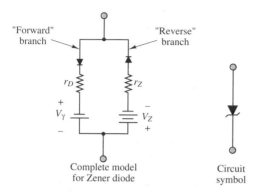

The Zener diode may be modeled as consisting of two parallel branches, representing forward and reverse operation.

Figure 7.51 Complete model for Zener diode

or not in simple voltage regulator problems, provided that the unregulated supply voltage is guaranteed to stay above V_Z (a problem arises, however, if the unregulated supply can drop below the Zener voltage). Assuming that the resistance r_Z is negligible with respect to R_S and R_L, we replace the Zener diode with the simplified circuit model of Figure 7.53, consisting of a battery of strength V_Z (the effects of the nonzero Zener resistance are explored in the examples and homework problems).

Figure 7.52 A Zener diode voltage regulator   **Figure 7.53** Simplified circuit for Zener regulator

Three simple observations are sufficient to explain the operation of this voltage regulator:

1. The load voltage must equal V_Z, as long as the Zener diode is in the reverse-breakdown mode. Then,

$$i_L = \frac{V_Z}{R_L} \tag{7.20}$$

2. The load current (which should be constant if the load voltage is to be regulated to sustain V_Z) is the difference between the unregulated supply current, i_S, and the diode current, i_Z:

$$i_L = i_S - i_Z \tag{7.21}$$

This second point explains intuitively how a Zener diode operates: any current in excess of that required to keep the load at the constant voltage V_Z is "dumped" to ground through the diode. Thus, the Zener diode acts as a sink to the undesired source current.

3. The source current is given by

$$i_S = \frac{v_S - V_Z}{R_S} \tag{7.22}$$

In the ideal case, the operation of a Zener voltage regulator can be explained very simply on the basis of this model. The examples and exercises will illustrate the effects of the practical limitations that arise in the design of a practical voltage regulator; the general principles will be discussed in the following paragraphs.

The Zener diode is usually rated in terms of its maximum allowable power dissipation. The power dissipated by the diode, P_Z, may be computed from

$$P_Z = i_Z V_Z \tag{7.23}$$

Thus, one needs to worry about the possibility that i_Z will become too large. This may occur either if the supply current is very large (perhaps because of an unexpected upward fluctuation of the unregulated supply), or if the load is suddenly removed and all of the supply current sinks through the diode. The latter case, of an open-circuit load, is an important design consideration.

Another significant limitation occurs when the load resistance is small, thus requiring large amounts of current from the unregulated supply. In this case, the Zener diode is hardly taxed at all in terms of power dissipation, but the unregulated supply may not be able to provide the current required to sustain the load voltage. In this case, regulation fails to take place. Thus, in practice, the range of load resistances for which load voltage regulation may be attained is constrained to a finite interval:

$$R_{L\ min} \leq R_L \leq R_{L\ max} \tag{7.24}$$

where $R_{L\ max}$ is typically limited by the Zener diode power dissipation and $R_{L\ min}$ by the maximum supply current. The following examples illustrate these concepts.

Example   7.8

In this example, we illustrate the selection of the power rating for a Zener diode. Consider the design of a regulator similar to that of Figure 7.52. Find the safe minimum power rating that will ensure that the Zener diode will not be damaged under worst-case conditions. Assume that $R_S = 50\ \Omega$, $R_L = 250\ \Omega$, the unregulated supply is 24 V, and the desired regulated voltage is $V_Z = 12$ V.

Solution:

Under normal operation, with a 250-Ω load, we may compute the source current to be

$$i_S = \frac{v_S - V_Z}{R_S} = \frac{12}{50} = 0.24 \text{ A}$$

and the load current to be

$$i_L = \frac{V_Z}{R_L} = \frac{12}{250} = 0.048 \text{ A}$$

Thus, the Zener current would be

$$i_Z = i_S - i_L = 0.192 \text{ A}$$

and the nominal Zener power dissipation could be computed to be

$$P_Z = i_Z V_Z = 0.192 \times 12 = 2.304 \text{ W}$$

However, if the load were accidentally disconnected from the regulator, all of the load current would be diverted to flow through the Zener diode. Thus, the worst-case Zener current is actually equal to i_S:

$$i_{Z \text{ max}} = i_S = \frac{v_S - V_Z}{R_S} = 0.24 \text{ A}$$

and therefore the maximum Zener diode power dissipation is

$$P_{Z \text{ max}} = i_{Z \text{ max}} V_Z = 2.88 \text{ W}$$

A safe design would include a Zener diode with a power rating at least equal to $P_{Z \text{ max}}$.

Example   7.9

This example illustrates the calculation of the range of allowable load resistances for a Zener regulator design. For the Zener regulator shown in Figure 7.54, we want

Figure 7.54

to maintain the load voltage at 14 V. Find the range of load resistances for which regulation can be obtained if the Zener diode is rated at 14 V, 5 W.

Solution:

The minimum load resistance for which a regulated load voltage of 14 V may be attained is found by requiring that the load voltage be 14 V and applying KVL subject to this constraint:

$$50\left(\frac{R_{L\ min}}{R_{L\ min} + 30}\right) = 14$$

$$R_{L\ min} = \frac{14}{50}(R_{L\ min} + 30)$$

$$= 11.7\ \Omega$$

The maximum current through the Zener diode that does not exceed the diode power rating may be computed by considering the 5-W power rating:

$$i_{Z\ max} = \frac{5}{14} = 0.357\ A$$

The current through the 20-Ω resistor will be

$$\frac{50 - 14}{30} = 1.2\ A$$

so that the maximum load resistance for which regulation occurs is

$$R_{L\ max} = \frac{14}{1.2 - 0.357}$$

$$= 16.6\ \Omega$$

Finally, the range of allowable load resistances is:

$$16.6\ \Omega \geq R_L \geq 11.7\ \Omega$$

Example   7.10

This example illustrates the effect of nonzero Zener resistance in a voltage regulator. We assume that, because of imperfect filtering, some ripple is present in the unregulated voltage source signal. This is usually the case in practice. The goal is to determine how much of the undesired ripple signal will appear at the output of the regulator (ideally, it should be completely removed by the regulator). Assume that 100 mV of ripple is present in addition to the unregulated 14-V (nominal) source. The circuit is sketched in Figure 7.55, and the source voltage has the appearance of the sketch of Figure 7.56. How much of the ripple will reach the load resistor? What is the actual DC load voltage? The Zener diode has $V_Z = 8$ V and a Zener resistance $r_Z = 10\ \Omega$. $R_S = 50\ \Omega$ and $R_L = 150\ \Omega$.

Figure 7.55

Figure 7.56

Solution:

Since the unregulated supply voltage is significantly larger than the Zener voltage, we are assured that the diode is operating in the Zener mode. To determine the load voltage, we can analyze the AC and DC behavior of the circuit separately, using the principle of superposition. Figure 7.57 depicts the AC and DC equivalent circuits, respectively.

The DC equivalent circuit shown in Figure 7.57, reveals that the actual DC load voltage is given by the sum of the contribution due to the DC portion of the unregulated supply and the contribution of the Zener diode:

AC equivalent circuit

$$V_L = 14 \left(\frac{10 \parallel 150}{(10 \parallel 150) + 50} \right) + 8 \left(\frac{50 \parallel 150}{(50 \parallel 150) + 10} \right)$$

$$= 2.21 + 6.32 = 8.53 \text{ V}$$

On the other hand, the AC component of the load voltage may be computed from the AC circuit to be

$$v_L = 0.1 \left(\frac{10 \parallel 150}{(10 \parallel 150) + 50} \right) = 0.016 \text{ V}$$

DC equivalent circuit

Figure 7.57

Thus, the load resistor will see approximately 16 mV of ripple, or about one sixth of what is present in the unregulated supply.

Note that accounting for the Zener resistance complicates the calculations to some extent, because the resistance introduces a coupling between the unregulated supply and the load, whereas in the case of the ideal Zener diode, such coupling does not exist. If the Zener resistance is significantly smaller than both R_S and R_L, these effects are less pronounced, and the behavior of the Zener regulator is much closer to the ideal case (see Drill Exercise 7.10).

DRILL EXERCISES

7.8   Show that the DC voltage output of the full-wave rectifier of Figure 7.42 is $2Nv_S/\pi$.

7.9   Compute the peak voltage output of the bridge rectifier of Figure 7.44, assuming diodes with 0.6-V offset voltage and a 110-V rms AC supply.

> Answer: 154.36 V

7.10   Compute the actual DC load voltage and the percentage of the ripple reaching the load (relative to the initial 100-mV ripple) for the circuit of Example 7.10 if $r_Z = 1 \ \Omega$.

> Answer: 8.06 V; 2%

Signal-Processing Applications

Among the numerous applications of diodes, there are a number of interesting signal-conditioning or signal-processing applications that are made possible by the nonlinear nature of the device. We explore three such applications here: the **diode limiter**, or **clipper**; the **diode clamp**; and the **peak detector.** Other applications are left for the homework problems.

The Diode Clipper (Limiter)

The *diode clipper* is a relatively simple diode circuit that is often employed to protect loads against excessive voltages. The objective of the clipper circuit is to keep the load voltage within a range—say, $-V_{max} \leq v_L(t) \leq V_{max}$—so that the maximum allowable load voltage (or power) is never exceeded. The circuit of Figure 7.58 accomplishes this goal.

Figure 7.58 Two-sided diode clipper

The circuit of Figure 7.58 is most easily analyzed by first considering just the branch containing D_1. This corresponds to clipping only the positive peak voltages; the analysis of the negative voltage limiter is left as a drill exercise. The circuit containing the D_1 branch is sketched in Figure 7.59;

Figure 7.59 Circuit model for the diode clipper

note that we have exchanged the location of the D_1 branch and that of the load branch for convenience. Further, the circuit is reduced to Thévenin equivalent form. Having reduced the circuit to a simpler form, we can now analyze its operation for two distinct cases: the ideal diode, and the piecewise linear diode.

1. Ideal diode model   For the ideal diode case, we see immediately that D_1 conducts if

$$\frac{R_L}{r_S + R_L} v_S(t) \geq V_{max} \tag{7.25}$$

and that if this condition occurs then (D_1 being a short circuit) the load voltage, v_L, becomes equal to V_{max}. The equivalent circuit for the "on" condition is shown in Figure 7.60.

If, on the other hand, the source voltage is such that

$$\frac{R_L}{r_S + R_L} v_S(t) < V_{max} \qquad (7.26)$$

then D_1 is an open circuit and the load voltage is simply

$$v_L(t) = \frac{R_L}{r_S + R_L} v_S(t) \qquad (7.27)$$

The equivalent circuit for this case is depicted in Figure 7.61.

Figure 7.60 Equivalent circuit for the one-sided limiter (diode on)

Figure 7.61 Equivalent circuit for the one-sided limiter (diode off)

The analysis for the negative branch of the circuit of Figure 7.58 can be conducted by analogy with the preceding derivation, resulting in the waveform for the two-sided clipper shown in Figure 7.62. Note how

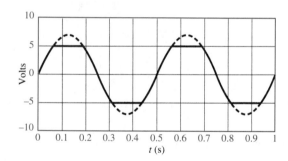

Figure 7.62 Two-sided (ideal diode) clipper input and output voltages

the load voltage is drastically "clipped" by the limiter in the waveform of Figure 7.62. In reality, such hard clipping does not occur, because the actual diode characteristic does not have the sharp on-off breakpoint the ideal diode model implies. One can develop a reasonable representation of the operation of a physical diode limiter by using the piecewise linear model.

2. Piecewise linear diode model   To avoid unnecessary complexity in the analysis, assume that V_{max} is much greater than the diode offset voltage, and therefore assume that $V_\gamma \approx 0$. We do, however, consider the finite diode resistance r_D. The circuit of Figure 7.59 still applies, and thus the determination of the diode on-off state is still based on whether $\frac{R_L}{r_S + R_L}v_S(t)$ is greater or less than V_{max}. When D_1 is open, the load voltage is still given by

$$v_L(t) = \frac{R_L}{r_S + R_L}v_S(t) \tag{7.28}$$

When D_1 is conducting, however, the corresponding circuit is as shown in Figure 7.63.

The effect of finite diode resistance on the limiter circuit.

Figure 7.63 Circuit model for the diode clipper (piecewise linear diode model)

The primary effect the diode resistance has on the load waveform is that some of the source voltage will reach the load even when the diode is conducting. This is most easily verified by applying superposition; it can be readily shown that the load voltage is now composed of two parts, one due to the voltage, and one proportional to $v_S(t)$:

$$v_L(t) = \frac{R_L \parallel r_S}{r_D + (R_L \parallel r_S)}v_{max} + \frac{r_D \parallel R_L}{r_S + (r_D \parallel R_L)}v_S(t) \tag{7.29}$$

It may easily be verified that as $r_D \to 0$, the expression for $v_L(t)$ is the same as for the ideal diode case. The effect of the diode resistance on the limiter circuit is depicted in Figure 7.64. Note how the clipping has a softer, more rounded appearance.

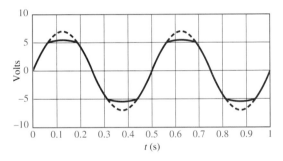

Figure 7.64 Voltages for the diode clipper (piecewise linear diode model)

The Diode Peak Detector

Another common application of semiconductor diodes, the *peak detector,* is very similar in appearance to the half-wave rectifier with capacitive filtering described in an earlier section. One of its more classic applications is in the demodulation of amplitude-modulated (AM) signals. We study this circuit in the following example.

Example   7.11   Peak Detector Circuit

In Chapter 4, a capacitive displacement transducer was introduced. It took the form of a parallel-plate capacitor composed of a fixed plate and a movable plate. The capacitance of this variable capacitor was shown to be a function of displacement; that is, it was shown that a movable-plate capacitor can serve as a linear transducer. Recall the expression derived in Example 4.17:

$$C = \frac{8.854A}{x}$$

where C is the capacitance in pF, A is the area of the plates in mm^2, and x is the (variable) distance in mm. If the capacitor is placed in an AC circuit, its impedance will be determined by the expression

$$Z_C = \frac{1}{j\omega C}$$

so that

$$Z_C = \frac{x}{j\omega 8.854A}$$

Thus, at a fixed frequency ω, the impedance of the capacitor will vary linearly with displacement. This property may be exploited in the bridge circuit of Figure 7.65, where a differential-pressure transducer is shown made of two movable-plate capacitors. If the capacitance of one of these capacitors increases as a consequence of a pressure differential across the transducer, the capacitance of the other must decrease by a corresponding amount, at least for small displacements (you may wish to refer to Example 4.4 for a picture of this transducer). The bridge is excited by a sinusoidal source.

Using phasor notation, Example 4.17 showed that the output voltage of the bridge circuit is given by

$$\mathbf{V}_{out}(j\omega) = \mathbf{V}_S(j\omega)\frac{x}{2d}$$

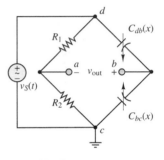

Figure 7.65 Bridge circuit for displacement transducer

provided that $R_1 = R_2$. Thus, the output voltage will vary as a scaled version of the input voltage in proportion to the displacement. A typical $v_{out}(t)$ is displayed in Figure 7.66 for a 0.05-mm "triangle" diaphragm displacement, with $d = 0.5$ mm and $\mathbf{V}_S$ a 100-Hz sinusoid with 1-V amplitude. Clearly, although the output voltage is a function of the displacement, x, it is not in a convenient form, since the displacement is proportional to the amplitude of the sinusoidal peaks.

The diode peak detector is a circuit capable of tracking the sinusoidal peaks without exhibiting the oscillations of the bridge output voltage. The peak detector operates by rectifying and filtering the bridge output in a manner similar to that of

Displacement input

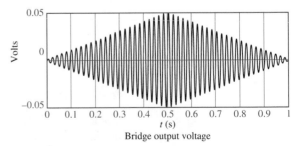

Bridge output voltage

Figure 7.66 Displacement and bridge output voltage waveforms the circuit of Figure 7.47. The ideal peak detector circuit is shown in Figure 7.67 and the response of a practical peak detector is shown in Figure 7.68. Its operation is based on the rectification property of the diode, coupled with the filtering effect of the shunt capacitor, which acts as a low-pass filter.

Peak detector circuit · Peak-detector output voltage

Figure 7.67 Peak detector circuit

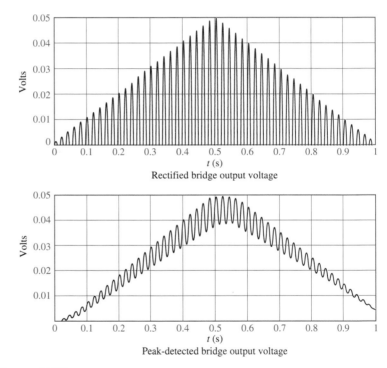

Figure 7.68 Rectified and peak-detected bridge output voltage waveforms

The Diode Clamp

Another circuit that finds common application is the *diode clamp,* which permits "clamping" a waveform to a fixed DC value. Figure 7.69 depicts two different types of clamp circuits.

The operation of the simple clamp circuit is based on the notion that the diode will conduct current only in the forward direction, and that

Figure 7.69 Diode clamp circuits

therefore the capacitor will charge during the positive half-cycle of $v_S(t)$ but
will not discharge during the negative half-cycle. Thus, the capacitor will
eventually charge up to the peak voltage of $v_S(t)$, V_{peak}. The DC voltage
across the capacitor has the effect of shifting the source waveform down by
V_{peak}, so that, after the initial transient period, the output voltage is

$$v_{out}(t) = v_S(t) - V_{peak} \qquad (7.30)$$

and the positive peaks of $v_S(t)$ are now clamped at 0 V. For equation 7.30
to be accurate, it is important that the RC time constant be greater than the
period, T, of $v_S(t)$:

$$RC \gg T \qquad (7.31)$$

Figure 7.70 depicts the behavior of the diode clamp for a sinusoidal input
waveform, where the dashed line is the source voltage, and the solid line
represents the clamped voltage.

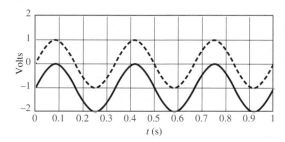

Figure 7.70 Ideal diode clamp input and output voltages

The clamp circuit can also work with the diode in the reverse direction;
the capacitor will charge to $-V_{peak}$ with the output voltage given by

$$v_{out}(t) = v_S(t) + V_{peak} \qquad (7.32)$$

Now the output voltage has its negative peaks clamped to zero, since the
entire waveform is shifted upward by V_{peak} volts. Note that in either case, the
diode clamp has the effect of introducing a DC component in a waveform that
does not originally have one. It is also possible to shift the input waveform
by a voltage different from V_{peak} by connecting a battery, V_{DC}, in series with

the diode, provided that

$$V_{DC} < V_{peak} \tag{7.33}$$

The resulting circuit is called a *biased diode clamp;* it is discussed in Example 7.12.

Example   7.12   Biased Diode Clamp

In this example, we design a biased diode clamp with the aim of clamping the signal

$$v_S(t) = 5 \sin(\omega t)$$

to a 3-VDC level (i.e., shifting the average value of the signal from 0 V to 3 V). Consider the biased clamp circuit of Figure 7.70. Once the capacitor has charged to V_{peak}, the output voltage will be given by

$$v_{out} = v_S(t) - V_{peak} + V_{DC}$$

But since V_{DC} must be smaller than V_{peak} (otherwise the diode would never conduct!), this circuit would never permit raising the DC level of v_{out}. To resolve this problem, we must invert both the diode and the battery, as shown in the circuit of Figure 7.71. Now the output voltage is given by

$$v_{out} = v_S(t) + V_{peak} - V_{DC}$$

and, to have a DC level of 3 V, we choose $V_{DC} = 2$ V. The resulting waveforms are shown in Figure 7.72.

Figure 7.71

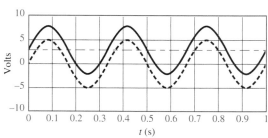

Figure 7.72

Photodiodes

Another property of semiconductor materials that finds common application in measurement systems is their response to light energy. In appropriately fabricated diodes, called **photodiodes,** when light reaches the depletion region of a *pn* junction, photons cause hole-electron pairs to be generated by a process called *photo-ionization*. This effect can be achieved by using a surface material that is transparent to light. As a consequence, the reverse saturation current depends on the light intensity (i.e., on the number of incident photons), in addition to the other factors mentioned earlier, in

Section 7.2. In a photodiode, the reverse current is given by $-(I_0 + I_p)$, where I_p is the additional current generated by photo-ionization. The result is depicted in the family of curves of Figure 7.73, where the diode charac-

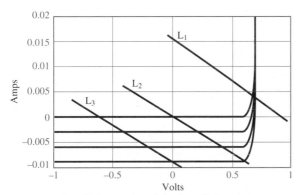

L$_1$: diode operation ; L$_2$: solar cell ; L$_3$: photosensor

Figure 7.73 Photodiode *i-v* curves

teristic is shifted downward by an amount related to the additional current generated by photo-ionization. Figure 7.73 depicts the appearance of the *i-v* characteristic of a photodiode for various values of I_p, where the *i-v* curve is shifted to lower values for progressively larger values of I_p. The circuit symbol is depicted in Figure 7.74.

Also displayed in Figure 7.73 are three load lines, which depict the three modes of operation of a photodiode. Curve L_1 represents normal diode operation, under forward bias. Note that the operating point of the device is in the positive *i*, positive *v* (first) quadrant of the *i-v* plane; thus, the diode dissipates positive power in this mode, and is therefore a passive device, as we already know. On the other hand, load line L_2 represents operation of the photodiode as a **solar cell;** in this mode, the operating point is in the negative *i*, positive *v*, or fourth, quadrant, and therefore the power dissipated by the diode is *negative*. In other words, the photodiode is generating power by converting light energy to electrical energy. Note further that the load line intersects the voltage axis at zero, meaning that no supply voltage is required to bias the photodiode in the solar-cell mode. Finally, load line L_3 represents the operation of the diode as a light sensor: when the diode is reverse-biased, the current flowing through the diode is determined by the light intensity; thus, the diode current changes in response to changes in the incident light intensity.

The operation of the photodiode can also be reversed by forward-biasing the diode and causing a significant level of recombination to take place in the depletion region. Some of the energy released is converted to light energy by emission of photons. Thus, a diode operating in this mode emits light when forward-biased. Photodiodes used in this way are called **light-emitting diodes (LEDs);** they exhibit a forward (offset) voltage of 1 to 2 volts. The circuit symbol for the LED is shown in Figure 7.75.

Figure 7.74 Photodiode circuit symbol

Figure 7.75 Light-emitting diode (LED) circuit symbol

Gallium arsenide (GaAs) is one of the more popular substrates for creating LEDs; gallium phosphide (GaP) and the alloy $GaAs_{1-x}P_x$ are also quite common. Table 7.1 lists combinations of materials and dopants used for common LEDs and the colors they emit. The dopants are used to create the necessary *pn* junction.

Table 7.1   LED materials and wavelengths

Material	Dopant	Wavelength	Color
GaAs	Zn	900 nm	Infrared
GaAs	Si	910–1,020 nm	Infrared
GaP	N	570 nm	Green
GaP	N, N	590 nm	Yellow
GaP	Zn, O	700 nm	Red
$GaAs_{0.6}P_{0.4}$		650 nm	Red
$GaAs_{0.35}P_{0.65}$	N	632 nm	Orange
$GaAs_{0.15}P_{0.85}$	N	589 nm	Yellow

The construction of a typical LED is shown in Figure 7.76, along with the schematic representation for an LED. A shallow *pn* junction is created with electrical contacts made to both *p* and *n* regions. As much of the upper surface of the *p* material is uncovered as possible, so that light can leave the device unimpeded. It is important to note that, actually, only a relatively small fraction of the emitted light leaves the device; the majority stays inside the semiconductor. A photon that stays inside the device will eventually collide with an electron in the valence band, and the collision will force the electron into the conduction band, emitting an electron-hole pair and absorbing the photon. To minimize the probability that a photon will be absorbed before it has an opportunity to leave the LED, the depth of the *p*-doped region is left very thin. Also, it is advantageous to have most of the recombinations that emit photons occur as close to the surface of the diode as possible. This is made possible by various doping schemes, but even so, of all of the carriers going through the diode, only a small fraction emit photons that are able to leave the semiconductor.

Figure 7.76 Light-emitting diode (LED)

A simple LED drive circuit is shown in Figure 7.77. From the standpoint of circuit analysis, LED characteristics are very similar to those of the silicon diode, except that the offset voltage is usually quite a bit larger. Typical values of V_γ can be in the range of 1.2 to 2 volts, and operating currents can range from 20 mA to 100 mA. Manufacturers usually specify an LED's characteristics by giving the rated operating-point current and voltage.

Figure 7.77 LED drive circuit and *i-v* characteristic

Example   7.13

The manufacturer of a certain LED suggests an operating point of 1.7 V and 40 mA.

1.   What is the total power consumption of the LED when operating?
2.   If the diode is to be used in the circuit of Figure 7.77 and $V_S = 5$ V, find the resistance R_S.
3.   How much power must the voltage source be able to provide?

Solution:

1.   The power consumption of the LED can be determined directly from the specifications:

$$P_{LED} = V_{LED}I_{LED} = (1.7)(40 \times 10^{-3}) = 68 \text{ mW}$$

2.   Writing KVL around the circuit of Figure 7.77, we obtain the equation

$$V_S = I_{LED}(R_S) + V_{LED}$$

or

$$R_S = \frac{V_S - V_{LED}}{I_{LED}} = \frac{(6 - 1.7)}{40 \times 10^{-3}}$$

from which the required value of R_S may be computed:

$$R_S = 82.5 \ \Omega$$

3.   To satisfy the power requirements of the circuit, the source must be able to provide 40 mA of current to both the resistor and the diode. Thus, the required power is

$$P_S = V_S I_{LED} = 5(40 \times 10^{-3})$$
$$= 200 \text{ mW}$$

Example   7.14   Opto-Isolators

Figure 7.78 Opto-isolator

One of the common applications of photodiodes and LEDs is the **opto-coupler,** or **opto-isolator.** This device, which is usually enclosed in a sealed package, uses the light to current and current to light conversion property of photodiodes and LEDs to provide signal connection between two circuits without any need for electrical connections. Figure 7.78 depicts the circuit symbol for the opto-isolator.

Because diodes are nonlinear devices, the opto-isolator is not used in transmitting analog signals: the signals would be distorted because of the nonlinear diode *i-v* characteristic. However, opto-isolators find a very important application when on-off signals need to be transmitted from high-power machinery to delicate computer control circuitry. The optical interface ensures that potentially damaging large currents cannot reach delicate instrumentation and computer circuits.

Example   7.15   Solar-Cell Array

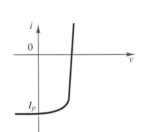

Figure 7.79 *i-v* curve of a solar cell

Figure 7.80 Circuit model for solar cell

As explained earlier in this section, a solar cell is a device that operates on the same principle as that of an LED. Instead of applying a forward bias to the terminals of an LED and generating a light in response to the diode current, the solar cell receives energy from light and converts this energy into an electrical current. The *i-v* characteristic of a photovoltaic cell is shown in Figure 7.79.

Note that the curve has the same shape as the diode curve but, instead of passing through the origin, is shifted down by the amount I_p. I_p is the strength of the current generated by the incident light on the cell. We can express the current and voltage relationship of the solar cell using the diode equation:

$$i_D = i_0(e^{qv_D/kT} - 1) - I_p$$

To increase the power output to a level sufficient for practical applications, solar cells can be connected in series (to increase output voltage) and in parallel (to increase output current). Such arrangements are commonly called **solar arrays.** The solar cell to be used in this example has the following characteristics when illuminated: $I_p = 20$ mA, $I_0 = 50$ pA, $R_S = 0.7$ Ω. The array has the minimum output requirements

$$V_{\text{out}} = 10 \text{ V} \qquad \text{when} \qquad I = 0.5 \text{ A}$$

Design an array that will meet these minimum requirements, using the circuit model for a solar cell shown in Figure 7.80.

Solution:

The array must be able to generate 0.5 A. Each photovoltaic cell can, at a maximum, generate 20 mA. This would seem to indicate that we must arrange the cells in parallel to increase the total current output. To find the minimum number, n_{min}, of branches needed to output the required current, we use the following expression:

$$n_{\text{min}} \times 20 \text{ mA} = 0.5 \text{ A}$$

$$n_{\text{min}} = \frac{0.5}{0.02} = 25$$

Since a total output voltage of 10 V is required, each parallel branch must consist of a series connection of m photovoltaic cells. The required number of series cells in each branch, m_{min}, is found by determining the open-circuit voltage of a cell from the diode equation:

$$I = I_0(e^{qv_D/kT} - 1) - I_p$$

Since the current is zero under open-circuit conditions,

$$I_p = I_0(e^{qv_{OC}/kT} - 1)$$

and

$$v_{OC} = \log_e\left(\frac{I_p}{I_0} + 1\right)\frac{kT}{q}$$

$$= \log_e\left(\frac{20 \text{ mA}}{50 \text{ pA}} + 1\right)0.026$$

Thus, we can compute the open-circuit voltage to be

$$V_{OC} = 0.515 \text{ V}$$

and since

$$m_{min}V_{OC} = 10 \text{ V}$$

we conclude that, at least nominally,

$$m_{min} = \frac{10 \text{ V}}{0.515 \text{ V}} = 19.4 \approx 20$$

That is, 20 solar cells will be needed in each branch.

The next step in the design process is to verify whether 25 branches of 20 cells each are sufficient to satisfy the original requirements of the problem. The appearance of the array is depicted in Figure 7.81. It should be pointed out that, because the open-circuit voltage and short-circuit current were used in the initial calculations, it is likely that the preliminary calculation is somewhat optimistic. Figure 7.82 depicts

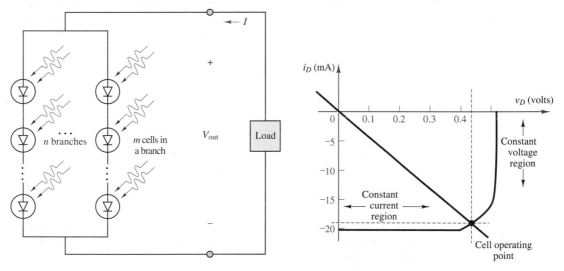

Figure 7.81 Solar array   **Figure 7.82** i-v curve of a solar cell

the i-v curve of one solar cell, clearly showing that the nominal values of short-circuit current (20 mA) and open-circuit voltage (0.515 V) cannot be achieved simultaneously, but that a compromise must be made, as represented by the operating point shown in the picture. To clarify this point further, consider that the cell output voltage can be at a maximum 0.4 V if the cell current is 20 mA. To determine an acceptable compromise, let us assume that one more branch is added to the array, so that now each branch must supply $\frac{1}{26}$ of the rated current, or

$$I_{branch} = \frac{0.5 \text{ A}}{n_{min} + 1} = \frac{0.5 \text{ A}}{26} = 19.23 \text{ mA}$$

The voltage supplied by a cell at this current level is then given by the expression

$$I_{cell} = I_0 (e^{q(v_{cell} - IR_S/kT)} - 1) - I_p$$

or

$$V_{cell} = IR_S + \frac{kT}{q} \left[\log_e \left(\frac{I_p + I}{I_0} + 1 \right) \right]$$

where I is referenced as entering the cell (I is a negative value, since the cell is supplying the power to the load).

Thus,

$$V_{cell} = (-19.23 \text{ mA})0.7 + 0.026 \log_e \left(\frac{20 \text{ mA} - 19.23 \text{ mA}}{50 \text{ pA}} + 1 \right) = 0.4168 \text{ V}$$

and the number of series cells required to obtain the desired 10-V output rating is given by

$$m = \frac{10}{0.4168} = 23.99 \approx 24$$

Thus, the array will consist of 26 branches, each containing 24 cells in series, or a total of 624 cells. Note that this is not the only solution to the problem. For example, it may be desirable to operate in either the constant-current or the constant-voltage region of the cell i-v curve, since in these regions the design is less sensitive to load variations. These designs are explored in the homework problems.

DRILL EXERCISES

7.11 Repeat the analysis of the diode clipper of Figure 7.58 for the branch containing D_2.

7.12 For the one-sided diode clipper of Figure 7.59, find the percentage of the source voltage that reaches the load if $R_L = 150 \text{ }\Omega$, $r_S = 50 \text{ }\Omega$, and $r_D = 5 \text{ }\Omega$. Assume that the diode is conducting, and use the circuit model of Figure 7.63.

> Answer: 8.8%

7.13 How would the diode clipper output waveform change if we used the offset diode model instead of the piecewise linear model in the analysis? [*Hint:* Compare Figures 7.23 and 7.35.]

7.14 Determine the value of load resistance that is compatible with the solar array design of Example 7.15. [*Hint:* Consider the characteristic curve of Figure 7.82.]

> Answer: 21.7 Ω

7.15 Compute the power output of each solar cell in Example 7.15. Would the cell generate more power if operated at the operating point shown, in the constant-voltage region (say, for $I = 12$ mA), or in the constant-current region (say, for $V = 0.3$ V)?

> Answer: $P_{cell} = 8$ mW; at the operating point shown.

CONCLUSION

- Semiconductor materials have conductive properties that fall between those of conductors and insulators. These properties make such materials useful in the construction of many electronic devices that exhibit nonlinear i-v characteristics. Of these devices, the diode is one of the most commonly employed.

- The semiconductor diode acts like a one-way current valve, permitting the flow of current only when biased in the forward direction. Although the behavior of the diode is described by an exponential equation, it is possible to approximate the operation of the diode by means of simple circuit models. The simplest model treats the diode as either a short circuit or an open circuit (the on-off, or ideal, model). The ideal model can be extended to include an offset voltage (usually 0.2 to 0.7 V), which represents the contact potential at the diode junction. A slightly more realistic model, the piecewise linear diode model, accounts for the effects of the diode forward resistance. With the aid of these circuit models it becomes possible to analyze diode circuits using the DC and AC circuit analysis techniques developed in earlier chapters.

- One of the most important properties of the semiconductor diode is that of rectification, which permits the conversion of AC voltages and currents to DC voltages and currents. Diode rectifiers can be of the half-wave type, or they can be full-wave. Full-wave rectifiers can be constructed in a conventional two-diode configuration, or in a bridge configuration. Diode rectifiers are an essential part of DC power supplies and are usually employed in conjunction with filter capacitors to obtain a relatively smooth DC voltage waveform. In addition to rectification and smoothing, it is also necessary to regulate the output of a DC power supply; Zener diodes accomplish this task by holding a constant voltage when reverse-biased above the Zener voltage.

• In addition to power supply applications, diodes find use in many signal-processing and signal-conditioning circuits. Of these, the diode limiter, peak detector, and clamp have been explored in the chapter. Further, since semiconductor material properties are also affected by light intensity, certain types of diodes, known as *photodiodes,* find application as light detectors, solar cells, or light-emitting diodes.

KEY TERMS

DEVICE DATA SHEETS

1N461A   General Purpose Diode

Typical Electrical Characteristic Curves (25°C Ambient Temperature unless otherwise noted)

Device No	V_{RRM} V Min	I_R nA Max	@	V_R V	V_F V Min	Max	@	I_F mA
1N461A	30	500		25		1.0		100

Forward Voltage vs Forward Current

Reverse Voltage vs Reverse Current

Dynamic Impedance vs Forward Current

1N746A   Zener Diode

Typical Electrical Characteristic Curves (25°C Ambient Temperature unless otherwise noted)

Device No.	V_Z V Nom	Tol. $\pm V_Z$ %	Z_Z Ω @ Max	I_Z mA	I_R μA @ Max	V_R V	T.C. %/°C Typ (Max)	P_D mW $T_A = 25°C$
1N746A	3.3	5.0	28.0	20	10	1.0	−0.070	500

Temperature Coefficient vs Zener Current

Dynamic Impedance vs Zener Current

Power Derating vs Ambient Temperature

HOMEWORK PROBLEMS

7.1   Consider the circuit of Figure P7.1. Determine whether the diode is conducting or not. Assume that the diode is an ideal diode.

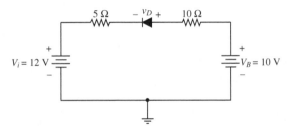

Figure P7.1

7.2   Repeat Problem 7.1 for $V_i = 12$ V and $V_B = 15$ V.

7.3   Consider the circuit of Figure P7.2. Determine whether the diode is conducting or not. Assume that the diode is an ideal diode.

Figure P7.2

7.4   Repeat Problem 7.3 for $V_B = 15$ V.

7.5   Repeat Problem 7.3 for $V_C = 15$ V.

7.6   Repeat Problem 7.3 for $V_C = 10$ V and $V_B = 15$ V.

7.7   For the circuit of Figure P7.3, find the range of V_{in} for which D_1 is forward-biased. Assume an ideal diode.

7.8   In the circuit of Figure P7.4, find the range of V_{in} for which D_1 is forward-biased. Assume ideal diodes.

Figure P7.3

Figure P7.4   **Figure P7.5**

7.9   For the circuit of Figure P7.5, assume that D_1 and D_2 are ideal diodes.

a.   For what range of values of V_1 is diode D_1 forward-biased?
b.   For what range of values of V_1 is diode D_2 forward-biased?
c.   What is V_o when $V_1 = 3$ V?

7.10   For the circuit of Figure P7.6, find the range of V_{in} for which D_1 is forward-biased. Assume an ideal diode.

7.11   Find V_o when $V_{in} = 8$ V in the circuit of Figure P7.6.

7.12   For the circuit of Figure P7.7, find the output voltage, V_{out}, if the input voltages V_1 and V_2 are as follows:

a. $V_1 = 0$ V, $V_2 = 0$ V
b. $V_1 = 5$ V, $V_2 = 5$ V
c. $V_1 = 5$ V, $V_2 = 0$ V

Figure P7.6

Figure P7.7

Assume ideal diodes.

*7.13   For the circuit of Figure P7.8, $V_{in} = V_0 \sin 2\pi 60t$, and for each diode, $V_\gamma = 0.7$ V when the diode is forward-biased. It is required that $V_o = 5$ V with less than 1 mV of ripple.

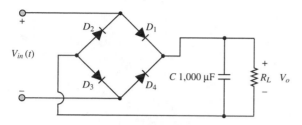

Figure P7.8

a.   What is the minimum value R_L can have?
b.   What is the value of V_{in} to achieve this specification?
c.   If D_1 is destroyed—that is, becomes an open circuit—what happens to V_o?

7.14   Find the operating point of the diode for the circuit of Figure P7.9 if the resistance R is

a.   1,000 Ω
b.   5,000 Ω

Find the power dissipated by the diode in each case.

Figure P7.9

7.15   A Zener diode's ideal i-v characteristic is shown in Figure P7.10(a). Given a Zener voltage, V_Z of 7.7 V, find the output voltage, V_{out}, for the circuit of Figure P7.10(b) if V_S is

a.   12 V
b.   20 V

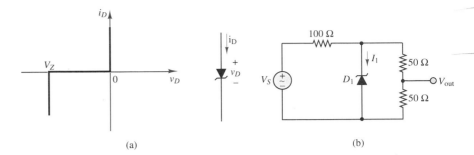

Figure P7.10

*7.16 A Zener diode is used in a switching circuit as shown in Figure P7.11. When V_1 reaches 7.2 V, switches S_2 and S_3 both close simultaneously and stay closed for 3.9 ms, and then both open.

a. Find the frequency for the voltage across the load resistor, R_L.

b. Find the power dissipated by R_L.

[*Hint:* At $t = 0$, S_1 closes and S_2 and S_3 are open. Find the waveforms for V_1 and V_L, and proceed from there.]

Figure P7.11

7.17 One of the more interesting applications of a diode, based on the diode equation, is an electronic thermometer. The concept is based on the empirical observation that if the current through a diode is nearly constant, the offset voltage is nearly a linear function of temperature, as shown in Figure P7.12(a).

a. Show that i_D in the circuit of Figure P7.12(b) is nearly constant in the face of variations in the diode voltage, v_D. This can be done by computing the percent change in i_D for a given percent change in v_D. Assume that v_D changes by 10 percent, from 0.6 to 0.66 V.

b. On the basis of the graph of Figure P7.12(a), write an equation for $v_D(T^\circ)$ of the form

$$v_D = \alpha T^\circ + \beta$$

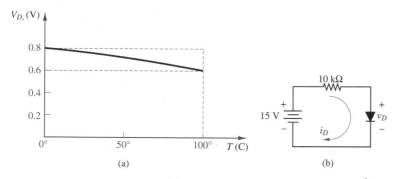

Figure P7.12

7.18 For the circuit of Figure P7.13, sketch $i_D(t)$ for the following conditions:

a. Use the ideal diode model.
b. Use the ideal diode model with offset ($V_\gamma = 0.6$ V).
c. Use the piecewise linear approximation with

$$r_D = 1 \text{ k}\Omega$$
$$V_\gamma = 0.6 \text{ V}$$

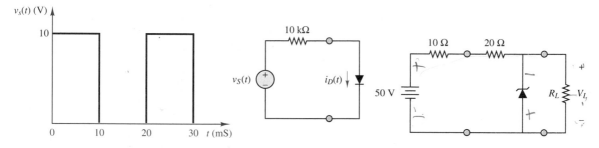

Figure P7.13   Figure P7.14

7.19 For the Zener regulator shown in the circuit of Figure P7.14, we desire to hold the load voltage to 14 V. Find the range of load resistances for which regulation can be obtained if the Zener diode is rated at 14 V, 5 W.

7.20 You have been asked to design a full-wave bridge rectifier for a power supply. A step-down transformer has already been chosen. It will supply 12 V rms to your rectifier. The full-wave rectifier is shown in the circuit of Figure P7.15.

a. If the diodes have an offset voltage of 0.6 V, sketch the input source voltage, $v_S(t)$, and the output voltage, $v_L(t)$, and state which diodes are on and which are off in the appropriate cycles of $v_S(t)$. The frequency of the source is 60 Hz.
b. If $R_L = 1,000$ Ω and a capacitor, placed across R_L to provide some filtering, has a value of 8 μF, sketch the output voltage, $v_L(t)$.
c. Repeat part b, with the capacitance equal to 100 μF.

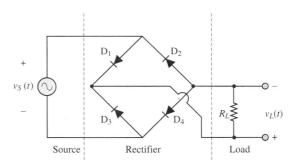

Figure P7.15

7.21 For the peak detector shown in Figure P7.16, find the charging and discharging time constants, given the diode i-v characteristic shown.

Figure P7.17

Figure P7.16

7.22 A piecewise linear model of a diode is shown in Figure P7.17. If $r_{D \text{ off}} = 100$ kΩ, $r_{D \text{ on}} = 100$ Ω, and $V_\gamma = 0.6$, sketch the i-v characteristics for -100 V $\leq v_D \leq 1$ V.

7.23 You have purchased an inexpensive diode that has the i-v characteristics shown in Figure P7.18. Determine the piecewise linear model, using the form shown in Figure P7.17.

Figure P7.19

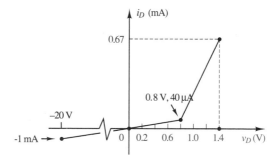

Figure P7.18

7.24 Sketch the output waveform for the circuit of Figure P7.19. Assume ideal diode characteristics, $v_S(t) = 10 \sin(2,000\pi t)$.

7.25 Sketch the output waveform for the circuit of Figure P7.20. Assume ideal diode characteristics, $v_S(t) = 10 \sin(2{,}000\pi t)$.

7.26 Sketch the output waveform for the circuit of Figure P7.21. Assume ideal diode characteristics, $v_S(t) = 10 \sin(2{,}000\pi t)$.

Figure P7.20

Figure P7.21

7.27 In the circuit of Figure P7.22, if the diodes D_1 and D_3 have an "on" resistance equal to 20 Ω and an offset voltage of 0.6 V, and if R_L is 3.3 kΩ, find v_{out} for the cases listed.

a. $v_1 = 0$ V, $v_2 = 0$ V
b. $v_1 = 5$ V, $v_2 = 5$ V
c. $v_1 = 0$ V, $v_2 = 5$ V
d. $v_1 = 5$ V, $v_2 = 0$ V

Figure P7.22

7.28 Find the output waveform for the circuit of Figure P7.23. The diode is ideal.

Figure P7.23   **Figure P7.24**

7.29 Find the output waveform for the circuit of Figure P7.24. The diode is ideal.

7.30 Find the output waveform for the circuit of Figure P7.25. The diode is ideal.

Figure P7.25

7.31   The circuit of Figure P7.26 forces the diode to operate at a certain voltage and current. If the diode has the given *i-v* characteristics, find the operating point of the diode.

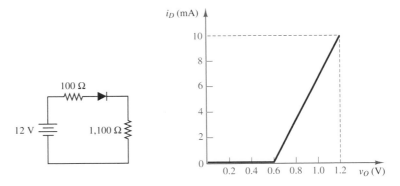

Figure P7.26

7.32   We can estimate the dynamic resistance in the forward direction of a silicon diode by differentiating the equation

$$i_D = I_0(e^{v_D/V_T} - 1) \approx I_0 e^{v_D/V_T}$$

$$\frac{di_D}{dv_D} = \frac{1}{V_T} I_0 e^{v_D/V_T} \approx \frac{i_D}{V_T}$$

where $V_T = \frac{kT}{q}$, and inverting the result and evaluating r_D at the operating point:

$$r_D \approx \frac{dv_D}{di_D}\bigg|_{Q \text{ point}} \approx \frac{V_T}{I_{DQpoint}} \approx \frac{25.6\text{mV}}{I_{DQpoint}}$$

Use this method to determine the dynamic resistance of the diode in the circuit of Figure P7.27 if the diode has the small-signal *i-v* characteristic shown in Figure 7.32.

Figure P7.27

*7.33   Assume that the operating point of diode in the circuit of Figure P7.28 is $I_Q = 7.0$ mA, $V_Q = 0.8$ V.

Figure P7.28

a.   Estimate the small-signal diode resistance, r_D.

b.   Estimate the peak values of the AC component of the diode current, using graphical analysis and circuit analysis methods.

Assume $C_C = \infty$ (open circuit for DC, short circuit for AC).

7.34 Assuming that the diodes are ideal, determine and sketch the *i-v* characteristics for the circuit of Figure P7.29. Consider the range $10 \geq v \geq 0$.

Figure P7.29   Figure P7.30

7.35 For the circuit of Figure P7.30, sketch the waveform of $v_L(t)$ if $v_S(t) = 15 \sin 2{,}000\pi t$, the Zener diode has $V_Z = 9$ V, $V_1 = 12$ V, $R_S = 50$ Ω, and $R_L = 150$ Ω. Consider the diodes to be ideal.

7.36 For the voltage limiter circuit of Figure P7.31, plot v_L versus v_S for $-20 < v_S < 20$ V. Assume that

$$R_S = 10 \ \Omega \quad V_1 = 15 \text{ V} \quad R_2 = 5 \ \Omega$$
$$R_1 = 10 \ \Omega \quad V_2 = 12 \text{ V} \quad R_L = 50 \ \Omega$$

Treat all diodes as ideal diodes.

Figure P7.31

7.37 For the voltage limiter circuit of Figure P7.32, plot the voltage across R_L versus v_S for $-20 < v_S < 20$ V. Assume that

$$R_S = 10 \ \Omega \quad V_{Z1} = 10 \text{ V} \quad R_2 = 10 \ \Omega \quad V_{Z4} = 5 \text{ V}$$
$$R_1 = 1 \ \Omega \qquad\qquad\quad R_L = 40 \ \Omega$$

Treat all diodes as ideal diodes.

7.38 The circuit of Figure P7.33(a) demonstrates the use of a diode as a path for current in an inductive circuit. At $t = 0$, the switch closes, causing the inductor to become a short circuit as $t \to \infty$. $R = 100$ Ω, $L = 100$ mH, and $V_S = 15$ V.

a. Assume that $v_L(t) = 0$ and the switch, S_1, has been closed for a long period of time. At $t = t_1$, the switch opens. Determine the voltage $v_L(t)$ for $t > t_1$.

b. Using the assumptions of part a, now consider the same circuit with a diode placed as in the circuit of Figure P7.33(b). Use the diode model of Figure

Figure P7.32

P7.33(b), with $r_{on} = r_D = 5\ \Omega$. At $t = t_1$, the switch opens. Determine the voltage $v_L(t)$ for $t \geq t_1$.

Figure P7.33(a)   **Figure P7.33(b)**

7.39 We are using a voltage source to charge an automotive battery as shown in the circuit of Figure P7.34(a). At $t = t_1$, the protective circuitry of the source causes switch S_1 to close, and the source voltage goes to zero. Find the currents, I_S, I_B, and I_{SW}, for the following conditions:

a. $t = t_1^-$

b. $t = t_1^+$

c.   What will happen to the battery after the switch closes?

Now we are going to charge the battery, using the circuit of Figure P7.34(b). Repeat parts a and b if the diode has an offset voltage of 0.6 V.

Figure P7.34(a)   **Figure P7.34(b)**

7.40 Repeat Drill Exercise 7.12 for $R_L = 1\ k\Omega$, $r_S = 100\ \Omega$, and $r_D = 10\ \Omega$.

7.41 Repeat Drill Exercise 7.12 for $R_L = 10\ k\Omega$, $r_S = 500\ \Omega$, and $r_D = 50\ \Omega$.

7.42 Repeat the design of the solar array of Example 7.15 under the assumption that the solar array will be operated in the constant-current region.

7.43 Repeat the design of the solar array of Example 7.15 under the assumption that the solar array will be operated in the constant-voltage region.

Bipolar Junction Transistors

C hapter 8 continues the discussion of electronic devices that began in Chapter 7 with the semiconductor diode. This chapter will not provide the kind of comprehensive treatment of electronic devices that would, for example, enable you to carry out the complete design of a transistor amplifier; but this chapter and the next will survey the two major families of electronic devices—bipolar and field-effect transistors—and the applications that are more likely to be encountered in the field, in part by providing a few application-oriented examples. Chapter 8 starts with a brief, qualitative discussion of the physics of the bipolar junction transistor (BJT), which naturally follows the discussion of the *pn* junction in Chapter 7. Circuit models for one type of BJT (the *npn* bipolar transistor) are then introduced, to illustrate how one can analyze transistor circuits using the techniques of Chapter 3. The circuit models are designed to facilitate analysis of device *i-v* characteristics, rather than discussion of the underlying physics. A few practical examples illustrate the use of the circuit models. Finally, a synopsis of the common bipolar transistor amplifier configurations is provided for reference at the end of the chapter.

By the end of Chapter 8, you should be able to:

- Describe the basic operation of bipolar junction transistors.
- Interpret the device characteristic curves and extract linear (small-signal) models for simple amplifiers from the device curves and data sheets.
- Recognize the more common amplifier configurations, and draw their linear (small-signal) equivalent circuits.
- Determine the voltage and current gains of an amplifier on the basis of its linear circuit model.
- Discuss the frequency response limitations of a transistor amplifier.

8.1   TRANSISTORS AS AMPLIFIERS AND SWITCHES

A transistor is a three-terminal semiconductor device that can provide two functions that are fundamental to the design of electronic circuits: **amplification** and **switching.** Amplification consists of magnifying a signal by transferring energy to it from an external source. A transistor switch is a device that enables the control of a relatively large current between or voltage across two terminals by means of a small control current or voltage, applied at a third terminal. In this chapter, we provide an introduction to the one of the two major families of transistors: *bipolar junction transistors,* or *BJTs.* The other family, *field-effect transistors,* or *FETs,* is discussed in Chapter 9.

The operation of the transistor as a linear amplifier can be explained qualitatively by the sketch of Figure 8.1, in which the four possible modes of operation of a transistor are illustrated by means of circuit models employing controlled sources (you may wish to review the section on controlled

(a) Current controlled current source   (b) Voltage controlled voltage source

(c) Voltage controlled current source   (d) Current controlled voltage source

Figure 8.1 Controlled-source models of linear amplifier transistor operation sources in Chapter 2). In Figure 8.1, controlled voltage and current sources are shown to generate an output proportional to an input current or voltage; the proportionality constant, μ, is called the internal *gain* of the transistor. As will be shown, the BJT acts essentially as a current-controlled device, while the FET behaves as a voltage-controlled device.

Transistors can also act in a nonlinear mode, as voltage- or current-controlled switches. When a transistor operates as a switch, a small voltage or current is used to control the flow of current between two of the transistor terminals in an on-off fashion. Figure 8.2 depicts the idealized operation of the transistor as a switch, suggesting that the switch is closed (on) whenever a control voltage or current is greater than zero and is open (off) otherwise. It will later become apparent that the conditions for the switch to be on or off need not necessarily be those depicted in Figure 8.2. The operation of transistors as switches will be discussed in more detail in Chapter 11.

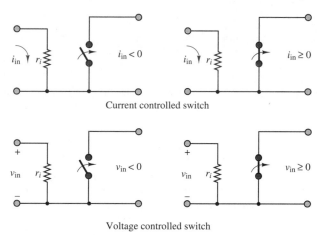

Figure 8.2 Models of ideal transistor switches

Example   8.1

Consider the simplified model of a transistor amplifier shown in Figure 8.3, where a signal source, $v_S(t)$, with internal resistance R_S, and a load resistance, R_L, have been connected to one of the four controlled-source models of Figure 8.1. The resistances

Figure 8.3

r_i and r_o and the gain, μ, are internal properties of the transistor, as will be shown in the next section. Determine the actual gain of this amplifier, $A_V = v_L/v_S$, on the basis of the circuit model of Figure 8.3.

Solution:

Since the output of the controlled voltage source is determined by the voltage v_{in}, we need to determine this quantity first:

$$v_{in} = v_S \frac{r_i}{r_i + R_S}$$

Thus, the output of the controlled current source will be

$$\mu v_{in} = \mu v_S \frac{r_i}{r_i + R_S}$$

and the output voltage, v_L, can be found by computing the voltage divider result:

$$v_L = \mu v_S \frac{r_i}{r_i + R_S} \frac{R_L}{r_o + R_L}$$

Finally, the amplifier voltage gain is found to be

$$A_V = \frac{v_L}{v_S} = \mu \frac{r_i}{r_i + R_S} \cdot \frac{R_L}{r_o + R_L}$$

Note that this gain, A_V, is necessarily less than the internal transistor voltage gain, μ! You can now show that if the conditions

$$r_i \gg R_S$$

$$r_o \gg R_L$$

apply, then the gain of the amplifier becomes approximately equal to the internal gain of the transistor. An important observation, then, is that, given a transistor with known internal parameters, r_i, r_o, and μ, the gain of the amplifier will depend on the internal resistance of the source and on the load resistance.

In the next sections, the physical operation of bipolar transistors will be described. Circuit models for bipolar transistors will be developed on the basis of their i-v characteristics. You will then be able to identify the models introduced in this section as simplified representations of the operations of actual transistor amplifiers.

DRILL EXERCISES

8.1   Repeat the analysis of Example 8.1 for the current-controlled voltage source model of Figure 8.1. What is the amplifier gain? Under what conditions would the gain A be equal to $\frac{\mu}{R_S}$?

Answer: $A = \mu \dfrac{1}{(r_i + R_S)} \dfrac{R_L}{(r_o + R_L)}$; $r_i \to 0$, $r_o \to 0$

8.2   Repeat the analysis of Example 8.1 for the current-controlled current source model of Figure 8.1. What is the amplifier voltage gain?

Answer: $A = \dfrac{1}{r_i + R_S} \dfrac{r_o R_L}{r_o + R_L} \mu$

8.3   Repeat the analysis of Example 8.1 for the voltage-controlled current source model of Figure 8.1. What is the amplifier voltage gain?

Answer: $A = \mu \dfrac{r_i}{(r_i + R_S)} \dfrac{r_o R_L}{(r_o + R_L)}$

8.2   THE BIPOLAR JUNCTION TRANSISTOR (BJT)

The *pn* junction studied in Chapter 7 forms the basis of a large number of semiconductor devices. The semiconductor diode, a two-terminal device, is the most direct application of the *pn* junction. In this section, we introduce the **bipolar junction transistor (BJT).** As we did in analyzing the diode, we will introduce the physics of transistor devices as intuitively as possible, resorting to an analysis of their *i-v* characteristics to discover important properties and applications.

A BJT is formed by joining three sections of semiconductor material, each with a different doping concentration. The three sections can be either a thin *n* region sandwiched between p^+ and *p* layers, or a *p* region between *n* and n^+ layers, where the superscript "plus" indicates more heavily doped material. The resulting BJTs are called *pnp* and *npn* transistors, respectively; we shall discuss only the latter in this chapter. Figure 8.4 illustrates the approximate construction, symbols, and nomenclature for the two types of BJTs.

The operation of the *npn* BJT may be explained by considering the transistor as consisting of two back-to-back *pn* junctions. The **base-emitter**

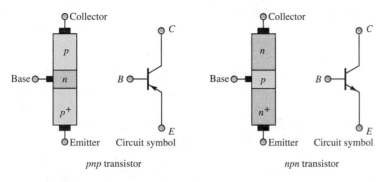

Figure 8.4 Bipolar junction transistors

(BE) junction acts very much like a diode when it is forward-biased; thus, one can picture the corresponding flow of hole and electron currents from base to emitter when the collector is open and the *BE* junction is forward-biased, as depicted in Figure 8.5. Note that the electron current has been

= Electron flow

= Hole flow

The *BE* junction acts very much as an ordinary diode when the collector is open. In this case, $I_B = I_E$.

Figure 8.5 Current flow in an *npn* BJT

When the *BC* junction is reverse-biased, the electrons from the emitter region are swept across the base into the collector.

= Electron flow

= Hole flow

Figure 8.6 Flow of emitter electrons into the collector in an *npn* BJT

shown larger than the hole current, because of the heavier doping of the *n* side of the junction. Some of the electron-hole pairs in the base will recombine; the remaining charge carriers will give rise to a net flow of current from base to emitter. It is also important to observe that the base is much narrower than the emitter section of the transistor.

Imagine, now, connecting the **base-collector (BC) junction** so that it is reverse-biased. In this case, an interesting phenomenon takes place: the electrons "emitted" by the emitter with the *BE* junction forward-biased reach the very narrow base region, and after a few are lost to recombination in the base, most of these electrons are "collected" by the collector. Figure 8.6 illustrates how the reverse bias across the *BC* junction is in such a direction as to sweep the electrons from the emitter into the collector. This phenomenon can take place because the base region is kept particularly narrow. Since the base is narrow, there is a high probability that the electrons will have gathered enough momentum from the electric field to cross the reversed-biased collector-base junction and make it into the collector. The result is that there is a net flow of current from collector to emitter (opposite in direction to the flow of electrons), in addition to the hole current from base to emitter. The electron current flowing into the collector through the base is substantially larger than that which flows into the base from the external circuit. One can see from Figure 8.6 that if KCL is to be satisfied, we must have

$$I_E = I_B + I_C \tag{8.1}$$

The most important property of the bipolar transistor is that the small base current controls the amount of the much larger collector current:

$$I_C = \beta I_B \qquad\qquad (8.2)$$

where β is a current amplification factor dependent on the physical properties of the transistor. Typical values of β range from 50 to 200. The operation of a *pnp* transistor is completely analogous to that of the *npn* device, with the roles of the charge carriers (and therefore the signs of the currents) reversed. The symbol for a *pnp* transistor was shown in Figure 8.4.

The detailed operation of bipolar transistors can be explained by resorting to a detailed physical analysis of the *npn* or *pnp* structure of these devices. The reader interested in such a discussion of transistors is referred to any·one of the number of excellent books on semiconductor electronics. The aim of this book, on the other hand, is to provide an introduction to the basic principles of transistor operation by means of simple linear circuit models based on the device *i-v* characteristic. Although it is certainly useful for the non–electrical engineer to understand the basic principles of operation of electronic devices, it is unlikely that most readers will engage in the design of high-performance electronic circuits or will need a detailed understanding of the operation of each device. The present chapter will therefore serve as a compendium of the basic ideas enabling an engineer to read and understand electronic circuit diagrams, and to specify the requirements of electronic instrumentation systems. The reader interested in furthering his or her knowledge of electronic circuits will find the fundamentals provided by this chapter very useful when undertaking the study of a more advanced textbook. The focus of this section will be on the analysis of the *i-v* characteristic of the *npn* BJT, based on the circuit notation defined in Figure 8.7. The device *i-v* characteristics will be presented qualitatively, without deriving the underlying equations, and will be utilized in constructing circuit models for the device.

The number of independent variables required to uniquely define the operation of the transistor may be determined by applying KVL and KCL to the circuit of Figure 8.7. It should be apparent that two voltages and two currents are sufficient to specify the operation of the device. Note that, since the BJT is a three-terminal device, it will not be sufficient to deal with a single *i-v* characteristic; it will soon become apparent that two such characteristics are required to explain the operation of this device. One of these characteristics relates the base current, i_B, to the base-emitter voltage, v_{BE}; the other relates the collector current, i_C, to the collector-emitter voltage, v_{CE}. As will be shown, the latter characteristic actually consists of a *family* of curves. To determine these *i-v* characteristics, consider the *i-v* curves of Figures 8.8 and 8.9, using the circuit notation of Figure 8.7. In Figure 8.8, the collector is open and the *BE* junction is shown to be very similar to a diode. The ideal current source, I_{BB}, injects a base current, which causes the junction to be forward-biased. By varying I_{BB}, one can obtain the open-collector *BE* junction *i-v* curve shown in the figure.

If a voltage source were now to be connected to the collector circuit, the voltage v_{CE} and, therefore, the collector current, i_C, could be varied,

The operation of the BJT is defined in terms of two currents and two voltages: i_B, i_C, v_{CE}, and v_{BE}.

KCL: $i_E = i_B + i_C$
KVL: $v_{CE} = v_{CB} + v_{BE}$

Figure 8.7 Definition of BJT voltages and currents

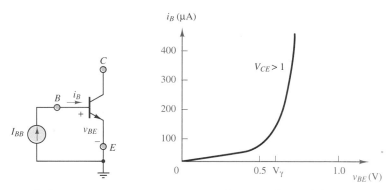

Figure 8.8 Determining the *BE* junction open-collector *i-v* characteristic

in addition to the base current, i_B. The resulting circuit is depicted in Figure 8.9. By varying both the base current and the collector-emitter voltage, one could then generate a plot of the device **collector characteristic.** This is also shown in Figure 8.9. Note that this figure depicts not just a single $i_C\text{-}v_{CE}$

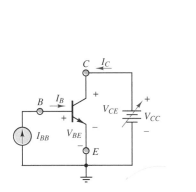

Figure 8.9(a) Ideal test circuit to determine the *i-v* characteristic of a BJT

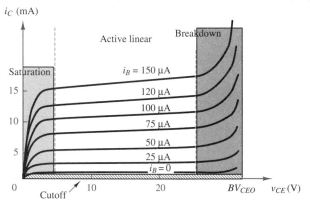

Figure 8.9(b) The collector-emitter output characteristics of a BJT

curve, but an entire family, since for each value of the base current, i_B, an $i_C\text{-}v_{CE}$ curve can be generated. Four regions are identified in the collector characteristic:

1. The **cutoff region,** where the base current is very small and essentially no collector current flows.
2. The **active linear region,** in which the transistor can act as a linear amplifier.
3. The **saturation region,** in which both junctions are forward-biased.
4. The **breakdown region,** which determines the physical limit of operation of the device.

From the curves of Figure 8.9, we note that as v_{CE} is increased, the collector current increases rapidly, until it reaches a nearly constant value;

this condition holds until the collector junction breakdown voltage, BV_{CEO}, is reached (for the purposes of this book, we shall not concern ourselves with the phenomenon of breakdown, except in noting that there are maximum allowable voltages and currents in a transistor). If we were to repeat the same measurement for a set of different values of i_B, the corresponding value of i_C would change in proportion; hence, the family of collector characteristic curves.

Determining the Operating State of a BJT

Before we discuss common circuit models for the BJT, it will be useful to consider the problem of determining the operating state of the transistor. A few simple voltage measurements permit a quick determination of the state of a transistor placed in a circuit. Consider, for example, the BJT described by the curves of Figure 8.9 if it is placed in the circuit of Figure 8.10. In this figure, voltmeters are used to measure the value of the collector, emitter, and base voltages. Can these simple measurements identify the operating state of the transistor? Assume that the measurements reveal the following conditions:

$$V_1 = 2 \text{ V}$$
$$V_2 = 1.3 \text{ V}$$
$$V_3 = 8 \text{ V}$$

What can be said about the operating state of the transistor?

The first observation is that knowing V_B and V_E permits determination of V_{BE}: $V_{BE} = 0.7$ V. Thus, we know that the BE junction is forward-biased. Another quick calculation permits determination of the relationship between base and collector current: the base current is equal to

$$I_B = \frac{4 - V_B}{40,000} \doteq 50 \ \mu A$$

while the collector current is

$$I_C = \frac{12 - V_C}{1,000} = 4 \text{ mA}$$

Thus, the current amplification (or gain) factor for the transistor is

$$\frac{I_C}{I_B} = 80$$

Such a value for the current gain suggests that the transistor is in the linear active region, because substantial current amplification is taking place. Finally, the collector-to-emitter voltage, V_{CE}, is found to be: $V_{CE} = V_C - V_E = 8 - 1.3 = 6.7$ V.

At this point, you should be able to locate the operating point of the transistor on the curves of Figures 8.8 and 8.9. These three variables uniquely determine the state of the transistor in the $I_C - V_{CE}$ and $I_B - V_{BE}$ characteristic curves. What would happen if the transistor were not in the linear active

Figure 8.10 Determination of the operating state of a BJT

region? The following examples answer this question and provide further insight into the operation of the bipolar transistor.

Example 8.2

The objective of this example is to determine the operating state of a BJT. The circuit is identical to the one in Figure 8.10, except that now the base voltage source, V_{BB}, is short-circuited.

Solution:

Since the base voltage source is short-circuited, the base will be at zero volts, the emitter-base junction will not conduct, and the emitter currents will be nearly zero.

From equation 8.1, it may also be concluded that the collector current must be zero. A glance at Figure 8.9 reveals that the transistor is in the cut-off state. The readings of the three voltmeters will be zero for V_B and V_E and $+12$ V for V_C, because there is no voltage drop across the 1-kΩ resistor.

In general, if the base supply voltage is not sufficient to forward-bias the base-emitter junction, the transistor will be in the cutoff region.

Example 8.3

Figure 8.11

For the circuit in Figure 8.11, the voltmeter readings are $V_1 = 2.7$ V, $V_2 = 2$ V, and $V_3 = 2.3$ V. Check the operating state of the transistor.

Solution:

One way to check the state of a transistor is to calculate V_{BC} and V_{BE} to see whether the *BE* and *CB* junctions are forward- or reverse-biased. Saturation is characterized by operation of both junctions with forward bias and with very small voltage drops, whereas active operation is characterized by a forward-biased *BE* junction and a reverse-biased *BC* junction. Thus, for an *npn* transistor in saturation, $V_{BE} > 0$ and $V_{BC} > 0$.

The circuit of Figure 8.11 has

$$V_{BE} = V_B - V_E = 2.7 - 2.0 = 0.7 \text{ V}$$

and

$$V_{BC} = V_B - V_C = 2.7 - 2.3 = 0.4 \text{ V}$$

Thus, the transistor is operating in the saturation region. The value $V_{CE} = V_C - V_E = 2.3 - 2 = 0.3$ V is very small. This is usually a good indication that a BJT is operating in saturation. You should now try to locate the operating point of this transistor in Figure 8.9, observing that $I_C = \frac{V_{CC} - V_{CE}}{R_C} = \frac{12 - 0.3}{1,000} = 11.7$ mA.

Selecting an Operating Point for a BJT

The family of curves shown in the collector *i-v* characteristic of Figure 8.9 reflects the dependence of the collector current on the base current. For each value of the base current, i_B, there exists a corresponding $i_C - v_{CE}$ curve. Thus, by appropriately selecting the base current and collector current (or collector-emitter voltage), we can determine the operating point, or **Q point,** of the transistor. The Q point of a device is defined in terms of the **quiescent** (or **idle**) **currents** and **voltages** that are present at the terminals of the device when DC supplies are connected to it. The circuit of Figure 8.12 illustrates an ideal **DC bias circuit,** used to set the Q point of the BJT in the approximate center of the collector characteristic. The circuit shown in Figure 8.12 is not a practical DC bias circuit for a BJT amplifier, but it is very useful for the purpose of introducing the relevant concepts. A practical bias circuit will be discussed later in this section.

> By appropriate choice of I_{BB}, R_C, and V_{CC}, the desired Q point may be selected.

Applying KVL around the base-emitter and collector circuits, we obtain the following equations:

$$I_B = I_{BB} \tag{8.3}$$

and

$$V_{CE} = V_{CC} - I_C R_C \tag{8.4}$$

which can be rewritten as

$$I_C = \frac{V_{CC}}{R_C} - \frac{V_{CE}}{R_C} \tag{8.5}$$

Figure 8.12 A simplified bias circuit for a BJT amplifier

Note the similarity of equation 8.5 to the load-line curves of Chapters 3 and 7. Equation 8.5 represents a line that intersects the I_C axis at $I_C = V_{CC}/R_C$ and the V_{CE} axis at $V_{CE} = V_{CC}$. The slope of the load line is $-1/R_C$. Since the base current, I_B, is equal to the source current, I_{BB}, the operating point may be determined by noting that the load line intersects the entire collector family of curves. The intersection point at the curve that corresponds to the base current $I_B = I_{BB}$ constitutes the operating, or Q, point. The load line corresponding to the circuit of Figure 8.12 is shown in Figure 8.13, superimposed on the collector curves for the 2N3904 transistor (data sheets for the 2N3904 transistor are included at the end of the chapter). In Figure 8.13, $V_{CC} = 15$ V, $\frac{V_{CC}}{R_C} = 40$ mA, $I_{BB} = 150$ μA.

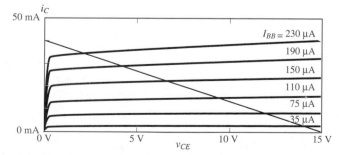

Figure 8.13 Load-line analysis of a simplified BJT amplifier

Once an operating point is established and DC currents I_{CQ} and I_{BQ} are flowing into the collector and base, respectively, the BJT can serve as a linear amplifier, as will be explained in Section 8.3. Example 8.4 serves as an illustration of the DC biasing procedures just described.

Example   8.4

The objective of this example is to illustrate numerically the DC biasing procedure for a BJT. For practical reasons, instead of an ideal current source, the base circuit shown in Figure 8.14 consists of a battery in series with a base resistance, R_B.

Assuming a base resistance $R_B = 62.7$ kΩ, a collector resistance $R_C = 375$ Ω, a collector DC voltage source $V_{CC} = 15$ V, and a base voltage supply $V_{BB} = 10$ V, we can sketch the load line shown in Figure 8.13. The equation for the load line is

$$V_{CE} = V_{CC} - R_C I_C$$

or

$$V_{CE} = 15 - 375 I_C$$

The value of the base current will then determine the Q point according to the approximate expression

$$I_B = \frac{V_{BB} - 0.6}{R_B}$$

in which we have assumed the BE junction is forward-biased and is therefore behaving like a diode with offset voltage equal to 0.6 V. This corresponds to a base current of 150 μA. The intersection of the load line with the collector i-v curve for a base current of 150 μA corresponds, approximately, to $V_{CE} = 7$ V and $I_C = 22$ mA. These values define the Q point of the transistor as

$$I_{BQ} = 150 \ \mu A; \ I_{CQ} = 22 \ mA; \ V_{CEQ} = 7 \ V$$

where the additional subscript Q is used to indicate the Q point. It is important to note that the quiescent DC base and collector currents flowing through the transistor at all times are necessary for amplification to occur. Thus, a transistor dissipates power continuously when subject to a DC bias, whether it is actually amplifying an input signal or not.

Figure 8.14

How can a transistor amplify a signal, then, given the $v_{BE} - i_B$ and $v_{CE} - i_C$ curves discussed in this section? The small-signal amplifier properties of the transistor are best illustrated by analyzing the effect of a small sinusoidal current superimposed on the DC current flowing into the base. The circuit of Figure 8.15 illustrates the idea, by including a small-signal AC source, of strength ΔV_B, in series with the base circuit. The effect of this AC source is to cause sinusoidal oscillations ΔI_B about the Q point, that is, around I_{BQ}. A quick study of the collector characteristic indicates that for a sinusoidal oscillation in i_B, a corresponding, but larger, oscillation will take place in the collector current. Figure 8.16 illustrates the concept. Note that the base current oscillates between 110 and 190 μA, causing the collector current to correspondingly fluctuate between 15.3 and 28.6 mA. The notation that will be used to differentiate between DC and AC (or fluctuating) components of transistor voltages and currents is as follows: DC (or quiescent) currents and voltages will be denoted by uppercase symbols; for example: I_B, I_C, V_{BE}, V_{CE}. AC components will be preceded by a "Δ": $\Delta I_B(t)$, $\Delta I_C(t)$, $\Delta V_{BE}(t)$, $\Delta V_{CE}(t)$. The complete expression for one of these quantities will therefore include both a DC term and a time-varying, or AC, term. For example, the collector current may be expressed by $i_C(t) = I_C + \Delta I_C(t)$.

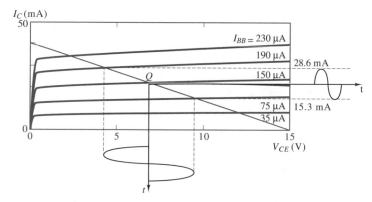

Figure 8.15 Circuit illustrating the amplification effect in a BJT

Figure 8.16 Amplification of sinusoidal oscillations in a BJT

The *i-v* characteristic of Figure 8.16 illustrates how this increase in collector current follows the same sinusoidal pattern of the base current but is greatly amplified. Thus, the BJT acts as a *current amplifier,* in the sense that any oscillations in the base current appear amplified at the collector. Since the voltage across the collector resistance, R_C is proportional to the collector current, one can see how the collector voltage is also affected by the amplification process. Example 8.5 illustrates numerically the effective amplification of the small AC signal that takes place in the circuit of Figure 8.15.

Example   8.5

The present example shows how amplification of small signals takes place in a BJT circuit. Let us assume that a DC operating point has been established—for example, as was illustrated in Example 8.4. In the circuit of Figure 8.17, the AC voltage source represents the signal to be amplified (e.g., the output of a transducer). The AC source induces an oscillation in the base current (think of this in terms of the principle of superposition). Correspondingly, a fluctuation in the collector current takes place, proportional to the base current oscillation. The current amplification factor may be deduced from the collector characteristic curve. Thus, the transistor acts as a current amplifier. In this particular configuration, voltage amplification takes place as well.

Figure 8.17

In the circuit of Figure 8.17, assume that $V_{CC} = 15$ V, $V_{BB} = 2.1$ V, $R_B = 10$ kΩ, and $R_C = 375$ Ω.

1. Determine the operating point of the transistor.
2. Determine the current gain $\beta = \frac{I_C}{I_B}$ at the quiescent point of the transistor.
3. Find the AC voltage gain of the circuit if the voltage gain A is defined as

$$A = \frac{\Delta V_o}{\Delta V_B}$$

Assume that each current and voltage consists of a DC and an AC component; for example,

$$i_C(t) = I_C + \Delta I_C(t)$$

where I_C is the DC component and $\Delta I_C(t) = \Delta I_C \cos(\omega t)$.

Solution:

1. The current into the base may be expressed as

$$i_B(t) = I_B + \Delta I_B(t)$$

where $\Delta I_B(t)$ is a small, time-dependent change in the base current given by

$$\Delta I_B(t) = \frac{\Delta V_B(t)}{10 \text{ k}\Omega}$$

and where it is assumed that the base-emitter voltage stays relatively constant and the *BE* junction forward resistance is negligible, relative to the 10-kΩ resistor. Both assumptions are quite reasonable (at least initially) on the basis of the circuit model for the diode presented in Chapter 7, since the *BE* junction acts, in effect, as a diode. Thus, we are in effect assuming that the BE junction acts as an ideal diode and that, therefore, $\Delta V_{BE} \approx 0$.

Consider, now, the DC base circuit. Assuming that the *BE* junction voltage is 0.6 V when the junction is forward-biased, and neglecting the AC source (in effect, this is equivalent to applying the principle of superposition to determine the voltages and currents due exclusively to the DC source), we may simplify the base circuit as shown in Figure 8.18. Application of KVL around the base circuit reveals that the DC base current is

Figure 8.18

$$I_B = \frac{2.1 - 0.6}{10,000} = 150 \ \mu\text{A}$$

The operating point can then be found by plotting the load line on the characteristic curves shown in Figure 8.19. The Q point is found to be

$$I_{BQ} = 150 \ \mu\text{A}; \ V_{CEQ} = 7.2 \text{ V}; \ I_{CQ} = 22 \text{ mA}$$

2. The current gain, β, is found by considering the effect of an oscillation in the base current on the collector current. If we follow the load line between, say, base currents of 110 and 190 μA, we note that at the intercepts with these two curves, the corresponding collector currents are 15.3 mA and 28.6 mA, respectively. Thus, for a change in base current of 80 μA, we see a corresponding change in collector current of 13.3 mA. The ratio between these two numbers is the current gain of the transistor: $\beta = 166.25$.

3. The total collector current is the sum of the DC current due to the DC bias
    circuit and the AC current caused by the amplified oscillations in the base
    current:

$$i_C(t) = 0.022 + 166.25\Delta I_B(t)$$

Thus, the voltage across the collector resistor, which may be viewed as the
output voltage in this example, also has corresponding DC and AC components:

$$v_o(t) = V_o + \Delta V_o(t)$$
$$= -i_C(t)R_C$$
$$= -I_{CQ}R_C - 166.25\Delta I_B(t)R_C$$

If we consider just the AC component of the output voltage, $\Delta V_o = \beta\Delta I_B R_C$,
we see that this is also an amplified version of the AC base voltage, ΔV_B. The
AC voltage gain is therefore given by

$$A_v = \frac{\Delta V_o}{\Delta V_B} = \frac{-\Delta I_C R_C}{\Delta I_B R_B} = \frac{-\beta\Delta I_B R_C}{\Delta I_B R_B} = -\beta\frac{R_C}{R_B} = -6.23$$

Note how the voltage gain is very much dependent on the base and collector
resistances in the circuit.

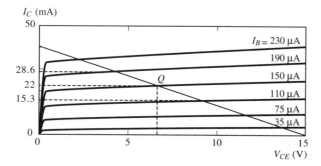

Figure 8.19 Operating point on the characteristic curve

In discussing the DC biasing procedure for the BJT, we pointed out
that the simple circuit of Figure 8.12 would not be a practical one to use
in an application circuit. In fact, the more realistic circuit of Example
8.4 is also not a practical biasing circuit. The reasons for this statement
are that two different supplies are required (V_{CC} and V_{BB})—a requirement
that is not very practical—and that the resulting DC bias (operating) point
is not very stable. This latter point may be made clearer by pointing out
that the location of the operating point could vary significantly if, say, the
current gain of the transistor, β, were to vary. A circuit that provides great
improvement on both counts is shown in Figure 8.20. Observe, first, that the
voltage supply, V_{CC}, appears across the pair of resistors R_1 and R_2, and that
therefore the base terminal for the transistor will see the Thévenin equivalent

Figure 8.20 Practical BJT
self-bias DC circuit

circuit composed of the equivalent voltage source.

$$V_{BB} = \frac{R_2}{R_1 + R_2} V_{CC} \qquad (8.6)$$

and of the equivalent resistance

$$R_B = R_1 \parallel R_2 \qquad (8.7)$$

Figure 8.21 shows a redrawn DC bias circuit that makes this observation more evident. The circuit to the left of the dashed line is now represented by the equivalent circuit composed of V_{BB} and R_B.

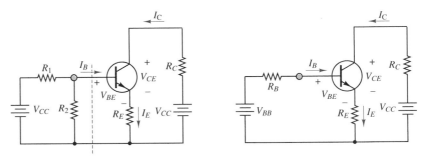

Figure 8.21 DC self-bias circuit represented in equivalent-circuit form

Recalling that the BE junction acts much as a diode, the following equations describe the DC operating point of the self-bias circuit. Around the base-emitter circuit,

$$V_{BB} = I_B R_B + V_{BE} + I_E R_E = [R_B + (\beta + 1)R_E]I_B + V_{BE} \qquad (8.8)$$

where V_{BE} is the BE junction voltage (diode forward voltage) and $I_E = (\beta + 1)I_B$. Around the collector circuit, on the other hand, the following equation applies:

$$V_{CC} = I_C R_C + V_{CE} + I_E R_E = I_C \left(R_C + \frac{\beta + 1}{\beta} R_E \right) + V_{CE} \qquad (8.9)$$

since $I_E = I_B + I_C = (1/\beta + 1)I_C$. These two equations may be solved to obtain an expression for the base current,

$$I_B = \frac{V_{BB} - V_{BE}}{R_B + (\beta + 1)R_E} \qquad (8.10)$$

from which the collector current can be determined as $I_C = \beta I_B$; and an expression for the collector-emitter voltage:

$$V_{CE} = V_{CC} - I_C \left(R_C + \frac{\beta + 1}{\beta} R_E \right) \qquad (8.11)$$

This last equation is the load-line equation for the bias circuit. Note that the effective load resistance seen by the DC collector circuit is no longer just R_C, but is now given by

$$R_C + \frac{\beta + 1}{\beta} R_E \approx R_C + R_E$$

The following example provides a numerical illustration of the analysis of a DC self-bias circuit for a BJT.

Example   8.6

Find the DC operating point of the circuit shown in Figure 8.20, given $R_1 = 100$ kΩ, $R_2 = 50$ kΩ, $R_C = 5$ kΩ, $R_E = 3$ kΩ, $V_{CC} = 15$ V, $\beta = 100$, and $V_{BE} = 0.7$ V.

Solution:

The first step is to determine the equivalent base supply voltage, V_{BB}, and the equivalent base resistance, R_B. From the expression

$$V_{BB} = \frac{R_2}{R_1 + R_2} V_{CC}$$

we obtain

$$V_{BB} = \frac{50,000}{100,000 + 50,000}(15 \text{ V}) = 5 \text{ V}$$

and the equivalent base resistance can be computed to be

$$R_B = R_1 \parallel R_2 = \frac{100,000 \times 50,000}{100,000 + 50,000} = 33.3 \text{ k}\Omega$$

Now, it is possible to compute the base current:

$$I_B = \frac{V_{BB} - V_{BE}}{R_B + (\beta + 1)R_E} = \frac{5 - 0.7}{33,333 + (100 + 1) \times 3,000} \text{ A}$$

$$= 12.8 \ \mu\text{A}$$

As a check, we can recompute the base voltage supply:

$$V_{BB} = V_{BE} + [R_B + (\beta + 1)R_E]I_B$$
$$= 0.7 + [33,333 + (100 + 1) \times 3,000] \times 12.8 \times 10^{-6}$$
$$= 5.0 \text{ V}$$

Knowing the current amplification factor, β, we can easily find the collector current:

$$I_C = \beta I_B = 1.28 \text{ mA}$$

Finally, the collector-to-emitter voltage can be computed from the expression

$$V_{CE} = V_{CC} - I_C \left(R_C + \frac{\beta + 1}{\beta} R_E \right)$$

$$= 15 - 1.28 \times 10^{-3} \times \left(5,000 + \frac{101}{100} \times 3,000 \right)$$

$$= 4.72 \text{ V}$$

Thus, the Q point of the transistor of Figure 8.20, given the component values of this example, is

$$I_{BQ} = 12.8 \ \mu\text{A} \qquad I_{CQ} = 1.28 \text{ mA} \qquad V_{CEQ} = 4.72 \text{ V}$$

The material presented in this section has illustrated the basic principles that underlie the operation of a BJT and the determination of its Q point. In the next two sections, we shall develop some simple circuit models for the BJT that will enable us to analyze the transistor amplifier in the linear active region using the familiar tools of linear circuit analysis.

DRILL EXERCISES

8.4 Describe the operation of a *pnp* transistor in the active region, by analogy with that of the *npn* transistor.

8.5 For the circuit given in Figure 8.11, the readings are $V_1 = 3$ V, $V_2 = 2.4$ V, and $V_3 = 2.7$ V. Determine the operating state of the transistor.

> Answer: Saturation

8.6 For the circuit given in Figure 8.21, assume that $R_B = 50$ kΩ, $R_E = 200$ Ω, $R_C = 1$ kΩ, $\beta = 100$, and $V_{CC} = 14$ V. Find the value of V_{BB} that yields a collector current $I_C = 6.3$ mA. What is the corresponding collector-emitter voltage (assume that $V_{BE} = 0.6$ V and that the transistor is in the active region)?

> Answer: $V_{BB} = 5$ V; $V_{CE} = 6.44$ V

8.7 What percent change in collector current would result if β were changed to 150 in Example 8.6? Why does the collector current increase by less than 50 percent?

> Answer: 3.74%; because R_E provides a negative feedback action that will keep I_C and I_E almost constant.

8.3   BJT CIRCUIT MODELS

The *i-v* characteristics and the simple circuits of the previous sections indicate that the BJT acts very much as a current-controlled current source: a small amount of current injected into the base can cause a much larger current to flow into the collector. This conceptual model, although somewhat idealized, is useful in describing a **large-signal model** for the BJT, that is, a model that describes the behavior of the BJT in the presence of relatively large base and collector currents, close to the limit of operation of the device. A more careful analysis of the collector curves will later reveal that it is also possible to generate a **small-signal model,** a model that describes the operation of the transistor as a linear amplifier of small AC signals. These models are certainly not a complete description of the properties of the BJT, nor do they accurately depict all of the effects that characterize the operation of such devices (for example, temperature effects, saturation, and cutoff); however, they are adequate for the intended objectives of this book, in that they provide a good, qualitative feel for the important features of transistor amplifiers.

Large-Signal Model of the BJT

The large-signal model for the BJT recognizes three basic operating modes of the transistor. When the *BE* junction is reversed-biased, no base current (and therefore no forward collector current) flows, and the transistor acts virtually as an open circuit; the transistor is said to be in the *cutoff region*. In practice, there is always a leakage current flowing through the collector, even when $V_{BE} = 0$ and $I_B = 0$. This leakage current is denoted by I_{CEO}. When the *BE* junction becomes forward-biased, the transistor is said to be in the *active region*, and the base current is amplified by a factor of β at the collector:

$$I_C = \beta I_B \tag{8.12}$$

Since the collector current is controlled by the base current, the controlled-source symbol is used to represent the collector current. Finally, when the base current becomes sufficiently large, the collector-emitter voltage, V_{CE}, reaches its saturation limit, and the collector current is no longer proportional to the base current; this is called the *saturation region*. The three conditions are described in Figure 8.22 in terms of simple circuit models. The corresponding collector curves are shown in Figure 8.23.

Example 8.7 illustrates the application of this large-signal model in a practical circuit and illustrates how to determine which of the three states is applicable, using relatively simple analysis.

Figure 8.22 BJT large-signal model

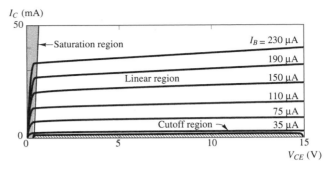

Figure 8.23 BJT collector characteristic

Example 8.7   LED Driver

Figure 8.24 LED driver circuit

The circuit shown in Figure 8.24 is being used to "drive" (i.e., provide power for) a light-emitting diode (LED) from a desktop computer. The signal available from the computer consists of a 5-V low-current output. The reason for using a BJT (rather than driving the LED directly with the signal available from the computer) is that the LED requires a substantial amount of current (at least 15 mA for the device used in this example) and the computer output is limited to 5 mA. Thus, some current amplification is required. The LED has an offset voltage of 1.4 V and a maximum power dissipation rating of 280 mW. The circuit has been designed so that the transistor will be either in cutoff, when the LED is to be turned off, or in saturation, when the LED is to be turned on.

Assume that the base-emitter junction of the transistor has an offset voltage of 0.7 V and that $R_B = 1$ kΩ and $R_S = 42.5$ Ω.

1. When the computer is supplying 5 V to the circuit, is the transistor in cutoff, in saturation, or in the linear region of operation?

2. Determine the current in the LED when "turned on," and, thus, state whether the LED will exceed its power rating.

Solution:

Figure 8.25 *BE* circuit for LED driver

1. The base-emitter circuit is considered first, to determine whether the *BE* junction is forward-biased. The equivalent circuit is shown in Figure 8.25.

   Writing KVL around the base circuit, we obtain

   $$(5 - 0.7) = I_B(1,000)$$

   or

   $$I_B = 4.3 \text{ mA}$$

   Since this current is greater than zero (i.e., since positive current flows into the base), the transistor is not in cutoff.

   Next, we need to determine whether the transistor is in the linear active or in the saturation region. One method that can be employed to determine whether the device is in the active region is to assume that it is and to solve for the circuit that results from the assumed condition. If the resulting equations are consistent, then the assumption is justified. Assuming that the transistor is in its active region, the following equations apply:

$$I_C = \beta I_B$$

or

$$I_C = 95(4.3 \text{ mA}) = 408.5 \text{ mA}$$

With reference to the circuit of Figure 8.26, KVL may be applied around the collector circuit, to obtain

$$5 = 1.4 + V_{CE} + I_C R_S$$

or

$$(5 - 1.4) - 408.5(42.5) = V_{CE}$$

or

$$V_{CE} = -13.76 \text{ V}$$

This result is clearly not possible, since the supply voltage is only 5 V! It must therefore be concluded that the transistor is not in the linear region and is therefore in saturation. To test this hypothesis, we can substitute the nominal saturation voltage for the BJT in place of V_{CE} (a typical saturation voltage would be $V_{CE \text{ sat}} = 0.2$ V) and verify that in this case we obtain a consistent solution:

$$I_C = \frac{(5 - 1.4 - 0.2)}{42.5} = 80 \text{ mA}$$

This is a reasonable result, stating that the collector current in the saturation region for the given circuit is of the order of 80 mA.

2. In part 1, it was determined that the transistor was in saturation. Using the circuit model for saturation, we may draw the equivalent circuit of Figure 8.27. Since $I_C = 80$ mA, the power dissipated by the LED may be calculated as follows:

$$P_{LED} = I_C V_{LED} = 80 \text{ mA} \times 1.4 \text{ V} = 112 \text{ mW}$$

Thus, the power limitation of the LED will not be exceeded.

Figure 8.26 Equivalent circuit of LED driver, assuming that BJT is in the linear active region

Figure 8.27 LED driver equivalent circuit, assuming that the BJT is in the saturation region

---

Example   8.8

This example demonstrates the use of the large-signal model for the BJT in a switching circuit. The input voltage, V_{in}, switches between two values, representing the circuit on and off conditions. In the example, we analyze the behavior of the transistor circuit of Figure 8.28 for different values of V_{in}, assuming a nominal transistor current gain $\beta = 60$. Find the output voltage, V_{out}, for: (1) $V_{in} = 0.2$ V; (2) $V_{in} = 0.6$ V; and (3) $V_{in} = 0.8$ V. The collector characteristic of transistor T is shown in Figure 8.9, and the base characteristic is shown in Figure 8.8.

Figure 8.28

Solution:

1. For $V_{in} = V_{BE} = 0.2$ V, the base-emitter junction is not forward-biased. Thus,

$$I_B \approx 0 \qquad I_C \approx 0 \qquad V_{out} = 15 \text{ V}$$

The transistor is in the cutoff state.

2. For $V_{in} = V_{BE} = 0.6$ V, the BE junction is forward-biased. From Figure 8.8, we find that $I_B \approx 125$ μA and consequently compute:

$$I_C = \beta I_B = 60 \times 0.125 = 7.5 \text{ mA}$$

Thus, the output voltage is

$$V_{out} = V_{CE} = 15 - 1,000 \times 0.0075 = 7.5 \text{ V}$$

3. For $V_{in} = V_{BE} = 0.8$ V, $I_B = 400$ μA. If the basic relation $I_C = \beta I_B$ were to hold here, we should have $I_C = 24$ mA and $V_{CE} = V_{CC} - I_C R_C = 15 - 0.024 \times 1,000 = -9$ V. This is an impossible situation. For an *npn* transistor, V_{CE} must be positive. Thus, the linear active region model does not apply, and the transistor is in saturation. Using the large-signal model of the BJT, we draw the equivalent circuit shown in Figure 8.29 and conclude that, for a nominal saturation voltage of 0.2 V, the output voltage is

$$V_{out} = 0.2 \text{ V}$$

The collector current is given by the expression

$$I_C = \frac{V_{CC} - V_{CE \, sat}}{R_C} = \frac{15 - 0.2}{1 \times 10^3} = 14.8 \text{ mA}$$

Note also that the effective β in this particular saturation condition is $I_C / I_B = 14.8/0.4 = 37$, which is smaller than the nominal value of 60, as is always the case for a transistor in saturation.

It is clear that a change as small as 0.6 V in the input voltage, from 0.2 V to 0.8 V, results in a corresponding change in output voltage from 15 V to 0.2 V, that is, from the cutoff state to the saturation state. This property is very useful in switching circuits. This type of circuit finds common application in digital logic devices, to be discussed in Chapter 12.

Figure 8.29 Transistor equivalent circuit using saturation model

The large-signal model of the BJT presented in this section treats the BE junction as an offset diode and assumes that the BJT in the linear active region acts as an ideal controlled current source. In reality, the BE junction is better modeled by considering the forward resistance of the *pn* junction; further, the BJT does not act quite like an ideal current-controlled current source. These phenomena are partially taken into account in the small-signal model introduced in the next section.

Small-Signal Models of the BJT

Small-signal models for the BJT take advantage of the relative linearity of the base and collector curves in the vicinity of an operating point. These linear circuit models work very effectively provided that the transistor voltages and currents remain within some region around the operating point. This condition is usually satisfied in small-signal amplifiers used to magnify low-level signals (e.g., outputs of transducers). For the purpose of our discussion, it is reasonable to use the **hybrid-parameter (h-parameter) small-signal model,** to be discussed presently.

Imagine the collector curves of Figure 8.19 magnified about the Q point. Figure 8.30 graphically depicts the interpretation of each of the h parameters relative to the operating point of the BJT.

The parameter h_{ie} is approximately equal to the ratio $\frac{\Delta V_{BE}}{\Delta I_B}$ in the neighborhood of the Q point; Figure 8.30(a) illustrates how this parameter is equal to the reciprocal of the slope of the $i_B - v_{BE}$ curve at the operating point. Physically, this parameter represents the forward resistance of the BE junction (a diode, in practice).

The parameter h_{re} is representative of the fact that the $i_B - v_{BE}$ curve is slightly dependent on the actual value of the collector-emitter voltage, ΔV_{CE}. However, this effect is virtually negligible in any applications of interest to us. Thus, we shall assume that $h_{re} \approx 0$. Figure 8.30(b) depicts the shift in the $i_B - v_{BE}$ curves represented by h_{re}. A typical value of h_{re} for $V_{CE} \geq 1$ V is around 10^{-2}.

Figure 8.30

The parameter h_{ie} is approximated in Figure 8.30(c) by the current ratio $\frac{\Delta I_C}{\Delta I_B}$. This parameter represents the current gain of the transistor and is approximately equivalent to the parameter β introduced earlier. For the purpose of our discussion, β and h_{fe} will be interchangeable, although they are not exactly identical.

The parameter h_{oe} may be calculated as $h_{oe} = \frac{\Delta I_C}{\Delta V_{CE}}$ from the collector characteristic curves, as shown in Figure 8.30(d). This parameter is a physical indication of the fact that the $i_c - v_{CE}$ curves in the linear active region are not exactly flat; h_{oe} represents the upward slope of these curves and therefore has units of conductance (S). Typical values of h_{oe} are around $10^{-5}S$. We shall often assume that this effect is negligible.

To be more precise, the h parameters are defined by the following set of equations:

$$h_{ie} = \left.\frac{\partial v_{BE}}{\partial i_B}\right|_{I_{BQ}} \quad (\Omega) \tag{8.13}$$

$$h_{oe} = \left.\frac{\partial i_C}{\partial v_{CE}}\right|_{V_{CEQ}} \quad (S) \tag{8.14}$$

$$h_{fe} = \left.\frac{\partial i_C}{\partial i_B}\right|_{I_{BQ}} \quad \left(\frac{A}{A}\right) \tag{8.15}$$

$$h_{re} = \left.\frac{\partial V_{BE}}{\partial V_{CE}}\right|_{V_{CEQ}} \quad \left(\frac{V}{V}\right) \tag{8.16}$$

The circuit of Figure 8.31 illustrates the small-signal model side by side with the BJT circuit symbol.

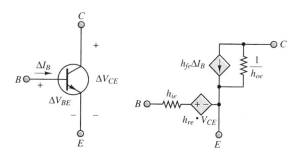

Figure 8.31 h-parameter small-signal model for BJT

Note that the small-signal model assumes that the DC bias point of the transistor has been established. The following convention will be used: each voltage and current is assumed to be the superposition of a DC component (the quiescent voltage or current) and a small-signal AC component. The former is denoted by an uppercase letter, and the latter by uppercase letter preceded by the symbol Δ. Thus

$$i_B = I_{BQ} + \Delta I_B \tag{8.17}$$

$$i_C = I_{CQ} + \Delta I_C \qquad (8.18)$$

$$v_{CE} = V_{CEQ} + \Delta V_{CE} \qquad (8.19)$$

Figure 8.32 depicts the appearance of the collector current $i_C(t)$ when $I_{CQ} = 15 \times 10^{-3}$ A and $\Delta I_C(t) = 0.1 \times 10^{-3} \sin \omega t$ A.

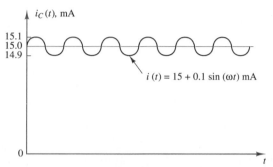

Figure 8.32 Superposition of AC and DC signals

To illustrate the application of the h-parameter small-signal model of Figure 8.31, consider the transistor amplifier circuit shown in Figure 8.33. This circuit is most readily analyzed if DC and AC equivalent circuits are treated separately. To obtain the DC circuit, the AC source is replaced by a short circuit. The resulting DC circuit is shown in Figure 8.34. The DC circuit may be employed to carry out a Q-point analysis similar to that of Examples 8.4 and 8.6. Since the objective of the present section is to illustrate the AC circuit model for the transistor amplifier, it will be assumed that the DC analysis (i.e., selection of the appropriate Q point) has already been carried out, and that a suitable operating point has been established.

$$i_B = I_{BQ} + \Delta I_B$$
$$i_C = I_{CQ} + \Delta I_C$$
$$v_{CE} = V_{CEQ} + \Delta V_{CE}$$

Figure 8.33 BJT amplifier

Figure 8.34 DC equivalent circuit for the BJT amplifier of Figure 8.33

Replacing the DC sources with short circuits, we obtain the AC equivalent circuit of Figure 8.35. The transistor may now be replaced by its h-parameter small-signal model, also shown in Figure 8.35. We may simplify the model by observing that h_{oe}^{-1} is a very large resistance and that if the load resistance R_L (in parallel with h_{oe}^{-1}) is small (i.e., if $R_L h_{oe} \le 0.1$),

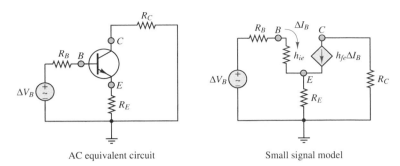

AC equivalent circuit   Small signal model

Figure 8.35 AC equivalent circuit and small-signal model for the amplifier of Figure 8.33

the resistor h_{oe}^{-1} in the model may be ignored. The linear AC equivalent circuit makes it possible to take advantage of the circuit analysis techniques developed in Chapters 2 and 3 to analyze the operation of the amplifier. For example, application of KVL around the base circuit loop yields the following equation:

$$\Delta V_B = \Delta I_B R_B + \Delta I_B h_{ie} + (h_{fe} + 1)\Delta I_B R_E \tag{8.20}$$

which can be solved to obtain

$$\frac{\Delta I_B}{\Delta V_B} = \frac{1}{R_B + h_{ie} + (h_{fe} + 1)R_E} \tag{8.21}$$

and

$$\Delta V_C = -\Delta I_C R_C$$
$$= -h_{fe}\Delta I_B R_C = \frac{-h_{fe}R_C}{R_B + h_{ie} + (h_{fe} + 1)R_E}\Delta V_B \tag{8.22}$$

Then, the **AC voltage gain** of the amplifier is given by the expression

$$A_V = \frac{\Delta V_C}{\Delta V_B} = \frac{-h_{fe}R_C}{R_B + h_{ie} + (h_{fe} + 1)R_E} \tag{8.23}$$

The small-signal model for the BJT will be further exploited in the next section.

Example   8.9

This example illustrates the computation of the AC equivalent circuit model for a typical *npn* transistor, the 2N5088. This transistor is designed for low-level, low-noise amplifier applications. Using the information $h_{fe} = 350$, $h_{ie} = 1.4$ kΩ, and $h_{oe} = 150$ μS, and the AC equivalent circuit of Figure 8.31, determine the nominal AC voltage gain of the transistor amplifier of Figure 8.33. Assume that $V_{CC} = 12$ V, $V_{BB} = 6$ V, $R_B = 100$ kΩ, $R_C = 500$ Ω, and $R_E = 100$ Ω.

Solution:

Before the AC model can be constructed, the operating point of the transistor must be determined. Using the procedure of Example 8.4, we establish the DC operating point as follows:

$$V_{CE} = V_{CC} - R_C I_C - R_E (I_C + I_B) \approx 12 - 600 I_C$$

With $h_{fe} = \beta = 350$, we obtain

$$I_B = \frac{V_{BB} - 0.6}{R_B + (1 + \beta)R_E} = 17.38 \ \mu A$$

$$I_C = \beta I_B = 6.08 \ mA$$

and can therefore solve for V_{CE}:

$$V_{CE} = 12 - 0.6 \times 6.08 = 8.35 \ V$$

Thus, the Q point for the circuit is

$$I_{BQ} = 17.58 \ \mu A \qquad I_{CQ} = 6.08 \ mA \qquad V_{CEQ} = 8.35 \ V$$

Finally, from equation 8.22, we have

$$A_V = \frac{-h_{fe}R_C}{R_B + h_{ie} + (h_{fe} + 1)R_E} = \frac{-175}{136.5} = -1.28$$

DRILL EXERCISES

8.8 Repeat the analysis of Example 8.7 if $R_S = 400 \ \Omega$. Which region is the transistor in? What is the collector current?

> Answer: Saturation; 8.5 mA

8.9 What is the power dissipated by the LED of Example 8.7 if $R_S = 30 \ \Omega$?

> Answer: 159 mW

8.10 Determine the AC voltage gain of a 2N3904 transistor, using the results of Example 8.5 and the data sheet provided at the end of the chapter. Use average values of the h parameters, and assume that $R_C = 1 \ k\Omega$, $R_B = 100 \ k\Omega$, and $R_E = 100 \ \Omega$.

8.4   BJT SMALL-SIGNAL AMPLIFIERS

The h-parameter model developed in the previous section is very useful in the small-signal analysis of various configurations of BJT amplifiers. In this section, a set of techniques will be developed to enable you to first establish the Q point of a transistor amplifier, then construct the small-signal model, and finally use the small-signal model for analyzing the small-signal behavior of the amplifier. A major portion of this section will be devoted to the analysis of the **common-emitter amplifier,** using this circuit as a case study to illustrate the analysis methods. At the end of the section, we will look briefly at two other common amplifier circuits: the **voltage follower,** or **common-collector amplifier;** and the **common-base amplifier.** A table will summarize the input and output resistance and gains of all three types of amplifiers. The discussion of the common-emitter amplifier will also provide an occasion to introduce, albeit qualitatively, the important issue of transistor amplifier frequency response. A detailed treatment of this last topic is beyond the intended scope of this book.

A complete common-emitter amplifier circuit is shown in Figure 8.36. The circuit may appear to be significantly different from the simple examples studied in the previous sections; however, it will soon become apparent that all the machinery necessary to understand the operation of a complete transistor amplifier is already available. The task at hand consists simply of piecing together the results obtained in the preceding sections.

Figure 8.36 The BJT common-emitter amplifier

The analysis of this circuit will be aimed at developing a simple small-signal linear AC equivalent-circuit model for the amplifier, to permit defining the open-circuit voltage and short-circuit current gains of the amplifier and its input and output resistances. To create this model, we shall develop a simple **two-port**[1] **equivalent circuit** for the amplifier; this equivalent circuit can then be used in connection with equivalent circuits for the load and source circuits to determine the actual gains of the amplifier as a function of the load and source impedances.

Figure 8.37 depicts the appearance of this simplified representation for a transistor amplifier, where r_i and r_o represent the input and output resistance

---

[1] A two-port circuit is a circuit that has an input and an output port, in contrast with the one-port circuits studied in Chapter 3, which had only a single port connecting the source to a load. The amplifier configuration shown in Figure 8.36 is representative of a general two-port circuit.

Figure 8.37 Equivalent-circuit model of voltage amplifier

of the amplifier, respectively, and μ is the open-loop voltage gain of the amplifier. Throughout this section, it will be shown how such a model can be obtained for a variety of amplifier configurations. You may wish to compare this model with that shown in Figure 8.3 in Section 8.1. Note that the model of Figure 8.37 makes use of the simple Thévenin equivalent circuit model developed in Chapter 3 in representing the input to the amplifier as a single equivalent resistance, r_i. Similarly, the circuit seen by the load consists of a Thévenin equivalent circuit. In the remainder of this section, it will be shown how the values of r_i, r_o, and μ may be computed, given a transistor amplifier design. It is also useful to define the overall voltage and current gains for the amplifier model of Figure 8.37 as follows. The amplifier voltage gain is defined as

$$A_V = \frac{v_{\text{out}}}{v_{\text{in}}} = \frac{R_L}{R_L + r_o}\mu \tag{8.24}$$

while the current gain is

$$A_I = \frac{i_{\text{out}}}{i_{\text{in}}} = \frac{r_i}{R_L}A_V \tag{8.25}$$

The first observation that can be made regarding the circuit of Figure 8.36 is that the AC input signal, $v_S(t)$, has been "coupled" to the remainder of the circuit through a capacitor, C_C (called a **coupling capacitor**). Similarly, the load resistance, R_L, has been connected to the circuit by means of an identical coupling capacitor. The reason for the use of coupling capacitors is that they provide separate paths for DC and AC currents in the circuit. In particular, *the quiescent DC currents do not reach the source or the load.* This is especially useful, since the aim of the circuit is to amplify the AC input signal only, and it would be undesirable to have DC currents flowing through the load. In fact, the presence of DC currents would cause unnecessary and undesired power consumption at the load. The operation of the coupling capacitors is best explained by observing that a capacitor acts as an open circuit to DC currents, while—if the capacitance is sufficiently large—it will act as a short circuit at the frequency of the input signal. Thus, in general, one wishes to make C_C as large as possible, within reason. The **emitter bypass capacitor,** C_E, serves a similar purpose, by "bypassing" the emitter resistance R_E *insofar as AC currents are concerned,* since the capacitor acts as a short circuit at the signal frequency. On the other hand,

C_E is an open circuit to DC currents, and therefore the quiescent current will flow through the emitter resistor, R_E. Thus, the emitter resistor can be chosen to select a given Q point, but it will not appear in the calculations of the AC gains. This dual role served by coupling and bypass capacitors in transistor amplifiers is of fundamental importance in their practical operation.

Figure 8.38 depicts the path taken by AC and DC currents in the circuit of Figure 8.36. Example 8.10 further explains the use of coupling capacitors.

DC current
——→

AC current
-----→

Figure 8.38 Effect of coupling capacitors on DC and AC current paths

Example   8.10

In this example, the numerical computation of the appropriate value for a coupling capacitor is illustrated for a signal $v_S(t)$ in the audio frequency range:

$$40\pi \leq \omega_S \leq 40,000\pi \text{ rad/s}$$

If we select a 1,000- μF coupling capacitor, its series impedance, Z_C, has magnitude

$$|Z_C| = \frac{1}{\omega C} = \frac{1}{40\pi \times 0.001} = 7.96 \ \Omega$$

at

$$\omega = 40\pi \text{ rad/s}$$

and magnitude 1,000 times smaller at the high end of the audio spectrum (i.e., $\omega = 40,000\pi$). On the other hand, the impedance of the coupling capacitor at the frequency $\omega = 0$ (DC) is infinite. Thus, the coupling capacitor offers negligible resistance to the flow of current from the source over the frequency range of interest, while completely isolating the AC source from the DC supply.

The amplifier of Figure 8.36 employs a single DC supply, V_{CC}, as does the DC self-bias circuit of Example 8.6. The resistors R_1 and R_2, in effect, act as a voltage divider that provides a suitable bias for the BE junction. This effect is most readily understood if separate DC and AC equivalent circuits for the common-emitter amplifier are portrayed as in Figure 8.39.

DC circuit   AC circuit

Figure 8.39 DC and AC circuits for the common-emitter amplifier

To properly interpret the DC and AC equivalent circuits of Figure 8.39, a few comments are in order. Consider, first, the DC circuit. As far as DC currents are concerned, the two coupling capacitors and the emitter bypass capacitor are open circuits. Further, note that the supply voltage, V_{CC}, appears across two branches, the first consisting of the emitter and collector resistors and the CE "junction," and of the base resistors. This DC equivalent circuit is used in determining the Q point of the amplifier—that is, the quantities I_{BQ}, V_{CEQ}, and I_{CQ}. These are the DC quiescent currents and voltage that define the operating point of the amplifier. In drawing the AC equivalent circuit, on the other hand, each of the capacitors has been replaced by a short circuit, as has the DC supply. The effect of the latter substitution (justified by the principle of superposition) is to create a direct path to ground for the resistors R_1 and R_C. Thus, R_1 appears in parallel with R_2. Note, also, that in the AC equivalent circuit, the collector resistance R_C appears in parallel with the load. This will have an important effect on the overall gain of the amplifier.

DC Analysis of the Common-Emitter Amplifier

It is useful in analyzing the DC circuit of Figure 8.39 to redraw it in a slightly different form, recognizing that the Thévenin equivalent circuit seen by the base consists of the equivalent voltage

$$V_{BB} = \frac{R_2}{R_1 + R_2} V_{CC} \tag{8.26}$$

and of the equivalent resistance

$$R_B = R_1 \parallel R_2 \tag{8.27}$$

Figure 8.40 DC bias circuit for the common-emitter amplifier

The resulting circuit is sketched in Figure 8.40. Application of KVL around the base and collector-emitter circuits yields the following equations, the solution of which determines the Q point of the transistor amplifier:

$$V_{CEQ} = V_{CC} - I_{CQ}R_C - \frac{\beta + 1}{\beta}I_{CQ}R_E \tag{8.28}$$

$$V_{BEQ} = V_{BB} - I_{BQ}R_B - \frac{\beta + 1}{\beta}I_{CQ}R_E \tag{8.29}$$

In these equations, the quiescent emitter current, I_{EQ}, has been expressed in terms of the collector current, I_{CQ}, according to the relation

$$I_{EQ} = \frac{\beta + 1}{\beta}I_{CQ} \tag{8.30}$$

The next two examples illustrate a number of practical issues in the choice of DC bias point, and in the determination of other important features of the common-emitter transistor amplifier.

Example   8.11

One of the important choices in the design of a transistor amplifier is the location of the operating point. Even when the cutoff and saturation limits of the transistor are taken into account, there is no unique design for a given transistor amplifier. In this example, we analyze two similar designs, leading to two rather different common-emitter transistor amplifiers. The two amplifier circuits are shown in Figure 8.41. In analyzing the designs, we will assume that $V_\gamma = 0.7$ V and $V_{CE\,sat} = 0.2$ V. Which is the preferred circuit, and why?

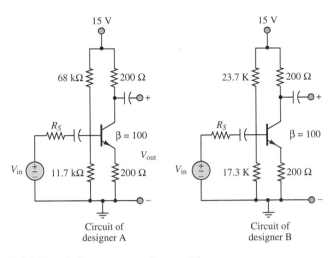

Figure 8.41 Two similar common-emitter amplifiers

Solution:

1. *Design A:* The equivalent circuit representing the base circuit of design A is shown in Figure 8.42, with

$$V_{BB} = \frac{11.7}{11.7 + 68}(15) = 2.2 \text{ V}$$

and

$$R_B = 68 \text{ k}\Omega \parallel 11.7 \text{ k}\Omega$$
$$\approx 10 \text{ k}\Omega$$

Next, we consider the base-emitter circuit, depicted in Figure 8.43, and apply KVL, to find that

$$V_{BB} - V_{BE} = I_B R_B + I_E R_E = I_B R_B + I_B(\beta + 1)R_E$$

or

$$2.2 - 0.7 = I_B[10,000 + 101(200)]$$

Solving for the base current yields

$$I_B = 50 \text{ } \mu\text{A}$$

and therefore

$$I_C = \beta I_B = 100(50 \text{ } \mu\text{A}) = 5 \text{ mA}$$

Now, writing KVL around the collector-emitter loop, shown in Figure 8.44, we can solve for V_{CE}:

$$V_{CE} = 15 - I_C R_C - I_E R_E$$
$$= 15 - I_C R_C - I_C \left(\frac{\beta + 1}{\beta}\right) R_E$$
$$= 15 - 0.005 \left[200 + \frac{101}{100}(200)\right]$$

so that

$$V_{CE} = 13 \text{ V}$$

Thus, we conclude that the Q point for design A is given by

$$I_{BQ} = 50 \text{ } \mu\text{A} \qquad V_{CEQ} = 13 \text{ V} \qquad I_{CQ} = 5 \text{ mA}$$

2. *Design B:* Repeating the analysis for design B and using the same form for the equivalent circuits, we can describe the base equivalent circuit by

$$V_{BB} = 6.33 \text{ V} \qquad R_B = 10 \text{ k}\Omega$$

Application of KVL around the base-emitter circuit reveals that

$$6.33 - 0.7 = I_B(10,000) + I_B(200)(101)$$

and therefore

$$I_B = 186.4 \text{ } \mu\text{A}$$

Figure 8.42 Equivalent base supply circuit of design A

Figure 8.43 Equivalent base-emitter circuit of design A

Figure 8.44 Collector-emitter circuit of design A

and

$$I_C = \beta I_B = 18.6 \text{ mA}$$

Finally, writing KVL around the collector-emitter loop yields

$$V_{CE} = 15 - I_C \left[200 + \frac{101}{100}(200) \right] = 7.5 \text{ V}$$

Thus, the Q point for circuit B is

$$I_{BQ} = 186 \ \mu\text{A} \qquad I_{CQ} = 18.6 \text{ mA} \qquad V_{CEQ} = 7.5 \text{ V}$$

Having completed the preliminary analysis, we plot the Q points on the load line for the circuit, to compare the results. The load-line equation for the collector circuit is given by the expression (obtained by applying KVL around the collector loop)

$$V_{CE} = V_{CC} - I_C(R_C + R_E)$$

and can most easily be sketched by finding the I_C and V_{CE} axis intercepts:

If $V_{CE} = 0$ V, then

$$I_C = \frac{15 \text{ V}}{R_C + R_E} = 37.5 \text{ mA}$$

If $I_C = 0$, then

$$V_{CE} = 15 \text{ V}$$

The results are displayed in Figure 8.45.

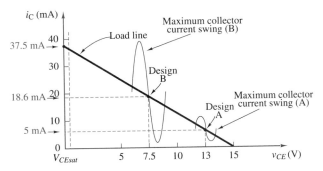

Figure 8.45 Operating points for designs A and B

We know that the AC voltage, v_{in}, will cause the base current—and therefore the collector current—to oscillate around the operating point along the load line. Clearly, design A is much too close to the cutoff region and allows the collector current to swing undistorted only over a range of less than 10 mA before entering the cutoff region. On the other hand, design B permits a much greater swing for the collector current, thus offering a much larger dynamic range for the output voltage. Thus, the clear choice is the circuit of design B, because of the large symmetric swing in i_C and v_{CE}.

Example   8.12

One of the practical limitations in the design of BJT amplifiers arises from the variability of the current gain, β, even among transistors of the same family. A typical
specification for a family of bipolar transistors—for example, the 2N3904—includes
a β ranging from 75 to 150! It is therefore important, in designing an amplifier, to be
able to compensate for such variability. It is especially important to obtain a stable
operating point, independent of the variation in current gain from device to device.
A common practice used to stabilize the operating point with respect to a range of β
for a given transistor consists of designing the equivalent base supply circuit so that

$$R_B = \frac{\beta_{min} R_E}{10}$$

where R_B is the equivalent base-circuit resistance and R_E is the emitter resistor.

1. Find the operating point of the transistor in circuit design B in Example 8.11 if:

   a. $\beta = \beta_{min} = 75$
   b. $\beta = \beta_{max} = 150$

2. Use the rule of thumb stated in this example to set the desired operating point
   at $I_C = 18.6$ mA, and thus find R_B, V_{BB}, and R_1 and R_2.

The DC circuit for this example is shown in Figure 8.46.

Figure 8.46 DC circuit for
Example 8.12

Solution:

1. a. If $\beta = \beta_{min} = 75$, we can derive the equivalent base circuit by repeating
      the analysis of Example 8.11. Then, by KVL, we find the base current from
      the expression

$$6.33 - 0.7 = I_B(10,000) + I_B(\beta_{min} + 1)200$$

   or

$$I_B = \frac{6.33 - 0.7}{10,000 + 76(200)} = 223 \ \mu A$$

   The corresponding collector current is then

$$I_C = \beta_{min} I_B = 16.725 \text{ mA}$$

   Applying KVL around the collector-emitter loop, we determine the
   collector-emitter voltage,

$$V_{CE} = 15 - I_C(200) - I_C \left(\frac{\beta_{min} + 1}{\beta_{min}} \right)(200) = 8.3 \text{ V}$$

   and the operating point:

$$I_{BQ} = 223 \ \mu A \qquad I_{CQ} = 16.7 \text{ mA} \qquad V_{CEQ} = 8.3 \text{ V}$$

   b. Now let us repeat the analysis for $\beta = \beta_{max}$. Using KVL around the base
      circuit, we obtain

$$6.33 - 0.7 = I_B[10,000 + (\beta_{max} + 1)200]$$

leading to

$$I_B = \frac{6.33 - 0.7}{10{,}000 + 151 \times 200} = 140 \ \mu A$$

and therefore

$$I_C = \beta_{max} I_B = 21 \ mA$$

Repeating the analysis for the collector circuit, we find that

$$V_{CE} = 15 - I_C \left[R_C + \left(\frac{\beta_{max} + 1}{\beta_{max}} \right) R_E \right]$$

$$= 15 - (21 \ mA) \left[200 + \frac{151}{150}(200) \right] = 6.57 \ V$$

so that the operating point is

$$I_{BQ} = 140 \ \mu A \qquad I_{CQ} = 21 \ mA \qquad V_{CEQ} = 6.57 \ V$$

The two Q points are shown in Figure 8.47, as they appear on the load line.

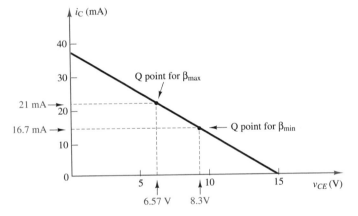

Figure 8.47 Variability in Q point due to change in β

The difference in operating point could then be as large as

$$\delta V_{CEQ} = 8.3 - 6.57 = 1.73 \ V$$

and

$$\delta I_{CQ} = 21 \ mA - 16.7 \ mA = 4.3 \ mA$$

for two identical amplifiers designed to operate at the same Q point.

2. Using the design rule of thumb, we find R_B to be

$$R_B = \frac{\beta_{min} R_E}{10} = \frac{75(200)}{10} = 1{,}500 \ \Omega$$

By writing KVL around the base-emitter loop and using the relationship

$$I_E = \frac{\beta_{min} + 1}{\beta_{min}} I_C$$

with $I_C =$ desired $I_C = 18.6$ mA, we arrive at the following equation:

$$V_{BB} = I_B R_B + 0.7 + I_E R_E$$

$$= \frac{18.6 \text{ mA}}{\beta_{min}}(1{,}500) + 0.7 + \frac{\beta_{min} + 1}{\beta_{min}}(18.6 \text{ mA})(200)$$

$$= 4.84 \text{ V}$$

Now, since

$$V_{BB} = \left(\frac{R_2}{R_1 + R_2}\right) 15 \text{ V}$$

and

$$R_B = \frac{R_2 R_1}{R_2 + R_1} = 1{,}500 \ \Omega$$

two equations in two unknowns may be obtained to solve for R_1 and R_2:

$$4.84 = \left(\frac{R_2}{R_1 + R_2}\right) 15$$

$$1{,}500 = \frac{R_1 R_2}{R_1 + R_2}$$

We obtain

$$R_1 = 4.65 \text{ k}\Omega \qquad R_2 = 2.21 \text{ k}\Omega$$

Thus, the final DC circuit will be as shown in Figure 8.48.

This design for the bias point will result in a far less sensitive Q point in the face of variations in the transistor current gain. The numerical answer is provided in Drill Exercise 8.12.

Figure 8.48 DC circuit resulting from application of rule of thumb

DRILL EXERCISES

8.11   Find the operating point of the circuit of design B in Example 8.11 if (a) $\beta = 90$ and (b) $\beta = 120$.

Answer: (a) $I_{BQ} = 199.6 \ \mu A$, $I_{CQ} = 17.96$ mA, $V_{CEQ} = 7.77$ V
(b) $I_{BQ} = 164.6 \ \mu A$, $I_{CQ} = 19.75$ mA, $V_{CEQ} = 7.065$ V

8.12   Verify that the operating point in design 2 of Example 8.12 is more stable than that of design 1.

Answer: $\delta V_{CEQ} = 0.47$ V $\delta I_{CQ} = 1.5$ mA

8.13   Repeat all parts of Example 8.12 if $\beta = \beta_{min} = 60$, $\beta = \beta_{max} = 200$, and $I_C = 10$ mA.

Answer: For $\beta = \beta_{min}$: $I_{BQ} = 254 \ \mu A$, $I_{CQ} = 15.2$ mA, $V_{CEQ} = 8.87$ V.
For $\beta = \beta_{max}$: $I_{BQ} = 112 \ \mu A$, $I_{CQ} = 22$ mA, $V_{CEQ} = 6.18$ V.
For $I_C = 10$ mA: $R_B = 1.2$ kΩ, $R_1 = 6.14$ kΩ, $R_2 = 1.49$ kΩ.

AC Analysis of the Common-Emitter Amplifier

To conduct the analysis of the AC circuit, we substitute the hybrid-parameter small-signal model of Figure 8.31 in the AC circuit of Figure 8.39, obtaining the linear AC equivalent circuit of Figure 8.49. Note how the emitter resis-

Figure 8.49 AC equivalent-circuit model for the common-emitter amplifier

tance is bypassed by the emitter bypass capacitor at AC frequencies; this, in turn, implies that the base to emitter (BE) junction appears in parallel with the equivalent resistance $R_B = R_1 \parallel R_2$. Similarly, the collector-to-emitter (CE) junction is replaced by the parallel combination of the controlled current source, $h_{fe}\Delta I_B$, with the resistance $1/h_{oe}$. Once again, since the DC supply provides a direct path to ground for the AC currents, the collector resistance appears in parallel with the load. It is important to understand why we have defined input and output voltages, v_{in} and v_{out}, rather than use the complete circuit containing the source, v_S, its internal resistance, R_S, and the load resistance, R_L. The AC circuit model is most useful if an equivalent input resistance and the equivalent output resistance and amplifier gain are defined as quantities independent of the signal source and load properties. This approach permits the computation of the parameters r_i, r_o, and μ shown in Figure 8.37, and therefore provides a circuit model that may be called upon for any source-and-load configuration. The equivalent circuit shown in Figure 8.49 allows viewing the transistor amplifier as a single equivalent circuit either from the source or from the load end, as will presently be illustrated.

The analysis of this circuit can easily be carried out by resorting to the DC circuit analysis tools of Chapters 2 and 3. First, compute the AC input current, i_{in},

$$i_{in} = \frac{v_{in}}{h_{ie} \parallel R_B} \tag{8.31}$$

and the AC base current,

$$\Delta I_B = \frac{v_{in}}{h_{ie}} \tag{8.32}$$

noting, further, that the input resistance of the circuit consists of the parallel combination of h_{ie} and R_B:

$$r_i = h_{ie} \parallel R_B \tag{8.33}$$

Next, observe that the AC base current is amplified by a factor of h_{fe} and that it flows through the parallel combination of the collector resistance, R_C, and the CE junction resistance, $1/h_{oe}$. The latter term is usually negligible with respect to load and collector resistances, since the slope of the collector i-v curves is very shallow; in the remainder of this chapter it will be assumed that h_{oe} is negligible and that the parallel resistance $1/h_{oe}$ is infinite. Thus, the AC short-circuit output current i_{out} is given by the expression

$$i_{out} = -h_{fe}\Delta I_B = -h_{fe}\frac{v_{in}}{h_{ie}} \tag{8.34}$$

and the AC open-circuit output voltage is given by the expression

$$v_{out} = -h_{fe}\Delta I_B R_C = -h_{fe}\frac{v_{in}}{h_{ie}}R_C \tag{8.35}$$

Note the negative sign, due to the direction of the controlled current source. This sign reversal is a typical characteristic of the common-emitter amplifier. Knowing the open-circuit output voltage and the short-circuit output current, we can find the output resistance of the amplifier as the ratio of these two quantities:

$$r_o = \frac{v_{out}}{i_{out}} = R_C \tag{8.36}$$

Next, if we define the **AC open-circuit voltage gain** of the amplifier, μ, by the expression

$$\mu = \frac{v_{out}}{v_{in}} \tag{8.37}$$

we find that

$$\mu = -h_{fe}\frac{R_C}{h_{ie}} \tag{8.38}$$

At this point, it is possible to take advantage of the equivalent-circuit representation of Figure 8.37, with the expressions for μ, r_o, and r_i just given. Figure 8.50 illustrates the equivalent circuit for the common-emitter amplifier. This circuit replaces the transistor amplifier and enables us to calculate the actual voltage and current gain of the amplifier for any given load-and-source pair. Referring to the circuit of Figure 8.38 and to the associated definition of the voltage and current gains, A_V and A_I, we can compute these gains to be

$$A_V = \frac{v_{out}}{v_{in}} = \frac{R_L}{R_L + r_o}\mu = -\frac{R_L}{R_L + r_o}h_{fe}\frac{R_C}{h_{ie}} \tag{8.39}$$

and

$$A_I = \frac{i_{out}}{i_{in}} = \frac{r_i}{R_L}A_V \tag{8.40}$$

Example 8.13 provides numerical values for a typical small-signal common-emitter amplifier.

Figure 8.50 Simplified equivalent circuit for the common-emitter amplifier

Example 8.13

A common-emitter amplifier is shown in Figure 8.51. The transistor has the following h parameters: $h_{ie} = 1,400\ \Omega$, $h_{fe} = 100$, and $h_{oe} = 125\ \mu\text{S}$. Resistors R_1 and R_2 provide the DC operating point, as discussed in Section 8.2. The source has an internal resistance of 500 Ω. Find the open-circuit voltage and current gain for this amplifier.

Figure 8.51 A common-emitter amplifier

Solution:

Replacing the BJT with its equivalent-circuit model results in the amplifier equivalent circuit shown in Figure 8.52, where

$$R_B = \frac{R_1 R_2}{R_1 + R_2} = \frac{20,000 \times 5,000}{20,000 + 5,000} = 4\ \text{k}\Omega$$

Figure 8.52

The input resistance r_i is determined from inspection of the circuit in Figure 8.52 to be

$$r_i = R_B \parallel h_{ie} = \frac{4,000 \times 1,400}{4,000 + 1,400} = \frac{56,000}{54} = 1,037\ \Omega$$

Figure 8.53

To determine the output resistance of the amplifier, we need to set $v_S = 0$. This results in $v_{in} = 0$ and the controlled-source input $\mu v_{in} = 0$. Thus,

$$r_o = R_C \parallel \frac{1}{h_{oe}} = \frac{4,000 \times 8,000}{4,000 + 8,000} = 2.67\ \text{k}\Omega$$

The complete AC model of the amplifier with source and load circuit is shown in Figure 8.53.

The open-circuit voltage gain, A_V, can be found by using equation 8.39:

$$A_V = \frac{v_{out}}{v_{in}} = \frac{-R_L}{R_L + r_o} h_{fe} \frac{R_C}{h_{ie}} = -\frac{2 \times 10^3}{4.67 \times 10^3} \times 100 \times \frac{4 \times 10^3}{1.4 \times 10^3} = -122.4$$

The current gain, A_I, can be found by using equation 8.40:

$$A_I = \frac{r_i}{R_L} A_V = \frac{1,037}{2,000} \times (-122.4) = -63.46$$

DRILL EXERCISES

8.14 Compute the actual voltage gain, v_{out}/v_S, of the amplifier of Example 8.13 for the following source-load pairs:

a. $R_S = 50 \ \Omega, \ R_L = 150 \ \Omega$
b. $R_S = 50 \ \Omega, \ R_L = 1,500 \ \Omega$
c. $R_S = 500 \ \Omega, \ R_L = 150 \ \Omega$

What conclusions can you draw from these results?

> Answer: (a) -14.5; (b) -98; (c) -10.25. The amplifier behaves more closely to the ideal model if the source resistance is small and the load resistance is large.

8.15 Calculate the current gains for the amplifier parameters of Drill Exercise 8.13.

> Answer: (a) -2.1; (b) -141.76; (c) -1.48

8.16 Repeat Example 8.13 for $h_{ie} = 2 \ k\Omega$ and $h_{fe} = 60$.

> Answer: $A_V = -51.39$; $A_I = -77.28$

Other BJT Amplifier Circuits

The common-emitter amplifier is a commonly employed configuration but is by no means the only type of BJT amplifier. Other amplifier configurations are also used, depending on the specific application requirements. Each type of amplifier is classified in terms of properties such as input and output resistance, and voltage and current gain. Rather than duplicate the detailed analysis just conducted for the common-emitter amplifier, we summarize the properties of the three more common BJT amplifier circuits in Table 8.1, which depicts the amplifier circuits, the corresponding small-signal models, and the expressions for the input and output resistance and open-circuit voltage gain. The methodology employed to derive these results is completely analogous to that surveyed in the previous section: the AC and DC circuits are identified first, then the transistor is replaced with its h-parameter model in the AC circuit, and standard circuit analysis techniques are used in deriving the desired expressions. We leave the derivation of the results of Table 8.1 to the student in a series of homework problems and recommend these exercises as an excellent means of practicing the newly acquired knowledge.

Table 8.1   BJT amplifier configurations

	Common emitter	Common collector	Common base
R_{in}	$h_{ie} \parallel R_B$	$R_B \parallel [h_{ie} + (h_{ie} + 1)R_E]$	$h_{ie}(h_{fe} + 1) + R_E$
R_o	$R_C \parallel \dfrac{1}{h_{oe}}$	$\dfrac{R_E \parallel \frac{1}{h_{oe}}(h_{ie} + R_E)}{R_E \parallel \frac{1}{h_{oe}}(h_{fe} + 1) + h_{ie} + R_B}$	$R_C \parallel \dfrac{1 + h_{fe}}{h_{oe}}$
$\lvert A_V \rvert$	$\dfrac{h_{fe}}{h_{ie}}\left(R_C \parallel \dfrac{1}{h_{oe}}\right)$	$\dfrac{(h_{fe} + 1)R_E \parallel \frac{1}{h_{oe}}}{(h_{fe} + 1)R_E \parallel \frac{1}{h_{oe}} + h_{ie}}$	$\dfrac{h_{fe}R_C(R_E \parallel h_{ie})}{R_E h_{ie}}$
$\lvert A_I \rvert$	$h_{fe}\dfrac{R_B}{R_B + h_{ie}}$	$\dfrac{h_{fe}R_B}{R_B + h_{ie}}$	$\dfrac{h_{fe}}{h_{fe} + 1}$
Amplifier circuit			
AC equivalent circuit			
Properties	Input impedance: medium Output impedance: medium Voltage gain: high Current gain: high Phase shift: 180°	Input impedance: high Output impedance: low Voltage gain: low Current gain: medium Phase shift: 0°	Input impedance: low Output impedance: high Voltage gain: high Current gain: low Phase shift: 0°

8.5   FREQUENCY RESPONSE OF SMALL-SIGNAL AMPLIFIERS

When the idea of a coupling capacitor was introduced earlier in this chapter, the observation was made that the capacitor C_C (see Figure 8.36) acts very nearly as a short circuit at AC frequencies. In fact, the presence of a series capacitor in the AC small-signal equivalent input circuit has very much the effect of a high-pass filter. Consider the small-signal equivalent in Figure 8.54, where a single coupling capacitor has been placed at the input, for

Figure 8.54 Small-signal equivalent circuit

simplicity. Let us compute the frequency response of the small-signal equivalent circuit. If we assume that $R_B \gg R_S$ and $R_B \gg h_{ie}$, then we can easily find that

$$\Delta I_B = \frac{v_{in}}{h_{ie}} \tag{8.41}$$

and

$$v_{in} = \frac{h_{ie}}{h_{ie} + R_S + 1/j\omega C_C} v_S \tag{8.42}$$

so that the output voltage is given by

$$v_{out} = -\left(R_C \parallel R_L\right) h_{fe} \Delta I_B = \left(R_C \parallel R_L\right) h_{fe} \frac{v_{in}}{h_{ie}}$$

$$= \left(R_C \parallel R_L\right) h_{fe} \frac{v_S}{h_{ie}} \frac{h_{ie}}{h_{ie} + R_S + 1/j\omega C_C} \tag{8.43}$$

This expression for v_{out} is clearly a function of frequency, approaching a constant as ω tends to infinity.

Figure 8.55 illustrates the general effect of a nonideal coupling capacitor on the frequency response of the common-emitter amplifier. Note that, in effect, the cutoff frequency is determined at the input circuit by the effective *RC* combination:

$$\omega_{3\,dB} = \frac{1}{(R_S + h_{ie})C_C} \tag{8.44}$$

Clearly, if we wished to design an *audio amplifier* (that is, an amplifier that can provide an undistorted frequency response over the range of frequencies audible to the human ear), we would select a coupling capacitor for which the lower cutoff frequency of Figure 8.55 was lower than $2\pi \cdot 20$ rad/s. There

Figure 8.55 Approximate frequency response of a common-emitter amplifier, assuming a finite coupling capacitor

is another phenomenon, however, that limits the frequency response of the amplifier from the high-frequency end: the so-called **parasitic capacitance** that is present between pairs of terminals at the transistor (don't forget that any two conductors separated by a dielectric—air, for example—form a capacitor!). Figure 8.56 illustrates the phenomenon of parasitic capacitances.

Figure 8.56 Effects of parasitic capacitance in a common-emitter amplifier

To simplify the analysis, assume that C_{BE} and C_{CB} are negligible in the equivalent circuit of Figure 8.56. The effect of an output coupling capacitor, C_C, would clearly be negligible (an open circuit) at the high frequencies, where parasitic capacitances are most likely to cause trouble. This permits considerable simplification in the analysis; the output voltage, v_{out}, becomes

$$v_{out} = -h_{fe}\Delta I_B \frac{R_C \parallel R_L}{1 + j\omega C_{CE}\left(R_C \parallel R_L\right)} \tag{8.45}$$

This function is sketched in Figure 8.57, and is clearly the response of a low-pass filter with $R_{eq} = R_C \parallel R_L$. Similarly, if we considered the input capacitance, C_{BE}, we would find that there is another cutoff frequency, determined by this parasitic capacitance and the input resistance of the amplifier.

Even though the present discussion is purely qualitative, you should by now have realized that any practical amplifier will have a band-pass response similar to that shown in Figure 8.58, where A_V is called the *mid-band gain* and is equal to the voltage gain computed in earlier sections. A complete discussion of the frequency response of transistor amplifiers is beyond the scope of this book.

Figure 8.57 Low-pass filter effect of parasitic capacitance in a common-emitter amplifier

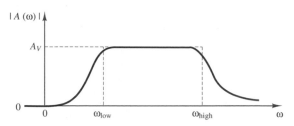

Figure 8.58 Frequency response of a common-emitter amplifier

CONCLUSION

- Transistors are three-terminal electronic semiconductor devices that can serve as linear amplifiers or switches.
- The bipolar junction transistor (BJT) acts as a current-controlled current source, amplifying a small base current by a factor ranging from 50 to 200. The operation of the BJT can be explained in terms of the device base-emitter and collector $i\text{-}v$ characteristics.
- Large- and small-signal linear circuit models for the BJT can be obtained by treating the transistor as a controlled current source. The small-signal, or h-parameter, model also takes into account the internal resistance of the BJT. Small-signal models can be used to analyze the operation of the BJT as a linear amplifier. Various amplifier configurations exist, each of which can be characterized by an equivalent input and output resistance and by an open-circuit voltage gain.
- Bipolar transistors are subject to the fundamental limitations of saturation, cutoff, and maximum power dissipation. Exceeding these limitations can introduce signal distortion and reduce the life of the device.

KEY TERMS

Amplification   *388*

Switching   *388*

Bipolar Junction Transistor (BJT)   *391*

Base-Emitter (*BE*) Junction   *391*

Base-Collector (*BC*) Junction   *392*

Collector Characteristic   *394*

Cutoff Region   *394*

Active Linear Region   *394*

Saturation Region   *394*

Breakdown Region   *394*

Q Point   *397*

Quiescent (Idle) Current or Voltage   *397*

DC Bias Circuit   *397*

Large-Signal Model of a BJT   *404*

Small-Signal Model of a BJT   *404*

Hybrid-Parameter (*h*-Parameter) Small-Signal Model   *409*

AC Voltage Gain   *412*

Common-Emitter Amplifier   *414*

Voltage Follower (Common-Collector Amplifier)   *414*

Common-Base Amplifier   *414*

Two-Port Circuit   *414*

Coupling Capacitor   *415*

Emitter Bypass Capacitor   *415*

AC Open-Circuit Voltage Gain   *425*

Parasitic Capacitance   *430*

DEVICE DATA SHEETS

2N5088   Low-Level Amplifier

Electrical Characteristics ($T_A = 25°C$)

Symbol	Conditions	Min	Typ	Max	Units
C_{ob}	$V_{CB} = 10V$, $f = 1$ MHz		3.0	4.0	pF
C_{ib}	$V_{EB} = 0.5V$, $f = 1$ MHz		16	25	pF
NF	$I_C = 100\mu A$, $V_{CE} = 5V$ $R_S = 2k\Omega$, $f = 1$ kHz		2.0		dB
f_T	$V_{CE} = 10V$, $I_C = 20$ mA	150	250		MHz
h_{FE}	$V_{CE} = 1.0V$, $I_C = 1$ mA $V_{CE} = 1.0V$, $I_C = 100$ mA $V_{CE} = 1.0V$, $I_C = 150$ mA	40 100 75	200	400	
$V_{CE(SAT)}$	$I_C = 150$ mA, $I_B = 15$ mA			0.5	V
$V_{BE(SAT)}$	$I_C = 150$ mA, $I_B = 15$ mA			1.0	V
BV_{CBO}	$I_C = 10\ \mu A$	80			
BV_{CEO}	$I_C = 10$ mA	65			
BV_{EBO}	$I_E = 10\ \mu A$	6.0			
I_{CBO}	$V_{CB} = 60V$			50	nA
I_{CES}	$V_{CE} = 60V$			50	nA
I_{EBO}	$V_{EB} = 4.0V$			50	nA
$P_{D(max)}$ TO-92 TO-236	$T_A = 25°C$ $T_C = 25°C$	600 350			mW mW

Collector Current vs Collector-Emitter Voltage

Collector Current vs Collector-Emitter Voltage

2N3904   General Purpose Amplifier and Switch

Electrical Characteristics ($T_A = 25°C$)

Symbol	Conditions	Min	Typ	Max	Units
t_{ON}	$I_C = 10$ mA, $I_{B1} = 1$ mA *(Figure 1)*		30	70	ne
t_{OFF}	$I_C = 10$ mA, $I_{B2} = 1$ mA *(Figure 2)*		150	250	ne
C_{ob}	$V_{CB} = 5V$, $f = 1$ MHz		2.7	4.0	pF
C_{ib}	$V_{EB} = 0.5V$, $f = 1$ MHz			8.0	pF
NF	$V_{CE} = 5V$, $I_C = 100\mu A$, $R_S = 1$ kΩ, $P_{BW} = 15.7$kHz		2.0		dB
h_{ie}	$I_C = 10$ mA, $V_{CE} = 20V$, $f = 100$ MHz	2.5	4.5		
h_{FE}	$I_C = 100$ μA, $V_{CE} = 5V$ $\quad$ $I_C = 1$ mA, $V_{CE} = 5V$ $\quad$ $I_C = 10$ mA, $V_{CE} = 5V$ $\quad$ $I_C = 50$ mA, $V_{CE} = 5V$ $\quad$ $I_C = 100$ mA, $V_{CE} = 5V$	40 90 60 40 20	150	360	
$V_{CE(SAT)}$	$I_C = 10$ mA, $I_B = 1$ mA			0.15	V
$V_{BE(SAT)}$	$I_C = 10$ mA, $I_B = 1$ mA			0.80	V
$V_{CE(SAT)}$	$I_C = 50$ mA, $I_B = 5$ mA			0.25	V
$V_{BE(SAT)}$	$I_C = 50$ mA, $I_B = 5$ mA			0.85	V
BV_{CBO}	$I_C = 10$ μA	60			V
BV_{CEO}	$I_C = 1$ mA	30			V
BV_{EBO}	$I_E = 10$ μA	6.0			V
I_{CBO}	$V_{CB} = 30V$			100	nA
I_{EBO}	$V_{EB} = 4V$			100	nA

DC Current Gain vs Collector Current

Base-Emitter ON Voltage vs Collector Current

Maximum Power Dissipation vs Ambient Temperature

Collector Saturation Voltage vs Collector Current

Base Saturation Voltage vs Collector Current

Collector Cutoff Current vs Ambient Temperature

Current Gain and Phase Angle vs Frequency

HOMEWORK PROBLEMS

8.1 Determine the region of operation for the following transistors:

a. *npn*, $V_{BE} = 0.8$ V, $V_{CE} = 0.4$ V

b. *npn*, $V_{CB} = 1.4$ V, $V_{CE} = 2.1$ V

c. *pnp*, $V_{CB} = 0.9$ V, $V_{CE} = 0.4$ V

d. *npn*, $V_{BE} = -1.2$ V, $V_{CB} = 0.6$ V

8.2 For the circuit in Figure 8.11, $R_C = 5$ kΩ, $R_E = 3$ kΩ, $R_B = 10$ kΩ, $V_{RB} = 4$ V, and $V_{CC} = 12$ V. Assume $V_{BE} = 0.7$ V. A single measurement indicates the reading of V_2 is 2.0 V. Find V_1, V_3, I_E, I_B, I_C, and β.

8.3 For the circuit in Figure 8.20, $V_{CC} = 20$ V, $R_C = 5$ kΩ, and $R_E = 1$ kΩ. Determine the region of operation of the transistor if:

a. $I_C = 1$ mA, $I_B = 20$ μA, $V_{BE} = 0.7$ V

b. $I_C = 3.2$ mA, $I_B = 0.3$ mA, $V_{BE} = 0.8$ V

c. $I_C = 3$ mA, $I_B = 1.5$ mA, $V_{BE} = 0.85$ V

Figure P8.1

8.4 For each transistor shown in Figure P8.1, determine whether the *BC* and *BE* junctions are forward- or reverse-biased, and also determine the operating state.

8.5   For the circuit in Figure 8.11, the potentials of the three terminals are measured as follows, with respect to the ground: $V_1 = 6.3$ V, $V_2 = 5.5$ V, and $V_3 = 6.0$ V. Determine the transistor operating state.

8.6   For the circuit shown in Figure P8.2, find I_E, I_B, I_C, and V_C, given $V_E = -0.7$ V.

8.7   An *npn* transistor is operated in the active region with $I_B = 0.2$ mA, $I_C = 12$ mA, $V_{BE} = 0.7$ V, and $V_{CE} = 14$ V. Find

a. I_C/I_E.

b. V_{CB}/V_{BE}.

c.   The total power dissipated by the transistor, defined here as $P = V_{CE}I_C + V_{BE}I_B$.

8.8   The magnitudes of a *pnp* transistor's emitter and base currents are 6 mA and 0.1 mA, respectively. The magnitudes of the voltages across the emitter-base and collector-base junctions are 0.65 and 7.3 V. Find

a. V_{CE}.

b. I_C.

c.   The total power dissipated in the transistor, defined here as $P = V_{CE}I_C + V_{BE}I_B$.

8.9   Repeat the calculations performed in Example 8.4, using $R_C = 0$, 100 Ω, 10 kΩ, and ∞.

8.10   For the circuit given in Figure 8.11, the following readings are measured:

a. $V_1 = 1.62$ V, $V_2 = 1$ V, $V_3 = 1.2$ V

b. $V_1 = 3.2$ V, $V_2 = 2.5$ V, $V_3 = 7$ V

c. $V_1 = 0.2$ V, $V_2 = 0$ V

What is the operating state of the transistor in each case?

8.11   For the circuit shown in Figure 8.20, $V_{CC} = 20$ V, $R_C = 5$ kΩ, and $R_E = 1$ kΩ. Determine the operating state of the transistor if

a. $I_C = 1$ mA, $I_B = 20$ mA, $V_{BE} = 0.7$ V

b. $I_C = 3.2$ mA, $I_B = 0.3$ mA, $V_{BE} = 0.8$ V

c. $I_C = 3$ mA, $I_B = 1.5$ mA, $V_{BE} = 0.85$ V

8.12   For the circuit given in Figure P8.3, verify that the transistor operates in the saturation region by computing the ratio of collector current to base current. (Hint: With reference to Figure 8.22, $V_\gamma = 0.6$ V, $V_{\text{sat}} = 0.2$ V.)

8.13   For the circuit shown in Figure P8.4, determine the base voltage V_{BB} required to saturate the transistor. Assume that $V_{CE\,\text{sat}} = 0.1$ V, $V_{BE\,\text{sat}} = 0.6$, and $\beta = 50$.

8.14   For the circuit shown in Figure 8.20, calculate the quiescent values of V_{BE}, I_B, V_{CE}, I_C for $V_{CC} = 15$ V, $R_1 = 10$ kΩ, $R_2 = 47$ kΩ, $R_C = 2$ kΩ, $R_E = 4$ kΩ, and $\beta = 75$.

Figure P8.2

Figure P8.3

Figure P8.4

8.15   An *npn* transistor is operated in the active region with the collector current 60 times the base current and with junction voltages of $V_{BE} = 0.6$ V and $V_{CB} = 7.2$ V. If $|I_E| = 4$ mA, find (a) I_B and (b) V_{CE}.

8.16   The collector characteristics for a certain transistor are shown in Figure P8.5.

Figure P8.5

a.   Find the ratio I_C/I_B for $V_{CE} = 10$ V and $I_B = 100\mu$A, 200 μA, and 600 μA.

b.   The maximum allowable collector power dissipation is 0.5 W for $I_B = 500\mu$A. Find V_{CE}.

Hint: A reasonable approximation for the power dissipated at the collector is the product of the collector voltage and current:

$$P = I_C V_{CE}$$

where P is the permissible power dissipation,
           I_C is the quiescent collector current,
           V_{CE} is the operating point collector-emitter voltage.

8.17   State whether each of the following conditions applies to active-region operation of an *npn* transistor, of a *pnp* transistor, of both, or of neither.

a. $V_{EB} = 0.6$ V
b. $I_C > I_E$
c. $|I_C| > |I_E|$
d. $I_E > I_B$
e. $V_{CE} > V_{CB}$

8.18   Use the collector characteristics of the 2N3904 *npn* transistor to determine the operating point (I_{CQ}, V_{CEQ}) of the transistor in the Figure P8.6, for each of the conditions listed. Also determine β.

a. $I_B = 0.1$ mA
b. $I_B = 0.19$ mA
c. $I_B = 0.04$ mA

Figure P8.6

8.19   Using the circuit shown in Figure P8.7 and the characteristic curves of a 2N3904 transistor,

a.   Determine the operating point of the transistor, Q, if $i_b(t) = 2.15$ mA $+ \Delta I_B$ and ΔI_B is very small.

b.   Determine β at this point.

c.   Estimate h_{fe} for this circuit at the operating point. Also estimate h_{ie}.

Figure P8.7

8.20   For the circuit in Figure 8.14, $R_C = 1$ kΩ, $V_{BB} = 5$ V, $\beta_{min} = 50$, and $V_{CC} = 10$ V. Find the range of R_B so that the transistor is in the saturation state.

8.21   For the circuit in Figure 8.14, $V_{CC} = 5$ V, $R_C = 1$ kΩ, $R_B = 10$ kΩ, and $\beta_{min} = 50$. Find the the range of values of V_{BB} so that the transistor is in saturation.

8.22   For the circuit of Figure 8.11, $V_{BB} = V_{CC} = 5$ V, $R_C = 1$ kΩ, $R_E = 1$ kΩ, and $\beta = 80$. Find the operating state of the transistor if

a. $R_B = 100$ kΩ

b. $R_B = 10$ kΩ

c. $R_B = 1$ kΩ

8.23   For the circuit in Figure 8.12, $I_{BB} = 20$ μA, $R_C = 2$ kΩ, $V_{CC} = 10$ V, and $\beta = 100$. Find I_C, I_E, V_{CE}, and V_{CB}.

8.24   For the circuit shown in Figure P8.8, find the maximum achievable voltage gain. Assume that $\beta = 100$.

Figure P8.8   **Figure P8.9**

*8.25   Find β for each of the transistors shown in Figure P8.9.

8.26   Using the h-parameters given in Example 8.9, construct the small-signal model of the common-emitter transistor amplifier in Figure 8.33.

8.27   The circuit shown in Figure P8.10 is known as a *common-base circuit*.

a.   Determine the operating point of the transistor circuit.

b.   Draw the h-parameter model for this circuit.

Figure P8.10

c. If the AC voltage gain is defined as $v_{out}(t)/v_S(t)$, find the voltage gain of this circuit for $h_{fe} = \beta$.

8.28 For the circuit shown in Figure P8.11, and given $R_E = 200\ \Omega$, $R_1 = 11.25\ k\Omega$, and $R_2 = 10\ k\Omega$,

Figure P8.11

a. Find the operating point of the transistor.
b. Draw the h-parameter circuit.
c. Determine the voltage gain v_{out}/v_{in} for $h_{fe} = \beta$, $h_{oe} = 10^{-4}$ S.

*8.29 Show that the voltage gain of the common-base amplifier is that given in Table 8.1. The effect of h_{oe} on A_V may be considered to be very small, so h_{oe} can be neglected in the calculation.

*8.30 Show that the output resistance of the common-base amplifier is that given in Table 8.1. Assume that the effect of h_{re} on the results is very small ($h_{re} \approx 0$).

*8.31 Show that the input resistance of the common-base amplifier is that given in Table 8.1. Assume that the effect of the h_{re} on the result is very small ($h_{re} \approx 0$).

8.32 Design the bias circuit of the common-emitter amplifier in Figure 8.20 to establish an emitter current $I_E = 1$ mA. Assume that the power supply V_{CC} is 15 V, and that $V_{BB} = \frac{1}{3}V_{CC}$, $V_{CB}(V_{CE}) = \frac{1}{3}V_{CC}$, and $I_C R_C = \frac{1}{3}V_{CC}$.

8.33 Consider a common emitter amplifier circuit with the following parameter values (see Figure 8.36): $V_{CC} = 10$ V, $R_1 = 1520\ \Omega$, $R_2 = 2920\ \Omega$, $R_S = 50\ \Omega$, $R_C = R_E = 100\ \Omega$, $R_L = 150\ \Omega$,

a. Find the operating point of the transistor.
b. Draw the AC equivalent circuit of the amplifier.
c. If $h_{oe} = 10^{-5}$ S and $h_{fe} = \beta$, determine the voltage gain as defined by v_{out}/v_{in}.
d. Find the input resistance, r_i.
e. Find the output resistance, r_o.

8.34 The transistor shown in the amplifier circuit in Figure P8.12 has $h_{ie} = 1.3\ k\Omega$, $h_{re} = 5 \times 10^{-4}$, $h_{fe} = 90$, and $h_{oe} = 120\ \mu$S. Find the voltage gain.

Figure P8.12

$R_E = 50\ \Omega$ $V_{CC} = 10$ V
$R_L = 100\ \Omega$ $R_{B1} = 513\ \Omega$
$\beta = 75 = h_{fe}$ $R_{B2} = 2730\ \Omega$

Figure P8.13

*8.35   The circuit shown in Figure P8.13 is a common-base amplifier. Using Table 8.1:

a.   Find the Q point of the transistor.
b.   Draw the AC equivalent circuit, using h parameters.
c.   Find the voltage gain v_L/v_{in}.
d.   Find the input resistance, r_i.
e.   Find the output resistance, r_o.

8.36   For the circuit shown in Figure P8.14, v_S is a small sine wave signal with zero average value. If $\beta = 100$,

a.   Find the value of R_E so that I_E is 1 mA.
b.   Find R_C so that V_C is 5 V.
c.   For $R_L = 5$ kΩ, find the small-signal equivalent circuit of the amplifier.
d.   Find the voltage gain.

*8.37   The circuit in Figure P8.15 is in a common-collector configuration. Using Table 8.1:

a.   Find the operating point of the transistor.
b.   Draw the AC equivalent circuit for $h_{oe} = 10^{-4}$ S.
c.   If the voltage gain is defined as v_{out}/v_{in}, find the voltage gain. If the current gain is defined as i_{out}/i_{in}, find the current gain.
d.   Find the input resistance, r_i.
e.   Find the output resistance, r_o.

Figure P8.14

$$R_E = 250 \ \Omega \qquad R_1 = 9221 \ \Omega$$
$$V_{CC} = 15 \ V \qquad C_B = \infty$$
$$R_2 = 6320 \ \Omega$$

Figure P8.15

*8.38   The circuit that supplies energy to an automobile's fuel injector is shown in Figure P8.16(a). The internal circuitry of the injector can be modeled as shown in Figure P8.16(b). The injector will inject gasoline into the intake manifold when $I_{inj} \geq 0.1$ A. The voltage V_{signal} is a pulse train whose shape is as shown in Figure P8.16(c). If the engine is cold and under start-up conditions, the signal duration, τ, is determined by the equation

$$\tau = BIT \times K_C + VCIT$$

where

$$BIT = \text{Basic injection time} = 1 \text{ ms}$$

$$K_C = \text{Compensation constant of temperature of coolant}(T_C)$$

$$VCIT = \text{Voltage-compensated injection time}$$

The characteristics of VCIT and K_C are shown in Figure P8.16(d).

Figure P8.16(a)   **Figure P8.16(b)**

If the transistor, Q_1, saturates at $V_{CE} = 0.3$ V and $V_{BE} = 0.9$ V, find the duration of the fuel injector pulse if

a. $V_{batt} = 13$ V, $T_C = 100°C$
b. $V_{batt} = 8.6$ V, $T_C = 20°C$

Figure P8.16(c)

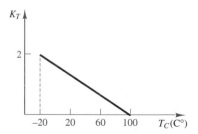

Figure P8.16(d)

*8.39 Some switching circuits require a higher voltage or current than may be available from a computer or another source. A power transistor drive circuit may be used to provide a "buffer" between the computer and load as shown in Figure P8.17(a).

Figure P8.17

If the waveform of the computer's output voltage, $v_{CO}(t)$, is as shown in Figure P8.17(b), determine the output voltage, $v_{LO}(t)$, given the following conditions.

$$V_{CE\,sat} = \pm0.2 \text{ V} \quad (+\text{for } npn, -\text{for } pnp)$$
$$V_\gamma = \pm0.8 \text{ V} \quad (+\text{for } npn, -\text{for } pnp)$$
$$R = 10 \text{ } \Omega \quad \beta_{sat} = 30$$
$$R_L = 50 \text{ } \Omega$$

Assume that the transistors are either in cutoff or in saturation.

*8.40 A DC relay coil can be modeled as an inductor with series resistance due to the windings, as shown in Figure P8.18(a). The relay in Figure P8.18(b) is being driven by a transistor circuit. When V_R is greater than 7.2 V_{DC}, the relay switch will close if it is in the open state. If the switch is closed, it will open when V_R is less

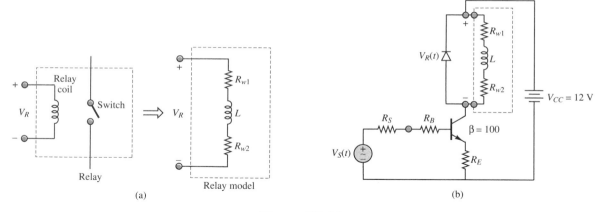

Figure P8.18

than 2.4 VDC. The diode "on" resistance is 10 Ω, and its off resistance is ∞. The relay is rated to dissipate 0.5 W when $V_R = 10$ V. If $R_E = 40$ Ω, $R_B = 450$ Ω, $R_S = 20$ Ω, $L = 100$ mH, $R_1 = R_2 = 2$ Ω, $V_{CE\,sat} = 0.3$ V, $V_\gamma = 0.75$ V, and $V_S(t)$ is a square wave that ranges from 0 volts to 4.8 volts and has a maximum current of 20 mA.

a.   Determine the maximum power dissipated by this circuit.

b.   Determine the time it takes for the relay to switch from the closed to the open position if V_S has been at 4.8 V for a very long time.

*8.41   The circuit shown in Figure P8.19 is used to switch a relay that turns a light off and on under the control of a computer. The relay dissipates 0.5 W at 5 VDC. It switches on at 3 VDC and off at 1.0 VDC. What is the maximum frequency with which the light can be switched? The inductance of the relay is 5 mH, and the transistor saturates at 0.2 V, $V_\gamma = 0.8$ V.

Figure P8.19

Figure P8.20

*8.42   A Darlington pair of transistors is connected as shown in Figure P8.20. The transistor parameters for small-signal operation are Q_1: $h_{ie} = 1.5$ kΩ, $h_{re} = 4 \times 10^{-4}$, $h_{oe} = 110$ μA/V, and $h_{fe} = 130$; Q_2: $h_{ie} = 200$ Ω, $h_{re} = 10^{-3}$, $h_{oe} = 500$ μA/V, and $h_{fe} = 70$. Calculate

a.   The overall current gain.

b.   The input impedance.

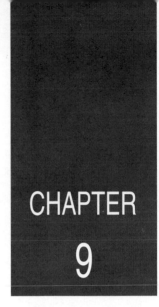

CHAPTER

9

Field-Effect Transistors
and Power Electronics

T his chapter introduces the second major family of transistors: the field-effect transistors, or FETs. Two representative devices are discussed, first by surveying their physical operation, then by examining their *i-v* characteristics. These are the junction field-effect transistor, or JFET, and the metal-oxide-semiconductor field-effect transistor, or MOSFET. Following a discussion of basic properties of these two devices, a section is devoted to the design of small-signal amplifiers using the circuit models for the devices.

The chapter ends with an introduction to power electronics, revisiting the voltage regulation problem introduced in Chapter 7, discussing some simple DC motor supplies, and analyzing a representative power electronic device: the silicon controlled rectifier, or SCR. By the end of Chapter 9, you should be able to:

- Describe the basic operation of junction field-effect transistors (JFETs) and insulated-gate field-effect transistors (IGFETs).
- Interpret the device characteristic curves and extract linear (small-signal) models for simple amplifiers from the device curves and data sheets.

- Recognize the more common JFET and MOSFET amplifier configurations, and draw their linear small-signal equivalent circuits.
- Compute the open-circuit voltage and current gains and the input and output resistance of the different amplifier configurations.
- Understand the operation of transistor power amplifiers, voltage regulators, and silicon controlled rectifiers.

9.1   THE JUNCTION FIELD-EFFECT TRANSISTOR (JFET)

The second family of transistor devices discussed in this book operates on a principle different from that of the *pn* junction introduced in Chapters 7 and 8. The concept that forms the basis of the operation of the **field-effect transistor,** or **FET,** is that the width of a conducting channel may be varied by the application of an external electric field. FETs form a very large family; a table of the different configurations of field-effect devices is provided in Section 9.2. To illustrate the operation of this class of devices, two types of field-effect transistor have been chosen: the *n*-channel **junction field-effect transistor,** or **JFET;** and the *n*-type enhancement **metal-oxide-semiconductor field-effect transistor,** or **MOSFET.** MOSFETs are part of the class of FETs known as **IGFETs,** which stands for **insulated-gate field-effect transistors.** In the next few sections, the physical operation of each of these devices will be outlined, and their use in different types of practical circuits will be discussed.

(a) JFET geometry, showing depletion layer

(b) *n*-channel JFET schematic symbols

(c) *p*-channel JFET schematic symbols

Figure 9.1 JFET geometry and circuit symbols

Basic Operation of the JFET

The physical structure and schematic symbols of the junction field-effect transistor are shown in Figure 9.1. JFETs are normally classified in two main groups: *n*-channel devices and *p*-channel devices. In many respects, these two types of JFETs differ from each other in a manner analogous to *npn* and *pnp* bipolar junction transistors. Figure 9.1 identifies three terminals for each device: the **source** (*S*), **gate** (*G*), and **drain** (*D*), which are similar in purpose to the BJT emitter, base, and collector terminals. The source and drain are connected by a **channel,** which is shown made of *n*-type material in the figure. The composition of the channel determines the type of JFET; for example, the device shown in Figure 9.1(a) is an *n*-channel JFET.

Figure 9.2 illustrates the operation of the (*n*-channel) JFET by showing a simplified schematic representation of the source, gate, and drain terminals, in relation to the channel; note that the channel in effect separates two *pn* junctions. In the figure, no connection is shown between the two gate regions; however, the two are connected inside the transistor. As a consequence of this geometry, two depletion regions (see Chapter 7) exist between each of the gate connections and the *n* channel. The width of these depletion regions (shown in white in Figure 9.2) determines the conduction characteristics of

Figure 9.2 Decreasing channel width with increasing drain-to-source voltage

the channel. To understand how it does so, consider the case where the gate-to-source voltage, v_{GS}, is kept equal to zero and the drain-to-source voltage, v_{DS}, is progressively increased. Initially, the depletion regions are small and a current flows from drain to source. When this condition is satisfied, small changes in v_{DS} result in a proportional change in the drain (or source[1]) current; thus, the JFET acts as a linear resistor in this region. As v_{DS} is further increased, the width of the depletion regions increases, and with it the resistance of the channel, since no current can flow through the depletion regions (remember that in these regions charge carriers have been *depleted*). Now Ohm's law is no longer satisfied: the drain current does not increase linearly with v_{DS}.

Eventually, a drain-to-source voltage is reached for which the width of the depletion region cannot increase any further, because the n channel has been *pinched off*. The voltage for which this condition occurs is called the **pinch-off voltage,** $-V_p$; note that the quantity V_p is by convention a negative voltage, and thus $-V_p$ is a positive voltage. When the drain-to-source voltage exceeds the pinch-off voltage, that is, when $v_{DS} > -V_p$, no further increase in drain current can take place. The drain current is now constant:

$$i_D = I_{DSS} = \text{ Drain (source) saturation current} \qquad v_{DS} > -V_p \quad \textbf{(9.1)}$$

Thus, for drain-to-source voltages greater than the pinch-off voltage, the

---

[1] Note that in a JFET, drain and source currents are virtually identical.

drain current is independent of v_{DS}. Eventually, as v_{DS} is further increased, a *breakdown* condition is reached, where the drain current sharply rises again.

Figure 9.3 summarizes the behavior of the JFET for $v_{GS} = 0$. Note that three regions may be identified in this *i-v* characteristic: the **ohmic** (or **triode**) **region,** the **saturation region,** and the **breakdown region.**

If the gate-to-source voltage is varied, a family of *i-v* curves may be generated. In a JFET, the gate-to-source voltage is always zero or negative, so that the *pn* junctions formed by each gate and the *n* channel are always reverse-biased. As v_{GS} is made more negative, i_D-v_{DS} curves similar to those of Figure 9.3 are generated, giving rise to a set of curves such as that shown in Figure 9.4. As v_{GS} is made sufficiently negative, the device enters the **cutoff region.** Note that a load line and an operating point are also shown on these curves, suggesting that the JFET can, in principle, be biased to serve as an amplifier.

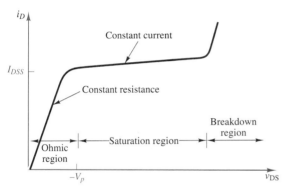

Figure 9.3 JFET i_D-v_{DS} characteristic for $v_{GS} = 0$

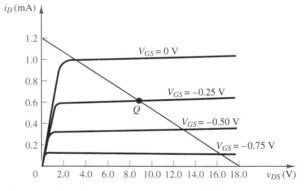

Figure 9.4 Typical drain curves for a JFET

Example   9.1

For the circuits in Figure 9.5, determine the operating state of the JFET, that is, whether it is in the ohmic or in the saturation region. Assume that $I_{DSS} = 10$ mA and $V_p = -4$ V.

Figure 9.5

Solution:

For Figure 9.5(a),

$$V_{DS} = V_D = V_{DD} - R_D I_D = 15 - 0.5 \times 10^3 (10^{-2}) = 10 \text{ V}$$

and since the condition for operation in the saturation region is $V_{DS} \geq V_{GS} - V_p$, the JFET is operating in the saturation region.

For Figure 9.5(b),

$$V_{DS} = V_D = 15 - 1.2 \times 10^3 (10^{-2}) = 3 \text{ V}$$

and since the condition for operation in the ohmic region is $V_{DS} \leq V_{GS} - V_p$, the JFET is operating in the ohmic region.

The JFET in the Ohmic Region

When operated in the ohmic region, the JFET operates very much as a **voltage-controlled resistor.** Figure 9.4 depicts a magnified plot of the JFET i_D-v_{DS} curves for v_{DS} below the pinch-off voltage. Note how the slope of the curves in this ohmic region depends on the gate-to-source voltage! By varying v_{GS} while keeping the transistor in the ohmic region, it is therefore possible to vary the resistance of the JFET. Since the resistance is equal to the inverse of the slope, it can be seen that the resistance of the JFET in the ohmic region increases as v_{GS} is made progressively more negative. In the ohmic region, the drain current is approximately given by the expression

$$i_D = I_{DSS} \left[2 \left(1 - \frac{v_{GS}}{V_p} \right) \left(\frac{v_{DS}}{-V_p} \right) - \left(\frac{v_{DS}}{V_p} \right)^2 \right] \tag{9.2}$$

For small values of v_{DS}, equation 9.2 may be approximated by

$$i_D \approx \frac{2 I_{DSS}}{-V_p} \left(1 - \frac{v_{GS}}{V_p} \right) v_{DS} = \frac{v_{DS}}{r_{DS}} \tag{9.3}$$

where r_{DS} is the equivalent resistance of the channel (for v_{DS} small) and is given by

$$r_{DS} = \frac{-V_p^2}{2 I_{DSS}(V_p - v_{GS})} \tag{9.4}$$

The JFET in the Saturation Region

In the saturation region, the JFET presents an essentially constant current characteristic for a given gate-to-source voltage. Comparing this characteristic with the operation of the bipolar junction transistor in the active region (see Chapter 8), we may observe a considerable similarity. The relatively flat family of curves for the JFET shows that the drain current can be controlled by relatively small changes in the gate-to-source voltage. In the operation of the bipolar transistor, on the other hand, changes in the base current result in modulation of the collector current. Thus, while the BJT in the forward active region acts as a current-controlled current source, the JFET in the saturation region is inherently a *voltage-controlled current source*.

In contrast with the current gain of the BJT, β, the JFET is characterized by a **transconductance, g_m,** which is equal to the change in drain current caused by a given change in gate-source voltage, for a given value of drain-source voltage:

$$g_m = \left. \frac{\partial i_D}{\partial v_{GS}} \right|_{V_{DS}} = \text{constant} \qquad (9.5)$$

A typical value of g_m is 10^{-3} S. For a JFET in the saturation region, the drain current is related to the gate-source voltage by the following relationship:

$$i_D \approx I_{DSS} \left(1 - \frac{v_{GS}}{V_p} \right)^2 \qquad (9.6)$$

As a final note, observe that saturation in an FET and in a BJT are two completely different concepts.

Small-Signal Model for JFET Amplifiers

The operation of the JFET in the saturation mode is simpler than the equivalent case of the BJT in the linear active region. This is because, while in a BJT the base circuit and the collector-emitter circuit are coupled to each other, in a JFET the gate circuit and the drain-source circuit are essentially isolated from each other, since the *pn* junctions between the gate and the channel are reverse-biased and therefore act as very large resistances (recall the discussion of the reverse-biased diode in Chapter 7). Figure 9.6 depicts the linear models of the operation of a BJT amplifier and of a JFET amplifier; note that in the small-signal model for the JFET, there is effectively no connection between the gate circuit and the drain-source circuit.

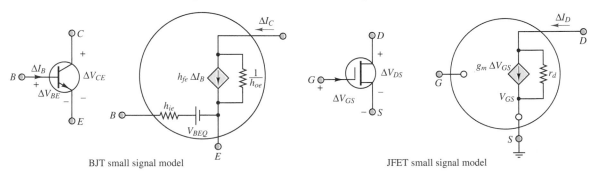

BJT small signal model   JFET small signal model

Figure 9.6 Comparison of BJT and JFET small-signal models

The operation of the JFET in the saturation region in response to small fluctuations in the gate-source voltage, ΔV_{GS}, is described by the controlled current source, $g_m \Delta V_{GS}$, and by the equivalent drain circuit resistance, r_{DS}, where

$$r_{DS} = \left. \frac{\partial v_{DS}}{\partial i_D} \right|_{V_{GS}=\text{constant}} \tag{9.7}$$

With reference to the drain curves of Figure 9.4, we note that a DC analysis similar to that carried out for the BJT leads to the determination of the Q point of the JFET amplifier. However, we can take advantage of the analytic relation between drain current and gate-source voltage, given in equation 9.6, to establish the Q point for the amplifier. In analogy with the discussion of bipolar transistors, we define the quiescent voltages V_{GSQ} and V_{DSQ} and the quiescent current I_{DQ}. The quadratic relationship of equation 9.6 allows us to determine the quiescent drain current, I_{DQ}, given that $V_{GS} = V_{GSQ}$ is known, and that the parameters I_{DSS} and V_p are known for any given transistor. Thus, the quiescent drain current is given by the expression

$$I_{DQ} = I_{DSS} \left(1 - \frac{V_{GSQ}}{V_p} \right)^2 \tag{9.8}$$

Consider, for example, the circuit of Figure 9.7, which depicts a simple JFET amplifier (after Figure 9.6) and its DC and AC equivalent-circuit models. Using the analysis methods of Chapter 8, we can determine the operating point from the DC circuit, using the load-line equation:

$$V_{DD} = V_{DSQ} + R_D I_{DQ} \tag{9.9}$$

The intersection of the load line with the quiescent gate-source voltage, V_{GSQ}, determines the operating point. For the amplifier of Figure 9.7, V_{GSQ} is equal to the battery voltage V_{GG}; note that the battery necessarily has negative polarity. Once the operating point is established, the AC circuit of Figure 9.7 may be used to determine the AC gains of the amplifier. In

Figure 9.7 Elementary JFET amplifier

particular, it may be seen by applying the current divider rule that

$$v_{\text{out}} = -g_m \Delta V_{GS} \frac{r_{DS}}{r_{DS} + R_D} R_D = -g_m v_{\text{in}} \frac{r_{DS}}{r_{DS} + R_D} R_D \qquad (9.10)$$

or that

$$A_V = -g_m \frac{r_{DS} R_D}{r_{DS} + R_D} \qquad (9.11)$$

That is, the magnitude of the voltage gain of the JFET amplifier is equal to the product of the transconductance and the parallel combination of the JFET resistance parameter, r_{DS}, with the drain resistor, R_D. Since, in general, r_{DS} is quite large, you see that, the larger R_D, the larger the voltage gain will be.

A more detailed analysis of FET amplifiers will be conducted in Section 9.3. The purpose of this discussion was to introduce the principal characteristics of FET amplifiers, in order to compare them with BJT amplifiers. Figure 9.8 depicts typical amplifier characteristics of the two types of devices, illustrating the voltage-controlled current source behavior of the JFET versus the current-controlled current source behavior of the BJT. It is important to note that the relationship between i_D and v_{GS} is nonlinear, even within the saturation (i.e., amplifier) region, whereas the corresponding BJT curve for the equivalent condition (i.e., the active region) is very nearly linear. This phenomenon may also be explained by observing that the spacing of the BJT collector curves is much more uniform than that of the JFET drain curves.

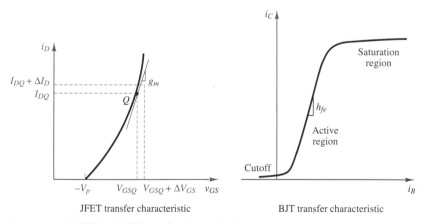

JFET transfer characteristic   BJT transfer characteristic

Figure 9.8 JFET and BJT transfer characteristics

Example   9.2

The characteristic curves and load line for a JFET were shown in Figure 9.4. Determine the voltage gain that can be achieved with the elementary JFET amplifier of Figure 9.7, assuming that the gate-source voltage is a shifted sine wave, given by

$$v_{GS} = -0.25 + 0.25 \sin \omega t$$

and that the Q point is $V_{DS} = 8.5$ V, $I_D = 0.6$ mA.

Solution:

The load resistance for this amplifier may be deduced from the curves of Figure 9.4 as the inverse of the slope of the load line:

$$R_D = \frac{\Delta V_{DS}}{\Delta I_D} = \frac{18 - 0}{1.2 \times 10^{-3} - 0} = 15 \text{ k}\Omega$$

For the stated oscillation in gate-source voltage (around the DC value -0.25 V), the peak-to-peak drain current may be found from the curves to be

$$\Delta I_D = (1.0 - 0.3) \times 10^{-3} = 0.7 \text{ mA}$$

This corresponds to a peak-to-peak drain-source voltage of

$$\Delta V_{DS} = 0.7 \times 10^{-3} \times 15 \times 10^3 = 10.5 \text{ V}$$

Note that the oscillation in drain-source voltage will take place around the quiescent value of 9.5 V. The (AC) voltage gain may therefore be computed as follows:

$$A_V = \frac{\Delta V_{DS}}{\Delta V_{GS}} = \frac{10.5}{2 \times 0.25} = 21$$

Example   9.3

Estimate the values of g_m and r_{DS} and find the open-circuit amplification factor μ for the amplifier of Example 9.2. Explain why the voltage gain, A_V, is much smaller than μ.

Solution:

Consider the two extremes reached by the gate-source voltage: $V_{GS} = 0$ V and $V_{GS} = -0.5$ V. In the first case, for a constant drain-source voltage $V_{DS} = 9.5$ V, the drain current is 1 mA. For $V_{GS} = -0.5$ V, the drain current is 0.3 mA, from the characteristic curves. Thus, the local value of the transconductance parameter is found as follows:

$$g_m = \left.\frac{\Delta I_D}{\Delta V_{GS}}\right|_{V_{DS}=\text{constant}} = \frac{(1 - 0.3) \times 10^{-3}}{0.5 - 0}$$

$$= 1.4 \times 10^{-3} \text{ S} = 1.4 \text{ mS}$$

The value of the parameter r_{DS} may be found by considering the curves of Figure 9.4 with $V_{GS} = \text{constant} = -0.25$ V. When $V_{DS} = 18$ V, the drain current is $I_D = 0.62$ mA. For $V_{DS} = 1$ V, $I_D = 0.60$ mA. Thus,

$$r_{DS} = \left.\frac{\Delta V_{DS}}{\Delta I_D}\right|_{V_{GS}=\text{constant}} = \frac{18 - 1}{0.62 \times 10^{-3} - 0.60 \times 10^{-3}} \times 10^3 = 0.85 \text{ M}\Omega$$

and the open loop voltage gain is

$$\mu = g_m r_{DS} = 1.4 \times 10^{-3} \times 0.85 \times 10^6 = 1,190$$

The equivalent circuit corresponding to these calculations is shown in Figure 9.7. Observe now that the actual voltage gain is calculated in the presence of a load resistance, R_D, which will necessarily appear in parallel with r_{DS} (if you are not convinced, examine the AC equivalent circuit of Figure 9.7). Thus, the actual voltage gain is

$$A_V = \frac{v_{out}}{\Delta V_{GS}} = -\left(\frac{r_{DS}}{r_{DS} + R_D}\right)\left(\frac{g_m \Delta V_{GS} R_D}{\Delta V_{GS}}\right)$$

$$= -\frac{850}{850 + 15} \times 1.4 \times 10^{-3} \times 15 \times 10^3 = -20.64$$

Why such a dramatic drop from the high open-loop voltage gain? The reason is that, to keep the JFET operating in the saturation region, it is necessary to employ a resistor $R_D \ll r_{DS}$, resulting in a relatively small output voltage.

Example   9.4

V_{CC} + 15 V

1 mA

R

+ 8 V

(a)

V_{CC} + 15 V

1 mA

R

+ 8 V

(b)

Figure 9.9

For the circuits shown in Figure 9.9,

1.   Find the voltage gain, by using a Q point of 1 mA at 8 V for the BJT circuit. The transconductance parameter, g_m, is 40 mS for a typical BJT.
2.   Find the voltage gain for the JFET circuit for $g_m = 1{,}000\ \mu$S, which is a typical value for a FET.
3.   Comment on the results.

Solution:

From the diagram for the BJT circuit, the collector resistor is computed to be

$$R = \frac{V_{CC} - V_{CE}}{I_C} = \frac{15 - 8}{1 \times 10^{-3}} = 7 \times 10^3\ \Omega$$

and since the transconductance is defined by the relation

$$g_m = \frac{i_{out}}{v_{in}} = -\frac{v_{out}}{v_{in}}\left(\frac{1}{R}\right)$$

and

$$v_{out} = -i_{out}R$$

the voltage gain is found to be

$$A_V = -\frac{v_{out}}{v_{in}} = -g_m R$$

For Figure 9.9(a), we have

$$A_V = -40 \times 10^{-3} \times 7 \times 10^3 = -280$$

The same calculation may be repeated for Figure 9.9(b), to obtain

$$A_V = -10^{-3} \times 7 \times 10^3 = -7$$

Comments

The transconductance of FETs is substantially smaller than that of BJTs; thus, the latter are preferred for use as amplifier stages. However, because of their high input impedance, FETs can serve as excellent input stages for multistage amplifiers.

Example   9.5   Current Sources

Since the drain current is essentially constant, provided that V_{DS} is more than a few volts, a JFET can be used as a current source, as shown in Figure 9.10, where the load is connected in series between the supply voltage and the drain. Although this circuit approximates an ideal current source quite well, the strength of the current is subject to some uncertainty, because of the spread in values of the transistor parameters, which is a consequence of the fabrication methods. For a given transistor design, the parameter I_{DSS} can vary by as much as a factor of 5. To adjust the operating point of the JFET and thus provide the desired source current, a source resistor is often employed, as shown in Figure 9.10(b). Typical ranges of current values that may be obtained by JFET current sources are between a few tenths of a milliampere and a few milliamperes.

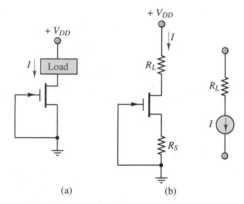

(a)   (b)

Figure 9.10 JFET current source

DRILL EXERCISES

9.1   Calculate I_D in Example 9.1.

> Answer: (a) -12.5 mA; (b) 9.4 mA

9.2   Calculate r_{DS} for the transistor of Example 9.1.

> Answer: 2 kΩ

9.3   Calculate the approximate value of I_D for Figure 9.5(b) in Example 9.1.

> Answer: 1.5 mA

9.4   Repeat Example 9.4 for $V_{CC} = 12$ V.

> Answer: BJT: -133.2; FET: -3.33

9.5   Draw the small-signal model for the circuit of Figure 9.9(b).

9.6   Find I_D as a function of V_{DS} for the circuit in Figure 9.10(a).

9.2   THE METAL-OXIDE-SEMICONDUCTOR FIELD-EFFECT TRANSISTOR (MOSFET)

The construction of the MOSFET we will be examining is shown in Figure 9.11. The device consists of a metal gate, insulated from the bulk of the *p*-type semiconductor material by an oxide layer (thus the terminology *metal-oxide-semiconductor*). Two n^+ regions on either side of the gate are called source and drain, respectively. The figure also illustrates how an electric field can be applied between the gate and the **bulk** of the semiconductor by connecting the gate to a voltage source. The effect of the electric field, the intensity of which depends on the strength of the gate voltage, is to push positive charge carriers away from the surface of the bulk material. As a consequence of this effect, the *p*-type material has a lower concentration of positive charge carriers near its surface, and it behaves progressively more like intrinsic semiconductor material and then, finally like *n*-type material as the electric field strength increases. Thus, in a narrow layer between n^+ regions, the *p*-type material is *inverted* and *n*-type charge carriers become available for conduction. This layer is called the *channel*. The width of the channel is, to a certain extent, controlled by the intensity of the field: the stronger the field, the farther into the bulk material the channel extends. Since the channel thus formed is *n*-type, the device of Figure 9.11 takes the name ***n*-channel MOSFET.** It is also called an **enhancement-mode** device, because the applied electric field "enhances" the conduction in the channel by attracting *n*-type charge carriers. There are also **depletion-mode** devices, in which the application of an electric field "depletes" the channel of charge

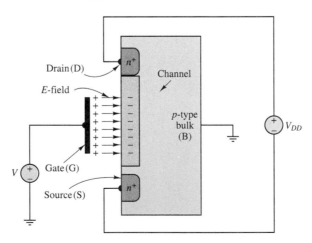

In a MOSFET, the electric field between the gate and the semiconductor can create a channel for conduction by "inverting" the *p*-type bulk material.

Figure 9.11 Metal-oxide-semiconductor field-effect transistor carriers, reducing the effective channel width. It is useful to think of enhancement-mode devices as being normally off: current cannot flow from drain to source unless a gate voltage is applied. On the other hand, depletion-mode devices are normally on; that is, the gate voltage is used to reduce the conduction of current from drain to source.

An analogous discussion could be carried out for **p-channel MOS-FETs,** in which the roles of the bulk and of the drain and source materials are reversed. Thus, in a *p*-channel device, conduction in the channel occurs through positive charge carriers. The correspondence between *n*-channel and *p*-channel devices is akin to that between *npn* and *pnp* bipolar devices.

Figure 9.12 illustrates the circuit symbols for the *n*-channel enhancement MOSFET of Figure 9.11.

If we now return to examine the *n*-channel enhancement MOSFET circuit of Figure 9.11, we see that when no gate voltage is applied, no conduction can take place, whether V_{DD} is a positive or a negative source, since the n^+-p-n^+ construction is such that one of the two *pn* junctions must be reversed at all times. Thus, the device is normally off, as already stated. One of the primary advantages of a MOSFET over the BJTs of Chapter 8 is that, because the gate is insulated, little or no current flows into the bulk from the gate. Thus, a MOSFET in practice draws little or no power from the circuit that controls the current flow from drain to source. A second point to be noted is that the current through the device is controlled by the gate voltage, while in the BJT it was a current that controlled the larger flow of current through the material. Other differences and similarities between these devices will be pointed out along the way.

At this stage, we are ready to examine the *i-v* characteristic curves of an *n*-channel enhancement MOSFET, shown in Figure 9.13. These characteristic curves could be obtained, for example, by varying the gate-to-source voltage, v_{GS}, and, for each value, measuring the drain current, i_D, versus the drain-to-source voltage, v_{DS}. The test circuit that would ideally be used to

Figure 9.12 Symbol for *n*-channel enhancement MOSFET

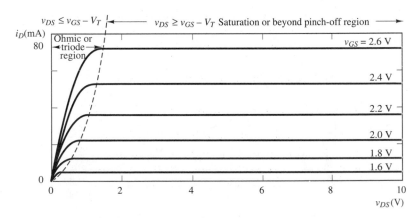

Figure 9.13 *n*-channel MOSFET drain characteristic

Figure 9.14 Test circuit to obtain the drain characteristic of Figure 9.13

generate such a characteristic is shown in Figure 9.14. Note that the bulk (*B*) forms two *pn* junctions, one with the source region and one with the drain region. For normal operation, these *pn* junctions are reverse-biased all the time. Reverse biasing can be achieved by connecting the bulk, *B*, to the source, that is by connecting the source to ground.

The first observation we make in comparing these curves with the BJT characteristic curves is that now the i_D-v_{DS} family of curves is parameterized in terms of a voltage, v_{GS}, whereas in a BJT, the collector family of curves was indexed by the base current. Second, we note that it is not necessary to portray two sets of relationships for FETs, as we did for BJTs by representing the base-emitter and the collector characteristics separately. Since the gate is insulated from the rest of the device, the gate current is zero, and we do not need to describe a gate characteristic; this is completely defined by the gate-source voltage alone. Thus in a MOSFET (as was the case with the JFET) we have only three variables: the gate-to-source voltage, v_{GS}, which controls conduction in the channel; the drain current, i_D (which is identically equal to the source current, i_S); and the drain-to-source voltage, v_{DS}.

In analyzing the drain curves more closely, we see that for values of v_{GS} less than or approximately equal to 1.4 V, no conduction occurs. As v_{GS} increases beyond this **threshold voltage**, V_T, some conduction starts to occur in the channel, as illustrated by the *i-v* characteristic. A load-line analysis, can be carried out to determine the operating point, as illustrated in Example 9.6.

Example   9.6

To illustrate the determination of the *Q* point for a MOSFET, we employ the circuit shown in Figure 9.15. Applying KVL around the drain circuit yields the familiar load-line equation:

$$10 = V_{DS} + 10,000 I_D$$

which is sketched in Figure 9.16, superimposed on the drain characteristic. The

Figure 9.15

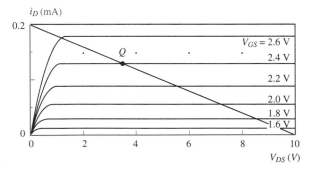

Figure 9.16

intersection of the load line with the i_D-v_{DS} curves for $V_{GS} = 2.4$ V occurs around

$$I_{DQ} = 0.125 \text{ mA} \qquad V_{DSQ} = 3.5 \text{ V}$$

Note the uneven spacing of the MOSFET curves with respect to v_{GS} (much more uneven than that of the BJT collector curves with respect to i_B).

It is important to note, though, that while the increase in collector current as a function of the base current (leading to amplification) was a very nearly linear function in the BJT ($i_C = \beta i_B$), resulting in equally spaced curves in the i_C-v_{CE} plots, a similar case does not hold for a MOSFET. The dependence of i_D on v_{GS} is nonlinear, taking the approximate form

$$i_D = K(v_{GS} - V_T)^2 \qquad v_{DS} > v_{GS} - V_T \qquad (9.12)$$

where the constant K depends on physical parameters and has the unit A/V^2, and where V_T is the threshold voltage. The value of V_T is controlled during device fabrication and typically lies in the range of 1 to 3 V. The nonlinearity in the relationship of equation 9.12 is reflected in the uneven spacing of the drain curves with respect to the gate voltage.

9.3   FET AMPLIFIERS

The discussion of FETs as amplifiers is analogous to that of BJT amplifiers. In particular, the **common-source amplifier** circuit is equivalent in structure to the common-emitter amplifier circuit studied earlier, and the **common-drain amplifier (source follower)** is analogous to the common-collector amplifier (emitter follower). In this section, we discuss the general features of FET amplifiers; to simplify the discussion, we have selected the n-channel enhancement MOSFET to represent the FET family. Although some of the details differ depending on the specific device, the discussion that follows applies in general to all FET amplifiers. Table 9.1 is a summary of FET symbols and will help you recognize a specific device in a circuit diagram.

Figure 9.17 depicts typical common-drain and common-source amplifiers, including coupling and bypass capacitors. One of the great features of FETs, and especially MOSFETs, is the high input impedance that can be achieved because the gate is effectively insulated from the substrate material. We shall illustrate this property in analyzing the source follower circuit. Before proceeding with the analysis of FET amplifiers, though, we shall discuss how one can construct a small-signal model analogous to the one that was obtained for the BJT.

In the case of MOSFETs, we can also use the analytic relation between drain current and gate-source voltage.

$$i_D = K(v_{GS} - V_T)^2 \qquad (9.13)$$

to establish the Q point for the transistor, which is defined by the quiescent voltages V_{GSQ}, V_{DSQ}, and by the quiescent current I_{DQ}. The quadratic

(a) Common drain amplifier

(b) Common source amplifier

Figure 9.17 Typical common-drain and common-source MOSFET amplifiers

Table 9.1   FET family circuit symbols

n-channel enhancement MOSFET	*p*-channel enhancement MOSFET
n-channel depletion MOSFET	*p*-channel depletion MOSFET
n-channel JFET	*p*-channel JFET

relationship allows us to determine the drain current, I_{DQ}, that will flow, given that $v_{GS} = V_{GSQ}$; the parameters K and V_T are a property of any given device. Thus,

$$I_{DQ} = K(V_{GSQ} - V_T)^2 \tag{9.14}$$

and it is now possible to determine the quiescent drain-to-source voltage from the load-line equation for the drain circuit:

$$V_{DD} = V_{DSQ} + R_D I_{DQ} \tag{9.15}$$

In a MOSFET, as in JFETs, it is possible to approximate the small-signal behavior of the device as a linear relationship by using the transconductance parameter g_m, where g_m is the slope of the i_D-v_{GS} curve at the Q point, as shown in Figure 9.8. Formally, we can write

$$g_m = \left. \frac{\partial i_D}{\partial v_{GS}} \right|_{I_{DQ}, V_{GSQ}} \tag{9.16}$$

Since an analytical expression for the drain current is known (equation 9.12), we can actually determine g_m as given by

$$g_m = \frac{\partial}{\partial v_{GS}}[K(v_{GS} - V_T)^2] \tag{9.17}$$

at the operating point (I_{DQ}, V_{CSQ}). Thus, evaluating the expression for the transconductance parameter, we obtain the expression

$$g_m = 2K(v_{GS} - V_T)\Big|_{I_{DQ}, V_{GSQ}} \tag{9.18}$$
$$= 2\sqrt{KI_{DQ}}$$

g_m is a function of the quiescent drain current, as expected, since the tangent to the i_D-v_{GS} curve of Figure 9.8 has a slope that is dependent on the value of I_{DQ}.

Example 9.7

Consider the Q point of the MOSFET of Example 9.6: $I_{DQ} = 55$ mA, $V_{DSQ} = 3.5$ V, $V_{GSQ} = 2.4$ V. Assume the values $V_T = 1.4$ V and $K = 0.125$ mA/V^2 for the MOSFET used in this example. Then, it follows that the transconductance parameter is given by

$$g_m = 2\sqrt{KI_{DQ}} = 2\sqrt{0.125 \times 10^{-3} \times 0.125 \times 10^{-3}}$$
$$= 0.25 \text{ mA/V}$$

The significance of this value of the transconductance g_m is that for every volt of incremental oscillation around the gate-to-source voltage, there will result a corresponding drain current fluctuation of 1 mA; for example, if $\Delta V_{GS} = \cos(\omega t)$, then $\Delta I_D = 0.001 \cos(\omega t)$. Thus, the MOSFET acts as a *voltage-controlled current source*. We shall see shortly how this behavior results in amplification.

Example 9.8

Determine all of the relevant transistor voltages and currents for the amplifier circuit shown in Figure 9.18. Assume that $V_T = 1$ V and $K = 0.5$ mA/V^2.

Solution:

Since the gate current is zero, the voltage between the gate and ground is

$$V_G = (10)\frac{10 \times 10^6}{10 \times 10^6 + 10 \times 10^6} = 5 \text{ V}$$

Let us assume for the moment that this gate voltage is such that the MOSFET is operating in the saturation region; we will proceed according to this assumption, and verify it once the analysis is complete. The voltage between the source and ground is $V_S = 6 \times 10^3 I_D$, so that the gate-source voltage is given by

$$V_{GS} = 5 - 6I_D$$

where I_D is in mA. The drain current may be obtained from equation 9.12.

$$I_D = K(V_{GS} - V_T)^2 = 0.5(5 - 6I_D - 1)^2$$

Figure 9.18

Solving this quadratic equation, we obtain two solutions:

$$I_{D1} = 0.89 \text{ mA} \qquad \text{and} \qquad I_{D2} = 0.5 \text{ mA}$$

Which of the two solutions is the correct one? One way of determining the correct solution is to solve for the gate-source voltage for each case. For $I_{D1} = 0.89$ mA, the gate-source voltage is $V_{GS} = 5 - 6I_D = -0.34$ V. For $I_{D2} = 0.5$ mA, $V_{GS} = 2$ V. Since V_{GS} needs to be $> V_T$ for the MOSFET to be in the saturation region, we must select

$$I_D = 0.5 \text{ mA}$$

to be consistent with the initial assumption. Correspondingly, the drain voltage is found to be

$$V_D = 10 - 6 \times 10^3 \times 0.5 \times 10^{-3} = 7 \text{ V}$$

and therefore $V_{DS} = V_D - V_S = 7 - 3 = 4$ V. Now it is possible to verify the initial assumption. Recall (equation 9.12) that the MOSFET is in the saturation region if $V_{DS} > V_{GS} - V_T$. From the calculations just carried out and the given value of V_T, $V_{DS} = 4$ V $> V_{GS} - V_T = 1$ V, and therefore the MOSFET is indeed in saturation. Thus, the initial assumption was correct.

The transconductance parameter allows us to define a very simple model for the MOSFET during small-signal operation: we replace the input circuit (gate) by an open circuit, since no current can flow into the insulated gate, and model the drain-to-source circuit by a controlled current source, $g_m \Delta V_{GS}$. This small-signal model is depicted in Figure 9.19.

It is important to appreciate the fact that the transconductance g_m is dependent on the quiescent value of the drain current, and therefore any MOSFET amplifier design is going to be very strongly dependent on the operating point.

MOSFET small signal model

MOSFET circuit symbol

Figure 9.19 MOSFET small-signal model

The MOSFET Common-Source Amplifier

It is useful at this stage to compare the performance of the common-source amplifier of Figure 9.17(b) with that of the BJT common-emitter amplifier. The DC equivalent circuit is shown in Figure 9.20; note the remarkable similarity with the BJT common-emitter amplifier DC circuit.

Figure 9.20 DC circuit for the common-source amplifier

The equations for the DC equivalent circuit are obtained most easily by reducing the gate circuit to a Thévenin equivalent with $V_{GG} = \frac{R_2}{R_1+R_2}V_{DD}$ and $R_G = R_1 \parallel R_2$. Then the gate circuit is described by the equation

$$V_{GG} = V_{GSQ} + I_{DQ}R_S \tag{9.19}$$

and the drain circuit by

$$V_{DD} = V_{DSQ} + I_{DQ}(R_D + R_S) \tag{9.20}$$

Note that, given V_{DD}, any value for V_{GG} may be achieved by appropriate selection of R_1 and R_2. Example 9.9 illustrates the computation of the Q point for a MOSFET amplifier.

Example   9.9

Given the curves for the MOSFET of Example 9.6, we wish to obtain the same Q point, $I_{DQ} = 55$ mA and $V_{DSQ} = 3.5$ V for $V_{GSQ} = 2.4$ V, with a single voltage supply, $V_{DD} = 10$ V, using the common-source MOSFET amplifier circuit of Figure 9.17.

Solution:

The gate equation, resulting from KVL, requires that

$$V_{GG} = V_{GSQ} + I_{DQ}R_S$$

or

$$V_{GG} = 2.4 + 0.055R_S$$

while the drain circuit imposes that

$$V_{DD} = V_{DSQ} + I_{DQ}(R_D + R_S)$$

or

$$10 = 3.5 + 0.055(R_D + R_S)$$

that is,

$$R_D + R_S = 118 \ \Omega$$

If we (arbitrarily) select $R_D = 100 \ \Omega$ and $R_S = 18 \ \Omega$ for this operating point, we solve for V_{GG} as follows:

$$V_{GG} = 3.4 \ \text{V}$$

To obtain $V_{GG} = 3.4$ V, we solve the equation

$$V_{GG} = \frac{R_2}{R_1 + R_2}V_{DD}$$

and choose

$$R_G = R_1 \parallel R_2 = 100 \ \text{k}\Omega$$

to obtain

$$(10\ v)\frac{R_2}{R_1 + R_2} = 3.4\ \text{V}$$

and

$$\frac{R_1 R_2}{R_1 + R_2} = 10^5\ \Omega$$

or

$$0.34 R_1 = 10^5 \Rightarrow R_1 = 294\ \text{k}\Omega$$

and therefore find that

$$R_2 = \frac{0.34}{0.66} R_1 = 151\ \text{k}\Omega$$

The amplifier circuit for the required Q point is shown in Figure 9.21.

Figure 9.21

Substituting the small-signal model in the common-source amplifier circuit of Figure 9.17, we obtain the small-signal AC equivalent circuit of Figure 9.22, where we have assumed the coupling and bypass capacitors to be short circuits at the frequency of the input signal $v(t)$. The circuit is

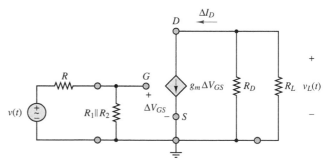

Figure 9.22 AC circuit for the common-source MOSFET amplifier

analyzed as follows. The load voltage, $v_L(t)$, is given by the expression

$$v_L(t) = -\Delta I_D \cdot (R_D \parallel R_L) \tag{9.21}$$

where

$$\Delta I_D = g_m \Delta V_{GS} \tag{9.22}$$

Thus, we need to determine ΔV_{GS} to write expressions for the voltage and current gains of the amplifier. Since the gate circuit is equivalent to an open circuit, we have:

$$\Delta V_{GS}(t) = \frac{(R_1 \parallel R_2)}{R + (R_1 \parallel R_2)} v(t) \tag{9.23}$$

and

$$\Delta V_{GS}(t) \approx v(t) \tag{9.24}$$

since we have purposely selected R_1 and R_2 to be quite large in the design of the DC bias circuit and therefore $R_1 \parallel R_2 \gg R$. Thus, we can express the load (output) voltage in terms of the input voltage $v(t)$ as

$$v_L(t) = -g_m \cdot (R_D \parallel R_L) \cdot v(t) \tag{9.25}$$

This expression corresponds to a voltage gain of

$$A_V = \frac{v_L(t)}{v(t)} = -g_m(R_D \parallel R_L) \tag{9.26}$$

Note that we have an upper bound on the voltage gain for this amplifier, $A_{V\,\text{max}}$, that is independent of the value of the load resistance:

$$|A_{V\,\text{max}}| \leq g_m R_D \tag{9.27}$$

This upper bound is achieved, of course, only when the load is an open circuit. This open-circuit voltage gain plays the same role in the analysis of the MOSFET amplifier as the parameter μ we defined in Chapter 8 for BJT amplifiers:

$$\mu = -g_m R_D \tag{9.28}$$

For the values of g_m and R_D quoted in Example 9.9, the magnitude of the voltage gain would be only $A_V = 0.1$. That is an attenuation! This result is very much in contrast with that obtained for the common-emitter BJT amplifier, where significant voltage amplification was easily attained. A useful comparison may be made here, in assessing typical values of the open-circuit voltage gain μ for small-signal BJT and FET amplifiers. For the MOSFET used in the examples of this section, a typical value of g_m would be 1,000 μA/V, or 10^{-3} A/V. Thus, for a design value of drain resistance of 100 Ω, we obtain the answer already noted: $\mu = -0.1$. If we were to calculate the value of μ for amplifier design B of Example 8.11, we would find that for a 2N3904 bipolar junction transistor, typical values of h_{fe} and h_{ie} are 100 A/A and 2 kΩ, respectively. Thus, the open-circuit voltage gain would be $\mu = -10$, giving a difference of two orders of magnitude between the two amplifiers!

Are MOSFET common-source amplifiers useless, then, since no voltage gain is provided? To answer this puzzling question, let us consider the AC current gain of the common-source amplifier. The current drawn from the input is very simply

$$i_{in} = \frac{v(t)}{R + R_G} \approx \frac{v(t)}{R_G} \tag{9.29}$$

while the output current is given by

$$i_L(t) = \frac{v_L(t)}{R_L} = \frac{A_V \cdot v(t)}{R_L} \tag{9.30}$$

Thus, the current gain is given by the following expression:

$$\begin{aligned} A_I &= \frac{i_L}{i_{in}} = \frac{A_V \cdot R_G}{R_L} = -g_m \frac{R_D R_L}{R_D + R_L} \frac{R_G}{R_L} \\ &= -g_m \frac{R_D R_G}{R_D + R_L} \end{aligned} \tag{9.31}$$

Note that we have purposely made R_G large, to limit the current required of the signal source, $v(t)$. Thus, the current gain for a common-source amplifier can be significant. For a 100-Ω load, we find that the effective voltage gain for the common-source amplifier is

$$A_V = -g_m R_D \parallel R_L = -0.05$$

and the corresponding current gain is

$$A_I = -g_m \frac{R_D R_G}{R_D + R_L} = -50$$

for the design values quoted in Example 9.9. Thus, the net **power gain** of the transistor, A_P, is greater than 1:

$$A_P = A_V \cdot A_I = 2.5$$

This brief discussion of the common-source amplifier has pointed to some important features of MOSFETs. First, MOSFETs do not have a very large transconductance, and therefore do not perform well by comparison with BJTs in voltage amplification. However, MOSFETs make up for this intrinsic lack of voltage gain by their inherently high input impedance, which requires very little current of a signal source and thus affords substantial current gains. Further, power MOSFETs can be constructed in a way that affords substantial voltage gains, as illustrated in Example 9.10.

Example   9.10

The n-channel enhancement-mode small-signal MOSFET BS170 is designed for high-voltage, high-speed power switching applications such as line drivers, relay drivers, microprocessor to high-voltage interface, and high-voltage display drivers. The typical value of the transconductance parameter for this type of device is much larger than for small-signal MOSFETs; for this device, $g_m = 200$ mS for $V_{DS} = 10$ V and $I_D = 250$ mA, and the minimum gate threshold voltage is 0.8 V at $V_{DS} = V_{GS}$ and $I_D = 1.0$ mA. Assume that the desired operating point is at $V_{DSQ} = 10$ V, $I_{DQ} = 250$ mA, $V_{GSQ} = 9$ V, and $V_{DD} = 25$ V. Complete the design of the common-source amplifier shown in Figure 9.17(b) for this particular transistor, and calculate the gains A_V and A_I.

Solution:

The equation for the gate circuit is

$$V_{GG} = V_{GSQ} + I_{DQ}R_S = 9 + 0.25R_S$$

while for the drain circuit, we obtain

$$V_{DD} = V_{DSQ} + I_{DQ}(R_D + R_S)$$

$$= 10 + 0.25(R_D + R_S) = 25 \text{ V}$$

Solving the two equations, we obtain a relationship between R_D and R_S

$$R_D + R_S = 60 \text{ }\Omega$$

If we select $R_S = 2$ Ω and $R_D = 58$ Ω, a gate supply voltage may be selected as follows:

$$V_{GG} = 9 + 0.25 \times 2$$

$$= 25\left(\frac{R_2}{R_1 + R_2}\right) = 9.5 \text{ V}$$

If we choose $R_G = R_1 \parallel R_2 = 100$ kΩ and solve the preceding equation, we get

$$R_1 \approx 263 \text{ k}\Omega \qquad R_2 \approx 161 \text{ k}\Omega$$

The voltage gain for an 80-Ω load may now be computed to be

$$A_V = -g_m(R_D \parallel R_L) = -200 \times 10^{-3} \times 33.6 = -6.7$$

and the current gain is

$$A_1 = -g_m\frac{R_D R_G}{R_D + R_L} = -8,400$$

You can see that this power MOSFET can afford substantial amplification of both voltage and current!

The MOSFET Source Follower

FET source followers are commonly used as **input stages** to many common instruments, because of their very high input impedance. A commonly used source follower is shown in Figure 9.23. The desirable feature of this amplifier configuration is that, as we shall soon verify, it provides a high input

Source follower

$R_G = R_1 \| R_2$

$V_{GG} = \dfrac{R_2}{R_1 + R_2} V_{DD}$

DC circuit

Figure 9.23 MOSFET source follower and DC circuit

Figure 9.24 AC equivalent circuit for the MOSFET source follower

impedance and a low output impedance. These features are both very useful and compensate for the fact that the voltage gain of the device is at most unity.

As usual, the Q point of the amplifier is found by writing KVL around the drain circuit:

$$V_{DSQ} = V_{DD} - I_{DQ}R_S \tag{9.32}$$

and by assuming that no current flows into the gate. The voltage V_{GSQ}, which controls the gate-to-source bias, is then given by the expression

$$V_{GSQ} = V_{GG} - I_{DQ}R_S \tag{9.33}$$

so that the desired Q point may be established by selecting appropriate values of R_1, R_2, and R_S (Example 9.11 will illustrate a typical design). The AC equivalent circuit is shown in Figure 9.24.

Since it is possible to select R_1 and R_2 arbitrarily, we see that the input resistance, R_G, of the source follower can be made quite large. Another observation is that, for this type of amplifier, the voltage gain is always less than 1. This fact may be verified by considering that

$$\Delta V_{GS} = v_{in}(t) - g_m \Delta V_{GS}R_S \tag{9.34}$$

which implies

$$\frac{\Delta V_{GS}}{v_{in}(t)} = \frac{1}{1 + g_m R_S} \tag{9.35}$$

Since the AC output voltage is related to ΔV_{GS} by

$$v_{out} = g_m \Delta V_{GS}R_S \tag{9.36}$$

it follows that

$$\frac{v_{out}}{\Delta V_{GS}} = g_m R_S \tag{9.37}$$

Thus, the open-circuit voltage gain, μ, may be obtained as follows:

$$\frac{v_{out}}{v_{in}} = \frac{v_{out}}{\Delta V_{GS}} = (g_m R_S) \left(\frac{1}{1 + g_m R_S} \right)$$

$$= \frac{g_m R_S}{g_m R_S + 1} < 1 \tag{9.38}$$

We can see that the open-circuit voltage gain will always be less than 1. The output resistance can be found by inspection from the circuit of Figure 9.24 to be equal to R_S.

In summary, the source follower amplifier has an input resistance, $R_G = R_1 \parallel R_2$, which can be made arbitrarily large; a voltage gain less than unity; and an output resistance, R_S, which can be made small by appropriate design.

Example   9.11

An enhancement MOSFET source follower is shown in Figure 9.25. Given $V_T = 1.5$ V, $K = 0.125$ mA/V^2, find the small-signal voltage gain and the input resistance.

Figure 9.25

Solution:

The DC operating point for the amplifier is found by using the equation relating the drain current to the gate-source voltage:

$$I_D = K(V_{GS} - V_T)^2$$
$$= 0.125 \times 10^{-3}(V_{GS} - 1.5)^2$$

Because the DC gate current is zero and no DC voltage will drop across R_G, it may be concluded that the gate-source voltage is equal to the drain voltage,

$$V_{GS} = V_D$$

Thus, the drain current is given by

$$I_D = 0.125 \times 10^{-3}(V_D - 1.5)^2$$

and the drain-source voltage is found from the load-line equation to be

$$V_{DS} = V_{DD} - R_D I_D = 15 - 10 \times 10^3 I_D$$

Solving the two preceding equations, we have

$$I_{DQ} = 1.06 \text{ mA} \qquad V_{DQ} = 4.4 \text{ V}$$

The transconductance parameter may also be found, once the operating point of the MOSFET is known:

$$g_m = 2K(V_{GS} - V_T) = 2 \times 0.125(4.4 - 1.5) = 0.725 \text{ mA/V}$$

Finally, the small-signal equivalent circuit is shown in Figure 9.26.

From the equivalent circuit, we may derive the following expression for the output voltage:

$$v_{\text{out}} = \Delta V_{DS} = -g_m \Delta V_{GS}(R_D \parallel R_L)$$

Since $\Delta V_{GS} = v_{in}(t)$, the voltage gain is

$$\frac{v_{out}}{v_{in}(t)} = -g_m(R_D \parallel R_L)$$

$$= -0.725 \times 10^{-3} \times (10 \times 10^3 \parallel 10 \times 10^3) = -3.625$$

Figure 9.26

The input current, i_{in}, is found to be

$$i_{in} = \frac{v(t) - v_{out}}{R_G} = \frac{v(t)}{R_G}\left(1 - \frac{v_{out}}{v(t)}\right)$$

$$= \frac{v(t)}{R_G}[1 - (-3.625)] = \frac{4.625v(t)}{R_G}$$

and from the expressions for the input current and input voltage, we can now compute the input resistance of the amplifier:

$$R_{in} = \frac{v_{in}(t)}{i_{in}} = \frac{R_G}{4.625} = 2.16 \text{ M}\Omega$$

You see that it is possible to design source followers with remarkably high input resistance.

DRILL EXERCISES

9.7 Compute the quiescent drain current and the small-signal transconductance parameter for a MOSFET with $V_T = 1.4$ V, $K = 2$ mA/V^2, for the following values of V_{GS}: 1.8 V, 2.0 V, 2.2 V, 2.4 V, 2.6 V, 2.8 V, 3.0 V.

> Answer: $I_D = $ 0.32, 0.72, 1.28, 2, 2.88, 3.92, 5.12 mA;
> $g_m = $ 1.6, 2.4, 3.2, 4, 4.8, 5.6, 6.4 mA/V

9.8 Find the open-circuit AC voltage gain, μ, for the amplifier design of Example 9.9. What is the effective voltage gain of the amplifier if the load resistance is 1 kΩ?

> Answer: $\mu = -0.1$; $A_V = -0.91$

9.9 Repeat Drill Exercise 9.8 for the values of g_m given in Drill Exercise 9.7.

> Answer: $\mu = $ -0.16, -0.24, -0.32, -0.4, -0.48, -0.56, -0.64

9.10 Design the circuit of Figure 9.10(b) so that the transistor operates at $i_D = I_{DQ} = 0.4$ mA, $V_{DQ} = 1$ V. Assume $V_{DD} = 5$ V, $V_p = -4$ V.

> Answer: $R_S = 0$, $R_L = 10$ kΩ

9.11 For the circuit in Figure 9.17(a), $V_{DD} = 15$ V, $R_1 = R_2 = 10$ MΩ, and $R_S = 6$ kΩ. Assume that $V_T = 1.5$ V and $K = 0.6$ mA/V^2, and find V_G, V_D, and V_S.

> Answer: $V_G = 7.5$ V, $V_D = 15$ V, $V_S = 4.84$ V

9.12 Calculate g_m for the circuit of Example 9.8 if $i_D = I_{DQ} - 0.5$ mA.

> Answer: 1 mA/V

9.4   POWER ELECTRONICS

So far, we have primarily considered low-power electronic devices, either in the form of small-signal linear amplifiers, or as switches and digital logic gates (the latter will be discussed in more detail in Chapter 12). There are many applications, however, in which it is desirable to provide a substantial amount of power to a load. Among the most common applications are loudspeakers (these can draw several amperes); electric motors and electromechanical actuators, which will be considered in greater detail in Chapters 14 through 16; and DC power supplies, which have already been analyzed to some extent in Chapter 7. In addition to such applications, the usage and control of electric power in industry requires electronic devices that can carry currents as high as hundreds of amperes, and voltages up to thousands of volts. Examples are found readily in the control of large motors and heavy industrial machinery.

The aim of the present section is to discuss qualitatively some of the more relevant issues in the design of *power amplifiers,* such as distortion and heat dissipation; to revisit the problem of voltage regulation in voltage supplies; and to introduce the *silicon controlled rectifier (SCR)* as a means of controlling the amount of power delivered to a load. This last topic, in particular, is likely to be useful to the non–electrical engineer dealing with electric motors and industrial loads.

Power Amplifiers

The brief discussion of power amplifiers in this section makes reference exclusively to the BJTs; this family of devices has traditionally dominated the field of **power amplifiers,** although in recent years semiconductor technology has made power FETs competitive with the performance of bipolar devices.

You may recall the notion of breakdown from the introductory discussion of BJTs. In practice, a bipolar transistor is limited in its operation by three factors: the maximum collector current, the maximum collector-emitter voltage, and the maximum power dissipation, which is the product of I_C and V_{CE}. Figure 9.27 illustrates graphically the power limitation of a BJT

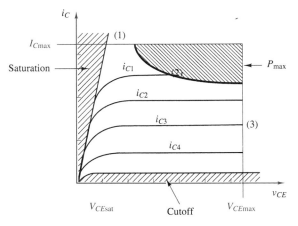

Figure 9.27 Limitations of a BJT amplifier

by showing the region where the maximum capabilities of the transistor are exceeded. (1) Exceeding the maximum allowable current $I_{C\,max}$ on a continuous basis will result in melting the wires that bond the device to the package terminals. (2) Maximum power dissipation is the locus of points for which $V_{CE}I_C = P_{max}$ at a case temperature of 25°C. The average power dissipation should not exceed P_{max}. (3) The instantaneous value of v_{CE} should not exceed $V_{CE\,max}$; otherwise, avalanche breakdown of the collector-base junction may occur. It is important to note that the linear operation of the transistor as an amplifier is also limited by the saturation and cutoff limits.

Thus, in effect, the operation of a BJT as a linear amplifier is rather severely limited by these factors. Consider, first, the effect of driving an amplifier beyond the limits of the linear active region, into saturation or cutoff. The result will be signal distortion. For example, a sinusoid amplified by a transistor amplifier that is forced into saturation, either by a large input or by an excessive gain, will be compressed around the peaks, because of the decreasing device gain in the extreme regions. Thus, to satisfy these limitations—and to fully take advantage of the relatively distortion-free linear active region of operation for a BJT—the Q point should be placed in the center of the device characteristic to obtain the **maximum symmetrical swing.** This point has already been discussed in Chapter 8 (see Example 8.11 and Figure 8.46, in particular).

The maximum power dissipation of the device, of course, presents a more drastic limitation on the performance of the amplifier, in that the transistor can be irreparably damaged if its power rating is exceeded. Values of the maximum allowable collector current, $I_{C\,max}$, of the maximum allowable transistor power dissipation, P_{max}, and of other relevant power BJT parameters are given in Table 9.2 for a few typical devices. Because of their large geometry and high operating currents, power transistors have typical parameters quite different from those of small-signal transistors.

From Table 9.2, we can find some of these differences:

1. β is low. It can be as low as 5; the typical value is 20 to 80.
2. $I_{C\,\text{max}}$ is typically in the ampere range; it can be as high as 100 A.
3. V_{CEO} is usually 40 to 100 V, but it can reach 500 V.

Table 9.2   Typical parameters for representative power BJTs

	2N4921	2N6306	2N5683
Type	*npn*	*npn*	*pnp*
Maximum I_C (continuous)	3 A	16 A	50 A
V_{CEO}	40 V	250 V	60 V
Power rating	30 W	125 W	300 W
β	10 @ I_C = 1 A	25 @ I_C = 3 A	20 @ I_C = 25 A
$V_{CE\,\text{sat}}$	0.6 V	0.8 V	1–5 V
$V_{BE\,\text{on}}$	1.3 V	1.3 V	2 V

Voltage Regulators

You will recall the discussion of the Zener diode as a voltage regulator in Chapter 7, where we introduced a voltage regulator as a three-terminal device that acts nearly as an ideal battery. Figure 9.28 depicts the appearance of a Zener diode *i-v* characteristic and shows a block diagram of a three-terminal regulator.

A simple Zener diode is often inadequate for practical voltage regulation. In some cases, the Zener resistance alone might cause excessive power dissipation in the Zener diode (especially when little current is required by the load). A more practical—and often-used—circuit is a regulator that includes a series pass transistor, shown in Figure 9.29. The operation of this voltage regulator is as follows. If the unregulated supply voltage, v_S, exceeds the Zener voltage, v_Z, by an amount sufficient to maintain the BJT in the

Figure 9.28 Zener diode

Figure 9.29 Practical voltage regulator

active region, then $v_{BE} \approx V_\gamma$ and $v_Z \approx V_Z$, the Zener voltage. Thus, the load voltage is equal to

$$v_L = V_Z - V_\gamma = \text{constant} \tag{9.39}$$

and is relatively independent of fluctuations in the unregulated source voltage, or in the required load current. The difference between the unregulated source voltage and the load voltage will appear across the CE "junction." Thus, the required power rating of the BJT may be determined by considering the largest unregulated voltage, $V_{S\,\text{max}}$:

$$\begin{aligned} P_{\text{BJT}} &= (V_{S\,\text{max}} - V_L)I_C \\ &\approx (V_{S\,\text{max}} - V_L)i_L \end{aligned} \tag{9.40}$$

The operation of a practical voltage regulator is discussed in more detail in Example 9.12.

Example   9.12

Use the large-signal model for the 2N4921 power BJT of Table 9.2 to compute the maximum load current that can be provided by a voltage regulator employing a 12.7-V Zener diode. What should the power rating of the Zener diode be? Assume that the nominal unregulated source voltage is $V_S = 20$ V and that $R_B = 47\ \Omega$ and $R_L = 10\ \Omega$.

Solution:

If we assume that the transistor is in the active region and the Zener diode is regulating, then, writing KVL for the circuit shown in Figure 9.30, we obtain

$$12.7 - V_\gamma - IR_L = 0$$

and

$$I = \frac{12.7 - 1.3}{10} = I_E = 1.14 \text{ A}$$

Applying KVL to the base circuit, as shown in Figure 9.31, we obtain an expression for the current through resistor R_B

$$I_{RB} = \frac{20 - 12.7}{R_B}$$

and

$$I_{RB} = \frac{20 - 12.7}{47} = 0.155 \text{ A}$$

We can now determine whether the transistor is in the active or the saturation region, and therefore verify the assumption made at the beginning of the problem. Consider the circuit of Figure 9.32. From the calculated value of the emitter current, the load voltage is computed to be

$$V_L = (1.14 \text{ A})(10\ \Omega) = 11.4 \text{ V}$$

Figure 9.30

Figure 9.31

Figure 9.32

The base voltage is fixed by the presence of the Zener diode:

$$V_B = 12.7 \text{ V}$$

so that

$$V_{CB} - 0.155 R_B = 0.155 \times 47 \ \Omega = 7.3 \text{ V}$$

Finally, the collector-emitter voltage may be found from the expression

$$V_{CE} = V_{CB} - V_{BE} = V_{CB} - (V_B - V_E) = 6 \text{ V}$$

Thus, we see that the transistor is in its linear region and therefore

$$I_B = I_E \frac{1}{\beta + 1} = \frac{I_L}{\beta + 1} = \frac{1.14}{10 + 1} = 103.6 \text{ mA}$$

Now, writing KCL at the base junction (Figure 9.33) suggests that

$$I_R = I_B + I_Z$$

and

$$I_Z = I_R - I_B = 0.155 - 0.1036 = 51.4 \text{ mA}$$

Thus, the power dissipated by the Zener diode is

$$P_Z = I_Z V_Z = (51.4 \text{ mA})(12.7 \text{ V})$$
$$= 652 \text{ mW}$$

Figure 9.33

The advantage of this configuration is that the Zener current is the base current going into the transistor, and therefore it is equal to only a fraction of the load current, thus reducing the power consumption of the Zener diode.

Three terminal voltage regulators are available in packaged form to include all the necessary circuitry (often including protection against excess heat dissipation). Regulators are rated in terms of the regulated voltage and power dissipation. Some types provide a variable regulated voltage by means of an external adjustment. Because of their requirement for relatively large power (and therefore heat) dissipation, voltage regulators often need to be attached to a **heat sink,** a thermally conductive assembly that aids in the cooling process. Figure 9.34 depicts the appearance of typical heat sinks. Heat sinking is a common procedure with many power electronic devices.

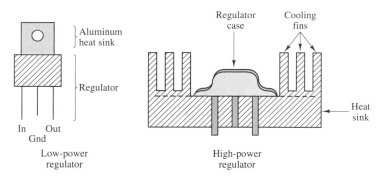

Figure 9.34 Heat-sink construction for voltage regulators

Inductive Loads and Electric Motor Supplies

One of the issues of practical importance in the design of a voltage supply is the nature of the load. In practice, loads are not always resistive, as will be seen in Chapters 14 through 16, where circuit models for electromechanical actuators and electric motors are introduced. A very common occurrence consists of a DC voltage supply providing current to a *DC motor*. For the purpose of the present discussion, it will suffice to state that a DC motor presents an inductive impedance to the voltage supply and requires a constant current from the supply to operate at a constant speed. The circuit of Figure 9.35 illustrates, as an example, a simple half-wave rectifier connected to an *RL* load.

Simple half-wave rectifier

The circuit on top in Figure 9.35, assuming an ideal diode, would present a serious problem during the negative half-cycle of the source voltage, since the requirement for continuity of current in the inductor (recall the discussion on continuity of inductor currents and capacitor voltages in Chapter 4) would be violated with D_1 off. Whenever the current flow through the inductor is interrupted (during the negative half-cycles of v_{AC}), the inductor attempts to build a **flyback voltage** proportional to di_L/dt. Since the rectifier does not provide any current during the negative half-cycle of the source voltage, the instantaneous inductor voltage could be very large and could lead to serious damage to either the motor or the rectifier.

Same arrangement with free-wheeling diode

Figure 9.35 Rectifier connected to an inductive load

The circuit shown on the bottom in Figure 9.35 contains a so-called **free-wheeling diode,** D_2. The role of D_2 is to provide continuity of current when D_1 is in the off state. D_2 is off during the positive half-cycle but turns on when D_1 ceases to conduct, because of the flyback voltage, $L\,di_L/dt$. Rather than build up a large voltage, the inductor now has a path for current to flow, through D_2, when D_1 is off. Thus, the energy stored by the inductor during the positive half-cycle of v_{AC} is utilized to preserve a continuous current through the inductor during the off period. Figure 9.36 depicts the load current for the circuit including the diode. Note that D_2 allows the energy-storage properties of the inductor to be utilized to smooth the pulselike supply current and to produce a nearly constant load current.

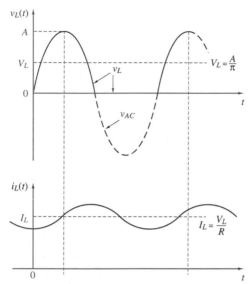

Figure 9.36 Operation of a free-wheeling diode

Analyzing the circuit on the right-hand side of Figure 9.35, with

$$v_{AC}(t) = A \sin(\omega t) \tag{9.41}$$

(and assuming that both D_1 and D_2 are ideal), we conclude that the DC component of the load voltage, V_L, must appear across the load resistor, R (no steady-state DC voltage can appear across the inductor, since $v_L = L\, di_L/dt$). Thus, the approximate DC current flowing through the load is given by

$$I_L = \frac{A}{\pi R} \tag{9.42}$$

since the average output voltage of a half-wave rectifier is A/π V for an AC source of peak amplitude A (see Chapter 7). The AC component of the load current (or "ripple" current) is not as simple to compute, since it is due to the AC component of v_L, which is not a pure sinusoid. The exact analysis would require the use of a Fourier series expansion. For the purposes of this discussion, it is not unreasonable to assume that most of the energy is at a frequency equal to that of the AC source:

$$i_L(t) \approx I_L + I_{AC} \cos(\omega t + \theta) \tag{9.43}$$

An acceptable approximation from which the amplitude of I_{AC} may be computed is

$$v_L(t) \approx \frac{A}{2} + \frac{A}{2} \sin \omega t \tag{9.44}$$

Figure 9.37 graphically illustrates the extent of this approximation.

A common alternative to the half-wave rectifier is the full-wave rectifier, which was discussed in Chapter 7.

Figure 9.37 Approximation of ripple voltage for a half-wave rectifier

Silicon Controlled Rectifiers

In a number of applications involving DC motors, it is useful to be able to externally control the amount of current flowing from an AC source to the load. A family of power semiconductor devices called **controlled rectifiers** allows for control of the rectifier state by means of a third input, called the **gate.** Figure 9.38 depicts the appearance of a **silicon controlled rectifier (SCR),** illustrating how the physical structure of this device consists of four layers, alternating p-type and n-type material. Note that the circuit symbol for the SCR suggests that this device acts as a diode, with provision for an additional external control signal.

The operation of the SCR can be explained in an intuitive fashion as follows. When the voltage v_{AK} is negative (i.e., providing reverse bias), the SCR acts just like a conventional pn junction in the off state. When v_{AK} is forward-biased *and* a small amount of current is injected into the gate, the SCR conducts forward current. The SCR then continues to conduct (even in the absence of gate current), provided that v_{AK} remains positive. Figure 9.39 depicts the i-v curve for the SCR. Note that the SCR has two stable states, determined by the bias v_{AK} and by the gate current. In summary, the SCR acts as a diode with a control gate that determines the time when conduction begins.

Figure 9.38 Silicon controlled rectifier (SCR)

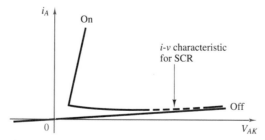

Switching occurs when i_G is injected into the gate.

Figure 9.39 SCR i-v characteristic

A somewhat more accurate description of SCR operation may be provided if we realize that the four-layer *pnpn* device can be modeled as a *pnp* transistor connected to an *npn* transistor. Figure 9.40 clearly shows that, physically, this is a realistic representation. Note that the anode current, i_A, is equal to the emitter current of the *pnp* transistor (labeled Q_p) and the base current of Q_p is equal to the collector current of the *npn* transistor, Q_n. Likewise, the base current of Q_n is the sum of the gate current and the collector current of Q_p. The behavior of this transistor model is explained as follows. Suppose, initially, i_G and i_{B_n} are both zero. Then it follows that Q_n is in cutoff, and therefore $i_{C_n} = 0$. But if $i_{C_n} = 0$, then the base current going into Q_p is also zero and Q_p is also in cutoff, and $i_{C_p} = 0$, consistent with our initial assumption. Thus, this is a stable state, in the sense that unless an external condition perturbs the SCR, it will remain off.

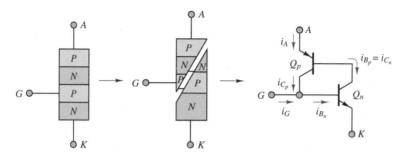

Figure 9.40 SCR two-transistor model

Now, suppose a small pulse of current is injected at the gate. Then $i_{B_n} > 0$, and Q_n starts to conduct, provided, of course, that $v_{AK} > 0$. At this point, i_{C_n}, and therefore i_{B_p}, must be greater than zero, so that Q_p conducts. It is important to note that once the gate current has turned Q_n on, Q_p also conducts, so that $i_{C_p} > 0$. Thus, even though i_G may cease, once this "on" state is reached, $i_{C_p} = i_{B_n}$ continues to drive Q_n so that the on state is also self-sustaining. The only condition that will cause the SCR to revert to the off state is the condition in which v_{AK} becomes negative; in this case, both transistors return to the cutoff state.

In a typical SCR application, the device is used as a half-wave rectifier that conducts only after a trigger pulse is applied to the gate. Without concerning ourselves with how the trigger pulse is generated, we can analyze the general waveforms for the circuit of Figure 9.41 as follows. Let the voltage $v_{trigger}$ be applied to the gate of the SCR at $t = \tau$. The voltage $v_{trigger}$ can be a short pulse, provided by a suitable trigger-timing circuit (Chapter 11 will discuss timing and switching circuits). At $t = \tau$, the SCR begins to conduct, and it continues to do so until the AC source enters its negative cycle. Figure 9.42 depicts the relevant waveforms.

Note how the DC load voltage is controlled by the firing time τ, according to the following expression:

$$< v_L >= V_L = \frac{1}{T} \int_{\tau}^{T/2} v_{AC}(t)\, dt \tag{9.45}$$

where T is the period of $v_{AC}(t)$. Now, if we let

$$v_{AC}(t) = A \sin \omega t \tag{9.46}$$

Figure 9.41 SCR rectifier circuit

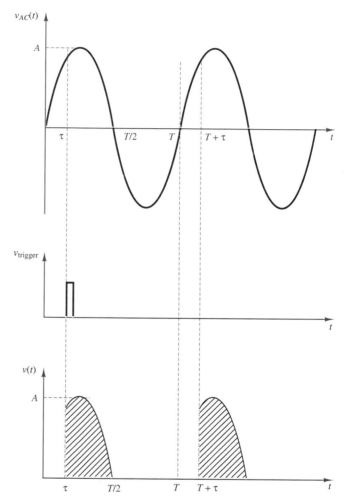

Figure 9.42 SCR rectifier waveforms

we can express the average (DC) value of the load voltage

$$V_L = \frac{1}{T} \int_{\tau}^{T/2} A \sin \omega t \, dt = (1 + \cos \omega t)\frac{A}{2\pi} \tag{9.47}$$

in terms of the **firing angle,** θ, defined as

$$\theta = \omega \tau \tag{9.48}$$

By evaluating the integral of equation 9.47, we can see that the (DC) load voltage amplitude depends on the firing angle, θ:

$$V_L = (1 + \cos \theta)\frac{A}{2\pi} \tag{9.49}$$

Example   9.13

SCRs are typically used in AC power control circuits. The circuit of Figure 9.43 is used to control power to a resistive load, such as a light bulb. The control circuitry will provide a pulse of current $i_G(t)$ to the SCR at $\omega t = \theta + n360$, where n is an integer 0, 1, 2, 3, ... based upon the position of the control knob. At the instant that $i_G(t)$ becomes nonzero, the SCR behaves like a forward-biased diode; that is, if $v_{AK}(t)$ is positive, then the SCR will allow current to pass in the forward direction. The gate current is pulsed with the timing shown in Figure 9.44.

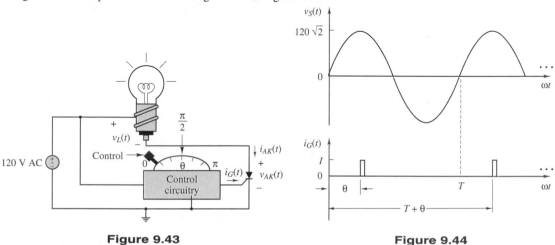

Figure 9.43   **Figure 9.44**

1. Find the rms load voltage as a function of the firing angle, θ, if $v_{AK\,on} = 0$ V.
2. Find the power supplied to the light bulb if its internal resistance is 240 Ω and $\theta = 0$.
3. Repeat part 2 for $\theta = \pi$.
4. Repeat part 2 for $\theta = \pi/2$.

Solution:

1. The load voltage will appear as shown in Figure 9.45, and the rms value of the load voltage may be computed as follows:

$$V_{L\,\mathrm{rms}} = \sqrt{\frac{(120\sqrt{2})^2}{2\pi} \int_{\theta}^{\pi} \sin^2 \omega t \, d(\omega t)}$$

$$= \sqrt{\frac{(120\sqrt{2})^2}{2\pi} \int_{\pi}^{\theta} \left(\frac{1}{2} - \frac{1}{2}\cos 2\omega t\right) d\omega t}$$

$$= \sqrt{\frac{(120\sqrt{2})^2}{4} - \frac{(120\sqrt{2})^2}{2\pi}\left(\frac{\theta}{2}\right) + \frac{(120\sqrt{2})^2}{4 \times 2\pi} \sin 2\theta}$$

After some simplification, the load voltage may be expressed as a function of

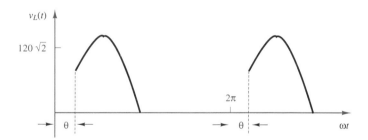

Figure 9.45

the firing angle, as follows:

$$V_L(\theta) = \frac{120\sqrt{2}}{2}\sqrt{1 - \frac{\theta}{\pi} + \sin 2\theta}$$

The load power may be now computed for the three cases:

2.
$$P = \frac{V_{\mathrm{rms}}^2}{R} = \frac{(120\sqrt{2}/2)^2}{240}$$
$$= 30 \text{ W}$$

3.
$$V_{\mathrm{rms}} = 0$$
$$P = 0$$

4.
$$P = \frac{\left(\frac{120\sqrt{2}}{2}\sqrt{1 - \frac{1}{2}}\right)^2}{240}$$
$$= 15 \text{ W}$$

Note that, since the SCR is off whenever power is not reaching the load, no power is wasted. This would not be the case if one simply used a resistive voltage divider to reduce the load voltage.

DRILL EXERCISES

9.13 Using the approximation given in equation 9.44, find the DC and AC load currents for the circuit of Figure 9.35 if $R = 10 \ \Omega$, $L = 0.3$ H, $A = 170$ V, and $\omega = 377$ rad/s.

> Answer: $I_L = 5.4$ A; $I_{\mathrm{AC}} = 0.75$ A; $\theta = 84.95°$

9.14 Repeat Example 9.12 using the 2N5683 transistor.

> Answer: $P = 632.8$ mW

9.15 Calculate the load voltage in Figure 9.41 for $A = 100$, $\theta = \pi/3$.

> Answer: $V_L = 23.87$ V

9.16 For the circuit in Example 9.13, the input AC voltage is 240 V. Find the rms value of the load voltage and the power at the firing angle $\theta = \pi/2$.

> Answer: 120 V, 60 W

CONCLUSION

- The field-effect transistor (FET) can be modeled as a voltage-controlled current source. It intrinsically offers lower voltage gain than the BJT, but it can provide much higher input resistance and therefore significant current gain. The FET i-v characteristic is more nonlinear than that of the BJT. FET amplifiers based on linear circuit models that are similar in structure to BJT amplifiers can be designed.

- When transistors are used in high-power electronic applications, a number of additional considerations must be made. The power dissipation of the device becomes very important, and it may become necessary to use heat sinks to allow for the necessary cooling. Further, it is often necessary to maximize the output voltage and current without incurring excessive distortion; this requires careful selection of the operating point to obtain maximum undistorted swing.

- Practical Zener diode voltage regulators usually employ a transistor to minimize the power dissipated by the Zener diode.

- Another commonly used three-terminal electronic device is the silicon controlled rectifier (SCR). The SCR acts as a diode with an additional control "gate," which can be used to determine when the diode is to turn on. The SCR is used mostly for controlled rectification, with application to DC motor supplies and other industrial loads that require AC to DC power conversion.

KEY TERMS

DEVICE DATA SHEETS
2N4856A N-Channel JFET

Electrical Characteristics ($T_A = 25°C$)

Symbol	Parameter	Conditions	Min	Typ	Max	Units
BV_{GSS}	Gate-Source Breakdown Voltage	$V_{DS} = OV, I_G = -1\mu A$	-30	-45		V
I_{DSS}	Zero Gate Voltage Drain Current	$V_{DS} = 2OV, V_{GS} = 0V$ Pulse Text	5.0	65	170	mA
I_{GSS}	Reverse Gate Leakage	$V_{GS} = -2OV, V_{DS} = 0V$		-15	-200	pA
$r_{DS(ON)}$	ON Resistance	$V_{DS} = 100$ mV, $V_{GS} = 0V$	20	35	100	Ω
g_{fs}	Forward Transconductance	$V_{DG} = 15V, I_D = 2$ mA			8.5	mmhos
$V_{GS(OFF)}$	Pinch Off Voltage	$V_{DS} = 2OV, I_D = 1$ nA	-0.5	-4.5	-9.0	V
$I_{D(OFF)}$	Drain OFF Current	$V_{DS} = 2OV, V_{GS} = -10V$		15	200	pA
C_{rss}	Feedback Capacitance	$V_{DG} = 15V, I_D = 5$ mA, $f = 1$ MHz		3.5	4.0	pF
C_{iss}	Input Capacitance	$V_{DG} = 15V, I_D = 5$ mA, $f = 1$ MHz		10	16	pF
o_n	Noise Voltage	$V_{DG} = 15V, I_D = 1$ mA, $f = 100$ Hz		6.0		$nV/\sqrt{Hz}$
t_{on}	Turn-On Time	$V_{DD} = 10V, I_D = 6.6$ mA		12	20	ns
t_{off}	Turn-Off Time	$V_{DD} = 10V, I_D = 6.6$ mA		40	80	ns

Transfer Characteristics

Transfer Characteristics

Parameter Interactions

Common Drain-Source Characteristics

Transfer Characteristics

Transfer Characteristics

Resistance vs Drain Current

HOMEWORK PROBLEMS

9.1 Determine the operating state for a JFET having:

a. $V_{DS} = -4$ V, $V_{GS} = 2$ V, $V_p = 1$ V
b. $V_{DS} = -4$ V, $V_{GS} = 1$ V, $V_p = 2$ V
c. $V_{DS} = 2$ V, $V_{GS} = 0$ V, $V_p = -3$ V
d. $V_{DS} = 4$ V, $V_{GS} = -2$ V, $V_p = -4$ V
e. $V_{DS} = 2$ V, $V_{GS} = -2$ V, $V_p = -1$ V
f. $V_{DS} = -3$ V, $V_{GS} = 1$ V, $V_p = 4$ V

9.2 Place the correct circuit symbol for a JFET in each circle in Figure P9.1, showing the proper connections to the external terminals. Identify the source and drain. Assume operation in the saturation region.

9.3 Repeat Example 9.1 for $I_{DSS} = 12$ mA and $V_p = -5$ V.

9.4 The transistors shown in Figure P9.2 have $|V_T| = 3$ V. Determine the operating state.

9.5 An n-channel JFET has $V_p = -2.8$ V. It is operating in the ohmic region, with $V_{GS} = -1$ V, $V_{DS} = 0.05$ V, and $I_D = 0.3$ mA. Find i_D if

a. $V_{GS} = -1$ V, $V_{DS} = 0.08$ V
b. $V_{GS} = 0$ V, $V_{DS} = 0.1$ V
c. $V_{GS} = -3.2$ V, $V_{DS} = 0.06$ V

9.6 An enhancement-type NMOS transistor with $V_T = 2$ V has its source grounded and a 3-VDC source connected to the gate. Determine the operating state if

a. $V_D = 0.5$ V
b. $V_D = 1$ V
c. $V_D = 5$ V

9.7 The characteristic curves of an enhancement-type MOSFET are shown in Figure 9.16. Assume that $K = 0.1$ mA/V^2 and $V_T = 1.5$ V and that the transistor is operated at $V_{GS} = 3.5$ V. Find the drain current obtained at

a. $v_{DS} = 2$ V
b. $v_{DS} = 10$ V

9.8 A n-channel enhancement MOSFET is found to have a measured drain current of 4 mA at $V_{GS} = V_{DS} = 10$ V and of 1 mA at $V_{GS} = V_{DS} = 6$ V. Find the parameters K and V_T.

9.9 An n-channel JFET has $V_p = -4$ V and $I_{DSS} = 10$ mA. Find the value of V_T and K.

9.10 In the circuit shown in Figure P9.3, the NMOS transistor has $V_T = 2$ V and $K = 10$ mA/V^2. Find R and V_D for $I_D = 0.4$ mA.

9.11 An enhancement-type NMOS transistor has $V_T = 2$ V and $I_D = 1$ mA when $V_{GS} = V_{DS} = 3$ V. Find the value of I_D for $V_{GS} = 4$ V.

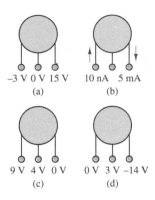

-3 V 0 V 15 V   10 nA 5 mA
       (a)   (b)

9 V 4 V 0 V   0 V 3 V −14 V
       (c)   (d)

Figure P9.1

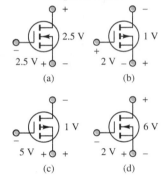

(a) 2.5 V, 2.5 V +, −
(b) 1 V, 2 V −, +
(c) 1 V, 5 V +, +
(d) 6 V, 2 V +, −

Figure P9.2

+20 V

R I_D

Figure P9.3

Figure P9.4

9.12   For $V_{DS} = 3$ V, $V_T = -4$ V, and $K = 0.625$ mA/V^2, find the change in I_D corresponding to a change in V_{GS} from 2.1 V to -1.5 V.

9.13   In the circuit shown in Figure P9.4, a drain voltage of 0.1 V is established. Find the current I_D for $V_T = 1$ V and $K = 0.5$ mA/V^2.

9.14   An n-channel enhancement-mode MOSFET is operated in the ohmic region, with $V_{DS} = 0.4$ V and $V_T = 3.2$ V. The effective resistance of the channel is given by $R_{DS} = 500/(V_{GS} - 3.2)$ Ω. Find I_D when $V_{GS} = 5$ V, $R_{DS} = 500$ Ω, and $V_{GD} = 4$ V.

9.15   In the circuit shown in Figure P9.5, the MOSFET operates in the saturation region, for $I_D = 0.5$ mA and $V_D = 3$ V. The enhancement-type PMOS has $V_T = -1$ V, $K = 0.5$ mA/V^2. Find

a. R_D.

b.   The largest allowable value of R_D for the MOSFET to remain in the saturation region.

Figure P9.5

9.16   Repeat Example 9.3 if the load line equation is given by $I_D = 1 - 0.0625V_{DS}$ (in mA).

9.17   Repeat Example 9.4 for a Q point of 2 mA at 6 V.

9.18   For the circuit in Figure 9.9, $R = 10$ kΩ, $g_m = 1,000$ μS, and $V_G = 2$ V. Calculate V_S and I_D.

9.19   For the circuit in Figure 9.20, $R_1 = 2$ MΩ, $R_2 = 3$ MΩ, $R_D = 6$ kΩ, $R_S = 6$ kΩ, $V_T = 1$ V, $V_{DD} = 10$ V, and $K = 0.5$ mA/V^2. Find V_G and I_D.

9.20   For the circuit shown in Figure 9.20, $V_T = 2$ V and $K = 0.25$ mA/V^2. Design the bias circuit so that $I_D = 1$ mA and $V_{DD} = 20$ V.

9.21   The three terminals of an n-channel enhancement-mode MOSFET are at potentials of 4 V, 5 V, and 10 V with respect to ground. Draw the circuit symbol, with the appropriate voltages at each terminal, if the device is operating

a.   In the ohmic region.

b.   Beyond pinch-off.

9.22   Repeat Example 9.6 for $V_{GS} = 2.2$ V.

9.23   Repeat Example 9.8 for $K = 0.05$ mA/V^2.

9.24   Repeat Example 9.9 for a Q point at $I_{DQ} = 40$ mA, $V_{DSQ} = 4.5$ V, for $V_{GSQ} = 2.3$ V.

9.25 An enhancement-type MOSFET has the parameters: $K = 0.5$ mA/V^2 and $V_T = 1.5$ V and the transistor is operated at $V_{GS} = 3.5$ V. Find the drain current obtained at

a. $V_{DS} = 3$ V
b. $V_{DS} = 10$ V

9.26 A n-channel enhancement MOSFET has $I_D = 6$ mA at $V_{GS} = V_{DS} = 12$ V and $I_D = 1.5$ mA at $V_{GS} = V_{DS} = 6$ V. Find parameters K and V_T.

9.27 An enhancement-type NMOS transistor with $V_T = 2.5$ V has its source grounded and a 4-VDC source connected to the gate. Find the operating region of the device if

a. $V_D = 0.5$ V
b. $V_D = 1.5$ V

9.28 An enhancement-type NMOS transistor has $V_T = 4$ V, $I_D = 1$ mA when $V_{GS} = V_{DS} = 6$ V. Neglecting the dependence of I_D on V_{DS} in saturation, find the value of I_D for $V_{GS} = 5$ V.

9.29 A NMOS transistor shown in Figure P9.6 has $V_T = 1.5$ V, $K = 0.4$ mA/V^2. Now v_G is a pulse with 0 V to 5 V, find the voltage levels of the pulse signal at the drain output.

9.30 A JFET having $V_p = -2$ V and $I_{DSS} = 8$ mA is operating at $V_{GS} = -1$ V and a very small V_{DS}. Find

a. r_{DS}
b. V_{GS} at which r_{DS} is half of its value in (a)

9.31 Analyze the circuit shown in Figure P9.7. Find drain current and drain voltage. The depletion MOSFET has $V_T = -1$ V, $K = 0.5$ mA/V^2.

Figure P9.6

Figure P9.7

Figure P9.8

9.32 The transistor shown in Figure P9.8 has $V_T = 2$ V, $K = 0.2$ mA/V^2. Find the DC voltage v_{out}.

9.33 If the equation of the dashed line in Figure 9.13 is $I_D = V_{DS}^2/800$ A for operation in the region beyond pinch-off, find I_D if

a. $V_{DS} = 1.5$ V, $V_{GS} = 2.9$ V
b. $V_{DS} = 5.2$ V, $V_{GS} = 2.9$ V

9.34 Repeat Example 9.12 for a 7-V Zener diode.

9.35 For the current regulator circuit shown in Figure P9.9, find the expression for R_S.

Figure P9.9   Figure P9.10

9.36 For the shunt-type voltage regulator shown in Figure P9.10, find the expression for the output voltage, V_{out}.

9.37 For the circuit shown in Figure 9.35, if the LR load is replaced by a capacitor, draw the output waveform and label the values.

9.38 Draw $v_L(t)$ and label the values for the circuit in Figure 9.35 if the diode forward resistance is 50 Ω, the forward bias voltage is 0.7 V, and the load resistor R is an open circuit.

9.39 For the circuit shown in Figure P9.11, v_{AC} is a sinusoid with 10 V peak amplitude, $R = 2$ kΩ, and the forward-conducting voltage of D is 0.7 V.

a. Sketch the waveform of $v_L(t)$.

b. Find the average value of $v_L(t)$.

Figure P9.11

9.40 A vehicle battery charge circuit is shown in Figure P9.12. Describe the circuit, and draw the output waveform (w_1 and w_2 represent the inductances of the windings of the alternator).

Figure P9.12

9.41 Repeat Example 9.13 for $\theta = \pi/3$ and $\pi/6$.

Operational Amplifiers

C hapter 10 introduces the operational amplifier as a building block of modern electronic instrumentation. As will be shown, electronic amplifiers are not limited to the amplification of electrical signals; they form the basis of many other instrumentation circuits—for example, active filters. In this chapter we will analyze the properties of the ideal amplifier and then carefully explore the features of a general purpose amplifier circuit known as the *operational amplifier*. Understanding the gain and frequency response properties of the operational amplifier is essential for the user of electronic instrumentation. Fortunately, the availability of operational amplifiers in integrated circuit form has made the task of analyzing such circuits quite simple. The models presented in this chapter are based on concepts that have already been explored at length in earlier chapters, namely, Thévenin and Norton equivalent circuits and frequency response.

Mastery of operational amplifier fundamentals is essential in any practical application of electronics. This chapter is aimed at developing your understanding of the fundamental properties of practical operational amplifiers. A number of useful applications are introduced in the examples and homework problems. Upon completion of the chapter, you should be able to:

- Analyze and design simple signal-conditioning circuits based on op-amps.
- Analyze and design simple active filters.
- Understand the operation of analog computers.
- Assess and understand the practical limitations of operational amplifiers.

10.1   AMPLIFIERS

One of the most important functions in electronic instrumentation is that of amplification. The need to amplify low-level electrical signals arises frequently in a number of applications. Perhaps the most familiar use of amplifiers arises in converting the low-voltage signal from a cassette tape player, a turntable, or a compact disc player to a level suitable for driving a pair of speakers. Figure 10.1 depicts a typical arrangement. Amplifiers have a number of applications of interest to the non–electrical engineer, such as the amplification of low-power signals from transducers (e.g., bioelectrodes, strain gauges, thermistors, and accelerometers) and other, less obvious functions that will be reviewed in this chapter—for example, filtering and impedance isolation. We turn first to the general features and characteristics of amplifiers, before delving into the analysis of the operational amplifier.

Cartridge

Phonograph

Source   Amplifier   Speakers
                                                 load

Figure 10.1

Ideal Amplifier Characteristics

The simplest model for an amplifier is depicted in Figure 10.2, where a signal, $v_S(t)$, is shown being amplified by a constant factor A, called the *gain* of the amplifier. Ideally, the load voltage should be given by the expression

$$v_L(t) = Av_S(t) \tag{10.1}$$

Note that the source has been modeled as a Thévenin equivalent, and the load as an equivalent resistance. Thévenin's theorem guarantees that this picture can be representative of more complex circuits. Hence, the equivalent source circuit is the circuit the amplifier "sees" from its input port, and R_L, the load, is the equivalent resistance seen from the output port of the amplifier.

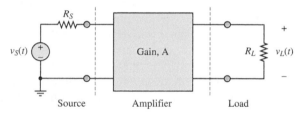

Figure 10.2 A voltage amplifier

What would happen if the roles were reversed? That is, what does the source see when it "looks" into the input port of the amplifier, and what does the load see when it "looks" into the output port of the amplifier? While it is not clear at this point how one might characterize the internal circuitry of an amplifier (which is rather complex), it can be presumed that the amplifier will act as an equivalent load with respect to the source, and as an equivalent source with respect to the load. After all, this is a direct application of Thévenin's theorem. Figure 10.3 provides a pictorial representation of this

Figure 10.3 Simple voltage amplifier model

simplified characterization of an amplifier. The "black box" of Figure 10.2 is now represented as an equivalent circuit with the following behavior. The input circuit has equivalent resistance R_{in}, so that the input voltage, v_{in}, is given by

$$v_{in} = \frac{R_{in}}{R_S + R_{in}} v_S \tag{10.2}$$

The equivalent input voltage seen by the amplifier is then amplified by a constant factor, A. This is represented by the controlled voltage source $A v_{in}$. The controlled source appears in series with an internal resistor, R_{out}, denoting the internal (output) resistance of the amplifier. Thus, the voltage presented to the load is

$$v_L = A v_{in} \frac{R_L}{R_{out} + R_L} \tag{10.3}$$

or, substituting the equation for v_{in},

$$v_L = \left(A \frac{R_{in}}{R_S + R_{in}} \frac{R_L}{R_{out} + R_L} \right) v_S \tag{10.4}$$

In other words, the load voltage is an amplified version of the source voltage.

Unfortunately, the amplification factor is now dependent on both the source and load impedances, and on the input and output resistance of the amplifier. Thus, a given amplifier would perform differently with different loads or sources. What are the desirable characteristics for a voltage amplifier that would make its performance relatively independent of source and load impedances? Consider, once again, the expression for v_{in}. If the input resistance of the amplifier, R_{in}, were very large, the source voltage, v_S, and the input voltage, v_{in}, would be approximately equal:

$$v_{in} \approx v_S \tag{10.5}$$

since

$$\lim_{R_{in} \to \infty} \left(\frac{R_{in}}{R_{in} + R_S} \right) = 1 \tag{10.6}$$

By an analogous argument, it can also be seen that the desired output resistance for the amplifier, R_{out}, should be very small, since for an amplifier with $R_{out} = 0$, the load voltage would be

$$v_L = A v_{in} \tag{10.7}$$

Combining these two results, we can see that as R_{in} approaches infinity and R_{out} approaches zero, the ideal amplifier magnifies the source voltage by a factor of A:

$$v_L = A v_S \tag{10.8}$$

just as was indicated in the "black box" amplifier of Figure 10.2.

Thus, two desirable characteristics for a general purpose voltage amplifier are a very large input impedance and a very small output impedance. In the next sections it will be shown how operational amplifiers provide these desired characteristics.

10.2   THE OPERATIONAL AMPLIFIER

An **operational amplifier** is an **integrated circuit,** that is, a large collection of individual electrical and electronic circuits integrated on a single silicon wafer. An operational amplifier—or op-amp—can perform a great number of operations, such as addition, filtering, or integration, which are all based on the properties of ideal amplifiers and of ideal circuit elements. The introduction of the operational amplifier in integrated circuit form marked the beginning of a new era in modern electronics. Since the introduction of the first IC op-amp, the trend in electronic instrumentation has been to move away from the discrete (individual-component) design of electronic circuits, toward the use of integrated circuits for a large number of applications. This statement is particularly true for applications of the type the non–electrical engineer is likely to encounter: op-amps are found in most measurement and instrumentation applications, serving as extremely versatile building blocks for any application that requires the processing of electrical signals.

In the following pages, simple circuit models of the op-amp will be introduced. The simplicity of the models will permit the use of the op-amp as a circuit element, or building block, without the need to describe its internal workings in detail. Integrated circuit technology has today reached such an advanced stage of development that it can be safely stated that for the purpose of many instrumentation applications, the op-amp can be treated as an ideal device.

The Open-Loop Model

The ideal operational amplifier behaves very much as an ideal **difference amplifier,** that is, a device that amplifies the difference between two input voltages. Operational amplifiers are characterized by near-infinite input resistance and very small output resistance. As shown in Figure 10.4, the output of the op-amp is an amplified version of the difference between the voltages present at the two inputs:[1]

$$v_{\text{out}} = A_{V(OL)}(v^+ - v^-) \tag{10.9}$$

The input denoted by a positive sign is called the **noninverting input** (or terminal), while that represented with a negative sign is termed the **inverting input** (or terminal). The amplification factor, or gain $A_{V(OL)}$ is called the **open-loop voltage gain** and is quite large by design, typically of the order of 10^5 to 10^7; it will soon become apparent why a large open-loop gain is a desirable characteristic. Together with the high input resistance and low output resistance, the effect of a large amplifier open-loop voltage gain, $A_{V(OL)}$, is such that op-amp circuits can be designed to perform very nearly as ideal voltage or current amplifiers. In effect, to analyze the performance of an op-amp circuit, only one assumption will be needed: that the current flowing into the input circuit of the amplifier is zero, or

$$i_{\text{in}} = 0 \tag{10.10}$$

This assumption is justified by the large input resistance and large open-loop gain of the operational amplifier. The model just introduced will be used to analyze three amplifier circuits in the next part of this section.

The Operational Amplifier in the Closed-Loop Mode

The Inverting Amplifier

One of the more popular circuit configurations of the op-amp, because of its simplicity, is the so-called inverting amplifier, shown in Figure 10.5. The

---

[1] The amplifier of Figure 10.4 is a *voltage amplifier;* another type of operational amplifier, called a *current* or *transconductance amplifier,* is described in the homework problems.

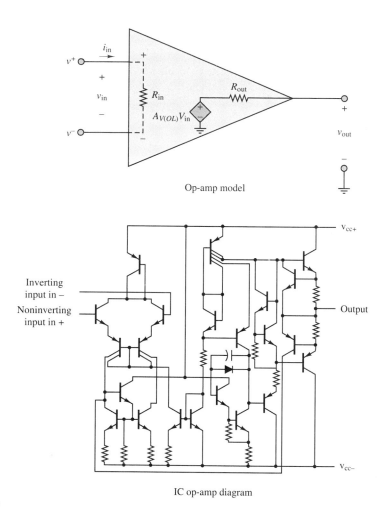

Op-amp model

IC op-amp diagram

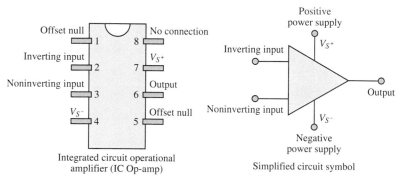

Integrated circuit operational
amplifier (IC Op-amp)

Simplified circuit symbol

Figure 10.4 Operational amplifier model symbols, and circuit diagram input signal to be amplified is connected to the inverting terminal, while the noninverting terminal is grounded. It will now be shown how it is possible to choose an (almost) arbitrary gain for this amplifier by selecting the ratio of two resistors. The analysis is begun by noting that at the inverting input node, KCL requires that

$$i_S + i_F = i_{in} \tag{10.11}$$

The current i_F, which flows back to the inverting terminal from the output, is appropriately termed **feedback current,** because it represents an input to the amplifier that is "fed back" from the output. Applying Ohm's law, we may determine each of the three currents shown in Figure 10.5:

$$i_S = \frac{v_S - v^-}{R_S} \qquad i_F = \frac{v_{out} - v^-}{R_F} \qquad i_{in} = 0 \tag{10.12}$$

Figure 10.5 Inverting amplifier

(the last by assumption, as stated earlier). The voltage at the noninverting input, v^+, is easily identified as zero, since it is directly connected to ground: $v^+ = 0$. Now, the open-loop model for the op-amp requires that

$$v_{out} = A_{V(OL)}(v^+ - v^-) = -A_{V(OL)}v^- \tag{10.13}$$

or

$$v^- = -\frac{v_{out}}{A_{V(OL)}} \tag{10.14}$$

Having solved for the voltage present at the inverting input, v^-, in terms of v_{out}, we may now compute an expression for the amplifier gain, v_{out}/v_S. This quantity is called the **closed-loop gain,** because the presence of a feedback connection between the output and the input constitutes a closed loop.[2] Combining equations 10.11 and 10.12, we can write

$$i_S = -i_F \tag{10.15}$$

and

$$\frac{v_S}{R_S} + \frac{v_{out}}{A_{V(OL)}R_S} = -\frac{v_{out}}{R_F} - \frac{v_{out}}{A_{V(OL)}R_F} \tag{10.16}$$

leading to the expression

$$\frac{v_S}{R_S} = -\frac{v_{out}}{R_F} - \frac{v_{out}}{A_{V(OL)}R_F} - \frac{v_{out}}{A_{V(OL)}R_S} \tag{10.17}$$

or

$$v_S = -v_{out}\left(\frac{1}{R_F/R_S} + \frac{1}{A_{V(OL)}R_F/R_S} + \frac{1}{A_{V(OL)}}\right) \tag{10.18}$$

[2] This terminology is borrowed from the field of automatic controls, of which the theory of closed-loop feedback systems forms the foundation.

If the open-loop gain of the amplifier, $A_{V(OL)}$, is sufficiently large, the terms $\frac{1}{A_{V(OL)}R_F/R_S}$ and $1/A_{V(OL)}$ are essentially negligible, compared with $1/(R_F/R_S)$. As stated earlier, typical values of $A_{V(OL)}$ range from 10^5 to 10^7, and thus it is reasonable to conclude that, to a close approximation, the following expression describes the closed-loop gain of the inverting amplifier:

$$v_{out} = -\frac{R_F}{R_S}v_S$$

Inverting amplifier closed-loop gain   **(10.19)**

That is, the closed-loop gain of an inverting amplifier may be selected simply by the appropriate choice of two externally connected resistors. The price for this extremely simple result is an inversion of the output with respect to the input, that is, a negative sign.

Next, we show that by making an additional assumption it is possible to simplify the analysis considerably. Consider that, as was shown for the inverting amplifier, the inverting terminal voltage is given by

$$v^- = -\frac{v_{out}}{A_{V(OL)}}$$

(10.20)

Clearly, as $A_{V(OL)}$ approaches infinity, the inverting terminal voltage is going to be very small (practically, of the order of microvolts). It may then be assumed that *in the inverting amplifier, v^- is virtually zero:*

$$v^- \approx 0$$

(10.21)

This fact prompts an interesting observation (which may not yet appear obvious at this point):

> The effect of the feedback connection from output to inverting input is to force the voltage at the inverting input to be equal to that at the noninverting input.

This is equivalent to stating that for an op-amp *with negative feedback,*

$$v^- \approx v^+$$

(10.22)

The analysis of the operational amplifier can now be greatly simplified if the following two assumptions are made:

1. $i_{in} = 0$
2. $v^- = v^+$ **(10.23)**

This technique will be tested in the next subsection by analyzing a noninverting amplifier configuration.

Why Feedback?

Why is such emphasis placed on the notion of an amplifier with a very large open-loop gain and with negative feedback? Why not just design an amplifier with a reasonable gain, say, $\times 10$, or $\times 100$, and just use it as such, without using feedback connections? In these paragraphs, we hope to answer these and other questions, introducing the concept of **negative feedback** in an intuitive fashion.

The fundamental reason for designing an amplifier with a very large open-loop gain is the flexibility it provides in the design of amplifiers with an (almost) arbitrary gain; it has already been shown that the gain of the inverting amplifier is determined by the choice of two external resistors—undoubtedly a convenient feature! It is important to appreciate that negative feedback is the mechanism that enables us to enjoy such flexibility in the design of linear amplifiers.

To understand the role of feedback in the operational amplifier, consider the internal structure of the op-amp shown in Figure 10.4. The large open-loop gain causes any difference in voltage at the input terminals to appear greatly amplified at the output. When a negative feedback connection is provided, as shown, for example, in the inverting amplifier of Figure 10.5, the output voltage, v_{out}, causes a current, i_F, to flow through the feedback resistance so that KCL is satisfied at the inverting node. Assume, for a moment, that the differential voltage

$$\Delta v = v^+ - v^-$$

is identically zero. Then, the output voltage will continue to be such that KCL is satisfied at the inverting node, that is, such that the current i_F is equal to the current i_S.

Suppose, now, that a small imbalance in voltage, Δv, is present at the input to the op-amp. Then the output voltage will be increased by an amount $A_{V(OL)}\Delta v$. Thus, an incremental current approximately equal to $A_{V(OL)}\Delta v/R_F$ will flow from output to input via the feedback resistor. The effect of this incremental current is to reduce the voltage difference Δv to zero, so as to restore the original balance in the circuit. One way of viewing negative feedback, then, is to consider it a self-balancing mechanism, which allows the amplifier to preserve zero potential difference between its input terminals.

A practical example that illustrates a common application of negative feedback is the thermostat. This simple temperature control system operates by comparing the desired ambient temperature and the temperature measured by a thermometer and turns a heat source on and off to maintain the difference between actual and desired temperature as close to zero as possible. An analogy may be made with the inverting amplifier if we consider that, in this case, negative feedback is used to keep the inverting-terminal voltage as close as possible to the noninverting-terminal voltage. The latter voltage is analogous to the desired ambient temperature in your home, while the former plays a role akin to that of the actual ambient temperature. The open-loop gain of the amplifier forces the two voltages to be close to each other, much the way the furnace raises the heat in the house to match the desired ambient temperature.

Example   10.1

Consider the inverting amplifier circuit of Figure 10.5. If $v_S(t) = 0.015 \cos(50t)$, $R_F = 10 \text{ k}\Omega$, and $R_S = 1 \text{ k}\Omega$, what is the output voltage, $v_{\text{out}}(t)$? Sketch both input and output voltage.

Solution:

$$v_{\text{out}} = -\left(\frac{R_F}{R_S}\right) v_S = -10 \times 0.015$$

$$= -150 \text{ mV}$$

$$v_{\text{out}}(t) = -0.15 \cos(50t)$$

The input and output voltage waveforms are shown in Figure 10.6.

Figure 10.6

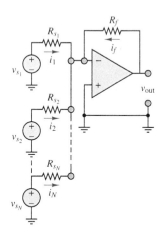

Figure 10.7 Summing
amplifier

A useful op-amp circuit that is based on the inverting amplifier is the **op-amp summer,** or **summing amplifier.** This circuit, shown in Figure 10.7, is used to add signal sources. The primary advantage of using the op-amp as a summer is that the summation occurs independently of load and source impedances, so that sources with different internal impedances will not interact with each other. The operation of the summing amplifier is best understood by application of KCL at the inverting node: the sum of the N source currents and the feedback current must equal zero, so that

$$i_1 + i_2 + \cdots + i_N = -i_F \tag{10.24}$$

But each of the source currents is given by the following expression:

$$i_n = \frac{v_{S_n}}{R_{S_n}} \qquad n = 1, 2, \ldots, N \tag{10.25}$$

while the feedback current is

$$i_F = \frac{v_{\text{out}}}{R_F} \tag{10.26}$$

Combining equations 10.25 and 10.26, and using equation 10.15, we obtain the following result:

$$\sum_{n=1}^{N} \frac{v_{S_n}}{R_{S_n}} = -\frac{v_{\text{out}}}{R_F} \tag{10.27}$$

or

$$v_{\text{out}} = \sum_{n=1}^{N} -\frac{R_F}{R_{S_n}} v_{S_n} \tag{10.28}$$

That is, the output consists of the weighted sum of N input signal sources, with the weighting factor for each source equal to the ratio of the feedback resistance to the source resistance.

The Noninverting Amplifier

To avoid the negative gain (i.e., phase inversion) introduced by the inverting amplifier, a **noninverting amplifier** configuration is often employed. A typical noninverting amplifier is shown in Figure 10.8; note that the input signal is applied to the noninverting terminal this time.

The noninverting amplifier can be analyzed in much the same way as the inverting amplifier. Writing KCL at the inverting node yields

$$i_F = i_S + i_{in} \approx i_S \tag{10.29}$$

where

$$i_F = \frac{v_{out} - v^-}{R_F} \tag{10.30}$$

$$i_S = \frac{v^-}{R_S} \tag{10.31}$$

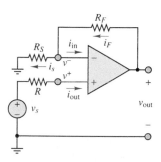

Figure 10.8 Noninverting amplifier

Now, since $i_{in} = 0$, the voltage drop across the source resistance, R_S, is equal to zero. Thus,

$$v^+ = v_S \tag{10.32}$$

and, using equation 10.22,

$$v^- = v^+ = v_S \tag{10.33}$$

Substituting this result in equations 10.29 and 10.30, we can easily show that

$$i_F = i_S \tag{10.34}$$

or

$$\frac{v_{out} - v_S}{R_F} = \frac{v_S}{R_S} \tag{10.35}$$

It is easy to manipulate equation 10.35 to obtain the result:

$$\boxed{\frac{v_{out}}{v_S} = 1 + \frac{R_F}{R_S}} \quad \begin{array}{l}\text{Noninverting amplifier} \\ \text{closed-loop gain}\end{array} \tag{10.36}$$

which is the closed-loop gain expression for a noninverting amplifier. Note that the gain of this type of amplifier is always positive and greater than (or equal to) 1.

The same result could have been obtained without making the assumption $v^+ = v^-$, at the expense of some additional work. The procedure one would follow in this latter case is analogous to the derivation carried out earlier for the inverting amplifier, and is left as an exercise.

In summary, in the preceding pages it has been shown that by constructing a nonideal amplifier with very large gain and near-infinite input resistance, it is possible to design amplifiers that have near-ideal performance and provide a variable range of gains, easily controlled by the selection of external resistors. The mechanism that allows this is negative feedback. From here on, unless otherwise noted, it will be reasonable and sufficient in analyzing new op-amp configurations to utilize the two assumptions

$$
\begin{array}{ll}
1. & i_{in} = 0 \\
2. & v^- = v^+
\end{array}
\qquad
\begin{array}{l}
\text{Approximations used for} \\
\text{ideal op-amps with} \\
\text{negative feedback}
\end{array}
\qquad \textbf{(10.37)}
$$

Example   10.2

The circuit of Figure 10.9 is called a **voltage follower** circuit. This terminology is due to the fact that the output voltage is always equal to the input voltage, that is, the output "follows" the input. Using the two assumptions just stated for op-amp analyses in this text, show that the output voltage does in fact equal the input voltage.

Figure 10.9 Voltage follower

Solution:

We have assumed that the current $i_{in} = 0$ and that $v^+ = v^-$. Therefore, with $v^- = v^+ = v_{in}$, and since v_{out} is short-circuited to v^-,

$$v_{out} = v_{in}$$

This amplifier circuit is commonly used to isolate circuits from each other. Homework problem 10.1 illustrates an application of this circuit.

The Differential Amplifier

The third closed-loop model examined in this chapter is a combination of the inverting and noninverting amplifiers; it finds frequent use in situations where the difference between two signals needs to be amplified. The basic **differential amplifier** circuit is shown in Figure 10.10, where the two

Figure 10.10 Differential amplifier

sources, v_1 and v_2, may be independent of each other, or may originate from the same process, as they are in Example 10.3.

The analysis of the differential amplifier may be approached by various methods; the one we select to use at this stage consists of:

1. Computing the noninverting- and inverting-terminal voltages, v^+ and v^-.
2. Equating the inverting and noninverting input voltages, $v^- = v^+$.
3. Applying KCL at the inverting node, where $i_2 = -i_1$.

Since it has been assumed that no current flows into the amplifier, the noninverting terminal voltage is given by the following voltage divider:

$$v^+ = \frac{R_2}{R_1 + R_2} v_2 \tag{10.38}$$

If the inverting-terminal voltage is assumed equal to v^+, then the currents i_1 and i_2 are found to be

$$i_1 = \frac{v_1 - v^+}{R_1} \tag{10.39}$$

and

$$i_2 = \frac{v_{out} - v^+}{R_2} \tag{10.40}$$

and since

$$i_2 = -i_1 \tag{10.41}$$

the following expression for the output voltage is obtained:

$$\begin{aligned} v_{out} &= R_2 \left[\frac{-v_1}{R_1} + \frac{1}{R_1 + R_2} v_2 + \frac{R_2}{R_1(R_1 + R_2)} v_2 \right] \\ &= \frac{R_2}{R_1}(v_2 - v_1) \end{aligned} \tag{10.42}$$

Thus, the differential amplifier magnifies the difference between the two input signals by the closed-loop gain R_2/R_1.

In practice, it is often necessary to amplify the difference between two signals that are both corrupted by noise or some other form of interference. In such cases, the differential amplifier provides an invaluable tool in amplifying the desired signal while rejecting the noise. Example 10.3 provides a realistic look at a very common application of the differential amplifier.

Example   10.3   An EKG Amplifier

This example illustrates the principle behind a two-lead electrocardiogram (EKG) measurement. The desired cardiac waveform is given by the difference between the potentials measured by two electrodes suitably placed on the patient's chest, as shown in Figure 10.11. A healthy, noise-free EKG waveform, $v_1 - v_2$, is shown in Figure 10.12.

Figure 10.11 Two-lead electrocardiogram   **Figure 10.12** EKG waveform

Unfortunately, the presence of electrical equipment powered by the 60-Hz, 110-VAC line current causes undesired interference at the electrode leads: the lead wires act as antennas and pick up some of the 60-Hz signal in addition to the desired EKG voltage. In effect, instead of recording the desired EKG signals, v_1 and v_2, the two electrodes provide the following inputs to the EKG amplifier, shown in Figure 10.13:

Lead 1:

$$v_1(t) + v_n(t) = v_1(t) + V_n \cos(377t + \phi_n)$$

Lead 2:

$$v_2(t) + v_n(t) = v_2(t) + V_n \cos(377t + \phi_n)$$

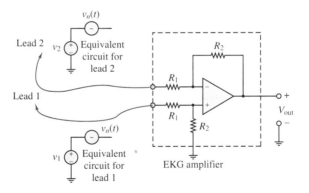

Figure 10.13 EKG amplifier

The interference signal, $V_n \cos(377t + \phi_n)$, is approximately the same at both leads, because the electrodes are chosen to be identical (e.g., they have the same lead lengths) and are in close proximity to each other. Further, the nature of the interference signal is such that it is common to both leads, since it is a property of the environment the EKG instrument is embedded in. On the basis of the analysis presented earlier, then,

$$v_{\text{out}} = \frac{R_2}{R_1} \left[(v_1 + v_n(t)) - (v_2 + v_n(t)) \right]$$

or

$$v_{\text{out}} = \frac{R_2}{R_1}(v_1 - v_2)$$

Thus, the differential amplifier nullifies the effect of the 60-Hz interference while amplifying the desired EKG waveform.

The preceding example introduced the concept of so-called **common-mode** and **differential-mode** signals. In the EKG example, the desired differential-mode EKG signal was amplified by the op-amp while the common-mode disturbance was canceled. Thus, the differential amplifier provides the ability to reject common-mode signal components (such as noise or undesired DC offsets) while amplifying the differential-mode components. This is a very desirable feature in instrumentation systems. In practice, rejection of the common-mode signal is not always complete: some of the common-mode signal component will always appear in the output. This fact gives rise to a figure of merit called the *common-mode rejection ratio,* which is discussed in Section 10.6.

Often, to provide impedance isolation between bridge transducers and the differential amplifier stage, the signals v_1 and v_2 are amplified separately. This technique gives rise to the so-called **instrumentation amplifier (IA),** shown in Figure 10.14. Example 10.4 illustrates the calculation of the closed-loop gain for a typical instrumentation amplifier.

Figure 10.14 Instrumentation amplifier

Example   10.4   Instrumentation Amplifier

In this example, we compute the closed-loop gain of the instrumentation amplifier of Figure 10.14.

Solution:

To carry out the desired analysis as easily as possible, it is helpful to observe that resistor R_1 is shared by the two input amplifiers. This corresponds to having each amplifier connected to a resistor equal to $R_1/2$, as shown in Figure 10.15(a). Because

Figure 10.15(a)

Figure 10.15(b)

of the symmetry of the circuit, one can view the shared resistor as being connected to ground in the center, so that the circuit takes the form of a noninverting amplifier, with closed-loop gain given by

$$A = 1 + \frac{2R_2}{R_1} \tag{10.43}$$

Thus, each of the input voltages is amplified by this gain, and the overall gain of the instrumentation amplifier can then be computed by considering that the voltage difference $(Av_1 - Av_2)$ is then amplified by the differential amplifier stage, with gain R_F/R, as shown in Figure 10.15(b):

$$v_{\text{out}} = \frac{R_F}{R}(Av_1 - Av_2) = \frac{R_F}{R}\left(1 + \frac{2R_2}{R_1}\right)(v_1 - v_2) \tag{10.44}$$

Because the instrumentation amplifier has widespread application, and in order to ensure the best possible match between resistors, the entire circuit of Figure 10.14 is often packaged as a single integrated circuit. The advantage of this configuration is that the resistors R_1 and R_2 can be matched much more precisely in an integrated circuit than would be possible using discrete components. A typical, commercially available integrated circuit package is the AD625. Data sheets for this device are provided at the end of Chapter 11.

Another simple op-amp circuit that finds widespread application in electronic instrumentation is the **level shifter.** The following two examples discuss its operation and its application in a transducer calibration circuit.

Example   10.5   Level Shifter

It is often desirable to add or subtract a DC voltage to or from a signal. A simple op-amp circuit called the *level shifter* accomplishes this function very elegantly. As an illustration, consider a sensor with output given by the voltage

$$v_{\text{sensor}}(t) = 1.8 + 0.1 \cos(\omega t)$$

To remove the 1.8-VDC offset, a level shifter circuit such as the one shown in Figure 10.16 may be employed. We can see that this circuit is somewhat similar to the differential amplifier in that it has two sources. If we consider the two input sources separately, we can see by superposition that the sensor output voltage is amplified by an inverting amplifier with gain $-R_F/R_S$, while the reference battery sees a noninverting amplifier with gain $(1 + R_F/R_S)$. Thus, the complete output is given by the expression

$$v_{\text{out}} = -\frac{R_F}{R_S}v_{\text{sensor}} + \left(1 + \frac{R_F}{R_S}\right)V_{\text{ref}}$$

and substituting the expression for v_{sensor}, we obtain

$$v_{\text{out}} = -\frac{R_F}{R_S}(1.8 + 0.1 \cos \omega t) + \left(1 + \frac{R_F}{R_S}\right)V_{\text{ref}}$$

$R_F = 200 \text{ k}\Omega; R_s = 10 \text{ k}\Omega$

Figure 10.16 Level shifter

We can rewrite v_{out} as

$$v_{\text{out}} = -\frac{R_F}{R_S}(0.1)\cos \omega t - \frac{R_F}{R_S}(1.8) + \left(1 + \frac{R_F}{R_S}\right)V_{\text{ref}}$$

Since the intent is to remove the DC offset, we require that

$$-\frac{R_F}{R_S}(1.8) + \left(1 + \frac{R_F}{R_S}\right)V_{\text{ref}} = 0$$

or

$$
\begin{aligned}
V_{\text{ref}} &= (1.8)\frac{R_F/R_S}{1 + R_F/R_S}\\
&= (1.8)\frac{20}{1 + 20}\\
&= 1.714 \text{ V}
\end{aligned}
$$

It would be rather inconvenient, however, to require a variable external voltage supply or a precision battery to be included in what is ideally to be a simple (and possibly adjustable) level-shifting circuit. To solve this problem, we introduce a common technique for obtaining a variable voltage supply, using no more than a potentiometer and the ordinary voltage supplies that would be used for the op-amp package. Figure 10.17 shows the potentiometer connection between the op-amp voltage supplies. It is a simple exercise to show that the voltage V_{ref}, referenced to ground, is given by the following expression:

$$V_{\text{ref}} = \left(\frac{R + \Delta R}{2R + R_P}\right)V_S^+ + \left(\frac{R + R_P - \Delta R}{2R + R_P}\right)V_S^-$$

or, if the voltage supplies are symmetrical (i.e., if $V_S^+ = -V_S^-$), then

$$V_{\text{ref}} = V_S^+ \left(\frac{2\Delta R - R_P}{2R + R_P}\right)$$

Figure 10.17

One can clearly see, then, that the potentiometer arrangement is capable of providing the desired reference voltage and may also be adjusted, if necessary. Thus, any undesired DC offset can be removed simply by adjusting a potentiometer setting.

Example   10.6   Transducer Calibration Circuit

In many practical instances, the output of a transducer is related to the physical variable we wish to measure in a form that requires some signal conditioning before it is completely useful. The most desirable form is one in which the electrical output of the transducer (for example, voltage) is related to the physical variable by a constant factor. Such a relationship is depicted in Figure 10.18(a), where k is the calibration constant relating voltage to temperature. Note that k is a positive number, and that the *calibration curve* passes through the (0, 0) point. On the other hand, the transducer characteristic of Figure 10.18(b) is best described by the following equation:

$$v_{\text{sensor}} = -\beta T + V_0$$

It is possible to modify the transducer calibration curve of Figure 10.18(b) to the more desirable one of Figure 10.18(a) by means of the simple circuit displayed in

(a)

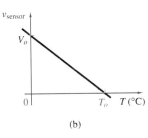

(b)

Figure 10.18 Transducer
calibration curves

Figure 10.19 Transducer
calibration circuit

Figure 10.19. This circuit provides the desired calibration constant k by a simple gain adjustment, while the zero (or bias) offset is adjusted by means of a potentiometer connected to the voltage supplies. The detailed operation of the circuit is described in the following paragraphs.

As noted before, nonideal characteristic can be described by the following equation:

$$v_{sensor} = -\beta T + V_0$$

Then, the output of the op-amp circuit of Figure 10.19 may be determined by using the principle of superposition:

$$v_{out} = -\frac{R_F}{R_S} v_{sensor} + \left(1 + \frac{R_F}{R_S}\right) V_{ref}$$

$$= -\frac{R_F}{R_S}(-\beta T + \beta) + \left(1 + \frac{R_F}{R_S}\right) V_{ref}$$

After substituting the expression for the transducer voltage and after some manipulation, we see that by suitable choice of resistors, and of the reference voltage source, we can compensate for the nonideal transducer characteristic. We want the following expression to hold:

$$v_{out} = \frac{R_F}{R_S}\beta T + \left(1 + \frac{R_F}{R_S}\right) V_{ref} - \frac{R_F}{R_S}V_0 = \alpha T$$

If we choose

$$\frac{R_F}{R_S}\beta = k$$

and

$$V_{ref} = \frac{R_F/R_S}{1 + R_F/R_S}V_0$$

then

$$V_{ref} \approx V_0 \qquad \text{if} \qquad \frac{R_F}{R_S} \gg 1$$

and we can directly convert the characteristic of Figure 10.18(b) to that of Figure 10.18(a). Clearly, the effect of selecting the gain resistors is to change the magnitude of the slope of the calibration curve. The fact that the sign of the slope changes is purely a consequence of the inverting configuration of the amplifier. The reference voltage source simply shifts the DC level of the characteristic, so that the curve passes through the origin.

DRILL EXERCISES

10.1 Consider an op-amp connected in the inverting configuration with a nominal closed-loop gain of $-R_F/R_S = -1,000$ (this would be the gain if the op-amp had an infinite open-loop gain). Determine the value of the closed-loop gain that includes the open-loop gain as a parameter, and compute the closed-loop gain for the following values of $A_{V(OL)}$: 10^7, 10^6, 10^5, and 10^4. How large do you think the open-loop gain should be for this op-amp, to achieve the desired closed-loop gain? [*Hint:* Do not assume that $A_{V(OL)}$ is negligible in equation 10.18.]

> Answer: $A_{V(CL)}$ = 999.9, 999.0, 990.1, 909.0
> $A_{V(OL)\,\min}$ = 10^6 for 0.1% error in $A_{V(CL)}$.

10.2 Repeat Drill Exercise 10.1 for $-R_F/R_S = -100$. What is the smallest value of $A_{V(OL)}$ you would recommend in this case?

> Answer: $A_{V(CL)}$ = 99.99, 99.99, 99.90, 99.00
> $A_{V(OL)\,\min}$ = 10^5 for 0.1% error in $A_{V(CL)}$.

10.3 Derive the result given for the differential amplifier by utilizing the principle of superposition. (Think of the differential amplifier as the combination of an inverting amplifier with input = v_2, plus a noninverting amplifier with input = v_1.)

10.4 For Example 10.5, find ΔR if the supply voltages are symmetrical at ±15 V and a 10-kΩ potentiometer is tied to two 10-kΩ resistors.

> Answer: ΔR = 6,714 Ω

10.5 For the circuit of Example 10.5, find the range of values of V_{ref} if the supply voltages are symmetrical at 15 V and a 1-kΩ potentiometer is tied to two 10-kΩ resistors.

> Answer: V_{ref} can have values between ±0.714 V.

10.6 Find the numerical values of R_F/R_S and V_{ref} if the temperature sensor of Example 10.6 has $\beta = 0.235$ and $V_0 = 0.7$ V and the desired relationship is $v_{\text{out}} = 10\,T$.

> Answer: R_F/R_S = 42.55; V_{ref} = 0.684 V

10.3 ACTIVE FILTERS

The range of useful applications of an operational amplifier is greatly expanded if energy-storage elements are introduced in the design; the frequency-dependent properties of these elements, studied in Chapter 4, will prove useful in the design of various types of op-amp circuits. In particular, it will be shown that it is possible to shape the frequency response of an operational amplifier by appropriate use of complex impedances in the input and feedback circuits. The class of filters one can obtain by means of op-amp designs is called **active filters,** because op-amps can provide amplification (gain) in addition to the filtering effects already studied in Chapter 6 for passive circuits (i.e., circuits comprising exclusively resistors, capacitors, and inductors).

The easiest way to see how the frequency response of an op-amp can be shaped (almost) arbitrarily is to replace the resistors R_F and R_S in Figures

Inverting

Noninverting

Figure 10.20 Op-amp circuits employing complex impedances

10.5 and 10.8 with impedances Z_F and Z_S, as shown in Figure 10.20. It is a straightforward matter to show that in the case of the inverting amplifier, the expression for the closed-loop gain is given by

$$\frac{V_{out}}{V_S}(\omega) = -\frac{Z_F}{Z_S} \tag{10.45}$$

whereas for the noninverting case, the gain is

$$\frac{V_{out}}{V_S}(\omega) = 1 + \frac{Z_F}{Z_S} \tag{10.46}$$

where Z_F and Z_S can be arbitrarily complex impedance functions and where V_S, V_{out}, I_F, and I_S are all phasors. Thus, it is possible to shape the frequency response of an ideal op-amp filter simply by selecting suitable ratios of feedback impedance to source impedance. By connecting a circuit similar to the low-pass filters studied in Chapter 6 in the feedback loop of an op-amp, the same filtering effect can be achieved and, in addition, the signal can be amplified.

The simplest op-amp low-pass filter is shown in Figure 10.21. Its analysis is quite simple if we take advantage of the fact that the closed-loop gain, as a function of frequency, is given by

$$A_{LP}(\omega) = -\frac{Z_F}{Z_S} \tag{10.47}$$

where

$$Z_F = R_Z \parallel \frac{1}{j\omega C_F} = \frac{R_F}{1 + j\omega C_F R_F} \tag{10.48}$$

and

$$Z_S = R_S \tag{10.49}$$

Note the similarity between Z_F and the low-pass characteristic of the passive *RC* circuit! The closed-loop gain $A_{LP}(\omega)$ is then computed to be

$$A_{LP}(\omega) = -\frac{Z_F}{Z_S} = -\frac{R_F/R_S}{1 + j\omega C_F R_F} \tag{10.50}$$

Figure 10.21 Active low-pass filter

This expression can be factored into two terms. The first is an amplification factor analogous to the amplification that would be obtained with a simple inverting amplifier (i.e., the same circuit as that of Figure 10.21 with the capacitor removed); the second is a low-pass filter, with a cutoff frequency dictated by the parallel combination of R_F and C_F in the feedback loop. The filtering effect is completely analogous to that which would be attained by the passive circuit shown in Figure 10.22. However, the op-amp filter also provides amplification by a factor of R_F/R_S.

It should be apparent that the response of this op-amp filter is just an amplified version of that of the passive filter. Figure 10.23 depicts the amplitude response of the active low-pass filter (in the figure $R_F/R_S = 10$ and $1/R_FC_F = 1$) in two different graphs; the first plots the amplitude ratio $\mathbf{V}_{out}/\mathbf{V}_S$ versus radian frequency, ω, on a logarithmic scale, while the second plots the amplitude ratio $20 \log_{10}(\mathbf{V}_{out}/\mathbf{V}_S)$ (in units of dB), also versus ω

Figure 10.22 Passive low-pass filter

Amplitude response of low-pass active filter

dB amplitude response of low-pass active filter

Figure 10.23 Normalized response of active low-pass filter

on a logarithmic scale. You will recall from Chapter 6 that dB frequency response plots are encountered very frequently. Note that in the dB plot, the slope of the filter response for frequencies significantly higher than the cutoff frequency,

$$\omega_0 = \frac{1}{R_FC_F} \tag{10.51}$$

is -20 dB/decade, while the slope for frequencies significantly lower than this cutoff frequency is equal to zero. The value of the response at the cutoff frequency is found to be, in units of dB,

$$|A_{LP}(\omega_0)|_{dB} = 20 \log_{10} \frac{R_F}{R_S} - 20 \log_{10} \sqrt{2} \tag{10.52}$$

where

$$-20 \log_{10} \sqrt{2} = -3 \text{ dB} \tag{10.53}$$

Thus, ω_0 is also called the **3-dB frequency.**

Among the advantages of such low-pass active filters is the ease with which the gain and the bandwidth can be adjusted by controlling the ratios R_F/R_S and $1/R_F C_F$, respectively.

It should be apparent at this point that it is possible to construct other types of filters by suitably connecting resistors and energy-storage elements to an op-amp. For example, a high-pass active filter can easily be obtained by using the circuit shown in Figure 10.24. Observe that the impedance of the input circuit is

$$Z_S = R_S + \frac{1}{j\omega C_S} \tag{10.54}$$

and that of the feedback circuit is

$$Z_F = R_F \tag{10.55}$$

Figure 10.24 Active high-pass filter

Then, the following gain function for the op-amp circuit can be derived:

$$A_{HP}(\omega) = -\frac{Z_F}{Z_S} = -\frac{R_F}{R_S + 1/j\omega C_S}$$
$$= -\frac{j\omega C_S R_F}{1 + j\omega R_S C_S} \tag{10.56}$$

As ω approaches zero, so does the response of the filter, whereas as ω approaches infinity, according to the gain expression of equation 10.56, the gain of the amplifier approaches a constant:

$$\lim_{\omega \to \infty} A_{HP}(\omega) = -\frac{R_F}{R_S} \tag{10.57}$$

That is, above a certain frequency range, the circuit acts as a linear amplifier. This is exactly the behavior one would expect of a high-pass filter. The high-pass response is depicted in Figure 10.25, in both linear and dB plots (in the figure, $R_F/R_S = 10$, $1/R_S C = 1$). Note that in the dB plot, the slope of the filter response for frequencies significantly lower than $\omega = 1/R_S C_S = 1$

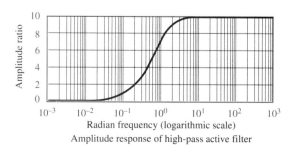

Amplitude response of high-pass active filter

dB amplitude response of high-pass active filter

Figure 10.25 Normalized response of active high-pass filter is +20 dB/decade, while the slope for frequencies significantly higher than this cutoff (or 3-dB) frequency is equal to zero.

As a final example of active filters, let us look at a simple active band-pass filter configuration. This type of response may be realized simply by combining the high- and low-pass filters we examined earlier. The circuit is shown in Figure 10.26.

The analysis of the band-pass circuit follows the same structure used in previous examples. First, we evaluate the feedback and input impedances:

$$Z_F = R_F \parallel \frac{1}{j\omega C_F} = \frac{R_F}{1 + j\omega C_F R_F} \tag{10.58}$$

$$Z_S = R_S + \frac{1}{j\omega C_S} = \frac{1 + j\omega C_S R_S}{j\omega C_S} \tag{10.59}$$

Next, we compute the closed-loop frequency response of the op-amp, as follows:

Figure 10.26 Active band-pass filter

$$A_{BP}(\omega) = -\frac{Z_F}{Z_S} = -\frac{j\omega C_S R_F}{(1 + j\omega C_F R_F)(1 + j\omega C_S R_S)} \tag{10.60}$$

The form of the op-amp response we just obtained should not appear as a surprise. It is very similar (although not identical) to the product of the low-pass and high-pass responses:

$$A_{LP}(\omega) = -\frac{R_F/R_S}{1 + j\omega C_F R_F} \tag{10.61}$$

$$A_{HP}(\omega) - \frac{j\omega C_S R_F}{1 + j\omega R_S C_S} \tag{10.62}$$

In particular, the denominator of $A_{BP}(\omega)$ is exactly the product of the denominators of $A_{LP}(\omega)$ and $A_{HP}(\omega)$. It is particularly enlightening to rewrite $A_{LP}(\omega)$ in a slightly different form, after making the observation that each RC product corresponds to some "critical" frequency:

$$\omega_1 = \frac{1}{R_F C_S} \qquad \omega_{HP} = \frac{1}{R_F C_F} \qquad \omega_{LP} = \frac{1}{R_S C_S} \tag{10.63}$$

It is easy to verify that for the case where

$$\omega_{HP} > \omega_{LP} \tag{10.64}$$

the response of the op-amp filter may be represented as shown in Figure 10.27 in both linear and dB plots (in the figure, $\omega_1 = 1$, $\omega_{HP} = 1,000$, $\omega_{LP} = 10$). The dB plot is very revealing, for it shows that, in effect, the band-pass response is the graphical superposition of the low-pass and high-pass responses shown earlier. The two 3-dB (or cutoff) frequencies are the same as in $A_{LP}(\omega)$, $1/R_F C_F$, and in $A_{HP}(\omega)$, $1/R_S C_S$. The third frequency, $\omega_1 = 1/R_F C_S$, represents the point where the response of the filter crosses the 0-dB axis (rising slope). Since 0 dB corresponds to a gain of 1, this frequency is called the **unity gain frequency.**

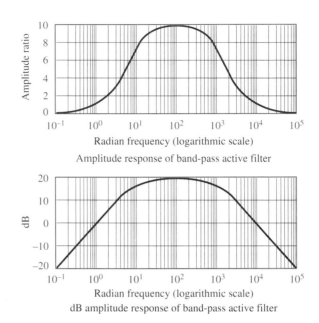

Figure 10.27 Normalized response of active band-pass filter

The ideas developed thus far can be employed to construct more complex functions of frequency. In fact, most active filters one encounters in practical applications are based on circuits involving more than one or two energy-storage elements. By constructing suitable functions for Z_F and Z_S, it is possible to realize filters with greater frequency selectivity (i.e., sharpness of cutoff), as well as flatter band-pass or band-rejection functions (that is, filters that either allow or reject signals in a limited band of frequencies). A few simple applications are investigated in the homework problems. One remark that should be made in passing though, pertains to the exclusive use of capacitors in the circuits analyzed thus far. One of the advantages of op-amp filters is that it is not necessary to use both capacitors and inductors to obtain a band-pass response. Suitable connections of capacitors can accomplish that task in an op-amp. This seemingly minor fact is of great importance in practice, because inductors are expensive to mass-produce to close tolerances and exact specifications and are often bulkier than capacitors with equivalent energy-storage capabilities. On the other hand, capacitors are easy to manufacture in a wide variety of tolerances and values, and in relatively compact packages, including in integrated circuit form.

Example 10.7 illustrates how it is possible to construct active filters with greater frequency selectivity by adding energy-storage elements to the design.

Example   10.7

The op-amp filter shown in Figure 10.28 provides more effective low-pass filtering than is obtained by using a single energy-storage element. In particular, if $R_2 C = R_1/L = 1/\omega_0$, one can obtain a relatively simple expression for $A(\omega)$. Find $A(\omega)$ for this special case.

Solution:

Recognizing that the gain of this active filter is given by

$$A(\omega) = \frac{\mathbf{V}_{out}}{\mathbf{V}_S} = -\frac{Z_F}{Z_S}$$

we can write

$$Z_F = R_2 \parallel \frac{1}{j\omega C} = \frac{R_2}{1 + j\omega C R_2} = \frac{R_2}{1 + j\omega/\omega_0}$$

and

$$Z_S = R_1 + j\omega L = R_1\left(1 + j\omega\frac{L}{R_1}\right) = R_1\left(1 + \frac{j\omega}{\omega_0}\right)$$

Figure 10.28

so that the closed-loop gain is given by the expression

$$A(\omega) = \frac{R_2/R_1}{(1 + j\omega/\omega_0)^2}$$

Note the similarity between this filter response and the response of the simple low-pass filter presented earlier in this section. The only difference is that the denominator is squared in this example. The linear and dB responses of this filter are compared in Figure 10.29 with those of the simple low-pass filter of Figure 10.21. Note that in the dB plot, the slope of the filter presented in this example is -40 dB/decade, that is, double that of the filter of Figure 10.21. Although the use of an inductor in the filter circuit is not practical, this example serves the purpose of illustrating the advantage of using multiple storage elements in a filter.

Comparison of low-pass active filters

Figure 10.29

Comparison of low-pass active filters (dB plot)

- - - - : Response of filter of figure 10.21
——— : Response of filter of example 10.7

DRILL EXERCISES

10.7 Design a low-pass filter with closed-loop gain of 100 and cutoff (3-dB) frequency equal to 800 Hz. Assume that only 0.01-μF capacitors are available. Find R_F and R_S.

> Answer: $R_F = 19.9$ kΩ, $R_S = 199$ Ω

10.8 Repeat the design of Drill Exercise 10.7 for a high-pass filter with cutoff frequency of 2,000 Hz. This time, however, assume that only standard values of resistors are available (see inside cover page for a table of standard values). Select the nearest component values, and calculate the percent error in gain and cutoff with respect to the desired values.

> Answer: $R_F = 820$ kΩ, $R_S = 8.2$ kΩ; error: gain $= 0\%$, $\omega_{3\,\text{dB}} = 2.95\%$

10.9 Find the frequency corresponding to attenuation of 1 dB (with respect to the maximum value of the amplitude response) for the filter of Drill Exercise 10.7.

> Answer: 407 Hz

10.10 What is the dB gain for the filter of Example 10.7 at the cutoff frequency, ω_0? Find the 3-dB frequency for this filter in terms of the cutoff frequency, ω_0, and note that the two are not the same.

> Answer: -6 dB; $\omega_{3\,\text{dB}} = 0.642\omega_0$

10.4   INTEGRATOR AND DIFFERENTIATOR CIRCUITS

In the preceding sections, we examined the frequency response of op-amp circuits for sinusoidal inputs. However, certain op-amp circuits containing energy-storage elements reveal some of their more general properties if we analyze their response to inputs that are time-varying but not necessarily sinusoidal. Among such circuits are the commonly used integrator and differentiator; the analysis of these circuits is presented in the following paragraphs.

The Ideal Integrator

Figure 10.30 Op-amp integrator

Consider the circuit of Figure 10.30, where $v_S(t)$ is an arbitrary function of time (e.g., a pulse train, a triangular wave, or a square wave). The op-amp circuit shown provides an output that is proportional to the integral of $v_S(t)$. The analysis of the integrator circuit is, as always, based on the observation that

$$i_S(t) = -i_F(t) \tag{10.65}$$

where

$$i_S(t) = \frac{v_S(t)}{R_S} \tag{10.66}$$

It is also known that

$$i_F(t) = C_F \frac{dv_{\text{out}}(t)}{dt} \tag{10.67}$$

from the fundamental definition of the capacitor. The source voltage can then be expressed as a function of the derivative of the output voltage:

$$\frac{1}{R_S C_F} v_S(t) = -\frac{dv_{\text{out}}(t)}{dt} \tag{10.68}$$

By integrating both sides of equation 10.68, we obtain the following result:

$$v_{\text{out}}(t) = -\frac{1}{R_S C_F} \int_{-\infty}^{t} v_S(t')\,dt' \tag{10.69}$$

This equation states that the output voltage is the integral of the input voltage.

There are numerous applications of the op-amp integrator, most notably the **analog computer,** which will be discussed in Section 10.5. The following example illustrates the operation of the op-amp integrator.

Example   10.8

Let $v_S(t)$ be a square wave of amplitude A and period $T = 10$ ms, as shown in Figure 10.31. Find $v_{\text{out}}(t)$ for an integrator with $C_F = 1\ \mu\text{F}$, $R_S = 10\ \text{k}\Omega$.

Solution:

The output of the ideal integrator is given by the expression

$$v_{\text{out}}(t) = -\frac{1}{R_S C_F} \int_{-\infty}^{t} v_S(t')\,dt' = -100 \int_{-\infty}^{t} v_S(t')\,dt'$$

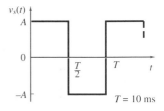

Figure 10.31

To integrate a square wave, we note that each period consists, in effect, of two segments, which are constant at $+A$ and $-A$, respectively. We further assume that the square wave starts at $t = 0$, and that no energy is stored in the capacitor for $t \le 0$. The technique we shall follow consists of integrating the waveform piecewise, taking into consideration the initial condition. For example,

$$\int_{-\infty}^{t} v_S(t')\,dt' = v_S(0) + \int_{0}^{t} v_S(t')\,dt'$$

Consider the first half-cycle of the square wave, starting at $t = 0$:

$$v_{\text{out}}(t) = v_{\text{out}}(0) - 100 \int_{0}^{T/2} A\,dt$$

$$= 0 - 100At \qquad 0 < t \le \frac{T}{2}$$

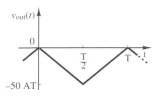

Figure 10.32

Now, to integrate over the second half-cycle, we must first evaluate $v_{out}(T/2)$ to obtain the initial condition for the next segment. Thus,

$$v_{out}(t) = v_{out}\left(\frac{T}{2}\right) - 100 \int_{T/2}^{T} -A\, dt$$

$$= -100A\frac{T}{2}\left[+100A\left(t - \frac{T}{2}\right)\right] \qquad \frac{T}{2} < t \leq T$$

Since the function is periodic, this waveform will repeat with period T, as shown in Figure 10.32.

Example   10.9   Charge Amplifiers

One of the most common families of transducers for the measurement of force, pressure, and acceleration is that of **piezoelectric transducers**. These transducers contain a piezoelectric crystal, a crystal that generates an electric charge in response to deformation. Thus, if a force is applied to the crystal (leading to a displacement), a charge is generated within the crystal. If the external force generates a displacement x_i, then the transducer will generate a charge q according to the expression

$$q = K_p x_i$$

Figure 10.33 depicts the basic structure of the piezoelectric transducer, and a simple circuit model. The model consists of a current source in parallel with a capacitor, where the current source represents the rate of change of the charge generated in response to an external force and the capacitance is a consequence of the structure of the transducer, which consists of a piezoelectric crystal (e.g., quartz or Rochelle salt) sandwiched between conducting electrodes (in effect, this is a parallel-plate capacitor).

Although it is possible, in principle, to employ a conventional voltage amplifier to amplify the transducer output voltage, v_0, given by

$$v_0 = \frac{1}{C}\int i\, dt = \frac{1}{C}\int \frac{dq}{dt}\, dt = \frac{q}{C} = \frac{K_p x_i}{C}$$

it is often advantageous to use a **charge amplifier.** The charge amplifier is essentially an integrator circuit, as shown in Figure 10.34, characterized by an extremely high

Figure 10.33 Piezoelectric transducer

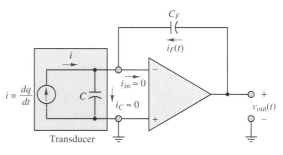

Figure 10.34 Charge amplifier

input impedance.[3] The high impedance is essential; otherwise, the charge generated by the transducer would leak to ground through the input resistance of the amplifier.

Because of the high input impedance, the input current into the amplifier is negligible; further, because of the high open-loop gain of the amplifier, the inverting-terminal voltage is essentially at ground potential. Thus, *the voltage across the transducer is effectively zero.* As a consequence, to satisfy KCL, the feedback current, $i_F(t)$, must be equal and opposite to the transducer current, i:

$$i_F(t) = -i$$

and since

$$v_{out}(t) = \frac{1}{C_F} \int i_F(t) \, dt$$

it follows that the output voltage is proportional to the charge generated by the transducer, and therefore to the displacement:

$$v_{out}(t) = \frac{1}{C_F} \int -i \, dt = \frac{1}{C_F} \int -\frac{dq}{dt} \, dt = -\frac{q}{C_F} = -\frac{K_p x_i}{C_F}$$

Since the displacement is caused by an external force or pressure, this sensing principle is widely adopted in the measurement of force and pressure.

The Ideal Differentiator

Using an argument similar to that employed for the integrator, we can derive a result for the ideal differentiator circuit of Figure 10.35. The relationship between input and output is obtained by observing that

$$i_S(t) = C_S \frac{dv_S(t)}{dt} \tag{10.70}$$

and

$$i_F(t) = \frac{v_{out}(t)}{R_F} \tag{10.71}$$

so that the output of the differentiator circuit is proportional to the derivative of the input:

$$v_{out}(t) = -R_F C_S \frac{dv_S(t)}{dt} \tag{10.72}$$

Although mathematically attractive, the differentiation property of this op-amp circuit is seldom used in practice, because differentiation tends to amplify any noise that may be present in a signal.

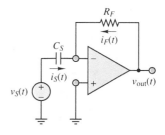

Figure 10.35 Op-amp differentiator

---

[3] Special op-amps are employed to achieve extremely high input impedance, through FET input circuits.

DRILL EXERCISES

10.11   Plot the frequency response of the ideal integrator in dB plot. Determine the slope of the curve in dB/decade. You may assume $R_S C_F = 10$.

> Answer: -20 dB/decade

10.12   Plot the frequency response of the ideal differentiator in a dB plot. Determine the slope of the curve in dB/decade. You may assume $R_F C_S = 100$.

> Answer: $+20$ dB/decade

10.13   Verify that if the triangular wave of Example 10.8 is the input to the ideal differentiator, the resulting output is the original square wave.

10.5   ANALOG COMPUTERS

Prior to the advent of digital computers, the solution of differential equations and the simulation of complex dynamic systems were conducted exclusively by means of analog computers. Analog computers still find application in engineering practice in the simulation of dynamic systems. The analog computer is a device that is based on three op-amp circuits introduced earlier in this chapter: the amplifier, the summer, and the integrator. These three building blocks permit the construction of circuits that can be used to solve differential equations and to simulate dynamic systems. Figure 10.36 depicts the three symbols that are typically employed to represent the principal functions of an analog computer.

   The simplest way to discuss the operation of the analog computer is to present an example. Consider the simple second-order mechanical system, shown in Figure 10.37, that represents, albeit in a greatly simplified fashion, one corner of an automobile suspension system. The mass, M, represents the mass of one quarter of the vehicle; the damper, B, represents the shock absorber; and the spring, K, represents the suspension spring (or strut). The differential equation of the system may be derived as follows:

$$M\frac{d^2 x_M}{dt^2} + B\left(\frac{dx_M}{dt} - \frac{dx_R}{dt}\right) + K(x_M - x_R) = 0 \tag{10.73}$$

Rearranging terms, we obtain the following equation, in which the terms related to the road displacement and velocity—x_R and dx_R/dt, respectively—are the forcing functions:

$$M\frac{d^2 x_M}{dt^2} + B\frac{dx_M}{dt} + Kx_M = B\frac{dx_R}{dt} + Kx_R \tag{10.74}$$

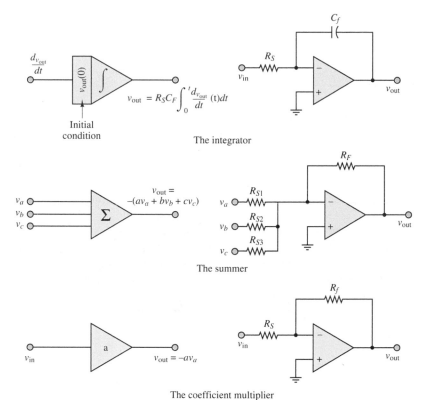

Figure 10.36 Elements of the analog computer

Assume that the car is traveling over a "washboard" surface on an unpaved road, such that the road profile is approximately described by the expression

$$x_R(t) = X \sin(\omega t) \tag{10.75}$$

It follows, then, that the vertical velocity input to the suspension is given by the expression

$$\frac{dx_R}{dt} = \omega X \cos(\omega t) \tag{10.76}$$

and we can write the equation for the suspension system in the form

$$M\frac{d^2x_M}{dt^2} + B\frac{dx_M}{dt} + Kx_M = B\omega X \cos(\omega t) + KX \sin(\omega t) \tag{10.77}$$

It would be desirable to solve the equation for the displacement x_M, which represents the motion of the vehicle mass in response to the road excitation. The solution can be used as an aid in designing the suspension system that best absorbs the road vibration, providing a comfortable ride for the

Figure 10.37 Model of automobile suspension

passengers. Equation 10.77 may be rearranged to obtain

$$\frac{d^2 x_M}{dt^2} = -\frac{B}{M}\frac{dx_M}{dt} - \frac{K}{M}x_M + \frac{B}{M}\omega X\cos(\omega t) + \frac{K}{M}X\sin(\omega t) \quad \textbf{(10.78)}$$

This equation is now in a form appropriate for solution by repeated integration, since we have isolated the highest derivative term; thus, it will be sufficient to integrate the right-hand side twice to obtain the solution for the displacement of the vehicle mass, x_M.

Figure 10.38 depicts the three basic operations that need to be performed to integrate the differential equation describing the motion of the mass, M. Note that in each of the three blocks—the summer and the two integrators—the inversion due to the inverting amplifier configuration used for the integrator is already accounted for. Finally, the basic summing and integrating blocks together with three coefficient multipliers (inverting amplifiers) are connected in the configuration that corresponds to the preceding differential equation (equation 10.78). You can easily verify that the analog computer circuit of Figure 10.39 does indeed solve the differential equation in x_M by repeated integration.

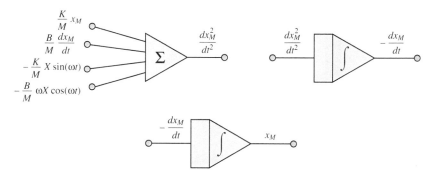

Figure 10.38 Solution by repeated integration

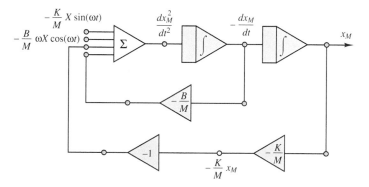

Figure 10.39 Analog computer simulation of suspension system

Scaling in Analog Computers

One of the important issues in analog computing is that of scaling. Since the analog computer implements an electrical analog of a physical system, there is no guarantee that the voltages and currents in the analog computer circuits will be of the same order of magnitude as the physical variables (e.g., velocity, displacement, temperature, or flow) they simulate. Further, it is not necessary that the computer simulate the physical system with the same time scale; it may be desirable in practice to speed up or slow down the simulation. Thus, the interest in time scaling and magnitude scaling.

Table 10.1

Physical system	Analog simulation
Physical variable, x	Voltage, v
Time variable, t	Simulated time, τ

Let Table 10.1 represent the physical and simulation variables. Considering time scaling first, let t denote real time and τ, the computer time variable. Then the time derivative of a physical variable can be expressed as

$$\frac{dx}{dt} = \frac{dx}{d\tau}\frac{d\tau}{dt} = \alpha\frac{dx}{d\tau} \tag{10.79}$$

where α is the scaling factor between real time and computer time:

$$\tau = \alpha t \tag{10.80}$$

For higher-order derivatives, the following relationship will hold:

$$\frac{d^n x}{dt^n} = \alpha^n \frac{d^n x}{d\tau^n} \tag{10.81}$$

While time scaling is likely to be prompted by a desire to speed up or slow down a computation, magnitude scaling is motivated by several different factors:

1. The relationship between physical variables and computer voltages (e.g., calibration constants).
2. Overloading of the op-amp circuits (we shall see in Section 10.6 that one of the fundamental limitations of the operational amplifier is its voltage range).
3. Loss of accuracy if voltages are too small (errors are usually expressed as a percentage of the full-scale range).

Thus, if the relationship between a physical variable and the computer voltage is $v = \beta x$, where β is a magnitude scaling factor, the derivative terms will be affected according to the relation

$$\frac{dx}{dt} = \frac{1}{\beta}\frac{dv}{dt} \tag{10.82}$$

Note that different scaling factors may be introduced at each point in the analog computer simulation, and so there is no general rule with regard to magnitude scaling. For example, if $v = \beta_0 x$, it is entirely possible to have $\frac{dx}{dt} = \frac{1}{\beta_1}\frac{dv}{dt}$.

Example 10.10

In this example we consider the implementation of the analog computer simulation of the suspension system described in Section 10.5. Let $M = 400$ kg, $B = 20{,}000$ N/m-s, and $K = 16 \times 10^5$ N/m. We wish to find the appropriate values of resistors and capacitors for each of the op-amp circuits.

Solution:

It is generally convenient to express resistances in $M\Omega$ and capacitances in μF; since the output of an integrator is scaled by the factor $1/RC$, this means that an ideal integrator has $RC = 1$, for $R = 1$ $M\Omega$ and $C = 1$ μF. Thus, we shall use these values for each integrator. The resulting circuit for each integrator will therefore be as shown in Figure 10.40(a). Similarly, the summer will employ 1-$M\Omega$ resistors everywhere, as shown in the figure. On the other hand, the two coefficient multipliers will require resistor ratios of $K/M = 4{,}000$ and $B/M = 50$. Possible solutions are shown in Figure 10.40(c) and (d). Finally, the unity gain inverting amplifier can be realized with two 1-$M\Omega$ resistors.

Note that it is not strictly necessary to use five op-amps in the simulation of the suspension system. For example, the summer and the first integrator could be combined into a single op-amp circuit, and one of the two coefficient multipliers could be eliminated. This idea is explored further in the homework problems.

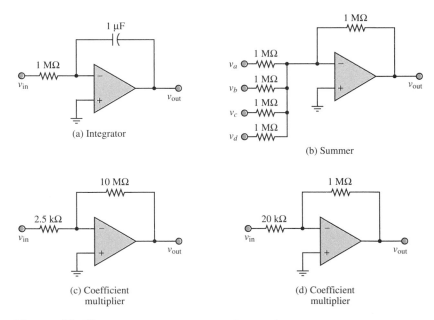

Figure 10.40 Analog computer simulation of suspension system

Example   10.11

In this example, we derive the differential equation of a system from the analog computer simulation circuit of Figure 10.41.

Figure 10.41 Analog computer simulation of unknown system

Solution:

We note first that the summing, scaling, and integrating functions have all been combined in the leftmost op-amp. Let the output of the first (leftmost) op-amp be denoted by y, that of the next by z, and the final output by x. Then, the following relationships hold:

$$x = -400z$$

and

$$y = -\frac{dz}{dt} = \frac{1}{400}\frac{dx}{dt}$$

Further, noting that

$$y = -\int [x + 2.5f(t)]\, dt$$

we can write

$$\frac{dy}{dt} = -x - 2.5f(t)$$

and finally obtain the equation

$$\frac{1}{400}\frac{d^2x}{dt^2} + x = -2.5f(t)$$

DRILL EXERCISES

10.14 Modify the gains of the coefficient multipliers in Example 10.10 if we wish to slow down the simulation by a factor of 10 (i.e., if $\alpha = 0.1$).

> Answer: $B/M = 5;\ K/M = 40$

10.15 For the simulation of Example 10.10, what will the largest magnitude of the voltage analog of x_M be for a road displacement $x_R = 0.01 \sin(100t)$? [*Hint:* Use phasor techniques to compute the frequency response $x_M(\omega)/F(\omega)$, where $f(t) = B\,dx_R/dt + Kx_R$, and evaluate the output voltage by multiplying the input by the magnitude of the frequency response at $\omega = 100$.]

> Answer: $x_{M\,max} = 0.0082$ m

10.16 Change the parameters of the analog computer simulation of Example 10.11 to simulate the differential equation $d^2x/dt^2 + 2x = -10f(t)$.

Answer:

10.6   PHYSICAL LIMITATIONS OF OP-AMPS

Thus far, the operational amplifier has been treated as an ideal device, characterized by infinite input resistance, zero output resistance, and infinite open-loop voltage gain. Although this model is adequate to represent the behavior of the op-amp in a large number of applications, it is important to realize that practical operational amplifiers are not ideal devices but exhibit a number of limitations that should be considered in the design of instrumentation. In particular, in dealing with relatively large voltages and currents, and in the presence of high-frequency signals, it is important to be aware of the nonideal properties of the op-amp. In the present section, we examine the principal limitations of the operational amplifier.

Voltage Supply Limits

As indicated in Figure 10.4, operational amplifiers (and all amplifiers, in general) are powered by external DC voltage supplies, V_S^+ and V_S^-, which are usually symmetrical and of the order of ± 10 V to ± 20 V. Some op-amps are especially designed to operate from a single voltage supply, but for the sake of simplicity we shall from here on consider only symmetrical supplies. The effect of limiting supply voltages is that amplifiers are capable of amplifying signals *only within the range of their supply voltages;* it would be physically impossible for an amplifier to generate a voltage greater than V_S^+ or less than V_S^-. This limitation may be stated as follows:

$$V_S^- < v_{\text{out}} < V_S^+ \tag{10.83}$$

For most op-amps, the limit is actually approximately 1.5 V less than the supply voltages. How does this practically affect the performance of an amplifier circuit? An example will best illustrate the idea.

Example   10.12

This example deals with the voltage supply limitations of a practical op-amp. Consider the inverting amplifier shown in Figure 10.42. If the input to the amplifier is $v_S(t) = 2\sin(2{,}000t)$, calculate and sketch the output.

Solution:

For an ideal op-amp, the output would be

$$v_{out}(t) = -20\sin 2{,}000t \text{ V}$$

However, the voltage supplies are ±15 V, and therefore $v_{out}(t)$ is limited to "swing" strictly between +15 V and −15 V. Figure 10.43 depicts the ideal and actual responses of the amplifier. For a practical op-amp, output saturation would most likely be reached somewhat before the voltage supply limits were reached, that is, around ±13.5 V.

Figure 10.42

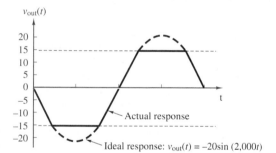

Figure 10.43 Op-amp output with voltage supply limit

Note how the voltage supply limit actually causes the peaks of the sine wave to be clipped in an abrupt fashion. This type of hard nonlinearity changes the characteristics of the signal quite radically, and could lead to significant errors if not taken into account. Just to give an intuitive idea of how such clipping can affect a signal, have you ever wondered why rock guitar has a characteristic sound that is very different from the sound of classical or jazz guitar? The reason is that the "rock sound" is obtained by overamplifying the signal, attempting to exceed the voltage supply limits, and causing clipping similar in quality to the distortion introduced by voltage supply limits in an op-amp. This clipping broadens the spectral content of each tone and causes the sound to be distorted.

One of the circuits most directly affected by supply voltage limitations is the op-amp integrator. The following example illustrates how saturation of an integrator circuit can lead to severe signal distortion.

Example   10.13

Consider the op-amp integrator of Figure 10.30, with $C_F = 20\ \mu\text{F}$ and $R_S = 10\ \text{k}\Omega$. Let the input to the integrator be a sinusoid with a small DC offset:

$$v_S(t) = 0.5 + 3\cos(10t)$$

Compute and sketch the output of the integrator, considering the voltage supply limitation of the op-amp.

Solution:

A noninverting ideal integrator would produce the following output:

$$v_{\text{out}}(t) = \frac{1}{R_S C_F}\int [0.5 + 3\cos(10t)]\,dt = 2.5t + 1.5\sin(10t)$$

Note that the presence of the DC offset in the input generates a term that increases linearly with time in the output and will eventually lead to saturation. Figure 10.44 depicts the op-amp input and output, the latter subject to ±15-V supply limits. Note that it is possible to eliminate this problem by modifying the design of the op-amp integrator to include a feedback resistor in parallel with the capacitor. This approach is discussed in homework problem 10.28.

Figure 10.44 Output of practical integrator with input DC offset

Frequency Response Limits

Another property of all amplifiers that may pose severe limitations to the op-amp is their finite bandwidth. We have so far assumed, in our ideal op-amp model, that the open-loop gain is a very large constant. In reality, $A_{V(OL)}$ is a function of frequency and is characterized by a low-pass response. For a typical op-amp,

$$A_{V(OL)}(\omega) = \frac{A_0}{1 + j\omega/\omega_0} \tag{10.84}$$

The cutoff frequency of the op-amp open-loop gain, ω_0, represents approximately the point where the amplifier response starts to drop off as a function of frequency, and is analogous to the cutoff frequencies of the RC and RL circuits of Chapter 6. Figure 10.45 depicts $A_{V(OL)}(\omega)$ in both linear and dB plots for the fairly typical values $A_0 = 10^6$ and $\omega_0 = 10\pi$. It should be apparent from this figure that the assumption of a very large open-loop gain becomes less and less accurate for increasing frequency. Recall the initial derivation of the closed-loop gain for the inverting amplifier: in obtaining the final result, $\mathbf{V}_{out}/\mathbf{V}_S = -R_F/R_S$, it was assumed that $A_{V(OL)} \to \infty$. This assumption is clearly inadequate at the higher frequencies.

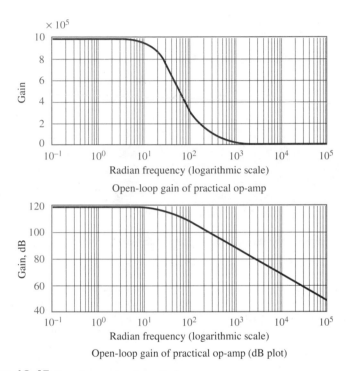

Figure 10.45 Open-loop gain of practical op-amp

The finite bandwidth of the practical op-amp results in a fixed **gain-bandwidth product** for any given amplifier. The effect of a constant gain-bandwidth product is that as the closed-loop gain of the amplifier is increased, its 3-dB bandwidth is proportionally reduced, until, in the limit, if the amplifier were used in the open-loop mode, its gain would be equal to A_0 and its 3-dB bandwidth would be equal to ω_0. The constant gain-bandwidth product is therefore equal to the product of the open-loop gain and the open-loop bandwidth of the amplifier: $A_0 \times \omega_0 = K$. When the amplifier is connected in a closed-loop configuration (e.g., as an inverting amplifier), its gain is typically much less than the open-loop gain and the 3-dB bandwidth of the amplifier is proportionally increased. To explain this further, Figure 10.46 depicts the case in which two different linear amplifiers (achieved through any two different negative feedback configurations) have been designed for the same op-amp. The first has closed-loop gain A_1, and the second has closed-loop gain A_2. The bold line in the figure indicates the open-loop frequency response, with gain A_0 and cutoff frequency ω_0. As the gain decreases from the open-loop gain, A_0, to A_1, we see that the cutoff frequency increases from ω_0 to ω_1. If we further reduce the gain to A_2, we can expect the bandwidth to increase to ω_2. Thus, *the product of gain and bandwidth in any given op-amp is constant.* That is,

$$A_0 \times \omega_0 = A_1 \times \omega_1 = A_2 \times \omega_2 = K \tag{10.85}$$

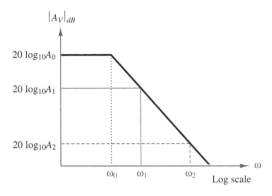

Figure 10.46

Example   10.14

The gain-bandwidth product is usually given in the manufacturer's specifications for a given op-amp. Suppose $A_0 = 10^6$ and $\omega_0 = 2\pi \times 5$ rad/s for a commercially available op-amp. If we wished to design an audio preamplifier with a bandwidth of 20 kHz, what is the maximum gain we should expect of the amplifier?

Solution:

The gain-bandwidth product for the amplifier is given by

$$A_0 \times \omega_0 = 2\pi \times 5 \times 10^6 = \pi \times 10^7 = K$$

Since the desired bandwidth is $\omega_1 = 2\pi \times 20{,}000$, the maximum allowable gain will be

$$A_1 = \frac{K}{\omega_1} = \frac{\pi \times 10^7}{2\pi \times 20{,}000} = 250$$

If we desire a gain larger than 250, we will have to purchase an op-amp with a greater gain-bandwidth product, or use two amplifiers in cascade, as shown in the next example.

Example   10.15   Cascade Amplifier

If a single amplifier is insufficient to obtain the desired amount of gain, because of its gain-bandwidth product limitation, one can resort to connecting two amplifiers in cascade, as shown in Figure 10.47. This example illustrates how a cascade amplifier can serve to increase the gain attainable for a given bandwidth. Let $K = 4\pi \times 10^6$ for both of the amplifiers in Figure 10.47. What is the 3-dB bandwidth of the cascade amplifier if $R_F/R_S = 100$ for each amplifier?

Figure 10.47 Cascade amplifier

Solution:

Let A_1 and ω_1 denote the gain and the 3-dB bandwidth of the first amplifier, and A_2 and ω_2 those of the second amplifier. Then the first amplifier has a 3-dB bandwidth given by

$$\omega_1 = \frac{A_0\omega_0}{A_1} = \frac{4\pi \times 10^6}{100} = 4\pi \times 10^4$$

Assuming an identical device, the second stage has the same approximate bandwidth:

$$\omega_2 = \frac{A_0\omega_0}{A_2} = \frac{4\pi \times 10^6}{100} = 4\pi \times 10^4$$

Thus, the *approximate* bandwidth of the cascade amplifier is $4\pi \times 10^4$. However, the gain of the cascade amplifier is equal to the product of the gains: $A_1 \times A_2 = 10{,}000$. Had we attempted to obtain the same amplification with a single op-amp by using resistors R_F' and R_S' such that

$$\frac{R_F'}{R_S'} = 10{,}000 = A'$$

we could have achieved a maximum bandwidth of only

$$\omega' = \frac{A_0\omega_0}{A'} = \frac{4\pi \times 10^6}{10{,}000} = 4\pi \times 10^2$$

In practice, the actual 3-dB bandwidth of the cascade amplifier is less than that of the single amplifiers, because the gain of each amplifier starts dropping off at frequencies somewhat lower than the nominal cutoff frequency. This problem is explored further in one of the Drill Exercises for Section 10.6.

Input Offset Voltage

Another limitation of practical op-amps results because even in the absence of any external inputs, it is possible that an **offset voltage** will be present at the input of an op-amp. This voltage is usually denoted by $\pm V_{os}$ and it is caused by mismatches in the internal circuitry of the op-amp. The offset voltage appears as a differential input voltage between the inverting and noninverting input terminals. The presence of an additional input voltage will cause a DC bias error in the amplifier output, as illustrated in Example 10.16. Typical and maximum values of V_{os} are quoted in manufacturers' data sheets. The worst-case effects due to the presence of offset voltages can therefore be predicted for any given application.

Example   10.16

Figure 10.48 Op-amp input offset voltage

To illustrate the phenomenon of DC offset, consider the case of a noninverting amplifier with a DC offset voltage $V_{os} = 1.5$ mV and a nominal (design) gain of $(1 + R_2/R_1) = 100$. The effect of the offset voltage is to cause a DC output even in the absence of any input. For the circuit shown in Figure 10.48, this DC error output voltage is

$$V_{\text{out},os} = \left(1 + \frac{R_2}{R_1}\right) V_{os} = 150 \text{ mV}$$

Thus, we can expect any AC output signal to be DC-shifted by 150 mV.

Input Bias Currents

Input bias currents I_{B-} and I_{B+} are always present at the input of an op-amp.

Figure 10.49

Another nonideal characteristic of op-amps results from the presence of small input bias currents at the inverting and noninverting terminals. Once again, these are due to the internal construction of the input stage of an operational amplifier. Figure 10.49 illustrates the presence of nonzero input bias currents (I_B) going into an op-amp.

Typical values of I_B depend on the semiconductor technology employed in the construction of the op-amp. Op-amps with bipolar transistor input stages may see input bias currents as large as 1 μA, while for FET input devices, the input bias currents are less than 1 nA. Since these currents depend on the internal design of the op-amp, they are not necessarily equal. One often designates the **input offset current** I_{os} as

$$I_{os} = I_{B+} - I_{B-} \tag{10.86}$$

The latter parameter is sometimes more convenient from the standpoint of analysis. The following example illustrates the effect of a nonzero input bias current on a practical amplifier design.

Example   10.17

The circuit shown in Figure 10.50 illustrates the effects of the input offset current I_{os}. Find the output voltage caused by the input offset currents in the absence of external inputs to the amplifier.

Solution:

Let the noninverting terminal resistance be $R_3 = R_1 \parallel R_2$. Then

$$v^+ = R_3 I_{B+}$$

and we can compute v_{out} by superposition

$$v_{out} = \left(1 + \frac{R_2}{R_1}\right) v^+ + R_2 I_{B-}$$

$$= \left(1 + \frac{R_2}{R_1}\right) \left(\frac{R_1 R_2}{R_1 + R_2}\right) I_{B+} + R_2 I_{B-}$$

$$= R_2 I_{os}$$

For a 100-kΩ feedback resistance and a (very worst case!) $I_{os} = 1\ \mu$A, the DC error voltage at the op-amp output could be as large as

$$v_{out}(I_{os}) = 10^5 \times 10^{-6} = 100\ \text{mV}$$

Figure 10.50

Output Offset Adjustment

Both the offset voltage and the input offset current contribute to an output offset voltage $V_{out,os}$. Some op-amps provide a means for minimizing $V_{out,os}$. For example, the μA741 op-amp provides a connection for this procedure. Figure 10.51 shows a typical pin configuration for an op-amp in an eight-pin dual-in-line package (DIP) and the circuit used for nulling the output offset voltage. The variable resistor is adjusted until v_{out} reaches a minimum (ideally, 0 volts). Nulling the output voltage in this manner removes the effect of both input offset voltage and current on the output.

Slew Rate Limit

Another important restriction in the performance of a practical op-amp is associated with rapid changes in voltage. The op-amp can produce only a finite rate of change at its output. This limit rate is called the **slew rate.** Consider an ideal step input, where at $t = 0$ the input voltage is switched from zero to V volts. Then we would expect the output to switch from 0 to $A \cdot V$ volts, where A is the amplifier gain. However, $v_{out}(t)$ can change at only a finite rate; thus,

$$\left| \frac{dv_{out}(t)}{dt} \right|_{max} = S_0 = \text{Slew rate} \qquad (10.87)$$

Figure 10.52 shows the response of an op-amp to an ideal step change in input voltage. Here, S_0, the slope of the boldface lines, represents the slew rate.

Figure 10.51 Output offset voltage adjustment

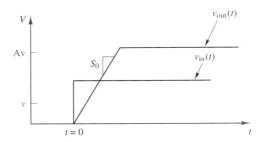

Figure 10.52 Slew rate limit in op-amps

The slew rate limitation can affect sinusoidal signals, as well as signals that display abrupt changes, as does the step voltage of Figure 10.52. This may not be obvious until we examine the sinusoidal response more closely. It should be apparent that the maximum rate of change for a sinusoid occurs at the zero crossing, as shown by Figure 10.52. To evaluate the slope of the waveform at the zero crossing, let

$$v(t) = A \sin \omega t \qquad (10.88)$$

so that

$$\frac{dv(t)}{dt} = \omega A \cos \omega t \qquad (10.89)$$

The maximum slope of the sinusoidal signal will therefore occur at $\omega t = 0, \pi, 2\pi, \ldots$, so that

$$\left| \frac{dv(t)}{dt} \right|_{\max} = \omega \times A = S_0 \qquad (10.90)$$

Thus, the maximum slope of a sinusoid is proportional to both the signal frequency and the amplitude. The curve shown by a dashed line in Figure 10.53 should indicate that as ω increases, so does the slope of $v(t)$ at the zero crossings. What is the direct consequence of this result, then? Example 10.18 gives an illustration of the effects of this slew rate limit.

Figure 10.53 The maximum slope of a sinusoidal signal varies with the signal frequency

Example   10.18

The manufacturer of an op-amp states that the slew rate of its device is

$$S_0 = 1V/\mu s$$

Would distortion be introduced in the output if we sought to amplify a 100,000-rad/s sinusoid of 1-V peak amplitude using an inverting amplifier configuration?

Solution:

Since

$$v_S(t) = 1 \sin(2\pi \times 10^5 t)$$

we can assume that the ideal output voltage would be

$$v_{out}(t) = -10 \sin(2\pi \times 10^5 t)$$

The maximum slope of $v_{out}(t)$ is

$$\left| \frac{dv(t)}{dt} \right|_{max} = A\omega = 10 \times 2\pi \times 10^5 = 6.28 \times 10^6 = 6.28 \ V/\mu s$$

This slope far exceeds the slew rate of the op-amp, $S_0 = 1 \ V/\mu s$. Figure 10.54 depicts the desired output (dashed line) and the actual output generated by the op-amp. Note how the distortion introduced by the slew rate limitation is so severe that the op-amp output resembles a triangular wave!

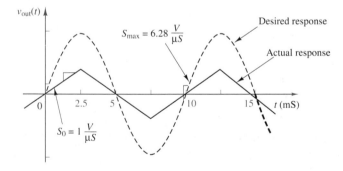

Figure 10.54 Distortion introduced by slew rate limit

Short-Circuit Output Current

Recall the model for the op-amp introduced in Section 10.2, which represented the internal circuits of the op-amp in terms of an equivalent input resistance, R_{in}, and a controlled voltage source, $A_V v_{in}$. In practice, the internal

Figure 10.55

source is not ideal, because it cannot provide an infinite amount of current (either to the load, or to the feedback connection, or both). The immediate consequence of this nonideal op-amp characteristic is that the maximum output current of the amplifier is limited by the so called short-circuit output current, I_{SC}:

$$|I_{out}| < I_{SC} \qquad (10.91)$$

To further explain this point, consider that the op-amp needs to provide current to the feedback path (in order to "zero" the voltage differential at the input) and to whatever load resistance, R_L, may be connected to the output. Figure 10.55 illustrates this idea for the case of an inverting amplifier, where I_{SC} is the load current that would be provided to a short-circuit load ($R_L = 0$).

Example   10.19

This example illustrates the effects of the short-circuit current limit. For an inverting amplifier, assume that $v_S = 0.05 \sin \omega t$ and $R_F/R_S = 100$. Then, we would expect to see the output voltage

$$v_{out}(t) = -5 \sin \omega t$$

If the manufacturer's data sheet rates the short-circuit output current at $I_{SC} = 50$ mA, what is the smallest load resistance, R_L, for which the op-amp will generate an undistorted output $v_{out}(t) = -5 \sin \omega t$ V?

Solution:

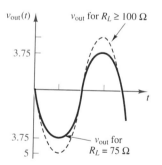

Figure 10.56 Distortion introduced by short-circuit current limit

Consider the peak values, for simplicity:

$$v_{out, peak} = 5 \text{ V}$$
$$I_{SC, peak} = 50 \text{ mA}$$
$$R_{L,min} = \frac{v_{out, peak}}{I_{SC, peak}} = \frac{5 \text{ V}}{50 \text{ mA}} = 100 \ \Omega$$

For any load resistance smaller than 100 Ω, the output of the op-amp will be distorted. For example, if $R_L = 75 \ \Omega$, we can calculate that the peak output voltage (which should ideally be 5 V) is only

$$v_{out, peak} = I_{SC} \times R_L = 50 \text{ mA} \times 75 \ \Omega = 3.75 \text{ V}$$

The effect of this distortion is to *compress* the extreme values of the waveform. Unlike the voltage supply limit, this distortion is not a hard clipping, but rather a soft rounding of the peak. Figure 10.56 illustrates this point.

Example   10.20

Consider an inverting amplifier with $R_S/R_F = 10$. If the input voltage is $v_S(t) = 1 \cos \omega t$ V, find the lowest value of R_L such that the short-circuit current limit is not exceeded, if $I_{SC} = 20$ mA.

Solution:

The nominal output voltage of the amplifier is

$$v_{\text{out}} = -10\cos\omega t$$

Thus, the maximum amplitude of the output voltage is

$$|v_{\text{out}}|_{\max} = 10\text{ V}$$

and the minimum load resistance that will not cause the output short-circuit current limit to be exceeded may be found as follows:

$$(R_L)_{\min} = \frac{|v_{\text{out}}|_{\max}}{I_{SC}} = \frac{10}{0.02}\ \Omega$$

Minimum $R_L = 500\ \Omega$

Common-Mode Rejection Ratio (CMRR)

Example 10.3 introduced the notion of differential-mode and common-mode signals. If we define A_{dm} as the **differential-mode gain** and A_{cm} as the **common-mode gain** of the op-amp, the output of an op-amp can then be expressed as follows:

$$v_{\text{out}} = A_{dm}(v_2 - v_1) + A_{cm}\left(\frac{v_2 + v_1}{2}\right) \tag{10.92}$$

Under ideal conditions, A_{cm} should be exactly zero, since the differential amplifier should completely reject common-mode signals. The departure from this ideal condition is a figure of merit for a differential amplifier and is measured by defining a quantity called the **common-mode rejection ratio (CMRR)**. The CMRR is defined as the ratio of the differential-mode gain to the common-mode gain and should ideally be infinite:

$$\text{CMRR} = \frac{A_{dm}}{A_{cm}}$$

The CMRR is often expressed in units of decibels (dB). The common-mode rejection ratio idea is explored further in the problems at the end of the chapter.

DRILL EXERCISES

10.17 In Example 10.15, we implicitly assumed that the gain of each amplifier was constant for frequencies up to the cutoff frequency. This is, in practice, not true, since the individual op-amp closed-loop gain starts dropping below the DC gain value according to the equation

$$A(\omega) = \frac{A_1}{1 + j\omega/\omega_1}$$

Thus, the calculations carried out in the example are only approximate. Find an expression for the closed-loop gain of the cascade amplifier. [*Hint:* The combined gain is equal to the product of the individual closed-loop gains.] What is the actual gain in dB at the cutoff frequency, ω_0, for the cascade amplifier?

Answer: 74 dB

10.18   What is the 3-dB bandwidth of the cascade amplifier of Example 10.15? [*Hint:* The gain of the cascade amplifier is the product of the individual op-amp frequency responses. Compute the magnitude of this product and set the magnitude of the product of the individual frequency responses equal to $(1/\sqrt{2}) \times 10{,}000$, and then solve for ω.]

Answer: $\Omega_{3dB} = 2\pi \times 12{,}800$ rad/s

Manufacturers generally supply values for the parameters discussed in this section in their device data specifications. Typical data sheets for common op-amps are given at the end of the chapter.

CONCLUSION

This chapter has described the fundamental properties and limitations of the operational amplifier.

- Ideal amplifiers represent fundamental building blocks of electronic instrumentation. With the concept of the ideal amplifiers in mind, one can design practical amplifiers, filters, integrators, and other useful signal-processing circuits.
- The operational amplifier closely approximates the characteristics of an ideal amplifier. The analysis of op-amp circuits may be carried out very easily, if it is assumed that the op-amp's input resistance and open-loop gain are very large. The inverting, noninverting, and differential amplifier configurations permit the design of useful electronic amplifiers simply by selecting a few external resistors.
- If energy-storage elements are used in the construction of op-amp circuits, it is possible to accomplish the functions of filtering, integration, and differentiation.
- The properties of summing amplifiers and integrators make it possible to build analog computers, which serve as an aid in the solution of differential equations, and in the simulation of dynamic systems.
- When op-amps are employed in more advanced applications, it is important to know that there are limitations on their performance that are not predicted by the simple op-amp model introduced at the beginning of the chapter. These include voltage supply limits, frequency response limits, offset voltages and currents, slew rate limits, and finite common-mode rejection ratio. In general, it is not difficult to compensate for these limitations in the design of op-amp circuits.

KEY TERMS

Operational Amplifier
(Op-Amp) *490*

Integrated Circuit (IC) *490*

Noninverting Input *491*

Inverting Input *491*

Open-Loop Voltage Gain *491*

Feedback Current *493*

Closed-Loop Gain *493*

Negative Feedback *495*

Op-Amp Summer (Summing
Amplifier) *496*

Noninverting Amplifier *497*

Voltage Follower *498*

Differential Amplifier *498*

Common-Mode Signal *501*

Differential-Mode Signal *501*

Instrumentation Amplifier
(IA) *501*

Level Shifter *502*

Active Filter *505*

3-db Frequency *508*

Unity Gain Frequency *509*

Analog Computer *513*

Piezoelectric Transducer *514*

Charge Amplifier *514*

Gain-Bandwidth Product *525*

Offset Voltage *527*

Input Offset Current *528*

Slew Rate *529*

Differential-Mode Gain *533*

Common-Mode Gain *533*

Common-Mode Rejection Ratio
(CMRR) *533*

DEVICE DATA SHEETS

LM741/LM741A/LM741C/LM741E   Operational Amplifier

Electrical Characteristics (Note 3)

Parameter	Conditions	LM741A / LM741E			LM741			LM741C			Units
		Min	Typ	Max	Min	Typ	Max	Min	Typ	Max	
Input Offset Voltage	$T_A = 25°C$ $R_S \le 10$ kΩ $R_S \le 50\Omega$		0.8	3.0		1.0	5.0		2.0	6.0	mV mV
	$T_{AMIN} \le T_A \le T_{AMAX}$ $R_S \le 50\Omega$ $R_S \le 10$ kΩ			4.0			6.0			7.5	mV mV
Average Input Offset Voltage Drift				15							μV/°C
Input Offset Voltage Adjustment Range	$T_A = 25°C$, $V_S = \pm 20$V	± 10				± 15			± 15		mV
Input Offset Current	$T_A = 25°C$		3.0	30		20	200		20	200	nA
	$T_{AMIN} \le T_A \le T_{AMAX}$			70		85	500			300	nA
Average Input Offset Current Drift				0.5							nA/°C
Input Bias Current	$T_A = 25°C$		30	80		80	500		80	500	nA
	$T_{AMIN} \le T_A \le T_{AMAX}$			0.210			1.5			0.8	μA
Input Resistance	$T_A = 25°C$, $V_S = \pm 20$V	1.0	6.0		0.3	2.0		0.3	2.0		MΩ
	$T_{AMIN} \le T_A \le T_{AMAX}$, $V_S = \pm 20$V	0.5									MΩ

Electrical Characteristics (Note 3) *(Continued)*

Parameter	Conditions	LM741A / LM741E Min	Typ	Max	LM741 Min	Typ	Max	LM741C Min	Typ	Max	Units
Input Voltage Range	$T_A = 25°C$							±12	±13		V
	$T_{AMIN} \leq T_A \leq T_{AMAX}$				±12	±13					V
Large Signal Voltage Gain	$T_A = 25°C$, $R_L \geq 2$ kΩ $V_S = ±20V$, $V_O = ±15V$	50									V/mV
	$V_S = ±15V$, $V_O = ±10V$				50	200		20	200		V/mV
	$T_{AMIN} \leq T_A \leq T_{AMAX}$, $R_L \geq 2$ kΩ, $V_S = ±20V$, $V_O = ±15V$	32									V/mV
	$V_S = ±15V$, $V_O = ±10V$				25			15			V/mV
	$V_S = ±5V$, $V_O = ±2V$	10									V/mV
Output Voltage Swing	$V_S = ±20V$ $R_L \geq 10$ kΩ	±16									V
	$R_L \geq 2$ kΩ	±15									V
	$V_S = ±15V$ $R_L \geq 10$ kΩ				±12	±14		±12	±14		V
	$R_L \geq 2$ kΩ				±10	±13		±10	±13		V
Output Short Circuit Current	$T_A = 25°C$	10	25	35		25			25		mA
	$T_{AMIN} \leq T_A \leq T_{AMAX}$	10		40							mA
Common-Mode Rejection Ratio	$T_{AMIN} \leq T_A \leq T_{AMAX}$ $R_S \leq 10$ kΩ, $V_{CM} = ±12V$				70	90		70	90		dB
	$R_S \leq 50\Omega$, $V_{CM} = ±12V$	80	95								dB
Supply Voltage Rejection Ratio	$T_{AMIN} \leq T_A \leq T_{AMAX}$, $V_S = ±20V$ to $V_S = ±5V$ $R_S \leq 50\Omega$	86	96								dB
	$R_S \leq 10\Omega$				77	96		77	96		dB
Bandwidth	$T_A = 25°C$	0.437	1.5								MHz
Slew Rate	$T_A = 25°C$, Unity Gain	0.3	0.7			0.5			0.5		V/μs
Supply Current	$T_A = 25°C$					1.7	2.8		1.7	2.8	mA

Dual-In-Line or S.O. Package

HOMEWORK PROBLEMS

10.1   Find v_1 in Figure P10.1(a) and (b). Note how the voltage follower holds v_1 in Figure P10.1(b) to $v_g/2$, while the 3-kΩ resistor "loads" the output in Figure P10.1(a).

(a)   (b)

Figure P10.1   **Figure P10.2**

10.2   Determine the output of the differential amplifier shown in Figure P10.2. Note that this amplifier is designed to amplify the difference between two signals.

10.3   In the circuit of Figure P10.3, find the voltage v.

Figure P10.3   **Figure P10.4**

10.4   In the circuit of Figure P10.4, find the current i.

10.5   In the circuit of Figure P10.5, find the current i.

Figure P10.5   **Figure P10.6**

10.6   In the circuit of Figure P10.6, find the voltage v_o.

Figure P10.7

10.7 In the circuit of Figure P10.7, find the current i.

10.8 In the circuit of Figure P10.8, find the voltage v.

Figure P10.8

10.9 In the circuit of Figure P10.9, find the current i.

Figure P10.9

10.10 Show that the circuit of Figure P10.10 is a noninverting summer.

Figure P10.10

Figure P10.11

10.11 For the circuit of Figure P10.11, find the voltage v and the current i.

10.12   For the circuit of Figure P10.12, find the current i.

Figure P10.12

10.13   For the circuit of Figure P10.13, determine the equivalent resistance of the inverting amplifier R_{eq} seen by the source. [*Hint:* Replace the signal source with an ideal 1-A current source, and track the path of the current.]

Figure P10.13

10.14   Differential amplifiers are often used in conjunction with the Wheatstone bridge. Consider the bridge shown in Figure P10.14, where each resistor is

Figure P10.14

a temperature-sensing element and the change in resistance is directly proportional to a change in temperature—that is, $\Delta R = \alpha(\pm\Delta T)$, where the sign is determined by the positive or negative temperature coefficient of the resistive element.

a. Find the Thévenin equivalent that the amplifier sees at point a and at point b. Assume that $|\Delta R|^2 \ll R_o$.

b. If $|\Delta R| = K\Delta T$, with K a numerical constant, find an expression for $v_{\text{out}}(\Delta T)$, that is, v_{out} as a function of change in temperature.

10.15   In the circuit of Figure P10.15, it is critical that the gain remain within 2 percent of its nominal value, 16. Find the resistor, R_S, that will accomplish the nominal gain requirement, and state what the maximum and minimum values of R_S can be. Will a *standard* 5 percent tolerance resistor be adequate to satisfy this requirement? (See Chapter 2 for resistor standard values.)

Figure P10.15

10.16   An inverting amplifier uses two 10 percent tolerance resistors: $R_F = 33$ kΩ, and $R_S = 1.1$ kΩ.

a. What is the nominal gain of the amplifier?

b. What is the maximum value of $|A_V|$?

c. What is the minimum value of $|A_V|$?

10.17   Find the voltage gain, $A_V = v_{\text{out}}/v_S$, for the op-amp circuit of Figure P10.16. $R_{S1} = 10$ kΩ, $R_{F1} = 47$ kΩ, $R_{F2} = 10$ kΩ, and $R_{S2} = 10$ kΩ.

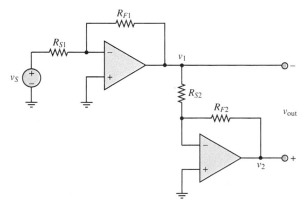

Figure P10.16

10.18 The circuit of Figure P10.17 includes a combination of resistors and op-amps. If $v_S(t) = 0.01 \cos \omega t$, find the output voltage.

Figure P10.17

10.19 The circuit of Figure P10.18 will remove the DC portion of the input voltage, $v_1(t)$, while amplifying the AC portion. Let $v_1(t) = 10 + 10^{-3} \sin \omega t$ V, $R_F = 10$ kΩ, and $V_{\text{batt}} = 20$ V.

a. Find R_S such that no DC voltage appears at the output.
b. What is $v_{\text{out}}(t)$, using R_S from part a?

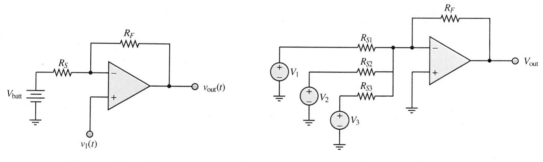

Figure P10.18   **Figure P10.19**

10.20 Use superposition to determine an expression for the output voltage of the circuit of Figure P10.19.

10.21 Use an inverting summing amplifier to obtain the following weighted sum of four different signal sources:

$$v_{\text{out}} = - \left(\frac{1}{4} \sin \omega_1 t + 5 \sin \omega_2 t + 2 \sin \omega_3 t + 16 \sin \omega_4 t \right)$$

Assume that $R_F = 10$ kΩ, and determine the required source resistors.

10.22 Assume that the op-amp in the circuit of Figure P10.20 is ideal, that is, that $A_{v(OL)} \rightarrow \infty$.

a. Use superposition to find v_{out} as a function of v_2.
b. Again using superposition, find v_{out} due to v_1.
c. If $v_2 = 5$ V and $v_1 = 4$ V, what is v_{out}?
d. Determine a general expression for v_{out}.

$R_1 = R_2 = R_3 = R_4$
$v_1 = 4$ V
$v_2 = 5$ V

Figure P10.20

Figure P10.21

10.23   Find the gain of the amplifier circuit in Figure P10.21. Assume that $R_1 = R_2 = R_3 = 3$ kΩ and $R_4 = 1$ kΩ.

10.24   Repeat Problem 10.23 if $R_4 = 100$ kΩ.

10.25   Consider the circuit of Figure P10.22.

Figure P10.22

a.   If $v_1 - v_2 = 1 \cos(1{,}000t)$ V, find the peak amplitude of v_{out}.
b.   Find the phase shift of v_{out}.

[Hint: use phasor analysis.]

10.26   The inverting amplifier shown in Figure P10.23 can be used as a high-pass filter.

a.   Derive frequency response of the circuit.
b.   If $R_1 = R_2 = 10$ kΩ and $C = 1$ μF, compute attenuation in dB at $\omega = 1000$ rad/s.
c.   Compute gain and phase at $\omega = 2000$ rad/s.
d.   Find range of frequencies over which the attenuation is less than 2 dB.

Figure P10.23

10.27   Repeat Problem 10.26 for $R_1 = 20$ kΩ, $R_2 = 10$ kΩ, and $C = 0.5$ μF.

10.28 The circuit shown in Figure P10.24 is an integrator. The capacitor is initially uncharged, and the source voltage is

$$v_{in}(t) = 10 \times 10^{-3} \sin(2{,}000\pi t) \text{ V}$$

a. At $t = 0$, the switch, S_1, is closed. How long does it take before clipping occurs at the output if $R_s = 10 \text{ k}\Omega$ and $C_F = 0.008 \ \mu\text{F}$?

b. At what times does the integration of the DC input cause the op-amp to saturate fully? Assume that the supply saturation voltage is 13.5 V.

Figure P10.24

*10.29 A practical integrator is shown in Figure 10.21 on p. 506. Note that the resistor in parallel with the feedback capacitor provides a path for the capacitor to discharge the DC voltage. Usually, the time constant $R_F C_F$ is chosen to be long enough not to interfere with the integration.

a. If $R_S = 10 \text{ k}\Omega$, $R_F = 2 \text{ M}\Omega$, $C_F = 0.008 \ \mu\text{F}$, and $v_S(t) = 10 \text{ V} + \sin(2{,}000\pi t)$ V, find $v_{out}(t)$ using phasor analysis.

b. Repeat part a if $R_F = 200 \text{ k}\Omega$, and if $R_F = 20 \text{ k}\Omega$.

c. Compare the time constants $R_F C_F$ with the period of the waveform for parts a and b. What can you say about the time constant and the ability of the circuit to integrate?

10.30 For the circuit of Figure P10.25,

a. Derive the frequency response of the circuit.

b. If $R_1 = 10 \text{ K}\Omega$, $R_2 = 20 \text{ k}\Omega$, $C = 1 \ \mu\text{F}$, and $C_2 = 2 \ \mu\text{F}$, compute the attenuation in dB at $\omega = 1{,}500$ rad/s.

c. Compute gain and phase shift at $\omega = 2{,}000$ rad/s.

10.31 Repeat Problem 10.30 for $R_1 = R_2 = 20 \text{ k}\Omega$, $C_1 = 1 \ \mu\text{F}$, and $C_2 = 1 \ \mu\text{F}$.

Figure P10.25

*10.32 Compute the frequency response of the circuit shown in Figure P10.26.

Figure P10.26

*10.33 The circuit of Figure 10.26 (on page 509) is a practical differentiator. Assuming an ideal op-amp with $v_S(t) = 10 \times 10^{-3} \sin(2{,}000\pi t)$ V, $C_S = 100\ \mu$F, $C_F = 0.008\ \mu$F, $R_F = 2$ MΩ, and $R_S = 10$ kΩ,

a. Determine the frequency response, $\frac{V_o}{V_S}(\omega)$.
b. Use superposition to find the actual output voltage (remember that DC = 0 Hz).

10.34 If a cosine wave of 2 V peak at 200 Hz is applied to the circuit of Figure 10.26, determine the output signal.

10.35 The inverting amplifier shown in Figure P10.27 can be used as a low-pass filter.

Figure P10.27

a. Derive frequency response of the circuit.
b. If $R_1 = R_2 = 100$ kΩ and $C = 0.1\ \mu$F, compute attenuation in dB at $\omega = 1000$ rad/s.
c. Compute gain and phase at $\omega = 2500$ rad/s.
d. Find range of frequencies over which the attenuation is less than 1 dB.

10.36 Repeat Problem 10.35 for $R_1 = 20$ kΩ, $R_2 = 100$ kΩ, and $C = 10\ \mu$F.

*10.37 Find an expression for the gain of the circuit of Figure P10.28.

Figure P10.28

10.38 Reduce the number of op-amps in the analog computer simulation of Section 10.5 by combining the summer and the first integrator into a single op-amp circuit, and by eliminating one of the two coefficient multipliers.

10.39 Derive the differential equation corresponding to the analog computer simulation circuit of Figure P10.29.

Figure P10.29

10.40   Construct the analog computer simulation corresponding to the following differential equation:

$$\frac{d^2x}{dt^2} + 100\frac{dx}{dt} + 10x = -5f(t)$$

*10.41   The ideal charge amplifier discussed in Example 10.9 will saturate in the presence of any DC offsets, as discussed in Section 10.6. The circuit of Figure P10.30 represents a practical charge amplifier, in which the user is provided with a choice of three time constants—$\tau_{\text{long}} = R_L C_F$, $\tau_{\text{medium}} = R_M C_F$, $\tau_{\text{short}} = R_S C_F$—which can be selected by means of a switch. Assume that $R_L = 10$ MΩ, $R_M = 1$ MΩ, $R_S = 0.1$ MΩ, and $C_F = 0.1$ μF. Analyze the frequency response of the practical charge amplifier for each case, and determine the lowest input signal frequency that can be amplified without excessive distortion for each case. Can this circuit amplify a DC signal?

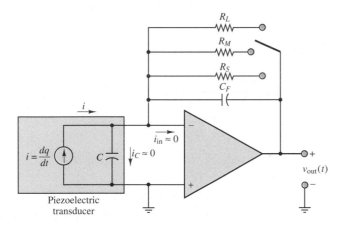

Figure P10.30

10.42   Consider a differential amplifier. We would desire the common-mode output to be less than 1 percent of the differential-mode output. Find the minimum dB common-mode rejection ratio (CMRR) to fulfill this requirement if the differential mode gain $A_{dm} = 1,000$. Let

$$v_1 = \sin(2,000\pi t) + 0.1\sin(120\pi t) \text{ V}$$

$$v_2 = \sin(2,000\pi t + 180°) + 0.1\sin(120\pi t) \text{ V}$$

$$v_o = A_{dm}(v_1 - v_2) + A_{cm}\left(\frac{v_1 + v_2}{2}\right)$$

10.43   Consider a standard inverting amplifier.

a.   Find a relationship showing that the closed-loop gain, $A_{V(CL)}$, is a function of R_F, R_S, and $A_{V(OL)}$; that is, find the closed-loop gain as a function of the resistors R_F and R_S and the open-loop gain.

b.   Plot $|A_{V(CL)}|$ versus $A_{V(OL)}$ on a range of $1 \leq A_{V(OL)} \leq 10^6$ if $R_F/R_S = 9$. Use a logarithmic scale for the open-loop gain axis.

c.   What conclusion can you draw from these results?

10.44   Use equation 10.92 to show that if $A_{dm} \gg A_{cm}$ and $A_{dm} \gg 1 + R_2/R_1$ in the circuit of Figure 10.10 (on page 498), then

$$v_{out} = \frac{R_2}{R_1}\left[(v_2 - v_1) + \frac{1}{CMRR}\frac{v_1 + v_2}{2}\right]$$

If $R_1 = R_2$, $v_2 = 5$ V, $v_1 = 3$ V, and CMRR $= 40$ dB, find v_{out}.

10.45   For the circuit of Figure P10.31,

a.   Derive frequency response of the circuit.

b.   If $R_1 = R_2 = 10$ kΩ and $C = 0.1$ μF, compute the attenuation in dB at $\omega = 1500$ rad/s.

c.   Compute the amplitude gain and phase shift at $\omega = 2000$ rad/s.

d.   Find range of frequencies over which the attenuation is less than 5 percent.

e.   What is the attenuation of part d in dB?
     [Hint: Find the transfer function in the form of $\frac{V_{out}}{(V_1 - V_2)}(\omega)$.]

Figure P10.31

10.46   Repeat Problem 10.45 if $R_1 = 5$ kΩ, $R_2 = 10$ kΩ, and $C = 0.4$ μF.

10.47   Repeat Problem 10.45 if $R_1 = 50$ kΩ, $R_2 = 100$ kΩ, and $C = 1$ μF.

10.48

a.   Given the voltage follower of Example 10.2, find an expression for the input impedance of the circuit in terms of R_{in}, $A_{V(OL)}$, and R_{out}.

b.   Using the parameters of the data sheet for the μA741 op-amp, determine the input resistance of the voltage follower.

10.49   For an inverting amplifier with $R_F = 47$ kΩ, $R_S = 3.3$ kΩ, and supply voltages $V_S^+ = 15$ V, $V_S^- = -12$ V,

a.   If the op-amp saturates at $0.9\,|V_S|$, plot the output voltage, $v_{out}(t)$, for

$$v_S(t) = 0.5 + 0.5 \sin 2\pi t \text{ V}$$

b.   Repeat part a if

$$v_S(t) = -1 + 0.5 \sin 2\pi t \text{ V}$$

c.   Assume that the slew rate of the op-amp is 2×10^6 V/s. If

$$v_S(t) = 0.5 \sin (4 \times 10^3 \pi t) \text{ V}$$

plot the output voltage $v_{out}(t)$ for the ideal case and for the slew rate–limited case.

10.50   Consider an inverting amplifier with open-loop gain 10^5. With reference to equation 10.18,

a.   If $R_S = 10$ kΩ and $R_F = 1$ MΩ, find the voltage gain $A_{V(CL)}$.
b.   Repeat part a if $R_S = 10$ kΩ and $R_F = 10$ MΩ.
c.   Repeat part a if $R_S = 10$ kΩ and $R_F = 100$ MΩ.
d.   Using the resistor values of part c, find $A_{V(CL)}$ if $A_{V(OL)} \to \infty$.

Figure P10.32

10.51

a.   If the op-amp of Figure P10.32 has an open-loop gain of 45×10^5, find the closed-loop gain for $R_F = R_S = 7.5$ kΩ, with reference to equation 10.18.
b.   Repeat part a if $R_F = 5(R_S) = 37,500$ Ω.

10.52

a.   Determine the closed-loop gain of the circuit shown in Figure P10.33 as a function of the resistors and of $A_{V(OL)}$, if $A_{V(OL)} = 10^6$.
b.   If $R_2 = R_3 = 642.8$ Ω, $R_1 = R_4 = 6,224$ Ω, and $v_{in}(t) = 4.68 \cos 27.6t$, find $v_{out}(t)$.

10.53   As discussed in Section 10.6, the open-loop gain, $A_{V(OL)}$, is actually a function of frequency, typically given by equation 10.84. If we substitute the equation for the open-loop frequency response in the closed-loop gain equation for an inverting amplifier (Problem 10.43), we can obtain a more complete expression for the closed-loop gain of the standard inverting amplifier, taking into account both the finite magnitude of $A_{V(OL)}$ and its limited frequency response.

Figure P10.33

a.   Determine an expression of the form

$$\frac{V_{out}}{V_{in}}(\omega) = \frac{K}{1 + j\omega/\omega^*}$$

where

$K =$ gain

$\omega^* =$ new 3-dB frequency

for the frequency response of the circuit shown, starting with the results of Problem 10.43. You may assume:

(1) $R_F + R_S \ll A_0 R_S$

or, equivalently,

$$\frac{R_F + R_S}{A_0 R_S} \ll 1$$

(2) $\dfrac{R_F + R_S}{R_S A_0} \gg 1$

b. Plot the response $|V_{out}/V_{in}|$ in dB versus frequency if

$$A_0 = 10^5 \quad \text{and} \quad f_0 = 5 = \frac{\omega_0}{2\pi}$$

Use the following cases and the equation you derived in part a.

Case	R_F	R_S
1	10 kΩ	10 kΩ
2	100 kΩ	10 kΩ
3	1 MΩ	10 kΩ
4	1 MΩ	1 kΩ

Plot all of these cases on the same set of axes with

$$1 \leq f \leq 1\text{MH}_3 \qquad -10 \leq |A_{CL}|_{dB} \leq 100 \text{ dB}$$

Also plot the open-loop gain $A_{V(OL)}(\omega)$ on these axes.

Integrated Circuit Electronics

C hapters 7, 8, and 9 served as an introduction to the operation of the most common electronic devices. In Chapter 10, the operational amplifier was introduced as the most common of the commercially available integrated circuit (IC) electronic devices, and its versatility was emphasized by a number of simple yet quite useful applications. In the present chapter, the discussion of electronic devices continues with an introduction to several useful *IC building blocks* that are commonly employed in instrumentation systems. The first two sections of the chapter discuss extensions of op-amp circuits introduced in Chapter 10: the IC instrumentation amplifier, and IC active filters. Section 10.3 introduces comparators and timer circuits, preparing the way for Chapters 12 and 13, which cover logic devices and digital systems, respectively. Switching circuits are first introduced by discussing op-amp–based switching and timing circuits (the comparator and multivibrators), and then a representative timer IC is described. Finally, BJT and MOSFET gates, which form the basis of all logic devices, are discussed.

Upon completion of Chapter 11, you should be able to:

- Specify and design an instrumentation amplifier, using either discrete op-amps or an integrated circuit package.
- Specify the pass-band and stop-band requirements of an active filter.

- Select the order of a Butterworth filter, given a set of specifications.
- Design a quadratic Sallen and Key active filter section.
- Understand the operation of op-amp comparators and Schmitt triggers, and select the noise protection level in a Schmitt trigger.
- Understand and design IC multivibrator circuits.
- Analyze diode, BJT, MOSFET, and CMOS gate circuits.

11.1   INSTRUMENTATION AMPLIFIERS

An **instrumentation amplifier (IA)** is a differential amplifier with very high input impedance, low bias current, and programmable gain that finds widespread application when low-level signals with large common-mode components are to be amplified in noisy environments. This situation occurs frequently when a low-level transducer signal needs to be preamplified, prior to further signal conditioning (e.g., filtering). Instrumentation amplifiers were briefly introduced in Chapter 10 (see Example 10.4), as an extension of the differential amplifier. You may recall that the IA introduced in Example 10.4 consisted of two stages, the first composed of two noninverting amplifiers, the second of a differential amplifier. Although the design in Chapter 10 is useful and is sometimes employed in practice, it suffers from a few drawbacks, most notably the requirement for very precisely matched resistors and source impedances to obtain the maximum possible cancellation of the common-mode signal. If the resistors are not matched exactly, the common-mode rejection ratio of the amplifier is significantly reduced, as the following will demonstrate.

The amplifier of Figure 11.1 has properly matched resistors ($R_2 = R_2'$, $R_F = R_F'$), except for resistors R and R', which differ by an amount ΔR such that $R' = R + \Delta R$. Let us compute the closed-loop gain for the amplifier.

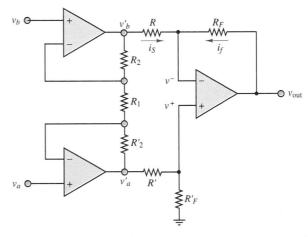

Figure 11.1 Discrete op-amp instrumentation amplifier

As shown in Example 10.4, the input-stage noninverting amplifiers have a closed-loop gain given by

$$A = \frac{v_b'}{v_b} = \frac{v_a'}{v_a} = 1 + \frac{2R_2}{R_1} \tag{11.1}$$

To compute the output voltage, we observe that the voltage at the noninverting terminal is

$$v^+ = \frac{R_F}{R_F + R + \Delta R} v_a' \tag{11.2}$$

and since the inverting-terminal voltage is $v^- = v^+$, the feedback current is given by

$$i_F = \frac{v_{out} - v^-}{R_F} = \frac{v_{out} - \frac{R_F}{R_F + R + \Delta R} v_a'}{R_F} \tag{11.3}$$

and the source current is

$$i_S = \frac{v_b' - v^-}{R} = \frac{v_b' - \frac{R_F}{R_F + R + \Delta R} v_a'}{R} \tag{11.4}$$

Applying KCL at the inverting node (under the usual assumption that the input current going into the op-amp is negligible), we set $i_F = -i_S$ and obtain the expression

$$\begin{aligned}
\frac{V_{out}}{R_F} &= \frac{v_a'}{R_F + R + \Delta R} - \frac{v_b'}{R} + \frac{R_F}{R} \frac{v_a'}{R_F + R + \Delta R} \\
&= \left(1 + \frac{R_F}{R}\right) \frac{v_a'}{R_F + R + \Delta R} - \frac{v_b'}{R}
\end{aligned}$$

so that the output voltage may be computed to be

$$v_{out} = R_F \left(\frac{R + R_F}{R}\right) \frac{v_a'}{R_F + R + \Delta R} - \frac{R_F}{R} v_b' \tag{11.5}$$

Note that if the term ΔR in the denominator were zero, the same result would be obtained as in Example 10.4: $v_{out} = (R_F/R)(v_a' - v_b')$; however, because of the resistor mismatch, there is a corresponding mismatch between the gains for the two differential signal components. Further—and more important—if the original signals, v_a and v_b, contained both differential-mode and common-mode components:

$$v_a = v_{a\,dif} + v_{com} \qquad v_b = v_{b\,dif} + v_{com} \tag{11.6}$$

such that

$$v_a' = A(v_{a\,dif} + v_{com}) \qquad v_b' = A(v_{b\,dif} + v_{com}) \tag{11.7}$$

then the common-mode components would not cancel out in the output of the amplifier, because of the gain mismatch, and the output of the amplifier would be given by

$$v_{\text{out}} = R_F \left(\frac{R + R_F}{R} \right) \frac{A(v_{a\,\text{dif}} + v_{\text{com}})}{R_F + R + \Delta R} - \frac{R_F}{R} A(v_{b\,\text{dif}} + v_{\text{com}}) \qquad \textbf{(11.8)}$$

resulting in the following output voltage:

$$v_{\text{out}} = v_{\text{out,dif}} + v_{\text{out,com}} \qquad \textbf{(11.9)}$$

with

$$v_{\text{out,dif}} = R_F \left(\frac{R + R_F}{R} \right) \frac{A v_{a\,\text{dif}}}{R_F + R + \Delta R} - \frac{R_F}{R} A v_{b\,\text{dif}} \qquad \textbf{(11.10)}$$

and

$$
\begin{aligned}
v_{\text{out,com}} &= R_F \left(\frac{R + R_F}{R} \right) \frac{A v_{\text{com}}}{R_F + R + \Delta R} - \frac{R_F}{R} A v_{\text{com}} \\
&= \frac{R_F}{R} \left(\frac{R + R_F}{R_F + R + \Delta R} - 1 \right) A v_{\text{com}}
\end{aligned}
\qquad \textbf{(11.11)}
$$

The common-mode rejection ratio (CMRR; see Section 10.6) is given in units of dB by

$$
\begin{aligned}
\text{CMRR}_{\text{dB}} &= \left| \frac{A_{\text{dif}}}{A_{\text{com}}} \right| = 20 \log_{10} \left| \frac{A_{\text{dif}}}{v_{\text{out,com}}/v_{\text{com}}} \right| \\
&= 20 \log_{10} \left| \frac{A_{\text{dif}}}{\frac{R_F}{R} \left(\frac{R+R_F}{R_F+R+\Delta R} - 1 \right) A} \right|
\end{aligned}
\qquad \textbf{(11.12)}
$$

where A_{dif} is the *differential gain* (which is usually assumed equal to the nominal design value). Since the common-mode gain, $v_{\text{out,com}}/v_{\text{com}}$, should ideally be zero, the theoretical CMRR for the instrumentation amplifier with perfectly matched resistors is infinite. In fact, even a small mismatch in the resistors used would dramatically reduce the CMRR, as the drill exercises at the end of the section illustrate. Even with resistors having 1 percent tolerance, the maximum CMRR that could be attained for typical values of resistors and an overall gain of 1,000 would be only 60 dB. In many practical applications, a requirement for a CMRR of 100 or 120 dB is not uncommon, and these would demand resistors of 0.01 percent tolerance (see Drill Exercise 11.3). It should be evident, then, that the "discrete" design of the IA, employing three op-amps and discrete resistors, will not be adequate for the more demanding instrumentation applications.

Example   11.1

Compute the common-mode gain and CMRR for the amplifier of Figure 11.1 if $A = 10$, $R = 1 \text{ k}\Omega$, $R_F = 10 \text{ k}\Omega$, and $\Delta R = 2$ percent.

Solution:

To compute the gain of the common-mode signal component, we need to compute the ratio $v_{\text{out,com}}/v_{\text{com}}$. Using equation 11.11 and the numerical values given above, we compute

$$\frac{v_{\text{out,com}}}{v_{\text{com}}} = 100\left(\frac{11}{11.02} - 1\right) = -0.1815$$

Now we can compute the CMRR in units of dB from equation 11.2, with $A_{\text{dif}} = A \times \frac{R_F}{R} = 100$

$$\text{CMRR}_{\text{dB}} = \left|\frac{A_{\text{dif}}}{A_{\text{com}}}\right| = 20\log_{10}\left|\frac{A_{\text{dif}}}{v_{\text{out,com}}/v_{\text{com}}}\right|$$

$$= 20\log_{10}\left|\frac{A_{\text{dif}}}{\frac{R_F}{R}\left(\frac{R+R_F}{R_F+R+\Delta R} - 1\right)A}\right|$$

$$= 20\log_{10}\left|\frac{100}{\frac{10}{1}\left(\frac{11}{11.02} - 1\right)10}\right| = 54.82\text{ dB}$$

The general expression for the CMRR of the instrumentation amplifier of Figure 11.1, without assuming any of the resistors are matched, is

$$\boxed{\text{CMRR} = \left|\frac{A_{\text{dif}}}{A_{\text{com}}}\right| = \left|\frac{(R_F/R)(1 + 2R_2/R_1)}{\frac{R_F}{R}\left[\frac{R_F'}{R_F}\left(\frac{R_F+R}{R_F'+R'}\right) - 1\right]}\right|} \qquad \textbf{(11.13)}$$

and it can easily be shown that the CMRR is infinite if the resistors are perfectly matched.

Example 11.1 illustrated some of the problems that are encountered in the design of instrumentation amplifiers using discrete components. Many of these problems can be dealt with very effectively if the entire instrumentation amplifier is designed into a single **monolithic integrated circuit,** where the resistors can be carefully matched by appropriate fabrication techniques and many other problems can also be avoided. The functional structure of an IC instrumentation amplifier is depicted in Figure 11.2. Specifications for a common IC instrumentation amplifier (and a more accurate circuit description) are shown in Figure 11.3. Among the features worth mentioning here are the programmable gains, which the user can set by suitably connecting one or more of the resistors labeled R_1 to the appropriate connection. Note that the user may also choose to connect additional resistors to control the amplifier gain, without adversely affecting the amplifier's performance, since R_1 requires no matching. In addition to the pin connection that permits programmable gains, two additional pins are provided, called **sense** and **reference.** These additional connections are provided to the user for the purpose of referencing the output voltage to a signal other than ground, by means of the reference terminal, or of further amplifying the output current (e.g., with a transistor stage), by connecting the sense terminal to the output of the current amplifier.

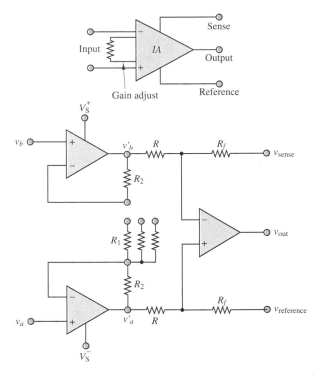

Figure 11.2 IC instrumentation amplifier

PIN CONFIGURATION   FUNCTIONAL BLOCK DIAGRAM

Figure 11.3 AD625 instrumentation amplifier data sheet

Example   11.2

Consider the IA of Figure 11.2, with the following resistor values: $R_F = R = 10 \text{ k}\Omega$, $R_2 = 20 \text{ k}\Omega$, and the following three choices for R_1: 80.2 Ω, 201 Ω, and 404 Ω. What are the possible gains that can be programmed with this amplifier?

Solution:

Recall that the gain of the input stage (for each of the differential inputs) is given by

$$A = 1 + \frac{2R_2}{R_1}$$

Thus, by connecting each of the three resistors, one can obtain the following gains:

$$A_1 = 1 + \frac{40,000}{80.2} = 500 \quad A_2 = 1 + \frac{40,000}{201} = 200 \quad A_3 = 1 + \frac{40,000}{404} = 100$$

It is also possible to obtain additional gains by connecting the resistors in parallel: 80.2 || 201 = 57.3 Ω ($A_4 \approx 700$), 80.2 || 404 = 66.9 Ω ($A_5 \approx 600$), 404 || 201 = 134.2 Ω ($A_6 \approx 300$).

DRILL EXERCISES

11.1   Use the definition of the common-mode rejection ratio (CMRR) given in Chapter 10 to compute the CMRR (in dB) of the amplifier of Example 11.1 if $R_F/R = 100$ and $A = 10$, and if $\Delta R = 5\%$ of R. Assume $R = 1 \text{ k}\Omega$, $R_F = 100 \text{ k}\Omega$.

> Answer: 66 dB

11.2   Repeat Drill Exercise 11.1 for a 1 percent variation in R.

> Answer: 80 dB

11.3   Repeat Drill Exercise 11.1 for a 0.01 percent variation in R.

> Answer: 120 dB

11.4   Calculate the mismatch in gains for the differential components for the 5 percent resistance mismatch of Drill Exercise 11.1.

> Answer: −6.1 dB

11.5   Calculate the mismatch in gains for the differential components for the 1 percent resistance mismatch of Exercise 11.2.

> Answer: −20.1 dB

11.6 What value of resistance R_1 would permit a gain of 1,000 for the IA of Example 11.2?

> Answer: 40 Ω

11.2   ACTIVE FILTERS

The need to filter sensor signals that may be corrupted by noise or other interfering or undesired inputs has already been approached in two earlier chapters. In Chapter 6, simple passive filters made of resistors, capacitors, and inductors were analyzed. It was shown that three types of filter frequency response characteristics can be achieved with these simple circuits: low-pass, high-pass, and band-pass. In Chapter 10, the concept of active filters was introduced, to suggest that it may be desirable to exploit the properties of operational amplifiers to simplify filter design, to more easily match source and load impedances, and to eliminate the need for inductors. The aim of this section is to discuss more advanced active filter designs, which find widespread application in instrumentation circuits.

Figure 11.4 depicts the general characteristics of a low-pass active filter, indicating that within the pass-band of the filter, a certain deviation from the nominal filter gain, A, is accepted, as indicated by the minimum and maximum pass-band gains, $A + \varepsilon$ and $A - \varepsilon$. The width of the pass-band is indicated by the cutoff frequency, ω_C. On the other hand, the stop-band, starting at the frequency ω_S, does not allow a gain greater than $A_{\min}$. Different types of filter designs achieve different types of frequency responses, which are typically characterized either by having a particularly flat pass-band frequency response (**Butterworth filters**), or by a very rapid transition between pass-band and stop-band (**Chebyshev filters,** and **Cauer,** or **elliptical, filters**), or by some other characteristic, such as a very linear phase response (**Bessel filters**). Achieving each of these properties usually involves trade-offs; for example, a very flat pass-band response will usually result in a relatively slow transition from pass-band to stop-band.

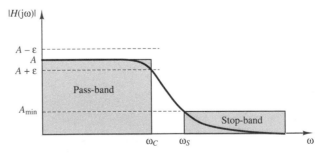

Figure 11.4 Prototype low-pass filter response

In addition to selecting a filter from a certain family, it is also possible to select the *order* of the filter; this is equal to the order of the differential equation that describes the input-output relationship of a given filter. In general, the higher the order, the faster the transition from pass-band to stop-band (at the cost of greater phase shifts and amplitude distortion, however). Although the frequency response of Figure 11.4 pertains to a low-pass filter, similar definitions also apply to the other types of filters.

Butterworth filters are characterized by a *maximally flat* pass-band frequency response characteristic; their response is defined by a magnitude-squared function of frequency:

$$|H(j\omega)|^2 = \frac{H_0^2}{1 + \varepsilon^2 \omega^{2n}} \tag{11.14}$$

where $\varepsilon = 1$ for maximally flat response and n is the order of the filter. Figure 11.5 depicts the frequency response (normalized to $\omega_C = 1$) of first-, second-, third-, and fourth-order Butterworth low-pass filters. The **Butterworth polynomials,** given in Table 11.1 in factored form, permit the design of the filter by specifying the denominator as a polynomial in s. For $s = j\omega$, one obtains the frequency response of the filter. Examples 11.4 and 11.5 illustrate filter design procedures that make use of these tables.

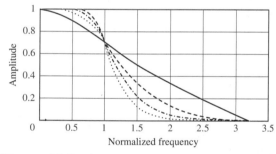

First order —————
Second order – – –
Third order –·–
Fourth order · · ·

Figure 11.5 Butterworth low-pass filter frequency response

Table 11.1   Butterworth polynomials in quadratic form

Order, n	Quadratic factors
1	$(s + 1)$
2	$(s^2 + \sqrt{2}s + 1)$
3	$(s + 1)(s^2 + s + 1)$
4	$(s^2 + 0.7654s + 1)(s^2 + 1.8478s + 1)$
5	$(s + 1)(s^2 + 0.6180s + 1)(s^2 + 1.6180s + 1)$

Figure 11.6 depicts the normalized frequency response of first- to fourth-order low-pass Chebyshev filters ($n = 1$ to 4), for $\varepsilon = 1.06$. Note that a certain amount of ripple is allowed in the pass-band; the amplitude of the ripple is defined by the parameter ε, and is constant throughout the pass-band. Thus, these filters are also called **equiripple filters.** Cauer, or

elliptical, filters are similar to Chebyshev filters, except for being characterized by equiripple both in the pass-band and in the stop-band. Design tables
exist to select the appropriate order of Butterworth, Chebyshev, or Cauer
filter for a specific application.

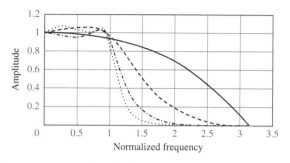

First order   ———
Second order – – –
Third order   –·–
Fourth order   · · ·

Figure 11.6 Chebyshev low-pass filter frequency response

Three common configurations of second-order active filters, which can
be used to implement **second-order** (or **quadratic**) **filter sections** using a
single op-amp, are shown in Figure 11.7. These filters are called **constant-K,**
or **Sallen and Key, filters** (after the name of the inventors). The analysis of
these active filters, although somewhat more involved than that of the active
filters presented in the preceding chapter, is based on the basic properties of
the ideal operational amplifier discussed earlier. Consider, for example, the

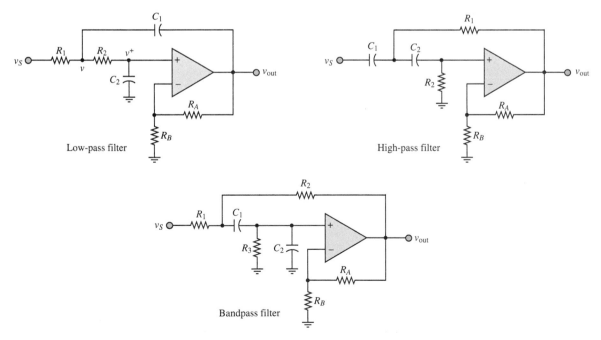

Figure 11.7 Sallen and Key active filters

low-pass filter of Figure 11.7. The first unusual aspect of the filter is the presence of both negative and **positive feedback;** that is, feedback connections are provided to both the inverting and the noninverting terminals of the op-amp. The analysis method consists of finding expressions for the input terminal voltages of the op-amp, v^+ and v^-, and using these expressions to derive the input-output relationship for the filter. This analysis is left as a homework problem. The frequency response of the low-pass filter is given by

$$H(j\omega) = \frac{K(1/R_1R_2C_1C_2)}{(j\omega)^2 + \left[\frac{1}{R_1C_1} + \frac{1}{R_2C_1} + \frac{1}{R_2C_2}(1-K)\right]j\omega + \frac{1}{R_1R_2C_1C_2}} \qquad (11.15)$$

This frequency response can be expressed in more general form as follows:

$$H(j\omega) = \frac{H_0\omega_C^2}{(j\omega)^2 + (\omega_C/Q)j\omega + \omega_C^2}$$

where

$$H_0 = K \qquad\qquad\qquad\qquad\qquad\qquad\qquad (11.16)$$

is the DC gain of the filter, and where

$$\omega_C = \sqrt{\frac{1}{R_1R_2C_1C_2}}$$

is the cutoff frequency, and where

$$\frac{1}{Q} = \sqrt{\frac{R_2C_2}{R_1C_1}} + \sqrt{\frac{R_1C_2}{R_2C_1}} + (1-K)\sqrt{\frac{R_1C_1}{R_2C_2}} \qquad (11.17)$$

is the inverse of the **quality factor,** Q. The Q of a filter is related to the overshoot in the transient response of the filter (recall the discussion of transient response in Chapter 6), and to the peaking (i.e., sharpness of the resonant peak) of the frequency response: a high-Q circuit will display more peaking, or overshoot, than a low-Q circuit.

Example   11.3

In this example, we shall determine the required order for a Butterworth filter, given a set of specifications. The filter is required to have 3 dB attenuation at $\omega_C = 1$ rad/s and an attenuation greater than 40 dB at the stop-band frequency, $\omega_S = 4\omega_C$. Find the order, n, of the Butterworth filter required to satisfy these constraints.

Solution:

To satisfy the first requirement, we substitute the value of the cutoff frequency in the expression for the magnitude-squared response

$$|H(j\omega)|^2 = \frac{H_0^2}{1 + \varepsilon^2\omega^{2n}}$$

with $\varepsilon = 1$, to obtain

$$|H(\omega = \omega_C)| = \frac{H_0}{\sqrt{1 + \varepsilon^2 \omega_C^{2n}}} = \frac{H_0}{\sqrt{1 + 1^{2n}}} = \frac{H_0}{\sqrt{2}}$$

which corresponds to 3 dB attenuation, since $20 \log_{10}(1/\sqrt{2}) = -3$ dB. Thus, the first requirement is automatically satisfied. To satisfy the second requirement, we set

$$\left| \frac{H_0}{H(\omega = \omega_S)} \right|_{dB} = 20 \log_{10} \left(\sqrt{1 + \omega_S^{2n}} \right) \geq 40 \text{ dB}$$

or, since $\omega_S = 4$,

$$1 + 4^{2n} \geq 10^4$$

and n can be computed from

$$2n \log_{10}(4) \geq \log_{10}(10^4 - 1)$$

or

$$n \geq 3.32$$

Thus, the required filter will be of the fourth order, $n = 4$, since n must be an integer. The actual attenuation in dB of the filter at the stop-band frequency will then be

$$A_{min,dB} = 20 \log_{10} \left(\sqrt{1 + \omega_S^{2n}} \right) = 48.1 \text{ dB}$$

which satisfies the original requirement.

Example   11.4

Compute the cutoff frequency, DC gain, and Q for a quadratic (order) low-pass filter section based on the Sallen and Key filter design of Figure 11.7. Assume that all the resistors are 500 Ω and all the capacitors are 2 μF.

Solution:

Recall the design equations for a second-order filter:

$$H(j\omega) = \frac{H_0 \omega_C^2}{(j\omega)^2 + (\omega_C/Q)j\omega + \omega_C^2}$$

$$H_0 = K$$

$$\omega_C = \sqrt{\frac{1}{R_1 R_2 C_1 C_2}}$$

$$\frac{1}{Q} = \sqrt{\frac{R_2 C_2}{R_1 C_1}} + \sqrt{\frac{R_1 C_2}{R_2 C_1}} + (1 - K)\sqrt{\frac{R_1 C_1}{R_2 C_2}}$$

Since the filter is connected in a noninverting configuration, the DC gain is given by

$$K = 1 + \frac{R_A}{R_B}$$

$$H_0 = K = 1 + \frac{500}{500} = 2$$

Thus, using the equations for a second-order filter:

$$\omega_C = \frac{1}{R_1 C_1} = 1{,}000 \text{ rad/s}$$

$$\frac{1}{Q} = 1 + 1 + (1 - 2) = 1$$

Example   11.5

This example illustrates the design of a Butterworth filter using the design method of Example 11.4. The Butterworth polynomial for a second-order filter is given in Table 11.1. With $s = j\omega$, the denominator polynomial becomes $(j\omega)^2 + \sqrt{2}(j\omega) + 1$. Since the general form of the frequency response is given by

$$H(j\omega) = \frac{H_0 \omega_C^2}{(j\omega)^2 + (\omega_C/Q)j\omega + \omega_C^2}$$

the following relationships must hold:

$$\frac{\omega_C}{Q} = \sqrt{2} \qquad \omega_C = 1 \qquad Q = \frac{1}{\sqrt{2}}$$

The gain corresponding to $Q = 1/\sqrt{2}$ is

$$K = 3 - \frac{1}{Q} = 3 - \sqrt{2} = 1.586$$

Thus,

$$R_A = 1.586 R_B$$

and the remaining components may be selected equal to those in Example 11.4.

DRILL EXERCISES

11.7 Determine the order of the filter required to satisfy the requirements of Example 11.3 if the stop-band frequency is moved to $\omega_S = 2\omega_C$.

> Answer: $n = 7$

11.8 What is the actual attenuation of the filter of Drill Exercise 11.7 at the stop-band frequency, ω_S?

> Answer: 42.1 dB

11.9 Design a quadratic filter section with $Q = 1$ and a cutoff frequency of 10 rad/s. Note that there can be many solutions, depending on your design.

> Answer: $R_1 = R_2 = 1 \text{ k}\Omega$; $C_1 = C_2 = 100 \text{ }\mu\text{F}$; $K = 2$

11.3   COMPARATORS AND TIMER ICs

Let us now explore some switching and timing circuits, first by looking further at the op-amp as a building block, and then by discussing a special-purpose timer IC that offers a number of attractive features in a single integrated circuit package. The switching applications illustrated in this section provide some background for the material of the last section of the chapter, which deals with analog and digital gate circuits.

The Op-Amp Comparator

The prototype of op-amp switching circuits is the op-amp comparator of Figure 11.8. This circuit, you will note, *does not employ feedback*. As a consequence of this,

$$v_{out} = A_{V(OL)}(v^+ - v^-) \tag{11.18}$$

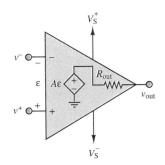

Because of the large gain that characterizes the open-loop performance of the op-amp ($A_{V(OL)} > 10^5$), any small difference between input voltages, ε, will cause large outputs. In particular, for ε of the order of a few tens of microvolts, the op-amp will go into saturation at either extreme, according to the voltage supply values and the polarity of the voltage difference (recall the discussion of the op-amp voltage supply limitations in Section 10.6). For example, if ε were a 1-mV potential difference, the op-amp output would ideally be equal to 100 V, for an open-loop gain $A_{V(OL)} = 10^5$ (and in practice the op-amp would saturate at the voltage supply limits). Clearly, any difference between input voltages will cause the output to saturate toward either supply voltage, depending on the polarity of ε.

Figure 11.8 Op-amp in open-loop mode

One can take advantage of this property to generate switching waveforms. Consider, for example, the circuit of Figure 11.9, in which a sinusoidal voltage source $v_{in}(t)$ of peak amplitude V is connected to the noninverting input. In this circuit, in which the inverting terminal has been connected to ground, the differential input voltage is given by

$$\varepsilon = V\cos(\omega t) \tag{11.19}$$

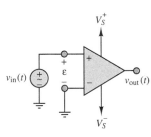

Figure 11.9 Noninverting op-amp comparator

and will be positive during the positive half-cycle of the sinusoid, and negative during the negative half-cycle. Thus, the output will saturate toward V_S^+ or V_S^-, depending on the polarity of ε: the circuit is, in effect, *comparing* $v_{in}(t)$ and ground, producing a positive v_{out} when $v_{in}(t)$ is positive, and a negative v_{out} when $v_{in}(t)$ is negative, independent of the amplitude of $v_{in}(t)$ (provided, of course, that the peak amplitude of the sinusoidal input is at least 1 mV, or so). The circuit just described is therefore called a **comparator,** and in effect performs a binary decision, determining whether $v_{in}(t) > 0$ or $v_{in}(t) < 0$. The comparator is perhaps the simplest form of an *analog-to-digital converter,* that is, a circuit that converts a continuous waveform into discrete values. The comparator output consists of only two discrete levels: "greater than" and "less than" a reference voltage. Analog-to-digital converters will be introduced more formally in Chapter 13.

The input and output waveforms of the comparator are shown in Figure 11.10, where it is assumed that $V = 1$ V and that the saturation voltage corresponding to the ±15-V supplies is approximately ±13.5 V. This circuit will be termed **noninverting comparator,** because a positive voltage differential, ε, gives rise to a positive output voltage. It should be evident that it is also possible to construct an inverting comparator by connecting the noninverting terminal to ground and connecting the input to the inverting terminal. Figure 11.11 depicts the waveforms for the **inverting comparator.** The analysis of any comparator circuit is greatly simplified if we observe that the output voltage is determined by the voltage difference present at the input terminals of the op-amp, according to the following relationship:

$$\boxed{\begin{aligned} \varepsilon > 0 &\Rightarrow V_{\text{out}} = V_{\text{sat}}^{+} \\ \varepsilon < 0 &\Rightarrow V_{\text{out}} = V_{\text{sat}}^{-} \end{aligned}} \quad \begin{array}{l} \text{Operation of} \\ \text{op-amp comparator} \end{array} \quad \begin{array}{r} \textbf{(11.20)} \\ \textbf{(11.21)} \end{array}$$

where V_{sat} is the saturation voltage for the op-amp (somewhat lower than the supply voltage, as discussed in Chapter 10). Typical values of supply voltages for practical op-amps are ±5 V to ±24 V.

A simple modification of the comparator circuit just described consists of connecting a fixed reference voltage to one of the input terminals; the effect of the reference voltage is to raise or lower the voltage level at which the comparator will switch from one extreme to the other. Example 11.6 describes one such circuit.

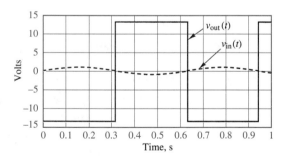

Figure 11.10 Input and output of noninverting comparator   **Figure 11.11** Input and output of inverting comparator

Example   11.6

This example considers the effect of a reference DC voltage applied to one of the input terminals of a comparator. The circuit of Figure 11.12 depicts a noninverting comparator with a DC source connected to its inverting terminal. Let $v_{\text{in}}(t)$ be a 1-V

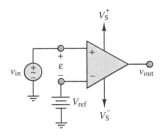

Figure 11.12

peak sinusoid, and let $V_{ref} = 0.6$ V. Sketch the input and output waveforms of the comparator.

Solution:

The analysis of the circuit is best conducted by analyzing the differential voltage across the op-amp input terminals, ε. An expression for ε may be obtained as follows:

$$\varepsilon = v_{in} - V_{ref}$$

and it follows that, since the output will saturate to the positive extreme whenever ε is positive, the following conditions hold:

$$v_{in} > V_{ref} \Rightarrow v_{out} = V_{sat}^+$$
$$v_{in} < V_{ref} \Rightarrow v_{out} = V_{sat}^-$$

Thus, the comparator will switch levels whenever the sinusoidal voltage (1 V peak, in this example) exceeds 0.6 V, or falls below this level. Figure 11.13 depicts the relevant waveforms, also showing the 0.6-V level. Note that now the comparator output voltage is asymmetric—that is, it is no longer a symmetrical square wave. This type of circuit is sometimes called a **comparator with offset.**

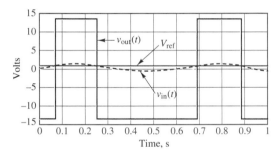

Figure 11.13 Waveforms of comparator with offset

Another useful interpretation of the op-amp comparator can be obtained by considering its **input-output transfer characteristic.** Figure 11.14 displays a plot of v_{out} versus v_{in} for a noninverting zero-reference (no offset) comparator. This circuit is often called a **zero-crossing comparator,** because

Figure 11.14 Transfer characteristic of zero-crossing comparator

the output voltage goes through a transition (V_{sat} to $-V_{sat}$, or vice versa) whenever the input voltage crosses the horizontal axis. You should be able to verify that Figure 11.15 displays the transfer characteristic for a comparator of the inverting type with a nonzero reference voltage.

Figure 11.15 Transfer characteristic of inverting comparator with offset

Very often, in converting an analog signal to a binary representation, one would like to use voltage levels other than $\pm V_{sat}$. Commonly used voltage levels in this type of switching circuit are 0 V and 5 V, respectively. This modified voltage transfer characteristic can be obtained by connecting a Zener diode between the output of the op-amp and the inverting input, in the configuration sometimes called a **level** or **Zener clamp.** The circuit shown in Figure 11.16 is based on the fact that a reverse-biased Zener diode will

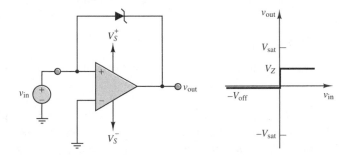

Figure 11.16 Level-clamped comparator

hold a constant voltage, V_Z, as was shown in Chapter 7. When the diode is forward-biased, on the other hand, the output voltage becomes the negative of the offset voltage, V_γ. An additional advantage of the level clamp is that it reduces the switching time. Input and output waveforms for a Zener-clamped comparator are shown in Figure 11.17, for the case of a sinusoidal $v_{in}(t)$ of peak amplitude 1 V and Zener voltage equal to 5 V.

Although the Zener-clamped circuit illustrates a specific issue of interest in the design of comparator circuits, namely, the need to establish desired reference output voltages other than the supply saturation voltages, this type of circuit is rarely employed in practice. Special-purpose integrated circuit (IC) packages are available that are designed specifically to serve as comparators. These can typically accept relatively large inputs and have provision for selecting the desired reference voltage levels (or, sometimes, are internally clamped to a specified voltage range). A representative product

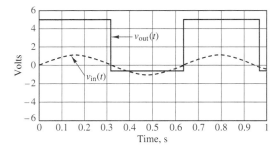

Figure 11.17 Zener-clamped comparator waveforms

is the LM311, which provides an open-collector output, as shown in Figure 11.18. The open-collector output allows the user to connect the output transistor to any supply voltage of choice by means of an external pull-up resistor, thus completing the output circuit. As we shall see in a later section of this chapter, where transistor switches are discussed, the actual value of the resistor is not critical, since the transistor is operated in the saturation mode; values between a few hundred and a few thousand ohms are typical. In the remainder of the chapter it will be assumed, unless otherwise noted, that the comparator output voltage will switch between 0 V and 5 V.

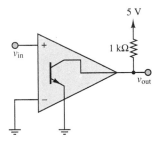

Figure 11.18 Open-collector comparator output with representative external supply connection

The Schmitt Trigger

One of the typical applications of the op-amp comparator is in detecting when an input voltage exceeds a present threshold level. The desired threshold is then represented by a DC reference, V_{ref}, connected to the noninverting input, and the input voltage source is connected to the inverting input, as in Figure 11.15. Under ideal conditions, for noise-free signals, and with an infinite slew rate for the op-amp, the operation of such a circuit would be as depicted in Figure 11.19. In practice, the presence of noise and the finite slew rate of practical op-amps will require special attention.

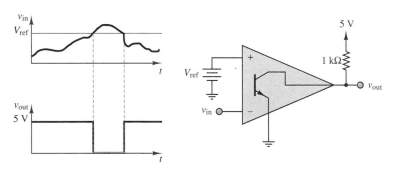

Figure 11.19 Waveforms for inverting comparator with offset

Two improvements of this circuit will be discussed in this section: how to improve the switching speed of the comparator, and how to design a circuit that can operate correctly even in the presence of noisy signals. If the input to the comparator is changing slowly, the comparator will not switch instantaneously, since its open-loop gain is not infinite and, more important, its slew rate further limits the switching speed. In fact, commercially available comparators have slew rates that are typically much lower than those of conventional op-amps. In this case, the comparator output would not switch very quickly at all. Further, in the presence of noisy inputs, a conventional comparator is inadequate, because the input signal could cross the reference voltage level repeatedly and cause multiple triggering. Figure 11.20 depicts the latter occurrence.

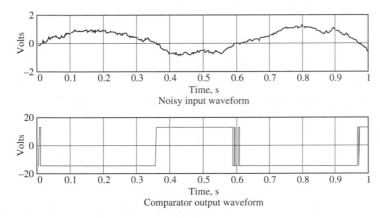

Figure 11.20 Comparator response to noisy inputs

One very effective way of improving the performance of the comparator is by introducing positive feedback. As will be explained shortly, positive feedback can increase the switching speed of the comparator and provide noise immunity at the same time. Figure 11.21 depicts a comparator circuit in which the output has been tied back to the *noninverting* input (thus the terminology *positive* feedback) by means of a resistive voltage divider. The effect of this positive feedback connection is to provide a reference voltage at the noninverting input equal to a fraction of the comparator output voltage; since the comparator output is equal to either the positive or the negative saturation voltage, $\pm V_{sat}$, the reference voltage at the noninverting input can be either positive or negative.

Consider, first, the case when the comparator output is $v_{out} = +V_{sat}$. It follows that

$$v^+ = \frac{R_2}{R_2 + R_1} V_{sat} \tag{11.22}$$

and therefore the differential input voltage is

$$\varepsilon = v^+ - v^- = \frac{R_2}{R_2 + R_1} V_{sat} - v_{in} \tag{11.23}$$

Figure 11.21 Transfer characteristic of the Schmitt trigger

For the comparator to switch from the positive to the negative saturation state, the differential voltage, ε, must then become negative; that is, the condition for the comparator to switch state becomes

$$v_{in} > \frac{R_2}{R_2 + R_1} V_{sat} \tag{11.24}$$

Since $\frac{R_2}{R_2 + R_1} V_{sat}$ is a positive voltage, the comparator will not switch when the input voltage crosses the zero level, but it will switch when the input voltage exceeds some positive voltage, which can be determined by appropriate choice of R_1 and R_2.

Consider, now, the case when the comparator output is $v_{out} = -V_{sat}$. Then

$$v^+ = -\frac{R_2}{R_2 + R_1} V_{sat} \tag{11.25}$$

and therefore

$$\varepsilon = v^+ - v^- = -\frac{R_2}{R_2 + R_1} V_{sat} - v_{in} \tag{11.26}$$

For the comparator to switch from the negative to the positive saturation state, the differential voltage, ε, must then become positive; the condition for the comparator to switch state is now

$$v_{in} < -\frac{R_2}{R_2 + R_1} V_{sat} \tag{11.27}$$

Thus, the comparator will not switch when the input voltage crosses the zero level (from the negative direction), but it will switch when the input voltage becomes more negative than a threshold voltage, determined by R_1 and R_2. Figure 11.22 depicts the effect of the different thresholds on the voltage transfer characteristic, showing the switching action by means of arrows.

The circuit just described finds frequent application and is called a **Schmitt trigger.**

If it is desired to switch about a voltage other than zero, a reference voltage can also be connected to the noninverting terminal, as shown in Figure 11.22. Now the expression for the noninverting terminal voltage is

$$v^+ = \frac{R_2}{R_2 + R_1} v_{out} + V_{ref} \frac{R_1}{R_2 + R_1} \tag{11.28}$$

and the switching levels for the Schmitt trigger are

$$v_{in} > \frac{R_2}{R_2 + R_1} V_{sat} + V_{ref} \frac{R_1}{R_2 + R_1} \tag{11.29}$$

for the positive-going transition, and

$$v_{in} < -\frac{R_2}{R_2 + R_1} V_{sat} + V_{ref} \frac{R_1}{R_2 + R_1} \tag{11.30}$$

Figure 11.22 Schmitt trigger (general circuit)

for the negative-going transition. In effect, the Schmitt trigger provides a noise-rejection range equal to $\pm \frac{R_2}{R_2+R_1}V_{\text{sat}}$ within which the comparator cannot switch. Thus, if the noise amplitude is contained within this range, the Schmitt trigger will prevent multiple triggering. Figure 11.23 depicts the response of a Schmitt trigger with appropriate switching thresholds to a noisy waveform. Example 11.7 provides a numerical illustration of this process.

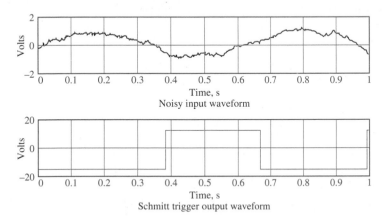

Figure 11.23 Schmitt trigger response to noisy waveforms

Example   11.7

In this example we analyze the practical design of a Schmitt trigger. Assume that the comparator has supply voltages equal to ± 18 V and saturation voltages equal to ± 16.5 V and it is desired to switch around the reference level $V_{\text{ref}} = 2$ V. Assume, further, that the noise corrupting the signal has a maximum excursion of ± 100 mV. Find the required values of the resistors in the circuit of Figure 11.24.

Solution:

Notice, first, that the reference voltage is obtained from the same (positive) voltage supply used for the op-amp. This is common practice and saves having a separate supply for the reference voltage. On the basis of the modified circuit, the expression for the noninverting-terminal voltage is

$$v^+ = \frac{R_2}{R_2 + R_1}v_{\text{out}} + \frac{R_2}{R_2 + R_3}V_S^+$$

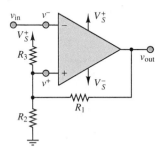

Figure 11.24

Since the required noise protection level is ± 100 mV, R_1 and R_2 can be computed from

$$\frac{\Delta V}{2} = \frac{R_2}{R_2 + R_1}V_{\text{sat}} = \frac{R_2}{R_2 + R_1}(16.5) = 0.1$$

where $\Delta V = 200$ mV is the noise protection range shown in Figure 11.25. Thus,

$$\frac{R_2}{R_2 + R_1} = \frac{0.1}{16.5} = 6.06 \times 10^{-3}$$

For $R_1 = 100$ kΩ, R_2 can be calculated to be approximately 610 Ω. Since the required reference voltage is 2 V, we can find R_3 by solving the equation

$$\frac{R_2}{R_2 + R_3}V_S^+ = \frac{R_2}{R_2 + R_3}(18) = 2$$

to obtain $R_3 = 4{,}880$ Ω.

It is instructive at this point to calculate the actual thresholds at which the comparator will switch; these are given by the expression

$$v^+ = \frac{R_2}{R_2 + R_1}v_{out} + \frac{R_2}{R_2 + R_3}V_S^+$$

where $v_{out} = \pm V_{sat}$. For a positive saturation voltage, the threshold will be

$$v^+ = \frac{R_2}{R_2 + R_1}(16.5) + 2 = 2.1 \text{ V}$$

while for the negative saturation voltage, the threshold is

$$v^+ = -\frac{R_2}{R_2 + R_1}(16.5) + 2 = 1.9 \text{ V}$$

as required. The transfer characteristic is shown in Figure 11.25.

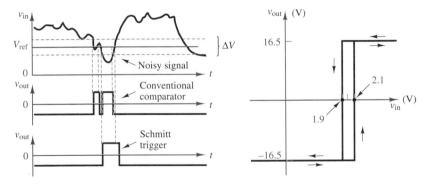

Figure 11.25

The Op-Amp Astable Multivibrator

This section describes an op-amp circuit useful in the generation of timing, or clock, waveforms. In the previous discussion, it was shown how it is possible to utilize positive feedback in a Schmitt trigger circuit. A small fraction of the large output saturation voltage was used to delay the switching of a comparator to make it less sensitive to random fluctuations in the input signal. A very similar circuit can be employed to generate a square-wave signal of fixed period and amplitude. This circuit is called an **astable multivibrator,** because it periodically switches between two states without ever reaching a stable state. The circuit is shown in Figure 11.26 in a somewhat simplified form.

The analysis of the circuit is similar to that of the Schmitt trigger, except for the presence of both negative and positive feedback connections. We start by postulating that the op-amp output is saturated at the positive supply saturation voltage. One can easily verify that, in this case, the noninverting terminal voltage is

$$v^+ = \frac{R_2}{R_2 + R_3} v_{\text{out}} \qquad (11.31)$$

and that the voltage at the inverting input is equal to the voltage across the capacitor, v_C. One can write an expression for the inverting terminal voltage, v^-, by noting that the capacitor charges at a rate determined by the feedback resistance and by the capacitance, according to the time constant:

$$\tau = R_1 C \qquad (11.32)$$

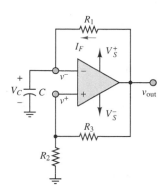

Figure 11.26 Simplified op-amp astable multivibrator

Since the current flowing into the inverting terminal is negligible, the inverting input voltage is in effect given by

$$v^-(t) = v_C(t) = V_{\text{sat}}(1 - e^{-t/\tau}) \qquad (11.33)$$

A sketch of the inverting input voltage as a function of time is shown in Figure 11.27.

The behavior of this op-amp multivibrator is best understood if the differential voltage, $\varepsilon = v^+ - v^-$, is considered. Assuming that the capacitor starts charging at $t = 0$ from zero initial charge and that V_{out} is saturated at $+V_{\text{sat}}$, one can easily show that this condition will be sustained for as long as ε remains positive. However, as the capacitor continues to charge, v^- will eventually grow larger than v^+, and ε will become negative. Since

$$v^+ = \frac{R_2}{R_2 + R_3} V_{\text{sat}} \qquad (11.34)$$

ε becomes negative as soon as the inverting-terminal voltage exceeds a threshold voltage

$$v^- > \frac{R_2}{R_2 + R_3} V_{\text{sat}} \qquad (11.35)$$

Figure 11.27

When the differential input voltage switches from a positive to a negative value, *the op-amp is forced to switch to the opposite extreme, $-V_{\text{sat}}$.* But when V_{out} switches to the negative saturation voltage, we see a sudden sign reversal in v^+:

$$v^+ = -\frac{R_2}{R_2 + R_3} V_{\text{sat}} \qquad (11.36)$$

Further, the capacitor, which has been charging toward $+V_{\text{sat}}$, now sees a negative voltage, $-V_{\text{sat}}$. This condition may be analyzed by resorting to the transient analysis methods of Chapter 6. The reversal of the output voltage causes the capacitor to discharge toward the new value of V_{out}, $-V_{\text{sat}}$, according to the function

$$v_C(t) = v_C(t_0)e^{-t/\tau} - V_{sat} \qquad (11.37)$$

where t_0 is the time at which the output voltage changes from $+V_{sat}$ to $-V_{sat}$. The discharging continues as long as the differential input voltage ε remains negative, since this guarantees that $v_{out} = -V_{sat}$. But this condition cannot last indefinitely, because at some point, as the capacitor discharges, $v_C(t) = v^-(t)$ will become more negative than $v^+(t)$ and ε will become positive again, causing the output of the op-amp to switch to $+V_{sat}$. This condition will occur when

$$v^- < -\frac{R_2}{R_2 + R_3}V_{sat} \qquad (11.38)$$

Figure 11.28 graphically summarizes the operation of the astable multivibrator. The corresponding op-amp output is depicted in Figure 11.29. The period,

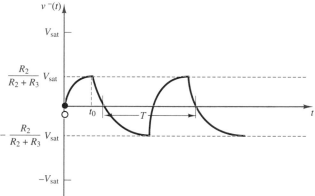

Figure 11.28 Astable multivibrator inverting-terminal voltage

T, of the waveform is determined by the charging and discharging cycle of the capacitor. It should be apparent that the introduction of a variable resistor R_1 gives rise to a circuit that can generate a variable-frequency square wave, since the frequency is directly determined by the time constant $\tau = R_1C$. It can be shown that the period of the square-wave waveform resulting from the astable multivibrator is given by

$$T = 2R_1C \log_e\left(\frac{2R_1}{R_2} + 1\right) \qquad (11.39)$$

Figure 11.29 Astable multivibrator output waveform

The Op-Amp Monostable Multivibrator (One-Shot)

An extension of the astable multivibrator is the **one-shot,** or **monostable multivibrator.** One-shots are available in integrated circuit form but can also be constructed from more general purpose op-amps. In this section, we first look at the op-amp one-shot, taking special note of the analogy with the astable multivibrator. Figure 11.30 depicts an op-amp monostable multivibrator. Its operation is summarized in the associated timing diagram, showing that if a negative voltage pulse is applied at the input, v_{in}, the output of the op-amp will switch from a stable state, $+V_{sat}$, to the unstable state, $-V_{sat}$, for a period of time T, which is determined by a charging time constant, after which the output will resume its stable value, $+V_{sat}$. In the analysis of the one-shot circuit, it will be assumed that the diode is an offset diode, with offset voltage V_{off}, and that $v_{out} = +V_{sat}$ for $t < t_0$.

For $t < t_0$, before the negative input pulse is applied, the input terminal voltages are given by

$$v^+ = V_{sat} \frac{R_2}{R_2 + R_3} \tag{11.40}$$

and

$$v^- = V_\gamma \tag{11.41}$$

Provided that $v^+ > v^-$, the op-amp will remain in its positive saturated state, then, until some external condition causes $\varepsilon = v^+ - v^-$ to become negative. This condition is brought about at $t = t_0$ by the "trigger" pulse v_{in}, which briefly lowers the noninverting-terminal voltage by δV volts. If δV is sufficiently large, the following condition will be satisfied:

$$v^+ = V_{sat} \frac{R_2}{R_2 + R_3} - \delta V < V_\gamma \tag{11.42}$$

and the op-amp will switch to the negative saturation state, as indicated in Figure 11.30. The noninverting terminal voltage may be expressed as a function of time for $t > t_0$, considering that v_{out} has switched to $-V_{sat}$ at $t = t_0$ and the diode now acts as an open circuit, since v^- is negative. With the diode out of the picture, then, the $R_1 C$ circuit will discharge, causing v^- to drop from its initial voltage (V_γ) toward $-V_{sat}$.

The circuit of Figure 11.31 depicts the equivalent switching circuit, illustrating that the capacitor discharges from the initial value of $v_C(t_0) = V_\gamma$ toward the final value, $-V_{sat}$. Recalling the analysis of transients in Chapter 6, we know that the capacitor voltage is given by the expression

$$v_C(t) = [v_C(t_0) + V_{sat}]e^{-(t-t_0)/R_1 C} - V_{sat} \tag{11.43}$$

As v^- becomes more negative, it eventually becomes smaller than v^+. Thus, when the condition

$$v_C(t) = [v_C(0) + V_{sat}]e^{-(t-t_0)/R_1 C} - V_{sat} = v^+ \tag{11.44}$$

is met, the comparator will switch back to the positive saturation state, $+V_{sat}$, and remain in that state until a new trigger pulse is provided. The value of the noninverting-terminal voltage during the time the output is in the negative saturated state is determined by the positive feedback circuit:

$$v^+ = -V_{sat} \frac{R_2}{R_2 + R_3} \tag{11.45}$$

Figure 11.30 Op-amp monostable multivibrator and typical waveforms

Figure 11.31 Equivalent charging circuit for monostable multivibrator

Clearly, the duration of the output pulse provided by the one-shot is determined by the time constant of the R_1C circuit, as well as by the value of the resistive voltage divider in the positive feedback network, as illustrated in Example 11.8.

Example   11.8

The op-amp monostable multivibrator of Figure 11.30 is assembled with the following component values: $R_1 = 20$ kΩ, $R_2 = 670$ Ω, $V_\gamma = 0.6$ V, $V_{sat} = 16$ V, $R_3 = 100$ kΩ, and $C = 4.7$ μF. Find the duration of the output pulse.

Solution:

Assume, without loss of generality, that $t_0 = 0$. Then the expression for the capacitor voltage (equal to the noninverting-terminal voltage) is

$$v_C(t) = [v_C(0) + V_{sat}]e^{-t/R_1C} - V_{sat}$$

The time at which $v^- = v^+$ may be found by setting this expression equal to the noninverting-terminal voltage

$$v^+ = -V_{sat}\frac{R_2}{R_2 + R_3} = -0.1065 \text{ V}$$

and thus obtaining the following relationship:

$$-0.1065 = 16.6e^{-10.64t} - 16$$

or

$$t = -\frac{1}{10.64} \log_e \frac{15.8935}{16.6} = 4.09 \text{ ms}$$

Thus, the duration of the output pulse is 4.09 ms.

Monostable multivibrators are usually employed in IC package form. An IC one-shot can generate voltage pulses when triggered by a **rising** or a **falling edge,** that is, by a transition in either direction in the input voltage. Thus, a one-shot IC offers the flexibility of external selection of the type of transition that will cause a pulse to be generated: a rising edge (from low voltage to high, typically zero volts to some threshold level), or a falling edge (high-to-low transition). Various input connections are usually provided for selecting the preferred triggering mode, and the time constant is usually set by selection of an external RC circuit. The output pulse that may be generated by the one-shot can also occur as a positive or a negative transition. Figure 11.32 shows the response of a one-shot to a triggering signal for the four conditions that may be attained with a typical one-shot.

A typical IC one-shot circuit based on the 74123 (see the data sheet at the end of the chapter) is displayed in Figure 11.33. The 74123 is a **dual one-shot,** meaning that the package contains two monostable multivibrators, which can be used independently. The outputs of the one-shot are indicated by the symbols Q_1, $\bar{Q}_1$, Q_2, and $\bar{Q}_2$, where the bar indicates the complement of the output. For example, if Q_1 corresponds to a positive-going output pulse, $\bar{Q}_1$ indicates a negative-going output pulse, of equal duration.

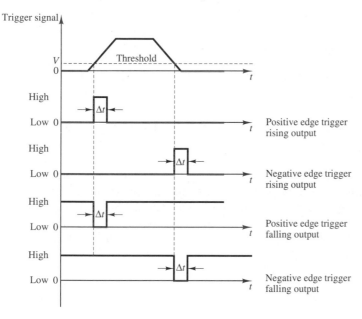

Figure 11.32 IC monostable multivibrator waveforms

Figure 11.33 Dual one-shot circuit

Timer ICs: The NE555

The discussion of op-amp–based timing circuits presented in the previous sections served the purpose of introducing a large family of integrated circuits that can provide flexible timing waveforms. These fall—for our purposes— into one of two classes: pulse generators, and clock waveform generators. Chapters 12 and 13 will delve into a more detailed analysis of digital timing circuits, a family to which the circuits of the previous sections belong. This section will now introduce a multipurpose integrated circuit that can perform both the monostable and astable multivibrator functions. The main advantage of the integrated circuit implementation of these circuits (as opposed to the discrete op-amp version previously discussed) lies in the greater accuracy and repeatability one can obtain with ICs, their ease of application, and the flexibility provided in the integrated circuit packages. The NE555 is a timer circuit capable of producing accurate time delays (pulses) or oscillation. In the time-delay, or monostable, mode, the time delay or pulse duration is controlled by an external RC network. In the astable, or clock generator, mode, the frequency is controlled by two external resistors and one capacitor. Figure 11.34 depicts typical circuits for monostable and astable operation of the NE555. Note that the threshold level and the trigger level can also be externally controlled. For the monostable circuit, the pulse width can be computed from the following equation:

$$T = 1.1R_1C \tag{11.46}$$

Figure 11.34 NE555 timer

For the astable circuit, the positive pulse width can be computed from the following equation:

$$T_+ = 0.69(R_1 + R_2) \tag{11.47}$$

and the negative pulse width can be computed from

$$T_- = 0.69R_2C \tag{11.48}$$

The use of the NE555 timer is illustrated in Example 11.9.

Example   11.9

The period of square wave generated by the monostable multivibrator of Figure 11.34 is 0.421 ms. Compute the values of R_1 and C.

Solution:

As was stated earlier, the period of the square wave generated by the monostable multivibrator of the NE555 can be computed from the following equation:

$$T = 1.1R_1C$$

By choosing $C = 1 \ \mu F$, we can obtain a value for R_1 from the following equation:

$$0.421 \ \text{ms} = 1.1R_1 \times 10^{-6}$$

Thus, $R_1 = 382.73 \ \Omega$. Note that this answer is not unique.

DRILL EXERCISES

11.10   Verify that the transfer characteristic of Figure 11.15 is correct.

11.11   For the comparator circuit shown in Figure 11.35, sketch the waveforms $v_{out}(t)$ and $v_S(t)$ if $v_S(t) = 0.1 \cos(\omega t)$ and $V_{ref} = 50 \ \text{mV}$.

11.12   Derive the expressions for the switching thresholds of the Schmitt trigger of Figure 11.21.

11.13   Explain why positive feedback increases the switching speed of a comparator.

11.14   Compute the period of the square wave generated by the multivibrator of Figure 11.26 if $C = 1 \ \mu F$, $R_1 = 10 \ \text{k}\Omega$, $R_3 = 100 \ \text{k}\Omega$, and $R_2 = 1 \ \text{k}\Omega$.

> Answer: 60.89 ms

11.15   Compute the value of R_3 that would increase the duration of the one-shot pulse in Example 11.8 to 10 ms.

> Answer: 9,300 Ω

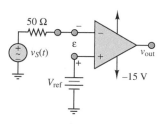

Figure 11.35

11.16 Compute the value of C that would increase the duration of the one-shot pulse in Example 11.8 to 10 ms.

Answer: 11.5 μF

11.4   TRANSISTOR SWITCHES AND GATES

In describing the properties of transistors in Chapters 8 and 9, it was suggested that, in addition to serving as amplifiers, three-terminal devices can be used as electronic switches in which one terminal controls the flow of current between the other two. It had also been hinted in Chapter 7 that diodes can act as on-off devices as well. In this section, we discuss the operation of diodes and transistors as electronic switches, illustrating the use of these electronic devices as the switching circuits that are at the heart of **analog** and **digital gates.** Transistor switching circuits form the basis of digital logic circuits, which will be discussed in more detail in the next chapter. The objective of this section is to discuss the internal operation of these circuits and to provide the reader interested in the internal workings of digital circuits with an adequate understanding of the basic principles.

An **electronic gate** is a device that, on the basis of one or more input signals, produces one of two or more prescribed outputs; as will be seen shortly, one can construct both digital and analog gates. A word of explanation is required, first, regarding the meaning of the words *analog* and *digital*. An analog voltage or current—or, more generally, an analog signal—is one that varies in a continuous fashion over time, in *analogy* (hence the expression *analog)* with a physical quantity. An example of an analog signal is the sensor voltage corresponding to ambient temperature on any given day, which may fluctuate between, say, 30° and 50°F. A digital signal, on the other hand, is a signal that can only take a finite number of values; in particular, a commonly encountered class of digital signals consists of **binary signals,** which can take only one of two values (for example, 1 and 0). A typical example of a binary signal would be the control signal for the furnace in a home heating system controlled by a conventional thermostat, where one can think of this signal as being "on" (or 1) if the temperature of the house has dropped below the thermostat setting (desired value), or "off" (or 0) if the house temperature is greater than or equal to the set temperature (say, 68°F). Figure 11.36 illustrates the appearance of the analog and digital signals in this furnace example.

The discussion of digital signals will be continued and expanded in Chapters 12 and 13. Digital circuits are an especially important topic, because a large part of today's industrial and consumer electronics is realized in digital form.

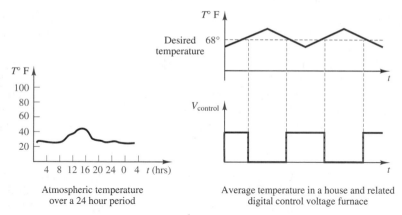

Atmospheric temperature
over a 24 hour period

Average temperature in a house and related
digital control voltage furnace

Figure 11.36 Illustration of analog and digital signals

Analog Gates

A common form of analog gate—probably the most important, in practice—employs an FET and takes advantage of the fact that current can flow in either direction in an FET biased in the ohmic region. Recall that the drain characteristic of the MOSFET discussed in Chapter 9 consists of three regions: ohmic, saturation, and breakdown. A MOSFET amplifier is operated in the saturation region, where the drain current is nearly constant for any given value of v_{GS}. On the other hand, a MOSFET biased in the ohmic state acts very much as a linear resistor. For example, for an n-channel enhancement MOSFET, the conditions for the transistor to be in the ohmic region are:

$$v_{GS} > V_T \qquad \text{and} \qquad |v_{DS}| \leq \frac{1}{4}(v_{GS} - V_T) \tag{11.49}$$

As long as the FET is biased within these conditions, it acts simply as a linear resistor, and it can conduct current in either direction (provided that v_{DS} does not exceed the limits stated in equation 11.49). In particular, the resistance of the channel in the ohmic region is found to be

$$r_{DS} = \frac{V_T^2}{2I_{DSS}(v_{GS} - V_T)} \tag{11.50}$$

so that the drain current is equal to

$$i_D = \frac{v_{DS}}{r_{DS}} \quad \text{for} \quad |v_{DS}| \leq \frac{1}{4}(v_{GS} - V_T) \quad \text{and} \quad v_{GS} > V_T \tag{11.51}$$

The most important feature of the MOSFET operating in the ohmic region, then, is that it acts as a voltage-controlled resistor, with the gate-source voltage, v_{GS}, controlling the channel resistance, r_{DS}. The use of the MOSFET

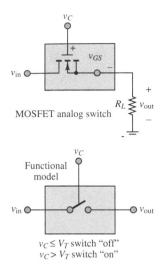

MOSFET analog switch

Functional model

$v_C \le V_T$ switch "off"
$v_C > V_T$ switch "on"

Figure 11.37 MOSFET analog switch

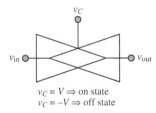

$v_C = V \Rightarrow$ on state
$v_C = -V \Rightarrow$ off state

Figure 11.38 Symbol for bilateral FET analog gate

as a switch in the ohmic region, then, consists of providing a gate-source voltage that can either hold the MOSFET in the cutoff region ($v_{GS} \le V_T$) or bring it into the ohmic region. In this fashion, v_{GS} acts as a control voltage for the transistor.

Consider the circuit shown in Figure 11.37, where we presume that v_C can be varied externally and that v_{in} is some analog signal source that we may wish to connect to the load R_L at some appropriate time. The operation of the switch is as follows. When $v_C \le V_T$, the FET is in the cutoff region and acts as an open circuit. When $v_C > V_T$ (with a value of v_{GS} such that the MOSFET is in the ohmic region), the transistor acts as a linear resistance, r_{DS}. If $r_{DS} \ll R_L$, then $v_{out} \approx v_{in}$. By using a pair of MOSFETs, it is possible to improve the dynamic range of signals one can transmit through this analog gate.

MOSFET analog switches are usually produced in integrated circuit (IC) form and denoted by the symbol shown in Figure 11.38.

Digital Gates

In this section, we explore the operation of diodes, BJTs, and FETs as digital gates. It will soon become apparent that, in general, digital circuits are simpler to analyze than analog circuits, because one is simply interested in whether a given device is conducting or not. Because of the common features shared by semiconductor diodes and bipolar junction transistors, these will be discussed together, while FETs will form the subject of a separate discussion.

Diode and BJT Gates

You will recall that a diode conducts current when it is forward-biased and otherwise acts very much as an open circuit. Thus, the diode can serve as a switch if properly employed. The circuit of Figure 11.39 is called an **OR gate**; it operates as follows. Let voltage levels greater than, say, 2 V

OR gate operation

$v_A = v_B = 0$ V $\longrightarrow$ diodes are off and $v_{out} = 0$

$\left.\begin{array}{l} v_A = 5\ \text{V} \\ v_B = 0\ \text{V} \end{array}\right\} \longrightarrow D_A$ is on, D_B is off

Equivalent circuit

$v_{out} = 5 - 0.6 = 4.4$ V

OR gate

Figure 11.39 Diode OR gate

correspond to a "logic 1" and voltages less than 2 V represent a "logic 0." Suppose, then, that the input voltages v_A and v_B can be equal to either 0 V or 5 V. If $v_A = 5$ V, diode D_A will conduct; if $v_A = 0$ V, D_A will act as an open circuit. The same argument holds for D_B. It should be apparent, then, that the voltage across the resistor R will be 0 V, or logic 0, if both v_A and v_B are 0. If either v_A or v_B is equal to 5 V, though the corresponding diode will conduct, and—assuming an offset model for the diode with $V_\gamma = 0.6$ V—we find that $v_{\text{out}} = 4.6$ V, or logic 1. Similar analysis yields an equivalent result if both v_A and v_B are equal to 5 V.

This type of gate is called an OR gate because v_{out} is equal to logic 1 (or "high") if either v_A *or* v_B (or both) is on, while it is logic 0 (or "low") if neither v_A nor v_B is on. Other functions can also be implemented; however, the discussion of diode gates will be limited to this simple introduction, because diode gate circuits, such as the one of Figure 11.39, are rarely if ever employed in practice. Most modern digital circuits employ transistors to implement switching and gate functions.

In discussing large-signal models for the BJT, we observed that the *i-v* characteristic of this family of devices includes a *cutoff* region, where virtually no current flows through the transistor. On the other hand, when a sufficient amount of current is injected into the base of the transistor, a bipolar transistor will reach *saturation,* and a substantial amount of collector current will flow. This behavior is quite well suited to the design of electronic gates and switches and can be visualized by superimposing a load line on the collector characteristic, as shown in Figure 11.40.

The operation of the simple BJT switch is illustrated in Figure 11.40, by means of load-line analysis. Writing the load-line equation at the collector circuit, we have

$$v_{CE} = V_{CC} - i_C R_C \tag{11.52}$$

and

$$v_{\text{out}} = v_{CE} \tag{11.53}$$

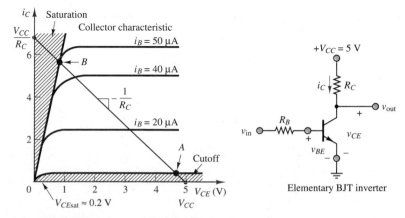

Figure 11.40 BJT switching characteristic

Thus, when the input voltage, V_{in}, is low (say, 0 V, for example) the transistor is in the cutoff region and little or no current flows, and

$$v_{out} = v_{CE} = V_{CC} \tag{11.54}$$

so that the output is "logic high."

When v_{in} is large enough to drive the transistor into the saturation region, a substantial amount of collector current will flow and the collector-emitter voltage will be reduced to the small saturation value, $V_{CE\,sat}$, which is typically a fraction of a volt. This corresponds to the point labeled B on the load line. For the input voltage v_{in} to drive the BJT of Figure 11.40 into saturation, a base current of approximately 50 μA will be required. Suppose, then, that the voltage v_{in} could take the values 0 V or 5 V. Then, if $v_{in} = 0$ V, v_{out} will be nearly equal to V_{CC}, or, again, 5 V. If, on the other hand, $v_{in} = 5$ V and R_B is, say, equal to 85 kΩ (so that the base current required for saturation flows into the base: $i_B \approx (5\ V - 0.7\ V)/R_B = 50.6\ \mu A$), we have the BJT in saturation, and $v_{out} = V_{CE\,sat} \approx 0.2$ V.

Thus, you see that whenever v_{in} corresponds to a logic high (or logic 1), v_{out} takes a value close to 0 V, or logic low (or 0); conversely, $v_{in} = $ "0" (logic "low") leads to $v_{out} = $ "1." The values of 5 V and 0 V for the two logic levels 1 and 0 are quite common in practice and are the standard values used in a family of logic circuits denoted by the acronym **TTL,** which stands for **transistor-transistor logic.** One of the more common TTL blocks is the inverter shown in Figure 11.40, so called because it "inverts" the input by providing a low output for a high input, and vice versa. This type of inverting, or "negative," logic behavior is quite typical of BJT gates (and of transistor gates in general).

In the following paragraphs, we introduce some elementary BJT logic gates, similar to the diode gates described previously; the theory and application of digital logic circuits is discussed in Chapter 12. Example 11.12 illustrates the operation of a **NAND gate,** that is, a logic gate that acts as an inverted AND gate (thus the prefix N in NAND, which stands for NOT).

Example   11.10   TTL NAND Gate

The transistors in the circuit of Figure 11.41 are connected in such a fashion as to achieve logic gate operation. Given values of V_1 and V_2, complete the following table:

v_1	v_2	State of Q_2	State of Q_3	v_{out}
0.2 V	0.2 V			
0.2 V	5.0 V			
5.0 V	0.2 V			
5.0 V	5.0 V			

Assume that transistors Q_1, Q_2, and Q_3 have a current gain $\beta = 50$, $R_1 = 5.7$ kΩ, $R_2 = 2$ kΩ, $R_3 = 2.4$ kΩ, $R_4 = 1.8$ kΩ, $V_\gamma = 0.6$ V, and $V_{CE\,sat} = 0.2$ V. Assume, further, that Q_1 can be modeled with the base-emitter and base-collector junctions as diodes, which may be considered ideal except for the offset voltage, assumed to be 0.6 V. Figure 11.42 depicts this model for transistor Q_1.

Figure 11.41 TTL NAND gate

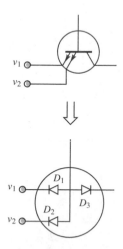

Figure 11.42

Solution:

Consider, first, the case $v_1 = v_2 = 0.2$ V. We will break the circuit into two parts to determine the voltages at different points. First, we consider the diode equivalent circuit of Q_1, shown in Figure 11.43; it should be apparent that both D_1 and D_2 will conduct. This means that

$$v_a = V_\gamma + 0.2 = 0.6 + 0.2 = 0.8 \text{ V}$$

Consider, next, the circuit including Q_2, which is shown in Figure 11.44. Writing KVL, as indicated in the figure, and assuming that Q_2 is turned on, we obtain

$$v_a - V_\gamma - V_{BE} = v_{E2}$$

Figure 11.43

Figure 11.44

Figure 11.45

Figure 11.46

or

$$v_{E2} = 0.8 - 0.6 - 0.6 = -0.4 \text{ V}$$

This result is inconsistent, since the collector voltage cannot be negative. Therefore, we conclude that Q_2 is actually in cutoff. With Q_2 in cutoff, the current through R_4 will be zero, causing v_{E2} to be zero as well.

Now consider the final stage of the circuit, shown in Figure 11.45. With $v_{E2} = 0$ V, Q_3 is clearly in cutoff, so that no current flows through R_3. Then

$$v_o = 5 - v_{R3} = 5 \text{ V}$$

Now consider the case where $V_1 = 0.2$ V and $V_2 = 5$ V. Again, consider the diode equivalent circuit of Q_1, shown in Figure 11.46. Once again, we can see that D_1 will conduct, being sufficiently forward-biased, while D_2 does not conduct, since it is reverse-biased. Thus,

$$v_a = v_1 + V_\gamma = 0.2 + 0.6 = 0.8 \text{ V}$$

which is exactly the same result obtained when we analyzed the case where $V_1 = v_2 = 0.2$ V. Thus, we conclude that the transistors Q_2 and Q_3 will be in the cutoff state, causing v_{out} to be 5 V. The same result is obtained for the case where the input voltages are $v_1 = 5$ V and $v_2 = 0.2$ V.

The final condition of interest is where $v_1 = v_2 = 5$ V, shown in Figure 11.47. Clearly, the diodes D_1 and D_2 are reverse-biased, and thus do not conduct. Writing

Figure 11.47

KVL around the circuit indicated in Figure 11.47 and assuming that Q_2 and Q_3 are turned on, we obtain

$$5 \text{ V} - I_{B2}R_1 - V_\gamma - v_{BE2} - v_{BE3} = 0$$

or

$$5 \text{ V} - I_{B2}5.7 \times 10^3 - 0.6 - 0.6 - 0.6 = 0$$

or

$$I_{B2} = \frac{3.2}{5.7 \times 10^3} = 561 \ \mu\text{A} = I_{B1}$$

Assume now that Q_3 is in saturation. Then, $v_{\text{out}} = 0.2$ V, implying that Q_2 might possibly be in saturation. Assuming that this is the case, the following expression holds:

$$v_{CEQ2} = 5 - I_{C2}R_2 - V_{BE3} = 0.2 \text{ V}$$
$$= 5 - I_{C2} \times 2 \times 10^3 - 0.6 = 0.2$$

leading to

$$I_{C2} = 2.1 \text{ mA}$$

Now, the validity of the assumptions regarding the state of Q_2 and Q_3 may be verified by checking the value of β. In fact,

$$\beta = \frac{I_{C2}}{I_{B2}} = \frac{2.1 \times 10^{-3}}{561 \times 10^{-6}} = 3.74$$

and we conclude that Q_2 is indeed in saturation, since $3.74 < 50$. Thus, the results obtained above hold.

Having completed the analysis, we may finally fill out the chart:

		State of		
v_1	v_2	Q_2	Q_3	v_{out}
0.2 V	0.2 V	Cutoff	Cutoff	5.0 V
0.2 V	5.0 V	Cutoff	Cutoff	5.0 V
5.0 V	0.2 V	Cutoff	Cutoff	5.0 V
5.0 V	5.0 V	Saturation	Saturation	0.2 V

You should have no difficulty in verifying that the output voltage is related to the inputs in the fashion of an inverted AND (or a NAND) gate.

The analysis method employed in Example 11.10 can be used to analyze any TTL gate. With a little practice, the calculations of this example will become familiar. The drill exercises and homework problems will reinforce the concepts developed in this section.

MOSFET Logic Gates

Having discussed the BJT as a switching element, we might quite naturally suspect that FETs may similarly serve as logic gates. In fact, in some respects, FETs are better suited to be employed as logic gates than BJTs. The n-channel enhancement MOSFET, discussed in Chapter 9, serves as an excellent illustration: because of its physical construction, it is normally off (that is, it is off until a sufficient gate voltage is provided), and therefore it does not require much current from the input signal source. Further, MOS devices offer the additional advantage of easy fabrication into integrated circuit form, making production economical in large volume. On the other hand, MOS devices cannot provide as much current as BJTs, and their switching speeds are not quite as fast—although these last statements may not hold true for long, because great improvements are taking place in MOS technology. Overall, it is certainly true that in recent years it has become increasingly common to design logic circuits based on MOS technology. In particular, a successful family of logic gates called **CMOS** (for **complementary metal-oxide semiconductor**) takes advantage of both *p*- and *n*-channel enhancement-mode MOSFETs to exploit the best features of both types of transistors. CMOS logic gates (and many other types of digital circuits constructed using the same technology) consume very little supply power, and have become the mainstay in pocket calculators, wristwatches, portable computers, and many other consumer electronics products. Without delving into the details of CMOS technology—which is beyond the intended scope of this book—we shall briefly illustrate the properties of MOSFET logic gates and of CMOS gates, in the remainder of this section.

Figure 11.48 depicts a MOSFET switch side by side with its drain *i-v* characteristic. Note the general similarity with the switching characteristic of the BJT shown in the previous section. When the input voltage, v_{in}, is zero, the MOSFET conducts virtually no current, and the output voltage, v_{out}, is equal to V_{DD}. When v_{in} is equal to 5 V, the MOSFET Q point moves from point A to point B along the load line, with $v_{DS} = 0.5$ V. Thus, the circuit acts as an inverter. Much as in the case of the BJT, the inverter forms the basis of all MOS logic gates.

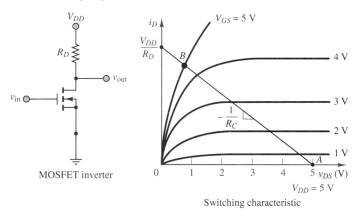

Figure 11.48 MOSFET switching characteristic

An elementary CMOS inverter is shown in Figure 11.49. Note first the simplicity of this configuration, which simply employs two enhancement-mode MOSFETs: *p*-channel at the top, denoted by the symbol Q_p, and *n*-channel at the bottom, denoted by Q_n.

Recall from Chapter 9 that when v_{in} is low, transistor Q_n is off. However, transistor Q_p sees a gate-to source voltage $v_{GS} = v_{in} - V_{DD} = -V_{DD}$; in a *p*-channel device, this condition is the counterpart of having $v_{GS} = V_{DD}$ for an *n*-channel MOSFET. Thus, when Q_n is off, Q_p is on and acts very much as a small resistance. In summary, when v_{in} is low, the output is $v_{out} \approx V_{DD}$. When v_{in} is high, the situation is reversed: Q_n is now on and acts nearly as a short circuit, while Q_p is open (since $v_{GS} = 0$ for Q_p). Thus, $v_{out} \approx 0$. The complementary MOS operation is depicted in Figure 11.49 in simplified form by showing each transistor as either a short or an open circuit, depending on its state. This simplified analysis is sufficient for the purpose of a qualitative analysis. The following examples illustrate methods for analyzing MOS switches and gates.

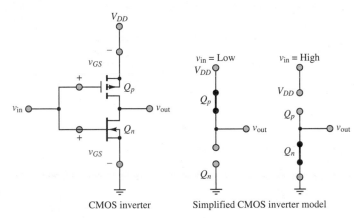

Figure 11.49 MOSFET inverter

Example   11.11

An n-channel MOSFET is used as a switch as shown in the circuit of Figure 11.50. The switching source signal is a 2.6-V signal applied to the gate at time t_1. The signal is zero for $t < t_1$.

1. Determine the operating point of the circuit for $t < t_1$.
2. Determine the operating point of the circuit for $t > t_1$.

Solution:

To determine the operating point, we draw the load line on the device characteristic curves, which are shown in Figure 11.51. The load line intercepts the v_{DS} axis at $v_{DS} = V_{DD} = 10$ V and the i_D axis at $v_{DS}/R_D = 10/125$, or 80 mA.

1. Next, to determine the operating point for $t < t_1$, we see that the gate voltage is 0 V and therefore the MOSFET is off. The operating point is then given by

$$v_{DS} = 10 \text{ V} \qquad i_D = 0 \text{ A}$$

2. For $t > t_1$, the gate voltage is 2.6 V, causing the transistor's operating point to move to a new position on the characteristic curves, determined by v_{GS}. The new operating point for the transistor is approximately

$$v_{DS} = 1.1 \text{ V} \qquad i_D = 66.5 \text{ mA}$$

Figure 11.50

Thus, the switching action at the gate causes the current through the MOSFET to switch from zero to approximately 60 mA. Consequently, the voltage across the drain resistor, $R_D = 125 \ \Omega$, is equal to 7.5 V, and the voltage across the MOSFET must be 2.5 V, by KVL.

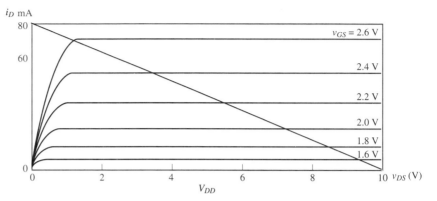

Figure 11.51

Example   11.12

The circuit shown in Figure 11.52 forms a CMOS gate. It is made of p-channel and n-channel MOSFETs. For analysis considerations, it may be assumed that the transistors are in either on or off, and that $V_T = 1.7$ V for the MOSFETs. Find the output voltage for each of the following cases:

The transistors in this circuit show the substrate for each transistor connected to its respective gate. In a true CMOS IC, the substrates for the p-channel transistors are connected to 5 V and the substrates of the n-channel transistors are connected to ground.

1. $v_1 = 0, \ v_2 = 0$
2. $v_1 = 5$ V, $v_2 = 0$ V
3. $v_1 = 0$ V, $v_2 = 5$ V
4. $v_1 = 5$ V, $v_2 = 5$ V

Solution:

To analyze each of these cases, we will represent the transistors that are on as linear resistors and those that are off as open circuits.

1. $v_1 = 0$ V, $v_2 = 0$ V

Figure 11.53 shows that no current will flow through the resistors representing M_1 and M_2, since M_3 and M_4 are both open circuits. This means that there will be no voltage drop across M_1 and M_2, and so

$$v_{\text{out}} = 5 \text{ V}$$

Figure 11.52

2. $v_1 = 5$ V, $v_2 = 0$ V

M_1 and M_4 will be turned off, since $v_{GS1,4} < V_T$, and M_2 and M_3 will be turned on, leading to the representation shown in Figure 11.54. Once again, no current will flow, since M_1 is an open circuit, causing $v_{\text{out}} = 0$ V.

With both v_1 and v_2 at zero volts, M_3 and M_4 will be turned off (in cutoff), since v_{GS} is less than V_T (0 V $<$ 1.7 V). M_1 and M_2 will be turned on, since the gate-to-source voltages will be greater than V_T.

Figure 11.53

Figure 11.54   **Figure 11.55**

3. $v_1 = 0$ V, $v_2 = 5$ V

Once again, we refer to Figure 11.52 and see that when we use the reasoning employed in part 2, M_1 and M_4 will be turned on and M_2 and M_3 will be in cutoff, as depicted in Figure 11.55.

4. $v_1 = v_2 = 5$ V

This time, we see from Figure 11.56 that M_3 and M_4 will both be turned on and M_1 and M_2 will be in cutoff. Clearly, $v_{out} = 0$ V.

A table may now be assembled, showing v_1, v_2, and v_{out}:

v_1	v_2	v_{out}
0	0	5
5	0	0
0	5	0
5	5	0

Figure 11.56

The table suggests that the circuit discussed in this example is a NOR gate—that is, an inverted OR gate.

DRILL EXERCISES

11.17 Show that both V_1 and V_2 must be high for the AND gate circuit shown in Figure 11.57 to give a logic high output.

11.18 Show that the circuit in Figure 11.58 acts as an AND gate, and construct a truth table as in Example 11.10.

Figure 11.57 Diode AND gate

(AND gate)

Figure 11.58 TTL AND gate

11.19 What value of R_D would ensure a drain-to-source voltage, v_{DS} of 5 V in the circuit of Example 11.11?

Answer: 250 Ω

11.20 Analyze the CMOS gate of Figure 11.59 and find the output voltages for the following conditions: (a) $v_1 = 0$, $v_2 = 0$; (b) $v_1 = 5$ V, $v_2 = 0$; (c) $v_1 = 0$, $v_2 = 5$ V; (d) $v_1 = 5$ V, $v_2 = 5$ V. Identify the logic function accomplished by the circuit.

Figure 11.59 CMOS AND gate

Answer:

v_1	v_2	v_{out}
0 V	0 V	5 V
5 V	0 V	5 V
0 V	5 V	5 V
5 V	5 V	0 V

NAND gate

11.21 Draw the schematic diagram of a four-input CMOS NAND gate.

11.22 Draw the schematic diagram of a four-input TTL NOR gate.

CONCLUSION

- Integrated circuits are common building blocks in the design of instrumentation systems. The relative simplicity of IC electronics brings the task of designing basic instrumentation circuits within reach of non-electrical engineers.

- IC instrumentation amplifiers with programmable gains, low bias current, and high CMRR are available commercially; these circuits require only the connection of a few external resistors for proper operation.

- Active filters can be designed using various prototype frequency response functions. Of these, Butterworth filters are possibly the simplest to design. Design tables are available to aid the engineer in the selection of the filter order and type for a given application.

- Op-amp comparators are the simplest form of analog-to-digital converter; various configurations exist, permitting the choice of reference voltage. The Schmitt trigger is an extension of the comparator that has the property of increasing the switching speed and of providing noise rejection. IC comparators are commercially available at very low cost.

- Multivibrators are switching circuits that can generate pulses or clock waveforms. Various types of such timer circuits exist in IC form, with easily programmable features, including selection of trigger and threshold levels, as well as polarities of input and output signals.

- Transistors form the basis of all digital and switching circuits. Transistor switching circuits can employ either BJT and FET circuits, giving rise to two very large families of digital logic circuits: TTL and CMOS. Each family is characterized by certain advantages; in particular, TTL circuits are faster and can provide greater load currents, while CMOS circuits have extremely low power consumption and are easily fabricated. Transistor switching circuits can be analyzed more easily than linear amplifier circuits, since one is usually concerned only with whether the device is on or off.

KEY TERMS

Instrumentation Amplifier (IA)   *550*

Monolithic Integrated Circuit   *553*

Sense and Reference Inputs   *553*

Butterworth Filter   *556*

Chebyshev Filter   *556*

Cauer (Elliptical) Filter   *556*

Bessel Filter   *556*

Butterworth Polynomial   *557*

Equiripple Filter   *557*

Second-Order (Quadratic) Filter   *558*

Constant-K (Sallen and Key) Filter   *558*

Positive Feedback   *559*

Quality Factor, Q *559*

Comparator   *562*

Noninverting Comparator   *563*

Inverting Comparator   *563*

Comparator with Offset   *564*

Input-Output Transfer Characteristic   *564*

Zero-Crossing Comparator   *564*

Level (Zener) Clamp   *565*

Schmitt Trigger   *568*

Astable Multivibrator   *570*

One-Shot (Monostable Multivibrator)   *572*

Rising Edge   *574*

DEVICE DATA SHEETS

LM311   Voltage Comparator

Features

- Operates from single 5V supply
- Input current: 150 nA max over temperature
- Offset current: 20 nA max over temperature
- Differential input voltage range: ±30V
- Power consumption: 135 mW at ±15V

Dual-In-Line Package

```
        NC   1          14  NC
        GND  2          13  NC
        INPUT 3         12  NC
        INPUT 4         11  V⁺
        NC   5          10  NC
        V⁻   6           9  OUTPUT
        BALANCE 7        8  BALANCE/
                            STROBE
```

TOP VIEW

Typical Applications**

Offset Balancing

Strobing

Increasing Input Stage Current*

Note: Do Not Ground Strobe Pin. Output is turned off when current is pulled from Strobe Pin.

**Note: Pin connections shown on schematic diagram and typical applications are for H08 metal can package.

*Increases typical common mode slew from 7.0V/μs to 18V/μs.

Detector for Magnetic Transducer

Relay Driver with Strobe

*Absorbs inductive kickback of relay and protects IC from severe voltage transients on V + + line.

Note: Do Not Ground Strobe Pin.

Electrical Characteristics for LM311

Parameter	Conditions	Min	Typ	Max	Units
Input Offset Voltage	$T_A = 25\,^\circ C$, $R_S \le 50k$		2.0	7.5	mV
Input Offset Current	$T_A = 25\,^\circ C$		6.0	50	nA
Input Bias Current	$T_A = 25\,^\circ C$		100	250	nA
Voltage Gain	$T_A = 25\,^\circ C$	40	200		V/mV
Response Time	$T_A = 25\,^\circ C$		200		ns
Saturation Voltage	$V_{IN} \le -10$ mV, $I_{OUT} = 50$ mA $T_A = 25\,^\circ C$		0.75	1.5	V
Strobe ON Current	$T_A = 25\,^\circ C$	1.5	3.0		mA
Output Leakage Current	$V_{IN} \ge 10$ mV, $V_{OUT} = 35$ V $T_A = 25\,^\circ C$, $I_{STROBE} = 3$ mA $V^- = V_{GRND} = -5V$		0.2	50	nA
Input Offset Voltage	$R_S \le 50K$			10	mV
Input Offset Current				70	nA
Input Bias Current				300	nA
Input Voltage Range		-14.5	$13.8, -14.7$	13.0	V
Saturation Voltage	$V^+ \ge 4.5V$, $V^- = 0$ $V_{IN} \le -10$ mV, $I_{OUT} \le 8$ mA		0.23	0.4	V
Positive Supply Current	$T_A = 25\,^\circ C$		5.1	7.5	mA
Negative Supply Current	$T_A = 25\,^\circ C$		4.1	5.0	mA

LM555/LM555C   Timer

Features

- Direct replacement for SE555/NE555
- Timing from microseconds through hours
- Operates in both astable and monostable modes
- Adjustable duty cycle
- Output can source or sink 200 mA
- Output and supply TTL compatible
- Temperature stability better than 0.005% per °C
- Normally on and normally off output

Applications

- Precision timing
- Pulse generation
- Sequential timing
- Time delay generation
- Pulse width modulation
- Pulse position modulation
- Linear ramp generator

Connection Diagram

Dual-In-Line and Small Outline Packages

Top View

Electrical Characteristics (T_A = 25°C, V_{CC} = +5V to +15V, unless otherwise specified)

Parameter	Conditions	Limits LM555			Limits LM555C			Units
		Min	Typ	Max	Min	Typ	Max	
Supply Voltage		4.5		18	4.5		16	V
Supply Current	V_{CC} = 5V, R_L = ∞		3	5		3	6	mA
	V_{CC} = 15V, R_L = ∞		10	12		10	15	mA
	(Low State) (Note 2)							
Timing Error, Monostable								
Initial Accuracy			0.5			1		%
Drift with Temperature	R_A = 1k to 100 kΩ		30			50		ppm/°C
	C = 0.1 μF, (Note 3)							
Accuracy over Temperature			1.5			1.5		%
Drift with Supply			0.05			0.1		%/V
Timing Error, Astable								
Initial Accuracy			1.5			2.25		%
Drift with Temperature	R_A, R_B = 1k to 100 kΩ		90			150		ppm/°C
	C = 0.1 μF, (Note 3)							
Accuracy over Temperature			2.5			3.0		%
Drift with Supply			0.15			0.30		%/V
Threshold Voltage			0.667			0.667		xV_{CC}
Trigger Voltage	V_{CC} = 15V	4.8	5	5.2		5		V
	V_{CC} = 5V	1.45	1.67	1.9		1.67		V
Trigger Current			0.01	0.5		0.5	0.9	μA
Reset Voltage		0.4	0.5	1	0.4	0.5	1	V
Reset Current			0.1	0.4		0.1	0.4	mA
Threshold Current	(Note 4)		0.1	0.25		0.1	0.25	μA
Control Voltage Level	V_{CC} = 15V	9.6	10	10.4	9	10	11	V
	V_{CC} = 5V	2.9	3.33	3.8	2.6	3.33	4	V
Pin 7 Leakage Output High			1	100		1	100	nA
Pin 7 Sat								
Output Low	V_{CC} = 15V, I_7 = 15 mA		150			180		mV
Output Low	V_{CC} = 4.5V, I_7 = 4.5 mA		70	100		80	200	mV
Output Voltage Drop (Low)	V_{CC} = 15V							
	I_{SINK} = 10 mA		0.1	0.15		0.1	0.25	V
	I_{SINK} = 50 mA		0.4	0.5		0.4	0.75	V
	I_{SINK} = 100 mA		2	2.2		2	2.5	V
	I_{SINK} = 200 mA		2.5			2.5		V
	V_{CC} = 5V							
	I_{SINK} = 8 mA		0.1	0.25				V
	I_{SINK} = 5 mA					0.25	0.35	V

Electrical Characteristics ($T_A = 25°C$, $V_{CC} = +5V$ to $+15V$, unless otherwise specified)
(continued)

Parameter	Conditions	Limits						Units
		LM555			LM555C			
		Min	Typ	Max	Min	Typ	Max	
Output Voltage Drop (High)	$I_{SOURCE} = 200$ mA, $V_{CC} = 15V$		12.5			12.5		V
	$I_{SOURCE} = 100$ mA, $V_{CC} = 15V$	13	13.3		12.75	13.3		V
	$V_{CC} = 5V$	3	3.3		2.75	3.3		V
Rise Time of Output			100			100		ns
Fall Time of Output			100			100		ns

DM74LS123   Dual Retriggerable One-Shot with Clear and Complementary Outputs

Features

- DC triggered from active-high transition of active-low transition inputs

- Retriggerable to 100% duty cycle

- Compensated for V_{CC} and temperature variations

- Triggerable from CLEAR input

- DTL, TTL, compatible

- Input clamp diodes

Connection Diagram

Dual-In-Line Package

Function Table

Inputs			Outputs	
CLEAR	A	B	Q	Q̄
L	X	X	L	H
X	H	X	L	H
X	X	L	L	H
H	L	↑	⊓	⊔
H	↓	H	⊓	⊔
↑	L	H	⊓	⊔

H = High Logic Level
L = Low Logic Level
X = Can Be Either Low or High
↑ = Positive Going Transition
↓ = Negative Going Transition
⊓ = A Positive Pulse
⊔ = A Negative Pulse

AD625   **Instrumentation Amplifier**

SPECIFICATIONS (typical @ $V_S = \pm\ 15V$, $R_L = 2k\Omega$ and $T_A = +25°C$ unless otherwise specified)

Model	AD625A/J/S Min	AD625A/J/S Typ	AD625A/J/S Max	AD625B/K Min	AD625B/K Typ	AD625B/K Max	AD625C Min	AD625C Typ	AD625C Max	Units
Gain										
Gain Equation		$\frac{2R_F}{R_G} + 1$			$\frac{2R_F}{R_G} + 1$			$\frac{2R_F}{R_G} + 1$		
Gain Range	1		10,000	1		10,000	1		10,000	
Gain Error[1]		$\pm.035$	±0.05		±0.02	±0.03		±0.01	±0.02	%
Nonlinearity, Gain = 1 − 256			±0.005			±0.002			±0.001	%
Gain > 256			±0.01			±0.008			±0.005	%
Gain vs. Temp. Gain < 1000[1]		5			5			5		ppm/°C
Input										
Input Impedance										
Differential Resistance		1			1			1		$G\Omega$
Differential Capacitance		4			4			4		pF
Common-Mode Resistance		1			1			1		$G\Omega$
Common-Mode Capacitance		4			4			4		pF
Input Voltage Range										
Differ. Input Linear (V_D)			±10			±10			±10	V
Common-Mode Linear (V_{CM})		$12V - \left(\frac{G}{2} \times V_D\right)$			$12V - \left(\frac{G}{2} \times V_D\right)$			$12V - \left(\frac{G}{2} \times V_D\right)$		
Common-Mode Rejection Ratio dc to 60Hz with 1 kΩ Source Imbalance										
G = 1	70	75		75	85		80	90		dB
G = 10	90	95		95	105		100	115		dB
G = 100	100	105		105	115		110	125		dB
G = 1000	110	115		115	125		120	140		dB
Output Rating		$\pm10V$ @ 5mA			$\pm10V$ @ 5mA			$\pm10V$ @ 5mA		
Dynamic Response										
Small Signal −3dB										
G = 1 ($R_F = 20k\Omega$)		650			650			650		kHz
G = 10		400			400			400		kHz
G = 100		150			150			150		kHz
G = 1000		25			25			25		kHz
Slew Rate		5.0			5.0			5.0		V/μs
Setting Time to 0.01%, 20V Step										
G = 1 to 200		15			15			15		μs
G = 500		35			35			35		μs
G = 1000		75			75			75		μs
Power Supply										
Power Supply Range		±5 to ±18			±5 to ±18			$\pm5 \pm18$		V
Quiescent Current		3.5	5		3.5	5		3.5	5	mA

HOMEWORK PROBLEMS

11.1 Repeat the derivation of Example 11.1 if the feedback resistance is $R_F + \Delta R$ and all other resistors are matched.

11.2 Repeat the derivation of Example 11.1 if the resistors R_2 and R_2' are mismatched by ΔR.

11.3 Derive the expression for the CMRR given in equation 11.12.

11.4 For the IA of Figure 11.1, find the gain of the input stage if $R_1 = 1$ kΩ and $R_2 = 10$ kΩ.

11.5 For the IA of Figure 11.1, find the gain of the input stage if $R_1 = 1.5$ kΩ and $R_2 = 80$ kΩ.

11.6 If $R_2 = 20$ KΩ for the IA of Figure 11.1, find R_1 so that an input-stage gain of 300 can be achieved.

11.7 Find the differential gain for the IA of Figure 11.1 if $R_2 = 5$ kΩ, $R_1 = R' = R = 1$ kΩ, and $R_F = 10$ kΩ.

11.8 Find the differential gain for the IA of Figure 11.1 if $R_1 = 2$ kΩ, $R_2 = R' = R = 10$ kΩ, and $R_F = 50$ k$\Omega = R_F'$.

11.9 Given $R_F = 10$ kΩ and $R_1 = 2$ kΩ for the IA of Figure 11.1, find R and R_2 so that a differential gain of 900 can be achieved.

11.10 Replace the cutoff frequency specification of Example 11.3 with $\omega_C = 10$ rad/s and determine the order of the filter required to achieve 40 dB attenuation at $\omega_S = 25$ rad/s.

11.11 Replace the cutoff frequency specification of Example 11.3 with $\omega_C = 10$ rad/s and determine the order of the filter required to achieve 60 dB attenuation at $\omega_S = 30$ rad/s.

11.12 Let $K = 1$ in a Sallen and Key low-pass filter prototype. If the cutoff frequency, ω_C, and the resistors $R' = R_1 = R_2$ and R are normalized to unity, derive relationships for C_1, C_2, and Q.

11.13 Derive the frequency response of the low-pass filter of Figure 11.7.

11.14 Derive the frequency response of the high-pass filter of Figure 11.7.

11.15 Derive the frequency response of the band-pass filter of Figure 11.7.

11.16 How would the components of Example 11.4 scale if the cutoff frequency were 500 rad/s?

11.17 Design a second-order Butterworth high-pass filter with a 10-kHz cutoff frequency, a DC gain of 10, $Q = 5$, and $V_S = \pm 15$ V.

11.18 Design a second-order Butterworth high-pass filter with a 25-kHz cutoff frequency, a DC gain of 15, $Q = 10$, and $V_S = \pm 15$ V.

11.19 Design a second-order Butterworth low-pass filter with a 10-kHz cutoff frequency, a DC gain of 10, $Q = 15$, and $V_S = \pm 15$ V.

11.20   Design a second-order Butterworth low-pass filter with a 15-kHz cutoff frequency, a DC gain of 15, $Q = 5$, and $V_S = \pm15$ V.

11.21   Design a band-pass filter with a low cutoff frequency of 200 Hz, a high cutoff frequency of 1 kHz, and a pass-band gain of 4. Calculate the value of Q for the filter. Also, draw the approximate frequency response of this filter.

11.22   Repeat Problem 11.21 for a low cutoff frequency of 500 Hz, a high cutoff frequency of 2 kHz, and a pass-band gain of 5.

11.23   Design two normalized low-pass quadratic filter sections with Q equal to 2 and 0.5, respectively. Plot the magnitude frequency response of each, making sure to plot enough points around the cutoff frequency. Explain what is meant by *peaking* in the frequency response.

11.24   The filter shown in Figure P11.1 is called an *infinite-gain multiple-feedback filter*. Derive the following expression for the filter's frequency response:

$$H(j\omega) = \frac{-(1/R_3R_2C_1C_2)R_3/R_1}{(j\omega)^2 + \left(\frac{1}{R_1C_1} + \frac{1}{R_2C_1} + \frac{1}{R_3C_1}\right)j\omega + \frac{1}{R_3R_2C_1C_2}}$$

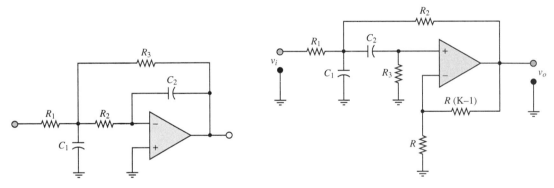

Figure P11.1   **Figure P11.2**

11.25   The filter shown in Figure P11.2 is a Sallen and Key band-pass filter circuit, where K is the DC gain of the filter. Derive the following expression for the filter's frequency response:

$$H(j\omega) = \frac{j\omega K/R_1C_1}{(j\omega)^2 + j\omega\left(\frac{1}{R_1C_1} + \frac{1}{R_3C_2} + \frac{1}{R_3C_1} + \frac{1-K}{R_2C_1}\right) + \frac{R_1+R_2}{R_1R_2R_3C_1C_2}}$$

11.26   Show that the expression for Q in the filter of Problem 11.24 is given by

$$\frac{1}{Q} = \sqrt{R_2R_3\frac{C_2}{C_1}}\left(\frac{1}{R_1} + \frac{1}{R_2} + \frac{1}{R_3}\right)$$

11.27   Design a Schmitt trigger to operate in the presence of noise with peak amplitude $= \pm150$ mV. The circuit is to switch around the reference value -1 V. Assume an op-amp with ±10-V supplies ($V_{\text{sat}} = 8.5$ V).

11.28   In the circuit of Figure P11.3, $R_1 = 100\ \Omega$, $R_2 = 56\ \text{k}\Omega$, $R_i = R_1 \parallel R_2$, and v_{in} is a 1-V peak-to-peak sine wave. Assuming that the supply voltages are ± 15 V, determine the threshold voltages (positive and negative v^+) and draw the output waveform.

Figure P11.3

11.29   Repeat Problem 11.28 for $R_1 = 1\ \text{k}\Omega$, $R_2 = 10\ \text{k}\Omega$, and v_{in} a 5-V peak-to-peak sine wave.

11.30   Repeat Problem 11.28 for $R_1 = 2\ \text{k}\Omega$, $R_2 = 15\ \text{k}\Omega$, and v_{in} a 2-V peak-to-peak sine wave.

11.31   For the circuit shown in Figure P11.4,

a.   Draw the output waveform for v_{in} a 4-V peak-to-peak sine wave at 100 Hz and $V_{\text{ref}} = 2$ V.

b.   Draw the output waveform for v_{in} a 4-V peak-to-peak sine wave at 100 Hz and $V_{\text{ref}} = -2$ V.

Note that the diodes placed at the input ensure that the differential voltage does not exceed the diode offset voltage.

11.32   For the circuit of Figure P11.5, v_{in} is a 100-mV peak sine wave at 5 kHz, $R = 10\ \text{k}\Omega$, D_1 is a 6.2-V Zener diode, and D_2 is a standart diode with $V_\gamma = 0.7$ V. Draw the output voltage waveform.

Figure P11.4

Figure P11.5

11.33   For the circuit of Figure P11.6, v_{in} is a 100-mV peak sine wave at 5 kHz, $R = 10\ \text{k}\Omega$, and D_1 and D_2 are 6.2-V Zener diodes. Draw the output voltage waveform.

11.34   Show that the period of oscillation of an op-amp astable multivibrator is given by the expression

$$T = 2R_1 C \log_e \left(\frac{2R_1}{R_2} + 1 \right)$$

11.35   Use the data sheets for the 74123 monostable multivibrator to analyze the connection shown in Figure 11.33. Draw a timing diagram indicating the approximate duration of each pulse, assuming that the trigger signal consists of a positive-going transition.

11.36   In the monostable multivibrator of Figure 11.34, $R_1 = 10\ \text{k}\Omega$ and the output pulse width $T = 10$ ms. Determine the value of C.

Figure P11.6

600   Chapter 11 Integrated Circuit Electronics

11.37 In the monostable multivibrator of Figure 11.34, R_1 = 2.2 kΩ, R_2 = 3.9 kΩ, and C = 0.1 μF. Determine the positive pulse width T_+ and the negative pulse width T_-.

11.38 Using the NE555 timer, design a monostable multivibrator having an output pulse of 10 ms.

11.39 Using the NE555 timer, design an astable multivibrator having an output pulse of 5 ms.

11.40 Using the astable multivibrator of Figure 11.34, design a circuit with the following characteristics:

> Pulse repetition rate: 500 pulses/s
> Duty cycle = 50%

[*Note:* Pulse repetition rate = 1/period.]

11.41 Using the astable multivibrator of Figure 11.34, design a circuit with the following characteristics:

> Pulse repetition rate: 700 pulses/s
> Duty cycle = 60%

Figure P11.7

11.42 In Figure P11.7, the minimum value of v_{in} for a high input is 2.0 V. Assume that transistor Q_1 has a β of at least 10. Find the range for resistor R_B that can guarantee that the transistor Q_1 is on.

11.43 Repeat Problem 11.42 if the minimum value of v_{in} for a high input is 2.7 V and transistor Q_1 has a β of at least 50.

11.44 Figure P11.8 shows a circuit with two transistor inverters connected in series, where $R_{1C} = R_{2C}$ = 10 kΩ and $R_{1B} = R_{2B}$ = 27 kΩ.

Figure P11.8

a. Find v_B, v_{out}, and the state of transistor Q_1 when v_{in} is low.
b. Find v_B, v_{out}, and the state of transistor Q_1 when v_{in} is high.

11.45 Repeat Problem 11.44 if $R_{1C} = R_{2C}$ = 5 kΩ and $R_{1B} = R_{2B}$ = 12 kΩ.

11.46 For the inverter of Figure P11.9, R_B = 5 kΩ and $R_{C1} = R_{C2}$ = 2 kΩ. Find the minimum values of β_1 and β_2 to ensure that Q_1 and Q_2 saturate when v_{in} is high.

Figure P11.9

11.47 For the inverter of Figure P11.9, $R_B = 4$ kΩ, $R_{C1} = 2.5$ kΩ, and $\beta_1 = \beta_2 = 4$. Show that Q_1 saturates when v_{in} is high. Find a condition for R_{C2} to ensure that Q_2 also saturates.

11.48 For the inverter of Figure P11.9, $R_{C1} = R_{C2} = 2$ kΩ and $\beta_1 = \beta_2 = 10$. Show that transistor Q_2 saturates, and find a condition for R_B so that Q_1 also saturates.

11.49 The basic circuit of a TTL gate is shown in the circuit of Figure P11.10. Determine the logic function performed by this circuit.

11.50 Figure P11.11 is a circuit diagram for a three-input TTL NAND gate. Assuming that all the input voltages are high, find v_{B1}, v_{B2}, v_{B3}, v_{C2}, and v_{out}. Also, indicate the state of each transistor.

Figure P11.10

Figure P11.11

11.51 Repeat Problem 11.50 for the case when v_1 and v_2 are high and v_3 is low.

11.52 For the circuit of Figure P11.11, discuss the purpose of transistor Q_4 and diode D_1.

11.53 Show that when two or more emitter-follower outputs are connected to a common load, as shown in the circuit of Figure P11.12, the OR operation results; that is, $v_o = v_1$ OR v_2.

11.54 For the CMOS NAND gate of Drill Exercise 11.20, identify the state of each transistor for $v_1 = v_2 = 5$ V.

11.55 Repeat Problem 11.54 for $v_1 = 5$ V and $v_2 = 0$ V.

11.56 Draw the schematic diagram of a two-input CMOS OR gate.

11.57 Draw the schematic diagram of a two-input CMOS AND gate.

11.58 Draw the schematic diagram of a two-input TTL OR gate.

11.59 Draw the schematic diagram of a two-input TTL AND gate.

Figure P11.12

CHAPTER
12

Digital Logic Circuits

Digital computers have taken a prominent place in engineering and science over the last two decades, performing a number of essential functions such as numerical computations and data acquisition. It is certainly not necessary to further stress the importance of these electronic systems in this book, since you have certainly already had some encounters with digital computers and programming languages. The objective of the chapter is to discuss the essential features of digital logic circuits, which are at the heart of digital computers, by presenting an introduction to *combinational* and *sequential logic circuits*.

The chapter starts with a discussion of the binary number system, and continues with an introduction to Boolean algebra. The self-contained treatment of Boolean algebra will enable you to design simple logic functions using the techniques of combinational logic, and several practical examples are provided to demonstrate that even simple combinations of logic gates can serve to implement useful circuits in engineering practice. In a later section, we introduce a number of logic modules which can be described using simple logic gates but which provide more advanced functions. Among these, we discuss flip-flops, read-only memories, multiplexers, and shift registers. The last section illustrates several design and application examples, to provide you with an idea of the capabilities afforded by simple digital logic circuits.

Chapter 12 provides the background needed to address the study of digital systems, which will be undertaken in Chapter 13. Upon completion of the chapter, you should be able to:

- Perform operations using the binary number system.
- Design simple combinational logic circuits using logic gates.
- Use Karnaugh maps to realize logical expressions.
- Interpret data sheets for multiplexers and EPROMs.
- Sketch the timing diagram for sequential circuits based on flip-flops.
- Describe the differences between different types of binary counters.
- Design simple logic circuits that include sequential elements.

12.1   ANALOG AND DIGITAL SIGNALS

One of the fundamental distinctions in the study of electronic circuits (and in the analysis of any signals derived from physical measurements) is that between analog and digital signals. As discussed in the preceding chapter, an **analog signal** is an electrical signal whose value varies in analogy with a physical quantity (e.g., temperature, force, or acceleration). For example, a voltage proportional to a measured variable pressure or to a vibration naturally varies in an analog fashion. Figure 12.1 depicts an arbitrary analog function of time, $f(t)$. We note immediately that for each value of time, t, $f(t)$ can take one value among any of the values in a given range. For example, in the case of the output voltage of an op-amp, we expect the signal to take any value between $+V_{sat}$ and $-V_{sat}$, where V_{sat} is the supply-imposed saturation voltage.

Figure 12.1 Analog signal

A **digital signal,** on the other hand, can take only a *finite number of values*. This is an extremely important distinction, as will be shown shortly. One example of a digital signal is the discrete approximation of an analog signal. Suppose, for example, that the output of a temperature sensor varies between 0 and 5 volts, with zero volts indicating a temperature of $0°$ and 10 volts corresponding, say, to $100°$. For a number of reasons (for example, if a computer is used to record the data), one might be interested only in detecting whether the measured temperature is within some range of temperatures—say, ΔT—rather than knowing exactly what the temperature is. If we divide the range of temperatures (or, equivalently, voltages) into 10 equal segments of 10° (or 0.5 V) each, we can then approximate the continuously varying function of time (either temperature or voltage) by a staircase function. Figure 12.2 illustrates the concept. Note that in Figure 12.2 it appears quite clearly that we need to make some decision regarding *when* we wish to make the approximation. Thus, timing becomes an important consideration in the representation of a digital signal.[1]

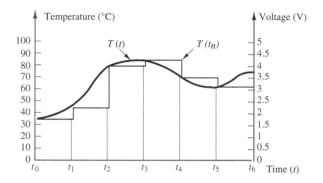

Figure 12.2 Discrete approximation of an analog signal

For the example of Figure 12.2, the "rules" employed in determining the approximation to the analog temperature, $T(t)$, are the following:

1. Measure the value of $T(t)$ at $t = t_n$, $n = 0, 1, 2, \ldots$; let this value be defined as $T(t_n)$.

2. Assume $T(t)$ to be constant until t_{n+1}; that is:

    $$T(t) = T(t_n) \qquad \text{for} \qquad t_n < t < t_{n+1}$$

3. At $t = t_{n+1}$, measure the temperature again, so that

    $$T(t) = T(t_{n+1}) \qquad \text{for} \qquad t_{n+1} < t < t_{n+2}$$

Note that we now have a set of time instants $\{t_n\}$ that is discrete (i.e., consists of a finite number of elements) and a set of voltages (analogous to temperatures) that is also discrete. In our example, the sets of times, t_n, voltages, V_m, and temperatures, T_m, are respectively given by the following expressions:

$$\{t_n\} = \{t_0, t_1, t_2, t_3, t_4, t_5, t_6\}$$
$$\{V_m\} = \{V_0, V_1, V_2, V_3, \ldots, V_{10}\} = \{0, 0.5, 1, 1.5, \ldots, 5\}$$
$$\{T_m\} = \{T_0, T_1, T_2, T_3, \ldots, T_{10}\} = \{0°, 10°, 20°, 30°, \ldots, 100°\}$$

The approximation shown in Figure 12.2 is clearly not very accurate. It should be apparent, however, that if the continuous function $T(t)$ were

---

[1] We are going to brush aside any timing issues in the introductory sections, where we deal exclusively with *combinational logic*, as will be explained soon. The issue of timing, and of *sequential logic*, will be addressed in the last section of this chapter, and in Chapter 13.

"sampled" at shorter intervals, and if the temperature (voltage) range were subdivided into a larger number of intervals (for example, $\Delta T = 1°$, corresponding to 0.05 V per increment), a more accurate *discrete approximation* to the continuous signal could be obtained. Why go through all this trouble? The fundamental reasons for the choice of a discrete signal representation are that discrete-valued signals are by their very nature compatible with digital computers, and that the discrete time representation is extremely useful, in that it allows the computer to perform necessary operations (e.g., mathematical calculations or storage in memory) between successive time samples, t_n and t_{n+1}. The discrete amplitude–discrete time representation is, in essence, the most natural representation for a digital computer. Most modern instrumentation employs a digital computer (or a microprocessor) to perform computations and manipulations, which might otherwise require exceedingly complex circuitry if they were implemented by analog devices.

We shall devote a section of the next chapter to the description of *analog-to-digital* and *digital-to-analog* conversion devices. An analog-to-digital converter is a device that enables a continuous measurement of a variable (such as temperature, pressure, or velocity) to be made compatible with the requirements of a digital computer, by converting it to a discrete approximation. A digital-to-analog converter converts the digital signals generated by a digital computer to a continuous voltage or current. In the remainder of this chapter, it will be tacitly assumed that all the relevant signals are already in digital form.

The most common digital signals are binary signals. A **binary signal** is a signal that can take only one of two discrete values and is therefore characterized by transitions between two states. Figure 12.3 displays a typical binary signal. In binary arithmetic (which we discuss in the next section), the two discrete values f_1 and f_0 are represented by the numbers 1 and 0. In binary voltage waveforms, these values are represented by two voltage levels. For example, in the TTL convention (see Chapter 11), these values are (nominally) 5 V and 0 V, respectively; while in CMOS circuits, these values can vary substantially. Other conventions are also used, including reversing the assignment—for example, by letting a 0-V level represent a logic 1 and a 5-V level represent a logic 0. Note that in a binary waveform, knowledge of the transition between one state and another (e.g., from f_0 to f_1

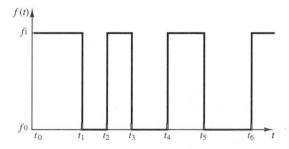

Figure 12.3 A binary signal

at $t = t_2$) is equivalent to knowledge of the state. Thus, digital logic circuits can operate by detecting transitions between voltage levels. The transitions are often called **edges,** and can be positive (f_0 to f_1) or negative (f_1 to f_0). Virtually all of the signals handled by a computer are binary. From here on, whenever we speak of digital signals, you may assume that the text is referring to signals of the binary type, unless otherwise indicated. The issue of how the discrete values in the example of Figure 12.3 are converted to binary signals will be approached in Chapter 13.

12.2   THE BINARY NUMBER SYSTEM

The binary number system is a natural choice for representing the behavior of circuits that operate in one of two states (on or off, 1 or 0, or the like). The diode and transistor gates and switches studied in Chapter 11 fall in this category. Table 12.1 shows the correspondence between decimal and binary number systems for decimal numbers up to 16.

Table 12.1   Conversion from decimal to binary

Decimal number, n_{10}	Binary number, n_2
0	0
1	1
2	10
3	11
4	100
5	101
6	110
7	111
8	1000
9	1001
10	1010
11	1011
12	1100
13	1101
14	1110
15	1111
16	10000

Binary numbers are based on powers of 2, whereas the decimal system is based on powers of 10. For example, the number 372 in the decimal system can be expressed as

$$372 = (3 \times 10^2) + (7 \times 10^1) + (2 \times 10^0)$$

while the binary number 10110 corresponds to the following combination of powers of 2:

$$10110 = (1 \times 2^4) + (0 \times 2^3) + (1 \times 2^2) + (1 \times 2^1) + (0 \times 2^0)$$

It is relatively simple to see the correspondence between the two number systems if we add the terms on the right-hand side of the previous expression. Let n_2 represent the number n **base 2** (i.e., in the binary system) and n_{10} the same number **base 10.** Then, our notation will be as follows:

$$10110_2 = 16 + 0 + 4 + 2 + 0 = 22_{10}$$

Note that a fractional number can also be similarly represented. For example, the number 3.25 in the decimal system may be represented as

$$3.25_{10} = 3 \times 10^0 + 2 \times 10^{-1} + 5 \times 10^{-2}$$

while in the binary system the number 10.011 corresponds to

$$10.011_2 = 1 \times 2^1 + 0 \times 2^0 + 0 \times 2^{-1} + 1 \times 2^{-2} + 1 \times 2^{-3}$$

$$= 2 + 0 + 0 + \frac{1}{4} + \frac{1}{8} = 2.375_{10}$$

Table 12.1 shows that it takes four binary digits, also called **bits,** to represent the decimal numbers up to 15. Usually, the rightmost bit is called the **least significant bit,** or **LSB,** and the leftmost bit is called the **most significant bit,** or **MSB.** Since binary numbers clearly require a larger number of digits than decimal numbers, the digits are usually grouped in sets of four, eight, or sixteen. Four bits are usually termed a **nibble,** eight bits are called a **byte,** and sixteen bits (or two bytes) form a **word.**

Addition and Subtraction

The operations of addition and subtraction are based on the simple rules shown in Table 12.2. Note that, just as is done in the decimal system, a carry is generated whenever the sum of two digits exceeds the largest single-digit number in the given number system, which is 1 in the binary system. The carry is treated exactly as in the decimal system. A few examples of binary addition are shown in Figure 12.4, with their decimal counterparts.

Table 12.2   Rules for addition

$0 + 0$	$= 0$
$0 + 1$	$= 1$
$1 + 0$	$= 1$
$1 + 1$	$= 0$ (with a carry of 1)

Decimal	Binary	Decimal	Binary	Decimal	Binary
5	101	15	1111	3.25	11.01
+6	+110	+20	+10100	+5.75	+101.11
11	1011	35	100011	9.00	1001.00

(Note that in this example, $3.25 = 3\frac{1}{4}$ and $5.75 = 5\frac{3}{4}$.)

Figure 12.4 Examples of binary addition

The procedure for subtracting binary numbers is based on the rules of Table 12.3. A few examples of binary subtraction are given in Figure 12.5, with their decimal counterparts.

Table 12.3   Rules for subtraction

$0 - 0$	$= 0$
$1 - 0$	$= 1$
$1 - 1$	$= 0$
$0 - 1$	$= 1$ (with a borrow of 1)

Decimal	Binary	Decimal	Binary	Decimal	Binary
9	1001	16	10000	6.25	110.01
-5	-101	-3	-11	-4.50	-100.10
4	0100	13	01101	1.75	001.11

Figure 12.5 Examples of binary subtraction

Multiplication and Division

Table 12.4   Rules for multiplication

$0 \times 0 = 0$
$0 \times 1 = 0$
$1 \times 0 = 0$
$1 \times 1 = 1$

Whereas in the decimal system the multiplication table consists of $10^2 = 100$ entries, in the binary system we only have $2^2 = 4$ entries. Table 12.4 represents the complete multiplication table for the binary number system.

Division in the binary system is also based on rules analogous to those of the decimal system, with the two basic laws given in Table 12.5. Once again, we need be concerned with only two cases, and just as in the decimal system, division by zero is not contemplated.

Table 12.5   Rules for division

$0 \div 1 = 0$
$1 \div 1 = 1$

Conversion from Decimal to Binary

The conversion of a decimal number to its binary equivalent is performed by successive division of the decimal number by 2, checking for the remainder each time. Figure 12.6 illustrates this idea with an example. The result

	Remainder
$37 \div 2 = 18 + 1$	
$18 \div 2 = 9 + 0$	
$9 \div 2 = 4 + 1$	
$4 \div 2 = 2 + 0$	
$2 \div 2 = 1 + 0$	
$1 \div 2 = 0 + 1$	
$37_{10} = 100101_2$	

	Remainder
$49 \div 2 = 24 + 1$	
$24 \div 2 = 12 + 0$	
$12 \div 2 = 6 + 0$	
$6 \div 2 = 3 + 0$	
$3 \div 2 = 1 + 1$	
$1 \div 2 = 0 + 1$	
$49_{10} = 110001_2$	

Figure 12.6 Example of conversion from decimal to binary

$2 \times 0.53 = 1.06 \rightarrow 1$
$2 \times 0.06 = 0.12 \rightarrow 0$
$2 \times 0.12 = 0.24 \rightarrow 0$
$2 \times 0.24 = 0.48 \rightarrow 0$
$2 \times 0.48 = 0.96 \rightarrow 0$
$2 \times 0.96 = 1.92 \rightarrow 1$
$2 \times 0.92 = 1.84 \rightarrow 1$
$2 \times 0.84 = 1.68 \rightarrow 1$
$2 \times 0.68 = 1.36 \rightarrow 1$
$2 \times 0.36 = 0.72 \rightarrow 0$
$2 \times 0.72 = 1.44 \rightarrow 1$

$0.53_{10} = 0.10000111101$

Figure 12.7 Conversion from decimal to binary

obtained in Figure 12.6 may be easily verified by performing the opposite conversion, from binary to decimal:

$$11001 = 2^5 + 2^4 + 2^0 = 32 + 16 + 1 = 49$$

The same technique can be used for converting decimal fractional numbers to their binary form, provided that the whole number is separated from the fractional part and each is converted to binary form (separately), with the results added at the end. Figure 12.7 outlines this procedure by converting the number 37.53 to binary form. The procedure is outlined in two steps. First, the integer part is converted; then, to convert the fractional part, one simple technique consists of multiplying the decimal fraction by 2 in successive stages. If the result exceeds 1, a 1 is needed to the right of the binary fraction being formed (100101 ..., in our example). Otherwise, a 0 is added. This procedure is continued until no fractional terms are left. In this case, the decimal part is 0.53_{10}, and Figure 12.7 illustrates the succession of calculations. Stopping the procedure outlined in Figure 12.7 after 11 digits results in the following approximation:

$$37.53_{10} = 100101.10000111101$$

Greater precision could be attained by continuing to add binary digits, at the expense of added complexity.

Complements and Negative Numbers

To simplify the operation of subtraction in digital computers, **complements** are used almost exclusively. In practice, this corresponds to replacing the operation $X - Y$ with the operation $X + (-Y)$. This procedure results in considerable simplification, since the computer hardware need include only adding circuitry. Two types of complements are used with binary numbers: the **one's complement** and the **two's complement.**

The one's complement of an n-bit binary number is obtained by subtracting the number itself from $(2^n - 1)$. Two examples are as follows:

$$a = 0101$$
$$\text{One's complement of } a = (2^4 - 1) - a$$
$$= (1111) - (0101)$$
$$= 1010$$

$$b = 101101$$
$$\text{One's complement of } b = (2^6 - 1) - b$$
$$= (111111) - (101101)$$
$$= 010010$$

The two's complement of an n-bit binary number is obtained by subtracting the number itself from 2^n. Two's complements of the same numbers a and b used in the preceding illustration are computed as follows:

$$a = 0101$$
$$\text{Two's complement of } a = 2^4 - a$$
$$= (10000) - (0101)$$
$$= 0011$$

$$b = 101101$$
$$\text{Two's complement of } b = 2^6 - b$$
$$= (1000000) - (101101)$$
$$= 010011$$

A simple rule that may be used to obtain the two's complement directly from a binary number is the following: starting at least significant (right-most) bit, copy each bit *until the first 1 has been copied,* and then replace each successive 1 by a 0 and each 0 by a 1. You may wish to try this rule on the two previous examples to verify that it is much easier to use than the subtraction from 2^n.

Different conventions exist in the binary system to represent whether a number is negative or positive. One convention, called the **sign-magnitude convention,** makes use of a *sign bit,* usually positioned at the beginning of the number, for which a value of 1 represents a minus sign and a value of 0, a plus sign. Thus, an eight-bit binary number would consist of a sign bit followed by seven *magnitude bits,* as shown in Figure 12.8(a). In a digital system that uses eight-bit signed integer words, we could represent integer numbers (decimal) in the range

$$-(2^7 - 1) \leq N \leq +(2^7 - 1)$$

or

$$-127 \leq N \leq +127$$

This is clearly an improvement over the example of Figure 12.2, where the temperature signal was represented by just 10 possible values. (How many bits would it take to represent the signal of Figure 12.2?)

A second convention uses the one's complement notation. In this convention, a sign bit is also used to indicate whether the number is positive (sign bit $= 0$) or negative (sign bit $= 1$). However, the magnitude of the binary number is represented by the true magnitude if the number is positive, and by its *one's complement* if the number is negative. Figure 12.8(b) illustrates the convention. For example, the number $(91)_{10}$ would be represented by the seven-bit binary number $(1011011)_2$ with a leading 0 (the sign bit): $(\mathbf{0}1011011)_2$. On the other hand, the number $(-91)_{10}$ would be represented by the seven-bit one's complement binary number $(100100)_2$ with a leading 1 (the sign bit): $(\mathbf{1}100100)_2$.

Most digital computers use the two's complement convention in performing integer arithmetic operations. The two's complement convention represents positive numbers by a sign bit of 0, followed by the true binary magnitude; negative numbers are represented by a sign bit of 1, *followed by the two's complement of the binary number,* as shown in Figure 12.8(c). The advantage of the two's complement convention is that the algebraic sum of two's complement binary numbers is carried out very simply by adding the two numbers *including the sign bit.* Example 12.1 illustrates two's complement addition.

Sign bit b_7	b_6	b_5	b_4	b_3	b_2	b_1	b_0
	$\leftarrow$ Actual magnitude of binary number $\rightarrow$						

Figure 12.8(a) Eight-bit sign-magnitude binary number

Sign bit b_7	b_6	b_5	b_4	b_3	b_2	b_1	b_0
	$\leftarrow$ Actual magnitude of binary number (if $b_7 = 0$) $\rightarrow$						
	$\leftarrow$ One's complement of binary number (if $b_7 = 1$) $\rightarrow$						

Figure 12.8(b) Eight-bit one's complement binary number

Sign bit b_7	b_6	b_5	b_4	b_3	b_2	b_1	b_0
	$\leftarrow$ Actual magnitude of binary number (if $b_7 = 0$) $\rightarrow$						
	$\leftarrow$ Two's complement of binary number (if $b_7 = 1$) $\rightarrow$						

Figure 12.8(c) Eight-bit two's complement binary number

Example   12.1   Two's Complement Operations

Perform the following subtractions in the binary system using two's complements:

1. $1011100 - 1110010$
2. $10101111 - 01110011$

Solution:

1.   As discussed in the preceding section, the operation $X - Y$ where $X = 1011100$ and $Y = 1110010$ can be replaced with the operation $X + (-Y)$. The next step is to find the two's complement of Y and add the result to X. Therefore,

$$1011100 - 1110010 = 1011100 + (2^7 - 1110010) = 1011100 + 0001110$$
$$= 1101010$$

2.   For this part, $X = 10101111$ and $Y = 01110011$. Using the same procedure as in part 1, $10101111 - 01110011 = 10101111 + (2^8 - 01110011) = 10101111 + 10001101 = 00111100$.

The Hexadecimal System

It should be apparent by now that representing numbers in base 2 and base 10 systems is purely a matter of convenience, given a specific application. Another base frequently used is the **hexadecimal system,** a direct derivation of the binary number system. In the hexadecimal (or hex) code, the bits in a binary number are subdivided into groups of four. Since there are 16 possible combinations for a four-bit number, the natural digits in the decimal system (0 through 9) are insufficient to represent a hex digit. To solve this problem, the first six letters of the alphabet are used, as shown in Table

Table 12.6 Hexadecimal code

0	0000
1	0001
2	0010
3	0011
4	0100
5	0101
6	0110
7	0111
8	1000
9	1001
A	1010
B	1011
C	1100
D	1101
E	1110
F	1111

12.6. Thus, in hex code, an eight-bit word corresponds to just two digits; for example:

$$1010\ 0111_2 = A7_{16}$$
$$0010\ 1001_2 = 29_{16}$$

Binary Codes

In this subsection, we describe two common binary codes that are often used for practical reasons. The first is a method of representing decimal numbers in digital logic circuits that is referred to as **binary-coded decimal,** or **BCD,** representation. In effect, the simplest BCD representation is just a sequence of four-bit binary numbers that stops after the first 10 entries, as shown in Table 12.7. There are also other BCD codes, all reflecting the same principle: that each decimal digit is represented by a fixed-length binary word. One should realize that although this method is attractive because of its direct correspondence with the decimal system, it is not efficient. Consider, for example, the decimal number 68. Its binary representation by direct conversion is the seven-bit number 1001000. On the other hand, the corresponding BCD representation would require eight bits:

$$68_{10} = 01101000_{BCD}$$

Another code that finds many applications is the **Gray code.** This is simply a reshuffling of the binary code with the property that any two consecutive numbers differ only by one bit. Table 12.8 illustrates the three-bit Gray code. The Gray code can be very useful in practical applications, because in counting up or down according to this code, the binary representation of a number changes only one bit at a time. The next example illustrates an application of the Gray code to a practical engineering problem.

Table 12.7 BCD code

0	0000
1	0001
2	0010
3	0011
4	0100
5	0101
6	0110
7	0111
8	1000
9	1001

Table 12.8 Three-bit Gray code

Binary	Gray
000	000
001	001
010	011
011	010
100	110
101	111
110	101
111	100

Example   12.2   Position Encoders

Position encoders are devices that output a digital signal proportional to their (linear or angular) position. These devices are very useful in measuring instantaneous position in *motion control* applications. Motion control is a technique that is used when it is necessary to accurately control the motion of a moving object; examples are found in robotics, machine tools, and servomechanisms. For example, in positioning the arm of a robot to pick up an object, it is very important to know its exact position at all times. Since one is usually interested in both rotational and translational motion, two types of encoders are discussed in this example: *linear* and *angular* position encoders.

An optical position encoder consists of an *encoder pad,* which is either a strip (for translational motion) or a disk (for rotational motion) with alternating black and white areas. These areas are arranged to reproduce some binary code, as shown in Figure 12.9, where both the conventional binary and Gray codes are depicted for a four-bit linear encoder pad. A fixed array of photodiodes (see Chapter 7) senses the reflected light from each of the cells across a row of the encoder path; depending on the amount of light reflected, each photodiode circuit will output a voltage corresponding to a binary 1 or 0. Thus, a different four-bit word is generated for each row of the encoder.

Suppose the encoder pad is 100 mm in length. Then its resolution can be computed as follows. The pad will be divided into $2^4 = 16$ segments, and each segment corresponds to an increment of $\frac{100}{16}$ mm $= 6.25$ mm. If greater resolution were necessary, more bits could be employed: an eight-bit pad of the same length would attain a resolution of $\frac{100}{256}$ mm $= 0.39$ mm.

A similar construction can be employed for the five-bit angular encoder of Figure 12.10. In this case, the angular resolution can be expressed in degrees of rotation, where $2^5 = 32$ sections correspond to $360°$. Thus, the resolution would be $360°/32 = 12.25°$. Once again, greater angular resolution could be obtained by employing a larger number of bits.

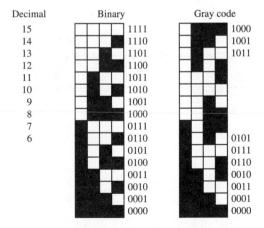

Figure 12.9 Binary and Gray code patterns for linear position encoders

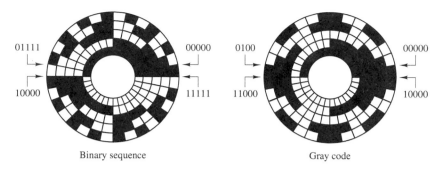

01111   00000   0100   00000

10000   11111   11000   10000

Binary sequence   Gray code

Figure 12.10 Binary and Gray code patterns for angular position encoders

Example   12.3

Convert the following binary numbers to hexadecimal numbers:

1.   100111
2.   1011101
3.   11001101

Solution:

The simplest method of conversion from a binary-base number to a hexadecimal-base number consists of grouping the binary number in subgroups of four starting from the right; that is, (1) 0010, 0111; (2) 0101, 1101; (3) 1100, 1101. Then, replacing each subgroup by its equivalent number in hexadecimal base:

1. 27_{16}
2. $5D_{16}$
3. CD_{16}

[*Note:* If you want to convert a hex number to a binary number, you need only replace each digit of the hex number by its equivalent four-bit binary number.]

Example   12.4

Convert the following binary numbers to hexadecimal numbers:

1.   101101111001
2.   100110110
3.   1101011011

Solution:

Replace each subgroup by its equivalent number in hexadecimal base:

1. $B79_{16}$

2. 136_{16}

3. $35B_{16}$

The hexadecimal base is widely used in digital computing because of its efficient representation of long strings of binary numbers.

DRILL EXERCISES

12.1   Convert the following decimal numbers to binary form:

a.   39

b.   59

c.   512

d.   0.4475

e. $\frac{25}{32}$

f.   0.796875

g.   256.75

h.   129.5625

i.   4,096.90625

Answer: (a) 100111; (b) 111011; (c) 1000000000;
(d) 0.011100; (e) 0.11001; (f) 0.110011;
(g) 100000000.11; (h) 10000001.1001; (i) 1000000000000.11101

12.2   Convert the following binary numbers to decimal:

a.   1101

b.   11011

c.   10111

d.   0.1011

e.   0.001101

f.   0.001101101

g.   111011.1011

h.   1011011.001101

i.   10110.0101011101

Answer: (a) 13; (b) 27; (c) 23; (d) 0.6875; (e) 0.203125;
(f) 0.21288065; (g) 59.6875; (h) 91.203125; (i) 22.340823125

12.3   Perform the following additions and subtractions. Express the answer in decimal form for problems (a)–(d) and in binary form for problems (e)–(h).

a. $1001.1_2 + 1011.01_2$

b. $100101_2 + 100101_2$

c. $0.1011_2 + 0.1101_2$

d. $1011.01_2 + 1001.11_2$

e. $64_{10} - 32_{10}$

f. $127_{10} - 63_{10}$

g. $93.5_{10} - 42.75_{10}$

h. $(84\frac{9}{32})_{10} - (48\frac{5}{16})_{10}$

Answer: (a) 20.75_{10}; (b) 74_{10}; (c) 1.5_{10}; (d) 21_{10};
(e) 100000_2; (f) 1000000_2; (g) 110010.11_2; (h) 100011.11111_2

12.4   How many possible numbers can be represented in a 12-bit word?

Answer: 4,096

12.5 If we use an eight-bit word with a sign bit (seven magnitude bits plus one sign bit) to represent voltages -5 V and $+5$ V, what is the smallest increment of voltage that can be represented?

Answer: 39 mV

12.6 Convert the following numbers from hex to binary or from binary to hex:

a. F83
b. 3C9
c. A6
d. 11010110_2
e. 10111001_2
f. 11011101101_2

Answer: (a) 111110000011; (b) 1111001001;
        (c) 10100110; (d) 1AE; (e) B9; (f) 6ED

12.7 Find the two's complement of the following binary numbers:

a. 11101001
b. 10010111
c. 1011110

Answer: (a) 00010111; (b) 01101001; (c) 10100010

12.8 Convert the following numbers from hex to binary, and find their two's complements:

a. F43
b. 2B9
c. A6

Answer: (a) 10111101; (b) 110101000111; (c) 1011010

12.3   BOOLEAN ALGEBRA

The mathematics associated with the binary number system (and with the more general field of logic) is called *Boolean,* in honor of the English mathematician George Boole, who published a treatise in 1854 entitled *An Investigation of the Laws of Thought, on Which Are Founded the Mathematical Theories of Logic and Probabilities.* The development of a *logical algebra,* as Boole called it, is one of the results of his investigations. The variables in a Boolean, or logic, expression can take only one of two values, usually represented by the numbers 0 and 1. These variables are sometimes referred to as true (1) and false (0).

AND and OR Gates

The basis of **Boolean algebra** lies in the operations of **logical addition,** or the **OR** operation, and **logical multiplication,** or the **AND** operation. Both of these find a correspondence in simple logic gates, as we shall presently illustrate. Logical addition, although represented by the symbol $+$, differs from conventional algebraic addition, as shown in the last rule listed in Table

OR gate

X	Y	Z
0	0	0
0	1	1
1	0	1
1	1	1

Truth table

Figure 12.11 Logical addition and the OR gate

Table 12.9   Rules of logical addition (OR)

$0 + 0 = 0$
$0 + 1 = 1$
$1 + 0 = 1$
$1 + 1 = 1$

12.9. Note that this rule differs from the last rule of binary addition studied in the previous section. Logical addition can be represented by the logic gate called an **OR gate,** whose symbol and whose inputs and outputs are shown in Figure 12.11. The OR gate represents the following logical statement:

If either X or Y is true (1), then Z is true (1).   **(12.1)**

This rule is embodied in the electronic gates discussed in Chapter 11, in which a logic 1 corresponds, say, to a 5-V signal and a logic 0 to a 0-V signal.

Logical multiplication is denoted by the center dot ($\cdot$), and is defined by the rules of Table 12.10. Figure 12.12 depicts the **AND gate,** which

AND gate

X	Y	Z
0	0	0
0	1	0
1	0	0
1	1	1

Truth table

Figure 12.12 Logical multiplication and the AND gate

Table 12.10   Rules of logic multiplication

$0 \cdot 0 = 0$
$0 \cdot 1 = 0$
$1 \cdot 0 = 0$
$1 \cdot 1 = 1$

corresponds to this operation. The AND gate corresponds to the following logical statement:

If both X *and* Y are true (1), then Z is true (1).   **(12.2)**

One can easily envision logic gates (AND and OR) with an arbitrary number of inputs; three- and four-input gates are not uncommon.

The rules that define a logic function are often represented in tabular form by means of a **truth table.** Truth tables for the AND and OR gates are shown in Figures 12.11 and 12.12. A truth table is nothing more than a tabular summary of all of the possible outputs of a logic gate, given all the possible input values. If the number of inputs is 3, the number of possible combinations grows from 4 to 8, but the basic idea is unchanged. Truth tables are very useful in defining logic functions. A typical logic design problem might specify requirements such as "the output Z shall be logic 1 only when

Logic gate realization of the statement "the output Z shall be logic 1 only when the condition ($X = 1$ AND $Y = 1$) OR ($W = 1$) occurs, and shall be logic 0 otherwise."

X	Y	W	Z
0	0	0	0
0	0	1	1
0	1	0	0
0	1	1	1
1	0	0	0
1	0	1	1
1	1	0	1
1	1	1	1

Truth table

Solution using logic gates

Figure 12.13 Example of logic function implementation with logic gates

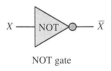

NOT gate

X	$\overline{X}$
1	0
0	1

Truth table for NOT gate

Figure 12.14 Complements and the NOT gate

X	Y	$(X+Y)$	$X \cdot (X+Y)$
0	0	0	0
0	1	1	0
1	0	1	1
1	1	1	1

Figure 12.16 Proof of rule 16 by perfect induction

the condition ($X = 1$ AND $Y = 1$) OR ($W = 1$) occurs, and shall be logic 0 otherwise." The truth table for this particular logic function is shown in Figure 12.13 as an illustration. The design consists, then, of determining the combination of logic gates that exactly implements the required logic function. Truth tables can greatly simplify this procedure.

The AND and OR gates form the basis of all logic design in conjunction with the **NOT gate.** The NOT gate is essentially an inverter (which can be constructed using bipolar or field-effect transistors, as discussed in Chapter 11), and it provides the complement of the logic variable connected to its input. The complement of a logic variable X is denoted by $\overline{X}$. The NOT gate has only one input, as shown in Figure 12.14.

To illustrate the use of the NOT gate, or inverter, we return to the design example of Figure 12.13, where we required that the output of a logic circuit be $Z = 1$ only if $X = 0$ AND $Y = 1$ OR if $W = 1$. We recognize that except for the requirement $X = 0$, this problem would be identical if we stated it as follows: "The output Z shall be logic 1 only when the condition ($\overline{X} = 1$ AND $Y = 1$) OR ($W = 1$) occurs, and shall be logic 0 otherwise." If we use an inverter to convert X to $\overline{X}$, we see that the required condition becomes ($\overline{X} = 1$ AND $Y = 1$) OR ($W = 1$). The formal solution to this elementary design exercise is illustrated in Figure 12.15.

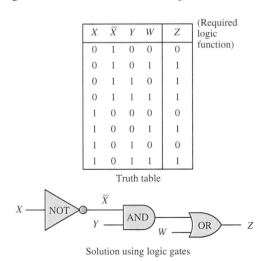

X	$\overline{X}$	Y	W	Z	(Required logic function)
0	1	0	0	0	
0	1	0	1	1	
0	1	1	0	1	
0	1	1	1	1	
1	0	0	0	0	
1	0	0	1	1	
1	0	1	0	0	
1	0	1	1	1	

Truth table

Solution using logic gates

Figure 12.15 Solution of a logic problem using logic gates

In the course of the discussion of logic gates, extensive use will be made of truth tables to evaluate logic expressions. A set of basic rules will facilitate this task. Table 12.11 lists some of the rules of Boolean algebra; each of these can be proven by using a truth table, as will be shown in examples and exercises. An example proof for rule 16 is given in Figure 12.16 in the form of a truth table. This technique can be employed to prove any of the laws of Table 12.11. From the simple truth table in Figure 12.16, which was obtained step by step, we can clearly see that indeed $X(X + Y) =$

X. This methodology for proving the validity of logical equations is called **proof by perfect induction.** The 19 rules of Table 12.11 can be used to simplify logic expressions.

Table 12.11   Rules of Boolean algebra

1. $0 + X = X$	
2. $1 + X = 1$	
3. $X + X = X$	
4. $X + \bar{X} = 1$	
5. $0 \cdot X = 0$	
6. $1 \cdot X = X$	
7. $X \cdot X = X$	
8. $X \cdot \bar{X} = 0$	
9. $\bar{\bar{X}} = X$	
10. $X + Y = Y + X$	Commutative law
11. $X \cdot Y = Y \cdot X$	
12. $X + (Y + Z) = (X + Y) + Z$	Associative law
13. $X \cdot (Y \cdot Z) = (X \cdot Y) \cdot Z$	
14. $X \cdot (Y + Z) = X \cdot Y + X \cdot Z$	Distributive law
15. $X + X \cdot Z = X$	Idempotent rule
16. $X \cdot (X + Y) = X$	
17. $(X + Y) \cdot (X + Z) = X + Y \cdot Z$	
18. $X + \bar{X} \cdot Y = X + Y$	
19. $X \cdot Y + Y \cdot Z + \bar{X} \cdot Z = X \cdot Y + \bar{X} \cdot Z$	

To complete the introductory material on Boolean algebra, which will enable you to design useful logic functions using gates, a few paragraphs need to be devoted to two very important theorems, called **De Morgan's theorems.** These are stated here in the form of logic functions:

$$\overline{(X + Y)} = \bar{X} \cdot \bar{Y} \qquad\qquad\qquad (12.3)$$

$$\overline{(X \cdot Y)} = \bar{X} + \bar{Y} \qquad\qquad\qquad (12.4)$$

These two laws state a very important property of logic functions:

> Any logic function can be implemented using only OR and NOT gates, or using only AND and NOT gates.

De Morgan's laws can easily be visualized in terms of logic gates, as shown in Figure 12.17.

The importance of De Morgan's laws is in the statement of the **duality** that exists between AND and OR operations: any function can be realized by just one of the two basic operations, plus the complement operation. This gives rise to two families of logic functions: **sums of products** and **products**

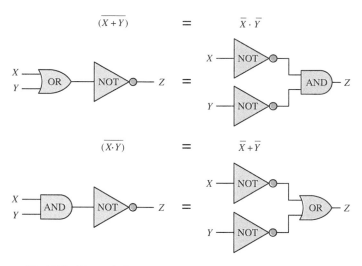

Figure 12.17 De Morgan's laws

of sums, as shown in Figure 12.18. Any logical expression can be reduced to either one of these two forms. Although the two forms are equivalent, it may well be true that one of the two has a simpler implementation (fewer gates). The following examples illustrate this point.

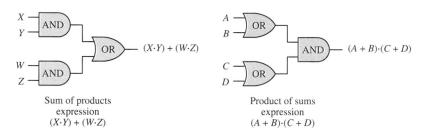

Figure 12.18 Sum-of-products and product-of-sums logic functions

Example   12.5

Using the rules of Boolean algebra, simplify the following logic function:

$$f(A,B,C,D) = \bar{A} \cdot \bar{B} \cdot D + \bar{A} \cdot B \cdot D + B \cdot C \cdot D + A \cdot C \cdot D$$

Solution:

$$f = \bar{A} \cdot \bar{B} \cdot D + \bar{A} \cdot B \cdot D + B \cdot C \cdot D + A \cdot C \cdot D$$
$$f = \bar{A} \cdot D \cdot (B + \bar{B}) + B \cdot C \cdot D + A \cdot C \cdot D \qquad \text{Rule 14}$$
$$f = \bar{A} \cdot D + B \cdot C \cdot D + A \cdot C \cdot D \qquad \text{Rule 4}$$
$$f = (\bar{A} + A \cdot C) \cdot D + B \cdot C \cdot D \qquad \text{Rule 14}$$
$$f = (\bar{A} + C) \cdot D + B \cdot C \cdot D \qquad \text{Rules 14 and 18}$$
$$f = \bar{A} \cdot D + C \cdot D + B \cdot C \cdot D \qquad \text{Rule 14}$$

$$f = \bar{A} \cdot D + C \cdot D \cdot (1 + B) \qquad\qquad \text{Rule 14}$$
$$f = \bar{A} \cdot D + C \cdot D \qquad\qquad\qquad \text{Rules 2 and 6}$$

Example   12.6

Use De Morgan's theorems to simplify the function

$$f(A,B,C,D) = \overline{[\bar{A} + B \cdot (\bar{C} + D)]}$$

Solution:

The point of this example is to show that the complexity of a circuit can be reduced by De Morgan's theorems (equations 12.3 and 12.4). It can be seen that the function f is in the form $\overline{X + Y}$, where $X = \bar{A}$ and $Y = B \cdot (\bar{C} + D)$. Therefore, using De Morgan's theorems, we have

$$f = \bar{\bar{A}} \cdot \overline{[B \cdot (\bar{C} + D)]} = A \cdot \overline{[B \cdot (\bar{C} + D)]}$$

It can now be seen that the function f is in the form $A \cdot \overline{(X \cdot Y)}$, where $X = B$ and $Y = (\bar{C} + D)$. Therefore, using De Morgan's theorems:

$$f = A \cdot [\bar{B} + \overline{(\bar{C} + D)}]$$

Applying De Morgan's theorems to $\overline{(\bar{C} + D)}$, we can simplify the function f to:

$$f = A \cdot [\bar{B} + \bar{\bar{C}} \cdot \bar{D}]$$

which can be simplified with the aid of rules 9 and 14 to

$$f = A \cdot \bar{B} + A \cdot C \cdot \bar{D}$$

Looking at the simplified function obtained using De Morgan's theorems, we can see that fewer gates are needed to realize the simplified function.

Example   12.7

Realize the function described by the following truth table:

A	B	C	y
0	0	0	0
0	0	1	1
0	1	0	0
0	1	1	1
1	0	0	1
1	0	1	1
1	1	0	1
1	1	1	1

Solution:

We may find a very simple solution to this problem by noting that the only occurrence of 0s in the output is when A and C are both 0, and that therefore the output y is 1 whenever either A and C or both are 1. But this is, in effect, the description of a two-input OR gate! Thus, the variable B is not necessary to realize the desired function. The solution is shown in Figure 12.19. Please note that in this problem, the function to be realized was simple, and the solution could be obtained by inspection. In general, however, formal techniques will be required to obtain the solution for more complex functions. You may wish to verify that in this example the logic function y is

$$y = \bar{A} \cdot \bar{B} \cdot C + \bar{A} \cdot B \cdot C + A \cdot \bar{B} \cdot \bar{C} + A \cdot \bar{B} \cdot C + A \cdot B \cdot \bar{C} + A \cdot B \cdot C$$

and that the expression can be simplified to $y = A + C$ by making use of the rules of Table 12.11. In the next sections, we shall develop formal techniques for the analysis and realization of more complex logic functions.

$A + C = y$ or

Figure 12.19

Example 12.8

Use De Morgan's theorem to realize the function $y = A + (B \cdot C)$ as a product-of-sums expression, and implement it using AND, OR, and NOT gates.

Solution:

Knowing that $\bar{\bar{y}} = y$, we can apply the first of De Morgan's laws to the complement of the function y to obtain the expression

$$\bar{y} = \overline{A + (B \cdot C)} = \bar{A} \cdot \overline{(B \cdot C)} = \bar{A} \cdot (\bar{B} + \bar{C})$$

Thus,

$$\bar{\bar{y}} = y = \overline{\bar{A} \cdot (\bar{B} + \bar{C})}$$

Using logic gates, we can then implement the function as shown in Figure 12.20.

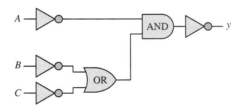

Figure 12.20

NAND and NOR Gates

In addition to the AND and OR gates we have just analyzed, the complementary forms of these gates, called NAND and NOR, are very commonly used in practice. In fact, NAND and NOR gates form the basis of most practical logic circuits. Figure 12.21 depicts these two gates, and illustrates how they can be easily interpreted in terms of AND, OR, and NOT gates by virtue of De Morgan's laws. You can readily verify that the logic function implemented by the NAND and NOR gates corresponds, respectively, to AND and OR gates followed by an inverter. It is very important to note that, by De Morgan's laws, the NAND gate performs a *logical addition* on the *complements* of the inputs, while the NOR gate performs a *logical multiplication* on the *complements* of the inputs. Functionally, then, any logic function could be implemented with either NOR or NAND gates only.

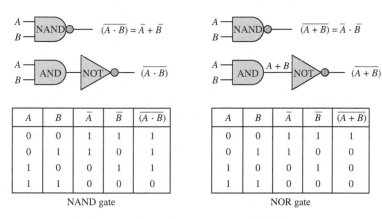

A	B	$\bar{A}$	$\bar{B}$	$\overline{(A \cdot B)}$
0	0	1	1	1
0	1	1	0	1
1	0	0	1	1
1	1	0	0	0

NAND gate

A	B	$\bar{A}$	$\bar{B}$	$\overline{(A + B)}$
0	0	1	1	1
0	1	1	0	0
1	0	0	1	0
1	1	0	0	0

NOR gate

Figure 12.21 Equivalence of NAND and NOR gates with AND and OR gates

In the next section we shall learn how to systematically approach the design of logic functions. First, we provide a few examples to illustrate logic design with NAND and NOR gates.

Example   12.9

Show that an AND gate can be obtained as a combination of NAND gates.

Solution:

Let us show this for two arbitrary inputs, A and B. The desired truth table is as follows:

		NAND	**AND**
A	B	$\overline{A \cdot B}$	$A \cdot B$
0	0	1	0
0	1	1	0
1	0	1	0
1	1	0	1

A	$B(=A)$	$A \cdot B$	$\overline{(A \cdot B)}$
0	0	0	1
1	1	1	0

Figure 12.22 NAND gate as an inverter

Looking at the truth table, we can see that we need only invert the output of the NAND gate. This is easily accomplished if we realize that a NAND gate can serve as an inverter when all of its inputs are tied together, as illustrated in Figure 12.22.

The desired logic circuit, which is functionally equivalent to an AND gate, is then shown in Figure 12.23.

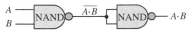

Figure 12.23

Example   12.10

Show that one can obtain an AND gate using a combination of NOR gates.

Solution:

A	$B(=A)$	$(A+B)$	$\overline{(A+B)}$
0	0	0	1
1	1	1	0

Figure 12.24 NOR gate as an inverter

The simplest way of solving this problem is by using De Morgan's theorems. We want to build an AND gate using only NOR gates. The output of an AND gate, f, for two arbitrary inputs A and B is $f = A \cdot B$. Knowing that $\overline{\overline{f}} = f$, we use De Morgan's law in the form $f = \overline{\overline{A} + \overline{B}}$ to obtain a function that can be implemented using a NOR gate with inputs $\overline{A}$ and $\overline{B}$. Also, we observe that a NOR gate can act as an inverter when its inputs are tied together. Figure 12.24 illustrates the behavior of a NOR gate as an inverter.

Next, we recall the statement made earlier in this section suggesting that a NOR gate performs a logical multiplication on the complements of its inputs. Upon a moment's reflection, we see that the logic circuit of Figure 12.25 provides the answer to our question.

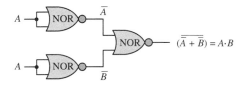

Figure 12.25

Example   12.11

Realize the following function using only NAND and NOR gates:

$$y = \overline{\overline{(A \cdot B)} + C}$$

Solution:

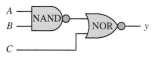

Figure 12.26

Since the term in parentheses appears as a complement product, it can be obtained by means of a NAND gate. Further, once the function $\overline{(A \cdot B)}$ has been realized, we can see that y is the complemented sum of two terms—that is, it can be obtained directly with a NOR gate. The resulting logic circuit is shown in Figure 12.26.

Can you find another solution to this problem that employs only two gates?

Example   12.12

Realize the logic function described by the following truth table, using only NAND and NOR gates.

A	B	C	y
0	0	0	1
0	0	1	0
0	1	0	1
0	1	1	0
1	0	0	0
1	0	1	0
1	1	0	0
1	1	1	0

Solution:

Note that the output has 1s occurring only when A and C are both 0, and that this corresponds to the function $\overline{A} \cdot \overline{C}$. Using De Morgan's theorems, we obtain:

$$y = \overline{A + C}$$

This function corresponds to a single NOR gate, as shown in Figure 12.27.

Figure 12.27

Example   12.13

Use De Morgan's theorems to implement the function $y = \overline{(A \cdot B) + C}$ using at least one NOR gate.

Solution:

To simplify the function, De Morgan's laws are applied to obtain

$$y = \overline{(A \cdot B) + C} = \overline{\overline{A \cdot B}} \cdot \overline{C}$$
$$= \overline{\overline{A} + \overline{B}} \cdot \overline{C}$$

resulting in the circuit shown in Figure 12.28.

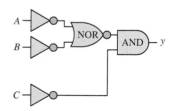

Figure 12.28

The XOR (Exclusive OR) Gate

It is a rather common practice for a manufacturer of integrated circuits to provide common combinations of logic circuits in a single integrated circuit package. We shall review many of these common **logic modules** in Section 12.5. An example of this idea is provided by the **exclusive OR (XOR)** gate, which provides a logic function similar, but not identical, to the OR gate

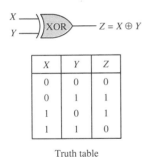

Truth table

Figure 12.29 XOR gate

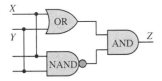

Figure 12.30 Realization of an XOR gate

we have already studied. The XOR gate acts as an OR gate, except when its inputs are all logic 1s; in this case, the output is a logic 0 (thus the term *exclusive*). Figure 12.29 shows the logic circuit symbol adopted for this gate, and the corresponding truth table. The logic function implemented by the XOR gate is the following: "either X or Y, but not both." This description can be extended to an arbitrary number of inputs.

The symbol adopted for the exclusive OR operation is $\oplus$, and so we shall write

$$Z = X \oplus Y$$

to denote this logic operation. The XOR gate can be obtained by a combination of the basic gates we are already familiar with. For example, if we observe that the XOR function corresponds to $Z = X \oplus Y = (X + Y) \cdot (\overline{X \cdot Y})$, we can realize the XOR gate by means of the circuit shown in Figure 12.30.

Common IC logic gate configurations and device numbers are given at the end of the chapter. These devices are typically available in both of the two more common device families, TTL and CMOS. The devices listed at the end of the chapter are available in CMOS technology under the numbers MM54HCXX or MM74HCXX. The same logic gate ICs are also available as TTL devices.

DRILL EXERCISES

12.9 Show that one can obtain an OR gate using NAND gates only. [*Hint:* Use three NAND gates.]

12.10 Show that one can obtain an AND gate using NOR gates only. [*Hint:* Use three NOR gates.]

12.11 Prepare a step-by-step truth table for the following logic expressions:

a. $\overline{(X + Y + Z)} + (X \cdot Y \cdot Z) \cdot \bar{X}$

b. $\bar{X} \cdot Y \cdot Z + Y \cdot (Z + W)$

c. $(X \cdot \bar{Y} + Z \cdot \overline{W}) \cdot (W \cdot X + \bar{Z} \cdot Y)$

[*Hint:* Your truth table must have 2^n entries, where n is the number of logic variables.]

12.12 Implement the logic functions of Drill Exercise 12.11 using NAND and NOR gates only. [*Hint:* Use De Morgan's theorems and the fact that $\bar{\bar{f}} = f$.]

12.13 Implement the logic functions of Drill Exercise 12.11 using AND, OR, and NOT gates only.

12.14 Show that the XOR function can also be expressed as $Z = X \cdot \bar{Y} + Y \cdot \bar{X}$. Realize the corresponding function using NOT, AND, and OR gates. [*Hint:* Use truth tables for the logic function Z (as defined in the exercise) and for the XOR function.]

12.4   KARNAUGH MAPS AND LOGIC DESIGN

In examining the design of logic functions by means of logic gates, we have discovered that more than one solution is usually available for the implementation of a given logic expression. It should also be clear by now that some combinations of gates can implement a given function more efficiently than others. How can we be assured of having chosen the most efficient realization? Fortunately, there is a procedure that utilizes a map describing all possible combinations of the variables present in the logic function of interest. This map is called a **Karnaugh map,** after its inventor. Figure 12.31 depicts the appearance of Karnaugh maps for two-, three-, and four-variable expressions in two different forms. As can be seen, the row and column assignments for two or more variables are arranged so that all adjacent terms change by only one bit. For example, in the two-variable map, the columns next to column 01 are columns 00 and 10. Also note that each map consists of 2^N **cells,** where N is the number of logic variables.

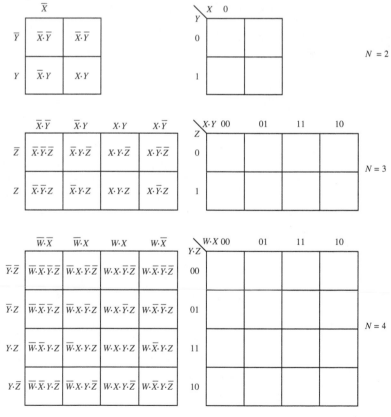

Figure 12.31 Two-, three-, and four-variable Karnaugh maps

Karnaugh map

	$\overline{X}\overline{Y}$	$\overline{X}Y$	XY	$X\overline{Y}$
$\overline{Z}$	0	1	1	0
Z	0	1	1	0

Karnaugh map

X	Y	Z	Desired Function
0	0	0	0
0	0	1	0
0	1	0	1
0	1	1	1
1	0	0	0
1	0	1	0
1	1	0	1
1	1	1	1

Truth table

Figure 12.32 Truth table and Karnaugh map representations of a logic function

Each cell in a Karnaugh map contains a **minterm,** that is, a product of the N variables that appear in our logic expression (in either uncomplemented or complemented form). For example, for the case of three variables ($N = 3$), there are $2^3 = 8$ such combinations, or minterms: $\overline{X}\cdot\overline{Y}\cdot\overline{Z}, \overline{X}\cdot\overline{Y}\cdot Z,$ $\overline{X}\cdot Y\cdot\overline{Z}, \overline{X}\cdot Y\cdot Z, X\cdot\overline{Y}\cdot\overline{Z}, X\cdot\overline{Y}\cdot Z, X\cdot Y\cdot\overline{Z},$ and $X\cdot Y\cdot Z$. The content of each cell—that is, the minterm—is the product of the variables appearing at the corresponding vertical and horizontal coordinates. For example, in the three-variable map, $X\cdot Y\cdot\overline{Z}$ appears at the intersection of $X\cdot Y$ and $\overline{Z}$. The map is filled by placing a value of 1 for any combination of variables for which the desired output is a 1. For example, consider the function of three variables for which we desire to have an output of 1 whenever the variables X, Y, and Z have the following values:

$X = 0$	$Y = 1$	$Z = 0$
$X = 0$	$Y = 1$	$Z = 1$
$X = 1$	$Y = 1$	$Z = 0$
$X = 1$	$Y = 1$	$Z = 1$

The same truth table is shown in Figure 12.32 together with the corresponding Karnaugh map.

The Karnaugh map provides an immediate view of the values of the function in graphical form. Further, the arrangement of the cells in the Karnaugh map is such that any two adjacent cells contain minterms that vary only in one variable. This property, as will be verified shortly, is quite useful in the design of logic functions by means of logic gates, especially if we consider the map to be continuously wrapping around itself, as if the top and bottom, and right and left, edges were touching each other. For the three-variable map given in Figure 12.31, for example, the cell $X\cdot\overline{Y}\cdot\overline{Z}$ is adjacent to $\overline{X}\cdot\overline{Y}\cdot\overline{Z}$ if we "roll" the map so that the right edge touches the left. Note that these two cells differ only in the variable X, a property we earlier claimed adjacent cells have.[2]

Shown in Figure 12.33 is a more complex, four-variable logic function, which will serve as an example in explaining how Karnaugh maps can be used directly to implement a logic function. First, we define a subcube as a set of 2^m adjacent cells, for $m = 1, 2, 3, \ldots, N$. Thus, a subcube can consist of 1, 2, 4, 8, 16, 32, ... cells. All the possible subcubes for the four-variable map of Figure 12.33 are shown in Figure 12.34. Note that there are no four-cell subcubes in this particular case. Note also that there is some overlap between subcubes. Examples of four-cell and eight-cell subcubes are shown in Figure 12.35 for some arbitrary expression.

[2] A useful rule to remember is that in a two-variable map there are two minterms adjacent to any given minterm; in a three-variable map, three minterms are adjacent to any given minterm; in a four-variable map, the number is four, and so on.

W	X	Y	Z	Desired Function
0	0	0	0	1
0	0	0	1	1
0	0	1	0	0
0	0	1	1	0
0	1	0	0	0
0	1	0	1	1
0	1	1	0	1
0	1	1	1	0
1	0	0	0	0
1	0	0	1	1
1	0	1	0	1
1	0	1	1	0
1	1	0	0	0
1	1	0	1	0
1	1	1	0	0
1	1	1	1	1

Truth table for four-variable expression

	$\overline{W}\cdot\overline{X}$	$\overline{W}\cdot X$	$W\cdot X$	$W\cdot\overline{X}$
$\overline{Y}\cdot\overline{Z}$	1	0	0	0
$\overline{Y}\cdot Z$	1	1	0	1
$Y\cdot Z$	0	0	1	0
$Y\cdot\overline{Z}$	0	1	0	1

Figure 12.33 Karnaugh map for a four-variable expression

1 cell subcubes

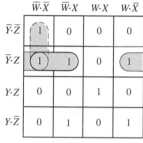

2 cell subcubes

Figure 12.34 One- and two-cell subcubes for Karnaugh map of Figure 12.33

(a)

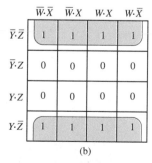

(b)

Figure 12.35 Four- and eight-cell subcubes for an arbitrary logic function

In general, one tries to find the largest possible subcubes to cover all of the "1" entries in the map. How do maps and subcubes help in the realization of logic functions, then? The use of maps and subcubes in minimizing logic expressions is best explained by considering the following rule of Boolean algebra:

$$Y \cdot X + Y \cdot \overline{X} = Y$$

where the variable Y could represent a product of logic variables (for example, we could similarly write $(Z \cdot W) \cdot X + (Z \cdot W) \cdot \overline{X} = Z \cdot W$ with $Y = Z \cdot W$). This rule is easily proven by factoring Y:

$$Y \cdot (X + \overline{X})$$

and observing that $X + \overline{X} = 1$, always. Then it should be clear that the variable X need not appear in the expression at all. Let us apply this rule to a more complex logic expression, to verify that it can also apply to this case. Consider the logic expression

$$\overline{W} \cdot X \cdot \overline{Y} \cdot Z + \overline{W} \cdot \overline{X} \cdot \overline{Y} \cdot Z + W \cdot \overline{X} \cdot \overline{Y} \cdot Z + W \cdot X \cdot \overline{Y} \cdot Z$$

and factor it as follows:

$$\overline{W} \cdot Z \cdot \overline{Y} \cdot (X + \overline{X}) + W \cdot \overline{Y} \cdot Z \cdot (\overline{X} + X) = \overline{W} \cdot Z \cdot \overline{Y} + W \cdot \overline{Y} \cdot Z$$
$$= \overline{Y} \cdot Z \cdot (\overline{W} + W) = \overline{Y} \cdot Z$$

Quite a simplification! If we consider, now, a map in which we place a 1 in the cells corresponding to the minterms $\overline{W} \cdot X \cdot \overline{Y} \cdot Z$, $\overline{W} \cdot \overline{X} \cdot \overline{Y} \cdot Z$, $W \cdot \overline{X} \cdot \overline{Y} \cdot Z$, and $W \cdot X \cdot \overline{Y} \cdot Z$, forming the previous expression, we obtain the Karnaugh map of Figure 12.36. It can easily be verified that the map of Figure 12.36 shows a single four-cell subcube corresponding to the term $\overline{Y} \cdot Z$.

We have not established formal rules yet, but it definitely appears that the map method for simplifying Boolean expressions is a convenient tool.

	$\overline{W}\cdot\overline{X}$	$\overline{W}\cdot X$	$W\cdot X$	$W\cdot\overline{X}$
$\overline{Y}\cdot\overline{Z}$	0	0	0	0
$\overline{Y}\cdot Z$	1	1	1	1
$Y\cdot Z$	0	0	0	0
$Y\cdot\overline{Z}$	0	0	0	0

Figure 12.36 Karnaugh map for the function $\overline{W}\cdot X\cdot\overline{Y}\cdot Z + \overline{W}\cdot\overline{X}\cdot\overline{Y}\cdot Z + W\cdot\overline{X}\cdot\overline{Y}\cdot Z + W\cdot X\cdot\overline{Y}\cdot Z$

	$\overline{W}\cdot\overline{X}$	$\overline{W}\cdot X$	$W\cdot X$	$W\cdot\overline{X}$
$\overline{Y}\cdot\overline{Z}$	1	1	1	1
$\overline{Y}\cdot Z$	1	1	1	1
$Y\cdot Z$	0	0	0	0
$Y\cdot\overline{Z}$	0	0	0	0

Figure 12.37

In effect, the map has performed the algebraic simplification automatically! We can see that in any subcube, one or more of the variables present will appear in both complemented *and* uncomplemented form in all their combinations with the other variables. These variables can be eliminated. As an illustration, in the *eight-cell* subcube case of Figure 12.37, the full-blown expression would be:

$$\overline{W}\cdot\overline{X}\cdot\overline{Y}\cdot\overline{Z} + \overline{W}\cdot X\cdot\overline{Y}\cdot\overline{Z} + W\cdot X\cdot\overline{Y}\cdot\overline{Z} + W\cdot\overline{X}\cdot\overline{Y}\cdot\overline{Z}$$

$$+ \overline{W}\cdot\overline{X}\cdot Y\cdot\overline{Z} + \overline{W}\cdot X\cdot Y\cdot\overline{Z} + W\cdot X\cdot Y\cdot\overline{Z} + W\cdot\overline{X}\cdot Y\cdot\overline{Z}$$

However, if we consider the eight-cell subcube, we note that the three variables X, W, and Y appear both in complemented and uncomplemented form in all their combinations with the other variables and thus can be removed from the expression. This reduces the seemingly unwieldy expression simply to $\overline{Z}$! In logic design terms, a simple inverter is sufficient to implement the expression.

The example just shown is a particularly simple one, but it illustrates how simple it can be to determine the minimal expression for a logic function. It should be apparent that the larger a subcube, the greater the simplification that will result. For subcubes that do not intersect, as in the previous example, the solution can be found easily, and is unique.

Sum-of-Products Realizations

Although not explicitly stated, the logic functions of the preceding section were all in sum-of-products form. As you know, it is also possible to realize logic functions in product-of-sums form. This section discusses the implementation of logic functions in sum-of-products form and gives a set of design rules. The next section will do the same for product-of-sums form logical expressions. The following rules are a useful aid in determining the minimal sum-of-products expression:

1. Begin with isolated cells. These must be used as they are, since no simplification is possible.
2. Find all cells that are adjacent to only one other cell, forming two-cell subcubes.
3. Find cells that form four-cell subcubes, eight-cell subcubes, and so forth.
4. The minimal expression is formed by the collection of the *smallest number of maximal subcubes*.

The following examples illustrate the application of these principles to a variety of problems.

Example   12.14

Design the logic circuit that implements the following truth table. Use only two-input gates.

A	B	C	D	y
0	0	0	0	1
0	0	0	1	1
0	0	1	0	1
0	0	1	1	0
0	1	0	0	0
0	1	0	1	1
0	1	1	0	0
0	1	1	1	0
1	0	0	0	1
1	0	0	1	1
1	0	1	0	0
1	0	1	1	1
1	1	0	0	0
1	1	0	1	1
1	1	1	0	0
1	1	1	1	1

Solution:

A common way of solving problems such as this is to use a Karnaugh map to simplify and implement the function, with the aim of minimizing the complexity of the resulting circuit. The function y may be implemented by covering the Karnaugh map shown in Figure 12.38. The map covering shown in the figure leads to the expression

$$y = A \cdot D + \bar{C} \cdot D + \bar{A} \cdot \bar{B} \cdot \bar{D} + \bar{B} \cdot \bar{C}$$

with the corresponding implementation using two-input gates shown in Figure 12.39.

Figure 12.38 Karnaugh map for Example 12.14

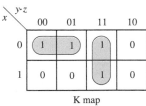

Figure 12.39

Example   12.15

Figure 12.40

Construct a truth table and derive the minimum sum-of-products equation from the circuit of Figure 12.40.

Solution:

The first step is to make a truth table. The second step is to use a Karnaugh map to simplify and implement the function. The truth table may be obtained by writing the function corresponding to the circuit:

$$f = \bar{x} \cdot \bar{y} + y \cdot z$$

and filling the table accordingly, as follows:

x	y	z	f
0	0	0	1
0	0	1	1
0	1	0	0
0	1	1	1
1	0	0	0
1	0	1	0
1	1	0	0
1	1	1	1

Correspondingly, we can select the appropriate entries in a three-variable Karnaugh map:

$y \cdot z$				
x	00	01	11	10
0	1	1	1	0
1	0	0	1	0

K map

The Karnaugh map verifies that the minimum sum-of-products expression is:

$$f = \bar{x} \cdot \bar{y} + y \cdot z$$

Example   12.16

Realize the following function using only two-input NAND gates:

$$f = (\bar{x} + \bar{y}) \cdot (\bar{z} + y)$$

Solution:

NAND gates naturally implement products, or serve as inverters, and thus there is no direct solution to this problem, unless the expression is manipulated into a different form. This can be done by applying De Morgan's laws:

1. $\bar{x} + \bar{y} = \overline{x \cdot y}$
2. $\bar{z} + y = \overline{z \cdot \bar{y}}$

Thus, the function f can be rewritten as $f = \overline{(x \cdot y)}\overline{(z \cdot \bar{y})}$ and consists only of product terms. The implementation with NAND gates requires the use of five gates, as shown in Figure 12.41.

Figure 12.41

Example   12.17

Prove that $x \cdot y + \bar{x} \cdot z + y \cdot z = x \cdot y + \bar{x} \cdot z$, thus showing that yz is a redundant term.

Solution:

One possible technique for simplifying a logic expression consists of using a Karnaugh map. In this case, the Karnaugh map is shown in Figure 12.42. The Karnaugh map immediately reveals that there is no need to circle the $y \cdot z$ term, thus making it redundant. Clearly, $f = x \cdot y + \bar{x} \cdot z$ is the minimum sum-of-products expression.

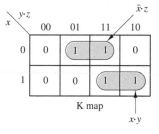

Figure 12.42

Example   12.18

Simplify the circuit of Figure 12.43 by reducing the number of gates to the minimum achievable number.

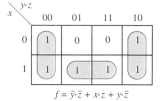

Figure 12.43   **Figure 12.44**

$$f = \bar{y} \cdot \bar{z} + x \cdot z + y \cdot \bar{z}$$

Solution:

To solve this problem, we need first to obtain a logic expression corresponding to the function of Figure 12.43. The second step is to use a Karnaugh map to simplify the function. The logic function corresponding to the logic circuit shown is

$$f = \bar{y} \cdot \bar{z} + x \cdot z + y \cdot \bar{z}$$

and the corresponding Karnaugh map is shown in Figure 12.44.

Inspection of the Karnaugh map, though, reveals that the map has been covered inefficiently, since two four-cell subcubes could have been used in place of three two-cell subcubes. Figure 12.45 depicts the alternative covering, which leads to the much simpler function

$$f = x + \bar{z}$$

The corresponding gate implementation is also shown in Figure 12.45.

This example emphasizes the importance of reducing logical expressions as much as possible prior to implementing a circuit. The Karnaugh map allows this simplification to be performed automatically, without having to explicitly reduce the logical expression.

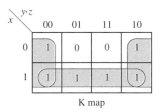

K map

Figure 12.45

Product-of-Sums Realizations

Thus far, we have exclusively worked with sum-of-products expressions, that is, logic functions of the form $A \cdot B + C \cdot D$. We know, however, that De Morgan's laws state that there is an equivalent form that appears as a product of sums, for example, $(W + Y) \cdot (Y + Z)$. The two forms are completely equivalent, logically, but one of the two forms may lead to a realization involving a smaller number of gates. When using Karnaugh maps, we may obtain the product-of-sums form very simply by following these rules:

1. Solve for the 0s exactly as for the 1s in sum-of-products expressions.

2. Complement the resulting expression.

The same principles stated earlier apply in covering the map with subcubes and determining the minimal expression. The following examples illustrate how one form may result in a more efficient solution than the other.

Example   12.19

This example illustrates the design of a logic function using both sum-of-products and product-of-sums implementations, thus showing that it may be possible to realize some savings by using one implementation rather than the other.

1. Realize the function f by a Karnaugh map using 0s.
2. Realize the function f by a Karnaugh map using 1s.

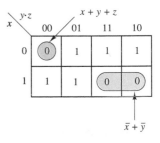

Figure 12.46

x	y	z	f
0	0	0	0
0	0	1	1
0	1	0	1
0	1	1	1
1	0	0	1
1	0	1	1
1	1	0	0
1	1	1	0

Solution:

1. Using 0s, we obtain the Karnaugh map of Figure 12.46, leading to the product-of-sums expression

$$f = (x + y + z) \cdot (\bar{x} + \bar{y})$$

2. If 1s are used, as shown in Figure 12.47, a sum-of-products expression is obtained, of the form

$$f = \bar{x} \cdot y + x \cdot \bar{y} + \bar{y} \cdot z$$

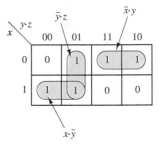

Figure 12.47

Example   12.20

Realize the logic function defined by the following truth table as the simplest product-of-sums form:

x	y	z	f
0	0	0	1
0	0	1	0
0	1	0	1
0	1	1	0
1	0	0	1
1	0	1	0
1	1	0	0
1	1	1	0

Solution:

To realize this function in product-of-sums form, the 0s are used in the map of Figure 12.48. The resulting expression is therefore given by

$$f = \bar{z} \cdot (\bar{x} + \bar{y})$$

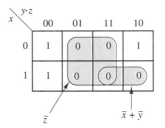

Figure 12.48

Example   12.21   Safety Circuit for Operation of a Stamping Press

In this example, the techniques illustrated in the preceding examples will be applied to a practical situation. A stamping press operator must press two buttons 1 meter apart to operate the press (this ensures that the operator's hands will not be caught in the press). These two logical inputs are ANDed together to ensure safe operation of the press. Table 12.12 summarizes additional logical conditions that must be satisfied for

Table 12.12   Conditions for operation of stamping press

(A) $b_1 \cdot b_2$	(B) Part A is in press	(C) Part B is in press	(D) Press is operable	Press operation 1 = pressing; 0 = not pressing
0	0	0	0	0
0	0	0	1	0
0	0	1	0	0
0	0	1	1	0
0	1	0	0	0
0	1	0	1	0
0	1	1	0	0
0	1	1	1	0
1	0	0	0	0
1	0	0	1	0
1	0	1	0	0
1	0	1	1	1
1	1	0	0	0
1	1	0	1	1
1	1	1	0	0
1	1	1	1	0

↑ Both buttons (b_1, b_2) must be pressed for this to be a 1.

the press to operate safely. Essentially, the press will operate if there is a workpiece in the press (but not two pieces), if the press is "ready," and only if both operator buttons are pressed. Using a Karnaugh map, realize the logic circuitry required for the safety circuit.

Solution:

Table 12.12 can be converted to a Karnaugh map, as shown in Figure 12.49. Since there are many more 0s than 1s in the table, the use of 0s in covering the map will

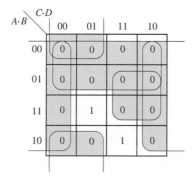

Figure 12.49

lead to greater simplification. This will result in a product-of-sums expression. The four subcubes shown in Figure 12.49 yield the equation

$$A \cdot D \cdot (C + B) \cdot (\overline{C} + \overline{B})$$

By De Morgan's law, this equation is equivalent to

$$A \cdot D \cdot (C + B) \cdot \overline{(C \cdot B)}$$

which can be realized by the circuit of Figure 12.50.

For the purpose of comparison, the corresponding sum-of-products circuit is shown in Figure 12.51. Note that this circuit employs a greater number of gates and will therefore lead to a more expensive design.

Figure 12.50

Figure 12.51

Don't Care Conditions

Another simplification technique may be employed whenever the value of the logic function to be implemented can be either a 1 or a 0. This condition may result from the specification of the problem and is not uncommon. Whenever it does not matter whether a position in the map is filled by a 1 or a 0, we use a so-called **don't care** entry, denoted by an **x**. Then the don't care can be used as either a 1 or a 0, depending on which results in a greater simplification (i.e., helps in forming the smallest number of maximal subcubes). The following examples illustrate the use of don't cares.

Example   12.22

When dealing with complex logic expressions involving several variables, it is often convenient to use a more compact form rather than writing out the entire function. For example, one can specify that a certain logic function be "true" (i.e., logic 1) for certain combinations, which are then expressed as binary numbers. In this example, we wish to find a minimum sum-of-products expression for the function

$$f(a,b,c,d) = \bar{a}\bar{b}\bar{c}\bar{d} + \bar{a}b\bar{c}\bar{d} + \bar{a}b\bar{c}d$$
$$+ a\bar{b}\bar{c}\bar{d} + a\bar{b}\bar{c}d + ab\bar{c}\bar{d}$$

Note that this could also be written as

$$f(a,b,c,d) = 0001 + 0010 + 0011 + 0101 + 1011 + 1100$$

if each uncomplemented variable is represented by a 1, and each complemented variable by a 0. Further, the binary numbers in the resulting expression may be converted to their decimal counterpart to obtain

$$f(a,b,c,d) = (1 + 2 + 3 + 5 + 11 + 12)_{10}$$
$$= \sum(1,2,3,5,11,12)_{10}$$

In this example we also assume that the following combinations may be treated as don't cares:

$$f(a,b,c,d) = \sum(4,6,10,14)_{10}$$

Solution:

The inputs that cannot occur may be treated as don't cares. If we use the symbol **x** to denote these inputs, the Karnaugh map corresponding to the problem specifications is as shown in Figure 12.52. If all **x** are given value of 1, the following expression can be obtained by the map covering of Figure 12.52:

$$f = b\bar{d} + \bar{b}c + \bar{a} \cdot \bar{c} \cdot d$$

Note that the **x**'s never occur, and so they may be assigned a 1 or a 0, whichever will best simplify the expression.

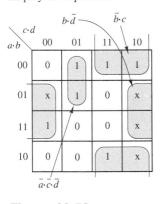

Figure 12.52

Example   12.23

Find a minimum sum-of-products expression for a function $f(a,b,c)$ whose output will be 1 for the combinations $(a,b,c)_{10} = 0, 2, 3$ and don't care for the combinations $(a,b,c)_{10} = 4, 5, 6$.

Solution:

The key is to use the don't care terms *if* they will simplify the expression. The simplest function that can be obtained with the proper use of the don't care terms is shown in Figure 12.53. Thus, the function f is

$$f = \bar{a} \cdot b + \bar{c}$$

Note that the don't care output for the combination $(a, b, c)_{10} = 5$ is not used. You should verify that using this additional don't care entry does not lead to any further simplification.

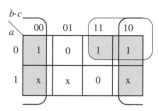

Figure 12.53

Example   12.24

Find a minimum sum-of-products expression for a function $f(a, b, c, d)$ whose output will be 1 for the combinations $(a, b, c, d)_{10} = 0, 3, 6, 9$ and don't care for the combinations $(a, b, c, d)_{10} = 10, 11, 13, 14, 15$.

Solution:

The Karnaugh map for the given function is shown in Figure 12.54(a). We know that for minimum sum of products, all the 1s have to be covered. With reference to Figure 12.54(a), the simplest function can be obtained by covering the map as shown in Figure 12.54(b). From this figure, it can be seen that the function is

$$f = \bar{a} \cdot \bar{b} \cdot \bar{c} \cdot \bar{d} + b \cdot c \cdot \bar{d} + a \cdot d + \bar{b} \cdot c \cdot d$$

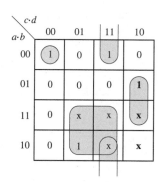

Figure 12.54(a)   **Figure 12.54(b)**

DRILL EXERCISES

12.15 Simplify the following expression to show that it corresponds to the function $\bar{Z}$:

$$\overline{W} \cdot \overline{X} \cdot \overline{Y} \cdot \overline{Z} + \overline{W} \cdot X \cdot \overline{Y} \cdot \overline{Z} + W \cdot X \cdot \overline{Y} \cdot \overline{Z} + W \cdot \overline{X} \cdot \overline{Y} \cdot \overline{Z} + \overline{W} \cdot \overline{X} \cdot Y \cdot \overline{Z} + \overline{W} \cdot X \cdot Y \cdot \overline{Z} + W \cdot X \cdot Y \cdot \overline{Z} + W \cdot \overline{X} \cdot Y \cdot \overline{Z}$$

12.16 Simplify the following expression, using a Karnaugh map:

$$\overline{W} \cdot \overline{X} \cdot \overline{Y} \cdot \overline{Z} + \overline{W} \cdot \overline{X} \cdot Y \cdot \overline{Z} + W \cdot X \cdot \overline{Y} \cdot \overline{Z} + W \cdot \overline{X} \cdot \overline{Y} \cdot \overline{Z} + W \cdot \overline{X} \cdot Y \cdot \overline{Z} + W \cdot X \cdot Y \cdot \overline{Z}$$

Answer: $W \cdot \bar{Z} + \bar{X} \cdot \bar{Z}$

12.17   Simplify the following expression, using a Karnaugh map:

$$\overline{W} \cdot \bar{X} \cdot \bar{Y} \cdot \bar{Z} + \overline{W} \cdot \bar{X} \cdot Y \cdot \bar{Z} + W \cdot X \cdot \bar{Y} \cdot \bar{Z} + W \cdot \bar{X} \cdot \bar{Y} \cdot \bar{Z} + W \cdot \bar{X} \cdot Y \cdot \bar{Z} + \overline{W} \cdot X \cdot \bar{Y} \cdot \bar{Z}$$

Answer: $\bar{Y} \cdot \bar{Z} + \bar{X} \cdot \bar{Z}$

12.18   The function y of Example 12.14 can be obtained with fewer gates if we use gates with three or four inputs. Find the minimum number of gates needed to obtain this function.

Answer: Nine gates

12.19   Verify that the product-of-sums expression for Example 12.19 can be realized with fewer gates.

12.20   Would a sum-of-products realization for Example 12.20 require fewer gates?

Answer: No

12.21   Prove that the circuit of Figure 12.51 can also be obtained from the sum of products.

12.22   In Example 12.22, assign a value of 0 to the don't care terms and derive the corresponding minimal expression. Is the new function simpler than the one obtained in Example 12.22?

Answer: $f = a \cdot b \cdot \bar{c} \cdot \bar{d} + \bar{a} \cdot \bar{c} \cdot d + \bar{a} \cdot \bar{b} \cdot c + \bar{b} \cdot c \cdot d$

12.23   In Example 12.23, assign a value of 0 to the don't care terms and derive the corresponding minimal expression. Is the new function simpler than the one obtained in Example 12.23?

Answer: $f = \bar{a} \cdot b + \bar{a} \cdot \bar{b} \cdot \bar{c}$

12.24   In Example 12.23, assign a value of 1 to all don't care terms and derive the corresponding minimal expression. Is the new function simpler than the one obtained in Example 12.23?

Answer: $f = \bar{a} \cdot b + a \cdot \bar{b} + \bar{c}$

12.25   In Example 12.24, assign a value of 0 to all don't care terms and derive the corresponding minimal expression. Is the new function simpler than the one obtained in Example 12.24?

Answer: $f = \bar{a} \cdot \bar{b} \cdot \bar{c} \cdot \bar{d} + \bar{a} \cdot \bar{b} \cdot c \cdot d + a \cdot \bar{b} \cdot \bar{c} \cdot d + \bar{a} \cdot b \cdot c \cdot \bar{d}$

12.26 In Example 12.24, assign a value of 1 to all don't care terms and derive the corresponding minimal expression. Is the new function simpler than the one obtained in Example 12.24?

Answer: $f = \bar{a} \cdot \bar{b} \cdot \bar{c} \cdot \bar{d} + b \cdot c \cdot \bar{d} + a \cdot d + \bar{b} \cdot c \cdot d + a \cdot c$

12.5   COMBINATIONAL LOGIC MODULES

The basic logic gates described in the previous section are used to implement more advanced functions and are often combined to form logic modules, which, thanks to modern technology, are available in compact integrated circuit (IC) packages. In this section and the next, we discuss a few of the more common logic modules, illustrating how these can be used to implement advanced logic functions. We shall consider the use of multiplexers, read-only memories, flip-flops, shift registers, and counters, preparing the way for the material on digital logic systems, to be introduced in Chapter 13.

Multiplexers

Multiplexers, or **data selectors,** are combinational logic circuits that permit the selection of one of many inputs. A typical multiplexer (MUX) has 2^n **data inputs,** n **control inputs,** and one output. Standard, commercially available MUXs allow for n up to 4; however, two or more MUXs can be combined if a greater range is needed. The MUX allows for one of 2^n inputs to be selected as the data output; the selection of which input is to appear at the output is made by way of the control inputs. Figure 12.55 depicts the block diagram of a four-input MUX. The input data lines are labeled $D_0, D_1, D_2,$ and D_3; the **data select,** or control, lines are labeled I_0 and I_1; and the output is available in both complemented and uncomplemented form, and is thus labeled F, or $\bar{F}$. Finally, an **enable** input, labeled E, is also provided, as a means of enabling or disabling the MUX: if $E = 1$, the MUX is disabled; if $E = 0$, it is enabled. The negative logic (MUX off when $E = 1$ and on when $E = 0$) is represented by the small "bubble" at the enable input, which represents a complement operation (just as at the output of NAND and NOR gates). The enable input is useful whenever one is interested in a cascade of MUXs; this would be of interest if we needed to select a line from a large number, say $2^8 = 256$. Then two 4-input MUXs could be used to provide the data selection of 1 of 8.

The material described in the previous sections is quite adequate to describe the internal workings of a multiplexer. Figure 12.56 shows the internal construction of a 1-of-4 MUX using exclusively NAND gates (inverters are also used, but the reader will recall that a NAND gate can act as an inverter if properly connected).

I_1	I_0	F
0	0	D_0
0	1	D_1
1	0	D_2
1	1	D_3

Truth table of
1-of-4 MUX

Figure 12.55 1-of-4 MUX

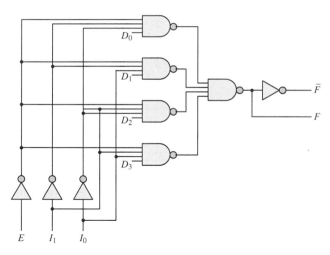

Figure 12.56 Internal structure of the 1-of-4 MUX

In the design of digital systems (for example, microcomputers), a single line is often required to carry two or more different digital signals. However, only one signal at a time can be placed on the line. A MUX will allow us to select, at different instants, the signal we wish to place on this single line. This property is shown here for a 1-of-4 MUX. It will be seen later, in Chapter 13, that this function is frequently used in data acquisition and communication systems. Figure 12.57 depicts the functional diagram of a 1-of-4 MUX, showing four data lines. D_0 through D_3, and two select lines, I_0 and I_1.

The data selector function of a MUX is best understood in terms of Table 12.13. In this truth table, the x's represent don't care entries. As can be seen from the truth table, the output selects one of the data lines depending on the values of I_1 and I_0, assuming that I_0 is the least significant bit. As an example, $I_1 I_0 = 10$ selects D_2, which means that the output, F, will select the value of the data line D_2. Therefore $F = 1$ if $D_2 = 1$ and $F = 0$ if $D_2 = 0$.

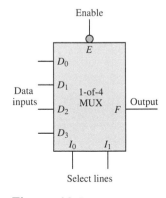

Figure 12.57 Functional diagram of four-input MUX

Table 12.13

I_1	I_0	D_3	D_2	D_1	D_0	F
0	0	x	x	x	0	0
0	0	x	x	x	1	1
0	1	x	x	0	x	0
0	1	x	x	1	x	1
1	0	x	0	x	x	0
1	0	x	1	x	x	1
1	1	0	x	x	x	0
1	1	1	x	x	x	1

Read-Only Memory (ROM)

Another common technique for implementing logic functions uses a **read-only memory,** or **ROM.** As the name implies, a ROM is a logic circuit that holds in storage ("memory") information—in the form of binary numbers—that cannot be altered but can be "read" by a logic circuit. A ROM is an array of memory cells, each of which can store either a 1 or a 0. The array consists of $2^m \times n$ cells, where n is the number of bits in each word stored in ROM. To access the information stored in ROM, m **address lines** are required. When an address is selected, in a fashion similar to the operation of the MUX, the binary word corresponding to the address selected appears at the output, which consists of n bits, that is, the same number of bits as the stored words. In some sense, a ROM can be thought of as a MUX that has an output consisting of a word instead of a single bit.

Figure 12.58 depicts the conceptual arrangement of a ROM with $n = 4$ and $m = 2$. The ROM table has been filled with arbitrary 4-bit words, just for the purpose of illustration. In Figure 12.58, if one were to select an enable input of 0 (i.e., on) and values for the address lines of $I_0 = 0$ and $I_1 = 1$, the output word would be $W_2 = 0110$, so that $b_0 = 0$, $b_1 = 1$, $b_2 = 1$, $b_3 = 0$. Depending on the content of the ROM and the number of address and output lines, one could implement an arbitrary logic function.

Unfortunately, the data stored in read-only memories must be entered during fabrication and cannot be altered later. A much more convenient type of read-only memory is the **erasable programmable read-only memory (EPROM),** the content of which can be easily programmed and stored and may be changed if needed. EPROMs find use in many practical applications, because of their flexibility in content and ease of programming. The following example illustrates the use of an EPROM to perform the linearization of a nonlinear function.

Figure 12.58 Read-only memory

Example   12.25   EPROM-Based Look-Up Table

One of the most common applications of EPROMs is the *arithmetic look-up table.* A look-up table is similar in concept to the familiar multiplication table and is used to store precomputed values of certain functions, eliminating the need for actually computing the function. A practical application of this concept is present in every automobile manufactured in the United States since the early 1980s, as part of the exhaust emission control system. In order for the catalytic converter to minimize the emissions of exhaust gasses (especially hydrocarbons, oxides of nitrogen, and carbon monoxide), it is necessary to maintain the *air-to-fuel ratio* (A/F) as close as possible to the stoichiometric value, that is, 14.7 parts of air for each part of fuel. Most modern-day engines are equipped with fuel injection systems that are capable of delivering accurate amounts of fuel to each individual cylinder: thus, the task of maintaining an accurate A/F amounts to measuring the mass of air that is aspirated into each cylinder and computing the corresponding mass of fuel. Many automobiles are equipped with a *mass air flow sensor,* capable of measuring the mass of air drawn into each cylinder during each engine cycle. Let the output of the mass air flow sensor be denoted by the variable M_A and let this variable represent the mass of air (in g) actually entering a cylinder during a particular stroke. It is then desired to compute the mass of fuel, M_F (also expressed in g) required to achieve an A/F of 14.7. This computation is simply:

$$M_F = \frac{M_A}{14.7} \text{ g}$$

Although the above computation is a simple division, its actual calculation in a low-cost digital computer (such as would be used on an automobile) is rather complicated. It would be much simpler to tabulate a number of values of M_A, to precompute the variable M_F, and then to store the result of this computation into an EPROM. If the EPROM address were made to correspond to the tabulated values of air mass, and the content at each address to the corresponding fuel mass (according to the precomputed values of the expression $M_F = \frac{M_A}{14.7}$), it would not be necessary to perform the division by 14.7. For each measurement of air mass into one cylinder, an EPROM address is specified and the corresponding content is read. The content at the specific address is the mass of fuel required by that particular cylinder.

In practice, the fuel mass needs to be converted into a time interval corresponding to the duration of time during which the fuel injector is open. This final conversion factor can also be accounted for in the table. Suppose for example that the fuel injector is capable of injecting K_F g of fuel per second; then the time duration, T_F, during which the injector should be open in order to inject M_F g of fuel into the cylinder is given by:

$$T_F = \frac{M_F}{K_F} \text{ s}$$

Therefore, the complete expression to be precomputed and stored in the EPROM is:

$$T_F = \frac{M_A}{14.7 \times K_F} \text{ s}$$

Figure 12.59 illustrates this process graphically.

To provide a numerical illustration consider a hypothetical engine capable of aspirating air in the range $0 < M_A < 0.55$ g and equipped with fuel injectors capable of injecting at the rate of 1.36 g/s. Thus, the relationship between T_F and M_A is:

$$T_F = 50 \times M_A \text{ ms}$$

If the digital value of M_A is expressed in dg (decigrams, tens of g), the look-up table of Figure 12.60 can be implemented, illustrating the conversion capabilities provided by the EPROM. Note that in order to represent the quantities of interest in an appropriate binary format compatible with the 8-bit EPROM, the units of air mass and of time have been scaled.

Figure 12.59 Use of EPROM look-up table in automotive fuel injection system

$M_A \times 10^{-2}$ (g)	Address (digital value of M_A)	Content (digital value of T_F)	$T_F \times 10^{-1}$ (ms)
0	00000000	00000000	0
1	00000001	00000101	5
2	00000010	00001010	10
3	00000011	00001111	15
4	00000100	00010100	20
5	00000101	00011001	25
.	.	.	.
.	.	.	.
.	.	.	.
.	.	.	.
51	00110011	11111111	255

Figure 12.60 Look-up table for automotive fuel injection application

Decoders and Read and Write Memory (RAM)

Decoders, which are commonly used for applications such as address decoding or memory expansion, are combinational logic circuits as well. Our reason for introducing decoders is to show some of the internal organization of semiconductor memory devices. An important application of decoders in the organization of a memory system will be discussed in Chapter 13.

Inputs			Outputs			
Enable	Select					
$\bar{G}$	A	B	Y_0	Y_1	Y_2	Y_3
1	X	X	1	1	1	1
0	0	0	0	1	1	1
0	0	1	1	0	1	1
0	1	0	1	1	0	1
0	1	1	1	1	1	0

Figure 12.61 2-to-4 decoder

Figure 12.61 shows the truth table for a 2-to-4 decoder. The decoder has an enable input, $\bar{G}$, and select inputs, B and A. It also has four outputs, Y_0 through Y_3. When the enable input is logic 1, all decoder outputs are forced to logic 1 regardless of the select inputs.

This simple description of decoders permits a brief discussion of the internal organization of an **SRAM (static read and write memory)**. SRAM is internally organized to provide memory with high speed (i.e., short access time), a large bit capacity, and low cost. The memory array in this memory device has a column length equal to the number of words W, and a row length equal to the number of bits per word, N. To select a word, a 1-of-W decoder is needed. Since the address inputs to the decoder select only one of the decoder's outputs, the decoder selects one word in the memory array. Figure 12.62 shows the internal organization of a typical SRAM.

Thus, to choose the desired word from the memory array, the proper address inputs are required. As an example, if the number of words in the memory array is 8, a 3-to-8 decoder is needed. Data sheets for 2-to-4 and 3-to-8 decoders from a CMOS family data book are provided at the end of the chapter.

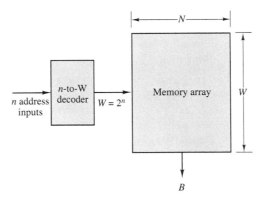

Figure 12.62 Internal organization of SRAM

DRILL EXERCISES

12.27   Which combination of the control lines will select the data line D_3 for a 1-of-4 MUX?

> Answer: $I_1I_0 = 11$

12.28   Show that a 3-to-8 MUX with eight data inputs (D_0 through D_7) and three control lines (I_0 through I_2) can be used as a data selector. Which combination of the control lines will select the data line D_5?

> Answer: For the first part, use the same method as in Drill Exercise 12.17, but for a 3-to-8 MUX. For the second part, $I_2I_1I_0 = 101$.

12.29   Which combination of the control lines will select the data line D_4 for a 3-to-8 MUX?

> Answer: $I_2I_1I_0 = 100$

12.30   How many address inputs do you need if the number of words in a memory array is 16?

> Answer: 4

12.6   SEQUENTIAL LOGIC MODULES

Thus far, the discussion of logic devices has focused on the general family of combinational logic devices. The feature that distinguishes combinational logic devices from the other major family—**sequential logic devices**—is that combinational logic circuits provide outputs that are based on a combination of present inputs only. On the other hand, sequential logic circuits depend on the present and past values of the inputs. Because of this "memory" property, sequential circuits can store information; this capability opens a whole new area of application for digital logic circuits.

Latches and Flip-Flops

The basic information-storage device in a digital circuit is called a **flip-flop.** There are many different varieties of flip-flops; however, all flip-flops share the following characteristics:

1. A flip-flop is a **bistable** device; that is, it can remain in one of two stable states (0 and 1) until appropriate conditions cause it to change state. Thus, a flip-flop can serve as a memory element.
2. A flip-flop has two outputs, one of which is the complement of the other.

RS Flip-Flop

It is customary to depict flip-flops by their block diagram and a name, such as A or X, representing the output variable. Figure 12.63 represents the so-called **RS flip-flop,** which has two inputs, denoted by S and R, and two outputs, A and $\bar{A}$. The value at A is called the *state* of the flip-flop. If $A = 1$, we refer to the device as *being in the 1 state*. Thus, we need define only one of the two outputs of the flip-flop. The two inputs, R and S, are used to change the state of the flip-flop, according to the following rules:

Figure 12.63 *RS* flip-flop

1. When $R = S = 0$, the flip-flop remains in its present state (whether 1 or 0).

2. When $S = 1$ and $R = 0$, the flip-flop is **set** to the 1 state (thus, the letter S, for "set").

3. When $S = 0$ and $R = 1$, the flip-flop is **reset** to the 0 state (thus, the letter R, for "reset").

4. It is not permitted for both S and R to be equal to 1. (This would correspond to requiring the flip-flop to set and reset at the same time.)

The rules just described are easily remembered by noting that 1s on the S and R inputs correspond to the set and reset commands, respectively.

A convenient means of describing the series of transitions that occur as the signals set to the flip-flop inputs change is the **timing diagram.** A timing diagram is a graph of the inputs and outputs of the RS flip-flop (or any other logic device) depicting the transitions that occur over time. In effect, one could also represent these transitions in tabular form; however, the timing diagram provides a convenient visual representation of the evolution of the state of the flip-flop. Figure 12.64 depicts a table of transitions for an RS flip-flop A, as well as the corresponding timing diagram.

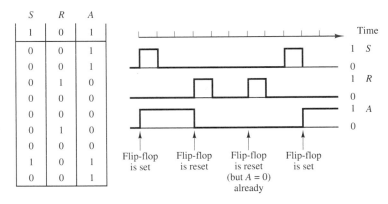

Figure 12.64 Timing diagram for the RS flip-flop

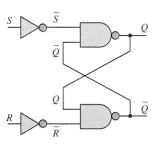

Figure 12.65 Logic gate implementation of the RS flip-flop

It is instructive to illustrate how an RS flip-flop can be constructed using simple logic gates. For example, Figure 12.65 depicts a realization of such a circuit consisting of four gates: two inverters and two NAND gates (actually, the same result could be achieved with four NAND gates). Consider the case in which the circuit is in the initial state $Q = 0$ (and therefore $\bar{Q} = 1$). If the input $S = 1$ is applied, the top NAND gate will see inputs $\bar{Q} = 1$ and $\bar{S} = 0$, so that $Q = \overline{(\bar{S} \cdot \bar{Q})} = \overline{(0 \cdot 1)} = 1$—that is, the flip-flop is set. Note that when Q is set to 1, $\bar{Q}$ becomes 0. This, however, does not affect the state of the Q output, since replacing $\bar{Q}$ with 0 in the expression

$$Q = \overline{(\bar{S} \cdot \bar{Q})}$$

does not change the result:

$$Q = \overline{(0 \cdot 0)} = 1$$

Thus, the cross-coupled feedback from outputs Q and $\bar{Q}$ to the input of the NAND gates is such that the set condition sustains itself. It is straightforward to show (by symmetry) that a 1 input on the R line causes the device to reset (i.e., causes $Q = 0$) and that this condition is also self-sustaining.

Example   12.26

What is the output of an RS flip-flop for the following series of inputs?

R	0	0	0	1	0	0	0
S	1	0	1	0	0	1	0

Solution:

On the basis of the rules stated earlier in this section, the state of the flip-flop can be described by means of the following truth table:

R	S	Q
0	1	1
0	0	1
0	1	1
1	0	0
0	0	0
0	1	1
0	0	1

A timing diagram can also be drawn, marking each transition:

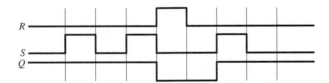

An extension of the RS flip-flop includes an additional enable input that is *gated* into each of the other two inputs. Figure 12.66 depicts an RS flip-flop consisting of two NOR gates. In addition, an enable input is connected through two AND gates to the RS flip-flop, so that an input to the R or S line will be effective only when the enable input is 1. Thus, any transitions will be controlled by the enable input, which acts as a synchronizing signal. The enable signal may consist of a **clock,** in which case the flip-flop is said to be **clocked** and its operation is said to be **synchronous.**

The same circuit of Figure 12.66 can be used to illustrate two additional features of flip-flops: the **preset** and **clear** functions, denoted by the inputs P and C, respectively. When P and C are 0, they do not affect the operation of the flip-flop. Setting $P = 1$ corresponds to setting $S = 1$, and therefore

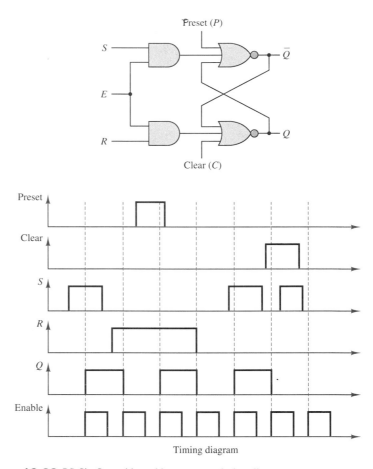

Figure 12.66 *RS* flip-flop with enable, preset, and clear lines

causes the flip-flop to go into the 1 state. Thus, the term *preset:* this function allows the user to preset the flip-flop to 1 at any time. When *C* is 1, the flip-flop is reset, or *cleared* (i.e., *Q* is made equal to 0). Note that these direct inputs are, in general, asynchronous; therefore, they allow the user to preset or clear the flip-flop at any time. A set of timing waveforms illustrating the function of the enable, preset, and clear inputs is also shown in Figure 12.66. Note how transitions occur only when the enable input goes high (unless the preset or clear inputs are used to overide the *RS* inputs).

Another extension of the *RS* flip-flop, called the **data latch,** is shown in Figure 12.67. In this circuit, the *R* input is always equal to the inverted *S* input, so that whenever the enable input is high, the flip-flop is set. This device has the dual advantage of avoiding the potential conflict that might arise if both *R* and *S* were high and reducing the number of input connections by eliminating the reset input. This circuit is called a data latch because once the enable input goes low, the flip-flop is latched to the previous value of the input. Thus, this device can serve as a basic memory element.

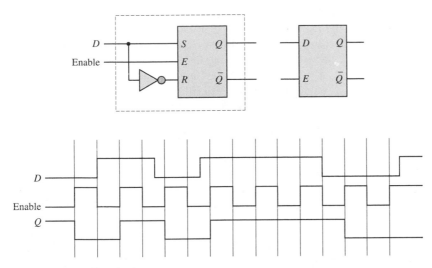

Figure 12.67 Data latch

D-Flip-Flop

The **D flip-flop** is an extension of the data latch that utilizes two *RS* flip-flops, as shown in Figure 12.68. In this circuit, a clock is connected to the enable

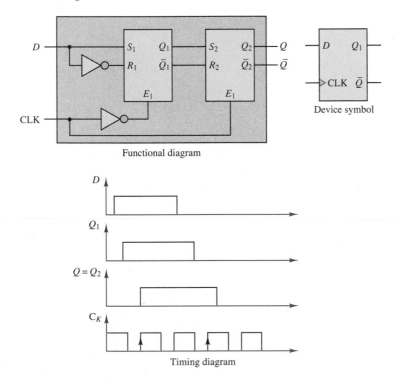

Figure 12.68 *D* flip-flop

input of each flip-flop. Since Q_1 sees an inverted clock signal, the latch is enabled when the clock waveform goes low. However, since Q_2 is disabled when the clock is low, the output of the D flip-flop will not switch to the 1 state until the clock goes high, enabling the second latch and transferring the state of Q_1 to Q_2. It is important to note that the D flip-flop changes state only on the positive edge of the clock waveform: Q_1 is set on the negative edge of the clock, and Q_2 (and therefore Q) is set on the positive edge of the clock, as shown in the timing diagram of Figure 12.68. This type of device is said to be **edge–triggered.** This feature is indicated by the "knife edge" drawing shown next to the CLK input in the device symbol. The particular device described here is said to be positive edge–triggered, or **leading edge–triggered,** since the final output of the flip-flop is set on a positive-going clock transition.

On the basis of the rules stated in this section, the state of the D flip-flop can be described by means of the following truth table:

D	CLK	Q
0	1	0
1	1	1

JK Flip-Flop

Another very common type of flip-flop is the **JK flip-flop,** shown in Figure 12.69. The JK flip-flop operates according to the following rules:

- When J and K are both low, no change occurs in the state of the flip-flop.
- When $J = 0$ and $K = 1$, the flip-flop is reset to 0.
- When $J = 1$ and $K = 0$, the flip-flop is set to 1.
- When both J and K are high, the flip-flop will toggle between states.

Functional diagram

Figure 12.69 JK flip-flop

Note that, functionally, the operation of the *JK* flip-flop can also be explained in terms of two *RS* flip-flops. When the clock waveform goes high, the "master" flip-flop is enabled; the "slave" receives the state of the master upon a negative clock transition. The "bubble" at the clock input signifies that the device is negative or **trailing edge–triggered.** This behavior is similar to that of an *RS* flip-flop, except for the $J = 1$, $K = 1$ condition, which corresponds to a toggle mode rather than to a disallowed combination of inputs.

Figure 12.70 depicts the truth table for the *JK* flip-flop. It is important to note that when both inputs are 0 the flip-flop remains in its previous state at the occurrence of a clock transition; when either input is high and the other is low, the *JK* flip-flop behaves like the *RS* flip-flop, whereas if both inputs are high, the output "toggles" between states every time the clock waveform undergoes a negative transition.

JK flip-flop

J_n	K_n	Q_{n+1}
0	0	Q_n
0	1	0 (reset)
1	0	1 (set)
1	1	$\bar{Q}_n$ (toggle)

Figure 12.70 Truth table for the *JK* flip-flop

Example   12.27   The *T* Flip-Flop

The **T flip-flop** is a *JK* flip-flop with the two inputs tied together (see Figure 12.71). The single input *T* flip-flop follows the transition rule $Q_{k+1} = Q_k$ if $T = 0$ and $Q_{k+1} = \bar{Q}_k$ if $T = 1$. As shown in the associated timing diagram, the *T* flip-flop toggles between the high and low states at a frequency half that of the clock when the *T* input is held high. It is easy to verify that since the *T* input is always high, the flip-flop acts so as to divide the input clock frequency by 2, as shown in the timing diagram.

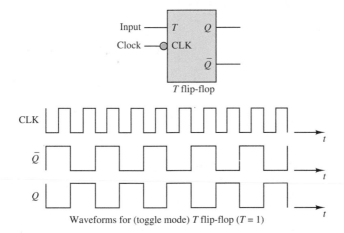

T flip-flop

Waveforms for (toggle mode) *T* flip-flop ($T = 1$)

Figure 12.71

Example   12.28

If the following inputs are applied (in the order shown) to a *JK* flip-flop, what is the output after each transition, and what is the final output ($Q_{\text{initial}} = 1$)?

Sequence:	J	0	1	0	1	0	0	1
	K	0	1	1	0	0	1	1

Solution:

Assume that the flip-flop is initially in the 1 state, that is, $Q_{init} = 1$. Then the output transition table is as follows:

J	K	Output (Q)
0	0	1
1	1	0
0	1	0
1	0	1
0	0	0
0	1	0
1	1	1

The timing diagram corresponding to these input/output combinations is as follows. Each line corresponds to a clock transition.

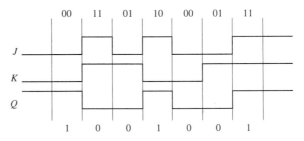

Digital Counters

One of the more immediate applications of flip-flops is in the design of **counters.** A counter is a sequential logic device that can take one of N possible states, stepping through these states in a sequential fashion. When the counter has reached its last state, it resets to zero and is ready to start counting again. For example, a three-bit binary counter would have $2^3 = 8$ possible states, and might appear as shown in the functional block of Figure 12.72. The input clock waveform causes the counter to step through the eight states, making one transition for each clock pulse. We shall shortly see that a string of JK flip-flops can accomplish this task exactly. The device shown in Figure 12.72 also displays a reset input, which forces the counter output to equal 0: $b_2 b_1 b_0 = 000$.

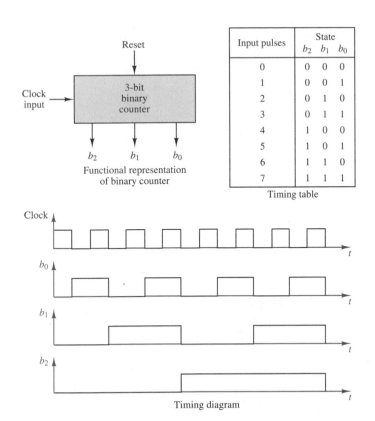

Input pulses	State		
	b_2	b_1	b_0
0	0	0	0
1	0	0	1
2	0	1	0
3	0	1	1
4	1	0	0
5	1	0	1
6	1	1	0
7	1	1	1

Timing table

Functional representation
of binary counter

Timing diagram

Figure 12.72 Binary counter

Although binary counters are very useful in many applications, one is often interested in a **decade counter,** that is, a counter that counts to 10 and then resets. A four-bit binary counter can easily be configured to provide this function by means of simple logic that resets the counter when it has reached the count $1010_2 = 10_{10}$. As shown in Figure 12.73, if we connect bits b_3 and b_1 to a four-input AND gate, along with $\overline{b_2}$ and $\overline{b_0}$, the output of the AND gate can be used to reset the counter after a count of 10. Additional logic can provide a "carry" bit whenever a reset condition is reached, which could be passed along to another decade counter, enabling counts up to 99. Decade counters can be cascaded so as to represent decimal digits in succession.

A simple implementation of the binary counter we have described in terms of its functional behavior is shown in Figure 12.74. The figure depicts a three-bit binary **ripple counter,** which is obtained from a cascade of three JK flip-flops. The transition table shown in the figure illustrates how the Q output of each stage becomes the clock input to the next stage, while each flip-flop is held in the toggle mode. The output transitions assume that the clock, CLK, is a simple square wave (all JKs are negative edge–triggered).

Input pulses	b_3	b_2	b_1	b_0
0	0	0	0	0
1	0	0	0	1
2	0	0	1	0
3	0	0	1	1
4	0	1	0	0
5	0	1	0	1
6	0	1	1	0
7	0	1	1	1
8	1	0	0	0
9	1	0	0	1
10	1	0	1	0

Figure 12.73 Decade counter

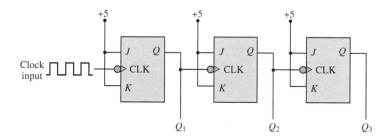

Input	Q_3	Q_2	Q_1
⎍	0	0	0
⎍	0	0	1
⎍	0	1	0
⎍	0	1	1
⎍	1	0	0
⎍	1	0	1
⎍	1	1	0
⎍	1	1	1

Figure 12.74 Ripple counter

This 3-bit ripple counter can easily be configured as a divide-by-8 mechanism, simply by adding an AND gate. To divide the input clock rate by 8, one output pulse should be generated for every eight clock pulses. If one were to output a pulse every time a binary 111 combination occurs, a simple AND gate would suffice to generate the required condition. This solution is shown in Figure 12.75. Note that the square wave is also included as an input to the AND gate; this ensures that the output is only as wide as the input signal. This application of ripple counters is further illustrated in the following example.

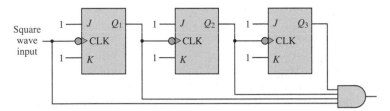

Figure 12.75 Divide-by-8 circuit

Example   12.29   **Divider Circuit**

A binary ripple counter provides a means to divide the fixed output rate of a clock by powers of 2. For example, the circuit of Figure 12.76 is divide-by-2 or a divide-by-4 counter. Draw the timing diagrams for the clock input, Q_0, and Q_1 to demonstrate these functions (assume positive edge–triggered flip-flops).

Figure 12.76

Solution:

To determine the operation of the divider circuit, a timing diagram is generated according to the rules for a positive edge–triggered JK flip-flop. From the diagram (page 658), it can be seen that Q_0 is switching at half the input clock frequency and Q_1 at half again the rate of Q_0, or one-fourth the original clock frequency.

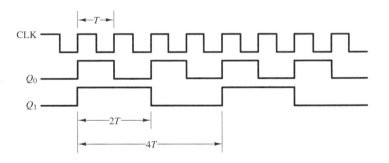

A slightly more complex version of the binary counter is the so-called **synchronous counter,** in which the input clock drives all of the flip-flops simultaneously. Figure 12.77 depicts a three-bit synchronous counter. In this figure, we have chosen to represent each flip-flop as a T flip-flop. The clocks to all the flip-flops are incremented simultaneously. The reader should verify that Q_0 toggles to 1 first and then Q_1 toggles to 1, and that the AND gate ensures that Q_2 will toggle only after Q_0 and Q_1 have both reached the 1 state ($Q_0 \cdot Q_1 = 1$).

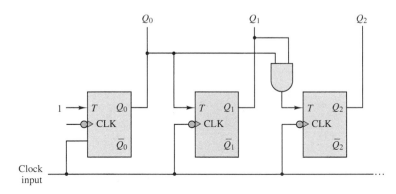

Figure 12.77 Three-bit synchronous counter

Other common counters are the **ring counter,** illustrated in Example 12.30, and the **up-down counter,** which has an additional select input that determines whether the counter counts up or down.

Example   12.30

The circuit shown in Figure 12.78 is called a **ring counter.** Analyze its operation by means of a timing diagram.

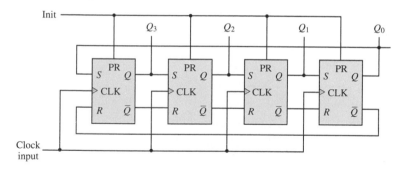

Figure 12.78 Ring counter

Solution:

The output from the first flip-flop (after initializing by means of a positive transition in the Init line) is $Q = 1$, $\bar{Q} = 0$. The complete sequence is as follows:

CLK	Q_3	Q_2	Q_1	Q_0
↑	1	0	0	0
↑	0	1	0	0
↑	0	0	1	0
↑	0	0	0	1
↑	1	0	0	0
↑	0	1	0	0
↑	0	0	1	0

This is called a ring counter because the logic 1 will propagate down the flip-flops and then rotate back to the first flip-flop in a circular fashion. This is how a shift register (introduced in the next section) would shift a value to the right at the rate of one bit per clock pulse.

Example   12.31

Another type of angular position encoder, besides the angular encoder discussed in Example 12.2, is the slotted encoder shown in Figure 12.79. This encoder can be used in conjunction with a pair of counters and a high-frequency clock to determine the speed of rotation of the slotted wheel. As shown in Figure 12.80, a clock of known frequency is connected to a counter while another counter records the number of slot pulses detected by an optical slot detector as the wheel rotates. Dividing the counter values, one could obtain the speed (in slots per second). For example, assume a clocking frequency of 1.2 kHz. If both counters are started at zero and at some instant the timer counter reads 2,850 and the encoder counter reads 3,050, then the speed of the rotating encoder is found to be:

1,200 Hz $\times$ (2,850 cycles/ 3,050 slots) = 1,284.2 slots per second

360 slots

360 slots; 1 increment = 1 degree

Figure 12.79

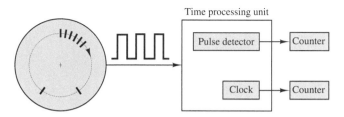

Time processing unit

Pulse detector → Counter

Clock → Counter

Figure 12.80 Calculating the speed of rotation of the slotted wheel

and

$$1{,}284.2 \text{ slots per second} \times 1° \text{ per slot} \times 2\pi/360 \text{ rad/degree} = 22.4 \text{ rad/s}$$

If this encoder is connected to a rotating shaft, it is possible to measure the angular position and velocity of the shaft. Such shaft encoders are used in measuring the speed of rotation of electric motors, machine tools, engines, and other rotating machinery. A typical application of the slotted encoder is to compute the ignition and injection timing in an automotive engine. In an automotive engine, information related to speed is obtained from the camshaft and the flywheel, which have known reference points. The reference points determine the timing for the ignition firing points and fuel injection pulses, and are identified by special slot patterns on the camshaft and crankshaft. Two methods are used to detect the special slots (reference points): *period measurement with additional transition detection (PMA)*, and *period measurement with missing transition detection (PMM)*. In the PMA method, an additional slot (reference point) determines a known reference position on the crankshaft or camshaft. In the PMM method, the reference position is determined by the absence of a slot. Figure 12.81 illustrates a typical PMA pulse sequence, showing the presence of an additional pulse. The additional slot may be used to determine the timing for the ignition pulses relative to a known position of the crankshaft. Figure 12.82 depicts a typical PMM pulse sequence. Because the period of the pulses is known, the additional slot or the missing slot can be easily detected and used as a reference position.

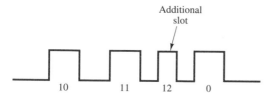

Figure 12.81 PMA pulse sequence

Figure 12.82 PMA pulse sequence

Registers

A register consists of a cascade of flip-flops that can store binary data, one bit in each flip-flop. The simplest type of register is the parallel input–parallel output register shown in Figure 12.83. In this register, the "load" input pulse, which acts on all clocks simultaneously, causes the parallel inputs $b_0 b_1 b_2 b_3$ to be transferred to the respective flip-flops. The D flip-flop employed in this register allows the transfer from b_n to Q_n to occur very directly. Thus, D flip-flops are very commonly used in this type of application. The binary word $b_3 b_2 b_1 b_0$ is now "stored," each bit being represented by the state of a flip-flop. Until the "load" input is applied again and a new word appears at the parallel inputs, the register will preserve the stored word.

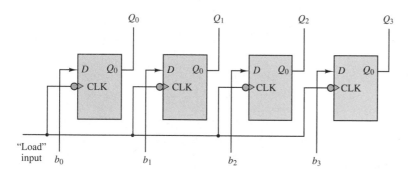

Figure 12.83 Four-bit parallel register

The construction of the parallel register presumes that the N-bit word to be stored is available in parallel form. However, it is often true that a binary word will arrive in serial form, that is, one bit at a time. A register that can accommodate this type of logic signal is called a **shift register.** Figure 12.84 illustrates how the same basic structure of the parallel register applies to the shift register, except that the input is now applied to the first flip-flop and shifted along at each clock pulse. Note that this type of register provides both a serial and a parallel output.

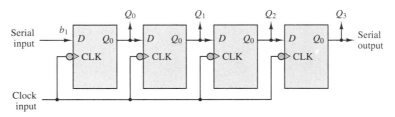

Figure 12.84 Four-bit shift register

Data sheets for some common flip-flops and registers are shown at the end of the chapter.

Example   12.32   Seven-Segment Displays

A **seven-segment display** is a very convenient device for displaying digital data. The display is shown in Figure 12.85. Operation of seven-segment display requires a decoder circuit to light the proper combinations of segments corresponding to the sixteen characters associated with the hexadecimal code.

This display, with the appropriate decoder driver, is capable of displaying values ranging from 0 to 15_{10}, or 0 to F in hex.

Figure 12.85 Seven-segment display

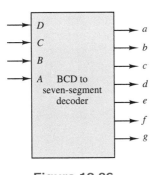

Figure 12.86

A typical BCD to seven-segment decoder function block is shown in Figure 12.86, where the lowercase letters correspond to the segments shown in Figure 12.85. The decoder features four data inputs (A,B,C,D), which are used to light the appropriate segment(s). The outputs of the decoder are connected to the seven-segment display. The decoder will light up the appropriate segments corresponding to the incoming value. A BCD to seven-segment decoder function is similar to the 2-to-4 decoder function described earlier and shown in Figure 12.61.

DRILL EXERCISES

12.31   The circuit shown in Figure 12.87 also serves as an *RS* flip-flop and requires only two NOR gates. Analyze the circuit to prove that it operates as an *RS* flip-flop. [*Hint:* Use a truth table with two variables, *S* and *R*.]

12.32   Derive the detailed truth table and draw a timing diagram for the *JK* flip-flop, using the two–flip-flop model of Figure 12.69.

Figure 12.87

12.33   The speed of the rotating encoder of Example 12.31 is found to be 9,425 rad/s. The encoder timer reads 10 and the clock counter reads 300. Assuming that both the timer counter and the encoder counter start at zero, find the clock frequency.

Answer: 1.5 kHz

CONCLUSION

- Digital logic circuits are at the basis of digital computers. Such circuits operate strictly on binary signals according to the laws of Boolean algebra.
- Logic circuits may be grouped into one of two families: combinational logic circuits, which implement logic functions without special consideration of timing issues; and sequential logic circuits, in which timing plays a major role.
- Part of the former group are all of the logic gates—AND, OR, NAND, NOT, and XOR—as well as logic modules such as multiplexers and read-only memory. These circuits can be used to implement logic functions.
- Sequential circuits, on the other hand, operate under timing constraints, typically provided by an external clock; the basic sequential element is the flip-flop, of which there exist several types, among them RS, D, JK, and T flip-flops. Flip-flops constitute the basic memory element in digital computers, and are also essential in the operation of digital counters, of registers, and of memories.
- The function of these logic circuits in a digital computer will become more evident when the architecture of digital computers is explored in the next chapter.

KEY TERMS

Analog Signal   *603*

Digital Signal   *603*

Binary Signal   *605*

Edge   *606*

Base 2   *607*

Base 10   *607*

Bit   *607*

Least Significant Bit (LSB)   *607*

Most Significant Bit (MSB)   *607*

Nibble   *607*

Byte   *607*

Word   *607*

Complement   *609*

One's Complement   *609*

Two's Complement   *609*

Sign-Magnitude Convention   *610*

Hexadecimal System   *611*

Binary-Coded Decimal (BCD) Representation   *612*

Gray Code   *612*

Position Encoder   *613*

Boolean Algebra   *617*

Logical Addition (OR)   *617*

Logical Multiplication (AND)   *617*

OR Gate   *617*

AND Gate   *617*

Truth Table   *617*

NOT Gate   *618*

DATA SHEETS FOR SELECTED LOGIC MODULES AND SEQUENTIAL LOGIC DEVICES

Common IC Logic Gates

TTL Name	CMOS Name	Description
SN74LS04	MM74HC04	Hex Inverter
SN74LS08	MM74HC08	Quad 2-Input AND Gate
SN74LS32	MM74HC32	Quad 2-Input OR Gate
SN74LS00	MM74HC00	Quad 2-Input NAND Gate
SN74LS02	MM74HC02	Quad 2-Input NOR Gate
SN74LS86	MM74HC86	Quad 2-Input Exclusive OR Gate
SN74LS266	MM74HC266	Quad 2-Input Exclusive NOR Gate
SN74LS10	MM74HC10	Triple 3-Input NAND Gate
SN74LS27	MM74HC27	Triple 3-Input NOR Gate
SN74LS20	MM74HC20	Dual 4-Input NAND Gate
SN74ALS25	CD4002C	Dual 4-Input NOR Gate
SN74LS30	MM74HC30	8-Input NAND Gate

MM54HC138/MM74HC138   3-to-8 Line Decoder

Features

- Typical propagation delay: 20 ns

- Wide power supply range: 2V–6V

- Low quiescent current: 80 μA maximum (74HC Series)

- Low input current: 1 μA maximum

- Fanout of 10 LS-TTL loads

Connection Diagram

Truth Table

Inputs					Outputs							
Enable		Select										
G1	$\overline{G2}$*	C	B	A	Y0	Y1	Y2	Y3	Y4	Y5	Y6	Y7
X	H	X	X	X	H	H	H	H	H	H	H	H
L	X	X	X	X	H	H	H	H	H	H	H	H
H	L	L	L	L	L	H	H	H	H	H	H	H
H	L	L	L	H	H	L	H	H	H	H	H	H
H	L	L	H	L	H	H	L	H	H	H	H	H
H	L	L	H	H	H	H	H	L	H	H	H	H
H	L	H	L	L	H	H	H	H	L	H	H	H
H	L	H	L	H	H	H	H	H	H	L	H	H
H	L	H	H	L	H	H	H	H	H	H	L	H
H	L	H	H	H	H	H	H	H	H	H	H	L

* $\overline{G2}$ = G2A + G2B

H = high level, L = low level, X = don't care

Operating Conditions

	Min	Max	Units
Supply Voltage (V_{CC})	2	6	V
DC Input or Output Voltage (V_{IN}, V_{OUT})	0	V_{CC}	V
Operating Temp. Range (T_A)			
MM74HC	−40	+85	°C
MM54HC	−55	+125	°C
Input Rise or Fall Times			
(t_r, t_f) V_{CC} = 2.0V		1000	ns
V_{CC} = 4.5V		500	ns
V_{CC} = 6.0V		400	ns

MM54HC139/MM74HC139   Dual 2-to-4 Line Decoder

Features

- Typical propagation delays—

  Select to outputs (4 delays): 18 ns

  Select to output (5 delays): 28 ns

  Enable to output: 20 ns

- Low power: 40 μW quiescent supply power

- Fanout of 10 LS-TTL devices

- Input current maximum 1 μA, typical 10 pA

Connection Diagram

Truth Table

'HC139

Inputs			Outputs			
Enable	Select					
G	B	A	Y0	Y1	Y2	Y3
H	X	X	H	H	H	H
L	L	L	L	H	H	H
L	L	H	H	L	H	H
L	H	L	H	H	L	H
L	H	H	H	H	H	L

H = high level, L = low level, X = don't care

Operating Conditions

	Min	Max	Units
Supply Voltage (V_{CC})	2	6	V
DC Input or Output Voltage ($V_{IN, OUT}$)	0	V_{CC}	V
Operating Temp. Range (T_A)			
MM74HC	−40	+85	°C
MM54HC	−55	+125	°C
Input Rise or Fall Times			
(t_r, t_f) V_{CC} = 2.0V		1000	ns
V_{CC} = 4.5V		500	ns
V_{CC} = 6.0V		400	ns

MM54HC153/MM74HC153   Dual 4-Input Multiplexer

Features

- Typical propagation delay: 24 ns

- Wide power supply range: 2V–6V

- Low quiescent current: 80 μA maximum (74HC Series)

- Low input current: 1 μA maximum

- Fanout of 10 LS-TTL loads

Connection Diagram

Dual-In-Line Package

Top View

Operating Conditions

	Min	Max	Units
Supply Voltage (V_{CC})	2	6	V
DC Input or Output Voltage (V_{IN}, V_{OUT})	0	V_{CC}	V
Operating Temp. Range (T_A)			
MM74HC	−40	+85	°C
MM54HC	−55	+125	°C
Input Rise or Fall Times			
(t_r, t_f) V_{CC} = 2.0V		1000	ns
V_{CC} = 4.5V		500	ns
V_{CC} = 6.0V		400	ns

Truth Table

Select Inputs		Data Inputs				Strobe	Output
B	A	C0	C1	C2	C3	G	Y
X	X	X	X	X	X	H	L
L	L	L	X	X	X	L	L
L	L	H	X	X	X	L	H
L	H	X	L	X	X	L	L
L	H	X	H	X	X	L	H
H	L	X	X	L	X	L	L
H	L	X	X	H	X	L	H
H	H	X	X	X	L	L	L
H	H	X	X	X	H	L	H

Select inputs A and B are common to both sections.

H = high level, L = low level, X = don't care.

MM54HC73/MM74HC73   Dual J-K Flip-Flops with Clear

Connection Diagram

Features

- Typical propagation delay: 16 ns
- Wide operating voltage range: 2–6V
- Low input current: 1 μA maximum
- Low quiescent current: 40 μA maximum (74HC Series)
- High output drive: 10 LS-TTL loads

Dual-In-Line Package

Top View

Truth Table

Inputs				Outputs	
CLR	CLK	J	K	Q	$\bar{Q}$
L	X	X	X	L	H
H	↓	L	L	Q0	$\bar{Q}$0
H	↓	H	L	H	L
H	↓	L	H	L	H
H	↓	H	H	TOGGLE	
H	H	X	X	Q0	$\bar{Q}$0

Operating Conditions

	Min	Max	Units
Supply Voltage (V_{CC})	2	6	V
DC Input or Output Voltage (V_{IN}, V_{OUT})	0	V_{CC}	V
Operating Temp. Range (T_A)			
MM74HC	−40	+85	°C
MM54HC	−55	+125	°C
Input Rise or Fall Times			
(t_r, t_f) V_{CC} = 2.0V		1000	ns
V_{CC} = 4.5V		500	ns
V_{CC} = 6.0V		400	ns

MM54HC76/MM74HC76   Dual J-K Flip-Flops with Preset and Clear

Connection Diagram

Features

- Typical propagation delay: 16 ns
- Wide operating voltage range
- Low input current: 1 μA maximum
- Low quiescent current: 40 μA maximum (74HC Series)
- High output drive: 10 LS-TTL loads

Dual-In-Line Package

Top View

Truth Table

Inputs					Outputs	
PR	CLR	CLK	J	L	Q	Q̄
L	H	X	X	X	H	L
H	L	X	X	X	L	H
L	L	X	X	X	L*	L*
H	H	↓	L	L	Q0	Q̄0
H	H	↓	H	L	H	L
H	H	↓	L	H	L	H
H	H	↓	H	H	TOGGLE	
H	H	H	X	X	Q0	Q̄0

*This is an unstable condition, and is not guaranteed

Operating Conditions

	Min	Max	Units
Supply Voltage (V_{CC})	2	6	V
DC Input or Output Voltage (V_{IN}, V_{OUT})	0	V_{CC}	V
Operating Temp. Range (T_A)			
MM74HC	−40	+85	°C
MM54HC	−55	+125	°C
Input Rise or Fall Times			
(t_r, t_f) V_{CC} = 2.0V		1000	ns
V_{CC} = 4.5V		500	ns
V_{CC} = 6.0V		400	ns

MM54HC74A/MM74HC74A   Dual D Flip-Flops with Preset and Clear

Features

- Typical propagation delay: 20 ns

- Wide power supply range: 2–6V

- Low quiescent current: 40 μA maximum (74HC Series)

- Low input current: 1 μA maximum

- Fanout of 10 LS-TTL loads

Connection Diagram

Dual-In-Line Package

TOP VIEW

Truth Table

Inputs				Outputs	
PR	CLR	CLK	D	Q	Q̄
L	H	X	X	H	L
H	L	X	X	L	H
L	L	X	X	H*	H*
H	H	↑	H	H	L
H	H	↑	L	L	H
H	H	L	X	Q0	Q̄0

Note: Q0 = the level of Q before the indicated input conditions were established.

* This configuration is nonstable; that is, it will not persist when preset and clear inputs return to their inactive (high) level.

Operating Conditions

	Min	Max	Units
Supply Voltage (V_{CC})	2	6	V
DC Input or Output Voltage (V_{IN}, V_{OUT})	0	V_{CC}	V
Operating Temp. Range (T_A)			
MM74HC	−40	+85	°C
MM54HC	−55	+125	°C
Input Rise or Fall Times			
(t_r, t_f) V_{CC} = 2.0V		1000	ns
V_{CC} = 4.5V		500	ns
V_{CC} = 6.0V		400	ns

MM54HCT193/MM74HCT193   Synchronous Binary Up/Down Counters

Features

- Low quiescent current: 80 μA maximum (74HCT Series)
- Low input current: 1 μA maximum
- TTL compatible inputs

Connection Diagram

Dual-In-Line Package

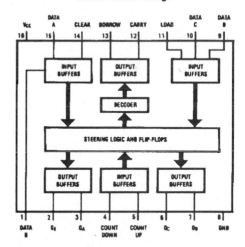

Truth Table

| Count | | Clear | Load | Function |
Up	Down			
↑	H	L	H	Count Up
H	↑	L	H	Count Down
X	X	H	X	Clear
X	X	L	L	Load

H = high level
L = low level
↑ = transition from low-to-high
X = don't care

Operating Conditions

	Min	Max	Units
Supply Voltage (V_{CC})	4.5	5.5	V
DC Input or Output Voltage			
(V_{IN}, V_{OUT})	0	V_{CC}	V
Operating Temp. Range (T_A)			
MM74HCT	−40	+85	°C
MM54HCT	−55	+125	°C
Input Rise or Fall Times			
(t_r, t_f)		500	ns

Logic Waveforms

'HCT193 Synchronous Binary Counters
Typical Clear, Load, and Count Sequences

Sequence:
(1) Clear outputs to zero.
(2) Load (preset) to binary thirteen
(3) Count up to fourteen, fifteen, carry, zero, one, and two.
(4) Count down to one, zero, borrow, fifteen, fourteen, and thirteen.
Note A: Clear overrides load data, and count inputs.
Note B: When counting up, count-down input must be high; when counting down, count-up input must be high.

HOMEWORK PROBLEMS

12.1 Convert the following base 10 numbers to hex and binary:

a.   401
b.   273
c.   15

d.   38
e.   56

12.2 Convert the following hex numbers to base 10 and binary:

a.   A
b.   66
c.   47

d.   21
e.   13

12.3 Convert the following base 10 numbers to binary:

a.   271.25
b.   53.375

c.   37.32
d.   54.27

12.4 Convert the following binary numbers to hex and base 10:

a. 1111
b. 1001101
c. 1100101

d. 1011100
e. 11101
f. 101000

12.5 Perform the following additions all in the binary system:

a. 11001011 + 101111
b. 10011001 + 1111011
c. 11101001 + 10011011

12.6 Perform the following subtractions all in the binary system:

a. 10001011 − 1101111
b. 10101001 − 111011
c. 11000011 − 10111011

12.7 Assuming that the most significant bit is the sign bit, find the decimal value of the following eight-bit binary numbers:

a. 11111000
b. 10011111
c. 01111001

12.8 Find the binary representation of the following decimal numbers:

a. 126
b. −126

c. 108
d. −98

12.9 Find the two's complement of the following binary numbers:

a. 1111
b. 1001101

c. 1011100
d. 11101

12.10 Using the method of proof by perfect induction, show the validity of rule 17 from Table 12.11.

12.11 Using the method of proof by perfect induction, show the validity of rule 18 from Table 12.11.

12.12 Using the method of proof by perfect induction, show that

$$(X + Y) \cdot (\bar{X} + X \cdot Y) = Y$$

12.13 Using De Morgan's theorems and the rules of Boolean algebra, simplify the following logic function:

$$F(X, Y, Z) = \bar{X} \cdot \bar{Y} \cdot \bar{Z} + \bar{X} \cdot Y \cdot Z + X \cdot (\overline{Y + Z})$$

12.14 Using De Morgan's theorems and the rules of Boolean algebra, simplify the following logic function:

$$F(X, Y, Z) = \overline{(X + Y)} \cdot \bar{Z} + \overline{X \cdot \bar{Y} \cdot \bar{Z}}$$

12.15 Simplify the logic function $F(A,B,C) = \bar{A} \cdot B \cdot \bar{C} + \bar{A} \cdot B \cdot C + A \cdot B \cdot \bar{C} + A \cdot B \cdot C$ using Boolean algebra.

12.16 Find the logic function defined by the truth table given in Figure P12.1.

12.17 What is the logic function of the circuit of Figure P12.2?

A	B	C	F
0	0	0	0
0	0	1	1
0	1	0	0
0	1	1	1
1	0	0	1
1	0	1	1
1	1	0	1
1	1	1	1

Figure P12.1

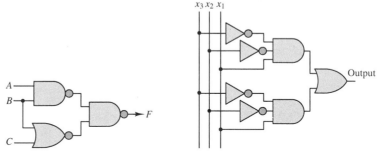

Figure P12.2   **Figure P12.3**

12.18 Use a truth table to show when the output of the circuit of Figure P12.3 is 1.

12.19 Find the logic function corresponding to the truth table of Figure P12.4.

A	B	C	F
0	0	0	1
0	0	1	0
0	1	0	0
0	1	1	0
1	0	0	1
1	0	1	0
1	1	0	1
1	1	1	1

Figure P12.4

12.20 Find the minimum expression for the output of the logic circuit shown in Figure P12.5.

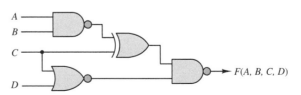

Figure P12.5

12.21 Using a Karnaugh map, find the minimum expression for the output of the logic circuit shown in Figure P12.6.

12.22 Determine the minimum expression for the following logic function, using a Karnaugh map:

$$f(A,B,C) = (A + B) \cdot A \cdot B + \bar{A} \cdot C + A \cdot \bar{B} \cdot C + \bar{B} \cdot \bar{C}$$

Figure P12.6

12.23

a. Build the Karnaugh map for the logic function defined by the truth table of Figure P12.7.

b. What is the minimum expression for this function?

c. Realize F using AND, OR, and NOT gates.

A	B	C	D	F
0	0	0	0	1
0	0	0	1	1
0	0	1	0	1
0	0	1	1	0
0	1	0	0	0
0	1	0	1	0
0	1	1	0	1
0	1	1	1	1
1	0	0	0	1
1	0	0	1	1
1	0	1	0	1
1	0	1	1	1
1	1	0	0	0
1	1	0	1	0
1	1	1	0	0
1	1	1	1	0

Figure P12.7

12.24 Fill in the Karnaugh map for the function defined by the truth table of Figure P12.8, and find the minimum expression for the function.

A	B	C	$f(A,B,C)$
0	0	0	0
0	0	1	1
0	1	0	1
0	1	1	0
1	0	0	1
1	0	1	0
1	1	0	0
1	1	1	1

Figure P12.8

12.25

a. Fill in the Karnaugh map for the logic function defined by the truth table of Figure P12.9.

b. What is the minimum expression for the function?

c. Realize the function, using only NAND gates.

12.26 Determine the minimum expression for the logic function

$$f(A,B,C,D) = A \cdot \bar{B} + A \cdot B \cdot \bar{C} + \bar{A} \cdot C \cdot D + A \cdot C \cdot \bar{D} + A \cdot B \cdot C \cdot D$$

A	B	C	D	F
0	0	0	0	1
0	0	0	1	1
0	0	1	0	1
0	0	1	1	1
0	1	0	0	0
0	1	0	1	1
0	1	1	0	0
0	1	1	1	1
1	0	0	0	1
1	0	0	1	1
1	0	1	0	0
1	0	1	1	0
1	1	0	0	0
1	1	0	1	1
1	1	1	0	1
1	1	1	1	0

Figure P12.9

12.27 What is the minimum expression for the output of the logic circuit shown in Figure P12.10?

Figure P12.10

12.28 Construct the Karnaugh map for the logic function defined by the truth table of Figure P12.11, and find the minimum expression for the function.

12.29 Simplify the following expression:

$$(C + \bar{A} \cdot B) \cdot [(\bar{A} \cdot \bar{B} + B) \cdot C + A]$$

12.30 Find the minimum output expression for the circuit of Figure P12.12.

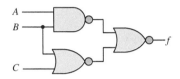

Figure P12.12

A	B	C	D	F
0	0	0	0	1
0	0	0	1	0
0	0	1	0	1
0	0	1	1	0
0	1	0	0	0
0	1	0	1	0
0	1	1	0	0
0	1	1	1	1
1	0	0	0	1
1	0	0	1	0
1	0	1	0	1
1	0	1	1	0
1	1	0	0	1
1	1	0	1	1
1	1	1	0	1
1	1	1	1	0

Figure P12.11

12.31 A logic circuit is to have three inputs, A, B, and C, and one output, F. The function F is to be a logic 1 if and only if an odd number of the input variables are at the logic 1 level. Make a truth table and a Karnaugh map for this function, and find the minimum expression for F.

12.32 Minimize the expression described in the truth table of Figure P12.13, and draw the circuit.

A	B	C	F
0	0	0	1
0	0	1	1
0	1	0	0
0	1	1	1
1	0	0	1
1	0	1	1
1	1	0	1
1	1	1	0

Figure P12.13

Figure P12.14

12.33 Find the minimum expression for the output of the logic circuit of Figure P12.14.

12.34   Simplify the following function (use a Karnaugh map) and draw a circuit:

$$F = \bar{A} \cdot C + A \cdot \bar{B} \cdot \bar{D} + \bar{A} \cdot \bar{C} \cdot \bar{D} + \bar{A} \cdot B \cdot \bar{C}$$

12.35   Find the minimum expression for the logic function at the output of the logic circuit of Figure P12.15.

12.36

a.   Complete the truth table for the circuit of Figure P12.16.

b.   What mathematical function does this circuit perform, and what do the outputs signify?

c.   How many standard 14-pin ICs would it take to construct this circuit?

Figure P12.15

x	y	C	S
0	0		
0	1		
1	0		
1	1		

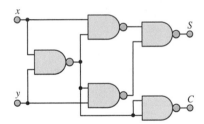

Figure P12.16

A	B	C	D	$f(A,B,C,D)$
0	0	0	0	1
0	0	0	1	0
0	0	1	0	1
0	0	1	1	1
0	1	0	0	0
0	1	0	1	1
0	1	1	0	0
0	1	1	1	0
1	0	0	0	0
1	0	0	1	1
1	0	1	0	0
1	0	1	1	0
1	1	0	0	1
1	1	0	1	0
1	1	1	0	1
1	1	1	1	1

Figure P12.17

12.37

a.   Fill in the Karnaugh map for the logic function defined by the truth table of Figure P12.17.

b.   What is the minimum expression for the function?

c.   Realize the function using a 1-of-8 multiplexer.

12.38

a.   Fill in the Karnaugh map for the function defined in the truth table of Figure P12.18.

b.   What is the minimum expression for the function?

c.   Draw the circuit, using AND, OR, and NOT gates.

A	B	C	$f(A,B,C)$
0	0	0	1
0	0	1	1
0	1	0	0
0	1	1	1
1	0	0	1
1	0	1	1
1	1	0	1
1	1	1	0

Figure P12.18

12.39

a.   Fill in the Karnaugh map for the logic function defined by the truth table of Figure P12.19.

b. What is the minimum expression for the function?

A	B	C	D	F
0	0	0	0	1
0	0	0	1	0
0	0	1	0	1
0	0	1	1	0
0	1	0	0	0
0	1	0	1	0
0	1	1	0	0
0	1	1	1	1
1	0	0	0	1
1	0	0	1	0
1	0	1	0	1
1	0	1	1	0
1	1	0	0	1
1	1	0	1	1
1	1	1	0	1
1	1	1	1	0

Figure P12.19

A	B	C	D	F
0	0	0	0	1
0	0	0	1	1
0	0	1	0	1
0	0	1	1	1
0	1	0	0	0
0	1	0	1	1
0	1	1	0	0
0	1	1	1	1
1	0	0	0	1
1	0	0	1	1
1	0	1	0	0
1	0	1	1	0
1	1	0	0	0
1	1	0	1	1
1	1	1	0	1
1	1	1	1	0

Figure P12.20

Figure P12.21

Figure P12.22

Figure P12.23

12.40

a. Fill in the Karnaugh map for the logic function defined by the truth table of Figure P12.20.

b. What is the minimum expression for the function?

c. Realize the function, using only NAND gates.

12.41   Design a circuit with a four-bit input representing the binary number $A_3A_2A_1A_0$. The output should be a 1 if the input value is divisible by 3. Assume that the circuit is to be used only for the digits 0 through 9 (thus, values for 10 to 15 can be don't cares).

a. Draw the Karnaugh map and truth table for the function.

b. Determine the minimum expression for the function.

c. Draw the circuit, using only AND, OR, and NOT gates.

12.42   Find the simplified sum-of-products representation of the function from the Karnaugh map shown in Figure P12.21. Note that **x** is the don't care term.

12.43   Find the simplified sum-of-products representation of the function from the Karnaugh map shown in Figure P12.22.

12.44   Find the simplified sum-of-products representation of the function from the Karnaugh map shown in Figure P12.23.

12.45

a. Fill in the truth table for the multiplexer circuit shown in Figure P12.24.

b. What binary function is performed by these multiplexers?

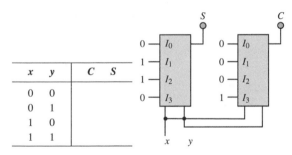

x	y	C	S
0	0		
0	1		
1	0		
1	1		

Figure P12.24

12.46 The circuit of Figure P12.25 can operate as a 4-to-16 decoder. Terminal EN denotes the enable input. Describe the operation of the 4-to-16 decoder. What is the role of logic variable A?

12.47 A 256×4 memory can be organized using linear selection. Draw a block diagram of the logical organization of the memory.

12.48 Draw a block diagram of the internal structure of a $1{,}024 \times 1$ SRAM using linear selection.

12.49 Suppose that you want to use a D flip-flop for a laboratory experiment. However, you have only T flip-flops. Assuming that you have all the logic gates available, make a D flip-flop using a T flip-flop and some logic gate(s).

12.50 Draw a timing diagram (four complete clock cycles) for A_0, A_1, and A_2 for the circuit of Figure P12.26. Assume that all initial values are 0. Note that all flip-flops are positive edge-triggered.

Figure P12.25

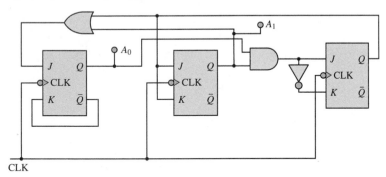

Figure P12.26

12.51 Assume that the slotted encoder shown in Figure P12.27 has a length of 1 meter and a total of 1,000 slots (i.e., there is one slot per millimeter). If a counter

Figure P12.27

is incremented by 1 each time a slot goes past a sensor, design a digital counting system that determines the speed of the moving encoder (in meters per second).

12.52 Find the output Q for the circuit of Figure P12.28.

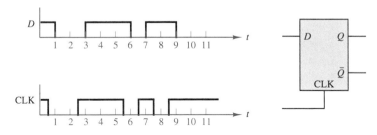

Figure P12.28

12.53 Describe how the ripple counter works. Why is it so named? What disadvantages can you think of for this counter?

12.54 Write the truth table for an RS flip-flop with enable (E), preset (P), and clear (C) lines.

12.55 A JK flip-flop is wired as shown in Figure P12.29 with a given input signal. Assuming that Q is at logic 0 initially and the trailing edge triggering is effective, sketch the output Q.

12.56 With reference to the JK flip-flop of Problem 12.55, assume that the output at the Q terminal is made to serve as the input to a second JK flip-flop wired exactly as the first. Sketch the Q output of the second flip-flop.

12.57 Assume that there is a flip-flop with the characteristic given in Figure P12.30, where A and B are the inputs to the flip-flop and Q is the next state output. Using necessary logic gates, make a T flip-flop from this flip-flop.

Figure P12.29

A	B	Q
0	0	$\bar{q}$
0	1	q
1	0	q
1	1	0

Figure P12.30

Digital Systems

T he objective of this chapter is to describe the application of modern microcomputer systems to data acquisition and control functions. Microcomputers have become a standard tool in the analysis of engineering data, in the design of experiments, and in the control of plants and processes. No longer a specialized electronic device to be used only by appropriately trained computer engineers, today's microcomputer—represented perhaps more commonly by the ubiquitous personal computer—is a basic tool in the engineering profession. The common thread in the application of microcomputers to various engineering fields is the use of the computer in the design of digital data acquisition instruments and of digital controllers.

Modern microcomputers are programmed with relative ease, have significant computing power and excellent memory storage capabilities, and are designed to be readily interfaced with other instruments and electronic devices, such as transducers, printers, or other computers. The basic functions performed by the microcomputer in a typical digital data acquisition or control application are easily described: input signals (often analog, sometimes already in digital form) are acquired by the computer and processed by means of suitable software (i.e., subjected to some mathematical manipulation) to produce the desired result, which is then outputted to either a display or a storage device, or is used in controlling a process, a plant, or an experiment.

This chapter will describe these various processes, with the aim of providing enough background information to enable you to understand the notation used in data books and instruction manuals. After an overview of digital data acquisition and control systems, the chapter describes the conversion of signals from analog to digital form, and back. The architecture of microcomputer systems is then discussed, with specific examples of practical systems. Finally, the issue of communications between computers is addressed.

13.1   DIGITAL SYSTEMS

To bring the broad range of applicability of microcomputers in engineering into perspective, it will be useful to stop for a minute to consider the possible application of microcomputer systems to different fields. The following list—by no means exhaustive—provides a few suggestions; it would be a useful exercise to imagine other likely applications in your own discipline.

Civil engineering	Measurement of stresses and vibration in structures
Chemical engineering	Process control
Industrial engineering	Control of manufacturing processes
Material and metallurgical engineering	Measurement of material properties
Marine engineering	Instrumentation to determine ship location, ship propulsion control
Aerospace engineering	Instrumentation for flight control and navigation
Mechanical engineering	Mechanical measurements, robotics, control of machine tools
Nuclear engineering	Radiation measurement, reactor instrumentation
Biomedical engineering	Measurement of physiological functions (e.g., electrocardiography and electroencephalography), control of experiments

The massive presence of microcomputers in engineering laboratories and in plants and production facilities can be explained by considering the numerous advantages the computer can afford over conventional instrumentation and control technologies. Consider, for example, the following points:

- A single microcomputer can perform computations and send signals from many different sensors measuring different parameters to many different display, storage, or control devices, under control of a single software program.
- The microcomputer is easily reprogrammed for any changes or adjustments to the measurement or control procedures, or in the computations.

• A permanent record of the activities performed by the microcomputer can be easily stored and retained.

It should be evident that the microcomputer can perform repetitive tasks, or tasks that require great accuracy and repeatability, far better than could be expected of human operators and analog instruments. What, then, does constitute a **digital data acquisition and control system?** Figure 13.1 depicts the basic blocks that form such a system. In the figure, the user of the microcomputer system is shown to interact with the microcomputer by means of software, often called **application software.** Application software is a collection of programs written either in **high-level languages,** such as FORTRAN, Pascal, or C, or in **assembly language** (a programming language very close to the internal code used by the microcomputer). The particular application software used may be commercially available or may be provided by the user; a combination of these two cases is the norm.

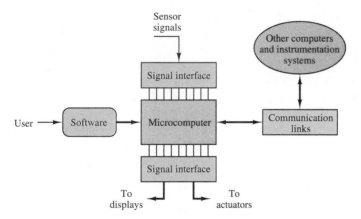

Figure 13.1 Structure of a digital data acquisition and control system

The signals that originate from real-world sensors—signals related to temperatures, vibration, or flow, for example—are interfaced to the micro-computer by means of specialized circuitry that converts analog signals to digital form and times the flow of information into the microcomputer using a clock reference, which may be internal to the microcomputer or externally provided. The heart of the signal interface unit is the *analog-to-digital con-verter,* or *ADC,* which will be discussed in some detail in this chapter. Not all sensor signals are analog, though. For example, the position of a switch or an on-off valve might be of interest; signals of this type are binary in nature, and the signal interface unit can route such signals directly to the microcomputer. Once the sensor data has been acquired and converted to digital form, the microcomputer can perform computations on the data and either display or store the results, or generate command outputs to actuators through another signal interface. Actuators are devices that can generate a physical output (e.g., force, heat, flow, pressure) from an electrical input.

Some actuators can be controlled directly by means of a digital signal (e.g., an on-off valve), but some require an analog input voltage or current, which can be obtained from the digital signal generated by the microcomputer by means of a *digital-to-analog converter,* or *DAC.*

In addition to the program control exercised by the user, the microcomputer may also respond to inputs originating from other computer and instrumentation systems through appropriate **communication links,** which also permit communication in the reverse direction. Thus, a microcomputer system dedicated to a complex task may consist of several microcomputers tied over a **communication network.**

13.2   MICROCOMPUTER ARCHITECTURE

Prior to delving into a description of how microcomputers interface with external devices (such as sensors and actuators) and communicate with the outside world, it will be useful to discuss the general architecture of a microcomputer, in order to establish a precise nomenclature and paint a clear picture of the major functions required in the operation of a typical microcomputer.

The general structure of a microcomputer is shown in Figure 13.2. It should be noted immediately that each of the blocks that are part of the microcomputer is connected with the **CPU bus,** which is the physical wire connection allowing each of the subsystems to communicate with the others. In effect, the CPU bus is simply a set of wires; note, however, that since only one set of signals can travel over the data bus at any one time, it is extremely important that the transmission of data between different parts of the microcomputer (e.g., from the A/D unit to memory) be managed properly, to prevent interference with other functions (e.g., the display of unwanted data

Figure 13.2 Microcomputer architecture

on a video terminal). As will be explained shortly, the task of managing the operation of the CPU bus resides within the **central processing unit,** or **CPU.** The CPU has the task of managing the flow of data and coordinating the different functions of the microcomputer, in addition to performing the data processing—in effect, the CPU is the heart and brains of the microcomputer. Some of the major functions of the CPU will be discussed in more detail shortly.

One of the important features of the microcomputer is its ability to store data. This is made possible by different types of **memory** elements: **read-only memory,** or **ROM; read-and-write memory,** or **RAM;** and **mass storage memory,** such as a hard drive or floppy disk, tape, or optical drives. ROM is *nonvolatile memory:* it will retain its data whether the operating power is on or off. ROM memory contains software programs that are used frequently by the microcomputer; one example is the *bootstrap program,* which is necessary to first start up the computer when power is turned on. RAM is memory that can be accessed very rapidly by the CPU; (data can be either read from or written into RAM very rapidly). RAM is therefore used primarily during the execution of programs to store partial or permanent results, as well as to store all of the software currently in use by the computer. The main difference between RAM and mass storage memory, such as a hard drive or a tape drive, is therefore in the speed of access: RAM can be accessed in under 1 microsecond, whereas a hard drive requires access time on the order of milliseconds, and tape drive, on the order of seconds. Another important distinction between RAM and mass memory is that the latter is far less expensive for an equivalent storage capability, the price typically being lower for longer access time.

A video terminal enables the user to enter programs and to display the data acquired by the microcomputer. The video terminal is one of many **peripherals** that enable the computer to communicate information to the outside world. Among these peripherals are printers, and devices that enable communication between computers, such as *modems* (a modem enables the computer to send and receive data over a telephone line). Finally, Figure 13.2 depicts real-time input/output (I/O) devices, such as analog-to-digital and digital-to-analog converters, and digital I/O devices. These are the devices that allow the microcomputer to read signals from external sensors, to output signals to actuators, and to exchange data with other computers.

13.3   INTRODUCTION TO MICROPROCESSORS

This section provides an overview of the structure of a general purpose microprocessor. The section builds on the foundation developed in Chapters 11 and 12 and requires no prior knowledge of digital computers. You may wish to explore this subject in greater depth, either by taking formal courses or through independent reading.

As explained earlier, the transfer of data from one location of the microprocessor to another (e.g., from the data registers to the ALU) occurs via a bus, which is a collection of data lines carrying individual bits that can form a word. Depending on the microprocessor structure, a word can be 8 bits, 16 bits, or 32 bits in length. An 8-bit word is referred to as a *byte*. The bits of a word can travel in parallel from one location to another. The width of the bus (number of data lines) is an important parameter in determining the processing speed and power of the microprocessor.

Figure 13.3 shows a simplified representation of a microprocessor's architecture. The two major functional units of a microprocessor are the **control unit** and the **arithmetic/logic unit (ALU).** In addition, a microprocessor has a program counter, an instruction register, general purpose registers, temporary registers, and a stack pointer. These components are briefly discussed here.

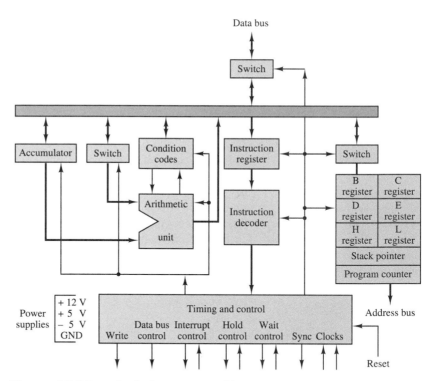

Figure 13.3 Example of microprocessor architecture

The control unit of a microprocessor controls and synchronizes all data transfers and manipulations in the microprocessor. The control unit uses inputs from a master clock to derive timing and control signals associated with each instruction. The ALU is where all computations take place; it is capable of performing the following operations on binary data:

- Binary addition and substraction
- Logical AND, OR, XOR
- Complement
- Rotation left or right

The ALU is made up of the logic devices studied in Chapter 12 and has the function of performing arithmetic operations (such as the two's complement arithmetic described in Chapter 12) and Boolean logic operations.

The **internal registers** include a **program counter,** the **instruction register, general purpose registers,** and **temporary registers.** The sequence of instructions constituting a program (often called **microinstructions**) is generated according to a programming language, which is in principle similar to FORTRAN, BASIC, or Pascal. To keep track of which instruction is to be executed next, the control unit has a dedicated register called the program counter (PC). The program counter holds the address of either the next instruction to be executed or the address of a multiword instruction. When the control unit requests that the memory transfer the data to the microprocessor, the data is transferred into the microprocessor through a data bus latch, and then into the instruction register. Registers inside the microprocessor are interconnected by an internal data bus. Temporary registers are used by the control unit to hold some information temporarily until the information is transferred to another register or used in a computation. To allow greater flexibility for the microprocessor, instructions can carry out transfers between general purpose registers. A microprocessor saves its register content in a storage register called the **stack** during subroutine calls and interrupts (an **interrupt** is a subroutine call initiated by external hardware). The stack consists of a group of specifically allocated read/write memory; a **stack pointer** is required to address a location or register in the stack.

Interrupts are an important part of the operation of a microprocessor. An interrupt is an electrical signal that is generated outside of the CPU and is connected to an input on the CPU. The interrupt causes the CPU to temporarily stop the execution of the program that is being executed to perform some operations on data coming from an external device. When an interrupt occurs, the processor automatically jumps to a designated program location and executes the interrupt service subroutine. When the interrupt subroutine is done, the computer returns to its place in the original program and resumes its operation. Example 13.1 illustrates the use of interrupts in a common application.

Example   13.1

In modern automotive instrumentation, a microprocessor performs all of the signal-processing operations for several measurements. A block diagram for such instrumentation is given in Figure 13.4. Depending on the technology used, the sensors' outputs can be either digital or analog. If the sensor signals are analog, they must be converted to digital format by means of an analog-to-digital converter (ADC), as shown in Figure 13.5. The analog-to-digital conversion process requires an amount of time that depends on the individual analog-to-digital converter (ADC), as will be explained in Section 13.4. After the conversion is completed, the ADC then signals the computer by changing the logic state on a separate line that sets its *interrupt request* flip-flop. This flip-flop stores the ADC's interrupt request until it is acknowledged (see Figure 13.6).

Figure 13.4 Automotive instrumentation

Figure 13.5 Sensor interface

Figure 13.6 Interrupt request in a microprocessor

When an interrupt occurs, the processor automatically jumps to a designated program location and executes the interrupt service subroutine. For the ADC, this would be a subroutine to read the conversion results and store them in some appropriate location, or to perform an operation on them. When the processor responds to the interrupt, the interrupt request flip-flop is cleared by a signal directly from the processor. To resume the execution of the program at the proper point upon completion of the ADC service subroutine, the program counter content is automatically saved before control is transferred to the service subroutine. The service subroutine saves in a stack the content of any registers it uses, and restores the registers' content before returning.

The interrupt may occur at any point in a program's execution, independent of the internal clock; it is therefore referred to as an *asynchronous* event.

The Intel 80X86 Microprocessor Family

In the remainder of this section, the properties and features of the Intel 80X86 microprocessor family will be explored. The reason for the choice of this device is that the 8086, 80186, 80286, 80386, and 80486 are commonly used processors used in personal computers (PCs); PCs are commonly employed in a variety of data acquisition and digital instrumentation applications in engineering plants and laboratories. We shall discuss the earliest (and simplest) version of this family of microprocessors, the 8086, in some detail, and then provide a summary of the characteristics of the more advanced versions.

The 8086 microprocessor is the first 16-bit processor from Intel Corporation. It is a 40-pin integrated circuit that was released in 1978. This processor is capable of addressing up to 1 Mbyte of memory and incorporates several hardware changes over previous products that make it function better and faster than its 8-bit predecessors, such as Intel's 8080 microprocessor. For example, the 8080 could address only up to 64 Kbytes of memory, versus the 1 Mbyte of the 8086. The 8086 can operate with clock speeds up to 8 MHz, a substantial increase over the 4-MHz maximum rating for the 8080. The 8086 can also be used as an 8-bit processor for simple applications; because of the greater simplicity of the 8-bit configuration, this is the configuration used in the problems and examples throughout this chapter.

The 8086 package has 16 pins dedicated to connections to the bus system, to be described shortly. For the 8-bit operations discussed in this book, a derivative of the 8086 is actually being used: the Intel 8088. These two processors are identical, except that the 8088 has only 8 pins for external bus connection. Although the 8086 has the advantage of being able to directly manipulate and transfer 16-bit words, the 8088 can in effect do the same by processing half of a 16-bit word, 8 bits at a time. The two bytes then form a 16-bit word in the CPU, where operations proceed as they would in the 16-bit version. The drawback of this simplified design is a decrease in processing speed.

Internal Organization of the 8086/8088

Data Registers

The organization of the registers within the CPU is shown in Figure 13.7. The data registers hold information vital to instructions, both past and present. Three data registers are available: B, C, and D. These registers can be accessed as 16-bits (BX, CX, DX) or, separately, as 8 bits (BL, CL, DL

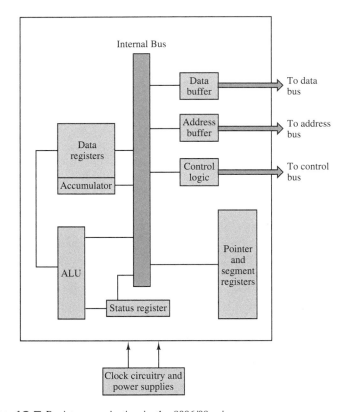

Figure 13.7 Register organization in the 8086/88 microprocessor.

for the lower 8 bits and BH, CH, DH for the upper 8 bits). Thus, the data register layout may look as follows:

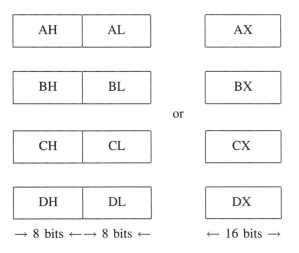

The **accumulator register** (AX) is also a data register, but it is used primarily for mathematical functions: results of mathematical operations appear in the accumulator and need to be transferred to other registers or into memory if they are to be saved. Also, the accumulator is used for I/O operations, storing data that is to be sent out to a port or holding information that has just been read. Thus, the AX register can be thought of as the main register, while the other data registers are used as temporary storage of variables and other data.

The reasons for using data registers are twofold. First, the registers are physically located on the microprocessor chip. Since off-chip operations (accessing memory in a floppy disk, reading from external input ports, or any other operation that uses the external buses) are quite slow, using these registers will speed up access to the variables considerably. Second, some of these registers serve a dual purpose: many are dedicated to particular operations, much as the accumulator is dedicated to math functions.

ALU and Status Register

Next on the list is the ALU, or arithmetic/logic unit. This is the heart of the processor. It is used to decode instructions, add numbers, and perform comparisons. Also, the ALU does not only send the outcomes of some operations to the accumulator, but it can also detect conditions and set flags in the *status register*. This is the register used in acting on comparison statements, such as "if greater than ..." and the like.

The status register is a 16-bit register; however, only 9 bits are used. The register looks as follows:

xxxxODITSZxAxPxC

Note that an "x" represents a bit that is not used, while the capital letters are bits that hold the following conditions:

O	Overflow flag
D	Direction
I	Interrupts enabled
T	Trap flag
S	Sign of operation
Z	Zero flag
A	Auxiliary carry
P	Parity flag
C	Carry flag

A detailed description of each of these flags is beyond the scope of this book. However, we shall give an example of how these flags are used. The easiest flag to understand is the zero flag, which is set when an arithmetic or logical operation results in accumulator contents equal to zero. Thus, when one compares two numbers and wishes to execute a *jump* to the location called *label* in the program, the operation that is in fact performed is the subtraction of the two numbers; a jump is executed if the zero flag is set (i.e., if Z = 1). You may compare this to a GO TO statement in FORTRAN or BASIC.

Pointer and Segment Registers

Next are the pointer and segment registers. These are registers that control the execution of the program and find the instructions to be executed. These hold the locations of instructions, the location of data, and the location of the *stack*.

The stack is simply an area of memory that is accessed using pop and push commands (described next); it provides for temporary storage of variables and values one does not wish to leave in the registers. By pushing a register onto the stack, the value is saved and the register can be used for another operation. Then, when the value is needed again, it is popped off the stack into its respective register. You may visualize the stack as a spring-loaded cylindrical container, into which we may push coins at the top, to pop them out again when we need them.

The need for the pointer and segment registers to track the areas where the data and code are located is vital to accessing the information correctly. The architecture of the 8086 microprocessor dictates that memory be organized as 64-Kbyte blocks, since this is a 16-bit machine (recall that $2^{16} = 65,536$, or 64 Kbytes). Only one block is available to the CPU at any time, effectively limiting the amount of data and code that can be accessed at one time to 64 Kbytes. However, the 8086 can access up to 1 Mbyte of memory through a clever addressing scheme that shifts address information by 4 bits, effectively making the address locations 20 bits long, or up to 1 Mbyte. Thus, since one cannot directly access the entire 1-Mbyte range, one must "point" to the 64-K block that is to be accessed. This is the job of the pointer and segment registers. These hold the information that is vital to the CPU for loading memory data and finding the next instruction. For example, if a JMP statement will jump to code that is not within the current 64-K block, the registers must change to point to the block in which the code is located. The same applies to accessing data in different blocks. Control of these registers and their contents entails an extensive understanding of memory allocation procedures and protocol, and is beyond the scope of this book.

Bus Buffers

The final group of functions found inside the CPU is a group of buffers. As you can see from Figure 13.7, buffers are connected to three buses that exit the CPU. These are the system buses that enable access to input/output (I/O) ports and to RAM and ROM memory chips and provide information vital to the functioning of the rest of the computer. These buffers are necessary to provide electrical isolation of subsystems that are not being used. Another function of the buffers is to drive the bus system connecting the CPU to a variety of external devices, taking into account the impedance of the bus. Thus, the bus buffers will condition the information through amplification and impedance isolation to ensure that loading does not occur.

External Support Circuitry

The CPU cannot function alone. It must be provided with power, and with a clock to drive its logic in an orderly fashion. The last functional block in Figure 13.7 is the circuitry required to accomplish these functions. For the 8086, the power supply will provide a 5-volt supply and two ground lines. The clock circuitry will vary with design; typically, one finds a crystal oscillator connected to a clock-generating IC. This IC is used to drive the oscillator as well as stabilize it, and to divide the oscillator's output down to the desired clock frequency for the CPU. Usually, the clock chip also provides outputs at various other frequencies for other applications, such as analog-to-digital conversion or communication.

Instruction Implementation in the 8086/8088

The basic instruction set in a microprocessor consists of a set of individual instructions, each of which consists of a simple sequence of elementary operations. Those sequences of operations that are performed routinely by the CPU are synthesized into instructions, which are then labeled with a *mnemonic,* that is, a simple abbreviation that suggests the basic function being performed. For example, JMP stands for the sequence of events that are necessary for the program to *jump* to a different point, MOV represents the operations required to *move* the content of a register to another register, and so forth. The essential instruction set for the 8086 is given in the accompanying box. The list is not complete, but it is representative of the complete instruction set.

In a microprocessor, an instruction (mnemonic) is decoded by an assembler into binary-coded information. For example, a "MOV AL,13H" mnemonic instruction (which means "move the value 13H into the lower 8 bits of the accumulator register") is converted into the hex code B013 by the assembler, where B0 is the MOV AL part and 13 is the number 13H, where the suffix "H" stands for hex. In binary form, this B013 becomes 1011 0000 0001 0011. This is the code the CPU will process through its hardware, either the ALU or operation-specific hardware (the 8086 microprocessor is capable of direct hardware multiplication and division, with a considerable increase in speed with respect to software operations). This sequence will run through the circuitry, producing outputs that accomplish the desired result. In this light, one can think of the 8086 as a very large logic circuit with 16 inputs capable of producing an enormous number of outputs. The operation

Abbreviated 8086 Instruction Set

S = source, D = destination, AX = accumulator (AL = low byte, AH = high byte)

Memory access instructions

MOV	D,S	Move byte/word from S to D
XCHG	D,S	Switches values in D and S
PUSH	S	Transfers S's value to the top of the stack
POP	D	Transfers the stack's top value to D

Mathematical instructions

ADD	D,S	Add two numbers—result in D
SUB	D,S	Subtract S from D—result in D
MUL	S	Multiply AX by S—result in AX
DIV	S	Divide AX by S—result in AX
INC	D	Add 1 to D—result in D
DEC	D	Subtract 1 from D—result in D
NEG	D	Negate D (subtract D from 0)

Logical instructions

AND	D,S	Bitwise logical AND of D and S—result in D
OR	D,S	Bitwise logical OR of D and S—result in D
XOR	D,S	Bitwise logical XOR of D and S—result in D
NOT	D	Invert the bits in D (form 1's complement)
TEST	D,S	Bitwise logical AND of D and S—result in Z flag

Compare and jump instructions

CMP	D,S	Compare D and S—results in flags
JE	*Label*	Jump if equal to *Label*
JNE	*Label*	Jump if not equal to *Label*
JG	*Label*	Jump if greater than *Label*
JGE	*Label*	Jump if greater than or equal to *Label*
JL	*Label*	Jump if less than *Label*
JLE	*Label*	Jump if less than or equal to *Label*
JMP	*Label*	Jump to *Label*

Subroutine and interrupt instructions

CALL	*Name*	Call the subroutine *Name*
RET		Return from subroutine entered with CALL
STI		Enable interrupt processing
INT	*Type*	Generate an interrupt of *Type* in software
IRET		Return from an interrupt

I/O instructions

IN	AX,*Port*	Read value of *Port* into AX
OUT	*Port*,AX	Write value in AX to *Port*

Miscellaneous instructions

NOP		No operation—delay instructions

is not quite as simple as this, but the general function of a microprocessor can be thought of as implementing large numbers of logic functions at very high speed.

Instructions are retrieved by the CPU using the pointer and segment registers, which tell the CPU where the next instruction is located. The binary form of the instruction is loaded into the instruction register (one can consider this register to be grouped with the segment and pointer registers), where it is then sent to the combinational circuitry to perform a specific function.

Example 13.2

In both large and small computers, it is desired to store the address of data in a main memory location, rather than have this address in an internal register. This procedure gives greater flexibility to the microprocessor. This example shows how this function is accomplished.

1. Assume that the desired data is 36H (the hexadecimal number 36), and use the 8086/8088 instruction set to write this data to I/O address A6H.

Solution:

The command to write to I/O ports is the OUT *Port*,AX command. Since this command uses the accumulator register and the 8088 system is an 8-bit system, the low byte of the accumulator must first be loaded with the value to be written. Thus, the code looks as follows:

```
MOV AL,36H       ;Load accumulator with 36 H
OUT 0A6H,AL      ;Write accumulator to I/O address A6 H
```

Note that a zero must be written in front of the address, since it begins with a letter. This enables the compiler to distinguish it as a value rather than a label or constant.

2. Assume that an input device is wired to an 8086 system at the address E6H. Write the code necessary to read a byte value from this input device.

Solution:

The command to read from an I/O port is in the IN AX,*Port* command. The values read are stored in the accumulator. Since the 8086 is an 8-bit system, you can load only a byte at a time into the accumulator.

To store the data in the low byte, the code is as follows:

```
IN AL,0E6H
```

To store the data in the high byte of the register, the code is

```
IN AH,0E6H
```

Note that a zero must be written in front of the address, since it begins with a letter. This enables the compiler to distinguish it as a value rather than a label or constant.

Example   13.3   Memory Access Command Examples

The microprocessor is often required to access the memory, either to receive data from memory or to store data there. Two methods for transferring data to and from memory are **immediate addressing** and **indirect addressing.** In immediate addressing, the data transferred is part of the instruction. In indirect addressing, the address contained in the instruction specifies the address of the register where data is. In minicomputers and large computers, both of the registers involved in indirect addressing are usually external registers in RAM. In the case of microprocessors, however, the register that contains the actual address of the data is usually an internal register. Figure 13.8 illustrates the two addressing modes.

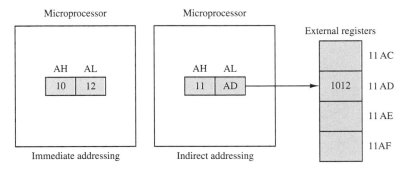

Figure 13.8 Addressing

For this example, the actual data is 1012H. As shown in Figure 13.8, the accumulator register holds the data for the case of immediate addressing. However, in the case of indirect addressing, the accumulator holds the address of the external register where the data is. Some commands and examples of memory access are given here.

Example of Immediate Addressing

Load register, immediate:

        MOV AL,12H   ;Loads AL register with 12 H
        MOV AH,10H   ;Loads AH register with 10 H

After the instructions are completed, the accumulator will hold the data (1012H).

Example of Indirect Addressing

Move the data to accumulator:

        LHLD 11ADH   ;loads data address into HL register pair
        MOV AX,11012   ;moves data from memory location pointed to by
                              HL to internal register AX.

After the instructions are completed, the accumulator will hold the address of the register where the data resides.

8086-Based Microcomputer Architecture

After the preceding introduction to the functions and architecture of the 8086 microprocessor, it will not be too difficult to provide an overview of a typical computer, viewed in its functional blocks. With the CPU we described in the previous section as its heart, a microcomputer consists of *memory* for data and instruction storage, *I/O* ports for communication with the external world, and *peripherals* to extend the microprocessor's capabilities. All of these components are connected to the CPU via a bus structure. For the 8086, this bus is 16 bits wide (physically, 16 wires lying side by side), while it is 8 bits wide for the 8088. Either bus runs the length of the computer, enabling each device to communicate with its host. Control over the devices is entirely up to the CPU's instructions, and thus to the programmer. Figure 13.9 depicts the structure of a general purpose computer based on the 8086 microprocessor.

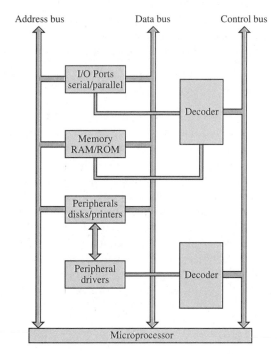

Figure 13.9 8086-based microcomputer

Bus System

In the block diagram of Figure 13.9, you can see three buses, each with a dedicated task. The **address bus,** which is a unidirectional bus, holds the address of the device the CPU wishes to communicate with. If one considers every address of each device a cell with a locked door, the values on the address bus would be the key to a particular door, and only that cell would be opened for access. The data bus, which is a bidirectional bus, plays its part by taking information to or from the device. Only information pointed to (or accessed by) the address bus will be available to the CPU through the data bus. The control bus carries the information concerning which operation is to be performed. For example, it selects whether data is to be read or written in memory or to I/O ports, and whether the peripherals being selected need to be controlled by drivers. The control bus carries the actual instruction, whereas the address bus carries the location information and the data bus contains either the result of the instruction or the command.

The two decoders shown are also an important part of the device selection process. Essentially, these decoders use the controller information to select a device, and they enable the device to become active. In the analogy of the locked cells, one can consider each device to be a building that contains cells. The decoders unlock the particular building and allow the address and data bus to pass through and unlock and react to the cell. All of this activity is controlled by the CPU and its instructions.

External Devices

The external devices shown in Figure 13.9 are divided into four groups, two of which are interconnected. These are the I/O ports, memory, the peripherals, and the peripheral drivers. The **I/O ports** serve an important function in making the computer a practical tool for real-life applications. These ports enable one to connect sensors, alarms, converters, and any other instrument to the computer. Instructions can be written to control these instruments and monitor any sensors. In this fashion, one can interface the machine to the outside world to perform dangerous tasks or repetitive analysis, or to capture actions too fast for humans to perceive. The computer can also control these processes by sending instructions to instruments that can be computer-controlled.

Memory serves a basic purpose: that of storing instructions or data. The distinction between read-and-write memory (RAM) and read-only memory (ROM) is important, since ROM is reserved for the storage of special instructions for use by the CPU. These are instructions that are vital to the CPU's life—instructions that must be available when the computer is turned on. RAM, on the other hand, is not so strictly controlled. This is the memory that allows programmers and users to temporarily store data, text, or measurements. The largest amount of memory the 8086 can access is 1 Mbyte; however, some of this is dedicated to program ROM, and so the user never has quite that much memory available.

The third and fourth devices, which are tied together, provide a connection with peripheral devices. Peripherals are devices that extend the computer's capabilities—for example, printers and disk drives. These, however, have additional hardware associated with them. While I/O ports and memories are directly accessible and controllable from the CPU, most peripherals are not. Disk drives must have their own controllers (or *drivers*), which spin the disk and read information from the magnetic surface; and printers must have their own drivers as well to select characters and advance the paper and ribbon. These drivers are under the control of the CPU, and are a necessary addition to the peripheral device's operation. With the knowledge of digital logic circuits from Chapter 12 and the material presented in this chapter, the following example is given to show how a memory system can be designed or expanded using auxiliary memory devices.

Example   13.4

This example illustrates the configuration of a memory system that is interfaced to an 8086 microprocessor. The memory system requires 8 Kbytes of ROM and 2 Kbytes of RAM. The ROM must start at location 0000H and must be immediately followed by the RAM. The following information is known for the processor: the system bus consists of a 16-bit address bus, A_0 through A_{15}, and an 8-bit bidirectional data bus, D_0 through D_7. Assume that the microprocessor has three control lines: $\overline{\text{WR}}$ (logic 0 indicates that the processor is writing), $\overline{\text{RD}}$ (logic 0 indicates that the processor is reading), and IO/$\overline{\text{M}}$ (logic 0 indicates that the address is that of the memory location as opposed to an I/O location).

Memory systems are designed to provide the amounts of ROM and RAM required for certain applications. The following design steps can be used:

1. Estimate the amount of ROM and RAM required for the given application.
2. Determine the address boundaries for ROM and RAM. Also, draw an initial memory map.
3. Select the types of memory devices that will be used for the design.
4. Determine the arrangement of memory devices needed to obtain the required word length and number of words.
5. Design the address-decoding logic.

These five steps will be applied to configure the required memory system. As the problem indicates, 8 Kbytes of ROM and 2 Kbytes of RAM are required. The next step is to draw the memory map for the system:

The storage requirements for this system can be met using only two devices: a 2764A 8-K × 8 EPROM, to provide the ROM; and an HM6116 2-K × 8 SRAM, to provide the RAM. Data sheets for these memory devices are given at the end of the chapter. Also, a 74ALS138 decoder is used for decoding. Address-decoding logic is required to generate the chip enable signal for the EPROM and the chip select signal for the SRAM. Memory devices must be selected one at a time, to preclude bus contention (i.e., the simultaneous occurrence of two signals on the bus). The EPROM must be enabled only when an address in the range from 0000H to 1FFFH is generated and IO/$\overline{\text{M}}$ is logic 0. The SRAM chip must be selected only when the address is in the range 2000H through 27FFH and IO/$\overline{\text{M}}$ is logic 0.

The following address bit map shows which address bits are decoded by the memory device and which are decoded by the address-decoding logic:

		$A_{15}-A_{12}$	$A_{11}-A_8$	A_7-A_4	A_3-A_0
ROM	0000H	0 0 0 0	0 0 0 0	0 0 0 0	0 0 0 0
	1FFFH	0 0 0 1	1 1 1 1	1 1 1 1	1 1 1 1
		Selects first 8-K block	Selects location within 8-K ROM		
		$A_{15}-A_{12}$	$A_{11}-A_8$	A_7-A_4	A_3-A_0
RAM	2000H	0 0 1 0	0 0 0 0	0 0 0 0	0 0 0 0
	27FFH	0 0 1 0	0 1 1 1	1 1 1 1	1 1 1 1
		Selects second 2-K block	Selects location with 2-K RAM		

Using the memory map and address bit map, we can configure the required memory system as shown in Figure 13.10.

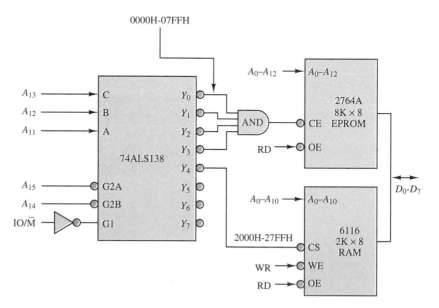

Figure 13.10 Memory system design

The implementation of Figure 13.10 has been designed to take advantage of the 74ALS138 1-of-8 decoder or its equivalent. The active high enable, G1, of the 74ALS138 is driven by the complement of $IO/\overline{M}$. Thus, the decoder is not enabled unless $IO/\overline{M}$ is logic 0. The active low enables, $\overline{G2A}$ and $\overline{G2B}$, are driven by address inputs A_{15} and A_{14}. This will ensure that the decoder is enabled only when both A_{15} and A_{14} are logic 0. The select inputs of the decoder, C, B, and A, are driven by A_{13}, A_{12}, and A_{11}. Therefore, each output of the decoder selects a 2-Kbyte block in the 64-Kbyte address space of the microprocessor. The first four 2-Kbyte blocks of the address space are occupied by the 2764A. The four-input AND will drive the chip enable low when any one of the first four 2-Kbyte blocks is selected. The output of the decoder that corresponds to the fifth 2-Kbyte block, 2000H–27FFH, drives the chip select of the HM6116. The memory device control inputs $\overline{WE}$ and $\overline{OE}$ are driven by $\overline{WR}$ and $\overline{RD}$, respectively. When an $\overline{RD}$ signal occurs, only the selected memory device will have its output buffers enabled and will be able to drive the data bus.

Evolution of the 8086 into the 80186, 80286, 80386, and 80486

In 1982, Intel introduced the 80186 and 80188. The 80186 microprocessor featured the processing power of the 8086 plus the support circuitry of 15 other chips. This "computer on a chip" contains two independent DMA (direct memory access) channels for high-speed peripherals such as disk drives, and a programmable interrupt controller. In terms of software, it provides the 8086 instruction set plus some additional instructions, valuable to people who are designing high-level language translators or compilers.

The same year, Intel announced the 80286, the microprocessor at the heart of the IBM PC/AT. The 80286 is an enhanced 80186 microprocessor that provides special features necessary for memory management and protection. These features are important for multi-user applications, in which several users share the same computer, usually through a local area network. They are also important for multi-tasking applications, in which several programs run at the same time.

In recent years, Intel has introduced the 80386 and 80486 microprocessors. The 80386 is a 32-bit microprocessor with 32-bit separate address and data buses. It has 4 gigabytes of physical memory and 64 terabytes of virtual memory. The 80486 is an enhanced 80386 microprocessor that includes an integrated floating-point processor.

The objective of Sections 13.2 and 13.3 has been to provide an overview of the structure of a general purpose microprocessor, and to motivate readers to pursue further study. The 8086 microprocessor was chosen for its simplicity and the fact that it is used in the IBM personal computer (PC), which is very commonly used in a variety of data acquisition applications.

DRILL EXERCISES

13.1   How many bits are in 256 bytes of data?

> Answer: 2,048

13.2   How many bits are in 4 Kbytes of data?

> Answer: 32,768

13.3   In Example 13.4, what is the output of the decoder that corresponds to the sixth 2-Kbyte block of the address?

> Answer: 2800H–2FFFH

13.4   Why are the address bits A_{15} and A_{14} connected to G2A and G2B in Example 13.4?

> Answer: To ensure that the decoder is active only when both A_{15} and A_{14} are logic 0.

13.4   SIGNAL INTERFACE

To take advantage of the capabilities of a microcomputer, it is necessary to suitably interface signals to and from external devices with the microcomputer. Such signals may be analog or digital. Depending on the nature of the signal, either an analog or a digital interface circuit will be required. The advantages in memory storage, programming flexibility, and computational power afforded by today's digital computers are such that the instrumentation designer often chooses to convert an analog signal to an equivalent digital representation, to exploit the capabilities of a microprocessor in processing the signal. In many cases, the data converted from analog to digital form remains in digital form for ease of storage, or for further processing. In some instances it is necessary to convert the data back to analog form. The latter condition arises frequently in the context of control system design, where an analog measurement is converted to digital form and processed by a digital computer to generate a control action (e.g., raising or lowering the temperature of a process, or exerting a force or a torque); in such cases, the output of the digital computer is converted back to analog form, so that a continuous signal becomes available to the actuators. Figure 13.11 illustrates the general appearance of a digital measuring instrument and of a digital controller acting on a plant or process.

The objective of this section is to describe how the digital-to-analog (D/A) and analog-to-digital (A/D) conversion blocks of Figure 13.11 function. After illustrating discrete circuits that can implement simple ADCs and DACs, we shall emphasize the use of ICs specially made for these tasks. Nowadays, it is uncommon (and impractical) to design such circuits using discrete components: the performance and ease of use of IC packages make them the preferred choice in virtually all applications.

Figure 13.11 Block diagrams of a digital measuring instrument and a digital control system

Digital-to-Analog Converters

We discuss digital-to-analog conversion first because it is a necessary part of analog-to-digital conversion in many A/D conversion schemes. A **digital-to-analog converter** will convert a binary word to an analog output voltage (or current). The binary word is represented in terms of 1s and 0s, where typically (but not necessarily), 1s correspond to a 5-volt level and 0s to a 0-volt signal. As an example, consider a four-bit binary word:

$$B = (b_3 b_2 b_1 b_0)_2 = (b_3 \cdot 2^3 + b_2 \cdot 2^2 + b_1 \cdot 2^1 + b_0 \cdot 2^0)_{10} \qquad \textbf{(13.1)}$$

The analog voltage corresponding to the digital word B would be

$$v_a = (8b_3 + 4b_2 + 2b_1 + b_0)\, \delta v \qquad \qquad \textbf{(13.2)}$$

where δv is the smallest *step size* by which v_a can increment. This least step size will occur whenever the LSB, b_0, changes from 0 to 1, and is the smallest increment the digital number can make. We shall also shortly see that the analog voltage obtained by the D/A conversion process has a "staircase" appearance because of the discrete nature of the binary signal.

The step size is determined on the basis of each given application, and is usually determined on the basis of the number of bits in the digital word to be converted to an analog voltage. We can see that, by extending the previous example for a four-bit word, the maximum value v_a can attain is

$$v_{a\,max} = (2^{n-1} + 2^{n-2} + \cdots + 2^1 + 2^0)\,\delta v$$
$$= (2^n - 1)\,\delta v \tag{13.3}$$

It is relatively simple to construct a DAC by taking advantage of the summing amplifier illustrated in Chapter 10. Consider the circuit shown in Figure 13.12, where each bit in the word to be converted is represented by means of a 5-V source and a switch. When the switch is closed, the bit takes a value of 1 (5 V); when the switch is not closed, the bit has value 0. Thus, the output of the DAC is proportional to the word $b_{n-1}b_{n-2}\ldots b_1 b_0$.

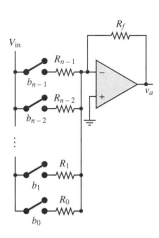

Figure 13.12 n-bit digital-to-analog converter (DAC)

You will recall that a property of the summing amplifier is that the sum of the currents at the inverting node is zero, yielding the relationship

$$v_a = -\left(\frac{R_f}{R_i} \cdot b_i \cdot 5\right) \qquad i = 0, 1, \ldots, n-1 \tag{13.4}$$

where R_i is the resistor associated with each bit and b_i is the decimal value of the ith bit (i.e., $b_0 = 2^0$, $b_1 = 2^1$, and so on). It is easy to verify that if we select

$$R_i = \frac{R_0}{2^i} \tag{13.5}$$

we can obtain weighted gains for each bit so that

$$v_a = -\frac{R_f}{R_0}(2^{n-1}b_{n-1} + \ldots + 2^1 b_1 + 2^0 b_0) \cdot 5 \text{ V} \tag{13.6}$$

and so that the analog output voltage is proportional to the decimal representation of the binary word. As an illustration, consider the case of a four-bit word; a reasonable choice for R_0 might be $R_0 = 10$ kΩ, yielding a resistor network consisting of 10-, 5-, 2.5-, and 1.25-kΩ resistors, as shown in Figure 13.13. The largest decimal value of a four-bit word is $2^4 - 1 = 15$, and so it is reasonable to divide this range into steps of 1 volt (i.e., $\delta v = 1$ V). Thus, the full-scale value of v_a is 15 V:

$$0 \le v_a \le 15 \text{ V}$$

and we select R_f according to the following expression:

$$R_f = \frac{\delta v\, R_0}{5} = 2 \text{ k}\Omega$$

Figure 13.13 Four-bit DAC

The corresponding four-bit DAC is shown in Figure 13.13.

The DAC transfer characteristic is such that the analog output voltage, v_a, has a steplike appearance, because of the discrete nature of the binary signal. The coarseness of the "staircase" can be adjusted by selecting the number of bits in the binary representation.

Example   13.5

Assume that the maximum analog voltage ($v_{a\,max}$) of an eight-bit DAC is 12 volts. Find the smallest step size (δv) by which v_a can increment.

Solution:

On the basis of the discussion in this section, we can use the following formula:

$$\delta v = \frac{v_{a\,max}}{2^8 - 1}$$

Thus, applying the formula, we obtain:

$$\delta v = \frac{12}{2^8 - 1} = 47.1 \text{ mV}$$

The practical design of a DAC is generally not carried out in terms of discrete components, because of problems such as the accuracy required of the resistor value. Many of the problems associated with this approach can be solved by designing the complete DAC circuit in integrated circuit (IC) form. The specifications stated by the IC manufacturer include the resolution, that is, the minimum nonzero voltage; the full-scale accuracy; the output range; the output settling time; the power supply requirements; and the power dissipation. The following example illustrates the use of a common integrated circuit DAC.

Example   13.6

Define the following concepts for a DAC, and show how they can be used to find the minimum number of bits required.

1. Range
2. Resolution

Solution:

1. The maximum output voltage is produced when all bits are logic 1s; it can be found from the following equation:

$$V_{max} = V_{in}\frac{R_f}{R_0}(2^n - 1)$$

    where V_{in} is the voltage corresponding to logic 1 and n is the number of bits. The minimum output voltage is produced when all bits are logic 0. In this case, $V_{in} = 0$. Thus, the range of the DAC is from zero to V_{max}.

2. Resolution is the minimum nonzero voltage that can be produced; it can be found by using the following equation:

$$\delta v = \frac{V_{max}}{2^n - 1} = V_{in}\frac{R_f}{R_0}$$

where δv is the resolution of the DAC.

If both the range and the resolution are specified, the minimum number of bits required in a DAC can be found by using the following equation:

$$n = \frac{\log\left(\frac{V_{max} - V_{min}}{\delta v} + 1\right)}{\log 2}$$

Because n must be an integer, any fractional part should be rounded up to the next higher integer.

Example   13.7

A typical DAC one would use in conjunction with the 8086 microprocessor is the AD558. This is an IC that can be used in a "stand-alone" configuration (without a microprocessor) or with a microprocessor interfaced to it. Using the data supplied in the data sheets at the end of the chapter, answer the following questions:

1. If one were to set up the AD558 for an output swing of 0 to 10 volts, what would be the smallest voltage output increment attainable?
2. On what is the maximum operating frequency (the largest frequency on which the DAC can perform conversion) of the AD558 dependent? Determine the maximum frequency attainable if the converter is to be run at full-scale input.

Solution:

1. Since this DAC is an eight-bit device, the total number of digital increments one could expect is 256. Thus, the finest voltage steps one could expect at the output would be:

$$\frac{10}{255} = 39.2 \text{ mV}$$

This means that for every increment of one bit, the output would jump (in a stepwise fashion) by 39.2 mV.

2. The maximum frequency at which a DAC can run is dependent on the settling time. This is defined as the time it takes for the output to settle to within one half of the least significant bit of its final value. Thus, only one transition can be made per settling time. The settling time for the AD558 depends on the voltage range and is defined for a positive going full scale step to $\pm\frac{1}{2}$ LSB. For an input voltage range of 0 to 2.56 V, the settling time is 0.8 μs, and the maximum conversion frequency is $\frac{1}{0.8\times10^{-6}} = 1.25$ MHz. For an input voltage range of 0 to 10 V, the settling time is 2 μs, and the corresponding maximum conversion frequency is 500 kHz.

Analog-to-Digital Converters

You should by now have developed an appreciation for the reasons why it is convenient to process data in digital form. The device that makes conversion of analog signals to digital form possible is the **analog-to-digital converter** (**ADC**), and, just like the DAC, it is also available as a single IC package. This section will illustrate the essential features of four types of ADCs: the tracking ADC, which utilizes a DAC to perform the conversion; the integrating ADC; the flash ADC; and (later) the successive-approximation ADC. In addition to discussing analog-to-digital conversion, we shall also introduce the *sample-and-hold amplifier.*

Quantization

The process of converting an analog voltage (or current) to digital form requires that the analog signal be quantized and encoded in binary form. The process of quantization consists of subdividing the range of the signal into a finite number of intervals; usually, one employs $2^n - 1$ intervals, where n is the number of bits available for the corresponding binary word. Following this quantization, a binary word is assigned to each interval (i.e., to each range of voltages or currents); the binary word is then the digital representation of any voltage (current) that falls within that interval. You will note that the smaller the interval, the more accurate the digital representation is. However, some error is necessarily always present in the conversion process; this error is usually referred to as *quantization error.* Let v_a represent the analog voltage and v_d its quantized counterpart, as shown in Figure 13.14 for an analog voltage in the range 0–16 V. In the figure, the analog voltage v_a takes on a value of $v_d = 0$ whenever it is in the range 0–1 V; for $1 \leq v_a < 2$, the corresponding value is $v_d = 1$; for $2 \leq v_a < 3$, $v_d = 2$; and so on, until, for $15 \leq v_a < 16$, we have $v_d = 15$. You see that if we now represent the quantized voltage v_d by its binary counterpart, as shown in the table of Figure 13.14, each 1-volt analog interval corresponds to a unique binary word. In this example, a four-bit word is sufficient to represent the analog voltage, although the representation is not very accurate. As the number of bits increases, the quantized voltage is closer and closer to the original analog signal; however, the number of bits required to represent the quantized value increases.

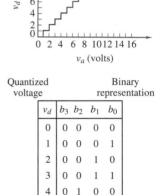

Quantized voltage	Binary representation			
v_d	b_3	b_2	b_1	b_0
0	0	0	0	0
1	0	0	0	1
2	0	0	1	0
3	0	0	1	1
4	0	1	0	0
⋮		⋮		
14	1	1	1	0
15	1	1	1	1

Figure 13.14 A digital voltage representation of an analog voltage

Tracking ADC

Although not the most efficient in all applications, the **tracking ADC** is an easy starting point to illustrate the operation of an ADC, in that it is based on the DAC presented in the previous section. The tracking ADC, shown in Figure 13.15, compares the analog input signal with the output of a DAC; the comparator output determines whether the DAC output is larger or smaller than the analog input to be converted to binary form. If the DAC output is smaller, then the comparator output will cause an up-down counter (see Chapter 12) to count up, until it reaches a level close to the analog signal; if the DAC output is larger than the analog signal, then the counter is forced to count down. Note that the rate at which the up-down counter is incremented is determined by the external clock, and that the binary counter output corresponds to the binary representation of the analog signal. A feature of the tracking ADC is that it follows ("tracks") the analog signal by changing one bit at a time.

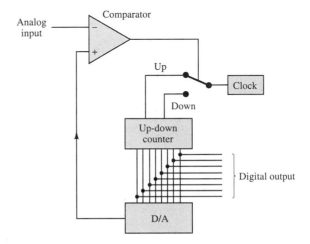

Figure 13.15 Tracking ADC

Integrating ADC

The **integrating ADC** operates by charging and discharging a capacitor, according to the following principle: if one can ensure that the capacitor charges (discharges) linearly, then the time it will take for the capacitor to discharge is linearly related to the amplitude of the voltage that has charged the capacitor. In practice, to limit the time it takes to perform a conversion, the capacitor is not required to charge fully. Rather, a clock is used to allow the input (analog) voltage to charge the capacitor for a short period of time, determined by a fixed number of clock pulses. Then the capacitor is allowed to discharge through a known circuit, and the corresponding clock count is incremented until the capacitor is fully discharged. The latter condition is verified by a comparator, as shown in Figure 13.16. The clock count accumulated during the discharge time is proportional to the analog voltage.

In the figure, the switch causes the counter to reset when it is connected to the reference voltage, V_{REF}. The reference voltage is used to provide a known, linear discharge characteristic through the capacitor (see the material on the op-amp integrator in Chapter 10). When the comparator detects that the output of the integrator is equal to zero, it switches state and disables the NAND gate, thus stopping the count. The binary counter output is now the digital counter part of the voltage v_a.

Other common types of ADC are the so-called successive-approximation ADC and the flash ADC.

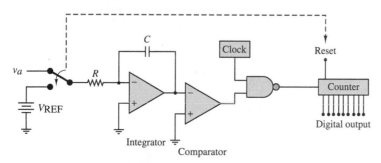

Figure 13.16 Integrating ADC

Flash ADC

The **flash ADC** is fully parallel and is used for high-speed conversion. A resistive divider network of 2^n resistors divides the known voltage range into that many equal increments. A network of $2^n - 1$ comparators then compares the unknown voltage with that array of test voltages. All comparators with inputs exceeding the unknown are "on"; all others are "off." This comparator code can be converted to conventional binary by a digital priority encoder circuit. For example, assume that the three-bit flash ADC of Figure 13.17 is set up with $V_{ref} = 8$ V. An input of 6.2 V is provided. If we number the comparators from the top of Figure 13.17, the state of each of the seven comparators is as given in Table 13.1.

Table 13.1   State of comparators in a 3-bit flash ADC

Comparator	Input on + Line	Input on − Line	Output
1	7 V	6.2 V	H
2	6 V	6.2 V	H
3	5 V	6.2 V	L
4	4 V	6.2 V	L
5	3 V	6.2 V	L
6	2 V	6.2 V	L
7	1 V	6.2 V	L

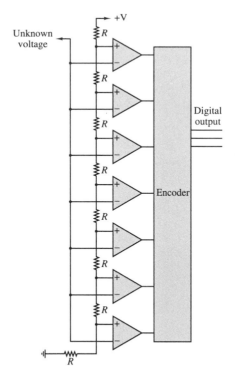

Figure 13.17 A three-bit flash ADC

Example   13.8

Answer the following questions for the flash ADC:

1. How many parallel comparators do you need for a four-bit flash ADC?
2. State the advantage and the disadvantage of using a flash ADC.

Solution:

1. The flash ADC converter uses $2^n - 1$ parallel comparators to determine simultaneously all n bits of the digital output. Thus, 15 comparators are needed.
2. The flash converter has the advantage of high speed because of its parallel comparators. The disadvantage is that it is costly to use this technique for a large number of bits, because of the number of comparators required.

In the preceding discussion, we explored a few different techniques for converting an analog voltage to its digital counterpart; these methods—and any others—require a certain amount of time to perform the A/D conversion. This is the ADC **conversion time,** and is usually quoted as one of the main specifications of an ADC device. A natural question at this point would be: If the analog voltage changes during the analog-to-digital conversion and the conversion process itself takes a finite time, how fast can the analog input signal change while still allowing the ADC to provide a meaningful digital representation of the analog input? To resolve the uncertainty generated by the finite ADC conversion time of any practical converter, it is necessary to use a sample-and-hold amplifier. The objective of such an amplifier is to "freeze" the value of the analog waveform for a time sufficient for the ADC to complete its task.

A typical sample-and-hold amplifier is shown in Figure 13.18. It operates as follows. An FET analog switch (see Chapter 11) is used to "sample" the analog waveform. Recall that when a voltage pulse is provided to the sample input of the FET switch (the gate), the FET enters the ohmic region and in effect becomes nearly a short circuit for the duration of the sampling pulse. While the FET conducts, the analog voltage, v_a, charges the "hold" capacitor, C, at a fast rate through the small "on" resistance of the FET. The duration of the sampling pulse is sufficient to charge C to the voltage v_a. Because the FET is virtually a short circuit for the duration of the sampling pulse, the charging (RC) time constant is very small, and the capacitor charges very quickly. When the sampling pulse is over, the FET returns to its nonconducting state, and the capacitor holds the sampled voltage without discharging, thanks to the extremely high input impedance of the voltage follower (buffer) stage. Thus, v_{SH} is the sampled-and-held value of v_a at any given sampling time.

Sample-and-hold amplifier

Functional representation of FET bilateral switch

Bilateral switch symbol for FET

Figure 13.18 Description of the sample-and-hold process

Example   13.9

Using the data sheet for the AD582 sample-and-hold amplifier, answer the following questions:

1.   What is the acquisition time?
2.   How can the acquisition time be reduced?
3.   What is the acquisition time for the AD582?

Solution:

1.   The acquisition time, T, is the time required for the output of the sample-and-hold amplifier to reach its final value (within a specified error bound) after it has been switched from the hold mode to the sample mode. It includes the switch delay time, the slewing interval, and the amplifier settling time.
2.   The acquisition time can be reduced by reducing the value of the hold capacitor, C_H.
3.   From the data sheet, $T = 6$ μs.

Figure 13.19 Sampled data

The appearance of the output of a typical sample-and-hold circuit is shown in Figure 13.19, together with the analog signal to be sampled. The time interval between samples, or **sampling interval,** $t_n - t_{n-1}$, allows the ADC to perform the conversion and make the digital version of the sampled signal available, say, to a computer or to another data acquisition and storage system. The sampling interval needs to be at least as long as the A/D conversion time, of course, but it is reasonable to ask how frequently one needs to sample a signal to preserve its fundamental properties (e.g., peaks and valleys, "ringing," fast transition). One might instinctively be tempted to respond that it is best to sample as frequently as possible, within the limitations of the ADC, so as to capture all the features of the analog signal. In fact, this is not necessarily the best strategy. How should we select the appropriate sampling frequency for a given application? Fortunately, an entire body of knowledge exists with regard to sampling theory, which enables the practicing engineer to select the best sampling rate for any given application. Given the scope of this chapter, we have chosen not to delve into the details of sampling theory, but, rather, to provide the student with a statement of the fundamental result: the **Nyquist sampling criterion.** The Nyquist criterion states that to prevent aliasing when sampling a signal, *the sample rate should be selected to be at least twice the highest-frequency component present in the signal.* Thus, if we were sampling an audio signal (say, music), we would have to sample at a frequency of at least 40 kHz (twice the highest audible frequency, 20 kHz). In practice, it is advisable to select sampling frequencies substantially greater than the Nyquist rate; a good rule of thumb is 5 to 10 times greater. The following example illustrates how the designer might take the Nyquist criterion into account in designing a practical A/D conversion circuit.

Example   13.10

A typical ADC one would use in conjunction with the 8086 microprocessor is the AD574. This is a successive-approximation converter. Using the data sheet supplied at the end of the chapter, answer the following questions:

1. What is the accuracy (in volts) the AD574 can provide if $V_{CC} = 15.0$ V and $0 \leq V_{in} \leq 15.0$ V?
2. On the basis of the data sheet, what is the highest-frequency signal you could convert using the AD574? (Assume that $V_{CC} = 15.0$ V.)
3. If the maximum conversion time available were 40 μs, what would be the highest-frequency signal you could expect to sample on the basis of the Nyquist criterion?

Solution:

1. According to the data sheet, the least significant bit (LSB) of this converter limits its accuracy, meaning that the output is accurate within ± 1 bit. For the 0- to 15-V swing, this gives a voltage accuracy of

$$\frac{V_{max} - V_{min}}{2^n - 1} \qquad \text{or} \qquad \frac{15}{2^{12} - 1} \times (\pm 1 \text{ bit}) = \pm 3.66 \text{ mV}$$

2. On the basis of the data sheet, the maximum conversion time is 35 μs. Therefore, the highest frequency of data conversion using the AD754 is

$$f_{max} = \frac{1}{35 \ \mu s} = 28.57 \ kHz$$

Thus, the highest signal frequency that could be represented, according to the Nyquist principle, is:

$$\frac{1}{2} f_{max} = \frac{28.57 \times 10^3}{2} = 14.285 \ kHz$$

This is the maximum theoretical signal frequency that can be represented without distortion, according to the Nyquist principle.

3. Following the same procedure discussed in part 2,

$$\frac{1}{2} f_{max} = \left(\frac{1}{40 \times 10^{-6}} \right) \left(\frac{1}{2} \right) = 12.5 \ kHz$$

Data Acquisition Systems

The structure of a data acquisition system, shown in Figure 13.20, can now be analyzed, at least qualitatively, since we have explored most of the basic building blocks. A typical data acquisition system often employs an *analog multiplexer,* to process several different input signals. A bank of bilateral analog FET switches, such as the one we described together with the sample-and-hold amplifier, provides a simple and effective means of selecting which of the input signals should be sampled and converted to digital form. Control logic, employing standard gates and counters, is used to select the desired *channel* (input signal), and to trigger the sampling circuit and the ADC. When the A/D conversion is completed, the ADC sends an appropriate *end of conversion* signal to the control logic, thereby enabling the next channel to be sampled.

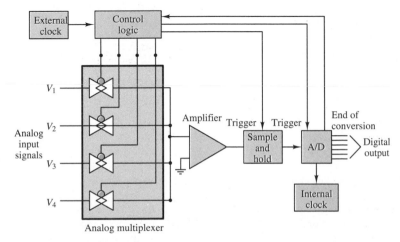

Data acquisition system

In the block diagram of Figure 13.20, four analog inputs are shown; if these were to be sampled at regular intervals, the sequence of events would appear as depicted in Figure 13.21. We notice, from a qualitative analysis of the figure, that the effective sampling rate for each channel is one fourth the actual external clock rate; thus, it is important to ensure that the sampling rate for each individual channel satisfies the Nyquist criterion. Further, although each sample is held for four consecutive cycles of the external clock, we must notice that the ADC can use only one cycle of the external clock to complete the conversion, since its services will be required by the next channel during the next clock cycle. Thus, the internal clock that times the ADC must be sufficiently fast to allow for a complete conversion of any sample within the design range. These and several other issues are discussed in the next example.

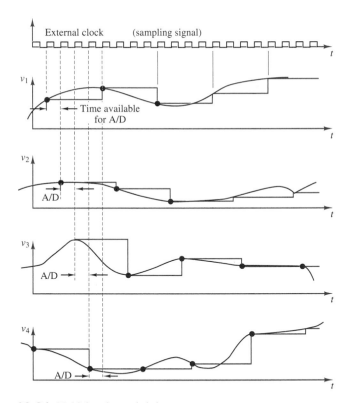

Figure 13.21 Multiplexed sampled data

Example   13.11   Data Acquisition System

This example discusses the internal structure of a typical data acquisition system, such as might be used in process monitoring and control, instrumentation, and test applications. The data acquisition system discussed in this example is the AT-MIO-

16 from National Instruments. The AT-MIO-16 is a high-performance, multifunction analog, digital, and timing input/output (I/O) board for the IBM PC/AT and compatibles. It contains a 12-bit ADC with up to 16 analog inputs, two 12-bit DACs with voltage outputs, eight lines of transistor-transistor-logic (TTL)–compatible digital I/O, and three 16-bit counter/timer channels for timing I/O. If additional analog inputs are required, the AMUX-64T analog multiplexer can be used. By cascading up to four AMUX-64T's, 256 single-ended or 128 differential inputs can be obtained. The AT-MIO-16 also uses the RTSI bus (real-time system interface bus) to synchronize multiboard analog, digital, and counter/timer operations by communicating system-level timing signals between boards.

Figure 13.22 is a block diagram of the AT-MIO-16 circuitry. Its major functions are described next.

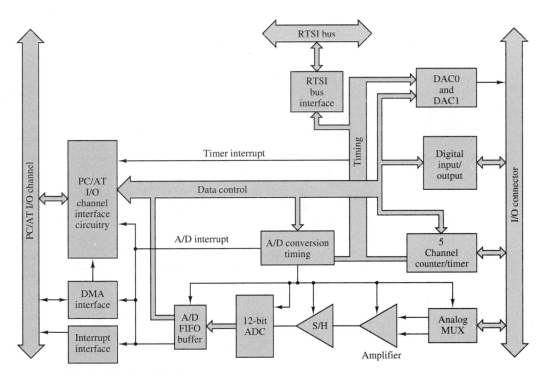

Figure 13.22 AT-MIO-16 block diagram

Analog Input: The AT-MIO-16 has two CMOS analog input multiplexers connected to 16 analog input channels. This data acquisition board has a software-programmable gain amplifier that can be used with voltage gains of 1, 10, 100, or 500 (the AT-MIO-16L) to accommodate low-level analog input signals, or with gains of 1, 2, 4, or 8 (the AT-MIO-16H) for high-level analog input signals. The AT-MIO-16 has a 12-bit ADC that gives an analog signal resolution of 4.88 mV with gain of 1 and an input range of ± 10 V. Finer resolutions up to 4.88 mV can be achieved by using gain and smaller input ranges. The board is available in three speeds: the AT-MIO-16(L/H)-9 contains a 9-μs ADC; the AT-MIO-16(L/H)-15 contains a 15-μs ADC; and the AT-MIO-16(L/H)-25 contains a 25-μs ADC. These conversions of the board have the following data acquisition sample rates on a single analog input channel:

Model number	Sampling rate	
	Typical case	Worst case
AT-MIO-16(L/H)-9	100 ksamples/s	91 ksamples/s
AT-MIO-16(L/H)-15	71 ksamples/s	59 ksamples/s
AT-MIO-16(L/H)-25	45 ksamples/s	37 ksamples/s

The timing of multiple A/D conversion is controlled either by the onboard counter/timer or by external timing signals. The onboard sample rate clock and sample counter control the onboard A/D timing. The AT-MIO-16 can generate both interrupts and DMA (direct memory access) requests on the PC/AT I/O channel. The interrupt can be generated when

1. An A/D conversion is available to be read from the A/D buffer.
2. The sample counter reaches its terminal count.
3. An error occurs.
4. One of the onboard timer clocks generates a pulse.

On the other hand, DMA requests can be generated whenever an A/D measurement is available from the A/D buffer.

Analog Output: The AT-MIO-16 has two double-buffered multiplying 12-bit DACs that are connected to two analog output channels. The resolution of the 12-bit DACs is 2.44 mV in the unipolar mode or 4.88 mV in the bipolar mode with the onboard 10-V reference. Finer resolutions can be achieved by using smaller voltages on the external reference. The analog output channels have an accuracy of ± 0.5 LSB and a differential linearity of ± 1 LSB. Voltage offset and gain error can be trimmed to zero.

Digital I/O: The AT-MIO-16 has eight digital I/O lines that are divided into two four-bit ports. The digital input circuitry has an eight-bit register that continuously reads the eight digital I/O lines, thus making read-back capability possible for the digital output ports, as well as reading incoming signals. The digital I/O lines are TTL-compatible.

Counter/Timer: The AT-MIO-16 uses the AM9513A counter/timer for time-related functions. The AM9513A contains five independent 16-bit counter/timers. A 1-MHz clock is the time baseline. Two of the AM9513A counter/timers are for multiple A/D conversion timing. The three remaining counters can be used for special data acquisition timing, such as expanding to a 32-bit sample counter or generating interrupts at user-programmable time intervals.

RTSI Bus Interface: The AT-MIO-16 is interfaced to the RTSI bus. You can send or receive the external analog input control signal; the waveform-generation timing signals; the output of counters 1, 2 and 5; the gate of counter 1; and the source of counter 5. You can send the RTSI bus the frequency output of the AM9513A.

PC/AT I/O Channel Interface: The PC/AT I/O channel interface circuitry includes address latches, address-decoding circuitry, data buffers, and interface timing and control signals.

I/O Connector: The I/O connector is a 50-pin male ribbon cable connector.

Software Support: The AT-MIO-16 also has software packages that control data acquisition functions on the PC-based data acquisition boards.

DRILL EXERCISES

13.5 Apply KCL at the inverting node of the summing amplifier of Figure 13.13 to show that equation 13.4 holds whenever $R_i = R_0/2^i$.

13.6 If the maximum analog voltage ($V_{a\,max}$) of a 12-bit DAC is 15 volts; find the smallest step size (δv) by which v_a can increment.

> Answer: 3.66 mV

13.7 Repeat Example 13.5 for the case of an eight-bit word with $R_0 = 10$ kΩ and the same range of v_a. Find the value of δv and R_f. Assume that ideal resistor values are available.

> Answer: $\delta v = 47.1$ mV; $R_f = 94.2$ Ω

13.8 For Figure 13.12, find V_{max} if $V_{in} = 4.5$ V.

> Answer: -13.5 V

13.9 For Figure 13.12, find the resolution if $V_{in} = 3.8$ V.

> Answer: 0.76 V

13.10 Find the minimum number of bits required in a DAC if the range of the DAC is from 0.5 to 15 V and the resolution of the DAC is 20 mV.

> Answer: 10

13.11 In Example 13.10, if the maximum conversion time available to you were 50 μs, what would be the highest-frequency signal you could expect to sample on the basis of the Nyquist criterion?

> Answer: $f_{max} = 10$ kHz

13.5   DATA TRANSMISSION IN DIGITAL INSTRUMENTS

One of the necessary aspects of data acquisition and control systems is the ability to transmit and receive data. Often, a microcomputer-based data acquisition system is interfaced to other digital devices, such as digital instruments or other microcomputers. In these cases it is necessary to transfer data directly in digital form. In fact, it is usually preferable to transmit data that is already in digital form, rather than analog voltages or currents. Among the chief reasons for the choice of digital over analog is that digital data is less sensitive to noise and interference than analog signals: in receiving a binary signal transmitted over a data line, the only decision to be made is whether the value of a bit is 0 or 1. Compared with the difficulty in obtaining a precise measurement of an analog voltage or current, either of which could be corrupted by noise or interference in a variety of ways, the probability of making an error in discerning between binary 0s and 1s is very small. Further, as will be shown shortly, digital data is often coded in such a way that many transmission errors may be detected and corrected for. Finally, storage and processing of digital data is much more readily accomplished than would be the case with analog signals. This section explores a few of the methods that are commonly employed in transmitting digital data; both parallel and serial interfaces are considered.

As was mentioned earlier, digital signals in a microcomputer are carried by a bus, consisting of a set of parallel wires each carrying one bit of information. In addition to the signal-carrying wires, there are also control lines that determine under what conditions transmission may occur. A typical computer data bus consists of eight parallel wires and therefore enables the transmission of one byte; digital data is encoded in binary according to one of a few standard codes, such as the BCD code described in Chapter 12, or the ASCII code, which is summarized in Table 13.2. This bus configuration is usually associated with **parallel transmission,** whereby all of the bits are transmitted simultaneously, along with some control bits.

Figure 13.23 depicts the general appearance of a parallel connection. Parallel data transmission can take place in one of two modes: **synchronous,** or **asynchronous.** In synchronous transmission, a timing clock pulse is transmitted along with the data over a control line. The arrival of the clock pulse indicates that valid data has also arrived. While parallel synchronous transmission can be very fast, it requires the added complexity of a synchronizing clock, and is typically employed only for internal computer data transmission. Further, this type of communication can take place only over short distances (approximately 4 m). Asynchronous data transmission, on the other hand, does not take place at a fixed clock rate, but requires a **handshake protocol** between sending and receiving ends. The handshake protocol consists of the transmission of *data ready* and *acknowledge* signals over two separate control wires. Whenever the sending device is ready to transmit data, it sends a pulse over the *data ready* line. When this signal reaches the receiver, and if the receiver is ready to receive the data, and *acknowledge* pulse is sent back, indicating that transmission may occur; at this point, the parallel data is transmitted.

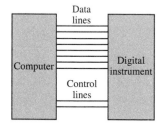

Figure 13.23 Parallel data transmission

Table 13.2   ASCII code

Graphic or control	ASCII (hex)	Graphic or control	ASCII (hex)	Graphic or control	ASCII (hex)
NUL	00	+	2B	V	56
SOH	01	,	2C	W	57
STX	02	-	2D	X	58
ETX	03	.	2E	Y	59
EOT	04	/	2F	Z	5A
ENQ	05	0	30	[	5B
ACK	06	1	31	\	5C
BEL	07	2	32	]	5D
BS	08	3	33	↑	5E
HT	09	4	34	←	5F
LF	0A	5	35	`	60
VT	0B	6	36	a	61
FF	0C	7	37	b	62
CR	0D	8	38	c	63
SO	0E	9	39	d	64
SI	0F	:	3A	e	65
DLE	10	;	3B	f	66
DC1	11	<	3C	g	67
DC2	12	=	3D	h	68
DC3	13	>	3E	i	69
DC4	14	?	3F	j	6A
NAK	15	@	40	k	6B
SYN	16	A	41	l	6C
ETB	17	B	42	m	6D
CAN	18	C	43	n	6E
EM	19	D	44	o	6F
SUB	1A	E	45	p	70
ESC	1B	F	46	q	71
FS	1C	G	47	r	72
GS	1D	H	48	s	73
RS	1E	I	49	t	74
US	1F	J	4A	u	75
SP	20	K	4B	v	76
!	21	L	4C	w	77
"	22	M	4D	x	78
#	23	N	4E	y	79
$	24	O	4F	z	7A
%	25	P	50	{	7B
&	26	Q	51	|	7C
'	27	R	52	}	7D
(	28	S	53	~	7E
)	29	T	54	DEL	7F
*	2A	U	55		

Perhaps the most common parallel interface is based on the **IEEE 488** standard, leading to the so called IEEE 488 bus, also referred to as **GPIB** (for general purpose instrument bus).

The IEEE 488 Bus

The IEEE 488 bus, shown in Figure 13.24, is an eight-bit parallel asynchronous interface that has found common application in digital instrumentation applications. The physical bus consists of 16 lines, of which 8 are used to carry the data, 3 for the handshaking protocol, and the rest to control the data flow. The bus permits connection of up to 15 instruments and data rates of up to 1 Mbyte/s. There is a limitation, however, in the maximum total length of the bus cable, which is 20 m. The signals transmitted are TTL-compatible and employ negative logic (see Chapter 12), whereby a logic 0 corresponds to a TTL high state (> 2 V) and a logic 1 to a TTL low state (< 0.8 V). Often, the eight-bit word transmitted over an IEEE 488 bus is coded in ASCII format (see Table 13.2), as illustrated in Example 13.12.

Data lines

```
0  ——————————————————————
1  ——————————————————————
2  ——————————————————————
3  ——————————————————————
4  ——————————————————————
5  ——————————————————————
6  ——————————————————————
7  ——————————————————————
```

Handshake lines

DAV	————————————————	Data valid
NRFD	————————————————	Not ready for data
NDAC	————————————————	Not data accepted

Control lines

IFC	————————————————	Interface clear
ATN	————————————————	Attention
REN	————————————————	Remote enable
SRQ	————————————————	Service request
EOI	————————————————	End or identify

Figure 13.24 IEEE 488 bus

Example   13.12

Determine the actual binary data sent by a digital voltmeter reading 3.405 V over an IEEE 488 bus if the data is encoded in ASCII format (see Table 13.2). Assume that the sequence is from most to least significant digit.

Solution:

From the ASCI conversion table (Table 13.2), we can operate the following conversions:

Control	ASCII (hex)
3	33
.	2E
4	34
0	30
5	35

Thus, the actual data sent, from most to least significant digit, is 33 2E 34 30 35 in hex code, or 110011 101110 110100 110000 110101 in binary.

In an IEEE 488 bus system, devices may play different roles and are typically classified as *controllers,* which manage the data flow; *talkers* (e.g., a digital voltmeter), which can only send data; *listeners* (e.g., a printer), which can only receive data; and *talkers/listeners* (e.g., a digital oscilloscope), which can receive as well as transmit data. The simplest system configuration might consist of just a talker and a listener. If more than two devices are present on the bus, a controller is necessary to determine when and how data transmission is to occur on the bus. For example, one of the key rules implemented by the controller is that only one talker can transmit at any one time; it is possible, however, for several listeners to be active on the bus simultaneously. If the data rates of the different listeners are different, the talker will have to transmit at the slowest rate, so that all of the listeners are assured of receiving the data correctly.

The set of rules by which the controller determines the order in which talking and listening are to take place is determined by a **protocol.** One aspect of the protocol is the handshake procedure, which enables the transmission of data. Since different devices (with different data rate capabilities) may be listening to the same talker, the handshake protocol must take into account these different capabilities. Let us discuss a typical handshake sequence that leads to transmission of data on an IEEE 488 bus. The three handshake lines used in the IEEE 488 have important characteristics that give the interface system wide flexibility, allowing interconnection of multiple devices that may operate at different speeds. The slowest active device controls the rate of data transfer, and more than one device can accept data simultaneously. The timing diagram of Figure 13.25 is used to illustrate the sequence in which the handshake and data transfer are performed:

1. All active listeners use the not ready for data (NRFD) line to indicate their state of readiness to accept a new piece of information. Nonreadiness to accept data is indicated if the NRFD line is held at

zero volts. If even one active listener is not ready, the NRFD line of the entire bus is kept at zero volts and the active talker will not transmit the next byte. When all active listeners are ready and they have released the NRFD line, it now goes high.

2. The designated talker drives all eight data input/output lines, causing valid data to be placed on them.

3. Two microseconds after putting valid data on the data lines, the active talker pulls the data valid (DAV) line to zero volts and thereby signals the active listeners to read the information on the data bus. The 2-μs interval is required to allow the data put on the data lines to reach (settle to) valid logic levels.

4. After the DAV is asserted, the listeners respond by pulling the NRFD line back down to zero. This prevents any additional data transfers from being initiated. The listeners also begin accepting the data byte at their own rates.

5. When each listener has accepted the data, it releases the not data accepted (NDAC) line. Only when the last active listener has released its hold on the NDAC line will that line go to its high-voltage-level state.

6. (a) When the active talker sees that NDAC has come up to its high state, it stops driving the data line. (b) At the same time, the talker releases the DAV line, ending the data transfer. The talker may now put the next byte on the data bus.

7. The listeners pull down the NDAC line back to zero volts and put the byte "away."

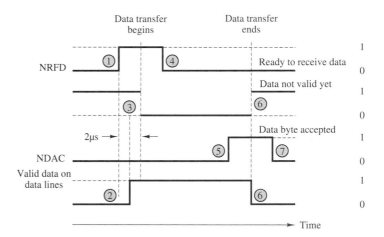

Figure 13.25 IEEE 488 data transmission protocol

Each of the instruments present on the data bus is distinguished by its own address, which is known to the controller; thus, the controller determines who the active talkers and listeners are on the bus by *addressing* them. To implement this and other functions, the controller uses the five control lines. Of these, ATN (attention) is used as a switch to indicate whether the controller is addressing or instructing the devices on the bus, or whether data transmission is taking place: when ATN is logic 1, the data lines contain either control information or addresses; with ATN = 1, only the controller is enabled to talk. When ATN = 0, only the devices that have been addressed can use the data lines. The IFC (interface clear) line is used to initialize the bus, or to clear it and reset it to a known condition in case of incorrect transmission. The REN (remote enable) line enables a remote instrument to be controlled by the bus; thus, any function that might normally be performed manually on the instrument (e.g., selecting a range or mode of operation) is now controlled by the bus via the data lines. The SRQ (service request) line is used by instruments on the bus whenever the instrument is ready to send or receive data; however, it is the controller who decides when to service the request. Finally, the EOI (end or identity) line can be used in two modes: when it is used by a talker, it signifies the end of a message; when it is used by the controller, it serves as a *polling* line, that is, a line used to interrogate the instrument about its data output.

Although it was mentioned earlier that the IEEE 488 bus can be used only over distances of up to 20 m, it is possible to extend its range of operation by connecting remote IEEE 488 bus systems over telephone communication lines. This can be accomplished by means of *bus extenders,* or by converting the parallel data to serial form (typically, in RS-232 format) and by transmitting the serial data over the phone lines by means of a modem. Serial communications and the RS-232 standard are discussed in the next section.

The RS-232 Standard

The primary reason why parallel transmission of data is not used exclusively is the limited distance range over which it is possible to transmit data on a parallel bus. Although there are techniques which permit extending the range for parallel transmission, these are complex and costly. Therefore, **serial transmission** is frequently used, whenever data is to be transmitted over a significant distance. Since serial data travels along one single path and is transmitted one bit at a time, the cabling costs for long distances are relatively low; further, the transmitting and receiving units are also limited to processing just one signal, and are also much simpler and less expensive. Two modes of operation exist for serial transmission: **simplex,** which corresponds to transmission in one direction only; and **duplex,** which permits transmission in either direction. Simplex transmission requires only one receiver and one transmitter, at each end of the link; on the other hand, duplex transmission can occur in one of two manners: **half-duplex** and **full-duplex.** In the former, although transmission can take place in both directions, it cannot occur simultaneously in both directions; in the latter case, both ends can simultaneously transmit and receive. Full-duplex transmission is usually implemented by means of four wires.

The data rate of a serial transmission line is measured in bits per second, since the data is transmitted one bit at a time. The unit of 1 bit/s is called a **baud;** thus, reference is often made to the baud rate of a serial transmission. The baud rate can be translated into a parallel transmission rate in words per second if the structure of the word is known; for example, if a word consists of 10 bits (start and stop bits plus an 8-bit data word) and the transmission takes place at 1,200 baud, 120 words are being transmitted every second. Typical data rates for serial transmission are standardized; the most common rates (familiar to the users of personal computer modem connections) are 300, 600, 1,200 and 2,400 baud. Baud rates can be as low as 50 baud or as high as 19,200 baud.

Like parallel transmission, serial transmission can also occur either synchronously or asynchronously. In the serial case, it is also true that asynchronous transmission is less costly but not as fast. A handshake protocol is also required for asynchronous serial transmission, as explained in the following. The most popular data-coding scheme for serial transmission is, once again, the ASCII code, consisting of a 7-bit word plus a **parity bit,** for a total of 8 bits per character. The role of the parity bit is to permit error detection in the event of erroneous reception (or transmission) of a bit. To see this, let us discuss the sequence of handshake events for asynchronous serial transmission and the use of parity bits to correct for errors. In serial asynchronous systems, handshaking is performed by using start and stop bits at the beginning and end of each character that is transmitted. The beginning of the transmission of a serial asynchronous word is announced by the "start" bit, which is always a 0 state bit. For the next five to eight successive bit times (depending on the code and the number of bits that specify the word length in that code), the line is switched to the 1 and 0 states required to represent the character being sent. Following the last bit of the data and the parity bit (which will be explained next), there is one bit or more in the 1 state, indicating "idle." The time period associated with this transmission is called the "stop" bit interval.

If noise pulses affect the transmission line, it is possible that a bit in the transmission could be misread. Thus, following the 5 to 8 transmitted data bits, there is a parity bit that is used for error detection. Here is how the parity bit works. If the transmitter keeps track of the number of 1s in the word being sent, it can send a parity bit, a 1 or a 0, to ensure that the total number of 1s sent is always even (even parity) or odd (odd parity). Similarly, the receiver can keep track of the 1s received to see whether there was an error with the transmission. If an error is detected, retransmission of the word can be requested.

Serial data transmission occurs most frequently according to the **RS-232 standard.** The RS-232 standard is based on the transmission of voltage pulses at a preselected baud rate; the voltage pulses are in the range -3 to -15 V for a logic 0 and in the range $+3$ to $+15$ V for a logic 1. It is important to note that this amounts to a negative logic convention and that the signals are *not* TTL-compatible. The distance over which such transmission can take place is up to approximately 17 m (50 ft). The RS-232 standard was designed to make the transmission of digital data compatible with existing telephone lines; since phone lines were originally designed to carry analog voice signals, it became necessary to establish some standard procedures to make digital transmission possible over them. The resulting standard describes the mechanical and electrical characteristics of the interface between *data terminal equipment* (DTE) and *data communication equipment* (DCE). DTE consists of computers, terminals, digital instruments, and related peripherals; DCE includes all of those devices that are used to encode digital data in a format that permits their transmission over telephone lines. Thus, the standard specifies how data should be presented by the DTE to the DCE so that digital data can be transmitted over voice lines.

A typical example of DCE is the **modem.** A modem converts digital data to audio signals that are suitable for transmission over a telephone line and is also capable of performing the reverse function, by converting the audio signals back to digital form. The term *modem* stands for *mod*ulate-*dem*odulate, because a modem modulates a sinusoidal carrier using digital pulses (for transmission) and demodulates the modulated sinusoidal signal to recover the digital pulses (at reception). Three methods are commonly used for converting digital pulses to an audio signal: **amplitude-shift keying, frequency-shift keying,** and **phase-shift keying,** depending on whether the amplitude, phase, or frequency of the sinusoid is modulated by the digital pulses. Figure 13.26 depicts the essential block of a data transmission system based on the RS-232 standard, as well as examples of digital data encoded for transmission over a voice line.

In addition to the function just described, however, the RS-232 standard also provides a very useful set of specifications for the direct transmission of digital data between computers and instruments. In other words, communication between digital terminal instruments may occur directly in digital form (i.e., without digital communication devices encoding the digital data in a form compatible with analog voice lines). Thus, this standard is also frequently used for direct digital communication.

The RS-232 standard can be summarized as follows:

- Data signals are encoded according to a negative logic convention using voltage levels of -3 to -15 V for logic 1 and $+3$ to $+15$ V for logic 0.
- Control signals use a positive logic convention (opposite to that of data signals).
- The maximum shunt capacitance of the load cannot exceed 2,500 pF; this, in effect, limits the maximum length of the cables used in the connection.
- The load resistance must be between 300 Ω and 3 kΩ.

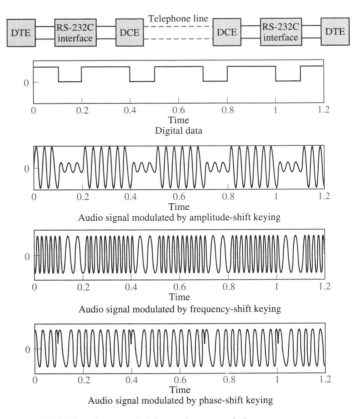

Figure 13.26 Digital data encoded for analog transmission

- Three wires are used for data transmission. One wire each is used for receiving and transmitting data; the third wire is a signal return line (signal ground). In addition, there are 22 wires that can be used for a variety of control purposes between the DTE and DCE.

- The male part of the connector is assigned to the DTE and the female part to the DCE. Figure 13.27 labels each of the wires in the 25-pin connector. Since each side of the connector has a *receive* and a *transmit* line, it has been decided by convention that the DCE transmits on the transmit line and receives on the receive line, while the DTE receives on the transmit line and transmits on the receive line.

- The baud rate is limited by the length of the cable; for a 50-ft length, any rate from 50 baud to 19.2 kbaud is allowed. If a longer cable connection is desired, the maximum baud rate will decrease according to the length of the cable, and **line drivers** can be used to amplify the signals, which are transmitted over twisted pair wires. Line drivers are simply signal amplifiers that are used directly on the digital signal, prior to encoding. For example, the signal generated by a DTE device (say a computer) may be transmitted over a distance of up to 10,000 ft (at a rate of 600 baud) prior to being encoded by the DCE.

• The serial data can be encoded according to any code, although the ASCII code is by far the most popular.

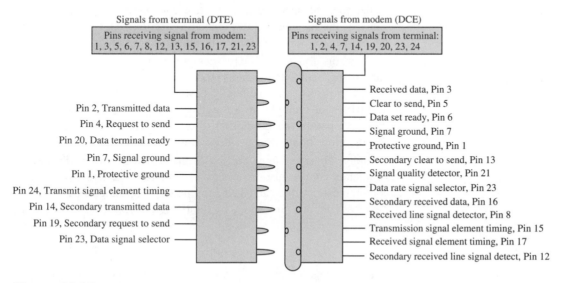

Figure 13.27 RS-232 connections

DRILL EXERCISE

13.12   Determine the actual binary data sent by a digital voltmeter reading 15.06 V over an IEEE 488 bus if the data is encoded in ASCII format (see Table 13.2). Assume that the sequence is from most to least significant digit.

Answer: 31 35 2E 30 36

CONCLUSION

• Digital systems play a prominent role in modern engineering. The microprocessor, in particular, has become an integral part of instrumentation and control systems. The microprocessor is much more flexible in its application than the logic circuits of Chapter 12, because it can be programmed to perform many different tasks. Depending on the computing power required, 8-, 16-, and 32-bit microprocessors are used for automating measurement and control functions in a variety of industrial applications.

- One of the important issues in the design of digital systems is the interface between microprocessors and external devices, in particular, sensors and actuators. Since these devices typically function in an analog fashion, it is necessary to convert sensor signals and actuator commands into digital form. These tasks are accomplished by analog-to-digital, and digital-to-analog converters, respectively. In defining the specifications of such converters, it is important to consider the number of bits (related to accuracy and resolution) and the conversion speed, among other factors.

- Data acquisition systems are special-purpose digital systems, comprising one or more microprocessors and ADC and DAC units; these computers are specifically designed to permit simultaneous acquisition of multiple sensor signals at variable rates. In addition, they also permit the generation of actuator command signals and can therefore be used to perform control functions.

- Complex digital systems consist of many instruments and micro-computers connected over a network. For these different devices to communicate, well-established communication protocols exist; two of the most common are defined by the IEEE 488 and RS-232 standards.

KEY TERMS

DEVICE DATA SHEETS

2764A Advanced 64K (8K × 8) Production and UV Erasable PROMs

2764A

27256	27128	2732A	2716					2716	2732A	27128	27256
V_{PP}	V_{PP}			V_{PP} ☐ 1		28 ☐ V_{CC}				V_{CC}	V_{CC}
A_{12}	A_{12}			A_{12} ☐ 2		27 ☐ $\overline{PGM}$				$\overline{PGM}$	A_{14}
A_7	A_7	A_7	A_7	A_7 ☐ 3		26 ☐ N.C.		V_{CC}	V_{CC}	A_{13}	A_{13}
A_6	A_6	A_6	A_6	A_6 ☐ 4		25 ☐ A_8		A_8	A_8	A_8	A_8
A_5	A_5	A_5	A_5	A_5 ☐ 5		24 ☐ A_9		A_9	A_9	A_9	A_9
A_4	A_4	A_4	A_4	A_4 ☐ 6		23 ☐ A_{11}		V_{PP}	A_{11}	A_{11}	A_{11}
A_3	A_3	A_3	A_3	A_3 ☐ 7		22 ☐ $\overline{OE}$		$\overline{OE}$	$\overline{OE}$ V_{PP}	$\overline{OE}$	$\overline{OE}$
A_2	A_2	A_2	A_2	A_2 ☐ 8		21 ☐ A_{10}		A_{10}	A_{10}	A_{10}	A_{10}
A_1	A_1	A_1	A_1	A_1 ☐ 9		20 ☐ $\overline{CE}$		$\overline{CE}$	$\overline{CE}$	$\overline{CE}$	$\overline{CE}$
A_0	A_0	A_0	A_0	A_0 ☐ 10		19 ☐ O_7		O_7	O_7	O_7	O_7
O_0	O_0	O_0	O_0	O_0 ☐ 11		18 ☐ O_6		O_6	O_6	O_6	O_6
O_1	O_1	O_1	O_1	O_1 ☐ 12		17 ☐ O_5		O_5	O_5	O_5	O_5
O_2	O_2	O_2	O_2	O_2 ☐ 13		16 ☐ O_4		O_4	O_4	O_4	O_4
Gnd	Gnd	Gnd	Gnd	GND ☐ 14		15 ☐ O_3		O_3	O_3	O_3	O_3

NOTE: INTEL "UNIVERSAL SITE"-COMPATIBLE EPROM PIN CONFIGURATIONS
ARE SHOWN IN THE BLOCKS ADJACENT TO THE 2764A PINS

Pin Configurations

PIN NAMES

A_0-A_{12}	ADDRESSES
$\overline{CE}$	CHIP ENABLE
$\overline{OE}$	OUTPUT ENABLE
O_0-O_7	OUTPUTS
$\overline{PGM}$	PROGRAM
N.C.	NO CONNECT

Note: Intel "Universal Site"-Compatible EPROM pin configurations are shown in the blocks adjacent to the P2764A pins.

HM6116 2048-word × 8-bit High Speed Static CMOS RAM

Absolute Maximum Ratings

Item	Symbol	Rating	Unit
Voltage on Any Pin Relative to GND	V_T	-0.5^* to $+7.0$	V
Operating Temperature	T_{opr}	0 to $+70$	°C
Storage Temperature (Plastic)	T_{stg}	-55 to $+125$	°C
Storage Temperature (Ceramic)	T_{stg}	-65 to $+150$	°C
Temperature Under Bias	T_{bias}	-10 to $+85$	°C
Power Dissipation	P_T	1.0	W

*Pulse Width 50 ns: -1.5 V

Pin Connections

240571-2

Truth Table

$\overline{CS}$	$\overline{OE}$	$\overline{WE}$	Mode	V_{CC} Current	I/O Pin	Ref. Cycle
H	×	×	Not Selected	I_{SB}, I_{SB1}	High Z	
L	L	H	Read	I_{CC}	Dout	Read Cycle (1) ∼ (3)
L	H	L	Write	I_{CC}	Din	Write Cycle (1)
L	L	L	Write	I_{CC}	Din	Write Cycle (2)

AD582   Low-Cost Sample/Hold Amplifier

Features

Suitable for 12-Bit Applications
High Sample/Hold Current Ratio: 10^7
Low Acquisition Time: 6μs to 0.1%
Low Charge Transfer: <2pC
High Input Impedance in Sample and Hold Modes
Connect in Any Op Amp Configuration
Differential Logic Inputs

SPECIFICATIONS (typical @ $+25°$C, $V_S = \pm15$V and $C_H = 1000$pF, A$= +1$ unless otherwise specified)

Model	AD582K	AD582S
Sample/hold characteristics		
Acquisition Time, 10V Step to 0.1%		
$C_H = 100$pF	6μs	*
Acquisition Time, 10V Step to 0.01%		
$C_H = 1000$pF	25μs	*
Aperture Time, 20V p-p Input,		
Hold 0V	200ns	*
Aperture Jitter, 20V p-p Input,		
Hold 0V	15ns	*
Setting Time, 20-V p-p Input		
Hold 0V, to 0.01%	0.5μs	*
Droop Current, Steady State, $\pm10V_{OUT}$	100pA max	*
Droop Current, T_{min} to T_{max}	1nA	150nA max
Charge Transfer	5pC max (1.5pC typ)	*
Sample to Hold Offset	0.5mV	*
Feedthrough Capacitance		
20V p-p, 10kHz Input	0.05pF	*
Transfer characteristics		
Open Loop Gain		
$V_{OUT} = 20$V p-p, $R_L = 2$k	25k min (50k typ)	*
Common Mode Rejection		
$V_{CM} = 20$V p-p	60dB min (70dB typ)	*
Small Signal Gain Bandwidth		
$V_{OUT} = 100$mV p-p, $C_H = 100$pF	1.5MHz	*
Full Power Bandwidth		
$V_{OUT} = 20$V p-p, $C_H = 100$pF	70kHz	*
Slew Rate		
$V_{OUT} = 20$V p-p, $C_H = 100$pF	3V/μs	*
Output Resistance		
Hold Mode, $I_{OUT} = \pm5$mA	12Ω	*
Linearity		
$V_{OUT} = 20$V p-p, $R_L = 2$k	±0.01%	*
Output Short Circuit Current	±25mA	*

*Specifications same as AD582K

SPECIFICATIONS (typical @ $+25°C$, $V_S = \pm15V$ and $C_H = 1000pF$, A$= +1$ unless otherwise specified) *(continued)*

Model	AD582K	AD582S
Analog input characteristics		
Offset Voltage	6mV max (2mV typ)	*
Offset Voltage, T_{min} to T_{max}	4mV	8mV max (5mV typ)
Bias Current	$3\mu A$ max ($1.5\mu A$ typ)	*
Offset Current	300nA max (75nA typ)	*
Offset Current T_{min} to T_{max}	100nA	400nA max (100nA typ)
Input Capacitance, f = 1MHz	2pF	*
Input Resistance, Sample or Hold		
20V p-p Input, A = +1	30MΩ	*
Absolute Max Diff Input Voltage	30V	*
Absolute Max Input Voltage, Either Input	$\pm V_S$	*
Digital input characteristics		
+Logic Input Voltage		
Hold Mode, T_{min} to T_{max}, −Logic @ 0V	+2V min	*
Sample Mode, T_{min} to T_{max}, −Logic @ 0V	+0.8V max	*
+Logic Input Current		
Hold Mode, +Logic @ +5V, −Logic @ 0V	$1.5\mu A$	*
Sample Mode, +Logic @ 0V, −Logic @ 0V	1nA	*
−Logic Input Current		
Hold Mode, +Logic @ +5V, −Logic @ 0V	$24\mu A$	*
Sample Mode, +Logic @ 0V, −Logic @ 0V	$4\mu A$	*
Absolute Max Diff Input Voltage, +L to −L	+15V/ − 6V	*
Absolute Max Input Voltage, Either Input	$\pm V_S$	*
Power supply characteristics		
Operating Voltage Range	$\pm9V$ to $\pm18V$	$\pm9V$ to $\pm22V$
Supply Current, $R_L = \infty$	4.5 mA max (3mA typ)	*
Power Supply Rejection,		
$\Delta V_S = 5V$, Sample Mode	60dB min (75dB typ)	*

*Specifications same as AD582K

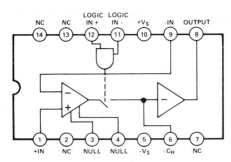

14 PIN DIP

AD558   DAC PORT Low-Cost Complete μP-Compatible 8-Bit DAC

Features

Complete 8-Bit DAC
Voltage Output—2 Calibrated Ranges
Internal Precision Band-Gap Reference
Single-Supply Operation: +5V to +15V
Full Microprocessor Interface
Fast: $1\mu s$ Voltage Setting to $\pm1/2$LSB
Low Power: 75mW
No User Trims
Guaranteed Monotonic Over Temperature
All Errors Specified T_{min} to T_{max}
Small 16-Pin DIP or PLCC Package
Single Laser-Wafer-Trimmed Chip for Hybrids
Low Cost

AD558 Pin Configuration (DIP)

SPECIFICATIONS (@ T_A = +25°C, V_{CC} = +5V to +15V unless otherwise specified)

Model	AD558J Min	AD558J Typ	AD558J Max	AD558K Min	AD558K Typ	AD558K Max	AD558S Min	AD558S Typ	AD558S Max	AD558T Min	AD558T Typ	AD558T Max	Units
Resolution			8			8			8			8	Bits
Relative accuracy[1]													
0 to +70 °C		$\pm1/2$			$\pm1/4$			$\pm1/2$			$\pm1/4$		LSB
−55 °C to 125 °C								$\pm3/4$			$\pm3/8$		LSB
Output													
Ranges[2]		0 to +2.56			0 to +2.56			0 to +2.56			0 to +2.56		V
		0 to +10			0 to +10			0 to +10			0 to +10		V
Output Settling Time[3]													
0 to 2.56 Volt Range		0.8	1.5		0.8	1.5		0.8	1.5		0.8	1.5	μs
0 to 10 Volt Range[4]		2.0	3.0		2.0	3.0		2.0	3.0		2.0	3.0	μs
Full scale accuracy[4]													
@ 25°C		±1.5			±0.5			±1.5			±0.5		LSB
T_{min} to T_{max}		±2.5			±1			±2.5			±1		LSB
Zero error													
@ 25°C		±1			$\pm1/2$			±1			$\pm1/2$		LSB
T_{min} to T_{max}		±2			±1			±2			±1		LSB
Digital inputs													
T_{min} to T_{max}													
Input Current		±100			±100			±100				100	μA
Data Inputs, Voltage													
Bit On −Logic"1"	2.0			2.0			2.0			2.0			V
Bit On −Logic"0"	0		0.8	0			0			0			V
Control Inputs, Voltage													
On −Logic"1"	2.0			2.0			2.0			2.0			V
On −Logic"0"	0		0.8	0		0.8	0		0.8	0		0.8	V
Input Capacitance		4			4			4			4		pF

SPECIFICATIONS (@ T_A = +25°C, V_{CC} = +5V to +15V unless otherwise specified)

(continued)

Model	AD558J			AD558K			AD558S			AD558T			Units
	Min	Typ	Max	Min	Typ	Max	Min	Typ	Max	Min	Typ	Max	
Power supply													
Operating Voltage Range													
(V_{CC})													
2.56 Volt Range	+4.5		+16.5	+4.5		+16.5	+4.5		+16.5	+4.5		+16.5	V
10 Volt Range	+11.4		+16.5	+11.4		+16.5	+11.4		+16.5	+11.4		+16.5	V
Current (I_{CC})		15	25		15	25		15	25		15	25	mA
Rejection Ratio			0.03			0.03			0.03			0.03	%%
Power dissipation,													
V_{CC} = 5V		75	125		75	125		75	125		75	125	mW
V_{CC} = 15V		225	375		225	375		225	375		225	375	mW
Operating temperature range	0		+70	0		+70	−55		+125	−55		+125	°C

[1] Relative Accuracy is defined as the deviation of the code transition points from the ideal transfer point on a straight line from the zero to the full scale of the device.

[2] Operation of the 0 to 10 volt output range requires a minimum supply voltage of +11.4 volts.

[3] Settling time is specified for a positive-going full-scale step to ± 1/2LSB. Negative-going steps to zero are slower, but can be improved with an external pull-down.

[4] The full range output voltage for the 2.56 range is 2.55V and is guaranteed with a +5V supply, for the 10V range, it is 9.960V guaranteed with a +15V supply.

Input Logic Coding

Digital Input Code			Output Voltage	
Binary	**Hexadecimal**	**Decimal**	**2.56V Range**	**10.00V Range**
0000 0000	00	0	0	0
0000 0001	01	1	0.010V	0.039V
0000 0010	02	2	0.020V	0.078V
0000 1111	0F	15	0.150V	0.586V
0001 0000	10	16	0.160V	0.625V
0111 1111	7F	127	1.270V	4.961V
1000 0000	80	128	1.280V	5.000V
1100 0000	CD	192	1.920V	7.500V
1111 1111	FF	255	2.55V	9.961V

AD574A   Complete 12-bit A/D Converter

Features

Complete 12-Bit A/D Converter with Reference and Clock
8- and 16-Bit Microprocessor Bus Interface
Guaranteed Linearity Over Temperature 0 to $+70°C$ – AD574AJ, K, L $-55°C$ to $+125°C$ – AD574AS, T, U
No Missing Codes Over Temperature
$35\mu s$ Maximum Conversion time
Buried Zener Reference for Long-Term Stability and Low Gain T.C. 10ppm/°C max AD574AL 12.5ppm/°C max AD574AU
Ceramic DIP, Plastic DIP or PLCC Package

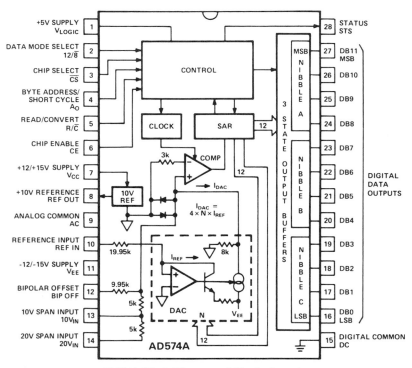

AD574A Block Diagram and Pin Configuration

SPECIFICATIONS (@ = 25°C with V_{CC} = +15V or +12V, V_{LOGIC} = +5V, V_{EE} = −15V or −12V unless otherwise indicated)

Model	AD574AJ			AD574AK			AD574AL			Units
	Min	Typ	Max	Min	Typ	Max	Min	Typ	Max	
Resolution			12			12			12	Bits
Linearity error @ $+25°C$			±1			$\pm1/2$			$\pm1/2$	LSB
T_{min} to T_{max}			±1			$\pm1/2$			$\pm1/2$	LSB
Differential linearity error (Minimum resolution for which no missing codes are guaranteed) T_{min} to T_{max}	11			12			12			Bits

SPECIFICATIONS (@ = 25°C with V$_{CC}$ = +15V or +12V, V$_{LOGIC}$ = +5V, V$_{EE}$ = −15V or −12V unless otherwise indicated) *(continued)*

Model	AD574AJ			AD574AK			AD574AL			Units
	Min	Typ	Max	Min	Typ	Max	Min	Typ	Max	
Unipolar offset (Adjustable to zero)			±2			±1			±1	LSB
Bipolar offset (Adjustable to zero)			±4			±4			±2	LSB
Full-scale calibration error (with fixed 50Ω resistor from REF OUT to REF IN) (Adjustable to zero)			0.25			0.25			0.125	% of F.S.
Temperature range	0		+70	0		+70	0		+70	°C
Analog input										
Input Ranges										
Bipolar	−5		+5	−5		+5	−5		+5	Volts
	−10		+10	−10		+10	−10		+10	Volts
Unipolar	0		+10	0		+10	0		+10	Volts
	0		+20	0		+20	0		+20	Volts
Input Impedance										
10 Volt Span	3	5	7	3	5	7	3	5	7	kΩ
20 Volt Span	6	10	14	6	10	14	6	10	14	kΩ
Digital characteristics										
Inputs (CE, $\overline{CS}$, R/$\overline{C}$, A$_0$)										
Logic "1" Voltage	+2.0		+5.5	+2.0		+5.5	+2.0		+5.5	Volts
Logic "0" Voltage	−0.5		+0.8	−0.5		+0.8	−0.5		+0.8	Volts
Current	−20		+20	−20		+20	−20		+20	μA
Capacitance		5			5			5		pF
Outputs (DB11–DB, STS)										
Logic "1" Voltage (I$_{SINK}$ ≤ 1.6mA)	+2.4			+2.4			+2.4			Volts
Logic "0" Voltage (I$_{SOURCE}$ ≤ 500μA)			+0.4			+0.4			+0.4	Volts
Leakage (DB11–DB0, High-Z State)	−20		+20	−20		+20	−20		+20	μA
Capacitance		5			5			5		pF
Power supplies										
Operating Range										
V$_{LOGIC}$	+4.5		+5.5	+4.5		+5.5	+4.5		+5.5	Volts
V$_{CC}$	+11.4		+16.5	+11.4		+16.5	+11.4		+16.5	Volts
V$_{EE}$	−11.4		−16.5	−11.4		−16.5	−11.4		−16.5	Volts
Power Dissipation		390	725		390	725		390	725	mW
Internal reference voltage	9.98	10.0	10.02	9.98	10.0	10.02	9.99	10.0	10.01	Volts
Output current (available for external loads) (External load should not change during conversion)			1.5			1.5			1.5	mA

HOMEWORK PROBLEMS

13.1   Draw a block diagram that shows the bus structure of a microprocessor-based system and the subsystems connected by the bus connects. For each subsystem, indicate the nature of the components that make up the system—that is, ROM or RAM, for example.

13.2   Explain the purpose of the ALU.

13.3   Name the internal registers of a microprocessor, and explain their functions.

13.4   Name the three different system buses, and explain their functions.

13.5   Suppose a microprocessor has n registers.

a.   How many control lines do you need to connect each register to all other registers?

b.   How many control lines do you need if a bus is used?

13.6   Explain the function of the status register (flag resister), and give an example.

13.7   What is the distinction between volatile and nonvolatile memory?

13.8   Draw a block diagram of a memory system that is interfaced to an 8086 microprocessor. The memory system requires 8 Kbytes of ROM and 8 Kbytes of RAM. The ROM must start at location 0000H and must be immediately followed by the RAM. The following information is known for the processor. The system bus consists of a 16-bit address bus, A_0 to A_{15}; and an 8-bit bidirectional data bus, D_0 to D_7. Assume that the microprocessor has three control lines: $\overline{WR}$ (logic 0 indicates that the processor is writing), $\overline{RD}$ (logic 0 indicates that the processor is reading), and IO/$\overline{M}$ (logic 0 indicates that the address is that of a memory location as opposed to an I/O location). The memory devices that are available are shown in Figure P13.1.

13.9   List two advantages of digital signal processing over analog signal processing.

13.10   Discuss the role of a multiplexer in a data acquisition system.

13.11   Discuss the purpose of using sample-and-hold circuits in data acquisition systems.

13.12   The unsigned decimal number 12_{10} is inputted to a four-bit DAC. Given that $R_f = R_0/15$, logic 0 corresponds to 0 V, and logic 1 corresponds to 4.5 V,

a.   What is the output of the DAC?

b.   What is the maximum voltage that can be outputted from the DAC?

c.   What is the resolution over the range 0 to 4.5 V?

d.   Find the number of bits required in the DAC if an improved resolution of 20 mV is desired.

13.13   The unsigned decimal number 215_{10} is inputted to an eight-bit DAC. Given that $R_f = R_0/255$, logic 0 corresponds to 0 V, and logic 1 corresponds to 10 V,

Figure P13.1

a. What is the output of the DAC?

b. What is the maximum voltage that can be outputted from the DAC?

c. What is the resolution over the range 0 to 10 V?

d. Find the number of bits required in the DAC if an improved resolution of 3 mV is desired.

13.14 Repeat Problem 13.13 for the unsigned decimal number 158_{10}.

13.15 The unsigned decimal number 98_{10} is inputted to an eight-bit DAC. Given that $R_f = R_0/255$, logic 0 corresponds to 0 V and logic 1 corresponds to 4.5 V,

a. What is the output of the DAC?

b. What is the maximum voltage that can be outputted from the DAC?

c. What is the resolution over the range 0 to 4.5 V?

d. Find the number of bits required in the DAC if an improved resolution of 0.5 mV is desired.

13.16 For the DAC circuit shown in Figure P13.2 (using an ideal op-amp), what value of R_f will give an output range of $-10 \leq V_0 \leq 0$ V? Assume that logic $0 = 0$ V and logic $1 = 5$V.

13.17 The unsigned decimal number 261_{10} is inputted to a 12-bit DAC. Given that $R_f = R_0/4,095$ and logic 1 corresponds to 10 V and logic 0 to 0 V,

Figure P13.2

a. What is the output of the DAC?

b. What is the maximum voltage that can be outputted from the DAC?

c. What is the resolution over the range 0 to 10 V?

d. Find the number of bits required in the DAC if an improved resolution of 1 mV is desired.

13.18 Repeat Problem 13.17 for logic $0 = 0$ V and logic $1 = 3.6$ V.

13.19 The unsigned decimal number 345_{10} is inputted to a 12-bit DAC. Given that $R_f = R_0/4,095$, logic 1 corresponds to 10 V, and logic 0 to 0 V,

a. What is the output of the DAC?

b. What is the maximum voltage that can be outputted from the DAC?

c. What is the resolution over the range 0 to 10 V?

d. Find the number of bits required in the DAC if an improved resolution of 0.5 mV is desired.

13.20 For the DAC circuit shown in Figure P13.2 (using an ideal op-amp), what value of R_f will give an output range of $-15 \leq V_0 \leq 5$ V?

13.21 Repeat Problem 13.19 for logic $0 = 0$ V and logic $1 = 4.5$ V.

13.22 A data acquisition system uses a DAC with a range of ± 15 V and a resolution of 0.01 V. How many bits must be present in the DAC?

13.23 A data acquisition system uses a DAC with a range of ± 10 V and a resolution of 0.04 V. How many bits must be present in the DAC?

13.24 A data acquisition system uses a DAC with a range of -10 to $+15$ V and a resolution of 0.004 V. How many bits must be present in the DAC?

13.25 A DAC is to be used to deliver velocity commands to a motor. The maximum velocity is to be 2,500 rev/min, and the minimum nonzero velocity is to be 1 rev/min. How many bits are required in the DAC? What will the resolution be?

13.26 The voltage range of a signal is ± 5 V, and a resolution of 0.005 percent of the voltage range is required. How many bits must be in the DAC to sense the signal to its full resolution?

13.27 The voltage range of feedback signal from a process is -5 V to $+15$ V, and a resolution of 0.05 percent of the voltage range is required. How many bits are required for the DAC?

13.28 Eight channels of analog information are being used by a computer to close eight control loops. Assume that all analog signals have identical frequency content and are multiplexed into a single ADC. The ADC requires 100 μs per conversion. The closed-loop software requires 500 μs of computation and output time for four of the loops, and for the other four it requires 250 μs. What is the maximum frequency content that the analog signal can have according to the Nyquist criterion?

13.29 A rotary potentiometer is to be used as a remote rotational displacement sensor. The maximum displacement to be measured is 180°, and the potentiometer is rated for 10 V and 270° of rotation.

a.   What voltage increment must be resolved by an ADC to resolve an angular displacement of 0.5°? How many bits would be required in the ADC for full-range detection?

b.   The ADC requires a 10-V input voltage for full-scale binary output. If an amplifier is placed between the potentiometer and the ADC, what amplifier gain should be used to take advantage of the full range of the ADC?

13.30 What is the maximum sampling time interval to ensure that the Nyquist principle is satisfied when a 75-Hz signal is being sampled?

13.31 A torque sensor has been mounted on a farm tractor engine. The voltage produced by the torque sensor is to be sampled by an ADC. The rotational speed of the crankshaft is 800 rev/min. Because of speed fluctuation caused by the reciprocating action of the engine, frequency content is present in the torque signal at twice the shaft rotation frequency. What is the minimum sampling period that can be used to ensure that the Nyquist principle is satisfied?

13.32 The output voltage of an aircraft altimeter is to be sampled using an ADC. The sensor outputs 0 V at 0 m altitude and outputs 10 V at 10,000 m altitude. If the allowable error in sensing is 10 m, find the minimum number of bits required for the ADC.

13.33 Consider a circuit that generates interrupts at fixed time intervals. Such a device is called a *real-time clock* and is used in control applications to establish the sample period as T seconds for control algorithms. Show how this can be done with a square wave (clock) that has a period equal to the desired time interval between interrupts.

13.34 Draw a block diagram of a circuit that will interface two interrupts, INT0 and INT1, to the INT input of a CPU so that INT1 has the higher priority and INT0 has the lower. In other words, a signal on INT1 is to be able to interrupt the CPU even when the CPU is currently handling an interrupt generated by INT0, but not vice versa.

13.35 Discuss the advantages of digital data transmission over analog data transmission.

13.36 Explain briefly the concept of interfacing as it is related to instrumentation systems.

13.37 Define the following terms:

a. Handshaking

b. Baud rate

13.38 Express the following decimal numbers in ASCII form:

a. 12

b. 345.2

c. 43.5

13.39 Express the following words in ASCII form:

a. Digital

b. Computer

c. Ascii

d. ASCII

13.40 Explain the method of handshaking used in serial transmission.

13.41 Explain why synchronous data transmission can be faster than asynchronous data transmission.

13.42 Explain why three wires are required for the handshaking technique employed by IEEE 488 bus systems.

13.43 Discuss the role of DTE and DCE in the RS 232 standard. Give examples of common DTE and DCE.

PART

III

ELECTROMECHANICS

739

Principles of Electromechanics

The objective of this chapter is to introduce the fundamental notions of electromechanical energy conversion, leading to an understanding of the operation of various electromechanical transducers. The chapter also serves as an introduction to the material on electric machines to be presented in Chapters 15 and 16.

The subject of electromechanical energy conversion is one that should be of particular interest to the non–electrical engineer, because it forms one of the important points of contact between electrical engineering and other engineering disciplines. Electromechanical transducers are commonly used in the design of industrial and aerospace control systems and in biomedical applications, and they form the basis of many common appliances. In the course of our exploration of electromechanics, we shall illustrate the operation of practical devices, such as loudspeakers, relays, solenoids, sensors for the measurement of position and velocity, and other devices of practical interest.

Upon completion of the chapter, you should be able to:

- Analyze simple magnetic circuits, to determine electrical and mechanical performance and energy requirements.
- Size a relay or solenoid for a given application.

- Describe the energy-conversion process in electromechanical systems.
- Perform a simplified linear analysis of electromechanical transducers.

14.1   ELECTRICITY AND MAGNETISM

The notion that the phenomena of electricity and magnetism are interconnected was first proposed in the early 1800s by H. C. Oersted, a Danish physicist. Oersted showed that an electric current produces magnetic effects (more specifically, a magnetic field). Soon after, the French scientist André Marie Ampère expressed this relationship by means of a precise formulation, known as *Ampère's law*. A few years later, the English scientist Faraday illustrated how the converse of Ampère's law also holds true, that is, that a magnetic field can generate an electric field; in short, *Faraday's law* states that a changing magnetic field gives rise to a voltage. We shall return to a more careful examination of both Ampère's and Faraday's laws later in this chapter.

As will be explained in the next few sections, the magnetic field forms a necessary connection between electrical and mechanical energy. Ampère's and Faraday's laws will formally illustrate the relationship between electric and magnetic fields, but it should already be evident from your own individual experience that the magnetic field can also convert magnetic energy to mechanical energy (for example, by lifting a piece of iron with a magnet). In effect, the devices we commonly refer to as *electromechanical* should more properly be referred to as electro*magneto*mechanical, since they almost invariably operate through a conversion from electrical to mechanical energy (or vice versa) by means of a magnetic field. Chapters 14 through 16 are concerned with the use of electricity and magnetic materials for the purpose of converting electrical energy to mechanical, and back.

The Magnetic Field and Faraday's Law

The quantities used to quantify the strength of a magnetic field are the **magnetic flux,** ϕ, in units of **webers** (Wb); and the **magnetic flux density, B,** in units of webers per square meters (Wb/m^2), or **teslas** (T). The latter quantity, as well as the associated **magnetic field intensity, H** (units of amperes per meter or A/m) are vectors.[1] Thus, the density of the magnetic flux and its intensity are in general described in vector form, in terms of the components present in each spatial direction (e.g., on the *x*, *y*, and *z* axes). In discussing magnetic flux density and field intensity in this chapter and the next, we shall almost always assume that the field is a *scalar field,* that is, that it lies in a single spatial direction. This will simplify many explanations.

---

[1] We'll use the boldface symbols **B** and **H** to denote the vector forms of *B* and *H*; the standard typeface will represent the scalar flux density or field intensity in a given direction.

It is customary to represent the magnetic field by means of the familiar *lines of force* (a concept also due to Faraday); we visualize the strength of a magnetic field by observing the density of these lines in space. You probably know from a previous course in physics that such lines are closed in a magnetic field, that is, that they form continuous loops exiting at a magnetic north pole (by definition) and entering at a magnetic south pole. The relative strengths of the magnetic fields generated by two magnets could be depicted as shown in Figure 14.1.

Magnetic fields are generated by electric charge in motion, and their effect is measured by the force they exert on a moving charge. As you may recall from previous physics courses, the vector force **f** exerted on a charge of q moving at velocity **u** in the presence of a magnetic field with flux density **B** is given by the equation

$$\mathbf{f} = q\mathbf{u} \times \mathbf{B} \tag{14.1}$$

where the symbol $\times$ denotes the (vector) cross product. If the charge is moving at a velocity **u** in a direction that makes an angle θ with the magnetic field, then the magnitude of the force is given by

$$f = quB \sin \theta \tag{14.2}$$

and the direction of this force is at right angles with the plane formed by the vectors **B** and **u.** This relationship is depicted in Figure 14.2.

The magnetic flux, ϕ, is then defined as the integral of the flux density over some surface area. For the simplified (but often useful) case of magnetic flux lines perpendicular to a cross-sectional area A, we can see that the flux is given by the following integral:

$$\phi = \int_A B \, dA \tag{14.3}$$

in webers (Wb), where the subscript A indicates that the integral is evaluated over the surface A. Furthermore, if the flux were to be uniform over the cross-sectional area A (a simplification that will be useful), the preceding integral could be approximated by the following expression:

$$\phi = B \cdot A \tag{14.4}$$

Figure 14.3 illustrates this idea, by showing hypothetical magnetic flux lines traversing a surface, delimited in the figure by a thin conducting wire.

Weaker magnetic field

Stronger magnetic field

Figure 14.1 Lines of force in a magnetic field

Figure 14.2 Charge moving in a constant magnetic field

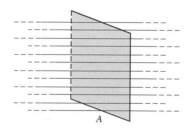

Figure 14.3 Magnetic flux lines crossing a surface

(a)

Current generating a magnetic flux
opposing the increase in flux due to **B**

(b)

Figure 14.4 Flux direction

Faraday's law states that if the imaginary surface A were bounded by a conductor—for example, the thin wire of Figure 14.3—then a *changing* magnetic field would induce a voltage, and therefore a current, in the conductor. More precisely, Faraday's law states that a time-varying flux causes an induced **electromotive force,** or **emf,** e, as follows:

$$e = -\frac{d\phi}{dt} \tag{14.5}$$

A little discussion is necessary at this point to explain the meaning of the $-$ sign in equation 14.5. Consider the one-turn coil of Figure 14.4, which forms a circular cross-sectional area, in the presence of a magnetic field with flux density **B** oriented in a direction perpendicular to the plane of the coil. If the magnetic field, and therefore the flux within the coil, is constant, no voltage will exist across terminals a and b; if, however, the flux were increasing and terminals a and b were connected—for example, by means of a resistor, as indicated in Figure 14.4(b)—current would flow in the coil in such a way that *the magnetic flux generated by the current would oppose the increasing flux*. Thus, the flux induced by such a current would be in the direction opposite to that of the original flux density vector, **B**. This principle is known as **Lenz's law.** The reaction flux would then point downward in Figure 14.4(a), or into the page in Figure 14.4(b). Now, by virtue of the **right-hand rule,** this reaction flux would induce a current flowing clockwise in Figure 14.4(b), that is, a current that flows out of terminal b and into terminal a. The resulting voltage across the hypothetical resistor R would then be negative. If, on the other hand, the original flux were decreasing, current would be induced in the coil so as to reestablish the initial flux; but this would mean that the current would have to generate a flux in the upward direction in Figure 14.4(a) (or out of the page in Figure 14.4(b)). Thus, the resulting voltage would change sign.

The polarity of the induced voltage can usually be determined from physical considerations; therefore the minus sign in equation 14.5 is usually left out. We will use this convention throughout the chapter.

In practical applications, the size of the voltages induced by the changing magnetic field can be significantly increased if the conducting wire is coiled many times around, so as to multiply the area crossed by the magnetic flux lines many times over. For an N-turn coil with cross-sectional area A, for example, we have the emf

$$e = N\frac{d\phi}{dt} \tag{14.6}$$

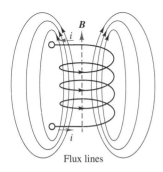

Right-hand rule

Flux lines

Figure 14.5 Concept of flux linkage

Figure 14.5 shows an N-turn coil *linking* a certain amount of magnetic flux; you can see that if N is very large and the coil is tightly wound (as is usually the case in the construction of practical devices), it is not unreasonable to presume that each turn of the coil links the same flux. It is convenient, in practice, to define the **flux linkage,** λ, as

$$\lambda = N\phi \tag{14.7}$$

so that

$$e = \frac{d\lambda}{dt} \tag{14.8}$$

Note that equation 14.8, relating the derivative of the flux linkage to the induced emf, is analogous to the equation describing current as the derivative of charge:

$$i = \frac{dq}{dt} \tag{14.9}$$

In other words, flux linkage can be viewed as the dual of charge in a circuit analysis sense, provided that we are aware of the simplifying assumptions just stated in the preceding paragraphs, namely, a uniform magnetic field perpendicular to the area delimited by a tightly wound coil. These assumptions are not at all unreasonable when applied to the inductor coils commonly employed in electric circuits.

What, then, are the physical mechanisms that can cause magnetic flux to change, and therefore to induce an electromotive force? Two such mechanisms are possible. The first consists of physically moving a permanent magnet in the vicinity of a coil—for example, so as to create a time-varying flux. The second requires that we first produce a magnetic field by means of an electric current (how this can be accomplished is discussed later in this section) and then vary the current, thus varying the associated magnetic field. The latter method is more practical in many circumstances, since it does not require the use of permanent magnets and allows variation of field strength by varying the applied current; however, the former method is conceptually simpler to visualize. The voltages induced by a moving magnetic field are called **motional voltages;** those generated by a time-varying magnetic field are termed **transformer voltages.** We shall be interested in both in this chapter, for different applications.

In the analysis of linear circuits, we had implicitly assumed that the relationship between flux linkage and current was a linear one:

$$\lambda = Li \tag{14.10}$$

so that the effect of a time-varying current was to induce a transformer voltage across an inductor coil, according to the expression

$$v = L\frac{di}{dt} \tag{14.11}$$

This is, in fact, the defining equation for the ideal **self-inductance, L.** In addition to self-inductance, however, it is also important to consider the **magnetic coupling** that can occur between neighboring circuits. Self-inductance measures the voltage induced in a circuit by the magnetic field generated by a current flowing in the same circuit. It is also possible that a second circuit in the vicinity of the first may experience an induced voltage as a consequence of the magnetic field generated in the first circuit. As we shall see in Section 14.4, this principle underlies the operation of all transformers.

Self and Mutual Inductance

Figure 14.6 depicts a pair of coils, one of which, L_1, is excited by a current, i_1, and therefore develops a magnetic field and a resulting induced voltage, v_1. The second coil, L_2, is not energized by a current, but links some of the flux generated by the current i_1 around L_1 because of its close proximity to the first coil. The magnetic coupling between the coils established by virtue of their proximity is described by a quantity called **mutual inductance** and defined by the symbol M. The mutual inductance is defined by the equation

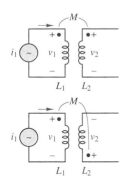

$$v_2 = M\frac{di_1}{dt} \tag{14.12}$$

The dots shown in the two figures indicate the polarity of the coupling between the coils. If the dots are at the same end of the coils, the voltage induced in coil 2 by a current in coil 1 has the same polarity as the voltage induced by the same current in coil 1; otherwise, the currents are in opposition, as shown in the lower part of Figure 14.6. Thus, the presence of such dots indicates that magnetic coupling is present between two coils. It should also be pointed out that if a current (and therefore a magnetic field) were present in the second coil, an additional voltage would be induced across coil 1. The voltage induced across a coil is, in general, equal to the sum of the voltages induced by self-inductance and mutual inductance.

Figure 14.6 Mutual inductance

Example   14.1   Linear Variable Differential Transformer (LVDT)

The linear variable differential transformer (LVDT) is a displacement transducer based on the mutual inductance concept just discussed. Figure 14.7 shows a simplified representation of an LVDT, which consists of a primary coil, subject to AC excitation, and of a pair of identical secondary coils, which are connected so as to result in the output voltage

$$v_{\text{out}} = v_1 - v_2$$

The ferromagnetic core between the primary and secondary coils can be displaced in proportion to some external motion, x, and determines the magnetic coupling between primary and secondary coils. Intuitively, as the core is displaced upward, greater coupling will occur between the primary coil and the top secondary coil, thus inducing a greater voltage in the top secondary coil. Hence, $v_{\text{out}} > 0$ for positive displacements. The converse is true for negative displacements. More formally, if the primary coil has resistance R_p and self-inductance L_p, we can write

$$iR_p + L_p\frac{di}{dt} = v_{\text{ex}}$$

Figure 14.7 Linear variable differential transformer

and the voltages induced in the secondary coils are given by

$$v_1 = M_1 \frac{di}{dt}$$

$$v_2 = M_2 \frac{di}{dt}$$

so that

$$v_{\text{out}} = (M_1 - M_2)\frac{di}{dt}$$

where M_1 and M_2 are the mutual inductances between the primary and the respective secondary coils. It should be apparent that each of the mutual inductances is dependent on the position of the iron core. For example, with the core at the *null position*, $M_1 = M_2$ and $v_{\text{out}} = 0$. The LVDT is typically designed so that $M_1 - M_2$ is linearly related to the displacement of the core, x.

Because the excitation is by necessity an AC signal (why?), the output voltage is actually given by the difference of two sinusoidal voltages at the same frequency, and is therefore itself a sinusoid, whose amplitude and phase depend on the displacement, x. Thus, v_{out} is an *amplitude-modulated (AM)* signal, similar to the one discussed in Example 4.17. To recover a signal proportional to the actual displacement, it is therefore necessary to use a demodulator circuit, such as the one discussed in Example 7.11.

In practical electromagnetic circuits, the self-inductance of a circuit is not necessarily constant; in particular, the inductance parameter, L, is not constant, in general, but depends on the strength of the magnetic field intensity, so that it will not be possible to use such a simple relationship as $v = L\,di/dt$, with L constant. If we revisit the definition of the transformer voltage,

$$e = N\frac{d\phi}{dt} \qquad (14.13)$$

we see that in an inductor coil, the inductance is given by

$$L = \frac{N\phi}{i} = \frac{\lambda}{i} \qquad (14.14)$$

This expression implies that the relationship between current and flux in a magnetic structure is linear (the inductance being the slope of the line). In fact, the properties of ferromagnetic materials are such that the flux-current relationship is nonlinear, as we shall see in the next subsection, so that the simple linear inductance parameter used in electric circuit analysis is not adequate to represent the behavior of the magnetic circuits of the present chapter. In any practical situation, the relationship between the flux linkage, λ, and the current is nonlinear, and might be described by a curve similar to that shown in Figure 14.8. Whenever the i-λ curve is not a straight line, it is more convenient to analyze the magnetic system in terms of energy calculations, since the corresponding circuit equation would be nonlinear.

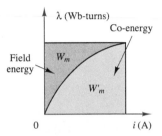

Figure 14.8

In a magnetic system, the energy stored in the magnetic field is equal to the integral of the instantaneous power, which is the product of voltage and current, just as in a conventional electrical circuit:

$$W_m = \int ei\,dt \tag{14.15}$$

However, in this case, the voltage corresponds to the induced emf, according to Faraday's law:

$$e = \frac{d\lambda}{dt} = N\frac{d\phi}{dt} \tag{14.16}$$

and is therefore related to the rate of change of the magnetic flux. The energy stored in the magnetic field could therefore be expressed in terms of the current by the integral

$$W_m = \int ei\,dt = \int \frac{d\lambda}{dt}i\,dt = \int i\,d\lambda \tag{14.17}$$

It should be straightforward to recognize that this energy is equal to the area above the λ-i curve of Figure 14.8. From the same figure, it is also possible to define a fictitious (but sometimes useful) quantity called **co-energy,** equal to the area under the curve and identified by the symbol W'_m. From the figure, it is also possible to see that the co-energy can be expressed in terms of the stored energy by means of the following relationship:

$$W'_m = i\lambda - W_m \tag{14.18}$$

Example 14.2 illustrates the calculation of energy, co-energy, and induced voltage using the concepts developed in these paragraphs.

Example   14.2

An iron-core inductor has the following empirically determined characteristic:

$$i = \lambda + 0.5\lambda^2$$

1. Determine the energy, the co-energy, and the incremental inductance for $\lambda = 0.5$ V·s.
2. Given that the coil resistance is 1 Ω and that the current through the coil is

    $$i(t) = 0.625 + 0.01\sin 400t$$

    determine the voltage across the terminals of the inductor.

Solution:

1. The energy and co-energy can be determined by integrating the function

    $$i = \lambda + 0.5\lambda^2$$

    over λ:

    $$W_m = \int_0^\lambda i(\lambda')\,d\lambda' = \frac{\lambda^2}{2} + 0.5\frac{\lambda^3}{3}$$

This function can be evaluated at $\lambda = 0.5$ to obtain

$$W_m = \frac{\lambda^2}{2} + 0.5\frac{\lambda^3}{3}\bigg|_{\lambda=0.5} = 0.1458 \text{ J}$$

and the corresponding co-energy is given by the expression

$$W'_m = i\lambda - W_m$$

To compute the current, i, required to generate a flux linkage $\lambda = 0.5$, we can use the following expression:

$$i = \lambda + 0.5\lambda^2 = 0.625 \text{ A}$$

and from the result, the co-energy may be computed as follows:

$$W'_m = (0.625)(0.5) - 0.1458 = 0.1667 \text{ J}$$

Further, since the *incremental inductance* (i.e., the inductance in the neighborhood of a particular point on the λ-i curve) is the reciprocal of the slope of the λ-i curve, it can be computed by differentiating the $i(\lambda)$ expression and evaluating it at $\lambda = 0.5$. Let the incremental inductance be L_Δ; then:

$$L_\Delta = \left(\frac{di}{d\lambda}\right)^{-1} = \frac{1}{1+\lambda} = 0.667 \text{ H}$$

Thus, for small variations of the current about the value $i = 0.625$ A, the nonlinear inductor will behave as if it were a linear inductor with incremental inductance $L_\Delta = 0.667$ H.

2. If the current consists of a small sinusoidal oscillation around the average current $i = 0.625$ A, it becomes possible to compute the corresponding voltage across the inductor using the well-known linear circuit laws:

$$v = iR + L_\Delta\frac{di}{dt}$$
$$= 0.625 + 0.01\sin(400t) \times 1 + (0.667)(4)\cos(400t)$$
$$= 0.625 + 0.01\sin(400t) + 2.668\cos(400t)$$
$$= 0.625 + 2.668\sin(400t + 89.8°)$$

The calculation of the energy stored in the magnetic field around a magnetic structure will be particularly useful later in the chapter, when the discussion turns to practical electromechanical transducers and it will be necessary to actually compute the forces generated in magnetic structures.

Ampère's Law

As explained in the previous section, Faraday's law is one of two fundamental laws relating electricity to magnetism. The second relationship, which forms a counterpart to Faraday's law, is **Ampère's law.** Qualitatively, Ampère's law states that the magnetic field intensity, **H,** is related to the current carried by the conductor in the vicinity of a conductor; thus Ampère's law establishes a dual relationship with Faraday's law.

In the previous section, we described the magnetic field in terms of its flux density, **B**, and flux ϕ. To explain Ampère's law and the behavior of magnetic materials, we need to define a relationship between the magnetic field intensity, **H**, and the flux density, **B**. These quantities are related by:

$$\begin{aligned} \mathbf{B} &= \mu \mathbf{H} \\ &= \mu_r \mu_0 \mathbf{H} \qquad \text{Wb/m}^2 \text{ or T} \end{aligned} \tag{14.19}$$

where the parameter μ is a scalar constant for a particular physical medium (at least, for the applications we consider here) and is called the **permeability** of the medium. The permeability of a material can be factored as the product of the permeability of free space, $\mu_0 = 4\pi \times 10^{-7}$ H/m, times the relative permeability, μ_r, which varies greatly according to the medium. For example, for air and for electrical conductors and insulators, μ_r is equal to 1. For ferromagnetic materials, the value of μ_r can take values in the hundreds or thousands. The size of μ_r represents a measure of the magnetic properties of the material. A consequence of Ampère's law is that, the larger the value of μ, the smaller the current required to produce a large flux density in an electromagnetic structure. Consequently, many electromechanical devices make use of ferromagnetic materials, called iron cores, to enhance their magnetic properties. Table 14.1 gives approximate values of μ_r for some common materials.

Table 14.1   Relative permeabilities for common materials

Material	μ_r
Air	1
Permalloy	100,000
Cast steel	1,000
Sheet steel	4,000
Iron	5,195

Conversely, the reason for introducing the magnetic field intensity is that it is independent of the properties of the materials employed in the construction of magnetic circuits. Thus, a given magnetic field intensity, **H,** will give rise to different flux densities in different materials. It will therefore be useful to define *sources* of magnetic energy in terms of the magnetic field intensity, so that different magnetic structures and materials can then be evaluated or compared for a given source. In analogy with electromotive force, this "source" will be termed **magnetomotive force.** As stated earlier, both the magnetic flux density and field intensity are vector quantities; however, for ease of analysis, scalar fields will be chosen by appropriately selecting the orientation of the fields, wherever possible.

Ampère's law states that the integral of the vector magnetic field intensity, **H,** around a closed path is equal to the total current linked by the closed path, i:

$$\oint \mathbf{H} \cdot d\mathbf{l} = \sum i \tag{14.20}$$

If the path is in the same direction as the direction of the magnetic field, we can use scalar quantities to state that

$$\oint \mathbf{H} \cdot d\mathbf{l} = \sum i \tag{14.21}$$

where dl is an increment in the direction of the closed path.

Figure 14.9 illustrates the case of a wire carrying a current i, and of a circular path of radius r surrounding the wire. In this simple case, you can see that the magnetic field intensity, **H,** is determined by the familiar right-hand rule. This rule states that if the direction of the current i points in the direction of the thumb of one's right hand, the resulting magnetic field encircles the conductor in the direction in which the other four fingers would encircle the thumb. Thus, in the case of Figure 14.9 the closed-path integral becomes equal to $H \cdot (2\pi r)$, since the path and the magnetic field are in the same direction, and therefore the magnitude of the magnetic field intensity is given by

$$H = \frac{i}{2\pi r} \tag{14.22}$$

Now, the magnetic field intensity is unaffected by the material surrounding the conductor, but the flux density depends on the material properties, since $B = \mu H$. Thus, the density of flux lines around the conductor would be far greater in the presence of a magnetic material than if the conductor were surrounded by air. The field generated by a single conducting wire is not very strong; however, if we arrange the wire into a tightly wound coil with many turns, we can greatly increase the strength of the magnetic field. For such a coil, with N turns, one can verify visually that the lines of force associated with the magnetic field link all of the turns of the conducting coil, so that we have effectively increased the current linked by the flux lines N-fold. The product $N \cdot i$ is a useful quantity in electromagnetic circuits, and is called the magnetomotive force,[2] $\mathcal{F}$ (often abbreviated mmf), in analogy with the electromotive force defined earlier:

$$\mathcal{F} = Ni \qquad \text{ampere-turns (A} \cdot \text{t)} \tag{14.23}$$

Figure 14.10 illustrates the magnetic flux lines in the vicinity of a coil. The magnetic field generated by the coil can be made to generate a much greater flux density if the coil encloses a magnetic material. The most common ferromagnetic materials are steel and iron; in addition to these, many alloys and oxides of iron—as well as nickel—and some artificial ceramic

Illustration of Ampère's law. By the right-hand rule, the current, i, generates a magnetic field intensity, **H,** in the direction shown.

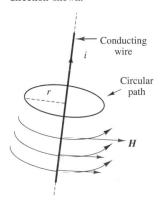

Figure 14.9 Illustration of Ampère's law

---

[2] Note that, although dimensionally equal to amperes, the units of magnetomotive force are ampere-turns.

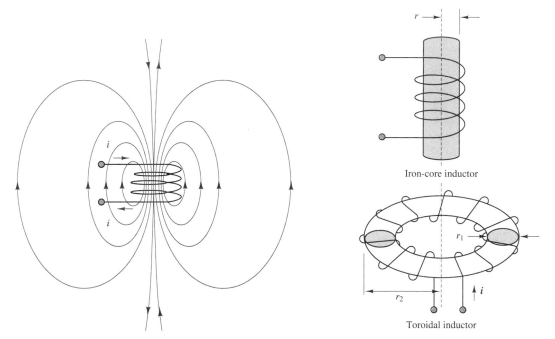

Figure 14.10 Magnetic field in the vicinity of a current-carrying coil   **Figure 14.11** Practical inductors

materials called **ferrites** also exhibit magnetic properties. Winding a coil around a ferromagnetic material accomplishes two useful tasks at once: it forces the magnetic flux to be concentrated near the coil and—if the shape of the magnetic material is appropriate—completely confines the flux within the magnetic material, thus forcing the closed path for the flux lines to be almost entirely enclosed within the ferromagnetic material. Typical arrangements are the iron-core inductor and the toroid of Figure 14.11. The flux densities for these inductors are given by the expressions

$$B = \frac{\mu N i}{2r} \qquad \text{flux density for tightly wound circular coil} \qquad \textbf{(14.24)}$$

$$B = \frac{\mu r_1^2 N i}{2r_2} \qquad \text{flux density for toroidal coil} \qquad \textbf{(14.25)}$$

Intuitively, the presence of a high-permeability material near a source of magnetic flux causes the flux to preferentially concentrate in the high-μ material, rather than in air, much as a conducting path concentrates the current produced by an electric field in an electric circuit. In the course of this section, we shall continue to develop this analogy between electric circuits and magnetic circuits. Figure 14.12 depicts an example of a simple electromagnetic structure, which, as we shall see shortly, forms the basis of the practical transformer.

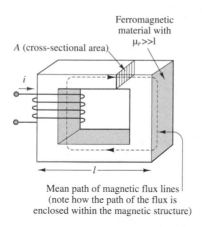

Figure 14.12 A simple electromagnetic structure

Table 14.2   Magnetic variables and units

Variable	Symbol	Units
Current	I	A
Magnetic flux density	B	$Wb/m^2 = T$
Magnetic flux	ϕ	Wb
Magnetic field intensity	H	A/m
Electromotive force	e	V
Magnetomotive force	$\mathscr{F}$	A · t
Flux linkage	λ	Wb · t

DRILL EXERCISES

14.1 A coil having 100 turns is immersed in a magnetic field that is varying uniformly from 80 mWb to 30 mWb in 2 seconds. Find the induced voltage in the coil.

> Answer: $e = -2.5$ V

14.2 The magnitude of **H** at a radius of 0.5 m from a long linear conductor is $1 \ A \cdot m^{-1}$. Find the current in the wire.

> Answer: $I = \pi$ A

14.3 The relation between the flux linkages and the current for a magnetic material is given by $\lambda = 6i/(2i + 1)$ Wb · t. Determine the energy stored in the magnetic field for $\lambda = 2$ Wb · t.

> Answer: $W_m = 0.648$ J

14.4   Verify that for the linear case, where the flux is proportional to the mmf, the energy stored in the magnetic field is $\frac{1}{2}Li^2$.

14.2   MAGNETIC CIRCUITS

It is possible to analyze the operation of electromagnetic devices such as the one depicted in Figure 14.12 by means of magnetic equivalent circuits, similar in many respects to the equivalent electrical circuits of the earlier chapters. Before we can present this technique, however, we need to make a few simplifying approximations. The first of these approximations assumes that there exists a **mean path** for the magnetic flux, and that the corresponding mean flux density is approximately constant over the cross-sectional area of the magnetic structure. Thus, a coil wound around a core with cross-sectional area A will have flux density

$$B = \frac{\phi}{A} \tag{14.26}$$

where A is assumed to be perpendicular to the direction of the flux lines. Figure 14.12 illustrates such a mean path and the cross-sectional area, A. Knowing the flux density, we obtain the field intensity:

$$H = \frac{B}{\mu} = \frac{\phi}{A\mu} \tag{14.27}$$

But then, knowing the field intensity, we can relate the mmf of the coil, $\mathcal{F}$, to the product of the magnetic field intensity, H, and the length of the magnetic (mean) path, l, for one leg of the structure:

$$\mathcal{F} = N \cdot i = H \cdot l \tag{14.28}$$

In summary, the mmf is equal to the magnetic flux times the length of the magnetic path, divided by the permeability of the material times the cross-sectional area:

$$\mathcal{F} = \phi \frac{l}{\mu A} \tag{14.29}$$

A review of this formula reveals that the magnetomotive force, $\mathcal{F}$, may be viewed as being analogous to the voltage source in a series electrical circuit, and that the flux, ϕ, is then equivalent to the electrical current in a series circuit and the term $l/\mu A$ to the *magnetic resistance* of one leg of the magnetic circuit. You will note that the term $l/\mu A$ is very similar to the term describing the resistance of a cylindrical conductor of length l and cross-sectional area A, where the permeability, μ, is analogous to the conductivity, σ. The term $l/\mu A$ occurs frequently enough to be assigned the name of **reluctance,** and the symbol $\mathcal{R}$.

In summary, when an N-turn coil carrying a current i is wound around a magnetic core such as the one indicated in Figure 14.12, the mmf, $\mathcal{F}$, generated by the coil produces a flux, ϕ, that is *mostly* concentrated within the core and is assumed to be uniform across the cross section. Within this simplified picture, then, the analysis of a magnetic circuit is analogous to that of resistive electrical circuits. This analogy is illustrated in Table 14.3 and in the examples in this section.

The usefulness of the magnetic circuit analogy can be emphasized by analyzing a magnetic core similar to that of Figure 14.12, but with a slightly modified geometry. Figure 14.13 depicts the magnetic structure and

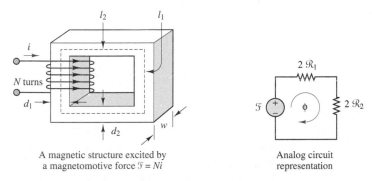

A magnetic structure excited by
a magnetomotive force $\mathcal{F} = Ni$

Analog circuit
representation

Figure 14.13 Analogy between magnetic and electric circuits

its equivalent circuit analogy. In the figure, we see that the mmf, $\mathcal{F} = Ni$, excites the magnetic circuit, which is composed of four legs: two of mean path length l_1 and cross-sectional area $A_1 = d_1 w$, and the other two of mean length l_2 and cross section $A_2 = d_2 w$. Thus, the reluctance encountered by the flux in its path around the magnetic core is given by the quantity $\mathcal{R}_{series}$, with

$$\mathcal{R}_{series} = 2\mathcal{R}_1 + 2\mathcal{R}_2$$

and

$$\mathcal{R}_1 = \frac{l_1}{\mu A_1}$$

$$\mathcal{R}_2 = \frac{l_2}{\mu A_2}$$

(14.30)

It is important at this stage to review the assumptions and simplifications made in analyzing the magnetic structure of Figure 14.13:

1. All of the magnetic flux is linked by all of the turns of the coil.
2. The flux is confined exclusively within the magnetic core.
3. The density of the flux is uniform across the cross-sectional area of the core.

You can probably see intuitively that the first of these assumptions might not hold true near the ends of the coil, but that it might be more reasonable if the coil is tightly wound. The second assumption is equivalent to stating that the relative permeability of the core is infinitely higher than that of air (presuming that this is the medium surrounding the core): if this were the case, the flux would indeed be confined within the core. It is worthwhile to note that we make a similar assumption when we treat wires in electric circuits as perfect conductors: the conductivity of copper is substantially greater than that of free space, by a factor of approximately 10^{15}. In the case of magnetic materials, however, even for the best alloys, we have a relative permeability only on the order of 2,000 to 6,000. Thus, an approximation that is quite appropriate for electric circuits is not nearly as good in the case of magnetic circuits. Some of the flux in a structure such as those of Figures 14.12 and 14.13 would thus not be confined within the core (this is usually referred to as **leakage flux**). Finally, the assumption that the flux is uniform across the core cannot hold for a finite-permeability medium, but it is very helpful in giving an approximate *mean* behavior of the magnetic circuit.

The magnetic circuit analogy is therefore far from being exact. However, short of employing the tools of electromagnetic field theory and of vector calculus, or advanced numerical simulation software, it is the most convenient tool at the engineer's disposal for the analysis of magnetic structures. In the remainder of this chapter, the approximate analysis based on the electric circuit analogy will be used to obtain approximate solutions to problems involving a variety of useful magnetic circuits, many of which you are already familiar with. Among these will be the loudspeaker, solenoids, automotive fuel injectors, sensors for the measurement of linear and angular velocity and position, and other interesting applications.

Table 14.3   Analogy between electric and magnetic circuits

Electrical quantity	Magnetic quantity
Electric field intensity, E, V/m	Magnetic field intensity, H, A · t/m
Voltage, v, V	Magnetomotive force, $\mathscr{F}$, A · t
Current, i, A	Magnetic flux, ϕ, Wb
Current density, J, A/m^2	Magnetic flux density, B, Wb/m^2
Resistance, R, Ω	Reluctance, $\mathscr{R} = l/\mu A$, A · t/Wb
Conductivity, σ, $1/\Omega$ · m	Permeability, μ, Wb/A · m

$i = .1$ A

N turns

h

l w

$l = 10$ cm, $h = 10$ cm, $w = 1$ cm

Figure 14.14

Example   14.3

Figure 14.14 shows a simple magnetic circuit. Calculate the flux, ϕ, the flux density, B, and the field intensity H, in the magnetic structure. Assume that $\mu_r = 1,000$, $N = 500$, and $i = 100$ mA. Also, assume that the structure's cross section is square, with dimension w.

Solution:

First we find the magnetomotive force, which is the source of the flux:

$$\mathscr{F} = \text{mmf} = Ni = (500)(0.1) = 50 \text{ A} \cdot \text{t}$$

Next, we estimate the mean path of the flux. Note that we assume that the flux is completely contained inside the structure (or, equivalently, that the permeability of the structure is significantly greater than that of free space). According to this assumption, then, the mean path would run through the geometric center of the structure, as shown in Figure 14.15.

The total circuit path length is computed to be

$$l_c = 9 \text{ cm} + 9 \text{ cm} + 9 \text{ cm} + 9 \text{ cm} = 36 \text{ cm}$$

Since the cross section is 1 cm $\times$ 1 cm in all parts of the structure, we can use the total length of the path to find the reluctance, $\mathcal{R}$, of the magnetic circuit, according to the following expression:

$$\mathcal{R} = \frac{l_c}{\mu A} = \frac{0.36}{(1,000)(4\pi \times 10^{-7})(0.01)(0.01)} = 2.865 \times 10^6 \text{ A} \cdot \text{t/Wb}$$

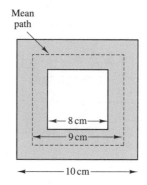

Figure 14.15

The corresponding equivalent magnetic circuit is shown in Figure 14.16.

A consequence of the simplifying approximations made in this section is that the mmf is linearly related to the flux:

$$\mathcal{F} = \phi \mathcal{R}$$

so that we can compute the magnetic flux in the structure to be

$$\phi = \frac{\mathcal{F}}{\mathcal{R}} = \frac{50 \text{ A} \cdot \text{t}}{2.865 \times 10^6 \text{ A} \cdot \text{t/Wb}}$$
$$= 1.75 \times 10^{-5} \text{ Wb}$$

Figure 14.16

The corresponding flux density is then computed as follows:

$$B = \frac{\phi}{A} = \frac{17.5 \times 10^{-6} \text{ Wb}}{0.0001 \text{ m}^2}$$
$$= 0.175 \text{ Wb/m}^2$$

and the magnetic field intensity is therefore given by

$$H = \frac{B}{\mu} = \frac{0.175 \text{ Wb/m}^2}{1,000(4\pi \times 10^{-7}) \text{ H/m}}$$
$$= 139.0 \text{ A} \cdot \text{t/m}$$

These calculations are useful, in practice, in determining practical parameters in the design or application of electromagnetic structures. For example, determining the magnetomotive force required to generate a certain amount of flux is necessary to determine the current that must be provided. Since, in practice, any coil has a finite resistance, this current will determine the required excitation voltage. As will be shown in a later section, these calculations can also be related to the determination of energy and force in electromagnetic transducers.

It will be useful, before proceeding further, to consider the analysis of the same simple magnetic structure when an **air gap** is present. Air gaps are very common in magnetic structures; in rotating machines, for example, air gaps are necessary to allow for free rotation of the inner core of the machine. The magnetic circuit of Figure 14.17 differs from the circuit analyzed in

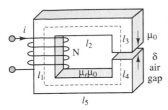

Figure 14.17 Magnetic circuit with air gap

Figure 14.18 Equivalent representation of magnetic circuit with an air gap

Example 14.3 simply because of the presence of an air gap; the effect of the gap is to break the continuity of the high-permeability path for the flux, adding a high-reluctance component to the equivalent circuit. The situation is analogous to adding a very large series resistance to a series electrical circuit. It should be evident from Figure 14.17 that the basic concept of reluctance still applies, although now two different permeabilities must be taken into account.

The equivalent circuit for the structure of Figure 14.17 may be drawn as shown in Figure 14.18, where $\mathcal{R}_n$ is the reluctance of path l_n, for $n = 1$, 2, ..., 5, and $\mathcal{R}_g$ is the reluctance of the air gap. The reluctances can be expressed as follows, if we assume that the magnetic structure has a uniform cross-sectional area, A:

$$
\mathcal{R}_1 = \frac{l_1}{\mu_r \mu_0 A} \qquad \mathcal{R}_2 = \frac{l_2}{\mu_r \mu_0 A}
$$

$$
\mathcal{R}_3 = \frac{l_3}{\mu_r \mu_0 A} \qquad \mathcal{R}_4 = \frac{l_4}{\mu_r \mu_0 A} \tag{14.31}
$$

$$
\mathcal{R}_5 = \frac{l_5}{\mu_r \mu_0 A} \qquad \mathcal{R}_g = \frac{\delta}{\mu_0 A_g}
$$

Note that in computing $\mathcal{R}_g$, the length of the gap is given by δ and the permeability is given by μ_0, as expected, but A_g is different from the cross-sectional area, A, of the structure. The reason is that the flux lines exhibit a phenomenon known as **fringing** as they cross an air gap. The flux lines actually *bow* out of the gap defined by the cross section, A, not being contained by the high-permeability material any longer. Thus, it is customary to define an area A_g that is greater than A, to account for this phenomenon. Example 14.4 describes the procedure for finding A_g in more detail and also discusses the phenomenon of fringing.

Example 14.4   Magnetic Structures with Air Gaps

In practical applications, magnetic circuits such as the ones discussed in this section are often used to generate forces or displacements as a result of the application of an electrical current. For example, the magnetic structure of Figure 14.19 might be used to displace a conducting bar in order to *make* or *break* an electrical contact. The techniques introduced in this section may be applied to understand the behavior of practical devices, such as solenoids and relays, which will be described shortly, and to carry out approximate calculations, for example, to compute the current required to energize a contact. This example illustrates the use of these techniques in a practical magnetic structure. The structure includes two air gaps that the flux must cross to establish a closed path. Assume that the relative permeability of the magnetic material is $\mu_r = 10,000$. Find:

1. The reluctance of each of the air gaps.
2. The total reluctance of the magnetic circuit.
3. The flux density in the bottom plate of the structure.

Figure 14.19 Electromagnetic structure with air gaps

Solution:

1. To find the reluctance of the air gaps, we must determine the effective cross-sectional area of the gaps. These air gaps will display the phenomenon known as *fringing,* which was briefly discussed earlier in this section; Figure 14.20 depicts the fringing of the flux lines across each gap.

   Notice how the flux lines tend to spread out in the air gap while in the magnetic structure they remain straight and completely contained. The assumption of constant flux density clearly breaks down in this case, but it is possible to partially compensate for this inaccuracy by increasing the effective area of the gap. A rule of thumb for this situation is to add the length of the air gap, l_g, to the gap's physical dimensions in both directions. For the structure in our example, the effective area in each gap will be found by the relation

   $$A_g = (1 + l_g)(1 + l_g) = (1 + 0.25)(1 + 0.25) = 1.5625 \text{ cm}^2$$

   The reluctance of the air gap is then found according to the usual expression, where now $\mu = \mu_0$:

   $$\mathcal{R}_g = \frac{l_g}{\mu_0 A_g}$$

   $$= \frac{0.0025 \text{ m}}{(4\pi \times 10^{-7} \text{ Wb/A} \cdot \text{m})(0.15625 \times 10^{-3} \text{ m}^2)}$$

   $$= 1.27 \times 10^7 \text{ A} \cdot \text{t/Wb}$$

Figure 14.20 Fringing effects in air gap

2. To find the total reluctance of the magnetic circuit, we must find the mean path followed by the flux. Figure 14.21 shows the mean path. The path indicates eight separate lengths that must be accounted for in finding the reluctance. The total length is given by

   $$l_c = l_1 + l_2 + l_3 + l_4 + l_5 + l_6 + l_g + l_g$$

   The total reluctance in this path is given by

   $$\mathcal{R} = \mathcal{R}_1 + \mathcal{R}_2 + \mathcal{R}_3 + \mathcal{R}_4 + \mathcal{R}_5 + \mathcal{R}_6 + \mathcal{R}_g + \mathcal{R}_g$$

   (where $\mathcal{R}_g$ has already been calculated in part 1). We also take note that the reluctance varies in proportion to the length of the path, and since l_5 and l_6 are

Figure 14.21

small compared with the total length of the path, we choose to ignore these contributions. Then:

$$\mathscr{R} \approx \mathscr{R}_1 + \mathscr{R}_2 + \mathscr{R}_3 + \mathscr{R}_4 + \mathscr{R}_g + \mathscr{R}_g$$

Now, to find the respective reluctances, we compute

$$\mathscr{R}_1 = \frac{l_2}{\mu A_1} = \frac{0.09}{(10,000)(4\pi \times 10^{-7})(0.0001)} = 71.6 \times 10^3 \text{ A} \cdot \text{t/Wb}$$

$$\mathscr{R}_3 = \mathscr{R}_2 = \frac{l_2}{\mu A_2} = \frac{0.045}{(10,000)(4\pi \times 10^{-7})(0.0001)}$$
$$= 35.8 \times 10^3 \text{ A} \cdot \text{t/Wb}$$

$$\mathscr{R}_4 = \frac{l_4}{\mu A_4} = \frac{0.09}{(10,000)(4\pi \times 10^{-7})(0.00005)} = 143.2 \times 10^3 \text{ A} \cdot \text{t/Wb}$$

and

$$\mathscr{R} = 71.6 \times 10^3 + 2 \times 35.8 \times 10^3 + 143.2 \times 10^3 + 2 \times 1.27 \times 10^7$$
$$= 25.686 \times 10^6 \text{ A} \cdot \text{t/Wb}$$

You can see that the air gaps, in spite of their small physical dimensions, contribute the greater part of the reluctance of the magnetic circuit.

3. To find the flux density, we must first determine the flux in the magnetic circuit from the expression

$$\phi = \frac{\mathscr{F}}{\mathscr{R}}$$

and since the mmf is given by

$$\mathscr{F} = Ni = (100)(1) = 100 \text{ A} \cdot \text{t}$$

the flux can be computed to be

$$\phi = \frac{100 \text{ A} \cdot \text{t}}{25.686 \times 10^6 \text{ A} \cdot \text{t/Wb}} = 3.89 \times 10^{-6} \text{ Wb}$$

Finally,

$$B = \frac{\phi}{A_4} = \frac{3.89 \times 10^{-6} \text{ Wb}}{0.00005 \text{ m}^2}$$
$$= 0.0778 \text{ Wb/m}^2$$

These calculations are essential in determining the specifications of electromechanical devices for a given application, since they can be used to determine the geometry and current required to accomplish a design objective.

Example   14.5   Magnetic Reluctance Position Sensor

A simple magnetic structure, very similar to those examined in the previous examples, finds very common application in the so-called variable-reluctance position sensor, which, in turn, finds widespread application in a variety of configurations for the measurement of linear and angular position and velocity. Figure 14.22 depicts one particular configuration that is used in many applications. In this structure, a permanent magnet with a coil of wire wound around it forms the sensor; a steel disk (typically connected to a rotating shaft) has a number of tabs that pass through the pole pieces of the sensor. The area of the tab is assumed equal to the area of the cross section of the pole pieces and is equal to a^2. The reason for the name *variable-reluctance sensor* is that the reluctance of the magnetic structure is variable, depending on whether or not a ferromagnetic tab lies between the pole pieces of the magnet.

The principle of operation of the sensor is that an electromotive force, e_S, is induced across the coil by the change in magnetic flux caused by the passage of the tab between the pole pieces when the disk is in motion. As the tab enters the volume between the pole pieces, the flux will increase, because of the lower reluctance of the configuration, until it reaches a maximum when the tab is centered between the poles of the magnet. Figure 14.23 depicts the approximate shape of the resulting voltage, which, according to Faraday's law, is given by

$$e_S = -\frac{d\phi}{dt}$$

The rate of change of flux is dictated by the geometry of the tab and of the pole pieces, and by the speed of rotation of the disk. It is important to note that, since the flux is changing only if the disk is rotating, this sensor cannot detect the static position of the disk.

One common application of this concept is in the measurement of the speed of rotation of rotating machines, including electric motors and internal combustion engines. In these applications, use is made of a *60-tooth wheel,* which permits the conversion of the speed rotation directly to units of revolutions per minute. The output of a variable-reluctance position sensor magnetically coupled to a rotating disk equipped with 60 tabs (teeth) is processed through a comparator or Schmitt trigger circuit (see Chapter 11). The voltage waveform generated by the sensor is nearly sinusoidal when the teeth are closely spaced, and it is characterized by one sinusoidal cycle for each tooth on the disk. If a negative zero-crossing detector (see Chapter 11) is employed, the trigger circuit will generate a pulse corresponding to the passage of each tooth, as shown in Figure 14.24. If the time between any two pulses is measured by means of a high-frequency clock, the speed of the engine can be directly determined in units of rev/min by means of a digital counter (see Chapter 12).

Figure 14.22 Variable-reluctance position sensor

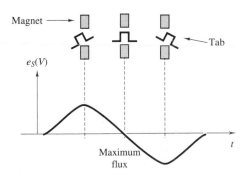

Figure 14.23 Variable-reluctance position sensor waveform

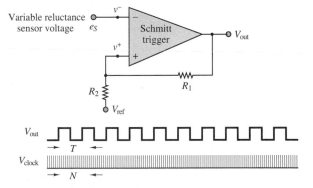

Figure 14.24 Signal processing for a 60-tooth wheel RPM sensor

Example   14.6

This example illustrates the calculation of the voltage induced in the magnetic reluctance sensor of Example 14.5 by a rotating toothed wheel. In particular, we will find an approximate expression for the reluctance and the induced voltage for the position sensor shown in Figure 14.25, and show that the induced voltage is speed-dependent. It will be assumed that the reluctance of the core and fringing at the air gaps are both negligible.

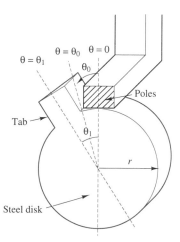

Figure 14.25 Reluctance sensor for measurement of angular position

Solution:

From the geometry of Example 14.5, the equivalent reluctance of the magnetic structure is twice that of one gap, since the permeability of the tab and magnetic structure are assumed infinite (i.e., they have negligible reluctance). When the tab and the poles are aligned, the angle θ is zero, as shown in Figure 14.25, and the area of the air gap is maximum. For angles greater than $2\theta_0$, the magnetic length of the air gaps is so large that the magnetic field may reasonably be taken as zero.

   To model the reluctance of the gaps, we assume the following simplified expression, where the area of overlap of the tab with the magnetic poles is assumed proportional to the angular displacement:

$$\mathcal{R} = \frac{2l_g}{\mu_0 A} = \frac{2l_g}{\mu_0 ar(\theta_1 - \theta)} \qquad \text{for} \qquad 0 < \theta < \theta_1$$

Naturally, this is an approximation; however, the approximation captures the essential idea of this transducer, namely, that the reluctance will decrease with increasing overlap area until it reaches a minimum, and then it will increase as the overlap area decreases. For $\theta = \theta_1$, that is, with the tab outside the magnetic pole pieces, we have $\mathcal{R}_{max} \rightarrow \infty$. For $\theta = 0$, that is, with the tab perfectly aligned with the pole pieces, we have $\mathcal{R}_{min} = 2l_g/\mu_0 ar\theta_1$. The flux ϕ may therefore be computed as follows:

$$\phi = \frac{Ni}{\mathcal{R}} = \frac{Ni\mu_0 ar(\theta_1 - \theta)}{2l_g}$$

The induced voltage e_S is found by

$$e_S = -\frac{d\phi}{dt} = -\frac{d\phi}{d\theta}\frac{d\theta}{dt} = \frac{Ni\mu_0 ar}{2l_g} \times \omega$$

where $\omega = d\theta/dt$ is the rotational speed of the steel disk. It should be evident that the induced voltage is speed-dependent. For $a = 1$ cm, $r = 10$ cm, $l_g = 0.1$ cm, $N = 100$ turns, $i = 10$ mA, $\theta_1 = 6° \approx 0.1$ rad, and $\omega = 400$ rad/s (approximately 3,800 rev/min), we have

$$\mathcal{R}_{max} = \frac{2 \times 0.1 \times 10^{-2}}{4\pi \times 10^{-7} \times 1 \times 10^{-2} \times 10 \times 10^{-2} \times 0.1} = 1.59 \times 10^7 \text{ At/Wb}$$

$$e_{S\,peak} = \frac{1000 \times 10 \times 10^{-3} \times 4\pi \times 10^{-7} \times 1 \times 10^{-2} \times 10^{-1}}{2 \times 0.1 \times 10^{-2}} \times 400$$

$$= 2.5 \text{ mV}$$

That is, the peak amplitude of e_s will be 2.5 mV.

Example   14.7

Figure 14.26 shows a coil of wire wrapped around an iron core. If the flux in the core is given by the equation $\phi = 4\sin(377t)$ mWb, and if the coil has 100 turns, what is the voltage produced at the terminals of the coil? What polarity is the voltage during the time when the flux is decreasing in the figure? (Neglect the flux leakage.)

Solution:

The direction of the voltage while the flux is decreasing in the reference direction must be positive to negative, as shown in the figure. The magnitude of the induced voltage is given by

$$e = N\frac{d\phi}{dt} = 100 \times \frac{d}{dt}(4 \times 10^{-3}\sin 377t)$$
$$= 150.8\cos(377t) = 150.8\sin(377t + 90°)$$

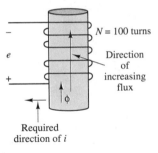

Figure 14.26

DRILL EXERCISES

14.5   If $\mathcal{R} = \mathcal{R}_g + \mathcal{R}_g$ in Example 14.4, calculate ϕ and B.

Answer: $\phi = 3.89 \times 10^{-6}$ Wb; $B = 0.0778$ Wb/m^2

14.6   Determine the equivalent reluctance of the structure of Figure 14.27 as seen

Figure 14.27

by the "source" if μ_r for the structure is 1,000, $l = 5$ cm, and all of the legs are 1 cm on a side.

> Answer: $\mathcal{R}_{eq} = 1.52 \times 10^6$ A·t/Wb

14.7   Find the equivalent reluctance of the magnetic circuit shown in Figure 14.28

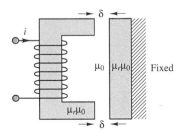

Figure 14.28

if μ_r of the structure is infinite, $\delta = 2$ mm, and the physical cross section of the core is 1 cm². Do not neglect fringing.

> Answer: $\mathcal{R}_{eq} = 22 \times 10^6$ A·t/Wb

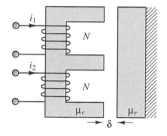

Figure 14.29

14.8   Find the equivalent magnetic circuit of the structure of Figure 14.29 is μ_r is infinite. Give expressions for each of the circuit values if the physical cross-sectional area of each of the legs is given by

$$A = l \times w$$

Do not neglect fringing.

> Answer: $\mathcal{R}_1 = \mathcal{R}_2 = \mathcal{R}_3 = \dfrac{\delta}{\mu_0(l+\delta)(w+\delta)}$; $\mathcal{F}_1 = Ni_1$; $\mathcal{F}_2 = Ni_2$

14.3   MAGNETIC MATERIALS AND B-H CURVES

In the analysis of magnetic circuits presented in the previous sections, the relative permeability, μ_r, was treated as a constant. In fact, the relationship between the magnetic flux density, **B** and the associated field intensity, **H,**

$$\mathbf{B} = \mu\mathbf{H} \tag{14.32}$$

is characterized by the fact that the relative permeability of magnetic materials is not a constant, but is a function of the magnetic field intensity. In effect, all magnetic materials exhibit a phenomenon called **saturation,** whereby the flux density increases in proportion to the field intensity until it cannot do so any longer. Figure 14.30 illustrates the general behavior of all magnetic materials. You will note that since the *B-H* curve shown in the

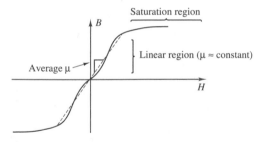

Figure 14.30 Permeability and magnetic saturation effects

figure is nonlinear, the value of μ (which is the slope of the curve) depends on the intensity of the magnetic field.

To understand the reasons for the saturation of a magnetic material, we need to briefly review the mechanism of magnetization. The basic idea behind magnetic materials is that the spin of electrons constitutes motion of charge, and therefore leads to magnetic effects, as explained in the introductory section of this chapter. In most materials, the electron spins cancel out, on the whole, and no net effect remains. In ferromagnetic materials, on the other hand, atoms can align so that the electron spins cause a net magnetic effect. In such materials, there exist small regions with strong magnetic properties (called **magnetic domains**), the effects of which are neutralized in unmagnetized material by other, similar regions that are oriented differently, in a random pattern. When the material is magnetized, the magnetic domains tend to align with each other, to a degree that is determined by the intensity of the applied magnetic field.

In effect, the large number of miniature magnets within the material are *polarized* by the external magnetic field. As the field increases, more and more domains become aligned. When all of the domains have become aligned, any further increase in magnetic field intensity does not yield an increase in flux density beyond the increase that would be caused in a nonmagnetic material. Thus, the relative permeability, μ_r, approaches 1 in the saturation region. It should be apparent that an exact value of μ_r cannot be determined; the value of μ_r used in the earlier examples is to be interpreted as an average permeability, for intermediate values of flux density. As a point of reference, commercial magnetic steels saturate at flux densities around a few teslas. Figure 14.33, shown later in this section, provides some actual *B-H* curves for common ferromagnetic materials.

The phenomenon of saturation carries some interesting implications with regard to the operation of magnetic circuits: the results of the previous section would seem to imply that an increase in the mmf (that is, an increase in the current driving the coil) would lead to a proportional increase in the magnetic flux. This is true in the *linear region* of Figure 14.30; however, as the material reaches saturation, further increases in the driving current (or, equivalently, in the mmf) do not yield further increases in the magnetic flux.

There are two more features that cause magnetic materials to further deviate from the ideal model of the linear *B-H* relationship: **eddy currents** and **hysteresis.** The first phenomenon consists of currents that are caused by any time-varying flux in the core material. As you know, a time-varying flux will induce a voltage, and therefore a current. When this happens inside the magnetic core, the induced voltage will cause "eddy" currents (the terminology should be self-explanatory) in the core, which depend on the resistivity of the core. Figure 14.31 illustrates the phenomenon of eddy currents. The effect of these currents is to dissipate energy in the form of heat. Eddy currents are reduced by selecting high-resistivity core materials, or by *laminating* the core, introducing tiny, discontinuous air gaps between core layers (see Figure 14.31). Lamination of the core reduces eddy currents greatly without affecting the magnetic properties of the core.

It is beyond the scope of this chapter to quantify the losses caused by induced eddy currents, but it will be important in Chapters 15 and 16 to be aware of this source of energy loss.

Hysteresis is another loss mechanism in magnetic materials; it displays a rather complex behavior, related to the magnetization properties of a material. The curve of Figure 14.32 reveals that the *B-H* curve for a magnetic material during magnetization (as *H* is increased) is displaced with respect to the curve that is measured when the material is demagnetized. To understand the hysteresis process, consider a core that has been energized for some time, with a field intensity of H_1 A · t/m. As the current required to sustain the mmf corresponding to H_1 is decreased, we follow the hysteresis curve from the point α to the point β. When the mmf is exactly zero, the material displays the **remanent (or residual) magnetization** B_r. To bring the flux density to zero, we must further decrease the mmf (i.e., produce a negative current), until the field intensity reaches the value $-H_0$ (point γ on the curve). As the mmf is made more negative, the curve eventually reaches the point α'. If the excitation current to the coil is now increased, the magnetization curve will follow the path α'-β'-γ'-α, eventually returning to the original point in the *B-H* plane, but via a different path.

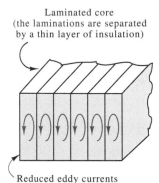

Laminated core
(the laminations are separated
by a thin layer of insulation)

Reduced eddy currents

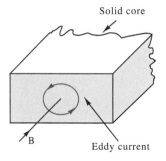

Solid core

B   Eddy current

Figure 14.31 Eddy currents in magnetic structures

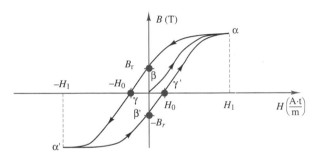

Figure 14.32 Hysteresis in magnetization curves

The result of this process, by which an *excess magnetomotive force* is required to magnetize or demagnetize the material, is a net energy loss. It is difficult to evaluate this loss exactly; however, it can be shown that it is related to the area between the curves of Figure 14.32. There are experimental techniques that enable the approximate measurement of these losses.

Figures 14.33 a–c depict magnetization curves for three very common ferromagnetic materials: cast iron, cast steel, and sheet steel. These curves will be useful in solving some of the homework problems.

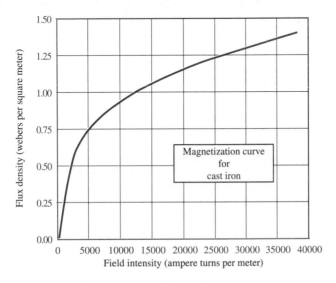

Figure 14.33(a) Magnetization curve for cast iron

Figure 14.33(b) Magnetization curve for cast steel

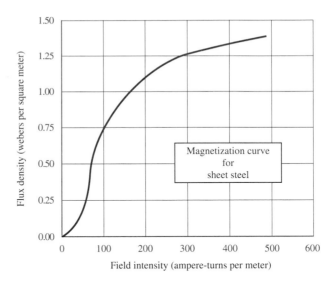

Figure 14.33(c) Magnetization curve for sheet steel

14.4   TRANSFORMERS

One of the more common magnetic structures in everyday applications is the transformer. The ideal transformer was introduced in Chapter 5 as a device that can step an AC voltage up or down by a fixed ratio, with a corresponding decrease or increase in current. The structure of a simple magnetic transformer is shown in Figure 14.34, which illustrates that a transformer is very similar to the magnetic circuits described earlier in this chapter. Coil L_1 represents the input side of the transformer, while coil L_2 is the output coil; both coils are wound around the same magnetic structure, which we show here to be similar to the "square doughnut" of the earlier examples.

The ideal transformer operates on the basis of the same set of assumptions we made in earlier sections: the flux is confined to the core, the flux links all turns of both coils, and the permeability of the core is infinite. The last assumption is equivalent to stating that an arbitrarily small mmf is sufficient to establish a flux in the core. In addition, we assume that the ideal

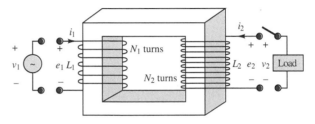

Figure 14.34 Structure of a transformer

transformer coils offer negligible resistance to current flow. Using this rather idealized picture, we can now resort to the principles developed earlier in this chapter to re-derive the already familiar results for the ideal transformer.

In essence, the operation of a transformer requires a time-varying current; if a time-varying voltage is applied to the primary side of the transformer, a corresponding current will flow in L_1; this current acts as an mmf and causes a (time-varying) flux in the structure. But the existence of a changing flux will induce an emf across the secondary coil! Without the need for a direct electrical connection, the transformer can couple a source voltage at the primary to the load; the coupling occurs by means of the magnetic field acting on both coils. Thus, a transformer operates by converting electric energy to magnetic, and then back to electric. The following derivation illustrates this viewpoint in the ideal case (no loss of energy), and compares the result with the definition of the ideal transformer in Chapter 5.

If a time-varying voltage source is connected to the input side, then by virtue of Faraday's law, a corresponding time-varying flux $d\phi/dt$ is established in coil L_1:

$$e_1 = N_1 \frac{d\phi}{dt} = v_1 \tag{14.33}$$

But since the flux thus produced also links coil L_2 an emf is induced across the output coil as well:

$$\tag{14.34}$$

This induced emf can be measured as the voltage v_2 at the output terminals, and one can readily see that the ratio of the open-circuit output voltage to input terminal voltage is

$$\frac{v_2}{v_1} = \frac{N_2}{N_1} \tag{14.35}$$

If a load current i_2 is now required by the connection of a load to the output circuit (by closing the switch in the figure), the corresponding mmf is $\mathscr{F}_2 = N_2 i_2$. This mmf, generated by the load current i_2, would cause the flux in the core to change; however, this is not possible, since a change in ϕ would cause a corresponding change in the voltage induced across the input coil. But this voltage is determined (fixed) by the source v_1 (and is therefore $d\phi/dt$), so that the input coil is forced to generate a **counter mmf** to oppose the mmf of the output coil; this is accomplished as the input coil draws a current i_1 from the source y_1 such that

$$i_1 N_1 = i_2 N_2 \tag{14.36}$$

or

$$\frac{i_2}{i_1} = \frac{N_1}{N_2} = \alpha \tag{14.37}$$

where α is the ratio of primary to secondary turns (the transformer ratio) and N_1 and N_2 are the primary and secondary turns, respectively. If there were any net difference between the input and output mmf, flux balance required by the input voltage source would not be satisfied. Thus, the two mmfs must be equal. As you can easily verify, these results are the same as in Chapter 5; in particular, the ideal transformer does not dissipate any power, since

$$v_1 i_1 = v_2 i_2 \tag{14.38}$$

Note the distinction we have made between the induced voltages (emf's), e, and the terminal voltages, v. In general, these are not the same.

The results obtained for the ideal case do not completely represent the physical nature of transformers. A number of loss mechanisms need to be included in a practical transformer model, to account for the effects of leakage flux, for various magnetic core losses (e.g., hysteresis), and for the unavoidable resistance of the wires that form the coils.

Commercial transformer ratings are usually given on the so-called nameplate, which indicates the normal operating conditions. The nameplate includes the following parameters:

> Primary-to-secondary voltage ratio.
>
> Design frequency of operation.
>
> (Apparent) rated output power.

For example, a typical nameplate might read 480:240 V, 60 Hz, 2 kVA. The voltage ratio can be used to determine the turns ratio, while the rated output power represents the continuous power level that can be sustained without overheating. It is important that this power be rated as the apparent power in kVA, rather than real power in kW, since a load with low power factor would still draw current and therefore operate near rated power. Another important performance characteristic of a transformer is its **power efficiency,** defined by:

$$\text{Power efficiency} = \eta = \frac{\text{Output power}}{\text{Input power}} \tag{14.39}$$

The following examples illustrate the use of the nameplate ratings and the calculation of efficiency in a practical transformer, in addition to demonstrating the application of the circuit models.

Example   14.8

A transformer has the following nameplate ratings:

120 V/480 V
48 kVA
60 Hz

If the transformer's primary side is the 480-V side, find

1.   The turns ratio $\alpha = N_2/N_1$.
2.   The current for which the device is rated.

Solution:

1.   From the ratio of primary to secondary voltages, the turns ratio is found to be

$$\alpha = \frac{N_2}{N_1} = \frac{120}{480} = \frac{1}{4}$$

2.   The kVA rating of the transformer can be used to find both primary current and secondary current: the primary current is 48 kVA/480 V = 100 A, while the secondary current is 48 kVA/120 V = 400 A. These calculations do not include the effects of the unavoidable losses that are always present in a transformer.

Example   14.9   Impedance Transformer

The ideal transformer of Figure 14.35 has a load Z_2 across its secondary output terminals. Find the equivalent reflected impedance from secondary to primary.

Solution:

For any value of load impedance Z_2 the secondary impedance looking into the secondary terminals from the load is

$$Z_1 = \frac{V_1}{I_1}$$

Figure 14.35 Ideal transformer

Since any changes in load at the secondary will be reflected as a change in current at the primary, it is convenient to simplify the transformer to a single equivalent circuit. From equation 14.35, we have

$$V_1 = \alpha V_2 \qquad I_1 = \frac{I_2}{\alpha}$$

Thus,

$$Z_1 = \frac{\alpha V_2}{I_2/\alpha} = \alpha^2 \frac{V_2}{I_2}$$

But V_2/I_2 is the secondary impedance, Z_2; therefore,

$$Z_1 = \alpha^2 Z_2$$

or

$$\frac{Z_1}{Z_2} = \alpha^2 = \left(\frac{N_1}{N_2}\right)^2$$

Figure 14.36 shows the impedance looking into the input terminals from the source when the secondary impedance has been reflected back to the primary.

Figure 14.36

DRILL EXERCISES

14.9   The high-voltage side of a transformer has 500 turns, and the low-voltage side has 100 turns. When the transformer is connected as a step-down transformer, the load current is 12 A. Calculate: (a) the turns ratio α; (b) the primary current.

> Answer: $\alpha = 5$; $I_1 = \frac{I_2}{\alpha} = 2.4$ A

14.10   Calculate the turns ratio if the transformer in Drill Exercise 14.9 is used as a step-up transformer.

> Answer: $\alpha = 0.2$

14.11   The output of a transformer under certain conditions is 12 kW. The copper losses are 189 W and the core losses are 52 W. Calculate the efficiency of this transformer.

> Answer: $\eta = 98\%$

14.12   The output impedance of a servo amplifier is 250 Ω. The servomotor that the amplifier must drive has am impedance of 2.5 Ω. Calculate the turns ratio of the transformer required to match these impedances.

> Answer: $\alpha = 10$

14.5   ELECTROMECHANICAL ENERGY CONVERSION

From the material developed thus far, it should be apparent that electromagnetomechanical devices are capable of converting mechanical forces and displacements to electromagnetic energy, and that the converse is also possible. The objective of this section is to formalize the basic principles of energy conversion in electromagnetomechanical systems, and to illustrate its usefulness and potential for application by presenting several examples of **energy transducers.** A transducer is a device that can convert electrical to mechanical energy (in this case, it is often called an **actuator),** or vice versa (in which case it is called a sensor).

Several physical mechanisms permit conversion of electrical to mechanical energy and back, the principal phenomena being the **piezoelectric effect,** consisting of the generation of a change in electric field in the presence of strain in certain crystals (e.g., quartz), and **electrostriction** and **magnetostriction,** in which changes in the dimension of certain materials lead to a change in their electrical (or magnetic) properties. Although these effects lead to some interesting applications, this chapter is concerned only with transducers in which electrical energy is converted to mechanical energy through the coupling of a magnetic field. It is important to note that all rotating machines (motors and generators) fit the basic definition of electromechanical transducers we have just given.

Forces in Magnetic Structures

It should be apparent by now that it is possible to convert mechanical forces to electrical signals, and vice versa, by means of the coupling provided by energy stored in the magnetic field. In this subsection, we discuss the computation of mechanical forces and of the corresponding electromagnetic quantities of interest; these calculations are of great practical importance in the design and application of electromechanical actuators. For example, a problem of interest is the computation of the current required to generate a given force in an electromechanical structure. This is the kind of application that is likely to be encountered by the engineer in the selection of an electromechanical device for a given task.

As already seen in this chapter, an electromechanical system includes an electrical system and a mechanical system, in addition to means through which the two can interact. The principal focus of this chapter has been the coupling that occurs through an electromagnetic field common to both the electrical and the mechanical system; to understand electromechanical energy conversion, it will be important to understand the various energy storage and loss mechanisms in the electromagnetic field. Figure 14.37 illustrates

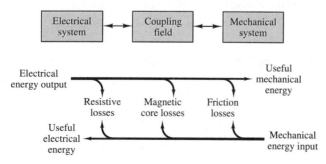

Figure 14.37

the coupling between the electrical and mechanical systems. In the mechanical system, energy loss can occur because of the heat developed as a consequence of *friction,* while in the electrical system, analogous losses are incurred because of *resistance*. Loss mechanisms are also present in the magnetic coupling medium, since *eddy current losses* and *hysteresis losses* are unavoidable in ferromagnetic materials. Either system can supply energy, and either system can store energy. Thus, the figure depicts the flow of energy from the electrical to the mechanical system, accounting for these various losses. The same flow could be reversed if mechanical energy were converted to electrical form.

The principle that will be applied throughout the section is that, for a mass to be displaced, some work needs to be done; this work corresponds to a change in the energy stored in the electromagnetic field. Let the displacement be given by the incremental quantity Δx, in meters, and let the corresponding change in stored energy (or work required) be given by ΔW, in joules. Then the corresponding force is

$$f = \frac{\Delta W}{\Delta x} \tag{14.40}$$

or, more generally,

$$f = \frac{dW}{dx} \tag{14.41}$$

Thus, to determine the forces present in a magnetic structure, it will be necessary to compute the energy stored in the magnetic field. To simplify the analysis, it will be assumed hereafter that the structures analyzed are magnetically linear. This is, of course, only an approximation, in that it neglects a number of practical aspects of electromechanical systems (for example, the nonlinear λ-i curves described earlier, and the core losses typical of magnetic materials), but it permits relatively simple analysis of many useful magnetic structures. Thus, although the analysis method presented in this section is only approximate, it will serve the purpose of providing a feeling for the direction and the magnitude of the forces and currents present in electromechanical devices.

On the basis of a linear approximation, it can be shown that the stored energy in a magnetic structure is given by the expression

$$W_m = \frac{\phi \mathscr{F}}{2} \tag{14.42}$$

and since the flux and the mmf are related by the expression

$$\phi = \frac{N_i}{\mathscr{R}} = \frac{\mathscr{F}}{\mathscr{R}} \tag{14.43}$$

the stored energy can be related to the reluctance of the structure according to the equation

$$W_m = \frac{\phi^2 \mathscr{R}(x)}{2} \tag{14.44}$$

where the reluctance has been explicitly shown to be a function of displacement, since these methods will be applied to electromechanical transducers with movable parts, which are therefore characterized by variable reluctance.

Now, equation 14.41 requires that a proper sign be selected, since the energy stored in the field decreases as the displacement, x, increases. Thus, *if the force acts in such a direction as to decrease the gap* (x in the negative direction), the correct expression is

$$f = -\frac{dW_m}{dx} = -\frac{\phi^2}{2} \frac{d\mathscr{R}}{dx} \tag{14.45}$$

One way to think of this is to visualize the force in equation 14.45 as a *pull force*.

The following examples illustrate the application of this approximate technique for the computation of forces and currents (the two problems of practical engineering interest to the user of such electromechanical systems) in some common devices.

Example   14.10   An Electromagnet

An electromagnet is used to support a solid piece of steel as shown in Figure 14.38. A force of 8,900 N is required to support the weight. The cross-sectional area of the magnet core (the fixed part) is 0.01 m^2. Determine the minimum current that can keep the weight from falling for $x = 1.5$ mm. Assume negligible reluctance for the steel parts, and negligible fringing in the air gap.

Figure 14.38

Solution:

We have already shown that in magnetic structures with air gaps, the reluctance is mostly due to the air gaps. This explains the assumption that the reluctance of the structure is negligible. For the structure of Figure 14.38, the reluctance is therefore given by

$$\mathcal{R}(x) = \frac{l}{\mu_0 A}$$

where $A = 0.01$ m^2 and $l = 2x$, and therefore

$$\mathcal{R}(x) = \frac{2x}{4\pi \times 10^{-7} \times 0.01} = \frac{x}{6.2832 \times 10^{-9}}$$

The magnitude of the force in the air gap is given by the expression

$$|f| = \frac{\phi^2}{2}\frac{d\mathcal{R}(x)}{dx} = \frac{N^2 i^2}{2\mathcal{R}^2}\frac{d\mathcal{R}(x)}{dx}$$

$$= \frac{i^2}{2}\frac{N^2}{\mathcal{R}^2}\frac{d\mathcal{R}}{dx} = \frac{i^2}{2}(700)^2\frac{6.2832 \times 10^{-9}}{x^2} = 8,900 \text{ N}$$

from which the current can be computed:

$$i^2 = 2 \times \frac{8,900(1.5 \times 10^{-3})}{(700)^2(6.2832 \times 10^{-9})} = 13.01$$

or

$$i = 3.607 \text{ A}$$

You should recognize the practical importance of these calculations in determining approximate current requirements and force-generation capabilities of electromechanical transducers.

Figure 14.39 Application
of the solenoid as a valve

One of the more common practical applications of the concepts discussed in this section is the **solenoid.** Solenoids find application in a variety of electrically controlled valves. The action of a solenoid valve is such that when it is energized, the plunger moves in such a direction as to permit the flow of a fluid through a conduit, as shown schematically in Figure 14.39.

The following example illustrates the calculations involved in the determination of forces and currents in a solenoid.

Example   14.11   A Solenoid

The magnetic structure shown in Figure 14.40 is a simplified representation of a solenoid. The presence of a flux in the air gap tends to pull the magnetic structures together, and serves the purpose of actuating the motion of the plunger.

1. Develop a general expression for the force exerted on the plunger as a function of the position, x.

2. Determine the mmf (and therefore the current) required to pull the plunger all the way into the field ($x = a$) if the spring constant is 1 N/m.

Assume that $a = 1$ cm, $l_g = 1$ mm, and μ_r is infinite. Neglect fringing, but assume that at $x = 0$, the plunger is still in the gap by the infinitesimal displacement ε.

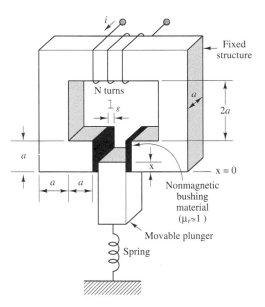

Figure 14.40 A solenoid

Solution:

1. As explained in this section, the force in the air gap can be expressed by

$$f_g = -\frac{1}{2}\phi^2 \frac{d\mathcal{R}_g}{dx}$$

Since most of the reluctance is due to the air gap, we approximate the total reluctance by

$$\mathcal{R} = 2\mathcal{R}_g$$

where the gap reluctance is the reluctance of the nonmagnetic bushing. As shown in Figure 14.41, the area of this gap is variable, depending on the position of the plunger.

$$\mathcal{R}_g = \frac{l_g}{\mu_0 A_g} = \frac{l_g}{\mu_0 ax}$$

$$\mathcal{R} = 2\mathcal{R}_g = \frac{2l_g}{\mu_0 ax}$$

Figure 14.41

The magnetic flux in the structure can then be found by the expression

$$\phi = \frac{\mathcal{F}}{\mathcal{R}} = \frac{\mu_0 Niax}{2l_g}$$

Now,

$$\frac{d\mathcal{R}_g}{dx} = \frac{d}{dx}\left(\frac{l_g}{\mu_0 ax}\right) = \frac{-l_g}{\mu_0 ax^2}$$

As discussed earlier, the force in the air gap is given by

$$f_g = -\frac{1}{2}\phi^2\left(\frac{d\mathcal{R}_g}{dx}\right)$$

Since we have two air gaps, we need to consider the force contribution due to both to compute the force:

$$f_g = -\frac{1}{2}\phi^2\left(2\frac{d\mathcal{R}_g}{dx}\right) = -\frac{1}{2}\left(\frac{\mu_0 Niax}{2l_g}\right)^2\left(-2\frac{l_g}{\mu_0 ax^2}\right)$$

$$= \frac{1}{4}\frac{\mu_0 a}{l_g}(Ni)^2$$

Thus, we see that the force has no dependence on position—that is, it is constant through the gap.

2. To compute the mmf required to sustain the force computed in part 1 at the position $x = 1$ cm, we evaluate the expression

$$f_g = \frac{1}{4}\frac{\mu_0 a}{l_g}(Ni)^2$$

at $x = 0.01$, to obtain

$$f_g = 1 \text{ (N/m)}(0.01 \text{ m}) = 0.01 \text{ N}$$

and compute the required mmf:

$$\text{mmf} = \mathscr{F} = Ni = \left(\frac{4f_g l_g}{\mu_0 a}\right)^{1/2} = \left[\frac{4(0.01)(0.001)}{4\pi \times 10^{-7}(0.01)}\right]^{1/2}$$

$$Ni = 56.42 \text{ A} \cdot \text{t}$$

This calculation permits the determination of two important variables: the current and number of turns required to produce the given force.

Another electromechanical device that finds common application in industrial practice is the **relay.** The relay is essentially an electromechanical switch that permits the opening and closing of electrical contacts by means of an electromagnetic structure similar to those discussed earlier in this section.

A relay such as would be used to start a high-voltage single-phase motor is shown in Figure 14.42. The magnetic structure has dimension equal to 1 cm on all sides, and the transverse dimension is 8 cm. The relay works as follows. When the push button is pressed, an electrical current flows through the coil and generates a field in the magnetic structure. The resulting force draws the movable part toward the fixed part, causing an electrical contact to be made. The advantage of the relay is that a relatively low-level current can be used to control the opening and closing of a circuit that can carry large currents. In this particular example, the relay is energized by a 120-VAC contact, establishing a connection in a 240-VAC circuit. Such relay circuits are commonly employed to remotely switch large industrial loads.

Circuit symbols for relays are shown in Figure 14.43. An example of the calculations that would typically be required in determining the mechanical and electrical characteristics of a simple relay are given in Example 14.12.

Figure 14.42 A relay

Figure 14.43 Circuit symbols for relays

Example   14.12   The Relay

For the simplified relay shown in Figure 14.44, determine the minimum amount of current required to keep the ferromagnetic plate at a distance of 0.5 cm from the electromagnet (which consists of a 10,000-turn coil) when the torque is 10 N·m. Assume that the ferromagnetic material is infinitely permeable and leakage and fringing effects are negligible.

Figure 14.44

Solution:

The force on the plate corresponding to the stated torque at a radius of 10 cm is

$$f' = \frac{10}{0.01} = 100 \text{ N}$$

To balance this force, the magnitude of the force developed by the electromagnet must be equal to 100 N. The total reluctance in the magnetic structure is due primarily to the two air gaps. Thus, the total reluctance is

$$\mathcal{R} = \frac{2x}{4\pi \times 10^{-7} \times 1 \times 10^{-4}}$$

The inductance can be found to be

$$L = \frac{N^2}{\mathcal{R}} = \frac{10,000^2}{1.59 \times 10^{10}x} = \frac{0.629 \times 10^{-2}}{x}$$

and the stored magnetic energy can be expressed (see Drill Exercise 14.4) as

$$W_m = \frac{1}{2}Li^2 = \frac{1}{2}\frac{\lambda^2}{L}$$

Note that L is a function of λ and x and represents the effective inductance of the entire magnetic circuit. From $f = -\partial W_m(\lambda, x)/\partial x$, the developed force is

$$f = \frac{1}{2}\frac{\lambda^2}{L^2}\frac{\partial L}{\partial x} = \frac{1}{2}i^2\frac{\partial L}{\partial x}$$

$$= \frac{1}{2}i^2\frac{dL}{dx} = -\frac{1}{2}i^2\frac{0.629 \times 10^{-2}}{x^2}$$

For $x = 0.5$ cm,

$$f = -125.8i^2$$

and since $f = -f'$, we have

$$125.8i^2 = 100$$

corresponding to:

$$i = 0.892 \text{ A}$$

The resulting magnetic circuit is depicted in Figure 14.45. The circuit assumes that the reluctance of the air gaps is far greater than that of the magnetic material.

Figure 14.45

Example   14.13   A Practical Solenoid Circuit

Figure 14.46

A common method to reduce the coil current required by a solenoid employs a normally closed (NC) switch in parallel with a *hold-in resistor*. In Figure 14.46, when the push button (PB) closes the circuit, full voltage is applied to the solenoid coil, bypassing the resistor through the NC switch. As the solenoid approaches the end of its stroke, a mechanical connection (it could also be electrical if the switch used is a relay) opens the switch, connecting the resistor in series with the coil. The additional voltage drop across the resistor is such that the solenoid voltage is reduced to the point where the voltage is just high enough to hold in, reducing the hold-in current to a minimum value, and thus preventing potential overheating.

Analysis of Electromechanical Transducers

The basic principle of operation of electromechanical transducers was presented in Section 14.1, where we stated that a magnetic field exerts a force on a charge moving though it. The equation describing this effect is the following:

$$\mathbf{f} = q\mathbf{u} \times \mathbf{B} \qquad (14.46)$$

This is a vector equation, as explained earlier. To correctly interpret equation 14.46, we must recall the right-hand rule, illustrated graphically in Figure 14.5. Figure 14.47 depicts a structure consisting of a sliding bar that makes

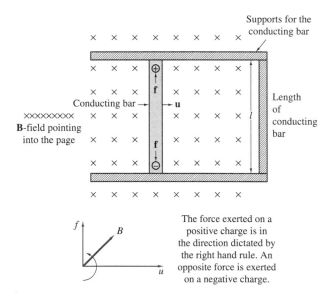

Figure 14.47 A simple electromechanical motion transducer

contact with a fixed electrically conducting frame. Although this structure does not represent a practical actuator, it will be a useful aid in explaining the operation of electromechanical motion devices. In Figure 14.47, and in all similar figures in this section, an "×" represents the "tail" of an arrow pointing into the page, while a dot (·) represents an arrow pointing out of the page; this convention will be useful in visualizing three-dimensional pictures.

It is important to observe that since positive and negative charges are forced in opposite directions, a potential difference will appear across the conducting bar; this potential difference is the electromotive force, or emf, already discussed. The emf offsets the force exerted by the magnetic field. In short, the electric field, e/l (corresponding to the electric force per unit charge), must equal the magnetic force per unit charge $f/q = Bu$. Thus, we have the relationship

$$e = Blu \qquad\qquad (14.47)$$

which holds whenever $\mathbf{B}$, $\mathbf{l}$, and $\mathbf{u}$ are mutually perpendicular, as in Figure 14.47. If equation 14.47 is analyzed in greater depth, it can be seen that the product lu (length times velocity) is the area crossed per unit time by the conductor. If one visualizes the conductor as "cutting" the flux lines going into the page in Figure 14.47, it may be concluded that the electromotive force is equal to the *rate at which the conductor "cuts" the magnetic lines of flux*. It will be useful for you to carefully absorb this notion of conductors cutting lines of flux, since this will greatly simplify your understanding of the material in this section and in the next chapter.

In general, $\mathbf{B}$, $\mathbf{l}$, and $\mathbf{u}$ are not necessarily perpendicular. In the case when they are not, one needs to consider the angles formed by the magnetic field with the normal to the plane containing $\mathbf{l}$ and $\mathbf{u}$, and the angle between $\mathbf{l}$ and $\mathbf{u}$. The former is the angle α of Figure 14.48, the latter, the angle β in the same figure. It should be apparent that the optimum values of α and β are $0°$ and $90°$, respectively. Thus, most practical devices are constructed with these values of α and β. Unless otherwise noted, it will be tacitly assumed that this is the case.

The *"Blu* law" just illustrated explains how a moving conductor in a magnetic field can generate an electromotive force. The counterpart of this relationship arises when one considers a current i flowing through the conductor. A path for current flow exists because the support of Figure 14.47 is made of conducting material, so that the conductor and the right-hand side of the support "rail" form a loop (in effect, a one-turn coil). To understand the effects of this current flow in the conductor, one must consider that a charge moving at a velocity u' (along the conductor and perpendicular to the velocity of the conducting bar, as shown in Figure 14.49) corresponds to a current $i = dq/dt$ along the length l of the conductor. This can be explained by considering the current i along a differential element dl and writing

$$i\, dl = \frac{dq}{dt} \cdot u'\, dt \qquad\qquad (14.48)$$

Figure 14.48

Figure 14.49

since the differential element dl would be traversed by the current in time dt at a velocity u'. Thus, we can write

$$i\,dl = dq\,u' \tag{14.49}$$

or

$$il = qu' \tag{14.50}$$

for the geometry of Figure 14.49.

From Section 14.1, the force developed by a charge moving in a magnetic field is, in general, given by

$$\mathbf{f}' = q\mathbf{u}' \times \mathbf{B} \tag{14.51}$$

we can replace the term $q\mathbf{u}'$ with $i\mathbf{l}$ to obtain

$$\mathbf{f}' = i\mathbf{l} \times \mathbf{B} \tag{14.52}$$

Using the right-hand rule, we determine that the force $\mathbf{f}'$ generated by the current i is in the direction that would push the conducting bar to the left. The magnitude of this force is $f' = Bli$ if the magnetic field and the direction of the current are perpendicular. If they are not, then we must consider the angle γ formed by $\mathbf{B}$ and $\mathbf{l}$; in the more general case,

$$f' = Bli\sin\gamma \tag{14.53}$$

The phenomenon we have just described is sometimes referred to as the "Bli law."

We must note now that the presence of a current in the loop formed by the conductor and the rail requires that the conductor move to the right at a velocity u (by the Blu law), thus cutting the lines of flux and generating the emf that gives rise to the current i. On the other hand, the same current causes a force f' to be exerted on the conductor (the Bli law) in the direction opposite to the movement of the conductor. Thus, it is necessary that an *externally applied force* f_{ext} exist to cause the conductor to move to the right with a velocity u. The external force must overcome the force f. This is the basis of electromechanical energy conversion.

An additional observation we must make at this point is that the current i flowing around a closed loop generates a magnetic field, as explained in Section 14.1. Since this additional field is generated by a one-turn coil in our illustration, it is reasonable to assume that it is negligible with respect to the field already present (perhaps established by a permanent magnet). Finally, we must consider that this coil links a certain amount of flux, which changes as the conductor moves from left to right. The area crossed by the moving conductor in time dt is

$$dA = lu\,dt \tag{14.54}$$

so that if the flux density, B, is uniform, the rate of change of the flux linked

by the one-turn coil is

$$\frac{d\phi}{dt} = B\frac{dA}{dt} = Blu \tag{14.55}$$

In other words, *the rate of change* of the flux linked by the conducting loop is equal to the emf generated in the conductor. You can see that this statement simply confirms Faraday's law.

Example   14.14

A set of parallel wires (shown in Figure 14.50) similar to those of common household appliance power cords each conduct 12 A. The wires are 2 m long, and they are separated 0.5 cm by insulation. How much force is exerted on each conductor?

Solution:

The magnetic field present will be due to the field generated by either of the wires. At a distance x from the axis of a wire, the flux density will be (by equation 14.22):

$$B = \mu H = \frac{\mu i}{2\pi x}$$

where μ is the permeability of the insulator, which is approximately μ_0. Then using the *Bli* law to calculate the force, we find:

$$f = Bli = \frac{\mu_0 i}{2\pi x} li = \frac{(4\pi \times 10^{-7})(12^2)(2)}{2\pi(0.005)}$$

$$= 11.52 \times 10^{-3} \text{ N}$$

This force corresponds to 11.52×10^{-3} N $\times$ 0.225 lb/N $= 2.59 \times 10^{-3}$ lb. Not exactly the kind of force that would tear the cable apart!

0.5 cm

Figure 14.50

It was briefly mentioned a few paragraphs back that the *Blu* and *Bli* laws indicate that, thanks to the coupling action of the magnetic field, a conversion of mechanical to electrical energy—or the converse—is possible. The simple structures of Figures 14.47 and 14.49 can, again, serve as an illustration of this energy-conversion process, although we have not yet indicated how these idealized structures can be converted into a practical device. In this section, we shall begin to introduce some physical considerations. Before we proceed any further, we should try to compute the power—electrical and mechanical—that is generated (or is required) by our ideal transducer. The electrical power is given by

$$P_E = ei = Blui \text{ (W)} \tag{14.56}$$

while the mechanical power required, say, to move the conductor from left to right is given by the product of force and velocity:

$$P_M = f_{ext}u = Bliu \text{ (W)} \tag{14.57}$$

The principle of conservation of energy thus states that in this ideal (lossless) transducer, we can convert a given amount of electrical energy to mechanical energy, or vice-versa.

Once again, we can utilize the same structure of Figure 14.47 to illustrate this reversible action. If the closed path that contains the moving conductor is now formed of a closed circuit containing a resistance R and a battery, V_B, as shown in Figure 14.51, the externally applied force, f_{ext}, generates a positive current, i, into the battery, provided that the emf is greater than V_B. When $e = Blu > V_B$, the ideal transducer acts as a *generator*. For any given set of values of B, l, R, and V_B, there will exist a velocity u for which the current i is positive. If the velocity is lower than this value, that is, if $e = Blu < V_B$—then the current i is negative and the conductor is forced to move to the right. In this case, the battery acts as source of energy, and the transducer acts as a *motor* (i.e., electrical energy drives the mechanical motion). These basic principles of electromechanical energy conversion will be applied to a practical transducer in the next example.

In practical transducers, we must be concerned with the inertia, friction, and elastic forces that are invariably present on the mechanical side. Similarly, on the electrical side, we must account for the inductance of the circuit, its resistance, and possibly some capacitance.

Consider the structure of Figure 14.52. In the figure, the conducting bar has been placed on a surface with coefficient of sliding friction D; it has a mass M and is attached to a fixed structure by means of a spring with spring constant K. The equivalent circuit representing the coil inductance and resistance is also shown. If we recognize that $u = dx/dt$ in the figure, we can write the equation of motion for the conductor as

$$m\frac{du}{dt} + du + k \int u\, dt = f' = Bli \qquad (14.58)$$

where the Bli term represents the driving input that causes the mass to move. The driving input in this case is provided by the electrical energy source,

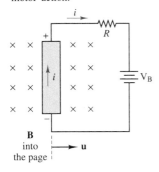

An ideal transducer: $i > 0 \Rightarrow$ generator action; $i < 0 \Rightarrow$ motor action.

Figure 14.51 Motor and generator action in an ideal transducer

A more realistic representation of the transducer of Figure 14.51

Figure 14.52

v_S. Thus, the transducer acts as a motor. On the electrical side, the circuit equation is:

$$v_S - L\frac{di}{dt} - Ri = e = Blu \tag{14.59}$$

The pair of equations just given could then be solved if we knew the excitation voltage, v_S, and the physical parameters of the mechanical and electrical circuits. For example, if the excitation voltage were sinusoidal, with

$$v_S(t) = V_S \cos \omega t \tag{14.60}$$

and the field density were constant:

$$B = B_0$$

we could postulate sinusoidal solutions for the transducer velocity, u, and current, i:

$$\begin{aligned} u &= U\cos(\omega t + \theta_u) \\ i &= I\cos(\omega t + \theta_i) \end{aligned} \tag{14.61}$$

and use phasor notation to solve for the unknowns (U, I, θ_u, θ_i).

The results obtained in the present section apply directly to transducers that are based on translational (linear) motion. It would be unreasonable to expect this type of transducer to operate in practice, except for very small displacement. Thus, the applications of this type of structure are somewhat limited, although quite common. The analysis methods developed in the section are illustrated in Example 14.15 by analyzing a very common transducer: the loudspeaker.

Example   14.15   The Loudspeaker

A loudspeaker, shown in Figure 14.53, uses a permanent magnet and a moving coil

Figure 14.53 Loudspeaker

to produce the vibrational motion that generates the pressure waves we perceive as sound. Vibration of the loudspeaker is caused by changes in the input current to a coil; the coil is, in turn, coupled to a magnetic structure that can produce time-varying forces on the speaker diaphragm. A simplified model for the mechanics of the speaker is also shown in Figure 14.53. The force exerted on the coil is also exerted on the mass of the speaker diaphragm, as shown in Figure 14.54, which depicts a free-body diagram of the forces acting on the loudspeaker diaphragm.

The force exerted on the mass, f_i, is the magnetic force due to current flow in the coil. The electrical circuit that describes the coil is shown in Figure 14.55, where L represents the inductance of the coil, R represents the resistance of the windings, and e is the emf induced by the coil moving through the magnetic field.

Determine the frequency response, $\frac{u}{V}(\omega)$, of the speaker using phasor analysis if the model parameters are:

$$L \approx 0\ \text{H} \qquad R = 8\ \Omega \qquad m = 0.001\ \text{kg} \qquad d = 2.75$$

$$k \approx 500{,}000\ \text{N/m} \qquad N = 47 \qquad B = 1\ T$$

Radius of coil $= 5$ cm

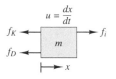

Figure 14.54 Forces acting on loudspeaker diaphragm

Figure 14.55 Model of transducer electrical side

Solution:

To determine the frequency response of the loudspeaker, we need to write the fundamental equations that describe the two subsystems that make up the loudspeaker. The electrical subsystem is described by the usual KVL relationship, applied to the circuit of Figure 14.55:

$$v = L\frac{di}{dt} + Ri + e$$

where e is the emf generated by the motion of the coil in the magnetic field. Next, according to Newton's law, we can write a force balance equation to describe the dynamics of the mechanical subsystem:

$$m\frac{du}{dt} = f_i - f_d - f_x = f_i - du - kx$$

Now, the coupling between the electrical and mechanical system is expressed in each of the two above equations by the terms e and f_i:

$$e = Bl\,u$$

$$f_i = Bl\,i$$

Since we desire the frequency response, we use phasor techniques to represent the electrical and mechanical subsystem equations:

$$\mathbf{V} = j\omega L\mathbf{I} + R\mathbf{I} + Bl\mathbf{U} \tag{14.62}$$

$$(j\omega m + d)\mathbf{U} + \frac{K}{j\omega}\mathbf{U} = Bl\mathbf{I} \tag{14.63}$$

Having assumed that the inductance of the coil is negligible, it is possible to simplify equation 14.62 and to solve for **I**:

$$\mathbf{I} = \frac{\mathbf{V} - Bl\mathbf{U}}{R} \tag{14.64}$$

Substituting (14.64) into (14.63) and accounting for the length of the coil, $l = 2\pi Nr$, the final expression for the frequency response of the loudspeaker is then given by:

$$\frac{U}{V} = \frac{2\pi NBr}{Rm} \times \frac{j\omega}{(j\omega)^2 + j\omega\left(\dfrac{d + \dfrac{(2\pi)^2 B^2 N^2 r^2}{R}}{m}\right) + \dfrac{k}{m}}$$

or, numerically,

$$\frac{U}{V} = \frac{2\pi \times 47 \times 1 \times 0.05}{8(0.001)}$$

$$\times \frac{j\omega}{(j\omega)^2 + j\omega\left(\dfrac{2.75 + 20 + \dfrac{(2\pi)^2(1)^2(47)^2(.05)^2}{8}}{0.001}\right) + \dfrac{500000}{0.001}}$$

$$\simeq \frac{1845 j\omega}{(j\omega + 13.8 \times 10^3)(j\omega + 36.2 \times 10^3)}$$

$$= \frac{0.051\left(\dfrac{j\omega}{13.8 \times 10^3}\right)}{\left(1 + \dfrac{j\omega}{36.2 \times 10^3}\right)\left(1 + \dfrac{j\omega}{13.8 \times 10^3}\right)}$$

This frequency response shows that the speaker has a lower cutoff frequency $f_{cl} = \frac{13,800}{2\pi} \approx 2,200$ Hz and an upper cutoff frequency of $f_{ch} = \frac{36,000}{2\pi} \approx 5,800$ Hz. In practice, a loudspeaker with such a frequency response would be useful only as a midrange speaker.

The methods introduced in this section will be applied in Chapters 15 and 16 to analyze rotating transducers, that is, electric motors and generators.

DRILL EXERCISES

14.13 The flux density of the earth's magnetic field is about 50 μT. Estimate the current required in a conductor of length 10 cm and mass 10 g to counteract the force of gravity if the wire is oriented in the optimum direction.

> Answer: $i = 196 \times 10^2$ A

14.14 In Example 14.15, we examined the frequency response of a loudspeaker. However, over a period of time, permanent magnets may become demagnetized. Find the frequency response of the same loudspeaker if the permanent magnet has lost its strength to a point where $B = 0.95$ T.

> Answer: $\dfrac{U}{V} = \dfrac{0.056(j\omega/15{,}902)}{(1 + j\omega/15{,}902)(1 + j\omega/31{,}443)}$

14.15   In Example 14.11, a solenoid is used to exert force on a spring. Estimate the position of the plunger if the number of turns in the solenoid winding is 1,000 and the current going into the winding is 40 mA.

Answer: $x = 0.5$ cm

14.16   Figure 14.50 shows a wire carrying a current in the presence of a magnetic field. The magnetic flux density is 0.2 Wb/m^2. If the wire is 1.0 m long and carries 0.8 A of current, find the magnetic force induced on the wire.

Answer: $f = 0.16$ N

14.17   For the circuit in Figure 14.50, the conducting bar is moving with a velocity of 6 m/s. The flux density is 0.5 Wb/m^2, and $l = 1.0$ m. Find the magnitude of the resulting induced voltage.

Answer: 3 V

CONCLUSION

- Magnetic fields form a coupling mechanism between electrical and mechanical systems, permitting the conversion of electrical energy to mechanical energy, and vice versa. The basic laws that govern such electromechanical energy conversion are Faraday's law, stating that a changing magnetic field can induce a voltage, and Ampère's law, stating that a current flowing through a conductor generates a magnetic field.

- The two fundamental variables in the analysis of magnetic structures are the magnetomotive force and the magnetic flux; if some simplifying approximations are made, these quantities are linearly related through the reluctance parameter, much in the same way as voltage and current are related through resistance according to Ohm's law. This simplified analysis permits approximate calculations of required forces and currents to be conducted with relative ease in magnetic structures.

- Magnetic materials are characterized by a number of nonideal properties, which should be considered in the detailed analysis of a magnetic structure. The most important phenomena are saturation, eddy currents, and hysteresis.

- Electromechanical transducers, which convert electrical signals to mechanical forces, or mechanical motion to electrical signals, can be analyzed according to the techniques presented in this chapter. Examples of such transducers are electromagnets, position and velocity sensors, relays, solenoids, and loudspeakers.

KEY TERMS

Magnetic Flux   742

Weber (Wb)   742

Magnetic Flux Density   742

Tesla (T)   742

Magnetic Field Intensity   742

Faraday's Law   744

Electromotive Force (emf)   744

Lenz's Law   744

Right-Hand Rule   744

Flux Linkage   744

Motional Voltages   745

Transformer Voltages   745

Self-Inductance   745

Magnetic Coupling   745

Mutual Inductance   746

Co-energy   748

Ampère's Law   749

Permeability   750

Magnetomotive Force
(mmf)   750

Ferrites   752

Mean Path   754

Reluctance   754

Leakage Flux   756

Air Gap   757

Fringing   758

Saturation   764

Magnetic Domains   765

Eddy Current   766

Hysteresis   766

Remanent (Residual)
Magnetization   766

Counter mmf   769

Power Efficiency   770

Energy Transducer   772

Actuator   772

Piezoelectric Effect   772

Electrostriction   772

Magnetostriction   772

Solenoid   776

Relay   778

HOMEWORK PROBLEMS

14.1   An iron-core inductor has the following characteristic:

$$i = \lambda + 0.5\lambda^2$$

a.   Determine the energy, co-energy, and incremental inductance for $\lambda = 0.5$ V·s.

b.   Given that the coil resistance is 1 Ω and that

$$i(t) = 0.625 + 0.01 \sin 400t \text{ A}$$

determine the voltage across the terminals on the inductor.

14.2   An iron-core inductor has the following characteristic:

$$\lambda = i - \frac{i^3}{3} + \frac{i^5}{5} \qquad -1 \text{ A} < i < 1 \text{ A}$$

a.   Determine the energy and the incremental inductance for $i = 0.75$ A.

b.   Given that the coil resistance is 1 Ω and that $i(t) = \sin t$, determine the voltage across the terminals of the inductor.

14.3   An iron-core inductor has the characteristic shown in Figure P14.1.

a.   Determine the energy and the incremental inductance for $i = 1.0$ A.

b.   Given that the coil resistance is 2 Ω and that $i(t) = 0.5 \sin 2\pi t$, determine the voltage across the terminals of the inductor.

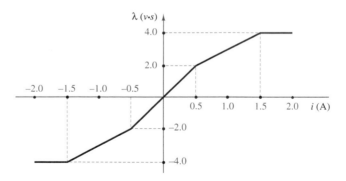

Figure P14.1

14.4 An iron-core inductor has the following characteristic:

$$\lambda = i - \frac{i^2}{6} + \frac{i^3}{120} \qquad -2\,A < i < 2\,A$$

a. Determine the energy, co-energy, and incremental inductance for $i = 1$ A.
b. Given that the coil resistance is 1 Ω and that $i(t) = \sin t$, determine the voltage across the terminals on the inductor.

14.5 For the circuit shown in Figure P14.2,

a. Determine the reluctance values and show the magnetic circuit, assuming that $\mu = 3{,}000\mu_0$.
b. Determine the inductance of the device.
c. The inductance of the device can be modified by cutting an air gap in the magnetic structure. If a gap of 0.1 mm is cut in the arm of length l_3, what is the new value of inductance?
d. As the gap is increased in size (length), what is the limiting value of inductance? Neglect leakage flux and fringing effects.

$N = 100$ turns	$A_2 = 25$ cm^2
$l_1 = 30$ cm	$l_3 = 30$ cm
$A_1 = 100$ cm^2	$A_3 = 100$ cm^2
$l_2 = 10$ cm	

Figure P14.2

14.6 The magnetic circuit shown in Figure P14.3 has two parallel paths. Find the flux and flux density in each of the legs of the magnetic circuit. Neglect fringing at the air gaps and any leakage fields. $N = 1{,}000$ turns, $i = 0.2$ A, $l_{g1} = 0.02$ cm, and $l_{g2} = 0.04$ cm.

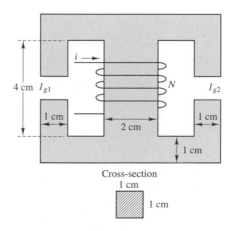

Figure P14.3

14.7 With reference to Figure 14.17, suppose that $A = A_g = 0.01$ m^2, $\delta = 0.005$ m, and $N = 1,000$ turns. Find the current if a flux of 0.01 Wb in the air gap is required, neglecting iron reluctance.

14.8 A cylindrical solenoid is shown in Figure P14.4. The plunger may move freely along its axis. The air gap between the shell and the plunger is uniform, and the diameter, d, is 25 mm. If the exciting coil carries a current of 7.5 A, find the force acting on the plunger when $x = 2$ mm. Assume $N = 200$ turns, and neglect the reluctance of the steel shell.

14.9 For the same magnetic structure of Problem 14.7, find the current that will produce a flux density in the air gap of 1.5 T.

14.10 For the transformer shown in Figure P14.5, $N = 1,000$ turns, $l_1 = 16$ cm, $A_1 = 4$ cm^2, $l_2 = 22$ cm, $A_2 = 4$ cm^2, $l_3 = 5$ cm, and $A_3 = 2$ cm^2. The relative permeability of the material is $\mu_r = 1,500$.

Figure P14.4

a.   Construct the equivalent magnetic circuit, and find the reluctance associated with each part of the circuit.
b.   Determine the self-inductance and mutual inductance for the pair of coils (i.e., L_{11}, L_{22}, and $M = L_{12} = L_{21}$).

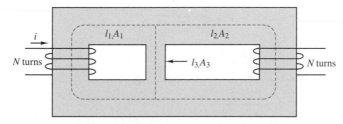

Figure P14.5

14.11 The cylindrical solenoid shown in Figure P14.4 carries a constant current of 5 A. Its armature is allowed to move slowly from $x = 5$ mm to 2 mm. Find

a.   The change in energy stored in the magnetic field
b.   The change in the magnetic co-energy.
c.   The change in the energy supplied by the source.

14.12 The double-excited electromechanical system shown in Figure P14.6 moves horizontally. Assuming that resistance, magnetic leakage, and fringing are negligible, the permeability of the core is very large, and the cross section of the structure is $w \times w$, find

a.   The reluctance of the magnetic circuit.
b.   The magnetic energy stored in the air gap.
c.   The force on the movable part as a function of its position.

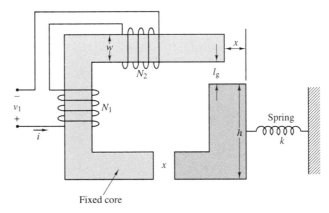

Figure P14.6

14.13 For the circuit shown in Figure P14.7:

a.   Determine the reluctance values and show the magnetic circuit, assuming that $\mu = \mu_r \mu_0$ and that $\mu_0 = 4\pi \times 10^{-7}$ H/m.
b.   Determine the inductance of the device.
c.   If an air gap of length 0.1 mm is cut in the magnetic structure directly across from the coil, what is the new value of inductance?

Neglect leakage flux and fringing effects.

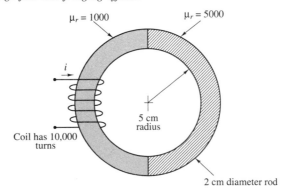

Figure P14.7

14.14

a. For the electromagnet shown in Figure P14.8, assume that the permeability of the magnetic material is infinite, so that the reluctance is all due to the air gap. For a force $f = 2,000$ N and a DC excitation current $I = 25$ A $(A_g = 20$ cm$^2)$, determine the number of turns N required to hold the air gap (i.e., x) at 0.1 cm.

b. What peak value of AC current would be required to generate the same average force?

14.15   The electromagnet of Figure P14.8 has reluctance given by $\mathcal{R}(x) = 7 \times 10^8(0.002 + x)$ H^{-1}, where x is the length of the variable gap in meters. The coil has 980 turns and 30 Ω resistance. For an applied voltage of 120 VDC, find:

a. The energy stored in the magnetic field for $x = 0.005$ m.

b. The magnetic force for $x = 0.005$ m.

14.16   For the DC relay shown in Figure P14.9, the resistance of the winding is $R = 25$ Ω, $N = 100$ turns, $x = 0.1$ cm, and the cross-sectional area of the gap is 0.01 m^2. At this position, the spring force is 2 N. Assume that the permeability of the magnetic material is infinite and that $\mu = \mu_0 = 4\pi \times 10^{-7}$ H/m for the air gap.

a. What current is required to hold the movable piece in this position, and what is the input voltage needed to get this current?

b. What are the inductance, L, and the energy stored in the magnetic field, W_m, at this position?

14.17   An electromagnet is used to support a solid piece of steel as shown in Example 14.10. A force of 10,000 N is required to support the weight. The cross-sectional area of the magnetic core (the fixed part) is 0.01 m^2. The coil has 1,000 turns. Determine the minimum current that can keep the weight from falling for $x = 1.0$ mm. Assume negligible reluctance for the steel parts and negligible fringing in the air gaps.

14.18   A core is shown in Figure P14.10, with $\mu_r = 1,200$ and $N = 100$. Find

a. The flux in each of the legs.

b. The flux density in each of the legs.

Neglect fringing effects at the air gap.

Figure P14.8

Figure P14.9

Figure P14.10

14.19 For the magnetic circuit shown in Figure P14.11, the self-inductance for the two coils is $L_{11} = L_{22} = 5 + (2x)^{-1}$ mH, and the mutual inductance between the two coils is $M_{12} = M_{21} = (4x)^{-1}$ mH, where x is in meters.

a. For $i_1 = 7.5$ A and $i_2 = 0$, what is the force between the coils for $x = 0.01$ m?

b. For the same currents as in part a, determine the voltage across the movable coil at $x = 0.01$ m when the movable coil is moving in the x direction at a constant speed of 10 m/s.

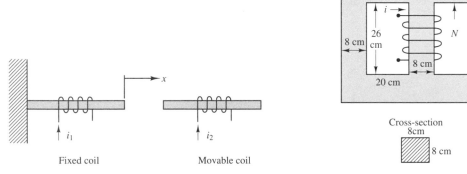

Fixed coil   Movable coil

Figure P14.11   **Figure P14.12**

14.20 A core is shown in Figure P14.12, with $\mu_r = 2,000$. Find:

a. The current needed to produce a flux density of 0.4 Wb/m² in the center leg.

b. The current needed to produce a flux density of 0.8 Wb/m² in the center leg.

14.21 A transformer is delivering power to a 300-Ω resistive load. To achieve the desired power transfer, the turns ratio is chosen so that the resistive load referred to the primary is 7,500 Ω. The parameter values, *referred to the secondary winding,* are:

$$r_1 = 20 \ \Omega \qquad L_1 = 1.0 \ \text{mH} \qquad L_m = 25 \ \text{mH}$$
$$r_2 = 20 \ \Omega \qquad L_2 = 1.0 \ \text{mH}$$

Core losses are negligible.

a. Determine the turns ratio.

b. Determine the input voltage, current, and power and the efficiency when this transformer is delivering 12 W to the 300-Ω load at a frequency $f = 10,000/2\pi$ Hz.

14.22 Derive and sketch the frequency response of the loudspeaker of Example 14.15 for (1) $k = 50,000\frac{N}{m}$ and (2) $k = 5 \times 10^6 \frac{N}{m}$. Describe qualitatively how the loudspeaker frequency response changes as the spring stiffness, k increases and decreases. What will the frequency response be in the limit as k approaches zero? What kind of speaker would this condition correspond to?

14.23 A 220/20-V transformer has 50 turns on its low-voltage side. Calculate

a. The number of turns on its high side.

b. The turns ratio α when it is used as a step-down transformer.

c. The turns ratio α when it is used as a step-up transformer.

14.24 The high-voltage side of a transformer has 750 turns, and the low-voltage side 50 turns. When the high side is connected to a rated voltage of 120 V, 60 Hz, a rated load of 40 A is connected to the low side. Calculate

a. The turns ratio.

b. The secondary voltage (assuming no internal transformer impedance voltage drops).

c. The resistance of the load.

14.25 A transformer is to be used to match an 8-Ω loudspeaker to a 500-Ω audio line. What is the turns ratio of the transformer, and what are the voltages at the primary and secondary terminals when 10 W of audio power is delivered to the speaker? Assume that the speaker is a resistive load and the transformer is ideal.

14.26 An electromechanical system is shown in Figure P14.13. Find the differential equations describing the system after the switch is closed. Neglect fringing and leakage and the friction.

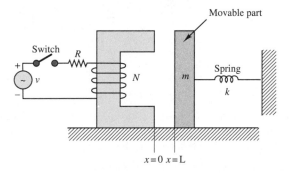

Figure P14.13

14.27 On the basis of Example 14.15, obtain an equivalent electrical circuit for the loudspeaker. [Hint: Eliminate velocity $\mathbf{U}$ and find the relation of $\mathbf{V} = f(\mathbf{I})$ first.]

14.28 A practical LVDT is typically connected to a resistive load. Derive the LVDT equations in the presence of a resistive load, R_L, connected across the output terminals, using the results of Example 14.1.

14.29 A relay is shown in Figure P14.14. Find the differential equations describing the system.

Figure P14.14

14.30 On the basis of the equations of Example 14.1, and of the results of Problem 14.28, derive the frequency response of the LVDT, and determine the range of frequencies for which the device will have maximum sensitivity for a given excitation. [*Hint:* Compute $dv_{\text{out}}/dv_{\text{ex}}$, and set the derivative equal to zero to determine the maximum sensitivity.]

14.31 The model of an electromechanical conversion system is shown in Figure P14.15. Find the differential equations describing the system.

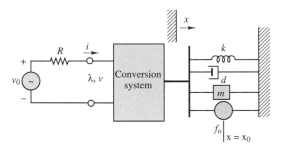

Figure P14.15

14.32 From the result of Problem 14.27, find the maximum mechanical impedance Z_m and the angular frequency where $Z_m = $ max. [Hint: The mechanical parameters of Problem 14.27 are defined as the mechanical impedance.]

14.33 The air gap inductor shown in Figure P14.16 has a height, h of 2.5 cm, width, w of 2.5 cm, air-gap dimension $l_g = 0.1$ cm, and $N = 400$ turns, and it is excited by a DC current, $I = 5$ A. Determine the force of attraction between sides of the gap, neglecting the reluctance of the core, leakage, and fringing.

14.34 The high-voltage side of a step-down transformer has 800 turns, and the low-voltage side has 100 turns. A voltage of 240 VAC is applied to the high side, and the load impedance is 3 Ω (low side). Find

a. The secondary voltage and current.

b. The primary current.

c. The primary input impedance from the ratio of primary voltage and current.

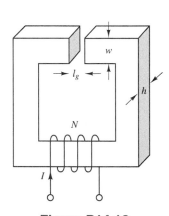

Figure P14.16

d. The primary input impedance.

14.35   Calculate the transformer ratio of the transformer in Problem 14.34 when it is used as a step-up transformer.

14.36   A 2,300/240-V, 60-Hz, 4.6-kVA transformer is designed to have an induced emf of 2.5 V/turn. Assuming an ideal transformer, Find

a.   The number of high-side turns, N_h, and low-side turns, N_1.

b.   The rated current of the high-voltage side, I_h.

c.   The transformer ratio when the device is used as a step-up transformer.

14.37   A wire of length 20 cm vibrates in one direction in a constant magnetic field with a flux density of 0.1 T; see Figure P14.17. The position of the wire as a function of time is given by $x(t) = 0.1 \sin 10t$ m. Find the induced emf across the length of the wire as a function of time.

14.38   A wire vibrating in a magnetic field induces a time-varying emf of

$$e_1(t) = 0.02 \cos 10t$$

A second wire is placed in the same magnetic field but has a length of 0.1 m, as shown in Figure P14.18. The position of this wire is given by $x(t) = 1 - 0.1 \sin 10t$. Find the induced emf $e(t)$ defined by the difference in emf's $e_1(t)$ and $e_2(t)$.

14.39   A solenoid having a cross section of 5 cm^2 is shown in Figure P14.19.

Figure P14.17

Figure P14.18

Figure P14.19

a.   Calculate the force exerted on the plunger when the distance x is 2 cm and the current in the coil (where $N = 100$ turns) is 5 A. Assume that the fringing and leakage effects are negligible. The relative permeabilities of the magnetic material and the nonmagnetic sleeve are 2,000 and 1.

b.   Develop a set of differential equations governing the behavior of the solenoid.

14.40 A conducting bar shown in Figure 14.47, is carrying 4 A of current in the presence of a magnetic field; $B = 0.3$ Wb/m^2. Find the magnitude and direction of the force induced on the conducting bar.

14.41 A wire, shown in Figure P14.20, is moving in the presence of a magnetic field, with $B = 0.4$ Wb/m^2. Find the magnitude and direction of the induced voltage in the wire.

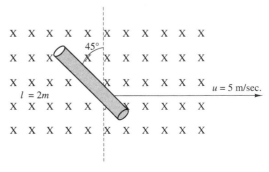

Figure P14.20

CHAPTER 15

Introduction to
Electric Machines

The objective of this chapter is to introduce the basic operation of rotating electric machines. The operation of the three major classes of electric machines—DC, synchronous, and induction—will first be described as intuitively as possible, building on the material presented in Chapter 14. The second part of the chapter will be devoted to a discussion of the applications and selection criteria for the different classes of machines.

The emphasis of this chapter will be on explaining the properties of each type of machine, with its advantages and disadvantages with regard to other types, and on classifying these machines in terms of their performance characteristics and preferred field of application. Chapter 16 will be devoted to a survey of special-purpose electric machines—many of which find common application in industry, especially for robotics and motion control applications—such as stepper motors, brushless DC motors, and single-phase induction motors. Selected examples and application notes will discuss some current issues of interest.

By the end of the chapter, you should be able to:

- Describe the principles of operation of DC and AC motors and generators.
- Interpret the nameplate data of an electric machine.
- Interpret the torque-speed characteristic of an electric machine.
- Select a machine for a given application on the basis of its torque-speed characteristic and other operating characteristics.
- Specify the requirements of a given application.

15.1   ROTATING ELECTRIC MACHINES

The range of sizes and power ratings and the different physical features of rotating machines are such that the task of explaining the operation of rotating machines in a single chapter may appear formidable at first. Some features of rotating machines, however, are common to all such devices. This introductory section is aimed at explaining the common properties of all rotating electric machines. We begin our discussion with reference to Figure 15.1, in which a hypothetical rotating machine is depicted in a cross-sectional view. In the figure, a circle with a cross inscribed in it indicates current flowing into the page, while a dot represents current out of the plane of the page.

In Figure 15.1, we identify a **stator,** of cylindrical shape, and a **rotor,** which, as the name indicates, rotates inside the stator, separated from the latter by means of an air gap. The rotor and stator each consist of a magnetic core, some electrical insulation, and the windings necessary to establish a magnetic flux (unless this is created by a permanent magnet). The rotor is mounted on a bearing-supported shaft, which can be connected to *mechanical loads* (if the machine is a motor) or to a *prime mover* (if the machine is a generator) by means of belts, pulleys, chains, or other mechanical couplings. The windings carry the electric currents that generate the magnetic fields and flow to the electrical loads, and also provide the closed loops in which voltages will be induced (by virtue of Faraday's law, as discussed in the previous chapter).

Figure 15.1 A rotating electric machine

Basic Classification of Electric Machines

An immediate distinction can be made between different types of windings characterized by the nature of the current they carry. If the current serves the sole purpose of providing a magnetic field and is independent of the load, it is called a *magnetizing,* or excitation, current, and the winding is termed a **field winding.** Field currents are nearly always DC and are of relatively low power, since their only purpose is to magnetize the core (recall the important role of high-permeability cores in generating large magnetic fluxes from relatively small currents). On the other hand, if the winding carries only the load current, it is called an **armature.** In DC and AC synchronous machines, separate windings exist to carry field and armature currents. In the induction motor, the magnetizing and load currents flow in the same winding, called the *input winding,* or *primary;* the output winding is then called the *secondary.* As we shall see, this terminology, which is reminiscent of transformers, is particularly appropriate for induction motors, which bear a significant analogy to the operation of the transformers studied in Chapters 5 and 14. Table 15.1 characterizes the principal machines in terms of their field and armature configuration.

It is also useful to classify electric machines in terms of their energy-conversion characteristics. A machine acts as a **generator** if it converts mechanical energy from a prime mover—e.g., an internal combustion engine—to electrical form. Examples of generators are the large machines used in power-generating plants, or the common automotive alternator. A machine is classified as a **motor** if it converts electrical energy to mechanical form. The latter class of machines is probably of more direct interest to you, because of its widespread application in engineering practice. Electric motors are used to provide forces and torques to generate motion in countless industrial applications. Machine tools, robotics, punches, presses, mills, and propulsion systems for electric vehicles are but a few examples of the application of electric machines in engineering.

Note that in Figure 15.1 we have explicitly shown the direction of two magnetic fields: that of the rotor, $\mathbf{B}_R$, and that of the stator, $\mathbf{B}_S$. Although these fields are generated by different means in different machines (e.g., permanent magnets, AC currents, DC currents), the presence of these fields is what causes a rotating machine to turn and enables the generation of electric power. In particular, we see that in Figure 15.1 the north pole of the rotor field will seek to align itself with the south pole of the stator field. It is this magnetic attraction force that permits the generation of torque in an electric motor; conversely, a generator exploits the laws of electromagnetic induction to convert a changing magnetic field to an electric current.

To simplify the discussion in later sections, we shall presently introduce some basic concepts that apply to all rotating electric machines. Referring to Figure 15.2, we note that for all machines the force on a wire is given by the expression

$$\mathbf{f} = i_w \mathbf{l} \times \mathbf{B} \tag{15.1}$$

where i_w is the current in the wire, $\mathbf{l}$ is a vector along the direction of the wire, and $\times$ denotes the cross product of two vectors. Then the torque for a multiturn coil becomes

$$T = KBi_w \sin \alpha \tag{15.2}$$

where

$B =$ magnetic flux density caused by the stator field

$K =$ constant depending on coil geometry

$\alpha =$ angle between $\mathbf{B}$ and the normal to the plane of the coil

In the hypothetical machine of Figure 15.2, there are two magnetic fields: one generated within the stator, the other within the rotor windings. Either (but not both) of these fields could be generated either by a current or by a permanent magnet. Thus, we could replace the permanent-magnet stator of Figure 15.2 with a suitably arranged winding to generate a stator field

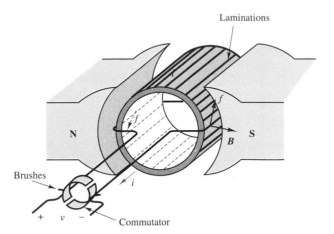

Figure 15.2 Stator and rotor fields and the force acting on a rotating machine

in the same direction. If the stator were made of a toroidal coil of radius R (see Chapter 14), then the magnetic field of the stator would generate a flux density B, where

$$B = \mu H = \mu \frac{Ni}{2\pi R} \tag{15.3}$$

and where N is the number of turns and i is the coil current. The direction of the torque is always the direction determined by the rotor and stator fields as they seek to align to each other (i.e., counterclockwise in the diagram of Figure 15.1).

It is important to note that Figure 15.2 is merely a general indication of the major features and characteristics of rotating machines. A variety of configurations exist, depending on whether each of the fields is generated by a current in a coil or by a permanent magnet, and on whether the load and magnetizing currents are direct or alternating. The type of excitation (AC or DC) provided to the windings permits a first classification of electric machines (see Table 15.1). According to this classification, one can define the following types of machines:

- *Direct-current machines:* DC current in both stators and rotor.
- *Synchronous machines:* AC current in one winding, DC in the other.
- *Induction machines:* AC current in both.

In most industrial applications, the induction machine is the preferred choice, because of the simplicity of its construction. However, the analysis of the performance of an induction machine is rather complex. On the other hand, DC machines are quite complex in their construction but can be analyzed relatively simply with the analytical tools we have already acquired. Therefore, the progression of this chapter will be as follows. We start with a section that discusses the physical construction of DC machines, both motors and generators. Then we continue with a discussion of synchronous machines, in which one of the currents is now alternating, since these can easily be understood as an extension of DC machines. Finally, we consider the case where both rotor and stator currents are alternating, and analyze the induction machine.

Table 15.1   Configurations of the three types of electric machines

Machine type	Winding	Winding type	Location	Current
DC	Input and output	Armature	Rotor	AC (winding)
				DC (at brushes)
	Magnetizing	Field	Stator	DC
Synchronous	Input and output	Armature	Stator	AC
	Magnetizing	Field	Rotor	DC
Induction	Input	Primary	Stator	AC
	Output	Secondary	Rotor	AC

Other Characteristics of Electric Machines

As already stated earlier in this chapter, electric machines are **energy-conversion devices,** and we are therefore interested in their energy-conversion **efficiency.** Typical applications of electric machines as motors or generators must take into consideration the energy losses associated with these devices. Figure 15.3 represents the various loss mechanisms you must consider in

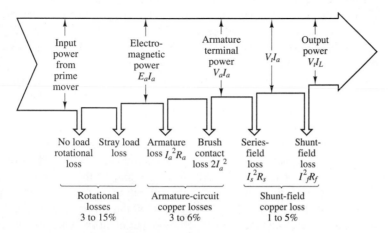

Figure 15.3(a) Generator losses, direct current

analyzing the efficiency of an electric machine for the case of direct current machines. It is important for you to keep in mind this conceptual flow of energy when analyzing electric machines. The sources of loss in a rotating machine can be separated into three fundamental groups: electrical (I^2R) losses, core losses, and mechanical losses.

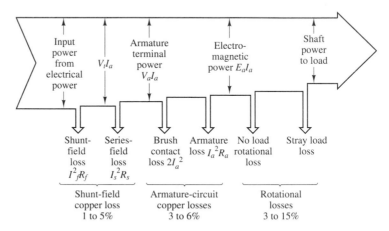

Figure 15.3(b) Motor losses, direct current

I^2R losses are usually computed on the basis of the DC resistance of the windings at 75°C; in practice, these losses vary with operating conditions. The difference between the nominal and actual I^2R loss is usually lumped under the category of *stray-load loss*. In direct current machines, it is also necessary to account for the *brush contact loss* associated with slip rings and commutators.

Mechanical losses are due to *friction* (mostly in the bearings) and *windage,* that is, the air drag force that opposes the motion of the rotor. In addition, if external devices (e.g., blowers) are required to circulate air through the machine for cooling purposes, the energy expended by these devices is also included in the mechanical losses.

Open-circuit core losses consist of hysteresis and eddy current losses, with only the excitation winding energized (see Chapter 14 for a discussion of hysteresis and eddy currents). Often these losses are summed to friction and windage losses to give rise to the *no-load rotational loss*. The latter quantity is useful if one simply wishes to compute efficiency. Since open-circuit core losses do not account for the changes in flux density caused by the presence of load currents, an additional magnetic loss is incurred that is not accounted for in this term. *Stray-load losses* are used to lump the effects of non-ideal current distribution in the windings and of the additional core losses just mentioned. Stray-load losses are difficult to determine exactly and are often assumed to be equal to 1.0 percent of the output power for DC machines; these losses can be determined by experiment in synchronous and induction machines.

The performance of an electric machine can be quantified in a number of ways. In the case of an electric motor, it is usually portrayed in the form of a graphical **torque-speed characteristic.** The torque-speed characteristic of a motor describes how the torque supplied by the machine varies as a function of the speed of rotation of the motor for steady speeds. As we shall see in later sections, the torque-speed curves vary in shape with the type of motor (DC, induction, synchronous) and are very useful in determining the performance of the motor when connected to a mechanical load. Figure 15.4 depicts the torque-speed curve of a hypothetical motor. We shall presently describe the essential elements of such a graphical representation of motor performance, and we shall later return to analyze the typical performance curve of each type of motor we encounter in our discussion. It is quite likely that in most engineering applications, the engineer is required to make a decision regarding the performance characteristics of the motor best suited to a specified task. However, it is probably not as likely that you will be involved in the design of an electric machine. Thus, we have chosen to present the material from a user's perspective, rather than from a designer's.

The first feature we note of the torque-speed characteristic is that it bears a strong resemblance to the i-v characteristics used in earlier chapters to represent the behavior of electrical sources. It should be clear that, according to this torque-speed curve, the motor is not an ideal source of torque (if it were, the curve would appear as a horizontal line across the speed range). One can readily see, for example, that the hypothetical motor represented by the curves of Figure 15.4 would produce maximum torque in the range of speeds between approximately 800 and 1,400 rev/min. What determines the actual speed of the motor (and therefore its output torque and power) is the torque-speed characteristic of the load connected to it, much as a resistive load determines the current drawn from a voltage source. In the figure, we display the torque-speed curve of a load, represented by the dashed line; the operating point of the motor-load pair is determined by the intersection of the two curves.

Another important observation pertains to the fact that the motor of Figure 15.4 produces a nonzero torque at zero speed. This fact implies that

Figure 15.4 Torque-speed curve for an electric motor

as soon as electric power is connected to the motor, the latter is capable of supplying a certain amount of torque; this zero-speed torque is called the **starting torque.** If the load the motor is connected to requires less than the starting torque the motor can provide, then the motor can accelerate the load, until the motor speed and torque settle to a stable value, at the operating point. The motor-load pair of Figure 15.4 would behave in the manner just described. However, there may well be circumstances in which a motor might not be able to provide a sufficient starting torque to overcome the static load torque that opposes its motion. Thus, we see that a torque-speed characteristic can offer valuable insight into the operation of a motor. As we proceed to discuss each type of machine in greater detail, we shall devote some time to the discussion of its torque-speed curve.

The most common means of conveying information regarding electric machines is the *nameplate*. Typical information conveyed by the nameplate is:

1. Type of device (e.g., DC motor, alternator)
2. Manufacturer
3. Rated voltage and frequency
4. Rated currents and volt-amperes
5. Rated speed and horsepower

The **rated voltage** is the voltage for which the machine was designed, and which will provide the desired magnetic flux. Operation at higher voltages will increase magnetic core losses, because of excessive core saturation. The **rated current** and **rated volt-amperes** are an indication of the typical current and power levels that will not cause undue overheating due to copper losses (I^2R losses) in the windings. These ratings are not absolutely precise, but they give an indication of the range of excitations for which the motor will perform without overheating. Peak power operation in a motor may exceed rated torque (horsepower) or currents by a substantial factor (up to as much as 6 or 7 times the rated value); however, continuous operation of the motor above the rated performance will cause the machine to overheat, and possibly to sustain damage. Thus, it is important to consider both peak and continuous power requirements when selecting a motor for a specific application. An analogous discussion is valid for the speed rating: while an electric machine may operate above rated speed for limited periods of time, the large centrifugal forces generated at high rotational speeds will eventually cause undesirable mechanical stresses, especially in the rotor windings, leading eventually even to self-destruction.

Another important feature of electric machines is the **regulation** of the machine speed or voltage, depending on whether it is used as a motor or as a generator, respectively. Regulation is the ability to maintain speed or voltage constant in the face of load variations. The ability to closely regulate speed in a motor or voltage in a generator is an important feature of electric machines; regulation is often improved by means of feedback control mechanisms, some of which will be briefly introduced in this chapter. We shall take the following definitions as being adequate for the intended purpose of this chapter:

$$\text{Speed regulation} = \frac{\text{Speed at no load} - \text{Speed at rated load}}{\text{Speed at rated load}} \quad \textbf{(15.4)}$$

$$\text{Voltage regulation} = \frac{\text{Voltage at no load} - \text{Voltage at rated load}}{\text{Voltage at rated load}} \quad \textbf{(15.5)}$$

Please note that the rated value is usually taken to be the nameplate value, and that the meaning of *load* changes depending on whether the machine is a motor, in which case the load is mechanical, or a generator, in which case the load is electrical.

Example 15.1 Regulation

The speed of a shunt DC motor drops from 1,800 rev/min at no load to 1,760 rev/min at rated load. Find the percentage of speed regulation.

Solution:

The percent speed regulation, SR%, is defined as follows:

$$\text{SR\%} = \frac{n_{nl} - n_{rl}}{n_{rl}} \times 100 = \frac{\omega_{nl} - \omega_{rl}}{\omega_{rl}} \times 100$$

where

$$n_{nl} = \text{speed at no load, rev/min}$$

$$\omega_{nl} = \text{speed at no load, rad/s}$$

$$n_{rl} = \text{speed at rated load, rev/min}$$

$$\omega_{rl} = \text{speed at rated load, rad/s}$$

The speed at no load is $n_{nl} = 1,800$ rev/min. The speed at rated load is $n_{rl} = 1,760$ rev/min. Thus,

$$\text{SR\%} = \frac{1,800 - 1,760}{1,760} \times 100 = 2.27\%$$

Example 15.2 Nameplate

The nameplate of a typical induction motor is shown in the table on page 808. The model number (sometimes abbreviated as MOD) uniquely identifies the motor to the manufacturer. It may be a style number, a model number, an identification number, or an instruction sheet.

The term *frame* (sometimes abbreviated as FR) refers principally to the physical size of the machine, as well as to certain construction features.

Ambient temperature (abbreviated as AMB, or MAX. AMB) refers to the maximum ambient temperature in which the motor is capable of operating. Operation of the motor in a higher ambient temperature may result in shortened motor life and reduced torque.

MODEL		19308 J-X				
TYPE		CJ4B		FRAME		324TS
VOLTS		230/460		°C AMB.		40
				INS. CL.		B
FRT. BRG		210SF		EXT. BRG		312SF
SERV FACT		1.0		OPER INSTR		C-517
PHASE	3	Hz	60	CODE	G	WDGS	1
H.P.		40				
R.P.M.		3565				
AMPS		106/53				
NEMA NOM.		EFF				
NOM. P.F.						
DUTY		CONT.		NEMA DESIGN		B

Insulation class (abbreviated as INS. CL.) refers to the type of insulation used in the motor. Most often used are class A (105°C) and class B (130°C).

The duty (DUTY), or time rating, denotes the length of time the motor is expected to be able to carry the rated load under usual service conditions. "CONT." means that the machine can be operated continuously.

The "CODE" letter sets the limits of starting kVA per horsepower for the machine. There are 19 levels, denoted by the letters A through V, excluding I, O, and Q.

Service factor (abbreviated as SERV FACT) is a term defined by NEMA (the National Electrical Manufacturers Association) as follows: "The service factor of a general-purpose alternating-current motor is a multiplier which, when applied to the rated horsepower, indicates a permissible horsepower loading which may be carried under the conditions specified for the service factor."

The voltage figure given on the nameplate refers to the voltage of the supply circuit to which the motor should be connected. Sometimes two voltages are given, for example, 230/460. In this case, the machine is intended for use on either a 230-V or a 460-V circuit. Special instructions will be provided for connecting the motor for each of the voltages.

The term "BRG" indicates the nature of the bearings supporting the motor shaft.

Example   15.3   Torque-Speed Curves

A variable-torque variable-speed motor has a torque output that varies directly with speed; hence, the horsepower output varies directly with the speed. Motors with this characteristic are commonly used with fans, blowers, and centrifugal pumps. Figure 15.5 shows typical torque-speed curves for this type of motor. Superimposed on the motor torque-speed curve is the torque-speed curve for a typical fan where the input power to the fan varies as the cube of the fan speed. Point A is the desired operating point, which could be determined graphically by plotting the load line and the motor torque-speed curve on the same graph, as illustrated in Figure 15.5.

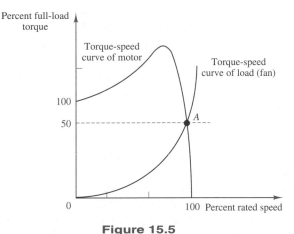

Figure 15.5

Basic Operation of All Rotating Machines

We have already seen in Chapter 14 how the magnetic field in electromechanical devices provides a form of coupling between electrical and mechanical systems. Intuitively, one can identify two aspects of this coupling, both of which play a role in the operation of electric machines: (1) magnetic attraction and repulsion forces generate mechanical torque, and (2) the magnetic field can induce a voltage in the machine windings (coils) by virtue of Faraday's law. Thus, we may think of the operation of an electric machine in terms of either a motor or a generator, depending on whether the input power is electrical and mechanical power is produced (motor action), or the input power is mechanical and the output power is electrical (generator action). Figure 15.6 illustrates the two cases graphically.

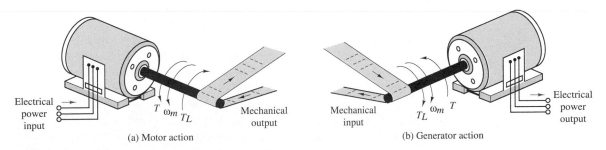

Figure 15.6 Generator and motor action in an electric machine

The coupling magnetic field performs a dual role, which may be explained as follows. When a current i flows through conductors placed in a magnetic field, a force is produced on each conductor, according to equation 15.1. If these conductors are attached to a cylindrical structure, a torque is generated, and if the structure is free to rotate, then it will rotate at an angular velocity, ω_m. As the conductors rotate, however, they move through a magnetic field and cut through flux lines, thus generating an electromotive force in opposition to the excitation. This emf is also called "counter" emf; it opposes the source of the current i. If, on the other hand, the rotating element of the machine is driven by a prime mover (for example, an internal combustion engine), then an emf is generated across the coil that is rotating in the magnetic field (the armature). If a load is connected to the armature, a current i will flow to the load, and this current flow will in turn cause a reaction torque on the armature that opposes the torque imposed by the prime mover.

You see, then, that for energy conversion to take place, two elements are required: (1) a coupling field, **B,** usually generated in the field winding; and (2) an armature winding that supports the load current, i, and the emf, e.

Magnetic Poles in Electric Machines

Before discussing the actual construction of a rotating machine, we should spend a few paragraphs to illustrate the significance of **magnetic poles** in an electric machine. In an electric machine, torque is developed as a consequence of magnetic forces of attraction and repulsion between magnetic poles on the stator and on the rotor; these poles produce a torque that accelerates the rotor and a reaction torque on the stator. Naturally, we would like a construction such that the torque generated as a consequence of the magnetic forces is continuous and in a constant direction. This can be accomplished if the number of rotor poles is equal to the number of stator poles. It is also important to observe that the number of poles must be even, since there have to be equal numbers of north and south poles.

Figure 15.7 depicts a two-pole machine in which the stator poles are constructed in such a way as to project closer to the rotor than to the stator structure. This type of construction is rather common, and poles constructed in this fashion are called **salient poles.** Note that the rotor could be constructed to have salient poles as well as the stator.

To understand magnetic polarity, we need to consider the direction of the magnetic field in a coil carrying current. Figure 15.8 shows how the *right-hand rule* can be employed to determine the direction of the magnetic flux. If one were to grasp the coil with the right hand, with the fingers curling in the direction of current flow, then the thumb would be pointing in the direction of the magnetic flux. As you may recall from the discussion in Chapter 14, magnetic flux is by convention viewed as entering the south pole and exiting from the north pole. Thus, to determine whether a magnetic

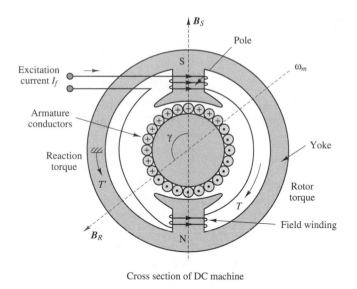

Cross section of DC machine

Figure 15.7 A two-pole machine

Figure 15.8 Right-hand rule

pole is north or south, we must consider the direction of the flux. Figure 15.9 shows a coil wound around a salient rotor pole. In this case, one can readily identify the direction of the magnetic flux and therefore the magnetic polarity of the pole by applying the right-hand rule, as illustrated in the figure.

Often, however, the coil windings are not arranged as simply as in the case of salient poles. In many machines, the windings are embedded in rectangular slots cut into the stator or rotor, so that the situation is similar to that of the stator depicted in Figure 15.10. This figure is a cross section in which the wire connections between "crosses" and "dots" have been cut away. In Figure 15.10, the dashed line indicates the axis of the stator flux according to the right-hand rule, indicating that the slotted stator in effect behaves like a pole pair. The north and south poles indicated in the figure are a consequence of the fact that the flux exits the bottom part of the structure (thus, the north pole indicated in the figure) and enters the top half of the structure (thus, the south pole). In particular, if you consider that the windings are arranged so that the current entering the right-hand side of the stator (to the right of the dashed line) flows through the back end of the stator and then flows outward from the left-hand side of the stator slots (left of the dashed line), you can visualize the windings in the slots as behaving in a manner similar to the coils of Figure 15.9, where the flux axis of Figure 15.10 corresponds to the flux axis of each of the coils of Figure 15.9. The actual circuit that permits current flow is completed by the front and back ends of the stator, where the wires are connected according to the pattern a-a', b-b', c-c', as depicted in the figure.

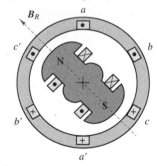

Figure 15.9 Magnetic field and currents in a salient rotor winding

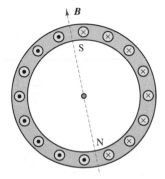

Figure 15.10 Magnetic field of stator

Another important consideration that facilitates understanding the operation of electric machines pertains to the use of AC currents. It should be apparent by now that if the current flowing into the slotted stator is alternating, the direction of the flux will also alternate, so that in effect the two poles will reverse polarity every time the current reverses direction, that is, every half-cycle of the sinusoidal current. Further, since the magnetic flux is approximately proportional to the current in the coil, as the amplitude of the current oscillates in a sinusoidal fashion, so will the flux density in the structure. Thus, *the magnetic field developed in the stator changes both spatially and in time*.

This property is typical of AC machines, where a *rotating magnetic field* is established by energizing the coil with an alternating current. As we shall see in the next section, the principles underlying the operation of DC and AC machines are quite different: in a direct-current machine, there is no rotating field, but a mechanical switching arrangement (the *commutator*) makes it possible for the rotor and stator magnetic fields to always align at right angles to each other.

DRILL EXERCISES

15.1   The percent speed regulation of a motor is 10 percent. If the full-load speed is 50π rad/s, find (*a*) the no-load speed in rad/s, and (*b*) the no-load speed in rev/min.

> Answer: (*a*) $\omega = 55\pi$ rad/s; (*b*) $n = 1650$ rev/min

15.2   The percent voltage regulation for a 250 V generator is 10 percent. Find the no-load voltage of the generator.

> Answer: 264 V

15.3   The nameplate of a three-phase induction motor indicates the following values:

H.P. = 10	Volt = 220 V
R.P.M. = 1750	Service factor = 1.15
Temperature rise = 60°C	Amp = 30 A

Find the rated torque, rated volt-amperes, and maximum continuous output power.

> Answer: $T = 40.7$ N-m; Volt-amperes = 11431 VA; $P_{max} = 11.5$ hp

15.4   A motor having the characteristic shown in Figure 15.4 is to drive a load; the load has a linear torque-speed curve and requires 150 percent of rated torque at 1500 rev/min. Find the operating point for this motor-load pair.

> Answer: 168.6% of rated torque; 1686 rev/min

15.2   DIRECT-CURRENT MACHINES

As explained in the introductory section, direct-current (DC) machines are easier to analyze than their AC counterparts, although their actual construction is made rather complex by the need to have a commutator, which reverses the direction of currents and fluxes to produce a net torque. The objective of this section is to describe the major construction features and the operation of direct-current machines, as well as to develop simple circuit models that are useful in analyzing the performance of this class of machines.

Physical Structure of DC Machines

A representative DC machine is depicted in Figure 15.7, with the magnetic poles clearly identified, for both the stator and the rotor. Figure 15.11 is a photograph of the same type of machine. Note the salient pole construction of the stator and the slotted rotor. As previously stated, the torque developed by the machine is a consequence of the magnetic forces between stator and rotor poles. This torque is maximum when the angle γ between the rotor and stator poles is 90°. Also, as you can see from the figure, in a DC machine the armature is usually on the rotor, and the field winding is on the stator.

Figure 15.11 Structure of a DC machine

To keep this torque angle constant as the rotor spins on its shaft, a mechanical switch, called a **commutator,** is configured so the current distribution in the rotor winding remains constant and therefore the rotor poles are consistently at 90°with respect to the fixed stator poles. In a DC machine, the magnetizing current is DC, so that there is no spatial alternation of the stator poles due to time-varying currents. To understand the operation of the commutator, consider the simplified diagram of Figure 15.12. In the figure, the brushes are fixed, and the rotor revolves at an angular velocity ω_m; the instantaneous position of the rotor is given by the expression $\theta = \omega_m t - \gamma$.

Figure 15.12 Rotor winding and commutator

The commutator is fixed to the rotor and is made up in this example of six segments that are made of electrically conducting material but are insulated from each other. Further, the rotor windings are configured so that they form six coils, connected to the commutator segments as shown in Figure 15.12.

As the commutator rotates counterclockwise, the rotor magnetic field rotates with it up to $\theta = 30°$. At that point, the direction of the current changes in coils L_3 and L_6 as the brushes make contact with the next segment. Now the direction of the magnetic field is $-30°$. As the commutator continues to rotate, the direction of the rotor field will again change from $-30°$ to $+30°$, and it will switch again when the brushes switch to the next pair of segments. In this machine, then, the torque angle, γ, is not always 90°, but can vary by as much as $\pm30°$; the actual torque produced by the machine would fluctuate by as much as ±14 percent, since the torque is proportional to $\sin \gamma$. As the number of segments increases, the torque fluctuation produced by the commutation is greatly reduced. In a practical machine, for example, one might have as many as 60 segments, and the variation of γ from 90° would be only $\pm3°$, with a torque fluctuation of less than 1 percent. Thus, the DC machine can produce a nearly constant torque (as a motor) or voltage (as a generator).

Configuration of DC Machines

In DC machines, the field excitation that provides the magnetizing current is occasionally provided by an external source, in which case the machine is said to be **separately excited** (Figure 15.13(a)). More often, the field excitation is derived from the armature voltage and the machine is said to be **self-excited.** The latter configuration does not require the use of a separate source for the field excitation and is therefore frequently preferred. If a machine is in the separately excited configuration, an additional source, V_f, is required. In the self-excited case, one method used to provide the field excitation is to connect the field in parallel with the armature; since the field winding typically has significantly higher resistance than the armature circuit (remember that it is the armature that carries the load current), this will not draw excessive current from the armature. Further, a series resistor can be added to the field circuit to provide the means for adjusting the field current independent of the armature voltage. This configuration is called a **shunt-connected** machine and is depicted in Figure 15.13(b). Another method for self-exciting a DC machine consists of connecting the field in series with the armature, leading to the **series-connected** machine, depicted in Figure 15.13(c); in this case, the field winding will support the entire armature current, and thus the field coil must have low resistance (and therefore relatively few turns). This configuration is rarely used for generators, since the generated voltage and the load voltage must always differ by the voltage drop across the field coil, which varies with the load current. Thus, a series generator would have poor (large) regulation. However, series-connected

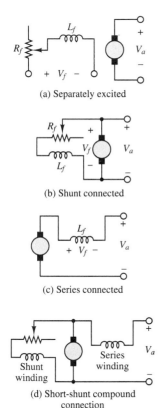

(a) Separately excited

(b) Shunt connected

(c) Series connected

(d) Short-shunt compound connection

(e) Long-shunt compound connection

Figure 15.13

motors are commonly used in certain applications, as will be discussed in a later section.

The third type of DC machine is the **compound-connected** machine, which consists of a combination of the shunt and series configurations. Figures 15.13(d) and (e) show the two types of connections, called the **short shunt** and the **long shunt,** respectively. Each of these configurations may be connected such that the series part of the field adds to the shunt part **(cumulative compounding)** or such that it subtracts **(differential compounding).**

DC Machine Models

As stated earlier, it is relatively easy to develop a simple model of a DC machine, which is well suited to performance analysis, without the need to resort to the details of the construction of the machine itself. This section will illustrate the development of such models in two steps. First, steady-state models relating field and armature currents and voltages to speed and torque are introduced; second, the differential equations describing the dynamic behavior of DC machines are derived.

When a field excitation is established, a magnetic flux, ϕ, is generated by the field current I_f. From equation 15.2, we know that the torque acting on the rotor is proportional to the product of the magnetic field and the current in the load-carrying wire; the latter current is the armature current, I_a (i_w, in equation 15.2). Assuming that, by virtue of the commutator, the torque angle, γ, is kept very close to 90°, and therefore $\sin \gamma = 1$, we obtain the following expression for the torque (in units of N-m) in a DC machine:

$$T = k_T \phi I_a \qquad \text{for } \gamma = 90° \tag{15.6}$$

You may recall that this is simply a consequence of the *Bli* law of Chapter 14. The mechanical power generated (or absorbed) is equal to the product of the machine torque and the mechanical speed of rotation, ω_m (in rad/s), and is therefore given by

$$P_m = \omega_m T = \omega_m k_T \phi I_a \tag{15.7}$$

Recall now that the rotation of the armature conductors in the field generated by the field excitation causes a **back emf,** E_b, in a direction that opposes the rotation of the armature. According to the *Blu* law (see Chapter 14), then, this back emf is given by the expression

$$E_b = k_a \phi \omega_m \tag{15.8}$$

where k_a is called the **armature constant** and is related to the geometry and magnetic properties of the structure. The voltage, E_b represents a countervoltage (opposing the DC excitation) in the case of a motor, and the generated voltage in the case of a generator. Thus, the electric power dissipated (or generated) by the machine is given by the product of the back emf and the armature current:

$$P_e = E_b I_a \tag{15.9}$$

The constants k_T and k_a in equations 15.6 and 15.8 are related to geometry factors, such as the dimension of the rotor and the number of turns in the armature winding; and to properties of materials such as the permeability of the magnetic materials. Note that in the ideal energy-conversion case, $P_m = P_e$, and therefore $k_a = k_T$. We shall in general assume such ideal conversion of electrical to mechanical energy (or vice versa) and will therefore treat the two constants as being identical: $k_a = k_T$. The constant k_a is given by

$$k_a = \frac{pN}{2\pi M} \tag{15.10}$$

where:

$p =$ number of magnetic poles

$N =$ number of conductors per coil

$M =$ number of parallel paths in armature winding

(a) Motor reference direction

An important observation concerning the units of angular speed must be made at this point. The equality between the constants k_a and k_T in equations 15.6 and 15.8 results from the choice of consistent units, namely, volts and amperes for the electrical quantities, and newton-meters and radians per second for the mechanical quantities. You should be aware that it is fairly common practice to refer to the speed of rotation of an electric machine in units of revolutions per minute (rev/min).[1] In this book, we shall uniformly use the symbol n to denote angular speed in rev/min; the following relationship should be committed to memory:

$$n \text{ (rev/min)} = \frac{60}{2\pi} \omega_m \text{ (rad/s)} \tag{15.11}$$

If the speed is expressed in rev/min, the armature constant changes as follows:

$$E_b = k'_a \phi n \tag{15.12}$$

where

$$k'_a = \frac{pN}{60M} \tag{15.13}$$

(b) Generator reference direction

Figure 15.14 Electrical circuit model of a separately excited DC machine

Having introduced the basic equations relating torque, speed, voltages, and currents in electric machines, we may now consider the interaction of these quantities in a DC machine at steady state, that is, operating at constant speed and field excitation. Figure 15.14 depicts the electrical circuit model of a separately excited DC machine, illustrating both motor and generator action. It is very important to note the reference direction of armature current flow, and of the developed torque, in order to make a distinction between the two modes of operation. The field excitation is shown as a voltage, V_f, generating the field current, I_f, that flows through a variable resistor, R_f,

---

[1] Note that the abbreviation RPM, although certainly familiar to the reader, is not a standard unit, and its use should be discouraged.

and through the field coil, L_f. The variable resistor permits adjustment of the field excitation. The armature circuit, on the other hand, consists of a voltage source representing the back emf, E_b, the armature resistance, R_a, and the armature voltage, V_a. This model is appropriate both for motor and for generator action. When $V_a < E_b$, the machine acts as a generator (I_a flows out of the machine). When $V_a > E_b$, the machine acts as a motor (I_a flows into the machine). Thus, according to the circuit model of Figure 15.14, the operation of a DC machine at steady state (i.e., with the inductors in the circuit replaced by short circuits) is described by the following equations:

$$I_f = \frac{V_f}{R_f} \qquad \text{and} \qquad V_a = R_a I_a + E_b \qquad \text{(motor action)}$$

$$I_f = \frac{V_f}{R_f} \qquad \text{and} \qquad V_a = -R_a I_a + E_b \qquad \text{(generator action)}$$

$$\text{(15.14)}$$

Equation pair 15.14 together with equations 15.6 and 15.8 may be used to determine the steady-state operating condition of a DC machine.

The circuit model of Figure 15.14 permits the derivation of a simple set of differential equations that describe the *dynamic* analysis of a DC machine. The dynamic equations describing the behavior of a separately excited DC machine are as follows:

$$V_a(t) = I_a(t)R_a + L_a\frac{dI_a(t)}{dt} + E_b(t) \qquad \text{(armature circuit)}$$

$$V_f(t) = I_f(t)R_f + L_f\frac{dI_f(t)}{dt} \qquad \text{(field circuit)}$$

$$\text{(15.15)}$$

These equations can be related to the operation of the machine in the presence of a load. If we assume that the motor is rigidly connected to an inertial load with moment of inertia J and that the friction losses in the load are represented by a viscous friction coefficient, d, then the torque developed by the machine (in the motor mode of operation) can be written as follows:

$$T(t) = T_L + d\omega_m(t) + J\frac{d\omega_m(t)}{dt} \qquad \text{(15.16)}$$

where T_L is the load torque. T_L is typically either constant or some function of speed, ω_m, in a motor. In the case of a generator, the load torque is replaced by the torque supplied by a prime mover, and the machine torque, $T(t)$, opposes the motion of the prime mover, as shown in Figure 15.14. Since the machine torque is related to the armature and field currents by equation 15.6, equations 15.15 and 15.16 are coupled to each other; this coupling may be expressed as follows:

$$T(t) = T_L + d\omega_m(t) + J\frac{d\omega_m(t)}{dt} \qquad \text{(15.17)}$$

or

$$k_a\phi I_a(t) = T_L + d\omega_m(t) + J\frac{d\omega_m(t)}{dt} \qquad \text{(15.18)}$$

The dynamic equations described in this section apply to any DC machine. In the case of a *separately excited* machine, a further simplification is possible, since the flux is established by virtue of a separate field excitation, and therefore

$$\phi = \frac{N_f}{\mathcal{R}} I_f = k_f I_f \qquad (15.19)$$

where N_f is the number of turns in the field coil, $\mathcal{R}$ is the reluctance of the structure, and I_f is the field current.

Magnetization Curve (Open-Circuit Characteristic)

To analyze the performance of a DC machine, it would be useful to obtain an open-circuit characteristic capable of predicting the voltage generated when the machine is driven at a constant speed ω_m by a prime mover. The common arrangement is to drive the machine at rated speed by means of a prime mover (or an electric motor). Then, with no load connected to the armature terminals, the armature voltage is recorded as the field current is increased from zero to some value sufficient to produce an armature voltage greater than the rated voltage. Since the load terminals are open-circuited, $I_a = 0$, and $E_b = V_a$, and since $k_a\phi = E_b/\omega_m$, the magnetization curve makes it possible to determine the value of $k_a\phi$ corresponding to a given field current, I_f, for the rated speed.

Figure 15.15 depicts a typical magnetization curve. Note that the armature voltage is nonzero even when no field current is present. This phenomenon is due to the *residual magnetization* of the iron core. The dashed lines in Figure 15.15 are called **field resistance curves** and are a plot of the voltage that appears across the field winding plus rheostat (variable resistor; see Figure 15.14) versus the field current, for various values of field winding plus rheostat resistance. Thus, the slope of the line is equal to the total field circuit resistance, R_f.

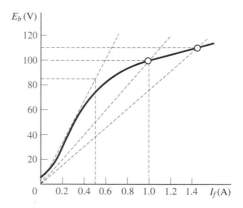

Figure 15.15 DC machine magnetization curve

Analysis of Direct-Current Generators

The operation of a DC generator may be readily understood with reference to the magnetization curve of Figure 15.15. As soon as the armature is connected across the shunt circuit consisting of the field winding and the rheostat, a current will flow through the winding, and this will in turn act to increase the emf across the armature. This **build-up** process continues until the two curves meet, that is, until the current flowing through the field winding is exactly that required to induce the emf. By changing the rheostat setting, the operating point at the intersection of the two curves can be displaced, as shown in Figure 15.15, and the generator can therefore be made to supply different voltages. The following examples illustrate the operation of the separately excited DC generator.

Example   15.4   Separately Excited DC Generator

A separately excited DC generator is rated at 100 V, 100 A, and 1,000 rev/min. The magnetization curve is shown in Figure 15.15. Some parameters of interest for the generator are as follows:

$$R_a = 0.14 \ \Omega \qquad V_f = 100 \ V \qquad R_f = 100 \ \Omega$$

1. If the prime mover is driving the generator at 800 rev/min, what is the no-load terminal voltage, V_a?
2. If a 1-Ω load is connected to the generator, what is the generated voltage?

Solution:

1. The field current in the machine is

$$I_f = \frac{V_f}{R_f} = \frac{100}{100} = 1 \ A$$

From the magnetization curve, it can be seen that this field current will produce 100 V at a speed of 1,000 rev/min. Since this generator is actually running at 800 rev/min, the induced emf may be found by assuming a linear relationship between speed and emf. This approximation is reasonable, provided that the departure from the nominal operating condition is small. Let n_0 and E_{b0} be the nominal speed and emf, respectively (i.e., 1,000 rev/min and 100 V); then,

$$\frac{E_b}{E_{bo}} = \frac{n}{n_o}$$

and therefore

$$E_b = \frac{n}{n_o} E_{bo} = \frac{800}{1,000} \times 100 = 80 \ V$$

The open-circuit (output) terminal voltage of the generator is equal to the emf from the circuit model of Figure 15.14; therefore:

$$v_a = E_b = 80 \ V$$

2.   When a load resistance is connected to the circuit (the practical situation), the terminal (or load) voltage is no longer equal to E_b, since there will be a voltage drop across the armature winding resistance. The armature (or load) current may be determined from the expression

$$i_a = i_L = \frac{E_b}{R_a + R_L} = \frac{80}{0.14 + 1} = 70.2 \text{ A}$$

where $R_L = 1 \ \Omega$ is the load resistance. The terminal (load) voltage is therefore given by

$$v_L = i_L R_L = 70.2 \times 1 = 70.2 \text{ V}$$

Example   15.5

A 1,000-kW, 2,000-V, 3,600-rev/min separately excited DC generator has an armature circuit resistance of 0.1 Ω. The flux per pole (ϕ) is 0.5 Wb. Find

1.   The induced voltage.
2.   The machine constant.
3.   The torque developed at the rated conditions.

Solution:

1.   The armature current may be found by observing that the rated power is equal to the product of the terminal (load) voltage and the armature (load) current; thus,

$$i_a = \frac{P_{\text{rated}}}{v_L} = \frac{1,000 \times 10^3}{2,000} = 500 \text{ A}$$

The generated voltage is equal to the sum of the terminal voltage and the voltage drop across the armature resistance (see Figure 15.14):

$$E_b = v_a + i_a R_a = 2,000 + 500 \times 0.1 = 2,050 \text{ V}$$

2.   The speed of rotation of the machine in units of rad/s is

$$\omega_m = \frac{2\pi n}{60} = \frac{2\pi \times 3,600}{60} = 377 \text{ rad/s}$$

Thus, the machine constant is found to be

$$k_a = \frac{E_b}{\phi \omega_m} = \frac{2,050}{0.5 \times 377} = 10.875 \frac{\text{V-s}}{\text{Wb-rad}}$$

3.   The torque developed is found from equation 15.6:

$$T = k_a \phi I_a = 10.875 \times 0.5 \times 500 = 2,718.8 \text{ N-m}$$

Since the compound-connected generator contains both a shunt and a series field winding, it is the most general configuration, and the most useful for developing a circuit model that is as general as possible. In the following discussion, we shall consider the so-called short-shunt, compound-connected generator, in which the flux produced by the series winding adds to that of the shunt winding. Figure 15.16 depicts the equivalent circuit

Figure 15.16 Compound generator circuit model

for the compound generator; circuit models for the shunt generator and for the rarely used series generator can be obtained by removing the shunt or series field winding element, respectively. In the circuit of Figure 15.16, the generator armature has been replaced by a voltage source corresponding to the induced emf and a series resistance, R_a, corresponding to the resistance of the armature windings. The equations describing the DC generator at steady state (i.e., with the inductors acting as short circuits) are:

$$E_b = k_a \phi \omega_m \text{ V} \tag{15.20}$$

$$T = \frac{P}{\omega_m} = \frac{E_b I_a}{\omega_m} = k_a \phi I_a \text{ N-m} \tag{15.21}$$

$$v_L = E_b - I_a R_a - I_S R_S \tag{15.22}$$

$$I_a = I_S + I_f \tag{15.23}$$

Note that in the circuit of Figure 15.16, the load and armature voltages are not equal, in general, because of the presence of a series field winding, represented by the resistor R_S and by the inductor L_S where the subscript "S" stands for "series." The expression for the armature emf is dependent on the air-gap flux, ϕ, to which the series and shunt windings in the compound generator both contribute, according to the expression

$$\phi = \phi_{\text{sh}} \pm \phi_S = \phi_{\text{sh}} \pm k_S I_a \tag{15.24}$$

Example   15.6

In many practical cases, it is not actually necessary to know the armature constant and the flux separately, but it is sufficient to know the value of the product $k_a \phi$. For example, suppose that the armature resistance of a DC machine is known and that, given a known field excitation, the armature current, load voltage, and speed of the machine can be measured. Then, the product $k_a \phi$ may be determined from equation 15.20, as follows:

$$k_a \phi = \frac{E_b}{\omega_m} = \frac{v_L - I_a(R_a + R_S)}{\omega_m}$$

where V_L, I_a, and ω_m are measured quantities for given operating conditions.

DRILL EXERCISES

15.5   A 24-coil, 2-pole DC generator has 16 turns per coil in its armature winding. The field excitation is 0.05 Wb per pole, and the armature angular velocity is 180 rad/s. Find the machine constant and the total induced voltage.

> Answer: $k_a = 122.23$; $E_b = 1100$ V

15.6   A 1000 kW, 1000 V, 2400 rev/min separately excited DC generator has an armature circuit resistance of 0.04 Ω. The flux per pole is 0.4 Wb. Find: (*a*) the induced voltage; (*b*) the machine constant; and (*c*) the torque developed at the rated conditions.

> Answer: (*a*) $E_b = 1040$ V; (*b*) $k_a = 10.34$; (*c*) $T = 4136$ N-m

15.7   A 100 kW, 250 V shunt generator has a field circuit resistance of 50 Ω and an armature circuit resistance of 0.05 Ω. Find: (*a*) the full load line current flowing to the load; (*b*) the field current; (*c*) the armature current; and (*d*) the full load generator voltage.

> Answer: (*a*) 400 A; :(*b*) 5 A; (*c*) 405 A; (*d*) 270.25 V

15.3   DIRECT-CURRENT MOTORS

DC motors are widely used in applications requiring accurate speed control—for example, in servo systems. Having developed a circuit model and analysis methods for the DC generator, we can rather straightforwardly extend these results to DC motors, since these are in effect DC generators with the roles of input and output reversed. Once again, we shall analyze the motor by means of both its magnetization curve and a circuit model. It will be useful to begin our discussion by referring to the schematic diagram of a cumulatively compounded motor, as shown in Figure 15.17. The choice of the compound-connected motor is the most convenient, since its model can be used to represent either a series or a shunt motor with minor modifications.

Figure 15.17 Equivalent circuit of a cumulatively compounded motor

The equations that govern the behavior of the DC motor follow and are similar to those used for the generator. Note that the only differences between these equations and those that describe the DC generator appear in the last two equations in the group, where the source voltage is equal to the *sum* of the emf and the voltage drop across the series field resistance and armature resistance, and where the source current now equals the *sum* of the field shunt and armature series currents.

$$E_b = k_a \phi \omega_m \tag{15.25}$$

$$T = k_a \phi I_a \tag{15.26}$$

$$V_s = E_b + I_a R_a + I_s R_s \tag{15.27}$$

$$I_s = I_f + I_a \tag{15.28}$$

Note that in these equations we have replaced the symbols V_L and I_L, used in the generator circuit model to represent the generator load current and voltage, with the symbols V_s and I_s, indicating the presence of an external source.

Speed-Torque and Dynamic Characteristics of DC Motors

The Shunt Motor

In a shunt motor (similar to the configuration of Figure 15.17, but with the series field short-circuited), the armature current is found by dividing the net voltage across the armature circuit (source voltage minus back emf) by the armature resistance:

$$I_a = \frac{V_s - k_a \phi \omega_m}{R_a} \tag{15.29}$$

An expression for the armature current may also be obtained from equation 15.26, as follows:

$$I_a = \frac{T}{k_a \phi} \tag{15.30}$$

It is then possible to relate the torque requirements to the speed of the motor by substituting equation 15.29 in equation 15.30:

$$\frac{T}{k_a \phi} = \frac{V_s - k_a \phi \omega_m}{R_a} \tag{15.31}$$

Equation 15.31 describes the steady-state torque-speed characteristic of the shunt motor. To understand this performance equation, we observe that if V_s, k_a, ϕ, and R_a are fixed in equation 15.31 (the flux is essentially constant in the shunt motor for a fixed V_s), then the speed of the motor is directly related to the armature current. Now consider the case where the load applied to the motor is suddenly increased, causing the speed of the motor to drop. As the speed decreases, the armature current increases, according to equation 15.29. The excess armature current causes the motor to develop additional torque, according to equation 15.30, until a new equilibrium is reached between the higher armature current and developed torque and the lower speed of rotation. The equilibrium point is dictated by the balance of mechanical and electrical power, in accordance with the relation

$$E_b I_a = T \omega_m \tag{15.32}$$

Thus, the shunt DC motor will adjust to variations in load by changing its speed to preserve this power balance. The torque-speed curves for the shunt motor may be obtained by rewriting the equation relating the speed to the armature current:

$$\omega_m = \frac{V_s - I_a R_a}{k_a \phi} \tag{15.33}$$

To interpret equation 15.33, one can start by considering the motor operating at rated speed and torque. As the load torque is reduced, the armature current will also decrease, causing the speed to increase in accordance with equation 15.33. The increase in speed depends on the extent of the voltage drop across the armature resistance, $I_a R_a$. The change in speed will be on the same order of magnitude as this drop; it typically takes values around 10 percent. This corresponds to a relatively good speed regulation, which is an attractive feature of the shunt DC motor (recall the discussion of regulation in Section 15.1). Normalized torque and speed curves for the shunt motor are shown in Figure 15.19. Note that, over a reasonably broad range of torques, up to rated value, the curve is relatively flat, indicating that the DC shunt motor acts as a reasonably constant-speed motor.

The dynamic behavior of the shunt motor is described by equations 15.15 through 15.18, with the additional relation

$$I_a(t) = I_s(t) - I_f(t) \tag{15.34}$$

Compound Motors

It is interesting to compare the performance of the shunt motor with that of the compound-connected motor; the comparison is easily made if we recall that a series field resistance appears in series with the armature resistance and that the flux is due to the contributions of both series and shunt fields. Thus, the speed equation becomes

$$\omega_m = \frac{V_s - I_a(R_a + R_S)}{k_a(\phi_{sh} \pm \phi_S)} \tag{15.35}$$

where

+ is for a cumulatively compounded motor.

− is for a differentially compounded motor.

ϕ_{sh} is the flux set up by the shunt field winding, assuming that it is constant.

ϕ_S is the flux set up by the series field winding, $\phi_S = k_S I_a$.

For the cumulatively compound motor, two effects are apparent: the flux is increased by the presence of a series component, ϕ_S; and the voltage drop due to I_a in the numerator term is increased by an amount proportional to the resistance of the series field winding, R_S. As a consequence, when the load to the motor is reduced, the numerator increases more dramatically than in the case of the shunt motor, because of the corresponding decrease in armature current, while at the same time the series flux decreases. Each of these effects causes the speed to increase; therefore, it stands to reason that the speed regulation of the compound-connected motor is poorer than that of the shunt motor. Normalized torque and speed curves for the compound motor (both differential and cumulative connections) are shown in Figure 15.19.

The differential equation describing the behavior of a compound motor differs from that for the shunt motor in having additional terms due to the series field component:

$$V_s = E_b(t) + I_a(t)R_a + L_a\frac{dI_a(t)}{dt} + I_s(t)R_S + L_S\frac{dI_s(t)}{dt}$$

$$= V_a(t) + I_s(t)R_S + L_S\frac{dI_s(t)}{dt} \tag{15.36}$$

The differential equation for the field circuit can be written as

$$V_a = I_f(t)(R_f + R_x) + L_f\frac{dI_f(t)}{dt} \tag{15.37}$$

We also have the following basic relations:

$$I_a(t) = I_s(t) - I_f(t) \tag{15.38}$$

and

$$E_b(t) = k_a I_a(t)\omega_m(t) \qquad \text{and} \qquad T(t) = k_a\phi I_a(t) \tag{15.39}$$

Series Motors

The series motor (see Figure 15.13(c)) behaves somewhat differently from the shunt and compound motors because the flux is established solely by virtue of the series current flowing through the armature. It is relatively simple to derive an expression for the emf and torque equations for the series motor if we approximate the relationship between flux and armature current by assuming that the motor operates in the linear region of its magnetization curve. Then we can write

$$\phi = k_S I_a \tag{15.40}$$

and the emf and torque equations become

$$E_b = k_a\omega_m\phi = k_a\omega_m k_S I_a \tag{15.41}$$

$$T = k_a\phi I_a = k_a k_S I_a^2 \tag{15.42}$$

The circuit equation for the series motor becomes

$$V_s = E_b + I_a(R_a + R_S) = (k_a\omega_m k_S + R_T)I_a \tag{15.43}$$

where R_a is the armature resistance, R_S is the series field winding resistance, and R_T is the total series resistance. From equation 15.43, we can solve for I_a and substitute in the torque expression (equation 15.42) to obtain the following torque-speed relationship:

$$T = k_a k_S \frac{V_s^2}{(k_a\omega_m k_S + R_T)^2} \tag{15.44}$$

which indicates the inverse squared relationship between torque and speed in the series motor. This expression describes a behavior that can, under certain conditions, become unstable. Since the speed increases when the load torque is reduced, one can readily see that if one were to disconnect the load altogether, the speed would tend to increase to dangerous values. To prevent excessive speeds, series motors are always mechanically coupled to the load. This feature is not necessarily a drawback, though, because series motors can develop very high torque at low speeds, and therefore can serve very well for traction-type loads (e.g., conveyor belts or vehicle propulsion systems). Torque and speed curves for the series motor are also shown in Figure 15.19.

The differential equation for the armature circuit of the motor can be given as

$$\begin{aligned}
V_s &= I_a(t)(R_a + R_S) + L_a\frac{dI_a(t)}{dt} + L_S\frac{dI_a(t)}{dt} + E_b \\
&= I_a(t)(R_a + R_S) + L_a\frac{dI_a(t)}{dt} + L_S\frac{dI_a(t)}{dt} + k_a k_S I_a \omega_m
\end{aligned} \tag{15.45}$$

Permanent-Magnet DC Motors

Permanent-magnet (PM) DC motors have become increasingly common in applications requiring relatively low torques and efficient use of space. The construction of PM direct-current motors differs from that of the motors considered thus far in that the magnetic field of the stator is produced by suitably located poles made of magnetic materials. Thus, the basic principle of operation, including the idea of commutation, is unchanged with respect to the wound-stator DC motor. What changes is that there is no need to provide a field excitation, whether separately or by means of the self-excitation techniques discussed in the preceding sections. Therefore, the PM motor is intrinsically simpler than its wound-stator counterpart.

The equations that describe the operation of the PM motor follow. The torque produced is related to the armature current by a torque constant, k_{PM}, which is determined by the geometry of the motor:

$$T = k_{TPM}I_a \tag{15.46}$$

As in the conventional DC motor, the speed rotation of the rotor produces the usual counter or back emf, E_b, which is linearly related to speed by a voltage constant, k_{aPM}:

$$E_a = k_{aPM}\omega_m \tag{15.47}$$

The equivalent circuit of the PM motor is particularly simple, since we need not model the effects of a field winding. Figure 15.18 shows the circuit model and the torque-speed curve of a PM motor.

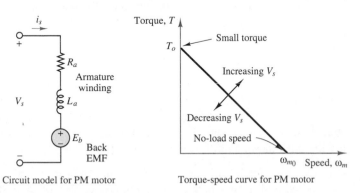

Circuit model for PM motor   Torque-speed curve for PM motor

Figure 15.18 Circuit model and torque-speed curve of PM motor

We can use the circuit model of Figure 15.18 to predict the torque-speed curve shown in the same figure, as follows. From the circuit model, for a constant speed (and therefore constant current), we may consider the inductor a short circuit and write the equation

$$
\begin{aligned}
V_s &= I_a R_a + E_b \\
&= I_a R_a + k_{aPM}\omega_m \\
&= \frac{T}{k_{TPM}}R_a + k_{aPM}\omega_m
\end{aligned} \tag{15.48}
$$

thus obtaining the equations relating speed and torque:

$$\omega_m = \frac{V_s}{k_{TPM}} - \frac{TR_a}{k_{aPM}k_{TPM}} \tag{15.49}$$

and

$$T = \frac{V_s}{R_a}k_{TPM} - \frac{\omega_m}{R_a}k_{aPM}k_{TPM} \tag{15.50}$$

From these equations, one can extract the stall torque, that is, the zero-speed torque:

$$T_0 = \frac{V_s}{R_a}k_{TPM} \tag{15.51}$$

and the no-load speed:

$$\omega_{m0} = \frac{V_s}{k_{aPM}} \tag{15.52}$$

Under dynamic conditions, assuming an inertia plus viscous friction load, the torque produced by the motor can be expressed as

$$k_{TPM}I_a(t) = T_{\text{load}}(t) + d\omega_m(t) + J\frac{d\omega_m(t)}{dt} \tag{15.53}$$

The differential equation for the armature circuit of the motor can be given as

$$\begin{aligned}
V_s &= I_a(t)R_a + L_a\frac{dI_a(t)}{dt} + E_b \\
&= I_a(t)R_a + L_a\frac{dI_a(t)}{dt} + k_{aPM}\omega_m(t)
\end{aligned} \tag{15.54}$$

The fact that the air-gap flux is constant in a permanent-magnet DC motor makes its characteristics somewhat different from those of the wound DC motor. A direct comparison of PM and wound-field DC motors reveals the following advantages and disadvantages of each configuration:

1. PM motors are smaller and lighter than wound motors for a given power rating, and have a high starting torque because of the reduced armature reaction effect.[2] Further, their efficiency is greater because there are no field winding losses.

2. An additional advantage of PM motors is their essentially linear speed-torque characteristic, which makes analysis (and control) much easier. Reversal of rotation is also accomplished easily, by reversing the polarity of the source.

3. A major disadvantage of PM motors is that they can become demagnetized by exposure to excessive magnetic fields, application of excessive voltage, or operation at excessively high or low temperatures.

4. A less obvious drawback of PM motors is that their performance is subject to greater variability from motor to motor than is the case for wound motors, because of variations in the magnetic materials.

In summary, four basic types of DC motors are commonly used. Their principal operating characteristics are summarized as follows, and their general torque-speed characteristics are depicted in Figure 15.19, assuming motors with identical voltage, power, and speed ratings.

>   *Shunt wound motor:* Field connected in parallel with the armature. With constant armature voltage and field excitation, the motor has good speed regulation (flat speed-torque characteristic).

---

[2] Armature current produces a field due to the armature winding that distorts the field produced by the excitation winding, leading to an effect known as *armature reaction*. This effect may be significant for high armature currents but will not be included in the analysis used here.

Compound wound motor: Field winding has both series and shunt components. This motor offers better starting torque than the shunt motor, but worse speed regulation.

Series wound motor: The field winding is in series with the armature. The motor has very high starting torque and poor speed regulation. It is useful for low-speed, high-torque applications.

Permanent-magnet motor: Field windings are replaced by permanent magnets. The motor has excellent starting torque, with speed regulation somewhat worse than that of the compound wound motor.

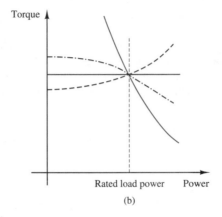

Figure 15.19 DC motor operating characteristics

Example   15.7   DC Shunt Motor

A four-pole, 3-hp, 240-VDC shunt motor has 1,000 conductors in the armature (i.e., $N = 1,000$, $M = 4$, as defined in equation 15.10). The load current is 30 A. The armature resistance is 0.6 Ω and the flux is 20 mWb. The field winding current is 1.4 A. Find the speed and torque.

Solution:

The output power at full load in watts may be computed using the conversion factor 1 hp = 746 W:

$$P_{\text{out}} = 3 \times 746 = 2,238 \text{ W}$$

The armature current may then be computed as the difference between the source current and the field current:

$$I_a = I_s - I_f = 30 - 1.4 = 28.6 \text{ A}$$

Thus, the no-load voltage is

$$E_b = V_s - I_a R_a = 240 - 28.6 \times 0.6 = 222.84 \text{ V}$$

and the armature constant is found from equation 15.10 to be

$$k_a = \frac{pN}{2\pi M} = \frac{4 \times 1{,}000}{2\pi \times 4} = \frac{500}{\pi} = 159.15$$

The speed may then be computed as follows:

$$\omega_m = \frac{E_a}{k_a \phi} = \frac{222.84}{159.15 \times 20 \times 10^{-3}} = 70 \text{ rad/s}$$

The torque developed by the motor is found as the ratio of output power to speed of rotation:

$$T = \frac{P_{\text{out}}}{\omega_m} = \frac{2{,}238}{70} = 31.97 \text{ N-m}$$

Example   15.8   DC Shunt Motor

A DC machine is used as a shunt motor. The magnetization curve for the machine used is shown in Figure 15.20. The size of the motor and its characteristics are typical of the small motors used in hand-held tools, such as power ratchets and screwdrivers. The motor is known to draw 8 A at full load and to have a speed of 120 rev/min. The armature winding resistance is estimated to be 0.2 Ω, and the battery voltage is 7.2 V. The field winding has 200 turns. The circuit configuration is shown in Figure 15.21.

Figure 15.20 Magnetization curve for a small DC motor

Figure 15.21 Shunt motor configuration

1. Estimate the current, i_f, required for full-load operation.
2. Determine the no-load speed.
3. Plot the speed-torque curve in the range from no-load torque to rated torque.
4. What is the power output at rated load?

Solution:

1. To find the field current, we must find the generated emf. Writing KVL around the armature circuit, we obtain

$$V_s = E_b + I_a R_a$$

$$E_b = V_s - I_a R_a = 7.2 - 8(0.2) = 5.6 \text{ V}$$

Having found the back emf, we can find the field current from the magnetization curve. At $E_b = 5.6$ V, we find that the field current is

$$I_f = 0.6 \text{ A}$$

2. To obtain the no-load speed, we use the equations

$$E_b = k_a \phi \frac{2\pi n}{60} \qquad T = k_a \phi I_a$$

leading to

$$V_s = I_a R_a + E_b = I_a R_a + k_a \phi \frac{2\pi}{60} n$$

or

$$n = \frac{V_s - I_a R_a}{k_a \phi (2\pi/60)}$$

At no load, the torque is zero, and we see that the current I_a must also be zero in the torque equation ($T = k_a \phi I_a$). Thus, the motor speed at no load is given by

$$n_{\text{no-load}} = \frac{V_s}{k_a \phi (2\pi/60)}$$

We can obtain an expression for $k_a \phi$ knowing that, at full load,

$$E_b = 5.6 \text{ V} = k_a \phi \frac{2\pi n}{60}$$

so that

$$k_a \phi = E_b \left(\frac{60}{2\pi n} \right) = 5.6 \left(\frac{60}{2\pi(120)} \right) = 0.44563 \frac{\text{V-s}}{\text{rad}}$$

Finally, we may solve for the no-load speed, in rev/min:

$$n_{\text{no-load}} = \frac{V_s}{k_a \phi (2\pi/60)} = \frac{7.2}{(0.44563)(2\pi/60)}$$

$$= 154.3 \text{ rev/min}$$

3. The torque at rated speed may be found as follows:

$$T_{\text{full-load}} = k_a \phi I_a = (0.44563)(8) = 3.565 \text{ N-m}$$

Now we have the two points necessary to construct the torque-speed curve for this motor, which is shown in Figure 15.22.

4. The power is related to the torque by the frequency of the shaft:

$$P = T\omega_m = (3.565)\left(\frac{120}{60} \right) 2\pi = 44.8 \text{ W}$$

or, equivalently,

$$P = 44.8 \text{ W} \times \frac{1}{746} \frac{\text{hp}}{\text{W}} = 0.06 \text{ hp}$$

Figure 15.22

Example   15.9   DC Series Motor

A 10-hp, 115-VDC series motor draws 40 A when operating at a full-load speed of 1,800 rev/min. If the supply current changes to 60 A, find the torque developed by the motor when it operates within its linear region.

Solution:

Within the linear region of operation, the flux per pole is directly proportional to the current in the field winding. That is,

$$\phi = k_S I_a$$

The full-load speed is

$$n = 1,800 \text{ rev/min}$$

or

$$\omega_m = \frac{2\pi n}{60} = 60\pi \text{ rad/s}$$

Rated output power is

$$P_{\text{out}} = 10 \times 746 = 7,460 \text{ W}$$

and full-load torque is

$$T_{40A} = \frac{7,460}{60\pi} = 39.58 \text{ N-m}$$

Thus, the machine constant may be computed from the torque equation for the series motor:

$$T = k_a k_S I_a^2 = K I_a^2$$

Thus, at full load,

$$K = k_a k_S = \frac{39.58}{40^2} = 0.0247$$

and we can compute the torque developed for a 60-A supply current to be

$$T_{60A} = 0.0247 \times 60^2 = 88.92 \text{ N-m}$$

Example   15.10   DC Motor Dynamic Response

The aim of this example is to illustrate the computation of a permanent-magnet DC motor dynamic response. Assume, for simplicity, that the mechanical friction parameter, d, in equation 15.17 is negligible and that the two torque constants for the PM motor are equal, that is, $k_{TPM} = k_{aPM} = k_{PM}$. Find the dynamic equations of a DC motor.

Solution:

To determine the dynamic response of the motor, two balance equations are needed; the first describes the mechanical behavior of the motor:

$$T = T_{\text{load}} + J\frac{d\omega_m(t)}{dt} \tag{15.55}$$

while the second is related to the electrical performance:

$$V_L = E_b + R_a I_a(t) + L_a \frac{dI_a(t)}{dt} \tag{15.56}$$

These two equations are related to each other by the characteristic equations for the permanent magnet DC motor:

$$E_b = k_{PM} \omega_m(t) \tag{15.57}$$

and

$$T = k_{PM} I_a(t) \tag{15.58}$$

When these two relations are substituted in the mechanical and electrical balance equations, the following differential equations are obtained:

$$k_{PM} I_a(t) = T_{\text{load}} + J \frac{d\omega_m(t)}{dt} \tag{15.59}$$

and

$$V_L = k_{PM} \omega_m(t) + R_a I_a(t) + L_a \frac{dI_a(t)}{dt} \tag{15.60}$$

Solving for $\omega_m(t)$ in equation 15.59 and substituting in equation 15.60, we obtain a second-order differential equation in the armature current:

$$k_{PM} I_a(t) = T_{\text{load}} + \frac{J}{k_{PM}} \left[\frac{dV_L}{dt} - R_a \frac{dI_a(t)}{dt} - L_a \frac{d^2 I_a(t)}{dt^2} \right] \tag{15.61}$$

A similar equation in $\omega_m(t)$ may also be obtained by substituting equation 15.60 in equation 15.59:

$$V_L = k_{PM} \omega_m(t) + \frac{R_a}{k_{PM}} \left[T_{\text{load}} + J \frac{d\omega_m(t)}{dt} \right]$$
$$+ \frac{L_a}{k_{PM}} \left[\frac{dT_{\text{load}}}{dt} + J \frac{d^2 \omega_m(t)}{dt^2} \right] \tag{15.62}$$

The two differential equations may be manipulated to obtain

$$\frac{L_a J}{k_{PM}} \frac{d^2 I_a(t)}{dt^2} + \frac{R_a J}{k_{PM}} \frac{dI_a(t)}{dt} + k_{PM} I_a(t) = \frac{J}{k_{PM}} \frac{dV_L}{dt} + T_{\text{load}} \tag{15.63}$$

and

$$\frac{L_a J}{k_{PM}} \frac{d^2 \omega_m(t)}{dt^2} + \frac{R_a J}{k_{PM}} \frac{d\omega_m(t)}{dt} + k_{PM} \omega_m(t)$$
$$= V_L - \frac{R_a}{k_{PM}} T_{\text{load}} - \frac{L_a}{k_{PM}} \frac{dT_{\text{load}}}{dt} \tag{15.64}$$

If we define the constants

$$\tau_m = \frac{R_a J}{k_{PM}^2} \quad \text{and} \quad \tau_a = \frac{L_a}{R_a} \tag{15.65}$$

the differential equations may be simplified as follows:

$$k_{PM} \tau_a \tau_m \frac{d^2 I_a(t)}{dt^2} + k_{PM} \tau_m \frac{dI_a(t)}{dt} + k_{PM} I_a(t) = \frac{J}{k_{PM}} \frac{dV_L}{dt} + T_{\text{load}} \tag{15.66}$$

and

$$k_{PM}\tau_a\tau_m \frac{d^2\omega_m(t)}{dt^2} + k_{PM}\tau_m \frac{d\omega_m(t)}{dt} + k_{PM}\omega_m(t)$$
$$= V_L - \frac{R_a}{k_{PM}}T_{\text{load}} - \frac{L_a}{k_{PM}}\frac{dT_{\text{load}}}{dt} \tag{15.67}$$

Speed Control of DC Motors

One of the features that have made DC motors popular in many applications is the ease with which their speed can be controlled. Revisiting the speed equation for the shunt DC motor, repeated here for convenience, we can see that the speed can be controlled by adjusting one of three parameters: the source voltage, V_s; the flux; or the armature resistance (and therefore the armature current):

$$\omega_m = \frac{V_s - I_a R_a}{k_a \phi_{\text{sh}}} \tag{15.68}$$

In the case of the shunt motor, the easiest adjustment is to control the field flux by means of a field rheostat, R_x, as illustrated in Figure 15.17 (neglect the series field connection in the figure). Another method involves the use of a rheostat, R_{am}, in series with the armature. The drawback of this method is that the armature rheostat needs to carry all of the armature current, and thus it must have a very high power rating; further, the presence of the additional series resistance also implies a power loss in the armature circuits, and thus reduced efficiency.

It is also possible to control the terminal voltage, V_s, by means of an external DC supply. Traditionally, this function was accomplished by means of a separate *motor-generator set*, in a configuration referred to as the **Ward-Leonard system**, shown in Figure 15.23, whereby a separate DC generator was used to produce the appropriate DC supply voltage to drive the motor. This type of speed control was used extensively in the past in applications requiring accurate speed control over a broad range. For example, it was (and to some extent still is) very popular in engine-dynamometer test cells,

Figure 15.23 Ward-Leonard DC motor speed-control system

where internal combustion engines are coupled to *electric dynamometers* (DC motors) that serve as a variable load for testing the engines under a wide variety of load and speed conditions. This application clearly requires a broad range of speeds, since an internal combustion engine (e.g., for automotive use) can range in speed from 500 to 5,000 rev/min.

The operation of the Ward-Leonard system can be briefly summarized as follows, with reference to Figure 15.23. An AC motor supplied by three-phase power (we shall study the characteristics of these motors in the second half of this chapter) drives both a *DC exciter* and a *DC generator*. The purpose of the exciter is to provide the field excitation current, acting as a separate excitation source. The generator, on the other hand, which is driven by the AC motor (the thick dashed line in the figure indicates a mechanical connection), converts the three-phase AC power to DC power, with variable DC voltage thanks to the generator field control rheostat, R_2, which controls the generator field current. The combination of AC motor and DC generator is called a **motor-generator (M-G) set.** The generator field control allows adjustment of the DC motor speed from zero up to rated speed with R_3 set for rated field current. Speeds greater than the rated speed can also be obtained by **field weakening,** that is, by reducing the motor field current by means of rheostat R_3.

One of the advantages of the Ward-Leonard system is that it allows for reversal of the power flow. Thus, if the DC motor is used as a brake (e.g., if it acts as a variable load coupled to another prime mover, such as an internal combustion engine), the generator can act as a motor, driving the AC machine, which in turn acts as a generator and can convert the energy that would be dissipated in braking to AC power, which is fed back to the grid. This process, which is called **regenerative braking,** results in obvious energy savings. However, such speed-control systems are much bulkier than what can be accomplished with today's solid-state electronics and are being supplanted by *solid-state electronic drives,* which can provide variable DC supplies from an AC source and, if properly designed, can also be regenerative in nature. An example of this type of electronic circuit was given in Chapter 9, where we studied the characteristics of SCRs and thyristors.

Another means of producing a variable DC supply is a **chopper drive.** A chopper supply operates by switching a constant DC voltage supply on and off at a variable rate, according to one of two principles: *pulse-width modulation,* or PWM; or *pulse-frequency modulation,* or PFM. In the case of PWM, the width of the "on" pulse is varied to vary the average (DC) voltage seen by the motor. Since the original DC supply is fixed, this method allows variation from zero up to the DC supply voltage. In the PFM method, the width of the pulse is kept fixed but the frequency of the switching is varied. Thus, at low frequency, the average voltage supplied to the motor is low; it increases at higher frequencies.

Which type of DC drive is best suited for a given application depends in part on the characteristics of the load. Thus, it will be necessary to delve into a brief discussion of the different types of loads that are commonly encountered. Typical applications of various types of DC motors will also be discussed. *Constant-torque loads* are quite common, and are characterized by a need for constant torque over the entire speed range. This need is usually due to friction; the load will demand increasing horsepower at higher speeds, since power is the product of speed and torque. Thus, the power required will increase linearly with speed. This type of loading is characteristic of conveyors, extruders, and surface winders.

Another type of load is one that requires *constant horsepower* over the speed range of the motor. Since torque is inversely proportional to speed with constant horsepower, this type of load will require higher torque at low speeds. Examples of constant-horsepower loads are machine tool spindles (e.g., lathes). This type of application requires very high starting torques.

Variable-torque loads are also common. In this case, the load torque is related to the speed in some fashion, either linearly or geometrically. For some loads, for example, torque is proportional to the speed (and thus horsepower is proportional to speed squared); examples of loads of this type are positive displacement pumps. More common than the linear relationship is the squared-speed dependence of inertial loads such as centrifugal pumps, some fans, and all loads in which a flywheel is used for energy storage.

To select the appropriate motor and adjustable speed drive for a given application, we need to examine how each method for speed adjustment operates on a DC motor. Armature voltage control serves to smoothly adjust speed from 0 to 100 percent of the nameplate rated value (i.e., base speed), provided that the field excitation is also equal to the rated value. Within this range, it is possible to fully control motor speed for a constant-torque load, thus providing a linear increase in horsepower, as shown in Figure 15.24. Field weakening allows for increases in speed of up to several times the base speed; however, field control changes the characteristics of the DC motor from constant torque to constant horsepower, and therefore the torque output

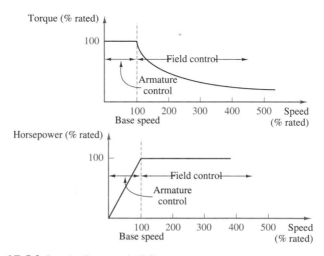

Figure 15.24 Speed adjustment in DC motors.

drops with speed, as shown in Figure 15.24. Operation above base speed requires special provision for field control, in addition to the circuitry required for armature voltage control, and is therefore more complex and costly.

DRILL EXERCISES

15.8   A series motor draws a current of 25 A and develops a torque of 100 N-m Find: (*a*) the torque when the current rises to 30 A if the field is unsaturated; and (*b*) the torque when the current rises to 30 A and the increase in current produces a 10 percent increase in flux.

> Answer: (*a*) 144 N-m; (*b*) 132 N-m

15.9   A 200 V DC shunt motor draws 10 A at 1800 rev/min. The armature circuit resistance is 0.15 Ω and the field winding resistance is 350 Ω. What is the torque developed by the motor?

> Answer: $T = \frac{P}{\omega_m} = 10.87$ N-m

15.10   Describe the cause-and-effect behavior of speed control method of changing armature voltage for a shunt DC motor.

> Answer: Increasing the armature voltage leads to an increase in armature current. Consequently, the motor torque increases until it exceeds the load torque, causing the speed to increase as well. The corresponding increase in back emf, however, causes the armature current to drop and the motor torque to decrease until a balance condition is reached between motor and load torque and the motor runs at constant speed.

15.4   AC MACHINES

From the previous sections, it should be apparent that it is possible to obtain a wide range of performance characteristics from DC machines, as both motors and generators. A logical question at this point should be: Would it not be more convenient in some cases to take advantage of the single- or multiphase AC power that is available virtually everywhere than to expend energy and use additional hardware to rectify and regulate the DC supplies required by direct-current motors? The answer to this very obvious question is certainly a resounding yes. In fact, the AC induction motor is the workhorse of many industrial applications, and synchronous generators are used almost exclusively for the generation of electric power worldwide. Thus, it is appropriate to devote a significant portion of this chapter to the study of AC machines, and of induction motors in particular. The objective of this section is to explain the basic operation of both synchronous and induction machines, and to outline their performance characteristics. In doing so, we shall also point out the relative advantages and disadvantages of these machines in comparison with direct-current machines.

Rotating Magnetic Fields

As mentioned in Section 15.1, the fundamental principle of operation of AC machines is the generation of a rotating magnetic field, which causes the rotor to turn at a speed that depends on the speed of rotation of the magnetic field. We shall now explain how a rotating magnetic field can be generated in the stator and air gap of an AC machine by means of AC currents.

Consider the stator shown in Figure 15.25, which supports windings a-a', b-b' and c-c'. The coils are geometrically spaced 120° apart, and a three-phase voltage is applied to the coils. As you may recall from the discussion of AC power in Chapter 5, the currents generated by a three-phase source are also spaced by 120°, as illustrated in Figure 15.26. The phase voltages would then be given by the expressions

$$v_a = A \cos \left(\omega_e t \right)$$

$$v_b = A \cos \left(\omega_e t - \frac{2\pi}{3} \right)$$

$$v_c = A \cos \left(\omega_e t + \frac{2\pi}{3} \right)$$

where ω_e is the frequency of the AC supply, or line frequency. The coils in each winding are arranged in such a way that the flux distribution generated by any one winding is approximately sinusoidal. Such a flux distribution may be obtained by appropriately arranging groups of coils for each winding over

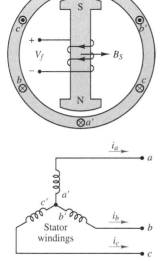

Figure 15.25 Two-pole three-phase stator

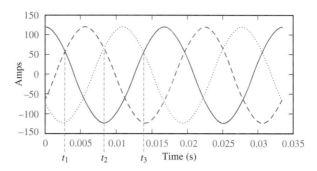

Figure 15.26 Three-phase stator winding currents

the stator surface. Since the coils are spaced 120°apart, the flux distribution resulting from the sum of the contributions of the three windings is the sum of the fluxes due to the separate windings, as shown in Figure 15.28. Thus, the flux in a three-phase machine rotates in space according to the vector diagram of Figure 15.27, and pulsates in amplitude, as shown in Figure 15.28.

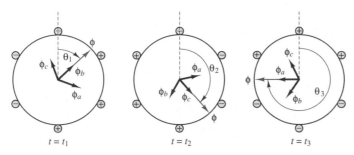

Figure 15.27 Rotating flux in a three-phase machine

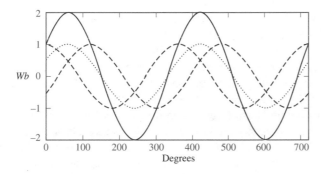

Figure 15.28 Flux distribution in three-phase stator winding

Since the resultant flux of Figure 15.28 is generated by the currents of Figure 15.26, the speed of rotation of the flux must be related to the frequency of the sinusoidal phase currents. In the case of the stator of Figure 15.25, the number of magnetic poles resulting from the winding configuration is two; however, it is also possible to configure the windings so that they have more poles. For example, Figure 15.29 depicts a simplified view of a four-pole stator.

In general, the speed of the rotating magnetic field is determined by the frequency of the excitation current, f, and by the number of poles present in the stator, p, according to the equation

$$n_s = \frac{120f}{p} \text{ rev/min}$$

or   (15.69)

$$\omega_s = \frac{2\pi n_S}{60} = \frac{2\pi \times 2f}{p}$$

where $n_s(\omega_s)$ is usually called the **synchronous speed.**

Now, the structure of the windings in the preceding discussion is the same whether the AC machine is a motor or a generator; the distinction between the two depends on the direction of power flow. In a generator, the electromagnetic torque is a reaction torque that opposes rotation of the machine; this is the torque against which the prime mover does work. In

Figure 15.29 Four-pole stator

a motor, on the other hand, the rotational (motional) voltage generated in the armature opposes the applied voltage; this voltage is the counter (or back) emf. Thus, the description of the rotating magnetic field given thus far applies to both motor and generator action in AC machines.

It is worthwhile to devote a few paragraphs to an additional discussion of magnetic poles in AC machines, since an intuitive understanding of how a rotating magnetic field is generated is essential. The motion and associated electromagnetic torque of an electric machine are the result of two magnetic fields that are trying to align with each other so that the south pole of one field attracts the north pole of the other. Figure 15.30 illustrates this action by analogy with two permanent magnets, one of which is allowed to rotate about its center of mass.

As described a few paragraphs earlier, the stator magnetic field rotates in an AC machine, and therefore the rotor cannot "catch up" with the stator field and is in constant pursuit of it. The speed of rotation of the rotor will therefore depend on the number of magnetic poles present in the stator and in the rotor. The magnitude of the torque produced in the machine is a function of the angle γ between the stator and rotor magnetic fields; precise expressions for this torque depend on how the magnetic fields are generated and will be given separately for the two cases of synchronous and induction machines. What is common to all rotating machines is that the number of stator and rotor poles must be identical if any torque is to be generated. Further, the number of poles must be even, since for each north pole there must be a corresponding south pole.

One important desired feature in an electric machine is an ability to generate a constant electromagnetic torque. With a constant-torque machine, one can avoid torque pulsations that could lead to undesired mechanical vibration in the motor itself and in other mechanical components attached to the motor (e.g., mechanical loads, such as spindles or belt drives). A constant torque may not always be achieved, although it will be shown that it is possible to accomplish this goal when the excitation currents are multiphase. A general rule of thumb, in this respect, is that it is desirable, insofar as possible, to produce a constant flux per pole.

Figure 15.30 Alignment action of poles

15.5   THE ALTERNATOR (SYNCHRONOUS GENERATOR)

One of the most common AC machines is the **synchronous generator,** or **alternator.** In this machine, the field winding is on the rotor, and the connection is made by means of brushes, in an arrangement similar to that of the DC machines studied earlier. The rotor field is obtained by means of a DC current provided to the rotor winding, or by permanent magnets. The rotor is then connected to a mechanical source of power and rotates at a speed that we will consider constant to simplify the analysis.

Figure 15.31 depicts a two-pole three-phase synchronous machine. Figure 15.32 depicts a four-pole three-phase alternator, in which the rotor poles

Figure 15.31 Two-pole synchronous machine

are generated by means of a wound salient pole configuration and the stator poles are the result of windings embedded in the stator according to the simplified arrangement shown in the figure, where each of the pairs a/a', b/b', and so on, contributes to the generation of the magnetic poles, as follows. The group a/a', b/b', c/c' produces a sinusoidally distributed flux (see Figure 15.28) corresponding to one of the pole pairs, while the group $-a/-a'$, $-b/-b'$, $-c/-c'$ contributes the other pole pair. The connections of the coils making up the windings are also shown in Figure 15.32. Note that the coils form a wye connection (see Chapter 5). The resulting flux distribution is such that the flux completes two sinusoidal cycles around the circumference of the air gap. Note also that each arm of the three-phase wye connection has been divided into two coils, wound in different locations, according to the schematic stator diagram of Figure 15.32. One could then envision analogous configurations with greater numbers of poles, obtained

Figure 15.32 A four-pole three-phase alternator

in the same fashion, that is, by dividing each arm of a wye connection into more windings.

The arrangement shown in Figure 15.32 requires that a further distinction be made between mechanical degrees θ_m, and electrical degrees, θ_e. In the four-pole alternator, the flux will see two complete cycles during one rotation of the rotor, and therefore the voltage that is generated in the coils will also oscillate at twice the frequency of rotation. In general, the electrical degrees (or radians) are related to the mechanical degrees by the expression

$$\theta_e = \frac{p}{2}\theta_m \qquad\qquad (15.70)$$

where p is the number of poles. In effect, the voltage across a coil of the machine goes through one cycle every time a pair of poles moves past the coil. Thus, the frequency of the voltage generator by a synchronous generator is

$$f = \frac{p}{2}\frac{n}{60} \text{ Hz} \qquad\qquad (15.71)$$

where n is the mechanical speed in rev/min. Alternatively, if the speed is expressed in rad/s, we have

$$\omega_e = \frac{p}{2}\omega_m \qquad\qquad (15.72)$$

where ω_m is the mechanical speed of rotation in rad/s. The number of poles employed in a synchronous generator is then determined by two factors: the frequency desired of the generated voltage (e.g., 60 Hz, if the generator is used to produce AC power), and the speed of rotation of the prime mover. In the latter respect, there is a significant difference, for example, between the speed of rotation of a steam turbine generator and that of a hydroelectric generator, the former being much greater.

A common application of the alternator is in automotive battery-charging systems—in which, however, the generated AC voltage is rectified to provide the DC current required for charging the battery.

15.6   THE SYNCHRONOUS MOTOR

Synchronous motors are virtually identical to synchronous generators with regard to their construction, except for an additional winding for helping start the motor and minimizing motor speed over- and undershoots. The principle of operation is, of course, the opposite: an AC excitation provided to the armature generates a magnetic field in the air gap between stator and rotor, resulting in a mechanical torque. To generate the rotor magnetic field, some DC current must be provided to the field windings; this is often accomplished by means of an **exciter**, which consists of a small DC generator propelled by the motor itself, and therefore mechanically connected to it. It was mentioned earlier that to obtain a constant torque in an electric motor, it is necessary to keep the rotor and stator magnetic fields constant relative to each other. This means that the electromagnetically rotating field in the stator and the mechanically rotating rotor field should be aligned at all times. The only condition for which this can occur is if both fields are rotating at the synchronous speed, $n_s = 120f/p$. Thus, synchronous motors are by their very nature constant-speed motors.

For a non–salient pole (cylindrical-rotor) synchronous machine, the torque can be written in terms of the AC stator current, $i_S(t)$, and of the DC rotor current, I_f:

$$T = k i_S(t) I_f \sin \gamma \tag{15.73}$$

where γ is the angle between the stator and rotor fields (see Figure 15.30). Let the angular speed of rotation be

$$\omega_m = \frac{d\theta_m}{dt} \text{ rad/s} \tag{15.74}$$

where $\omega_m = 2\pi n/60$, and let ω_e be the electrical frequency of $i_S(t)$, where $i_S(t) = \sqrt{2} I_S \sin \omega_e t$. Then the torque may be expressed as follows:

$$T = k\sqrt{2} I_S \sin \omega_e t I_f \sin \gamma \tag{15.75}$$

where k is a machine constant, I_S is the rms value of the stator current, and I_f is the DC rotor current. Now, the rotor angle γ can be expressed as a function of time by

$$\gamma = \gamma_0 + \omega_m t \tag{15.76}$$

where γ_0 is the angular position of the rotor at $t = 0$; the torque expression then becomes

$$\begin{aligned} T &= k\sqrt{2} I_S I_f \sin \omega_e t \sin (\omega_m t + \gamma_0) \\ &= k\frac{\sqrt{2}}{2} I_S I_f \cos [(\omega_m - \omega_e)t - \gamma_0] - \cos [(\omega_m + \omega_e)t + \gamma_0] \end{aligned} \tag{15.77}$$

It is a straightforward matter to show that the average value of this torque, $<T>$, is different from zero only if $\omega_m = \pm\omega_e$, that is, only if the motor is turning at the synchronous speed. The resulting average torque is then given by

$$<T> = k\sqrt{2} I_S I_f \cos \gamma_0 \tag{15.78}$$

Note that equation 15.77 corresponds to the sum of an average torque plus a fluctuating component at twice the original electrical (or mechanical) frequency. The fluctuating component results because in the foregoing derivation, a single-phase current was assumed. The use of multiphase currents reduces the torque fluctuation to zero and permits the generation of a constant torque.

A per-phase circuit model describing the synchronous motor is shown in Figure 15.33, where the rotor circuit is represented by a field winding equivalent resistance and inductance, R_f and L_f, respectively, and the stator circuit is represented by equivalent stator winding inductance and resistance, L_S and R_S, respectively, and by the induced emf, E_b. From the exact

Figure 15.33 Per-phase circuit model

equivalent circuit as given in Figure 15.34, we have

$$V_s = E_b + I_s(R_s + jX_s) \tag{15.79}$$

Where X_s is known as the synchronous reactance and includes magnetizing reactance.

The motor power is

$$P_{out} = \omega_s T = |V_s|\,|I_s|\cos\theta \tag{15.80}$$

for each phase, where T is the developed torque and θ is the angle between V_s and I_s.

When the phase winding resistance R_s is neglected, the circuit model of a synchronous machine can be redrawn as shown in Figure 15.34. The input power (per phase) is equal to the output power in this circuit, since no power is dissipated in the circuit; that is:

$$P_\phi = P_{in} = P_{out} = |\mathbf{V}_S|\,|\mathbf{I}_S|\cos\theta \tag{15.81}$$

Also by inspection of Figure 15.35, we have

$$d = |\mathbf{E}_b|\sin\delta = |\mathbf{I}_S|\,X_S\cos\theta \tag{15.82}$$

Then

$$|\mathbf{E}_b|\,|\mathbf{V}_S|\sin\delta = |\mathbf{V}_S|\,|\mathbf{I}_S|\,X_S\cos\theta = X_S P_\phi \tag{15.83}$$

The total power of a three-phase synchronous machine is then given by

$$P = (3)\frac{|\mathbf{V}_S|\,|\mathbf{E}_b|}{X_S}\sin\delta \tag{15.84}$$

Because of the dependence of the power upon the angle δ, this angle has come to be called the **power angle**. If δ is zero, the synchronous machine cannot develop useful power. The developed power has its maximum value at δ equal to 90°. If we assumed that $|\mathbf{E}_b|$ and $|\mathbf{V}_S|$ are constant, we can draw the curve shown in Figure 15.35, relating the power and power angle in a synchronous machine.

A synchronous generator is usually operated at a power angle varying from 15° to 25°. For synchronous motors and small loads, δ is close to 0°, and the motor torque is just sufficient to overcome its own windage and friction losses; as the load increases, the rotor field falls further out of phase with the stator field (although the two are still rotating at the same speed), until δ reaches a maximum at 90°. If the load torque exceeds the maximum torque, which is produced for $\delta = 90°$, the motor is forced to slow down below synchronous speed. This condition is undesirable, and provisions are usually made to shut the motor down automatically whenever synchronism is lost. The maximum torque is called the **pull-out torque** and is an important measure of the performance of the synchronous motor.

Figure 15.34

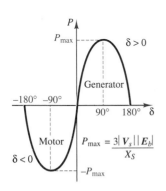

Figure 15.35 Power versus power angle for a synchronous machine

Accounting for each of the phases, the total torque is given by

$$T = \frac{m}{\omega_s} |\mathbf{E}_b| |\mathbf{I}_S| \cos \theta \tag{15.85}$$

where m is the number of phases. From figure 15.35, we have $E_b \sin \delta = X_S I_S \cos \theta$. Therefore, for a three-phase machine, the developed torque is

$$T = \frac{P}{\omega_s} = \frac{3}{\omega_s} \frac{|\mathbf{V}_S| |\mathbf{E}_b|}{X_S} \sin \delta \qquad \text{N-m} \tag{15.86}$$

Typically, analysis of multiphase motors is performed on a per-phase basis, as illustrated in the examples that follow.

Example   15.11

A 460-V, three-phase, wye-connected synchronous motor has a full-load stator winding current of 12.5 A. The synchronous impedance of the motor is $1.0 + j\,12\ \Omega$. Find:

1. The kVA rating of the motor.
2. The induced voltage and the power angle of the rotor when the motor is fully loaded at 0.707 power factor lagging.

Solution:

The circuit model for the motor is shown in Figure 15.36.

1. The per-phase current in the wye-connected stator winding is

    $$I_S = |\mathbf{I}_S| = 12.5\ \text{A}$$

    The per-phase voltage is

    $$V_S = |\mathbf{V}_S|\,\frac{460}{\sqrt{3}} = 265.58\ \text{V}$$

    The kVA rating of the motor is expressed in terms of the apparent power, S (see Chapter 5):

    $$S = 3V_S I_S = 3 \times 265.58 \times 12.5 = 9{,}959\ \text{W}$$

2. From the equivalent circuit, we have

    $$\mathbf{E}_b = \mathbf{V}_S - \mathbf{I}_S(R_S + jX_S)$$
    $$= 265.58 - (12.5\angle{-45°}) \times (1 + j\,12) = 179.31\angle{-32.83°}$$

    The induced line voltage is defined to be

    $$V_{\text{line}} = \sqrt{3}E_b = \sqrt{3} \times 179.31 = 310.57\ \text{V}$$

    From the expression for $\mathbf{E}_b$, we can find the power angle:

    $$\delta = -32.83°$$

    The minus sign indicates that the machine is in motor mode.

Figure 15.36

Example   15.12

A 208-V, 45-kVA, 0.8 power factor leading, Δ-connected 60-Hz synchronous machine has a synchronous reactance of 2.5 Ω and a negligible armature resistance. Its friction and windage losses are 1.5 kW, and the core losses are 1.0 kW. Initially, the shaft is supplying a 15-hp load. Find $\mathbf{I}_S$, $\mathbf{I}_{line}$, and $\mathbf{E}_b$.

Solution:

The output power of the motor is 15 hp; that is

$$P_{out} = 15 \text{ hp} \times 0.746 \text{ kW/hp} = 11.19 \text{ kW}$$

The electric power supplied to the machine is

$$P_{in} = P_{out} + P_{mech} + P_{core\text{-}loss} + P_{elec\text{-}loss}$$
$$= 11.19 \text{ kW} + 1.5 \text{ kW} + 1.0 \text{ kW} + 0 \text{ kW} = 13.69 \text{ kW}$$

As discussed in Chapter 5, the resulting line current is

$$I_{line} = \frac{P_{in}}{\sqrt{3} \text{ V} \cos \theta} = \frac{13.69}{\sqrt{3} \times 208 \times 0.8} = 47.5 \text{ A}$$

Because of the Δ connection, the armature current is

$$\mathbf{I}_S = \frac{1}{\sqrt{3}} \mathbf{I}_{line} = 27.4 \angle 36.87° \text{ A}$$

The emf may be found from the equivalent circuit and KVL:

$$\mathbf{E}_b = \mathbf{V}_S - jX_s \mathbf{I}_S$$
$$= 208 \angle 0° - j2.5(27.4 \angle 36.87°) = 255 \angle -12.4° \text{ V}$$

The power angle is

$$\delta = -12.4°$$

Synchronous motors are not very commonly used in practice, for various reasons, among which are that they are essentially required to operate at constant speed (unless a variable-frequency AC supply is available) and that they are not self-starting. Further, separate AC and DC supplies are required. It will be seen shortly that the induction motor overcomes most of these drawbacks.

DRILL EXERCISES

15.11 A synchronous generator has a multipolar construction that permits changing its synchronous speed. If only two poles are energized, at 50 Hz, the speed is 3000 rev/min. If the number of poles is progressively increased to 4, 6, 8, 10, and 12, find the synchronous speed for each configuration.

> Answer: 1500 rev/min; 1000 rev/min; 750 rev/min; 600 rev/min; 500
> rev/min; 428 rev/min

15.12 Draw the complete equivalent circuit of a synchronous generator and its phasor diagram.

15.13 Find the maximum of pullout torque of the synchronous motor.

> Answer: $T_{\max} = \dfrac{3V_S E_b}{\omega_m X_S}$

15.7   THE INDUCTION MOTOR

The induction motor is the most widely used electric machine, because of its relative simplicity of construction. The stator winding of an induction machine is similar to that of a synchronous machine; thus, the description of the three-phase winding of Figure 15.25 also applies to induction machines. The primary advantage of the induction machine, which is almost exclusively used as a motor (its performance as a generator is not very good), is that no separate excitation is required for the rotor. The rotor typically consists of one of two arrangements: a **squirrel cage,** or a **wound rotor.** The former contains conducting bars short-circuited at the end and embedded within it; the latter consists of a multiphase winding similar to those used for synchronous machines, but electrically short-circuited.

In either case, the induction motor operates by virtue of currents induced from the stator field in the rotor. In this respect, its operation is similar to that of a transformer, in that currents in the stator (which acts as a primary coil) induce currents in the rotor (acting as a secondary coil). In most induction motors, no external electrical connection is required for the rotor, thus permitting a simple, rugged construction, without the need for slip rings or brushes. Unlike the synchronous motor, the induction motor does not operate at synchronous speed, but at a somewhat lower speed, which is dependent on the load. Figure 15.37 illustrates the appearance of a squirrel-cage induction motor. The following discussion will focus mainly on this very common configuration.

You are acquainted with the notion of a rotating stator magnetic field. Imagine now that a squirrel-cage rotor is inserted in a stator in which such a rotating magnetic field is present. The stator field will induce voltages in the cage conductors, and if the stator field is generated by a three-phase source, the resulting rotor currents—which circulate in the bars of the squirrel cage, with the conducting path completed by the shorting rings at the end of the cage—are also three-phase, and are determined by the magnitude of the induced voltages and by the impedance of the rotor. Since the rotor currents are induced by the stator field, the number of poles and the speed of rotation

Figure 15.37 Squirrel-cage induction motor

of the induced magnetic field are the same as those of the stator field, *if the rotor is at rest*. Thus, when a stator field is initially applied, the rotor field is synchronous with it, and the fields are stationary with respect to each other. Thus, according to the earlier discussion, a *starting torque* is generated.

If the starting torque is sufficient to cause the rotor to start spinning, the rotor will accelerate up to its operating speed. However, an induction motor can never reach synchronous speed; if it did, the rotor would appear to be stationary with respect to the rotating stator field, since it would be rotating at the same speed. But in the absence of relative motion between the stator and rotor fields, no voltage would be induced in the rotor. Thus, an induction motor is limited to speeds somewhere below the synchronous speed, n_s. Let the speed of rotation of the rotor be n; then, the rotor is losing ground with respect to the rotation of the stator field at a speed $(n_s - n)$. In effect, this is equivalent to backward motion of the rotor at the **slip speed,** defined by $(n_s - n)$. The **slip,** s, is usually defined as a fraction of n_s:

$$s = \frac{n_s - n}{n_s} \tag{15.87}$$

which leads to the following expression for the rotor speed:

$$n = n_s(1 - s) \tag{15.88}$$

The slip, s, is a function of the load, and the amount of slip in a given motor is dependent on its construction and rotor type (squirrel cage or wound rotor). Since there is a relative motion between the stator and rotor fields, voltages will be induced in the rotor at a frequency called the **slip frequency,** related to the relative speed of the two fields. This gives rise to an interesting phenomenon: the rotor field travels relative to the rotor at the slip speed sn_s, but the rotor is mechanically traveling at the speed $(1 - s)n_s$, so that the net effect is that the rotor field travels at the speed

$$sn_s + (1 - s)n_s = n_s \tag{15.89}$$

that is, at synchronous speed. The fact that the rotor field rotates at synchronous speed—although the rotor itself does not—is extremely important, because it means that the stator and rotor fields will continue to be stationary with respect to each other, and therefore a net torque can be produced.

As in the case of DC and synchronous motors, important characteristics of induction motors are the starting torque, the maximum torque, and the torque-speed curve. These will be discussed shortly, after some analysis of the induction motor is performed in the next few paragraphs.

Example 15.13

A four-pole, three-phase, 230-V, 60-Hz induction motor runs at a speed of 1,725 rev/min under full-load conditions. Find the slip and the frequency of the voltage induced in the rotor at the rated speed.

Solution:

The synchronous speed of the motor is

$$n_s = \frac{120f}{p} = \frac{120 \times 60}{4} = 1{,}800 \text{ rev/min}$$

The slip is

$$s = \frac{n_s - n}{n_s} = \frac{1{,}800 - 1{,}725}{1{,}800} = 0.0417$$

The rotor frequency, f_R, is

$$f_R = sf = 0.0417 \times 60 = 2.5 \text{ Hz}$$

The induction motor can be described relatively simply by means of an equivalent circuit, which is essentially that of a rotating transformer. (See Chapter 14 for a circuit model of the transformer.) Figure 15.38 depicts such a circuit model, where:

R_S = stator resistance

X_S = stator reactance

X_m = magnetizing (mutual) reactance

R_C = equivalent core-loss resistance

E_S = per-phase induced voltage in stator windings

E_R = per-phase induced voltage in rotor windings

The primary internal stator voltage, $\mathbf{E}_S$, is coupled to the secondary rotor voltage, $\mathbf{E}_R$, by an ideal transformer with an effective turns ratio α. For the rotor circuit, the induced voltage at any slip will be

$$\mathbf{E}_R = s\mathbf{E}_{R0} \qquad (15.90)$$

where E_{R0} is the induced rotor voltage at the condition in which the rotor is stationary. Also, $X_R = \omega_R L_R = 2\pi f_R L_R = 2\pi s f L_R = s X_{R0}$, where $X_{R0} = 2\pi f L_R$ is the reactance when the rotor is stationary. The rotor current is given by the expression

$$\mathbf{I}_R = \frac{\mathbf{E}_R}{R_R + jX_R} = \frac{s\mathbf{E}_{R0}}{R_R + jsX_{R0}} = \frac{\mathbf{E}_{R0}}{R_R/s + jX_{R0}} \qquad (15.91)$$

The result rotor equivalent circuit is shown in Figure 15.39.

Figure 15.38 Circuit model for induction machine

Figure 15.39 Rotor circuit

The voltages, currents, and impedances on the secondary (rotor) side can be reflected to the primary (stator) by means of the effective turns ratio. When this transformation is effected, the transformed rotor voltage is given by

$$\mathbf{E}_2 = \mathbf{E}'_R = \alpha \mathbf{E}_{R0} \tag{15.92}$$

The transformed (reflected) rotor current is

$$\mathbf{I}_2 = \frac{\mathbf{I}_R}{\alpha} \tag{15.93}$$

The transformed rotor resistance can be defined as

$$R_2 = \alpha^2 R_R \tag{15.94}$$

and the transformed rotor reactance can be defined by

$$X_2 = \alpha^2 X_{R0} \tag{15.95}$$

The final per-phase equivalent circuit of the induction motor is shown in Figure 15.40.

Figure 15.40 Equivalent circuit of an induction machine

The following examples illustrate the use of the circuit model in determining the performance of the induction motor.

Example   15.14

A 460-V, 25-hp, 60-Hz, four-pole, wye-connected induction machine has the following parameters (per phase), referred to the stator circuit:

$$R_S = 0.641 \ \Omega \qquad R_2 = 0.332 \ \Omega$$
$$X_S = 1.106 \ \Omega \qquad X_2 = 0.464 \ \Omega \qquad X_m = 26.3 \ \Omega$$

For $s = 0.022$ at the rated voltage and frequency and an output power of 14 hp, find

1. The speed.
2. The stator current.
3. The power factor.
4. The output torque.

You may neglect core losses ($R_c = 0$).

Solution:

1. The per-phase equivalent circuit is shown in Figure 15.40. The synchronous speed is found to be

$$n_s = \frac{120f}{p} = \frac{120 \times 60}{4} = 1,800 \text{ rev/min}$$

or

$$\omega_s = 1,800 \times \frac{2\pi}{60} = 188.5 \text{ rad/s}$$

The rotor mechanical speed is

$$n = (1 - s)n_s = 1,760 \text{ rev/min}$$

or

$$\omega_m = (1 - s)\omega_s = 184.4 \text{ rad/s}$$

2. The reflected rotor impedance is found from the parameters of the per-phase circuit to be

$$Z_2 = \frac{R_2}{s} + jX_2 = \frac{0.332}{0.022} + j0.464$$
$$= 15.09 + j0.464 \ \Omega$$

The combined magnetization plus rotor impedance is therefore equal to

$$Z = \frac{1}{1/jX_m + 1/Z_2} = \frac{1}{-j0.038 + 0.0662\angle-1.76°} = 12.94\angle31.1°$$

and the total impedance is

$$Z_{\text{total}} = Z_S + Z = 0.641 + j1.106 + 11.08 + j6.68$$
$$= 11.72 + j7.79 = 14.07\angle33.6° \ \Omega$$

Finally, the stator current is given by

$$I_S = \frac{V_s}{Z_{\text{total}}} = \frac{460/\sqrt{3}\angle0°}{14.07\angle33.6°} = 18.88\angle-33.6° \text{ A}$$

3. The power factor is

$$pf = \cos 33.6° = 0.833 \qquad \text{lagging}$$

4. The output power, P_{out}, is

$$P_{\text{out}} = 14 \times 746 = 10.444 \text{ kW}$$

and the output torque is

$$T = \frac{P_{\text{out}}}{\omega_m} = \frac{10,444 \text{ W}}{184.4 \text{ rad/s}} = 56.64 \text{ N-m}$$

Example   15.15

A 500-V, three-phase, 50-Hz, eight-pole, wye-connected induction motor has the following equivalent circuit parameters:

$$R_S = 0.13 \ \Omega \qquad R_2' = 0.32 \ \Omega \qquad X_S = 0.6 \ \Omega \qquad X_2' = 1.48 \ \Omega$$

The magnetic branch admittance is

$$Y_m = G_c + jB_m = 0.004 - j0.05 \ \Omega^{-1}$$

referred to the primary side, where G_c is the conductance and B_m the susceptance associated with the core loss and mutual inductance, respectively. The full-load slip is 5 percent. Find:

1. The full-load electromagnetic torque
2. The stator input current
3. The power factor

using the approximate equivalent circuit. The effective stator-to-rotor turns ratio per phase is 1/1.57. Neglect mechanical losses.

Solution:

The approximate equivalent circuit of the three-phase induction motor on a per-phase basis is shown in Figure 15.41. The parameters of the model are calculated as follows:

$$R_2 = 0.32 \times \left(\frac{1}{1.57}\right)^2 = 0.13 \ \Omega$$

$$X_2 = 1.48 \times \left(\frac{1}{1.57}\right)^2 = 0.6 \ \Omega$$

$$Z = R_S + \frac{R_2}{s} + j(X_S + X_2)$$

$$= 0.13 + \frac{0.13}{0.05} + j(0.6 + 0.6) = 2.73 + j1.2$$

Using the approximate circuit,

$$\mathbf{I}_2 = \frac{\mathbf{V}_s}{Z} = \frac{(500/\sqrt{3})\angle 0°}{2.73 + j1.2} = 88.8 - j39$$

$$\mathbf{I}_R = \mathbf{V}_s G_c = 288.7 \times 0.004 = 1.15 \ A$$

$$\mathbf{I}_m = -j\mathbf{V}_s B_m = 288.7 \times (-j0.05) = -j14.4 \ A$$

$$\mathbf{I}_1 = \mathbf{I}_2 + \mathbf{I}_R + \mathbf{I}_m = 89.95 - j53.4$$

$$\text{Input power factor} = \frac{89.95}{105} = 0.856 = 85.6\%$$

$$\text{Torque} = \frac{3P}{\omega_s} = \frac{3I_2^2 R_2/s}{2\pi f/p} = 935 \ \text{N-m}$$

Figure 15.41

Performance of Induction Motors

The performance of induction motors can be described by torque-speed curves similar to those already described for DC motors. Figure 15.42 depicts an induction motor torque-speed curve, with five torque ratings marked *a* through *e*. Point *a* is the *starting torque,* also called **breakaway torque,** and is the torque available with the rotor "locked," that is, in a stationary position. At this condition, the frequency of the voltage induced in the rotor is highest, since it is equal to the frequency of rotation of the stator field; consequently, the inductive reactance of the rotor is greatest. As the rotor accelerates, the torque drops off, reaching a minimum value called the **pull-up torque** (point *b*); this typically occurs somewhere between 25 and 40

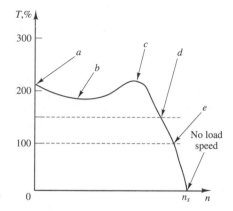

Figure 15.42 Performance curve for induction motor

percent of synchronous speed. As the rotor speed continues to increase, the rotor reactance decreases further (since the frequency of the induced voltage is determined by the relative speed of rotation of the rotor with respect to the stator field). The torque becomes a maximum when the rotor inductive reactance is equal to the rotor resistance; maximum torque is also called **breakdown torque** (point *c*). Beyond this point, the torque drops off, until it is zero at synchronous speed, as discussed earlier. Also marked on the curve are the *150 percent torque* (point *d*), and the *rated torque* (point *e*).

A general formula for the computation of the induction motor steady-state torque-speed characteristic is

$$T = \frac{1}{\omega_e} \frac{mV_S^2}{2\left[R_S + \sqrt{R_S^2 + (X_S + X_R)^2}\right]} \tag{15.96}$$

where *m* is the number of phases.

Different construction arrangements permit the design of induction motors with different torque-speed curves, thus permitting the user to select the motor that best suits a given application. Figure 15.43 depicts the four basic classifications, classes A, B, C, and D, as defined by NEMA. The determining

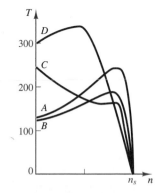

Figure 15.43 Induction motor classification

features in the classification are the locked-rotor torque and current, the breakdown torque, the pull-up torque, and the percent slip. Class A motors have a higher breakdown torque than class B motors, and a slip of 5 percent or less. Motors in this class are often designed for a specific application. Class B motors are general purpose motors; this is the most commonly used type of induction motor, with typical values of slip of 3 to 5 percent. Class C motors have a high starting torque for a given starting current, and a low slip. These motors are typically used in applications demanding high starting torque but having relatively normal running loads, once running speed has been reached. Class D motors are characterized by high starting torque, high slip, low starting current, and low full-load speed. A typical value of slip is around 13 percent.

Factors that should be considered in the selection of an AC motor for a given application are the *speed range,* both minimum and maximum, and the speed variation. For example, it is important to determine whether constant speed is required; what variation might be allowed, either in speed or in torque; or whether variable speed operation is required, in which case a variable-speed drive will be needed. The torque requirements are obviously important as well. The starting and running torque should be considered; they depend on the type of load. Starting torque can vary from a small percentage of full load to several times full-load torque. Furthermore, the excess torque available at start-up determines the *acceleration characteristics* of the motor. Similarly, *deceleration characteristics* should be considered, to determine whether external braking might be required.

Another factor to be considered is the *duty cycle* of the motor. The duty cycle, which depends on the nature of the application, is an important consideration when the motor is used in repetitive, noncontinuous operation, such as is encountered in some types of machine tools. If the motor operates at zero or reduced load for periods of time, the duty cycle—that is, the percentage of the time the motor is loaded—is an important selection criterion. Last, but by no means least, are the *heating properties* of a motor. Motor temperature is determined by internal losses and by ventilation; motors operating at a reduced speed may not generate sufficient cooling, and forced ventilation may be required.

Thus far, we have not considered the dynamic characteristics of induction motors. Among the integral-horsepower induction motors, the most common dynamic problems are associated with starting and stopping and with the ability of the motor to continue operation during supply system transient disturbances. Dynamic analysis methods for induction motors depend to a considerable extent on the nature and complexity of the problem and the associated precision requirements. When the electrical transients in the motor are to be included as well as the motional transients, and especially when the motor is an important element in a large network, the simple transient equivalent circuit of Figure 15.44 provides a good starting approximation. In the circuit model of Figure 15.44, X'_S is called the *transient reactance*. The voltage E'_S is called the *voltage behind the transient reactance* and is assumed

Figure 15.44 Simplified induction motor dynamic model

to be equal to the initial value of the induced voltage, at the start of the transient. R_S is the stator resistance. The dynamic analysis problem consists of selecting a sufficiently simple but reasonably realistic representation that will not unduly complicate the dynamic analysis, particularly through the introduction of nonlinearities.

It should be remarked that the basic equations of the induction machine, as derived from first principles, are quite nonlinear. Thus, an accurate dynamic analysis of the induction motor, without any linearizing approximations, requires the use of computer simulation.

DRILL EXERCISES

15.14 A three-phase induction motor has six poles. (*a*) If the line frequency is 60 Hz, calculate the speed of the magnetic field in rev/min. (*b*) Repeat the calculation if the frequency is changed to 50 Hz.

> Answer: (*a*) $n = 1200$ rev/min (*b*) $n = 1000$ rev/min

15.15 A four-pole induction motor operating at a frequency of 60 Hz has a full-load slip of 4 percent. Find the frequency of the voltage induced in the rotor (*a*) at the instant of starting and (*b*) at the full load

> Answer: (*a*) $f_R = 60$ Hz (*b*) $f_R = 2.4$ Hz

15.16 A four-pole, 1746 rev/min, 220 V, 3-phase, 60 Hz, 10 hp, Y-connected induction machine has the following parameters: $R_S = 0.4$ Ω, $R_2 = 0.14$ Ω, $jX_m = 16$ Ω, $jX_S = 0.35$ Ω, $jX_2 = 0.35$ Ω, $R_C = 0$. Using Figure 15.40, find: (*a*) the stator current; (*b*) the rotor current; (*c*) the motor power factor; and (*d*) the stator power input.

> Answer: (*a*) $25.88\angle - 22.45°$ A (*b*) $24.355\angle - 6.51°$ A (*c*) 0.9242 (*d*) 3083 W

CONCLUSION

The principles developed in Chapter 14 can be applied to rotating electric machines, to explain how mechanical energy can be converted to electrical energy, and vice versa. The former function is performed by electric generators, while the latter is provided by electric motors.

- Electric machines are described in terms of their mechanical characteristics, their torque-speed curves, and their electrical characteristics, including current and voltage requirements. Losses and efficiency are an important part of the operation of electric machines, and it should be recognized that there will be electrical losses (due to the resistance of the windings), mechanical losses (friction and windage), and magnetic core losses (eddy currents, hysteresis). The main mechanical components of an electric machine are the stator, rotor, and air gap. Electrically, the important parameters are the armature (load current–carrying) circuit, and the field (magnetizing) circuit. Magnetic fields establish the coupling between the electrical and mechanical systems.

• Electric machines are broadly classified into DC and AC machines; the former use DC excitation for both the field and armature circuits, while the latter may be further subdivided into two classes: synchronous machines, and induction motors. AC synchronous machines are characterized by DC field and AC armature excitation. Induction machines (of the squirrel-cage type), on the other hand, do not require a field excitation, since this is provided by electromagnetic induction. Typically, DC machines have the armature windings on the rotor, while AC machines have it on the stator.

• The performance of electric machines can be approximately predicted with the use of circuit models, or of performance curves. The selection of a particular machine for a given application is driven by many factors, including the availability of suitable electrical supplies (or prime movers), the type of load, and various other concerns, of which heat dissipation and thermal characteristics are probably the most important.

KEY TERMS

HOMEWORK PROBLEMS

15.1 A DC shunt motor has a no-load saturation characteristic given by

$$E_b = \frac{300I_f}{1 + I_f} \qquad \text{(at 1,000 rev/min)}$$

where E_b is the open-circuit voltage and I_f the shunt field current. The armature resistance is 0.12 Ω, and the shunt field resistance is 247 Ω. The motor is operating from a 230-volt DC source.

a. If the load is adjusted to make the armature current 80 A, find the speed in revolutions per minute.

b. If the no-load line current is 3 A, determine the value of the torque necessary to make the armature current 80 A (assume that the starting torque is zero).

15.2 In a DC machine, the air-gap flux density is 4 W_b/m^2. The area of the pole face is 2 cm $\times$ 4 cm. Find the flux per pole in the machine.

15.3 A separately excited generator produces 400 V at 1,500 rev/min. Find the induced emf if the speed changes to 2,000 rev/min. Assume that the flux is constant.

15.4 A 120-V, 10-A shunt generator has an armature resistance of 0.6 Ω. The shunt field current is 2 A. Determine the voltage regulation of the generator.

15.5 The voltage regulation of a 220-V shunt generator is 10 percent. Find the terminal voltage when the load is reduced to zero.

15.6 A 20-kW, 230-V separately excited generator has an armature resistance of 0.2 Ω and a load current of 100 A. Find:

a. The terminal voltage.

b. The output power.

15.7 A 10-kW, 120-VDC series generator has an armature resistance of 0.1 Ω and a series field resistance of 0.05 Ω. Assuming that it is delivering rated current at rated speed, find (a) the armature current and (b) the armature voltage.

15.8 The armature resistance of a 30-kW, 440-V shunt generator is 0.1 Ω. Its shunt field resistance is 200 Ω. Find

a. The power developed at rated load.

b. The load, field, and armature currents.

c. The electrical power loss.

15.9 A four-pole, 450-kW, 4.6-kV shunt generator has armature and field resistances of 2 and 333 Ω. The generator is operating at the rated speed of 3,600 rev/min. Find the no-load voltage of the generator and terminal voltage at half load.

15.10 A shunt DC motor has a shunt field resistance of 400 Ω and an armature resistance of 0.2 Ω. The motor nameplate rating values are 440 V, 1,200 rev/min, 100 hp, and full-load efficiency of 90 percent. Find:

a.   The motor line current.

b.   The field and armature currents.

c.   The counter emf at rated speed.

d.   The output torque.

15.11   A 30-kW, 240-V generator is running at half load at 1,800 rev/min with efficiency of 85 percent. Find the total losses and input power.

15.12   A 400-V, 100-kW, long-shunt compound generator has a series field resistance of 0.04 Ω, a shunt field resistance of 100 Ω, and an armature resistance of 0.02 Ω. The rated current is delivered at rated speed of 100 rev/min. Find

a.   The armature current.

b.   The armature voltage.

15.13   A 220-VDC shunt motor has an armature resistance of 0.2 Ω and a rated armature current of 50 A. Find

a.   The voltage generated in the armature.

b.   The power developed.

15.14   The open-circuit characteristic for a DC shunt generator operating at 1,500 rev/min is given by

$$E = \frac{367I_f}{0.233 + I_f}$$

where E is the open-circuit voltage and I_f is the shunt field current. The other parameters are

$$R_a = 0.14 \ \Omega \qquad R_f = 459 \ \Omega$$

a.   Determine the no-load terminal voltage when this machine is operating as a self-excited shunt generator.

b.   What is the terminal voltage for an armature current of 50 A?

c.   What is the terminal voltage for the compound generator for an armature current of 50 A?

15.15   A 200-VDC shunt motor has the following parameters:

$$R_a = 0.1 \ \Omega \qquad R_f = 100 \ \Omega$$

When running at 1,100 rev/min with no load connected to the shaft, the motor draws 4 A from the line. Find E and the rotational losses at 1,100 rev/min (assuming that the stray-load losses can be neglected).

15.16   A 230-VDC shunt motor has the following parameters:

$$R_a = 0.5 \ \Omega \qquad R_f = 75 \ \Omega$$
$$P_{\text{rot}} = 500 \ \text{W (at 1,120 rev/min)}$$

When loaded, the motor draws 46 A from the line. Find

a.   The speed, P_{dev}, and T_{sh}.

b.   If $L_f = 25$ H, $L_a = 0.008$ H, and the terminal voltage has a 115-V change, find $i_a(t)$ and $\omega_m(t)$.

15.17   A 200-VDC shunt motor with an armature resistance of 0.1 Ω and a field resistance of 100 Ω draws a line current of 5 A when running with no load at 955 rev/min. Determine the motor speed, the motor efficiency, the total losses (i.e., rotational and stray-load losses), and the load torque (i.e., T_{sh}) that will result when the motor draws 40 A from the line.

15.18   A self-excited DC shunt generator is delivering 20 A to a 100-V line when it is driven at 200 rad/s. The magnetization characteristic is shown in Figure P15.1. It is known that $R_a = 1.0$ Ω and $R_f = 100$ Ω. When the generator is disconnected from the line, the drive motor speeds up to 220 rad/s. What is the terminal voltage?

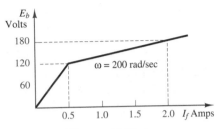

Figure P15.1

15.19   A 230-V DC shunt motor has $R_a = 0.05$ Ω and $R_f = 75$ Ω.

a.   At no load, the motor draws 7 A from the line and runs at a speed of 1,120 rev/min. Determine the copper losses and the rotational loss (assuming the stray losses to be negligible) for this machine operating at no load.

b.   The machine is loaded so that the line current becomes 46 A. What is the speed (in rev/min) at this load, and what are the developed torque and the shaft torque?

15.20   A DC shunt motor has a no-load saturation characteristic given by

$$E_b = \frac{1,020 I_f}{0.9 + I_f} \qquad \text{at 1,200 rev/min}$$

where E_b is the generated voltage and I_f is the shunt field current. The armature resistance is 0.48 Ω, the field resistance is 368 Ω, and the armature current is 50 A at a terminal voltage of 460 V.

a.   What is the speed (in rev/min)?

b.   What is the developed torque?

15.21   The machine of example 15.8 is being used in a series connection. That is, the field coil is connected in series with the armature, as shown in Figure P15.2. The machine is to be operated under the same conditions as in the previous example, that is, $n = 120$ rev/min, $I_a = 8$ A. In the operating region, $\phi = kI_f$, and $k = 200$. The armature resistance is 0.2 Ω, and the resistance of the field winding is negligible.

Figure P15.2

a.   Find the number of field winding turns necessary for full-load operation.

b.   Find the torque output for the following speeds:

     1. $n' = 2n$

     2. $n' = 3n$

     3. $n' = \dfrac{n}{2}$

     4. $n' = \dfrac{n}{4}$

c.   Plot the speed-torque characteristic for the conditions of part b.

15.22   A non–salient pole, Y-connected, three-phase, two-pole synchronous machine has a synchronous reactance of 7 Ω and negligible resistance and rotational losses. One point on the open-circuit characteristic is given by $V_o = 400$ V (phase voltage) for a field current of 3.32 A. The machine is to be operated as a motor, with a terminal voltage of 400 V (phase voltage). The armature current is 50 A, with power factor 0.85, leading.

    Determine E_b, field current, torque developed, and power angle δ.

15.23   A non–salient pole, Y-connected, three-phase, two-pole synchronous machine has $X_S = 1.9$ Ω and negligible phase resistance and rotational losses. The field current is adjusted to give an excitation voltage of 380 V per phase (open-circuit voltage when operated as a generator). It is to be connected as a motor to a 254-V (per phase) line and will drive a load such that $|\delta| = 30°$. Find the armature current and power factor. Is the current capacitive or inductive?

15.24   A non–salient pole, Y-connected, three-phase, two-pole synchronous generator is connected to a 400-V (line to line), 60-Hz, three-phase line. The stator impedance is $0.5 + j\,1.6$ Ω (per phase). The generator is delivering rated current (36 A) at unity power factor to the line. Determine the power angle for this load and the value of E_b for this condition. Sketch the phasor diagram, showing $\mathbf{E}_b$, $\mathbf{I}_S$, and $\mathbf{V}_S$.

15.25   A non–salient pole, three-phase, two-pole synchronous motor is connected in parallel with a three-phase, Y-connected load so that the per-phase equivalent circuit is as shown in Figure P15.3. The parallel combination is connected to a 220-V (line to line), 60-Hz, three-phase line. The load current $\mathbf{I}_L$ is 25 A at a power factor of 0.866 inductive. The motor has $X_S = 2$ Ω and is operating with $I_f = 1$ A and $T = 50$ N-m at a power angle of $-30°$. (Neglect all losses for the motor.)

    Find $\mathbf{I}_S$, P_{in} (to the motor), the overall power factor (i.e., angle between $\mathbf{I}_1$ and $\mathbf{V}_S$), and the total power drawn from the line.

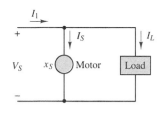

Figure P15.3

15.26   A non-salient pole, Y-connected, three-phase, two-pole synchronous motor is connected to a 220-V (line to line) 60-Hz source. The motor is driving a load such that $T_{load} = 85$ N-m and is known to have total rotational and stray-load losses of 942.5 W. The total stator copper loss at this load is 600 W. The measured phase parameters are $R_S = 0.02$ Ω and $X_S = 1.269$ Ω.

    Find $I_S E_b \delta$, the power factor, and the total power into the machine.

15.27   A four-pole, three-phase, Y-connected, non–salient pole synchronous motor has a synchronous reactance of 10 Ω. This motor is connected to a $230\sqrt{3}$ V (line to line), 60-Hz, three-phase line and is driving a load such that $T_{shaft} = 30$ N-m. The line current is 15 A, leading the phase voltage. Assuming that all losses can be neglected, determine the power angle δ and E for this condition. If the load is removed, what is the line current, and is it leading or lagging the voltage?

15.28   A 10-hp, 230-V, 60-Hz, three-phase wye-connected synchronous motor delivers full load at a power factor of 0.8 leading. The synchronous reactance is 6 Ω, the rotational loss is 230-V, and the field loss is 50 W. Find

a.   The armature currents.
b.   The motor efficiency.
c.   The power angle.

Neglect the stator winding resistance.

15.29   A synchronous motor with a leading power factor can be used to correct power factor. Assume that the power of a load is S' with a lagging power factor angle θ. The apparent power requirement of a synchronous motor is S'_m at a leading power factor angle θ_m. The connection is shown in Figure P15.4. Find the new power factor, and explain how the synchronous motor can be used to improve the power angle.

Figure P15.4

15.30   Draw the phasor diagram of the synchronous motor of Problem 15.29.

15.31   A 2,000-hp, unity power factor, three-phase, Y-connected, 2,300-V, 30-pole, 60-Hz synchronous motor has a synchronous reactance of 1.95 Ω per phase. Neglect all losses. Find the maximum power and torque.

15.32   A 1,200-V, three-phase, wye-connected synchronous motor takes 110 kW (exclusive of field winding loss) when operated under a certain load at 1,200 rev/min. The back emf of the motor is 2,000 V. The synchronous reactance is 10 Ω per phase, with negligible winding resistance. Find the line current and the torque developed by the motor.

15.33   The per-phase impedance of a 600-V, three-phase, Y-connected synchronous motor is $5 + j50$ Ω. The motor takes 24 kW at a leading power factor of 0.707. Determine the induced voltage and the power angle of the motor.

15.34   A 74.6-kW, three-phase, 440-V (line to line), four-pole, 60-Hz induction motor has the following (per-phase) parameters referred to the stator circuit:

$$R_S = 0.06\ \Omega \quad X_S = 0.3\ \Omega \quad X_m = 5\ \Omega \quad R_R = 0.08\ \Omega \quad X_R = 0.3\ \Omega$$

The no-load power input is 3,240 W at a current of 45 A. Determine the line current, the input power, the developed torque, the shaft torque, and the efficiency at $s = 0.02$.

15.35   A 60-Hz, four-pole, Y-connected induction motor is connected to a 400-V (line to line), three-phase, 60-Hz line. The equivalent circuit parameters are:

$$R_S = 0.2 \ \Omega \qquad R_R = 0.1 \ \Omega$$
$$X_S = 0.5 \ \Omega \qquad X_R = 0.2 \ \Omega$$
$$X_m = 20 \ \Omega$$

When the machine is running at 1,755 rev/min, the total rotational and stray-load losses are 800 W. Determine the slip, input current, total input power, mechanical power developed, shaft torque, and efficiency.

15.36 A three-phase, 60-Hz induction motor has eight poles and operates with a slip of 0.05 for a certain load. Determine

a.   The speed of the rotor with respect to the stator.
b.   The speed of the rotor with respect to the stator magnetic field.
c.   The speed of the rotor magnetic field with respect to the rotor.
d.   The speed of the rotor magnetic field with respect to the stator magnetic field.

15.37 A three-phase, two-pole, 400-V (per phase), 60-Hz induction motor develops 37 kW (total) of mechanical power (P_m) at a certain speed. The rotational loss at this speed is 800 W (total). (Stray-load loss is negligible.)

a.   If the total power transferred to the rotor is 40 kW, determine the slip and the output torque.
b.   If the total power into the motor (P_{in}) is 45 kW and R_S is 0.5 Ω find I_S and the power factor.

15.38 A four-pole induction motor operating at a frequency of 600 Hz has a full-load rotor slip of 5 percent. Calculate the rotor frequency

a.   At the instant of starting.
b.   At full load.

15.39 Calculate the full-load rotor speed if in Problem 15.38 the slip is

a.   5 percent when the line frequency is 60 Hz.
b.   7 percent when the line frequency is 50 Hz.

15.40 A motor will be used to drive an 18 kW load working intermittently. Let the overload capability of a motor, λ, be the ratio of the maximum intermittent power the motor can produce to the nominal motor power rating. Select one of the following two motors based on overload capability.

   Motor 1: $P_e = 10$ kW, $n = 1460$ rpm, $\lambda = 2.5$.
   Motor 2: $P_e = 14$ kW, $n = 1460$ rpm, $\lambda = 2.8$.

15.41 A 60-Hz, four-pole, Y-connected induction motor is connected to a 200-V (line to line), three-phase, 60 Hz line. The equivalent circuit parameters are:

$$R_S = 0.48 \ \Omega \qquad \text{Rotational loss torque} = 3.5 \text{ N-m}$$
$$X_S = 0.8 \ \Omega \qquad R_R = 0.42 \ \Omega \ (\text{referred to the stator})$$
$$X_m = 30 \ \Omega \qquad X_R = 0.8 \ \Omega \ (\text{referred to the stator})$$

The motor is operating at slip $s = 0.04$. Determine the input current, input power, mechanical power, and shaft torque (assuming that stray-load losses are negligible).

15.42

a. A three-phase, 220-V, 60-Hz induction motor runs at 1,140 rev/min. Determine the number of poles (for minimum slip), the slip, and the frequency of the rotor currents.

b. To reduce the starting current, a three-phase squirrel-cage induction motor is started by reducing the line voltage to $V_S/2$. By what factor are the starting torque and the starting current reduced?

15.43 The power rating of a motor can be modified to account for different ambient temperature, according to the following table:

Ambient temperature	30°C	35°C	40°C	45°C	50°C	55°C
Variation of rated power	+8%	+5%	0	−5%	−12.5%	−25%

A motor with $P_e = 10$ kW is rated up to 85°C. Find the actual power for each of the following conditions:

a. Ambient temperature is 50°C.

b. Ambient temperature is 25°C.

15.44 The speed-torque characteristic of an induction motor has been empirically determined as follows:

Speed	1470	1440	1410	1300	1100	900	750	350	0	rev/min
Torque	3	6	9	13	15	13	11	7	5	N-m

The motor will drive a load requiring a starting torque of 4 N-m and increase linearly with speed to 8 N-m at 1500 rpm.

a. Find the steady-state operating point of the motor.

b. Equation 15.96 predicts that the motor speed can be regulated in the face of changes in load torque by adjusting the stator voltage. Find the change in voltage required to maintain the speed at the operating point of part a if the load torque increases to 10 N-m.

15.45 An induction motor, operating from a 60-Hz line, develops 10 hp at 1,745 rev/min. Find the speed and the horsepower if the load torque is reduced by one half.

15.46 A 10-hp, 230-V, three-phase induction motor is marked with code letter G. Its starting kVA per horsepower will be less than 6.3. Find the upper limit of the starting current.

15.47 Repeat Problem 15.46 for the induction motor in Example 15.2.

15.48 Find the speed of the rotating field of a six-pole, three-phase motor connected to (a) a 60-Hz line and (b) a 50-Hz line, in rev/min and rad/s.

15.49 A six-pole, three-phase, 440-V, 60-Hz induction motor has the following model impedances:

$$R_S = 0.8 \ \Omega \qquad X_S = 0.7 \ \Omega$$
$$R_R = 0.3 \ \Omega \qquad X_R = 0.7 \ \Omega$$
$$X_m = 35 \ \Omega$$

Calculate the input current and power factor of the motor for a speed of 1,200 rev/min.

15.50 An eight-pole, three-phase, 220-V, 60-Hz induction motor has the following model impedances:

$$R_S = 0.78 \ \Omega \qquad X_S = 0.56 \ \Omega \qquad X_m = 32 \ \Omega$$
$$R_R = 0.28 \ \Omega \qquad X_R = 0.84 \ \Omega$$

Find the input current and power factor of this motor for $s = 0.02$.

15.51 A nameplate is given in Example 15.2. Find the rated torque, rated volt amperes, and maximum continuous output power for this motor.

15.52 Find the Thevénin equivalent of the circuit of Figure 15.41.

15.53 Explain the disadvantages of

a. Using a 20 hp motor to drive a 10 hp load continuously.
b. Using a 5 hp motor to drive a 10 hp load continuously.

15.54 In addition to the factors of rated voltage, rated current, frequency, temperature rise, speed, and duty cycle, give some other considerations affecting the selection of

a. Generators.
b. Motors.

CHAPTER

16

Special-Purpose
Electric Machines

T he objective of this chapter is to introduce the operating principles and
performance characteristics of a number of special-purpose electric
machines that find widespread engineering application in a variety of fields,
ranging from robotics to vehicle propulsion, aerospace, and automotive control. In Chapters 14 and 15, you were introduced to the operating principles of
the major classes of electric machines: DC machines, synchronous machines,
and induction motors. The machines discussed in Chapter 16 operate according to the essential principles described earlier, but are also characterized by
unique features that set them apart from the machines described in Chapter
15. The first of these special-purpose machines is the brushless DC motor.
Next, we discuss stepping motors, illustrating a very natural match between
electromechanical devices and digital logic. A discussion of universal motors
and single-phase induction motors follows, with a brief description of the
types of electronic drives used to supply power to these machines. The
discussion of the electronic drives ties the electromechanics material with
the subject of power electronics introduced in Chapter 9.

The machines introduced in Chapter 16 are used in many applications requiring fractional horsepower, or the ability to accurately control position, velocity, or torque. Among the applications presented in the chapter are electric vehicles and motors for control and robotics.

By the end of this chapter, you should be able to:

- Describe the principles of operation of brushless DC motors, stepping motors, universal motors, and single-phase motors.
- Select an appropriate motor and determine the required performance characteristics for a given application.
- Describe the characteristics of the electronic drives required for each of the important motors discussed in the chapter.

16.1   BRUSHLESS DC MOTORS

In spite of its name, the **brushless DC motor** is actually not a DC motor, but (typically) a permanent-magnet synchronous machine; the name is actually due not to the construction of the machine, but to the fact that its operating characteristics resemble those of a shunt DC motor with constant field current. This characteristic can be obtained by providing the motor with a power supply whose electrical frequency is always identical to the mechanical frequency of rotation of the rotor. To generate a source of variable frequency, use is made of DC-to-AC inverters, consisting of banks of transistors that are switched on and off at a frequency corresponding to the rotor speed; thus, the inverter converts a DC source to an AC source of variable frequency. As far as the user is concerned, then, the source of excitation of a brushless DC motor is DC, although the current that actually flows through the motor windings is AC. (The operation of the inverters will be explained shortly.) In effect, the brushless DC motor is a synchronous motor in which the torque angle, δ, is kept constant by an appropriate excitation current.

Brushless DC motors also require measurement of the position of the rotor to determine its speed of rotation, and to generate a supply current at the same frequency. This function is accomplished by means of a position-sensing arrangement that usually consists either of a magnetic Hall-effect position sensor, which senses the passage of each pole in the rotor, or of an optical encoder similar to the encoders discussed in Chapter 12.

Figure 16.1 depicts the appearance of a brushless DC motor. Note how the multiphase winding is similar to that of the synchronous motor of Chapter 15. The brushless motor consists of a stator with a multiphase winding, usually three-phase; a permanent-magnet rotor; and a rotor position sensor. It is interesting to observe that since the commutation is performed electronically by switching the current to the motor—rather than by brushes, as in DC motors—the brushless motor can be produced in many different configurations, including, for example, very flat ("pancake") motors. Figure 16.1 shows the classical configuration of inside rotor; outside stator. For simple machines, it is also possible to resort to an outside rotor, with greater

Figure 16.1 Two-pole brushless DC motor with three-phase stator winding

ease of magnet attachment and inherently smoother rotation, but with infe-
rior thermal characteristics, since a stator encased within the rotor structure
cannot be cooled efficiently.

In conventional DC motors, the supply voltage is limited by brush wear
and sparking that can occur at the commutator, often resulting in the need
for transformers to step down the supply voltage. In brushless DC motors,
on the other hand, such a concern does not exist, because the commutation
is performed electronically without the need for brushes. Further, since, in
general, the armature (load-carrying winding) is on the stator and thus the
losses are concentrated in the stator, liquid cooling (if required) is feasible
and does not involve excessive complexity. You will recall that in a conventional DC motor the armature is on the rotor, and therefore auxiliary liquid
cooling is very difficult to implement.

Another important advantage of brushless DC motors is that by sealing
the stator, submersible units can be built. In addition to these operational
advantages, it should be noted that these motors are also characterized by
easier construction: the construction of the stator in a brushless DC motor
is similar to that in traditional induction motors, and is therefore suitable
for automated production. The windings may also be fitted with temperature
sensors, providing the possibility of additional thermal protection.

The permanent-magnet rotor is typically made either of rare earth
magnets (Sm-Co) or of ceramic magnets (ferrites). Rare earth magnets have
outstanding magnetic properties, but they are expensive and in limited supply,
and therefore the more commonly employed materials are ceramic magnets.
Rare earth magnet motors can be a cost-effective solution—since they allow
much greater fluxes to be generated by a given supply current—in applications where high speed, high efficiency, and small size are important.
Brushless DC motors can be rated up to 250 kW at 50,000 rev/min. The
rotor position sensor must be designed for operation inside the motor, and
must withstand the backlash, vibrations, and temperature range typical of
motor operation.

Brushless DC motors do require a position-sensing device, though,
to permit proper switching of the supply current. Recall that the brushless
DC motor replaces the cumbersome mechanical commutation arrangement
with electronic switching of the supply current. The most commonly used
position-sensing devices are *position encoders* and **resolvers.** The resolver,
shown in Figure 16.2, is a rotating machine that is mechanically coupled to
the rotor of the brushless motor and consists of two stator and two rotor
windings; the stator windings are excited by an AC signal, and the resulting
rotor voltages are proportional to the sine and cosine of the angle of rotation
of the rotor, thus providing a signal that can be directly related to the instantaneous position of the rotor. The resolver has two major disadvantages:
first, it requires a separate AC supply; second, the resolver output must be
appropriately decoded to obtain a usable position signal. For these reasons,
angular position encoders (see Chapter 12) are often used. You will recall
that such encoders provide a digital signal directly related to the position

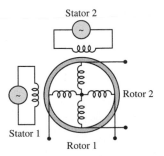

Figure 16.2 Resolver

of a rotating shaft. Their output can therefore be directly used to drive the current supply for a brushless motor.

To understand the operation of the brushless DC motor, it will be useful to make an analogy with the operation of a permanent-magnet (PM) DC motor. As discussed in Chapter 15, in a permanent-magnet DC motor, a fixed magnetic field generated by the permanent magnets interacts with the perpendicular field induced by the currents in the rotor windings, thus creating a mechanical torque. As the rotor turns in response to this torque, however, the angle between the stator and rotor fields is reduced, so that the torque would be nullified within a rotation of 90 electrical degrees. To sustain the torque acting on the rotor, permanent-magnet DC motors incorporate a commutator, fixed to the rotor shaft. The commutator switches the supply current to the stator so as to maintain a constant angle, $\delta = 90°$, between interacting fields. Because the current is continually switched between windings as the rotor turns, the current in each stator winding is actually alternating, at a frequency proportional to the number of motor magnetic poles and the speed.

The basic principle of operation of the brushless DC motor is essentially the same, with the important difference that the supply current switching takes place electronically, instead of mechanically. Figure 16.3

Transistor supply for brushless DC motor

SCR supply for brushless DC motor

Figure 16.3 Transistor and SCR drives for a brushless DC motor

depicts a transistor switching circuit capable of switching a DC supply so as to provide the appropriate currents to a three-phase rotor winding. The electronic switching device consists of a rotor position sensor, fixed on the motor shaft, and an electronic switching module that can supply each stator winding. Diagrams of the phase-to-phase back emf's and the switching sequence of the inverter are shown in Figure 16.4. The back emf waveforms shown in Figure

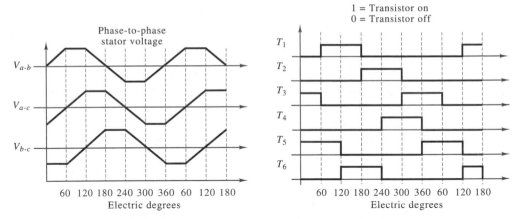

Figure 16.4 Phase voltages and transistor (SCR) switching sequence for the brushless DC motor drive of Figure 16.3

16.4 are called *trapezoidal;* the total back emf of the inverter is obtained by piecewise addition of the motor phase voltages and is a constant voltage, proportional to motor speed. You should visually verify that the addition of the three phase voltages of Figure 16.4 leads to a constant voltage. The brushless DC motor is therefore similar to a standard permanent-magnet DC motor, and can be described by the following simplified equations:

$$V = k_a \omega_m + R_w I \tag{16.1}$$

$$T = k_T I \tag{16.2}$$

where

$$k_a = k_T$$

and where

$$V = \text{motor voltage}$$
$$k_a = \text{armature constant}$$
$$\omega_m = \text{mechanical speed}$$
$$R_w = \text{winding resistance}$$
$$T = \text{motor torque}$$
$$k_T = \text{torque constant}$$
$$I = \text{motor (armature) current}$$

The speed and torque of a brushless DC motor can therefore be controlled with any variable-speed DC supply, such as one of the supplies briefly discussed in Chapter 15. Further, since the brushless motor has intrinsically higher torque and lower inertia than its DC counterpart, its response speed is superior to that obtained from traditional DC motors.

One important difference between the conventional DC and the brushless motor, however, is due to the coarseness of the electronic switching compared with the mechanical switching of the brush-type DC motor (recall the discussion of torque ripple due to the commutation effect in DC motors in Chapter 15). In practice, one cannot obtain the exact trapezoidal emf of Figure 16.4 by means of the transistor switching circuit of Figure 16.3, and a voltage ripple results as a consequence, leading to a torque ripple in the motor. Additional phase windings on the stator could solve the problem, at the expense of further complexity in the drive electronics, since the switching sequence would be more complex. Thus, brushless motors suffer from an inherent trade-off between torque ripple and drive complexity.

Among other applications, brushless DC motors find use in the design of servo loops in control systems—for example, in computer disk drives, and in propulsion systems for electric vehicles. The comparisons between the conventional DC motor and the brushless DC motor are summarized in the following table:

Conventional DC motors
___

Advantages:
1. Controllability over a wide range of speeds
2. Capability of rapid acceleration and deceleration
3. Convenient control of shaft speed and position by servo amplifiers

Disadvantages:
1. Commutation (through brushes) causes wear, electrical noise, and sparking
___

Brushless DC motors
___

Advantages:
1. Controllability over a wide range of speeds
2. Capability of rapid acceleration and deceleration
3. Convenient control of shaft speed and position
4. No mechanical wear or sparking problem due to commutation
5. Better heat dissipation capabilities

Disadvantages:
1. Need for more complex power electronics than the brush-type DC motor for equivalent power rating and control range
___

Example   16.1

This example explains a sinusoidal torque-generation scheme for brushless motors, in contrast to the system that produces the trapezoidal waveforms of Figure 16.4. The torque-generation scheme uses a magnet and field coil arrangement that produces a torque function resulting from the sine-cosine relationship between the currents in two coils, as shown in Figure 16.5.

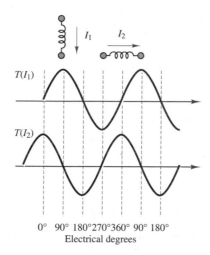

Figure 16.5 Sinusoidal torque-generation circuit and current waveforms for a brushless DC motor

The torques produced by the respective coils are

Coil 1: $T_1 = I_{m1} k_T \sin \theta$

Coil 2: $T_2 = I_{m2} k_T \cos \theta$

where T_1 and T_2 are instantaneous torque values and k_T is the torque constant of each winding. The current in each coil is controlled as a function of shaft angle by the switching electronics, as follows:

$I_{m1} = I_m \sin \theta$

$I_{m2} = I_m \cos \theta$

where I_m is the amplitude of the desired motor current. Thus, the total output torque is

$$T = T_1 + T_2$$
$$= I_m k_T \sin \theta \times \sin \theta + I_m k_T \cos \theta \times \cos \theta$$
$$= k_T I_m (\sin^2 \theta + \cos^2 \theta) = k_T I_m$$

Thus, the torque generated is directly proportional to the supply current amplitude:

$$T = k_T I_m$$

Example   16.2

A three-phase brushless DC motor is running at 500 rev/min. Four phases conduct in one cycle. The motor parameters are

$k_T = 0.037$ N-m/A (per phase) $I = 3.6$ A (per phase)

Find the torque generated by the motor.

Solution:

Since four phases conduct during each cycle, the torque is given by

$$T = 4k_T I = 4 \times 0.037 \times 3.6 = 0.53 \text{ N-m}$$

DRILL EXERCISES

16.1   List four features of the brushless DC motor that differ from a conventional shunt-type DC motor.

16.2   Repeat Example 16.2 for $I = 7.2$ A.

> Answer: $T = 1$ N-m

16.3   Convert the result of Drill Exercise 16.2 to oz-in.

> Answer: $T = 149.76$ oz-in

16.4   Verify that the transistor and SCR switching circuits of Figure 16.3 perform the same function.

16.2   STEPPING MOTORS

Stepping, or **stepper, motors** are motors that convert digital information to mechanical motion. The principles of operation of stepping motors have been known since the 1920s; however, their application has seen a dramatic rise with the increased use of digital computers. Stepping motors, as the name suggests, rotate in distinct steps, and their position can be controlled by means of logic signals. Typical applications of stepping motors are line printers, positioning of heads in magnetic disk drives, and any other situation where continuous or stepwise displacements are required.

Stepping motors can generally be classified in one of three categories: variable-reluctance, permanent-magnet, and hybrid types. It will soon be shown that the principles of operation of each of these devices bear a definite resemblance to those of devices already encountered in this book. Stepping motors have a number of special features that make them particularly useful in practical applications. Perhaps the most important feature of a stepping motor is that the angle of rotation of the motor is directly proportional to the number of input pulses; further, the angle error per step is very small and does not accumulate. Stepping motors are also capable of rapid response to starting, stopping, and reversing commands, and can be driven directly by digital signals. Another important feature is a self-holding capability that makes it possible for the rotor to be held in the stopped position without the use of brakes. Finally, a wide range of rotating speeds—proportional to the frequency of the pulse signal—may be attained in these motors.

Figure 16.6 depicts the general appearance of three types of stepping motors. The **permanent-magnet-rotor stepping motor,** Figure 16.6(a), permits a nonzero holding torque when the motor is not energized. Depending on the construction of the motor, it is typically possible to obtain step angles of 7.5, 11.25, 15, 18, 45, or 90 degrees. The angle of rotation is determined by the number of stator poles, as will be illustrated shortly in an example. The **variable-reluctance stepping motor,** Figure 16.6(b), has an iron multipole rotor and a laminated wound stator, and rotates when the teeth on the rotor are attracted to the electromagnetically energized stator teeth. The rotor inertia of a variable-reluctance stepping motor is low, and the response is very quick, but the allowable load inertia is small. When the windings are not energized, the static torque of this type of motor is zero. Generally, the step angle of the variable-reluctance stepping motor is 15 degrees.

The **hybrid stepping motor,** Figure 16.6(c), is characterized by multitoothed stator and rotor, the rotor having an axially magnetized concentric magnet around its shaft. It can be seen that this configuration is a mixture of the variable-reluctance and permanent-magnet types. This type of motor generally has high accuracy and high torque and can be configured to provide a step angle as small as 1.8 degrees.

For any of these configurations, the principle of operation is essentially the same: when the coils are energized, magnetic poles are generated in the stator, and the rotor will align in accordance with the direction of the magnetic field developed in the stator. By reversing the phase of the currents in the coils, or by energizing only some of the coils (this is possible in motors with more than two stator poles), the alignment of the stator magnetic field can take one of a discrete number of positions; if the currents in the coils are pulsed in the appropriate sequence, the rotor will advance in a step-by-step fashion. Thus, this type of motor can be very useful whenever precise incremental motion must be attained. As mentioned earlier, typical applications are printer wheels, computer disk drives, and plotters. Other applications are found in the control of the position of valves (e.g., control of the throttle valve in an engine, or of a hydraulic valve in a fluid power system), and in drug-dispensing apparatus for clinical applications.

The following examples illustrate the operation of a four-pole, two-phase permanent-magnet stepping motor, and of a similar motor of the variable-reluctance type. The operation of these motors is representative of all stepping motors.

(a) Permanent-magnet
stepping motor

(b) Variable-reluctance
stepping motor

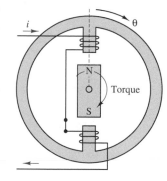

(c) Hybrid stepping motor

Figure 16.6 Stepping
motor configurations

Example   16.3

The PM stepper motor shown in Figure 16.7 can be operated by applying either or both of the currents i_1 and i_2, and by selecting the polarity of the currents. The

Figure 16.7 Two-phase four-pole PM stepper motor

currents are provided by a bipolar source (i.e., both positive and negative currents can be provided).

Initially, it will be assumed that the rotor is at rest; this condition can be attained with $i_1 > 0$ and $i_2 = 0$. The simplest mode of operation is the full-step, single-phase sequence, in which current is supplied to only one of the two coils at any given time while the other coil is deenergized. The current sequence shown in the following table will then cause the rotor to advance in 90° increments. Thus, if current is supplied to the coils with the appropriate switching sequence, the motor will rotate by a 90° step with every current pulse.

Full-step, single-phase sequence

i_1	i_2	θ
+	0	0°
0	+	90°
−	0	180°
0	−	270°
+	0	0°

It is also possible to obtain intermediate steps by supplying current to both coils. In this case, the magnetic stator poles will cause the rotor to align halfway between the stator teeth, also in increments of 90°, but shifted by 45° with respect to the singe-phase stepping sequence. The accompanying table illustrates the stepping sequence for this full-step, two-phase mode of operation:

Full-step, two-phase sequence

i_1	i_2	θ
+	+	45°
−	+	135°
−	−	225°
+	−	315°
+	+	45°

Finally, the two sequences may be combined, since the current commands required by each are distinct, to obtain the half-step sequence described in the next table. It should be apparent that for this particular motor, it is not possible to obtain finer increments in position than 45°. However, by increasing the number of windings and stator teeth, greater resolution can be obtained.

Half-step sequence

i_1	i_2	θ
+	0	0°
+	+	45°
0	+	90°
−	+	135°
−	0	180°
−	−	225°
0	−	270°
+	−	315°
+	0	0°

Example   16.4

Variable-reluctance (VR) stepper motors are characterized by the presence of salient poles in both the rotor and the stator, as shown in Figure 16.8, where a two-pole rotor is depicted. The operation of the VR motor is slightly different from that of the PM type, because the rotor is not magnetically polarized and therefore it is not necessary to have a bipolar supply to cause the rotor to rotate through a full revolution. Figure 16.8 depicts an ideal arrangement in which a DC source and four switches (controlled by a logic circuit, such as one of the circuits discussed in Chapter 12) are used to step the motor through its sequence of positions.

Figure 16.9 illustrates how the rotor would be positioned for three different positions of the switches; you should be able to infer the complete switching sequence for the motor. Note that, just as in the preceding example, the smallest increment in position that can be attained by this motor is 45°.

Figure 16.8 Two-phase four-pole VR stepping motor

Figure 16.9 Two-phase four-pole VR motor positioning sequence

Example   16.5

The step angle of a VR stepping motor is determined by the number of teeth on the rotor and the stator, as well as the number of phases. The following equations give the relationship between the number of steps per revolution, the number of phases, the number of teeth in each phase, and the step angle (resolution):

$$N = tm$$

where

$$N = \text{number of steps per revolution}$$

$$t = \text{number of teeth per phase}$$

$$m = \text{number of phases}$$

and

$$\Delta\theta = \frac{360°}{tm}$$

where $\Delta\theta$ = step angle, in degrees, or resolution.

For example, if a VR stepping motor has 12 teeth on the rotor and three phases, we can compute the steps per revolution as follows:

$$N = tm = 12 \times 3 = 36$$

It also follows that the step angle is given by

$$\Delta\theta = \frac{360°}{tm} = \frac{360°}{36} = 10°$$

Example   16.6

The torque equation of a stepping motor may be written as

$$T = 0.314 \times 10^{-6} \frac{tL(r + g/2)\mathscr{F}^2}{g} \text{ N-m}$$

where

$$T = \text{torque, N-m}$$

$$t = \text{number of teeth per phase}$$

$$L = \text{axial length of rotor, m}$$

$$g = \text{rotor-to-stator radial air gap, m}$$

$$r = \text{rotor radius, m}$$

$$\mathscr{F} = \text{magnetomotive force developed across the two air gaps in series through which a line of flux must pass in one phase}$$

For a stepping motor having the following data:

$$t = 16 \text{ teeth per phase (48 steps per revolution, for three-phase excitation)}$$

$$g = 6.35 \times 10^{-5} \text{ m}$$

$r = 1.29 \times 10^{-2}$ m

$\mathscr{F} = 720$ A-t

$L = 6.35 \times 10^{-3}$ m

we may compute the torque T using the preceding relationship, as follows:

$$T = \frac{0.314 \times 10^{-6} \times 16 \times 6.35 \times 10^{-3} (1.29 \times 10^{-2} + \frac{1}{2} \times 6.35 \times 10^{-5}) \times 720^2}{6.35 \times 10^{-5}}$$

$$= 3.37 \text{ N-m}$$

From the preceding examples, you should now have a feeling for the operation of variable-reluctance and PM stepping motors. The hybrid configuration is characterized by multitooth rotors that are made of magnetic materials, thus providing a variable-reluctance geometry in conjunction with a permanent-magnet rotor.

An ideal torque-speed characteristic for a stepper motor is shown in Figure 16.10. Two distinct modes of operation are marked on the curve: the **locked-step mode,** and the **slewing mode.** In the first mode, the rotor comes to rest (or at least decelerates) between steps; this is the mode commonly used to achieve a given rotor position. In the locked-step mode, the rotor can be started, stopped, and reversed. The slewing mode, on the other hand, does not allow stopping or reversal of the rotor, although the rotor still advances in synchronism with the stepping sequence, as described in the preceding examples. This second mode could be used, for example, in rewinding or fast-forwarding a tape drive.

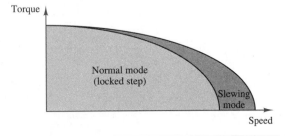

Figure 16.10 Ideal torque-speed characteristic of a stepping motor

The power supply, or driver, required by a stepping motor is shown in block diagram form in Figure 16.11; it includes a DC power supply, to provide the required current to drive the motor, in addition to logic and switching circuits to provide the appropriate inputs at the right time. One of the important considerations in driving a stepping motor is the excitation mode, which can be one-phase or two-phase. The driver is the circuit that arranges, distributes, and amplifies pulse trains from the logic circuit determining the stepping sequence; the driver excites each winding of the stepping motor at specified turns. In the **one-phase excitation** mode, current

Figure 16.11 Power supply for stepping motor

is supplied to one phase at a time, with the advantages of low power consumption and good step-angle accuracy. Input signal pulses and the change in the condition of each phase excitation are shown in Figure 16.12. In the **two-phase excitation** mode, current is simultaneously provided to two phases. Input signal pulses and the change in the condition of each phase excitation are also shown in Figure 16.12.

Input signal pulses and the change in phase excitation of the one-phase excitation mode

Input signal pulses and the change in phase excitation of the two-phase excitation mode

Figure 16.12 One- and two-phase excitation waveforms for stepper motors

In addition to the classification of the excitation by phase, stepping motor drives are also classified according to whether the drive supplies are unipolar or bipolar, that is, whether they can supply current in one or both directions. Unipolar excitation is clearly simpler, although in the case of the two-phase excitation mode, only half of the motor windings are used, with an obvious decrease in performance. Figure 16.13 shows a circuit diagram of the unipolar drive and the sequence of phase excitation.

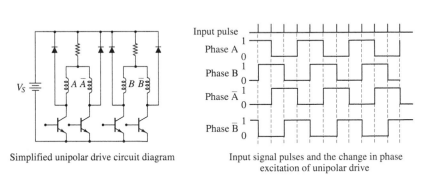

Simplified unipolar drive circuit diagram

Input signal pulses and the change in phase excitation of unipolar drive

Figure 16.13 Unipolar drive for stepper motor

When a bipolar drive is used, motor windings are used effectively, because of the bidirectional exciting current; when operated in this mode, a stepping motor can generate a large output torque at low speed compared with the unipolar drive. Figure 16.14 shows two versions of the bipolar drive. The first requires two power supplies, one for each polarity, while the second requires only one power supply but needs four switching transistors per phase to reverse the polarity.

Simplified two-power-supply bipolar drive circuit diagram

Simplified single-power-supply bipolar drive circuit diagram

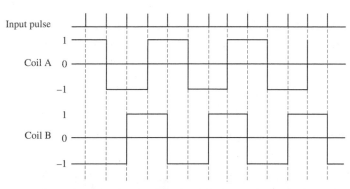

Input pulse signals and the change in phase excitation of bipolar drive

Figure 16.14 Bipolar drive for stepper motors

DRILL EXERCISES

16.5   Explain why the term *variable-reluctance* is used to describe the class of stepping motors so named.

16.6   Determine the smallest increment in angular position that can be obtained with a PM stepper motor with six stator teeth and three-phase current excitation.

> Answer: $\Delta\theta = 20°$

16.7   Express the stepping sequence of the variable-reluctance stepping motor of Example 16.4 as a four-digit binary sequence.

16.8   Derive the excitation waveforms corresponding to the direction of rotation opposite to that caused by the stepping sequence shown in Figure 16.12.

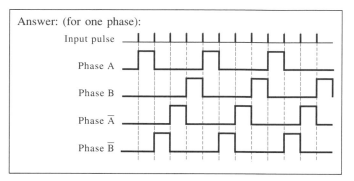

Answer: (for one phase):

16.9 For Example 16.6, express the torque in units of lb-in.

Answer: 30 lb-in

16.3   SINGLE-PHASE AC MOTORS

In Chapter 15, two types of AC machines were discussed: synchronous, and induction. In the discussion of these devices, especially in their motor applications, three-phase excitation was assumed; however, in many practical applications—and especially in small household appliances and small industrial motors—three-phase sources are not readily available, and it would be desirable to use single-phase excitation. Unfortunately, single-phase power does not lend itself to the generation of a rotating magnetic field: single-phase currents in the winding of an AC machine lead to a magnetic field that pulsates in amplitude but does not rotate in space. Thus, it would not be possible to use the AC machines described in Chapter 15 if only single-phase power were available. The aim of this section is to discuss the construction and the operating and performance characteristics of single-phase AC motors. The discussion will focus mainly on the *universal motor* and on **single-phase induction motors.**

   Fractional-horsepower (as opposed to **integral-horsepower**) **motors** represent by far the major share of the industrial production of electric motors. Many fractional-horsepower motors are designed for single-phase use, since single-phase AC power is readily available practically anywhere. Many applications are related to household appliances: refrigerator compressors, air conditioners, fans, electric tools, washer and dryer motors, and others. For the rest of this chapter, we shall examine qualitatively the principle of operation of single-phase motors and look at a few applications. The variety of designs for practical single-phase motors is such that it would not be possible to present the detailed principles of operation for all common types. However, it is hoped that the introduction provided in this chapter will help you in decoding the manufacturer's specifications for a given motor, and in making a preliminary selection for a given application.

Example   16.7

A small motor, as defined by the American Standards Association (ASA) and the National Electrical Manufacturers Association (NEMA; see Chapter 15) is a "motor built in a frame smaller than that having a continuous rating of 1 hp, open type, at 1700 to 1800 rev/min." Small motors are generally considered *fractional-horsepower motors*. However, since the determination is based on frame size and on a given speed range, the classification of a motor is not always obvious. Let us give two examples.

1. Consider a $\frac{3}{4}$-hp, 1,200-rev/min motor. This motor is not considered a fractional-horsepower motor, because of its frame size. If the same frame size were used for an 1,800-rev/min motor, it would produce a rating of more than 1 hp. Thus, it is considered an integral-horsepower motor of

$$0.75 \text{ hp} \times \frac{1,800}{1,200} = 1.125 \text{ hp}$$

   In other words, since the motor is capable of integral-horsepower performance at speeds of 1,700 to 1,800 rev/min, it is classified as an integral-horsepower motor.

2. Consider now a 1.25-hp, 3,600-rev/min motor. This motor is classified as a fractional-horsepower motor, in spite of the fact that its power output is actually greater than 1 hp. If the same motor were used at a speed of 1,800 rev/min, it would produce a rating of less than 1 hp:

$$1.25 \times \frac{1,800}{3,600} = 0.625 \text{ hp}$$

Thus, we see once again that some attention must be paid to the speed of operation of the motor in determining its classification. The term *fractional horsepower* relates more to the physical size of the machine than to the actual power output rating.

The Universal Motor

If it were possible to operate a DC motor from a single-phase AC supply, a wide range of simple applications would become readily available. Recall that the direction of the torque produced by a DC machine is determined by the direction of current flow in the armature conductors and by the polarity of the field; torque is developed in a DC machine because the commutator arrangement permits the field and armature currents to remain in phase, thus producing torque in a constant direction. A similar result can be obtained by using an AC supply, and by connecting the armature and field windings in series, as shown in Figure 16.15. A series DC motor connected in this con-figuration can therefore operate on a single-phase AC supply, and is referred to as a **universal motor.** An additional consideration is that, because of the AC excitation, it is necessary to reduce AC core losses by laminating the

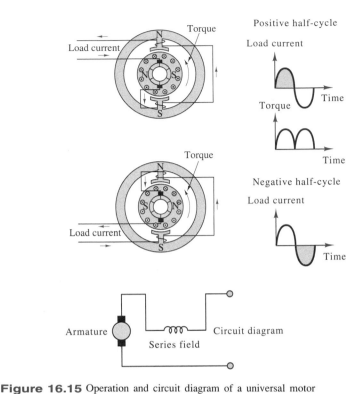

Figure 16.15 Operation and circuit diagram of a universal motor

Figure 16.16 Torque-speed curve of a universal motor

Figure 16.17 Torque-speed characteristics of a universal motor

stator; thus, the universal motor differs from the series DC motor discussed in Chapter 15 in its construction features. Typical torque-speed curves for AC and DC operation of a universal motor are shown in Figure 16.16. As shown in Figure 16.15, the load current is sinusoidal and therefore reverses direction each half-cycle; however, the torque generated by the motor is always in the same direction, resulting in a pulsating torque, with nonzero average value.

As in the case of a DC series motor, the best method for controlling the speed of a universal motor is to change its (rms) input voltage. The higher the rms input voltage, the greater the resulting speed of the motor. Approximate torque-speed characteristics of a universal motor as a function of voltage are shown in Figure 16.17.

Example   16.8

A 120-V, 60-Hz, two-pole universal motor operates at a speed of 800 rev/min at full load and draws a current of 17.85 A at a lagging power factor of 0.912. The parameters of the windings are:

Series field winding:

$$R_f = 0.65 \ \Omega \qquad X_f = 1.2 \ \Omega$$

Armature winding:

$$R_a = 1.36 \ \Omega \qquad X_a = 1.6 \ \Omega$$

Find:

1. The generated voltage.
2. The power output.
3. The shaft torque.
4. The efficiency of the motor (assume that the rotational loss is 80 W).

Solution:

Using the circuit model for the series motor discussed in Chapter 15, we can obtain the circuit shown in Figure 16.18.

1. The impedance angle may be found from the power factor:

$$\cos^{-1} \theta = 0.912 \qquad \theta = -24.22°$$

and therefore the expression for the back emf can be evaluated as follows:

$$\mathbf{E}_b = \mathbf{V}_S - \mathbf{I}_a(R_f + R_a + jX_f + jX_a)$$
$$= 120 - (17.58\angle-24.22°)(0.65 + 1.36 + j\,1.2 + j\,1.6)$$
$$= 74.1\angle-24.22°$$

Figure 16.18 Equivalent circuit of a universal motor

2. The total power developed by the motor is given by the product of the induced emf and the series current:

$$P_{\text{total}} = E_b \times I_s = 74.1 \times 17.85 = 1{,}322.69 \ \text{W}$$

The useful output power is the difference between the total power developed by the motor and the rotational losses:

$$P_{\text{out}} = P_{\text{total}} - P_{\text{rot}} = 1{,}322.69 - 80 = 1{,}242.69 \ \text{W}$$

3. The angular speed in units of rad/s is

$$\omega_m = \frac{2\pi \times 800}{60} = 83.776 \ \text{rad/s}$$

and therefore the output torque, T_{out} is

$$T_{\text{out}} = \frac{P_{\text{out}}}{\omega_m} = \frac{1{,}242.69}{837.76} = 14.83 \ \text{N-m}$$

4. The input power (or power supplied to the motor) is found by computing the phasor product of the supply voltage and current:

$$P_{\text{in}} = V_s I_s \cos \theta = 120 \times 17.58 \times 0.912 = 1{,}923.96 \ \text{W}$$

Knowing both the input and the output power, we can compute the efficiency of the motor according to the following expression:

$$\text{Efficiency} = \frac{1{,}242.69}{1{,}923.96} = 0.6459 = 64.59\%$$

Example   16.9

Compute an expression for the average torque of an AC motor based on the circuit diagram of Figure 16.18. Assume that the magnetic flux is such that the motor operates in the linear region of its magnetization curve.

Solution:

The instantaneous magnetic flux ϕ of an AC series motor produced by the motor current i_a in the field is

$$\phi = k_S i_a$$

The instantaneous torque is given by

$$T(t) = k_T \phi(t) i_a(t)$$

Thus, the time-average torque, T_{ave}, over one period of the supply waveform, τ, is

$$T_{\text{ave}} = \frac{1}{\tau\omega} \int_0^{\tau\omega} k_T k_S i_a^2 \, dt$$

When operated on AC supplies, the universal motor develops a pulsating torque, one pulse for each half-cycle, as shown in Figure 16.15. The average value of this torque, based on the preceding expression, may then be computed to be

$$T_{\text{ave}} = \frac{1}{\tau/\omega} \int_0^{\tau/\omega} k_T k_S i_a^2 \, d\alpha = k_T k_S I_a^2 = K I_a^2$$

where $K = k_T k_S$ and I_a is the rms value of the armature current, i_a.

Single-Phase Induction Motors

Figure 16.19 Single-phase induction motor

A typical single-phase induction motor bears close resemblance to the polyphase squirrel-cage induction motor discussed in Chapter 15, the major difference being in the configuration of the stator winding. A simplified schematic diagram of such a motor, with a single winding, is shown in Figure 16.19; the winding is typically distributed around the stator so as to produce an approximately sinusoidal mmf.

Assume that the mmf for a practical motor can be generated so as to approximate the following function:

$$\mathscr{F} = F_{\max} \cos(\omega t) \cos(\theta_m)$$

This function can be written as the sum of two components, as follows:

$$\mathscr{F}^+ = \frac{1}{2} F_{\max} \cos(\theta_m - \omega t)$$

$$\mathscr{F}^- = \frac{1}{2} F_{\max} \cos(\theta_m + \omega t)$$

These two components may be interpreted as representing two mmf waves traveling in opposite directions around the stator. Each of these mmf's produces torque according to the induction principles described in Chapter 15; however, the two components are equal and opposite, and no net torque results if the rotor is at rest. The resulting mmf is pulsating (i.e., changing in amplitude), but not rotating in space, as it would be in a polyphase stator. If the rotor is made to turn in either direction, however, the two mmf's will not be equal any longer, because the motion of the rotor will induce an additional mmf, which will add to one of the two mmf's and subtract from the other. Thus, a net torque will be established, causing the motor to continue its rotation in the same direction in which it was started. In particular, if the rotor is started in the forward direction, the forward mmf, $\mathscr{F}^+$, will be greater than the backward mmf, and the motor will continue to rotate in the forward direction.

Figure 16.20 depicts an equivalent circuit for the single-phase induction motor *with stationary rotor,* where the parameters in the circuit are defined as follows:

V_S = supply voltage

R_S = resistance of stator winding

X_S = leakage reactance of stator winding

X_m = magnetizing reactance of stator winding

X_R = leakage reactance of rotor referred to stator at standstill

R_R = leakage resistance of rotor referred to stator at standstill

E_b = voltage induced in the stator winding by the (stationary) pulsating flux in air gap

Figure 16.20 Circuit model for single-phase induction motor with rotor at standstill

Figure 16.21 depicts the equivalent circuit for the same motor with the rotor rotating with slip s. Note that the circuit is asymmetrical, because of the different air-gap flux forward and backward components, E_f and E_b, respectively. The factors of 0.5 come from the resolution of the pulsating stator mmf into forward and backward components. Note further that the reflected rotor impedance is asymmetrical because of the presence of the slip parameter in the expression for the reflected rotor resistance. Further,

Figure 16.21 Circuit model for single-phase induction motor with rotor in motion

the circuit model also confirms that the forward induced voltage, E_f, must be greater than the backward voltage, E_b, since the slip is always less than 1.

It can be shown that the torque components in the forward and backward directions are given by the expressions

$$T_f = \frac{P_f}{\omega_s} \tag{16.3}$$

and

$$T_b = \frac{P_b}{\omega_s} \tag{16.4}$$

where ω_s is the synchronous speed and

$$P_f = I_1^2 R_f \tag{16.5}$$

Here, R_f is the resistive component of the forward field impedance; also,

$$P_b = I_1^2 R_b \tag{16.6}$$

where R_b is the resistive component of the backward field impedance. Since the torque produced by the backward field is in the opposite direction to that produced by the forward field, the net torque will consist of the difference between the two:

$$T = T_f - T_b = \frac{I_1^2(R_f - R_b)}{\omega_s} \tag{16.7}$$

The mechanical power developed by the motor is

$$\begin{aligned} P_{mech} = T\omega_m &= T\omega_s(1 - s) = (P_f - P_b)(1 - s) \\ &= I_1^2(R_f - R_b)(1 - s) \end{aligned} \tag{16.8}$$

Example   16.10

A 115-V, 60-Hz, four-pole, single-phase induction motor is rotating at a speed of 1,710 rev/min. Find the slip of the field in

1. The direction of rotation.
2. The backward direction.

Solution:

1. First, we find the synchronous speed of the motor:

$$n_s = \frac{120f}{p} = \frac{120 \times 60}{4} = 1{,}800 \text{ rev/min}$$

The slip in the direction of rotation can now be computed, since we know the mechanical speed of rotation of the motor:

$$s_f = \frac{n_s - n_m}{n_s} = \frac{1{,}800 - 1710}{1{,}800} = 0.05 = 5\%$$

2. The slip in the backward direction is simply computed as:

$$s_b = 2 - s_f = 2 - 0.05 = 1.95$$

Example 16.11

A four-pole single-phase induction motor is rated at $\frac{1}{4}$ hp, 110 V, 60 Hz. The parameters of the circuit model are:

$$R_S = 1.5 \ \Omega \qquad X_S = 2 \ \Omega$$

$$R_R = 3 \ \Omega \qquad X_R = 2 \ \Omega$$

$$X_m = 50 \ \Omega \qquad s = 0.05$$

The motor is operating at rated voltage and rated frequency. Find the input current and electromagnetic torque.

Solution:

Using the equivalent circuit shown in Figure 16.21, we find that for the given numerical values, the impedance seen by the emf $E_b Z_b$, is much smaller than that seen by E_f. This observation permits the following approximation:

$$0.5Z_b = (0.5)\frac{jX_m[R_R/(2 - s) + jX_R]}{R_R/(2 - s) + j(X_m + X_R)} \approx 0.5\left(\frac{R_R}{2 - s} + jX_R\right)$$

This approximation is valid for values of slip less than 0.15. The corresponding equivalent circuit is shown in Figure 16.22.

Figure 16.22 Approximate circuit model for single-phase induction motor

Using the approximate circuit, we find that

$$0.5Z_b = 0.5(1.538 + j2) \ \Omega = 0.5R_b + j0.5X_b$$

and, for the forward direction of rotation,

$$0.5Z_f = \frac{j25(30 + j1)}{30 + j26} = 18.9\angle 51°$$

$$= 11.9 + j14.69 = 0.5R_f + j0.5X_f$$

Let $Z_S = R_S + jX_S$; then,

$$Z_S = 1.5 + j2\ \Omega$$

and the total impedance is

$$Z = Z_S + 0.5Z_f + 0.5Z_b = 14.169 + j\,17.69 = 22.66\angle 51.3°\ \Omega$$

The stator current may then be found as follows:

$$I_S = \frac{V_S}{Z} = \frac{110}{22.66\angle 51.3°} = 4.85\angle -51.3°\ A$$

The power factor is $pf = \cos(51.3°) = 0.625$. Such a low value for the line power factor is typical of single-phase motors.

The forward field air-gap power is

$$P_f = I_S^2 0.5R_f = (4.85)^2 \times 11.9 = 279.9\ W$$

The backward field air-gap power is

$$P_b = I_S^2 0.5R_b = (4.85)^2 \times 1 = 18.1\ W$$

The synchronous speed is

$$\omega_s = \frac{4\pi f}{P} = 179.1\ \text{rad/s}$$

The torque is

$$T = \frac{P_f - P_b}{\omega_s} = 1.38\ \text{N-m}$$

Example   16.12

The machine in Example 16.11 is operated at rated voltage and frequency and with slip $s = 0.05$. The combined mechanical and core losses are 30 W. Find

1.   The output torque.
2.   The output power.
3.   The efficiency.

Solution:

For the same operating conditions as in Example 16.11, the mechanical power is

$$P_{\text{mech}} = (1 - s)(P_f - P_b) = 0.95(279.9 - 18.1) = 248.71\ W$$

Subtract the losses to find the output power:

$$P_{\text{out}} = P_{\text{mech}} - P_{\text{loss}} = 248.71 - 30 = 218.71\ W$$

The shaft speed is

$$\omega_m = (1 - s)\omega_s = 179\ \text{rad/s}$$

The output torque is

$$T_{\text{out}} = \frac{P_{\text{out}}}{\omega_m} = \frac{218.71}{179} = 1.22\ \text{N-m}$$

The electrical losses are

$$P_{R_S} = I_S^2 R_S = (4.85)^2 \times 2 = 47 \text{ W}$$

$$P_{R_{R(f)}} = sP_f = 0.05 \times 279.9 = 14 \text{ W}$$

$$P_{R_{R(b)}} = (2 - s)P_b = 1.95 \times 18.1 = 35.3 \text{ W}$$

and the total loss is

$$P_{\text{loss}} = P_{R_S} + P_{R_{R(f)}} + P_{R_{R(b)}} + P_{\text{rot}} = 126.3 \text{ W}$$

Finally, the efficiency may be computed to be:

$$\text{Efficiency} = 1 - \frac{\sum \text{losses}}{P_{\text{out}} + \sum \text{losses}} = 1 - \frac{126.3}{218.71 + 126.3} = 0.634 = 63.4\%$$

The equations and circuit models in the preceding examples suggest that a single-phase induction motor is capable of sustaining a torque, and of reaching its operating speed, once it is started by external means. However, because the magnetic field in a single-phase winding is stationary, a single-phase motor is not self-starting. The speed-torque characteristic of a typical single-phase induction motor shown in Figure 16.23 clearly shows that the starting torque for this motor is zero. The curve also shows that the motor can operate in either direction, depending on the direction of the initial starting torque, which must be provided by separate means.

Figure 16.23 Torque-speed curve of a single-phase induction motor

Classification of Single-Phase Induction Motors

Thus far, we have not mentioned how the initial starting torque can be provided to a single-phase motor. In practice, single-phase motors are classified by their starting and running characteristics, and several methods exist to provide nonzero starting torque. The aim of this section is to classify single-phase motors by describing their configuration on the basis of the method of starting. For each class of motor, a torque-speed characteristic will also be described.

Split-Phase Motors

Split-phase motors are constructed with two separate stator windings, called **main** and **auxiliary winding;** the axes of the two windings are actually at 90° with respect to each other, as shown in Figure 16.24. The auxiliary winding current is designed to be out of phase with the main winding current, as a result of different reactances of the two windings. Different winding reactances can be attained by having a different ratio of resistance to inductance—for example, by increasing the resistance of the auxiliary winding. In particular, the auxiliary winding current, $\mathbf{I}_{\text{aux}}$, leads the main winding current, $\mathbf{I}_{\text{main}}$. The net effect is that the motor sees a two-phase (unbalanced) current that results in a rotating magnetic field, as in any polyphase stator arrangement. Thus, the motor has a nonzero starting torque, as shown in Figure 16.25.

Figure 16.24 Split-phase motor

Figure 16.25 Torque-speed curve of split-phase motor

Once the motor has started, a centrifugal switch is used to disconnect the auxiliary winding, since a single winding is sufficient to sustain the motion of the rotor. The switching action permits the use of relatively high-resistance windings, since these are not used during normal operation and therefore one need not be concerned with the losses associated with a higher-resistance winding. Figure 16.25 also depicts the combined effect of the two modes of operation of the split-phase motor.

Split-phase motors have appropriate characteristics (at very low cost) for fans, blowers, centrifugal pumps, and other applications in the range of $\frac{1}{20}$ to $\frac{1}{2}$ hp.

Capacitor-Type Motors

Another method for obtaining a phase difference between two currents that will give rise to a rotating magnetic field is by the addition of a capacitor. Motors that use this arrangement are termed **capacitor-type motors.** These motors make different use of capacitors to provide starting or running capabilities, or a combination of the two. The **capacitor-start motor** is essentially identical to the split-phase motor, except for the addition of a capacitor in series with the auxiliary winding, as shown in Figure 16.26. The addition of

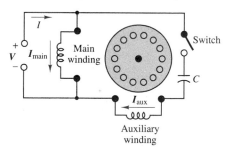

Figure 16.26 Capacitor-start motor

the capacitor changes the reactance of the auxiliary circuit in such a way as to cause the auxiliary current to lead the main current. The advantage of using the capacitor as a means for achieving a phase split is that greater starting torque may be obtained than with the split-phase arrangement. A centrifugal switching arrangement is used to disconnect the auxiliary winding above a certain speed, in the neighborhood of 75 percent of synchronous speed.

Figure 16.27 depicts the torque-speed characteristic of a capacitor-start motor. Because of their higher starting torque, these motors are very useful in connection with loads that present a high static torque. Examples of such loads are compressors, pumps, and refrigeration and air-conditioning equipment.

It is also possible to use the capacitor-start motor without the centrifugal switch, leading to a simpler design. Motors with this design are called **permanent split-capacitor motors;** they offer a compromise between running and starting characteristics. A typical torque-speed curve is shown in Figure 16.28.

A further compromise can be achieved by using two capacitors, one to obtain a permanent phase split and the resulting improvement in running characteristics, the other to improve the starting torque. A small capacitance is sufficient to improve the running performance, while a much larger capacitor provides the temporary improvement in starting torque. A motor with this design is called a **capacitor-start capacitor-run motor;** its schematic diagram is shown in Figure 16.29. Its torque-speed characteristic is similar to that of a capacitor-start motor.

Figure 16.27 Torque-speed curve for a capacitor-start motor

Figure 16.28 Torque-speed curve for a permanent split-capacitor motor

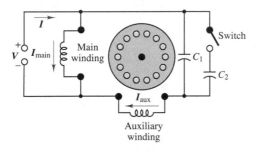

Figure 16.29 Capacitor-start capacitor-run motor

Example   16.13

A $\frac{1}{3}$-hp, 120-V, 60-Hz capacitor-start motor has the following parameters:

For the main winding,

$$R_m = 4.5 \ \Omega \qquad X_m = 3.7 \ \Omega$$

For the auxiliary winding,

$$R_a = 9.5 \ \Omega \qquad X_a = 3.5 \ \Omega$$

Find the required starting capacitance of C in Figure 16.29.

Solution:

Capacitors can be used to improve the starting performance, running performance, or both. By means of a starting capacitor of appropriate value, the auxiliary-winding

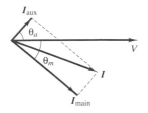

Figure 16.30 Starting phasor diagram for capacitor-start motor

current $\mathbf{I}_{aux}$ at standstill could be made to lead the main-winding current $\mathbf{I}_{main}$ by 90 degrees, as shown in Figure 16.30. The 90° shift will provide the maximum possible starting torque.

The impedance angle of the main winding is

$$\theta_m = \tan^{-1}\frac{X_m}{R_m} = \tan^{-1}\frac{3.7}{4.5} = 39.4°$$

The impedance angle of the auxiliary winding is

$$\theta_a = 39.6° - 90° = -50.6°$$

The minus sign means that $\mathbf{I}_{aux}$ leads the terminal voltage. The required capacitive reactance can be determined from the relationship

$$\tan^{-1}\frac{3.5 - X_C}{9.5} = -50.6°$$

or

$$X_C = 1.21 \times 9.5 + 3.5 = 15.07\ \Omega$$

Thus,

$$C = \frac{1}{X_C\omega} = \frac{1}{377 \times 15.07} = 176 \times 10^{-6}\ \text{F} = 176\mu\text{F}$$

Shaded-Pole Motors

Figure 16.31 Shaded-pole motor

The last type of single-phase induction motor discussed in this chapter is the **shaded-pole motor.** This type of motor operates on a different principle from the motors discussed thus far. The stator of a shaded-pole motor has a salient pole construction, as shown in Figure 16.31, that includes a shading coil consisting of a copper band wound around part of each pole. The flux in the shaded portion of the pole lags behind the flux in the unshaded part, achieving an effect similar to a rotation of the flux in the direction of the shaded part of the pole. This flux rotation in effect produces a rotating field that enables the motor to have a starting torque. This construction technique is rather inexpensive and is used in motors up to about $\frac{1}{20}$ hp.

A typical torque-speed characteristic for a shaded-pole motor is given in Figure 16.32.

Figure 16.32 Torque-speed curve of a shaded-pole motor

Example   16.14

A typical nameplate of a split-phase motor is represented in the accompanying table. This example will explain the significance of the most important nameplate parameters. We shall also calculate the rated slip s, the synchronous speed, and the rated torque from the nameplate data.

THERMAL PROTECTED				SPLIT-PHASE FAN & BLOWER MOTOR			
MOD	4k800	HP	1/3	RPM	1725	KVA CODE	N
V	115			A.	5.5		
FR	48Y	Hz	60	PH	1	DUTY	CONT.
INS.CL.	B	MAX. AMB	40 C	S.F.	1.35	BRG	SLEEVE

Solution:

An explanation of the nameplate for a typical electric motor was given in Chapter 15. This example focuses on a few specific items of interest in the case of a split-phase motor. As you can see, the nameplate directly indicates the split-phase motor classification. Following the Hz designation is the phase information. AC systems may have one, two, or three phases. Single-phase and three-phase systems are the most common.

The code letter following "KVA CODE" indicates the locked-rotor kilovolt-amperes per horsepower, as explained in *NEMA Motor and Generator Standards,* NEMA Publication Number MG1-10.37. The symbol "N" means that this motor has a maximum locked-rotor kilovolt-amperes per horsepower of 12.5. Since the motor is rated at $\frac{1}{3}$ hp, the maximum locked-rotor kilovolt-amperes is 12.5/3 = 4.167. The maximum locked-rotor amperes at 115 V will be 4.167 kVA/115 V = 36.23 A.

A large percentage of fractional-horsepower motors are now provided with built-in thermal protection. The use of such protection will also be indicated in the motor nameplate—here for example, by "THERMAL PROTECTED."

Bearing is abbreviated as "BRG." Fractional-horsepower motors normally use one of two types of bearings: sleeve or ball.

A variety of additional information may appear on the nameplate. This may include instructions for connecting the motor to a source of supply, reversing the direction of rotation, lubricating the motor, or operating it safely.

For the machine in this example, the synchronous speed is

$$n_s = 1,800 \text{ rev/min}$$

The slip at rated speed is

$$s = \frac{n_s - n}{n_s} = \frac{1,800 - 1,725}{1,800} = 0.042$$

The power is

$$P = \frac{1}{3} \text{ hp} = \frac{1}{3} \times 746 = 248.7 \text{ W}$$

The rated torque is

$$T = \frac{K \times \text{power}}{\text{rev/min}}$$

where the constant K is given by

$\quad K = 0.97376$ when T is expressed in meter-kilograms

$\quad K = 9.549$ when T is expressed in newton-meters

$$T = 9.549 \times \frac{248.7}{1,725} = 1.377 \text{ N-m}$$

Summary of Single-Phase Motor Characteristics

Four basic classes of single-phase motors are commonly used:

1. Single-phase induction motors are used for the larger home and small business tasks, such as furnace oil burner pumps, or hot water or hot air circulators. Refrigerator compressors, lathes, and bench-mounted circular saws are also powered with induction motors.

2. Shaded-pole motors are used in the smaller sizes for quiet, low-cost applications. The size range is from $\frac{1}{30}$ hp (24.9 W) to $\frac{1}{2}$ hp (373 W), particularly for fans and similar drives in which the starting torque is low.

3. Universal motors will operate on any household AC frequency or on DC without modification or adjustment. They can develop very high speed while loaded, and very high power for their size. Vacuum cleaners, sewing machines, kitchen food mixers, portable electric drills, portable circular saws, and home motion-picture projectors are examples of applications of universal motors.

4. The capacitor-type motor finds its widest field of application at low speeds (below 900 rev/min) and in ratings from $\frac{3}{4}$ hp (0.5595 kW) to 3 hp (2.238 kW) at all speeds, especially in fan drives.

DRILL EXERCISES

16.10 Prove that the mmf for the single-phase induction motor can be expressed as the sum of two components traveling in opposite directions.

16.11 For the circuit shown in Figure 16.29, draw the starting phasor diagram relating $\mathbf{V}$, $\mathbf{I}$, $\mathbf{I}_{\text{main}}$, and $\mathbf{I}_{\text{aux}}$.

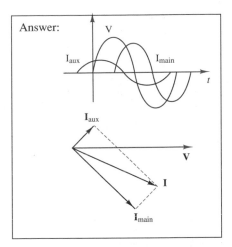

Answer:

16.12

a. What is the zero-speed torque for a single-phase induction motor?

b. If external torque is applied to the machine, what is its terminal speed?

16.13 For the motor of Example 16.14, calculate the rated torque in meter-kilograms.

> Answer: 0.14 m-kg

CONCLUSION

- The most common engineering applications of electric motors make use of a number of special-purpose motors, with operating characteristics that may be derived from the fundamental principles presented in Chapters 14 and 15.

- The brushless DC motor is a PM synchronous motor in which the mechanical commutation of conventional DC motors is replaced by electronic commutation. Brushless DC motors can be made quite compact and find application in electric vehicle propulsion and motion control.

- Stepping motors—of the variable-reluctance, PM, or hybrid type—permit fine angular displacement control by moving in fixed, discrete steps. Typical applications are in robotics and control systems.

- The universal motor is very similar in construction to a series DC motor but can operate on AC supplies; its speed can be controlled by means of electronic circuitry of modest complexity. Thus, the universal motor finds common application in both variable- and fixed-speed appliances, such as power drills and vacuum cleaners, respectively.

• Squirrel-cage induction motors can be operated on a single-phase AC supply if a means is provided for establishing a starting torque. Various techniques are commonly employed, such as split-phase, capacitor-start, and shaded-pole construction. The different types are characterized by differing torque-speed characteristics that make the single-phase induction motor a very versatile device. This is the most commonly employed electric machine.

KEY TERMS

HOMEWORK PROBLEMS

16.1 From a mechanical standpoint, the brushless motor consists of how many parts?

16.2 Discuss and analyze the discrepancy between the ideal trapezoidal waveforms of the brushless DC drive of Figure 16.4 and the actual waveforms that would result if winding resistance and practical transistor switching characteristics were taken into account. Explain how this leads to torque ripple in brushless DC motors.

16.3 Derive the dynamic equation for a stepping motor coupled to a load. The motor moment of inertia is J_m, the load moment of inertia is J_L, the viscous damping coefficient is D, and motor friction torque is T_f.

16.4 Sketch the rotor-stator configuration of a hybrid stepper motor capable of 18° steps. [*Hint:* The rotor will have five teeth.]

16.5 Use a binary counter and logic gates to implement the stepping motor binary sequence of Drill Exercise 16.7.

16.6 The half-step excitation scheme of a stepper motor is a combination of the single-phase and two-phase excitations. This scheme reduces the step angle by half. Derive the excitation sequence for a three-phase motor and verify the result.

16.7 The schematic diagram of a four-phase, two-pole PM stepper motor is shown in Figure P16.1. The phase coils are excited in sequence by means of a logic circuit. Find

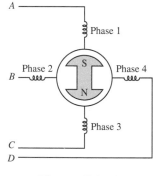

Figure P16.1

a. The logic schedule for full-stepping of this motor.

b. The displacement angle of the full step.

16.8 A PM stepper motor is designed to provide a full-step angle of 15°. Find the number of stator and rotor poles.

16.9 A bridge driver scheme for a two-phase stepping motor is shown in Figure P16.2. Find the excitation sequences of the bridge operation (fill in the blanks of the table).

Figure P16.2

Clock state	Reset	1	2	3	4	5	6	7	8
S_1									
S_2									
S_3									
S_4									
S_5									
S_6									
S_7									
S_8									

16.10 For a two-phase, two-pole-rotor PM stepper motor, derive a switching sequence for continuous rotation and draw the waveform of on-off periods of circuits in each of the two phases.

16.11 Determine whether the following motors are integral- or fractional-horse-power motors:

a. $\frac{3}{4}$ hp, 900 rev/min

b. $1\frac{1}{2}$ hp, 3,600 rev/min

c. $\frac{3}{4}$ hp, 1,800 rev/min

d. $1\frac{1}{2}$ hp, 6,000 rev/min

16.12   The spatial fluctuation of the stator mmf $\mathscr{F}_1$ is expressed as

$$\mathscr{F}_1 = F_{1(peak)} \cos \theta$$

where θ is the electrical angle measured from the stator coil axis and $F_{1(peak)}$ is the instantaneous value of the mmf wave at the coil axis and is proportional to the instantaneous stator current. If the stator current is a cosine function of time, the instantaneous value of the spatial peak of the pulsating mmf wave is

$$F_{1(peak)} = F_{1(max)} \cos \omega t$$

where $F_{1(max)}$ is the peak value corresponding to maximum instantaneous current. Derive the expression for $\mathscr{F}_1$, and verify that for a single-phase winding, both forward and backward components are present.

16.13   A 200-V, 60-Hz, 10-hp single-phase induction motor operates at an efficiency of 0.86 and a power factor of 0.9. What capacitor should be placed in parallel with the motor so that the feeder supplying the motor will operate at unity power factor?

16.14   A 230-V, 50-Hz, two-pole single-phase induction motor is designed to run at 3 percent slip. Find the slip in the opposite direction of rotation. What is the speed of the motor in the normal direction of rotation?

16.15   A sine-cosine resolver is shown in Figure P16.3. The stator winding $S_1 S_2$ is supplied by the excitation voltage $v_t = V_t \sin \omega t$. When $\theta = 0$, that is, when the axis of winding ab is in alignment with the axis of winding $S_1 S_2$, the induced voltage in the rotor winding ab will be $E_{ab} = K V_t \sin \omega t$ and $cd = 0$, where K is the effective turns ratio of the rotor winding to the stator winding. If the rotor is turned counterclockwise through an angle of θ, then the induced voltage produced by the forward field is $E_{ab}^+ = \frac{1}{2} K V_t \sin(\omega t - \theta)$ and the induced voltage produced by the backward field is $E_{ab} = \frac{1}{2} V_t \sin(\omega t + \theta)$. Find the total voltage in winding cd. Verify that the magnitude of the voltage E_{cd} is the cosine function of the angle θ.

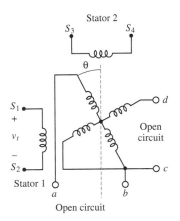

Figure P16.3

16.16 A $\frac{1}{4}$-hp, 110-V, 60-Hz, four-pole capacitor-start motor has the following parameters:

$$R_S = 2.02 \ \Omega \qquad X_S = 2.8 \ \Omega$$
$$R_R = 4.12 \ \Omega \qquad X_R = 2.12 \ \Omega$$
$$X_m = 66.8 \ \Omega \qquad s = 0.05$$

Find

a. The stator current.
b. The mechanical power.
c. The rotor speed.

16.17 A $\frac{1}{4}$-hp, four-pole, 110-V, 60-Hz single-phase induction motor has the following data:

$$R_S = 1.86 \ \Omega \qquad X_S = 2.56 \ \Omega$$
$$R_R = 3.56 \ \Omega \qquad X_R = 2.56 \ \Omega$$
$$X_m = 53.5 \ \Omega \qquad s = 0.05$$

Find the mechanical power output.

16.18 A one-phase, 115-V, 60-Hz, four-pole induction motor has the following parameters:

$$R_S = 0.5 \ \Omega \qquad X_S = 0.4 \ \Omega$$
$$R_R = 0.25 \ \Omega \qquad X_R = 0.4 \ \Omega$$
$$X_m = 35 \ \Omega$$

Find the input current and developed torque when the motor speed is 1,730 rev/min.

16.19 The no-load test of a single-phase induction motor is made by running the motor without load at rated voltage and rated frequency. Derive the equivalent circuit of a single-phase induction motor for the no-load test. [*Hint:* The no-load slip is very small.]

16.20 Derive the equivalent circuit of a single-phase induction motor for the locked-rotor test. Neglect the magnetizing current.

16.21 The design for a $\frac{1}{8}$-hp, two-pole, 115-V universal motor gives the effective resistances of the armature and series field as 4 Ω and 6 Ω, respectively. The output torque is 0.17 N-m when the motor is drawing rated current of 1.5 A (rms) at a power factor of 0.88 at rated speed. Find:

a. The full-load efficiency.
b. The rated speed.
c. The full-load copper losses.
d. The combined windage, friction, and iron losses.
e. The motor speed when the rms current is 0.5 A, neglecting phase differences and saturation.

16.22   A 240-V, 60-Hz, two-pole universal motor operates at a speed of 12,000 rev/min on full load and draws a current of 6.5 A at 0.94 power factor lagging. The series field-winding impedance is 4.55 $+j\,3.2$ ohms, and the armature circuit impedance is 6.15 $+j\,9.4$ ohms. Find

a.   The back emf of the motor.
b.   The mechanical power developed by the motor.
c.   The power output if the rotational loss is 65 W.
d.   The efficiency of the motor.

16.23   A single-phase motor is drawing 20 A from a 400-V, 50-Hz supply. The power factor is 0.8 lagging. What value of capacitor connected across the circuit will be necessary to raise the power factor to unity?

16.24   A three-phase induction motor is required to operate from a single-phase source. One possible connection is shown in Figure P16.4. Will the motor work? Explain why or why not.

16.25   What type of motor would you select to perform the following tasks? Justify your selection.

a.   Vacuum cleaner
b.   Refrigerator
c.   Air conditioner compressor
d.   Air conditioner fan
e.   Variable-speed sewing machine
f.   Clock
g.   Electric drill
h.   Tape drive
i.   *X-Y* plotter

Figure P16.4

Linear Algebra, Complex Numbers, Laplace Transform

I. SOLVING SIMULTANEOUS LINEAR EQUATIONS, CRAMER'S RULE AND MATRIX EQUATION

The solution of simultaneous equations, such as those that are often seen in circuit theory, may be obtained relatively easily by using Cramer's rule. This method applies to 2×2 or higher systems of equations. Cramer's rule requires the use of the concept of determinant. The method of determinants is valuable because it is systematic, general, and useful in solving complicated problems. A determinant is defined as a square array of numbers having a numerical value, such as

$$det = \begin{vmatrix} a_{11} & a_{12} \\ a_{21} & a_{22} \end{vmatrix} \tag{A.1}$$

In this case the determinant is 2×2 array, with two rows and two columns and its determinant defined as

$$det = a_{11}a_{22} - a_{12}a_{21} \tag{A.2}$$

A third-order, or 3×3, determinant such as

$$det = \begin{vmatrix} a_{11} & a_{12} & a_{13} \\ a_{21} & a_{22} & a_{23} \\ a_{31} & a_{32} & a_{33} \end{vmatrix} \tag{A.3}$$

has three rows and three columns and its determinant is defined as

$$det = a_{11}(a_{22}a_{33} - a_{23}a_{32}) - a_{12}(a_{21}a_{33} - a_{23}a_{31}) \\ + a_{13}(a_{21}a_{32} - a_{22}a_{31}) \tag{A.4}$$

For higher-order determinants, you may refer to a linear algebra book. To illustrate Cramer's method, a set of two equations in general form will be solved here. A set of two linear simultaneous algebraic equations in two unknowns can be written in the form

$$a_{11}x_1 + a_{12}x_2 = b_1 \\ a_{21}x_1 + a_{22}x_2 = b_2 \tag{A.5}$$

where x_1 and x_2 are the two unknowns to be solved for. The coefficients a_{11}, a_{12}, a_{21} and a_{22} are known quantities. The two quantities on the right-hand side, b_1 and b_2, are also known (these are typically the source currents and voltages in a circuit problem). The set of equations can be arranged in matrix form, as shown in equation A.6.

$$\begin{bmatrix} a_{11} & a_{12} \\ a_{21} & a_{22} \end{bmatrix} \begin{bmatrix} x_1 \\ x_2 \end{bmatrix} = \begin{bmatrix} b_1 \\ b_2 \end{bmatrix} \tag{A.6}$$

In equation A.6, a coefficient matrix multiplied by a vector of unknown variables is equated to a solution vector. Cramer's rule can then be applied to find x_1 and x_2 using the following formulas:

$$x_1 = \frac{\begin{vmatrix} b_1 & a_{12} \\ b_2 & a_{22} \end{vmatrix}}{\begin{vmatrix} a_{11} & a_{12} \\ a_{21} & a_{22} \end{vmatrix}}$$

$$x_2 = \frac{\begin{vmatrix} a_{11} & b_1 \\ a_{21} & b_2 \end{vmatrix}}{\begin{vmatrix} a_{11} & a_{12} \\ a_{21} & a_{22} \end{vmatrix}} \tag{A.7}$$

Thus, the solution is given by the ratio of two determinants: the denominator is the determinant of the matrix of coefficients, while the numerator is the determinant of the same matrix with the solution vector ($\begin{bmatrix} b_1 \\ b_2 \end{bmatrix}$ in this case) substituted in place of the column of the coefficient matrix corresponding to the desired variable (i.e., first column for x_1, second column for x_2, etc.). In a circuit analysis problem, the coefficient matrix is formed by the resistance (or conductance) values, the vector of unknowns is composed of the mesh currents (or node voltages), and the solution vector contains the source currents or voltages.

In practice, many calculations involve solving higher-order systems of linear equations. Therefore, a variety of computer software packages are often used to solve higher-order systems of linear equations.

DRILL EXERCISES

A.1 Use Cramer's rule to solve the system

$$5v_1 + 4v_2 = 6$$
$$3v_1 + 2v_2 = 4$$

Answer: $v_1 = 2$; $v_2 = -1$

A.2 Use Cramer's rule to solve the system

$$i_1 + 2i_2 + i_3 = 6$$
$$i_1 + i_2 - 2i_3 = 1$$
$$i_1 - i_2 + i_3 = 0$$

Answer: $i_1 = 1$; $i_2 = 2$; $i_3 = 1$

A.3 Convert the following system of linear equations into a matrix equation as shown in equation A.6 and find matrices A and b.

$$2i_1 - 2i_2 + 3i_3 = -10$$
$$-3i_1 + 3i_2 - 2i_3 + i_4 = -2$$
$$5i_1 - i_2 + 4i_3 - 4i_4 = 4$$
$$i_1 - 4i_2 + i_3 + 2i_4 = 0$$

Answer: $A = \begin{bmatrix} 2 & -2 & 3 & 0 \\ -3 & 3 & -2 & 1 \\ 5 & -1 & 4 & -4 \\ 1 & -4 & 1 & 2 \end{bmatrix}$ and $b = \begin{bmatrix} -10 \\ -2 \\ 4 \\ 0 \end{bmatrix}$

II.   INTRODUCTION TO COMPLEX ALGEBRA

From your earliest training in arithmetic, you have dealt with real numbers such as 4, -2, $\frac{5}{9}$, π, e, etc., which may be used to measure distances in one direction or another from a fixed point. However, a number such as x that satisfies the equation

$$x^2 + 9 = 0 \qquad\qquad\qquad (A.8)$$

is not a real number. Imaginary numbers were introduced to solve equations such as equation A.8. Imaginary numbers add a new dimension to our numbering system. To deal with imaginary numbers, a new element, j, is added to the number system having the property

$$j^2 = -1$$

or $\qquad\qquad\qquad\qquad\qquad\qquad\qquad\qquad\qquad (A.9)$

$$j = \sqrt{-1}$$

Thus, we have $j^3 = -1$, $j^4 = 1$, $j^5 = j$, etc. Using equation A.9, we can see that the solutions to equation A.8 are $\pm j\,3$. In mathematics, the symbol i is used for the imaginary unit, but this might be confused with current in electrical engineering. Therefore, the symbol j is used in this book.

A complex number (indicated in boldface notation) is an expression of the form

$$A = a + jb \qquad\qquad\qquad (A.10)$$

where a and b are real numbers. The complex number A has a real part, a, and an imaginary part, b, which can be expressed as

$$a = \operatorname{Re} A$$
$$b = \operatorname{Im} A \qquad\qquad\qquad (A.11)$$

It is important to note that a and b are both real numbers. The complex number $a + jb$ can be represented on a rectangular coordinate plane, called the *complex plane,* by interpreting it as a point (a, b). That is, the horizontal coordinate is a in real axis and the vertical coordinate is b in imaginary axis, as shown in Figure A.1. The complex number $A = a + jb$ can also be uniquely located in the complex plane by specifying the distance r along a straight line from the origin and the angle θ, which this line makes with the real axis, as shown in Figure A.1. From the right triangle of Figure A.1, we can see that:

$$r = \sqrt{a^2 + b^2}$$
$$\theta = \tan^{-1}\left(\frac{b}{a}\right) \qquad\qquad\qquad (A.12)$$
$$a = r \cos \theta$$
$$b = r \sin \theta$$

Then, we can represent a complex number by the expression:

$$A = r e^{j\theta} = r\angle\theta \qquad\qquad\qquad (A.13)$$

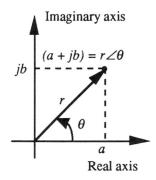

Figure A.1 Polar form representation of complex numbers

which is called the polar form of the complex number. The number r is called the magnitude (or amplitude) and the number θ is called the angle (or argument). The two numbers are usually denoted by: $r = |A|$ and $\theta = \arg A = \angle A$.

Given a complex number $A = a + jb$, the complex conjugate of A, denoted by the symbol A^*, is defined by the following equalities:

$$\begin{aligned} \operatorname{Re} A^* &= \operatorname{Re} A \\ \operatorname{Im} A^* &= -\operatorname{Im} A \end{aligned} \tag{A.14}$$

That is, the sign of the imaginary part is reversed in the complex conjugate.

Finally, we should remark that two complex numbers are equal *if and only if* the real parts are equal and the imaginary parts are equal. This is equivalent to stating that two complex numbers are equal only if their magnitudes are equal and their arguments are equal.

The following examples and drill exercises should help clarify these explanations.

Example   A.1

Convert the complex number $A = 3 + jb$ to its polar form.

Solution:

$$r = \sqrt{3^2 + 4^2} = 5 \quad \theta = \tan^{-1}\left(\frac{4}{3}\right) = 53.13°$$

$$A = 5\angle 53.13°$$

Example   A.2

Convert the number $A = 4\angle -60°$ to its complex form.

Solution:

$$\begin{aligned} a &= 4\cos(-60°) = 4\cos(60°) = 2 \\ b &= 4\sin(-60°) = -4\sin(60°) = -2\sqrt{3} \end{aligned}$$

Thus, $A = 2 - j2\sqrt{3}$.

Addition and *subtraction* of complex numbers take place according to the following rules:

$$\begin{aligned} (a_1 + jb_1) + (a_2 + jb_2) &= (a_1 + a_2) + j(b_1 + b_2) \\ (a_1 + jb_1) - (a_2 + jb_2) &= (a_1 - a_2) + j(b_1 - b_2) \end{aligned} \tag{A.15}$$

Multiplication of complex numbers in polar form follows the law of exponents. That is, the magnitude of the product is the product of the individual magnitudes, and the angle of the product is the sum of the individual angles, as shown below.

$$(A)(B) = (Ae^{j\theta})(Be^{j\phi}) = ABe^{j(\theta + \phi)} = AB\angle(\theta + \phi) \tag{A.16}$$

If the numbers are given in rectangular form and the product is desired in rectangular form, it may be more convenient to perform the multiplication directly, using the rule that $j^2 = -1$, as illustrated in equation A.17.

$$
\begin{aligned}
(a_1 + jb_1)(a_2 + jb_2) &= a_1 a_2 + ja_1 b_2 + ja_2 b_1 + j^2 b_1 b_2 \\
&= (a_1 a_2 + j^2 b_1 b_2) + j(a_1 b_2 + a_2 b_1) \\
&= (a_1 a_2 - b_1 b_2) + j(a_1 b_2 + a_2 b_1)
\end{aligned} \tag{A.17}
$$

Division of complex numbers in polar form follows the law of exponents. That is, the magnitude of the quotient is the quotient of the magnitudes, and the angle of the quotient is the difference of the angles, as shown in equation A.18.

$$\frac{A}{B} = \frac{Ae^{j\theta}}{Be^{j\phi}} = \frac{A\angle\theta}{B\angle\phi} = \frac{A}{B}\angle(\theta - \phi) \tag{A.18}$$

Division in the rectangular form can be accomplished by multiplying the numerator and denominator by the complex conjugate of the denominator. Multiplying the denominator by its complex conjugate converts the denominator to a real number and simplifies division. This is shown in Example A.4. Powers and roots of a complex number in polar form follow the laws of exponents, as shown in equations A.19 and A.20.

$$A^n = (Ae^{j\theta})^n = A^n e^{jn\theta} = A^n\angle n\theta \tag{A.19}$$

$$A^{1/n} = (Ae^{j\theta})^{1/n} = A^{1/n} e^{j1/n\theta}$$

$$= \sqrt[n]{A}\angle\left(\frac{\theta + k2\pi}{n}\right) \qquad k = 0, \pm1, \pm2, \ldots \tag{A.20}$$

Example   A.3

Perform the following operations given that $A = 2 + j3$ and $B = 5 - j4$.

    (a) $A + B$ (b) $A - B$ (c) $2A + 3B$

Solution:

$$A + B = (2 + 5) + j(3 + (-4)) = 7 - j$$

$$A - B = (2 - 5) + j(3 - (-4)) = -3 + j7$$

For part c, $2A = 4 + j6$ and $3B = 15 - j12$. Thus, $2A + 3B = (4 + 15) + j(6 + (-12)) = 19 - j6$

Example   A.4

Perform the following operations both in rectangular and polar form, given that $A = 3 + j3$ and $B = 1 + j\sqrt{3}$.

   (a) AB (b) $A \div B$

Solution:

a.   In rectangular form:

$$AB = (3 + j3)(1 + j\sqrt{3}) = 3 + j3\sqrt{3} + j3 + j^2 3\sqrt{3}$$
$$= (3 + j^2 3\sqrt{3}) + j(3 + j3\sqrt{3})$$
$$= (3 - 3\sqrt{3}) + j(3 + j3\sqrt{3})$$

To obtain the answer in polar form, we need to convert A and B to their polar forms:

$$A = 3\sqrt{2}e^{j45°} = 3\sqrt{2}\angle 45°$$
$$B = \sqrt{4}e^{j60°} = 2\angle 60°$$

Then,

$$AB = (3\sqrt{2}e^{j45°})(\sqrt{4}e^{j60°}) = 6\sqrt{2}\angle 105°$$

b.   To find $A \div B$ in rectangular form, we can multiply A and B by B^*.

$$\frac{A}{B} = \frac{3 + j3}{1 + j\sqrt{3}} \frac{1 - j\sqrt{3}}{1 - j\sqrt{3}}$$

Then,

$$\frac{A}{B} = \frac{(3 + 3\sqrt{3}) + j(3 - 3\sqrt{3})}{4}$$

In polar form, the same operation may be performed as follows:

$$A \div B = \frac{3\sqrt{2}\angle 45°}{2\angle 60°} = \frac{3\sqrt{2}}{2}\angle\left(45° - 60°\right) = \frac{3\sqrt{2}}{2}\angle -15°$$

Euler's Identity

This formula extends the usual definition of the exponential function to allow for complex numbers as arguments. Euler's identity states that

$$e^{j\theta} = \cos\theta + j\sin\theta \tag{A.21}$$

All the standard trigonometry formulas in the complex plane are direct consequences of Euler's identity. The two important formulas are:

$$\cos\theta = \frac{e^{j\theta} + e^{-j\theta}}{2} \qquad \sin\theta = \frac{e^{j\theta} - e^{-j\theta}}{2j} \tag{A.22}$$

Example   A.5

Using Euler's formula, show that

$$\cos \theta = \frac{e^{j\theta} + e^{-j\theta}}{2}$$

Solution:

Using Euler's formula

$$e^{j\theta} = \cos \theta + j \sin \theta$$

Extending the above formula, we can obtain

$$e^{-j\theta} = \cos(-\theta) + j \sin(-\theta) = \cos \theta - j \sin \theta$$

Thus,

$$\cos \theta = \frac{e^{j\theta} + e^{-j\theta}}{2}$$

DRILL EXERCISES

A.4 In a certain AC circuit, $V = ZI$ where $Z = 7.75\angle 90°$ and $I = 2\angle -45°$. Find V.

> Answer: $15.5\angle 45°$

A.5 In a certain AC circuit, $V = ZI$ where $Z = 5\angle 82°$ and $V = 30\angle 45°$. Find I.

> Answer: $6\angle -37°$

A.6 Show that the polar form of AB in Example A.4 is equivalent to its rectangular form.

A.7 Show that the polar form of $A \div B$ in Example A.4 is equivalent to its rectangular form.

A.8 Using Euler's formula, show that $\sin \theta = \frac{e^{j\theta} - e^{-j\theta}}{2j}$

III.   COMPLEX FREQUENCY AND THE LAPLACE TRANSFORM

Complex Frequency

In Chapter 4, we considered circuits with sinusoidal excitations such as

$$v(t) = A \cos(\omega t + \phi) \tag{A.23}$$

which we also wrote in the equivalent phasor form

$$V(j\omega) = Ae^{j\phi} = A\angle\phi \qquad \text{(A.24)}$$

The two expressions just given are related by

$$v(t) = \text{Re}\,(Ve^{j\omega t}) \qquad \text{(A.25)}$$

As was shown in Chapter 4, the phasor notation is extremely useful in solving AC steady-state circuits, in which the voltages and currents are *steady-state sinusoids* of the form of equation A.23. We now consider a different class of waveforms, useful in the transient analysis of circuits, namely, *damped sinusoids*. The most general form of a damped sinusoid is

$$v(t) = Ae^{\sigma t}\cos{(\omega t + \phi)} \qquad \text{(A.26)}$$

As one can see, a damped sinusoid is a sinusoid multiplied by a real exponential, $e^{\sigma t}$. The constant σ is real and is usually zero or negative in most practical circuits. Figures A.2a and b depict the case of a damped

Figure A.2a Damped sinusoid for negative σ

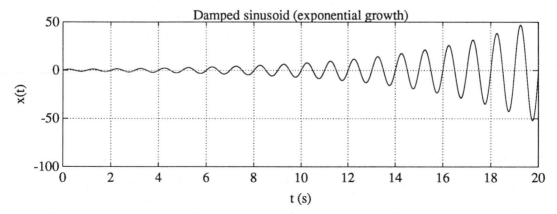

Figure A.2b Damped sinusoid for positive σ

sinusoid with negative σ and with positive σ, respectively. Note that the case $\sigma = 0$ corresponds exactly to a sinusoidal waveform. The definition of phasor voltages and currents given in Chapter 4 can be easily extended to account for the case of damped sinusoidal waveforms by defining a new variable, s, called the *complex frequency:*

$$s = \sigma + j\omega \tag{A.27}$$

Note that the special case $\sigma = 0$ corresponds to $s = j\omega$, that is, the familiar steady-state sinusoidal (phasor) case. We shall now refer to the complex variable $V(s)$ as the **complex frequency domain** representation of $v(t)$. It should be observed that from the viewpoint of circuit analysis, the use of the Laplace transform is in all analogous to phasor analysis; that is, substituting the variable s wherever $j\omega$ was used is the only step required to describe a circuit using the new notation.

DRILL EXERCISES

A.9 Find the complex frequencies that are associated with

a. $5e^{-4t}$,
b. $\cos 2\omega t$,
c. $\sin(\omega t + 2\theta)$,
d. $4e^{-2t}\sin(3t - 50°)$,
e. $e^{-3t}(2 + \cos 4t)$.

> Answer: a. -4; b. $\pm j2\omega$; c. $\pm j\omega$; d. $-2 \pm j3$; e. -3 and $-3 \pm j4$

A.10 Find s and $V(s)$ if $v(t)$ is given by

a. $5e^{-2t}$
b. $5e^{-2t}\cos(4t + 10°)$
c. $4\cos(2t - 20°)$

> Answer: a. -2, $5\angle 0°$; b. $-2 + j4$, $5\angle 10°$; c. $j2$, $4\angle -20°$

A.11 Find $v(t)$ if

a. $s = -2$, $V = 2\angle 0°$
b. $s = j2$, $V = 12\angle -30°$
c. $s = -4 + j3$, $V = 6\angle 10°$

> Answer: a. $2e^{-2t}$; b. $12\cos(2t - 30°)$; c. $6e^{-4t}\cos(3t + 10°)$

All the concepts and rules used in AC network analysis (see Chapter 4) such as impedance, admittance, KVL, KCL, Thévenin's and Norton's theorems, etc., carry over to the damped sinusoid case exactly. In the complex frequency domain, the current $I(s)$ and voltage $V(s)$ are related by the expression

$$V(s) = Z(s)I(s) \qquad\qquad \text{(A.28)}$$

where $Z(s)$ is the familiar impedance, with s replacing $j\omega$. We may obtain $Z(s)$ from $Z(j\omega)$ by simply replacing $j\omega$ by s. For a resistance, R, the impedance is

$$Z_R(s) = R \qquad\qquad \text{(A.29)}$$

for an inductance, L, the impedance is

$$Z_L(s) = sL \qquad\qquad \text{(A.30)}$$

and for a capacitance C, it is

$$Z_C(s) = \frac{1}{sC} \qquad\qquad \text{(A.31)}$$

Impedances in series or parallel are combined in exactly the same way as in the AC steady-state case since we only replace $j\omega$ by s.

Example A.6

Consider a series R-L circuit with $R = 4\ \Omega$ and $L = 2\ H$, excited by a voltage source $v(t) = 10e^{-2t}\cos 5t$ V. Find the impedance seen by the source, and determine a time domain expression for the series current flowing in the circuit.

Solution:

The impedance seen by the source terminal is given by

$$Z(s) = 4 + 2s$$

Thus, since the input voltage phasor is

$$V(s) = 10\angle 0°\ \text{V}$$

we have

$$I(s) = \frac{V(s)}{Z(s)} = \frac{10\angle 0°}{2s + 4} = \frac{10\angle 0°}{2(-2 + j5) + 4} = 1\angle -90°\ \text{A}$$

Therefore, the time domain expression for the current is

$$i(t) = e^{-2t}\cos(5t - 90°)\ \text{A}$$

Just as frequency response functions $H(j\omega)$ were defined in Chapter 4, it is possible to define a **transfer function,** $H(s)$. This can be a ratio of a voltage to a current, a ratio of a voltage to a voltage, a ratio of a current to a current, or a ratio of a current to a voltage. The transfer function $H(s)$ is a function of network elements and their interconnections. Using the transfer function and knowing the input (voltage or current) to a circuit, we can find an expression for the output either in the complex frequency domain, or in the time domain. As an example suppose $V_i(s)$ and $V_o(s)$ are the input and output voltages to a circuit, respectively, in complex frequency notation. Then

$$H(s) = \frac{V_o(s)}{V_i(s)} \tag{A.32}$$

from which we can obtain the output in the complex frequency domain by computing

$$V_o(s) = H(s)V_i(s) \tag{A.33}$$

If $V_i(s)$ is a known damped sinusoid, we can then proceed to determine $v_o(t)$ by means of the method illustrated earlier.

DRILL EXERCISES

A.12   Given the transfer function $H(s) = \frac{3(s+2)}{s^2+2s+3}$ and the input $V_i(s) = 4\angle0°$, find the forced response $v_o(t)$ if

a. $s = -1$
b. $s = -1 + j1$
c. $s = -2 + j1$

Answer: a. $6e^{-t}$; b. $12\sqrt{2}e^{-t}\cos(t + 45°)$; c. $6e^{-2t}\cos(t + 135°)$

A.13   Given the transfer function $H(s) = \frac{2(s+4)}{s^2+4s+5}$ and the input $V_i(s) = 6\angle30°$, find the forced response $v_o(t)$ if

a. $s = -4 + j1$
b. $s = -2 + j2$

Answer: a. $3e^{-4t}\cos(t + 165°)$; b. $8\sqrt{2}e^{-2t}\cos(2t - 105°)$

The Laplace Transform

The Laplace transform, named after the French mathematician and astronomer Pierre Simon de Laplace, is defined by

$$\mathcal{L}[f(t)] = F(s) = \int_0^\infty f(t)e^{-st}\, dt \tag{A.34}$$

The function $F(s)$ is the Laplace transform of $f(t)$ and is a function of the complex frequency, $s = \sigma + j\omega$, considered earlier. Note that the function

$f(t)$ is defined only for $t \geq 0$. This definition of the Laplace transform is referred to as the **one-sided** or **unilateral Laplace transform** since $f(t)$ is evaluated only for positive t. In order to conveniently express arbitrary functions only for positive time, we introduce a special function called the **unit step function,** $u(t)$, defined by the expression

$$
\begin{aligned}
u(t) &= 0, \quad t < 0 \\
&= 1, \quad t > 0
\end{aligned}
\tag{A.35}
$$

Example   A.7

Find the Laplace transform of $f(t) = e^{-at} u(t)$.

Solution:

The Laplace transform of $f(t)$ can be obtained by applying equation A.34:

$$
F(s) = \int_0^\infty e^{-at} e^{-st} \, dt = \int_0^\infty e^{-(s+a)t} \, dt = \frac{1}{s+a} e^{-(s+a)t} \Big|_0^\infty = \frac{1}{s+a}
$$

Thus,

$$
\mathcal{L}[e^{-at}] = \frac{1}{s+a}
$$

Example   A.8

Find the Laplace transform of $f(t) = \cos(\omega t)\, u(t)$.

Solution:

Using Euler's formula, we can expand $f(t)$ into: $f(t) = \frac{1}{2}(e^{j\omega t} + e^{-j\omega t}) u(t)$. We can now apply equation A.34 to find the Laplace transform of $f(t)$.

$$
F(s) = \int_0^\infty \frac{1}{2}(e^{j\omega t} + e^{-j\omega t}) e^{-st} \, dt = \frac{1}{2} \left(\int_0^\infty e^{-(s+j\omega)t} \, dt + \int_0^\infty e^{-(s-j\omega)t} \, dt \right)
$$

Evaluating the integrals, we find the Laplace transform of $f(t)$ to be

$$
\mathcal{L}[\cos \omega t] = \frac{s}{s^2 + \omega^2}
$$

DRILL EXERCISES

A.14   Find the Laplace transform of the following functions:

a. $u(t)$

b. $\sin(\omega t)\, u(t)$

c. $t u(t)$

> Answer: a. $\dfrac{1}{s}$; b. $\dfrac{\omega}{s^2 + \omega^2}$; c. $\dfrac{1}{s^2}$

A.15 Find the Laplace transform of the following functions:

a. $e^{-at} \sin \omega t \, u(t)$

b. $e^{-at} \cos \omega t \, u(t)$

Answer: a. $\dfrac{\omega}{(s+a)^2 + \omega^2}$; b. $\dfrac{(s+a)}{(s+a)^2 + \omega^2}$

From what we discussed so far on Laplace transform, it is obvious that we may compile a lengthy table of functions and their Laplace transforms by repeated application of equation A.34 for various cases of $f(t)$. Then, we could obtain a wide variety of inverse transforms by matching entries in the table. Table A.1 lists some of the more common **Laplace transform pairs.** The computation of the **inverse Laplace transform** is in general rather complex if one wishes to consider arbitrary functions of s. In many practical cases, however, it is possible to use combinations of known transform pairs to obtain the desired result.

Table A.1   Laplace transform pairs

$f(t)$	$E(s)$
$\delta(t)$	1
$u(t)$	$\dfrac{1}{s}$
$e^{-at} u(t)$	$\dfrac{1}{s+a}$
$\sin \omega t \, u(t)$	$\dfrac{\omega}{s^2 + \omega^2}$
$\cos \omega t \, u(t)$	$\dfrac{s}{s^2 + \omega^2}$
$e^{-at} \sin \omega t \, u(t)$	$\dfrac{\omega}{(s+a)^2 + \omega^2}$
$e^{-at} \cos \omega t \, u(t)$	$\dfrac{s+a}{(s+a)^2 + \omega^2}$
$tu(t)$	$\dfrac{1}{s^2}$

Example   A.9

Find the inverse of the transform

$$F(s) = \frac{2}{s+3} + \frac{4}{s^2 + 4} - \frac{4}{s}$$

Solution:

We need to find the inverse transform of $F_1(s)$, $F_2(s)$, and $F_3(s)$ where

$$F_1(s) = \frac{2}{s+3}, F_2(s) = \frac{4}{s^2+4}, \text{ and } F_3(s) = \frac{4}{s}$$

Looking at Table A.1, we can see that

$$f_1(t) = 2\mathcal{L}^{-1}\left(\frac{1}{s+3}\right) = 2e^{-3t}u(t)$$

$$f_2(t) = F_2(s) = 2\mathcal{L}^{-1}\left(\frac{2}{s^2+2^2}\right) = 2\sin 2t\, u(t)$$

$$f_3(t) = 4\mathcal{L}^{-1}\left(\frac{1}{s}\right) = 4u(t)$$

Thus,

$$f(t) = f_1(t) + f_2(t) + f_3(t) = (2e^{-3t} + 2\sin 2t - 4)u(t)$$

Example   A.10

Find the inverse of the transform

$$F(s) = \frac{2s+5}{s^2+5s+6}$$

Solution:

As you can see, there is no direct entry in Table A.1 that we can use to obtain $f(t)$. However, we can obtain a partial fraction expansion of $F(s)$ and apply the table to each term in the expansion. Obtaining a partial fraction expansion is the opposite operation of obtaining a common denominator. We illustrate this by obtaining the partial fraction expansion of $F(s)$.

$$F(s) = \frac{2s+5}{s^2+5s+6} = \frac{A}{s+2} + \frac{B}{s+3}$$

The next step is to determine the constants A and B. There are many ways of determining A and B. However, the simplest means of determining A and B is to note that

$$(s+2)F(s) = \frac{2s+5}{s+3} = A + \frac{B(s+2)}{s+3}$$

and

$$(s+3)F(s) = \frac{2s+5}{s+2} = \frac{A(s+3)}{s+2} + B$$

Therefore,

$$A = (s+2)F(s) = \left.\frac{2s+5}{s+3}\right|_{s=-2} = 1$$

$$B = (s+3)F(s) = \left.\frac{2s+5}{s+2}\right|_{s=-3} = 1$$

Thus,

$$F(s) = \frac{2s + 5}{s^2 + 5s + 6} = \frac{1}{s + 2} = \frac{1}{s + 3}$$

We can now use Table A.1 to find the inverse transform of $F(s)$. Using Table A.1:

$$f(t) = (e^{-2t} + e^{-3t})u(t)$$

DRILL EXERCISE

A.16 Find the inverse Laplace transform of each of the following functions:

a. $F(s) = \frac{1}{s^2+5s+6}$

b. $F(s) = \frac{s-1}{s(s+2)}$

c. $F(s) = \frac{3s}{(s^2+1)(s^2+4)}$

d. $F(s) = \frac{1}{(s+2)(s+1)^2}$

Answer: a. $f(t) = (e^{-2t} - e^{-3t})u(t)$; b. $f(t) = \left(\frac{3}{2}e^{-2t} - \frac{1}{2}\right)u(t)$; c. $f(t) = (\cos t - \cos 2t)u(t)$; d. $f(t) = (e^{-2t} + te^{-t} - e^{-t})u(t)$

EIT Exam Review

INTRODUCTION

Each of the 50 states regulates the engineering profession by requiring individuals who intend to practice the profession to become registered professional engineers (PE). To become a professional engineer, it is necessary to satisfy four requirements. The first is the completion of a B.S. degree in engineering from an accredited college or university (although it is theoretically possible to be registered without having completed a degree). The second is the successful completion of the Engineer-in-Training (Engineering Fundamentals) Examination. This is an eight-hour exam that covers general engineering undergraduate education. The third requirement is two to four years of engineering experience after passing the Engineer-in-Training (EIT) exam. Finally, the fourth requirement is successful completion of the Principles and Practice of Engineering or Professional Engineer (PE) Examination.

The EIT exam is a two-part national examination given twice a year (in April and October). The exam is divided into two four-hour sessions. The morning session consists of 140 multiple choice questions (five possible answers are given); the afternoon session consists of 70 questions. The exam is prepared by the National Council of Examiners for Engineering and Surveying.[1]

The aim of this appendix is to assist you in preparing for one part of the EIT exam, entitled Electrical Circuits. This part of the EIT consists of a total of 14 questions in the morning session, and 10 questions in the afternoon session. The examination topics for the electrical circuits part are the following:

DC Circuits
AC Circuits
Three-Phase Circuits
Capacitance and Inductance
Transients
Diode Applications
Operational Amplifiers (ideal)
Electric and Magnetic Fields
Electric Machinery

Solved sample problems for each of these areas are given in the next few pages, together with a summary of useful examples you may refer to in the text. You will find this appendix invaluable in preparing for the Electrical Circuits part of the EIT exam.

The detailed topics covered in the electrical circuits part of the EIT exam are listed below, along with the relevant chapter in the text.

DC Circuits
  Relationship between current and charge (Chapter 2)
  Current and voltage laws (Chapter 2)
  Power and energy (Chapter 2)
  Equivalent series and parallel elements (Chapter 2)
  Maximum power transfer (Chapter 3)
  Thévenin and Norton theorems (Chapter 3)
AC Circuits
  Algebra of complex numbers (Appendix A)
  Phasor representation (Chapter 4)
  Concept of time domain and frequency domain (Chapter 4)
  Impedance concepts (Chapter 4)
  Complex power, apparent power, and power factor (Chapter 5)
  Ideal transformers (Chapter 5)

---

[1] P. O. Box 1686 (1820 Seneca Creek Road), Clemson, SC, 29633-1686

Three-Phase Circuits
    Computation of power (Chapter 5)
    Line and phase currents for delta and wye circuits (Chapter 5)
Capacitance and Inductance
    Energy storage (Chapter 4)
    RL or RC (first-order) transients (Chapter 6)
Transients (Chapter 6)
Diode Applications (Chapter 7)
Operational Amplifiers (ideal) (Chapter 10)
Electric and Magnetic Fields
    Work done in moving a charge in an electric field (Chapter 2)
    Relationship between voltage and work (Chapter 2)
    Faraday's and Ampere's laws (Chapter 14)
    Magnetic reluctance concepts (Chapter 14)
Electric Machinery
    AC and DC motor fundamentals (Chapters 15, 16)

1.   DC CIRCUITS

DC circuit analysis is the foundation of electrical engineering. The aim of
EIT exam problems in this subject area is to verify the basic understanding
of circuit analysis. Chapters 2 and 3 of the text cover all the relevant review
topics.

DRILL EXERCISES

DC.1   Assuming the connecting wires and the battery have negligible resistance,
the voltage across the 25 Ω resistance in Figure B.1 is

a. 25 V   b. 60 V   c. 50 V   d. 15 V   e. 12.5 V

Solution:

This problem calls for application of the voltage divider rule, discussed in Section
2.6. Applying the voltage divider rule to the circuit of Figure B.1 we have

$$v_{25\ \Omega} = 60\frac{25}{3 + 2 + 25} = 50 \text{ V}$$

Thus, the answer is c.

DC.2   Assuming the connecting wires and the battery have negligible resistance,
the voltage across the 6 Ω resistor in Figure B.2 is

a. 6 V   b. 3.5 V   c. 12 V   d. 4 V   e. 3 V

Figure B.1

Figure B.2

Solution:

This problem can be solved most readily by applying nodal analysis (Section 3.1), since one of the node voltages is already known. Applying KCL at the node v, we obtain

$$\frac{12 - v}{2} = \frac{v}{6} + \frac{v}{12}$$

This equation can be solved to show that $v = 4$ V. Note that it is also possible to solve this problem by mesh analysis (Section 3.2). You are encouraged to try this method as well.

DC.3 A 125 V battery charger is used to charge a 75 V battery with internal resistance of 1.5 Ω. If the charging current is not to exceed 5 A, the minimum resistance in series with the charger must be

a. 10 Ω b. 5 Ω c. 38.5 Ω d. 41.5 Ω e. 8.5 Ω

Solution:

The circuit of Figure B.3 describes the charging arrangement. Applying KVL to the circuit of Figure B.3, we obtain

$$i_{max}R + 1.5i_{max} - 125 + 75 = 0$$

and using $i = i_{max} = 5$ A, we can find R from the following equation:

$$5R + 7.5 - 125 + 75 = 0$$

$$R = 8.5 \ \Omega$$

Thus, e is the correct answer.

Figure B.3

If you wish to reinforce the concepts illustrated in the sample problems given in this section, you may review some of the examples in Chapters 2 and 3. In particular, the following examples illustrate the fundamental techniques of circuit analysis:

Chapter 2: Examples 2.6, 2.9, 2.10, 2.11
Chapter 3: Examples 3.1, 3.5, 3.11, 3.12, 3.13, 3.14, 3.15

2.   AC CIRCUITS

One of the most important applications of electricity in the engineering profession is the analysis of AC electric power systems. These topics are a very important part of the EIT exam and are discussed in the next two sections. You should conduct a careful review of Chapters 4 and 5 in preparing for this section of the exam.

DRILL EXERCISES

AC.1 A voltage sine wave of peak value 100 V is in time phase with a current sine wave of peak value 4 A. When the phase angle is 60° later than a time at which the voltage and the current are both zero, the instantaneous power is most nearly

a. 250 W   b. 200 W   c. 400 W   d. 150 W   e. 100 W

Solution:

As discussed in Section 5.1, the instantaneous AC power $p(t)$ is

$$p(t) = \frac{VI}{2}\cos\theta + \frac{VI}{2}\cos(2\omega t + \theta_V + \theta_I)$$

In this problem, when the phase angle is 60° later than a "zero crossing," we have $\theta_V = \theta_I = 0$, $\theta = \theta_V - \theta_I = 0$, $2\omega t = 120°$. Thus, we can compute the power at this instant as

$$p = \frac{\left(\dfrac{100}{\sqrt{2}}\dfrac{4}{\sqrt{2}}\right)}{2} + \frac{\left(\dfrac{100}{\sqrt{2}}\dfrac{4}{\sqrt{2}}\right)}{2}\cos 120° = 250 \text{ W}$$

The correct answer is a.

AC.2 A sinusoidal voltage whose amplitude is $20\sqrt{2}$ V is applied to a 5-Ω resistor. The root-mean-square value of the current is

a. 5.66 A   b. 4 A   c. 7.07 A   d. 8 A   e. 10 A

Solution:

From Section 4.2, we know that

$$V_{\text{rms}} = \frac{V}{\sqrt{2}} = \frac{20\sqrt{2}}{\sqrt{2}} = 20 \text{ V}$$

Thus, $I_{\text{rms}} = \frac{20}{5} = 4$ A. Therefore, b is the correct answer.

AC.3 The magnitude of the steady-state, root-mean-square voltage across the capacitor in the circuit of Figure B.4 is

a. 30 V   b. 15 V   c. 10 V   d. 45 V   e. 60 V

Solution:

This problem requires the use of impedances (Section 4.4). Using the voltage divider rule for impedances, we write the voltage across the capacitor as

$$\mathbf{V} = 30\angle 0° \times \frac{-j10}{10 - j10 + j10}$$
$$= 30\angle 0° \times -j1 = 30\angle 0° \times 1\angle -90° = 30\angle -90°$$

Thus, the rms amplitude of the voltage across the capacitor is 30 V, and a is the correct answer. Note the importance of the phase angle in this kind of problem.

Figure B.4

10 Ω
-j10 Ω
+j10 Ω
30 Vrms

The next set of questions (AC.4 to AC.8) pertain to single-phase AC power calculations and refer to the single-phase electrical network shown in Figure B.5. In this figure, $\mathbf{E}_S = 480\angle 0°$ V; $I_S = 100\angle -15°$ A; $\omega = 120\pi$ rad/s. Further, Load A is a bank of single-phase induction machines. The

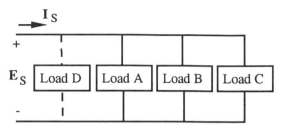

Figure B.5

bank has an efficiency of 80 percent, a power factor 0.70 lagging and a load of 20 hp. Load B is a bank of overexcited single-phase synchronous machines. The machines draw 15 kVA and the load current leads the line voltage by 30 degrees. Load C is a lighting (resistive) load and absorbs 10 kW. Load D is a proposed single-phase synchronous condenser which will correct the source power factor to unity. This material is covered in Sections 5.1 and 5.2.

DRILL EXERCISES

AC.4 The root-mean-square magnitude of Load A current, I_A, is most nearly

a. 44.4 A   b. 31.08 A   c. 60 A   d. 38.85 A   e. 55.5 A

Solution:

The output power P_o of the single-phase induction motor is: $P_o = 20 \times 746 = 14{,}920$ W. The input electric power P_{in} is:

$$P_{in} = \frac{P_o}{\eta} = \frac{14920}{0.80} = 18{,}650 \text{ W}$$

P_{in} can be expressed as:

$$P_{in} = E_S I_A \cos\theta_A$$

Therefore, the rms magnitude of the current $\mathbf{I}_A$ is found as

$$I_A = \frac{P_{in}}{E_S \cos\theta_A} = \frac{18650}{480 \times 0.70} = 55.5015 = 55.5 \text{ A}$$

Thus, the correct answer is e.

AC.5 The phase angle of $\mathbf{I}_A$ with respect to the line voltage E_S is most nearly

a. 36.87°   b. 60°   c. 45.6°   d. 30°   e. 48°

Solution:

The phase angle between $\mathbf{I}_A$ and $\mathbf{E}_S$ is:

$$\theta = \cos^{-1} 0.70 = 45.57° = 45.6°$$

The correct answer is c.

AC.6 The power absorbed by synchronous machines is most nearly

a. 20000 W   b. 7500 W   c. 13000 W   d. 12990 W   e. 15000 W

Solution:

The apparent power, S, is known to be 15 kVA, and θ is 30°. From the power triangle, we have

$$P = S \cos \theta$$

Therefore, the power drawn by the bank of synchronous motors is:

$$P = 15000 \times \cos 30° = 12990.38 \approx 12.99 \text{ kW}$$

The answer is d.

AC.7 The power factor of the system before load D is installed is most nearly

a. 0.70 lagging   b. 0.866 leading   c. 0.866 lagging   d. 0.966 leading   e. 0.966 lagging

Solution:

From the expression for the current $\mathbf{I}_S$, we have

$$pf = \cos \theta = \cos(0° - (-15°)) = \cos 15° = 0.966 \text{ lagging}$$

The correct answer is e.

AC.8 The capacitance of the synchronous condenser which will give a unity power factor of the system is most nearly

a. 219 μF   b. 187 μF   c. 132.7 μF   d. 240 μF   e. 132.7 pF

Solution:

The reactive power Q_A in load A is:

$$Q_A = P_A \times \tan \theta_A$$
$$\theta_A = \cos^{-1} 0.70 = 45.57°$$

Therefore,

$$Q_A = 18650 \times \tan 45.57° = 19024.82 \text{ VAR}$$

The reactive power Q_B in load B is:

$$Q_B = S \times \sin \theta_B = 15000 \times \sin(-30°) = -7500 \text{ VAR}$$

The total reactive power Q is:

$$Q = Q_A + Q_B = 19024.82 - 7500 = 11524.82 \text{ VAR}$$

To cancel this reactive power, we set

$$Q_C = -Q = -11524.82 \text{ VAR}$$

and

$$Q_C = -\frac{E_S^2}{X_C} \text{ and } X_C = -\frac{1}{\omega C}$$

Therefore, the capacitance required to obtain a power factor of unity is:

$$C = -\frac{Q_C}{\omega E_S^2} = \frac{11524.82}{120\pi \times 480^2} = 132.7 \ \mu\text{F}$$

The correct answer is c.

In addition to these sample problems, you may also find it useful to review the following examples in the text: Chapter 4, Examples 4.8, 4.9, 4.11, 4.12, 4.13, 4.14, 4.16, 4.18, 4.20, 4.21; Chapter 5, Examples 5.1, 5.2, 5.3, 5.4, 5.5, 5.6, 5.7, 5.8, 5.9, 5.12.

3. CAPACITANCE AND INDUCTANCE

Problems that pertain to this topic are typically oriented toward the calculation of voltages, current, and energy storage in these elements. The following two problems are based on the waveform of Figure B.6.

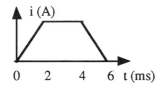

Figure B.6

DRILL EXERCISES

CL.1   A coil with inductance of 1 H and negligible resistance carries the current shown. The maximum energy stored in the inductor is:

a. 2 J   b. 0.5 J   c. 0.25 J   d. 1 J   e. 0.2 J

Solution:

The energy stored in an inductor is $W = \frac{1}{2}Li^2$ (see Section 4.1). Since the maximum current is 1 A, the maximum energy will be $W_{max} = \frac{1}{2}Li_{max}^2 = \frac{1}{2}$ J. Thus b is the correct answer.

CL.2   The maximum voltage that will appear across the coil is:

a. 5 V   b. 100 V   c. 250 V   d. 500 V   e. 5000 V

Solution:

Since the voltage across an inductor is given by $v = L\frac{di}{dt}$, we need to find the maximum (positive) value of $\frac{di}{dt}$. This will occur anywhere between $t = 0$ and $t = 2$ ms. The corresponding slope is: $\frac{di}{dt}\Big|_{max} = \frac{1}{2 \times 10^{-3}} = 500$. Therefore $v_{max} = 1 \times 500 = 500$ V, and the correct answer is d.

In addition to these sample problems, you may also find it useful to review the following examples in the text: Chapter 4, Examples 4.1, 4.2, 4.3, 4.5, 4.6, 4.7; Chapter 6, Examples 6.11, 6.12, 6.13, 6.14.

4.   THREE-PHASE CIRCUITS

Another aspect of circuit analysis that plays a prominent role in engineering practice is the use of three-phase AC circuits for power distribution and industrial applications. Three-phase power is therefore an important section of the EIT exam. This material is covered primarily in Section 5.4 of the text. The following sample problems are representative of the type of questions that might appear in the exam.

DRILL EXERCISES

3–PH.1   A 3-phase circuit is shown in Figure B.7. Load resistors (66 Ω) are connected in delta and supplied by a 220-volt balanced three-phase source through three lines of 2-ohm resistance. The magnitude of the root-mean-square, line-to-line voltage across each 66-ohm resistor is most nearly

a.198 V   b. 110 V   c. 201 V   d. 220 V   e. 120 V

Figure B.7

Solution:

Since the load in this problem is Δ-connected, it must first be converted to an equivalent Y-form. The phase impedance of the Δ-connected load is 66 Ω, so the equivalent phase impedance of the corresponding Y-form is:

$$Z_Y = \frac{Z_\Delta}{3} = \frac{66}{3} = 22 \ \Omega$$

The phase voltage is:

$$\mathbf{V}_\phi = \frac{208}{\sqrt{3}} \angle 0° \ \text{V}$$

The per-phase equivalent circuit of this problem is shown in Figure B.8. The load voltage is obtained by the voltage divider rule:

$$\mathbf{V}_L = \frac{22}{2+22} \times \mathbf{V}_\phi = \frac{208 \times 22}{\sqrt{3} \times 24} = \frac{201.67}{\sqrt{3}} \angle 0° \ \text{V}$$

The rms line-to-line voltage across the 66 Ω resistor therefore is:

$$\mathbf{V}_{66\Omega} = \sqrt{3}\,\mathbf{V}_L = 201.67 \angle 0° \ \text{V}$$

Figure B.8

The correct answer is c.

3–PH.2   A three-phase load is composed of three impedance of $9.0 + j9.0$ ohms and connected in wye. The balanced three-phase source is 208 volt (line to line). The current in each line is most nearly

a. 40 A   b. 16.3 A   c. 13.3 A   d. 9 A   e. 6 A

Solution:

The phase voltage is $\frac{208}{\sqrt{3}}$. Therefore, the magnitude of phase current I_{an} is

$$I_{an} = \frac{\left(\frac{208}{\sqrt{3}}\right)}{9 + j9} = \frac{208}{\sqrt{3}} \frac{1}{12.73} = 9.43 \text{ A}$$

The line current is equal to the phase current in a Y-connected system. The answer is d.

The next four questions refer to a three-phase system with line-to-line voltage of 220 V rms, with ABC phase sequence, and with phase reference V_{AB} shown in the phase diagram of Figure B.9. The load is a balanced delta connection, shown in Figure B.10, with branch impedances $Z = 30 - j40$ ohms, $j = \sqrt{-1}$.

Figure B.9   Figure B.10

DRILL EXERCISES

3–PH.3 The phase current is most nearly

a. $4.4\angle 53.13°$ A   b. $2.4\angle 53.13°$ A   c. $4.4\angle 0°$ A   d. $4.4\angle -53.13°$ A
e. $2.4\angle -53.13°$ A

Solution:

The load current $\mathbf{I}_{AB}$ is given by

$$\mathbf{I}_{AB} = \frac{\mathbf{V}_{AB}}{Z} = \frac{220\angle 0°}{30 - j40} = \frac{220\angle 0°}{50\angle -53.13°} = 4.4\angle 53.13° \text{ A}$$

The correct answer is a.

3-PH.4 The line current I_A (in amperes) is most nearly

a. $4.4\angle -186.87°$ b. $4.4\angle 23°$ c. 7   d. $7.6\angle 23°$ e. $7\angle -186.87°$

Solution:

The current phasor diagram for this Δ-connected system is shown in Figure B.11. From the relationship among three-phase currents, we have

$$\mathbf{I}_{CA} = \mathbf{I}_{AB}\angle -240° = 4.4\angle(53.13°-240°) = 4.4\angle -186.87° \text{ A}$$

From the phasor diagram, we have

$$\mathbf{I}_A = \mathbf{I}_{AB} - \mathbf{I}_{CA} = 4.4\angle 53.13° - 4.4\angle -186.87°$$
$$= 2.64 + j3.52 - (-4.37 + j0.53) = 7.62\angle 23.1° \text{ A}$$

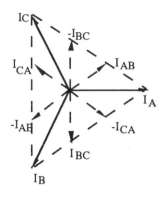

Figure B.11

Therefore, the correct answer is d.

3–PH.5 The power factor is most nearly

a. 1.0   b. 0.6 leading   c. 0.866 leading   d. 0   e. 0.8 lagging

Solution:

The impedance angle θ is:

$$\theta = \tan^{-1}\frac{40}{30} = 53.13°$$

Therefore, the power factor is:

$$pf = \cos\theta = 0.6, \text{ leading}$$

We can also get the answer directly from the expression for the current $\mathbf{I}_{AB}$. The correct answer is b.

3–PH.6 The total real power delivered from the source to the load is most nearly

a. 1496 W   b. 580 W   c. 1742 W   d. 2904 W   e. 850 W

Solution:

The total power P delivered to the balanced load is:

$$P = 3V_{AB}I_{AB}\cos\theta = 3 \times 220 \times 4.4 \times 0.6 = 1742.4 \approx 1742 \text{ W}$$

The answer is c.

Examples 5.15 and 5.16 will be useful in supplementing the ideas reviewed in these sample problems.

5.   DIODE APPLICATIONS

Diodes are used in many practical circuits in electric power systems and in signal processing applications. Chapter 7 is devoted entirely to diodes; Section 7.3, in particular, discusses circuit models that may be useful in the analysis of diode circuits, while Section 7.4 presents several applications. Questions DA.1 and DA.2 refer to the circuit of Figure B.12.

Figure B.12

DRILL EXERCISES

DA.1   The circuit of Figure B.11 is

a.   peak detector
b.   half-wave rectifier
c.   bridge rectifier
d.   voltage doubler
e.   full-wave rectifier

Solution:

The correct answer is e (see Chapter 7).

DA.2   The inductor L and the capacitor C serve the function of

a.   converting the AC input to DC output
b.   increasing the peak value of the output voltage
c.   protecting the diodes
d.   a high-pass filter
e.   reducing the ripple component of the output voltage

Solution:

The correct answer is e.

In addition to the two sample problems given above, you may wish to review the following examples in Chapter 7: 7.1, 7.2, 7.3, 7.7, 7.9, 7.12.

6. OPERATIONAL AMPLIFIERS

The content of Chapter 10, Sections 1 through 5, matches quite closely the requirements of the EIT exam in this area; Section 10.6 does not apply because EIT exam questions are targeted at ideal op-amps. For a review of this material we recommend revisiting Examples 10.1 and 10.2 and Homework Problems 10.2 to 10.13.

7.   TRANSIENTS

The analysis of transient circuits is discussed in the second half of Chapter 6 (Sections 6.3, 6.4, 6.5). One of the fundamental topics in transient analysis is the solution of first-order circuits. The following problem is a representative sample.

DRILL EXERCISE

TR.1   The expression for the current in the 2-ohm resistor in Figure B.13 for time greater than zero is:

a. $-3e^{-0.5t} + 3$ A   b. $3e^{-0.5t} + 3$ A   c. $-3e^{0.5t} + 3$ A   d. $-6e^{-0.5t} + 6$ A   e. $6e^{-0.5t} + 6$ A

Applying KVL to the circuit when the switch is closed ($t \geq 0$), we have

$$4\frac{di}{dt} + 2i = 12$$

Solving the differential equation:

$$i(t) = k_1 e^{-0.5t} + 6 \text{ A} \quad t \geq 0$$

$$i(0^-) = i(0^+) = 0$$

Therefore,

$$i(t) = -6e^{-0.5t} + 6 \text{ A} \quad t \geq 0$$

Therefore, the correct answer is d.

Figure B.13

Additional solved review problems related to transient analysis may be found in Examples 6.11, 6.12, 6.13, 6.18, 6.19, 6.20, 6.21.

8.   ELECTRIC AND MAGNETIC FIELDS

The aim of this section is to test knowledge of the fundamental relationships relating the electric and magnetic fields. Most of the useful relations may be found in Chapter 14. For a more complete review, you may wish to consult a physics text.

DRILL EXERCISES

EM.1   Which of the following is a true characteristic of magnetic flux lines?

a.   They cross each other.
b.   They begin and end on electric charges.
c.   They are parabolic.

d.   They are continuous.

e.   None of the above.

Solution:

As discussed in Chapter 14, magnetic flux lines are continuous. Thus, d is the correct answer.

EM.2   For the circuit of Figure B.14 where $i = 2$ A, $\phi = 1 \times 10^{-3}$ Wb, cross-sectional area $= 5$ in^2, and the mean flux path length $= 2$ in, the total reluctance $\mathcal{R}$ of the magnetic circuit is

Figure B.14

a. $1 \times 10^5 \frac{A\text{-}t\text{-}in^2}{Wb}$

b. $2 \times 10^5 \frac{A\text{-}t\text{-}in^2}{Wb}$

c. $1.5 \times 10^5 \frac{A\text{-}t\text{-}in^2}{Wb}$

d. $3.5 \times 10^4 \frac{A\text{-}t}{Wb\text{-}in}$

e. $2 \times 10^5 \frac{A\text{-}t}{Wb\text{-}in^2}$

Solution:

From Chapter 14, the relationship between magnetomotive force and flux is

$$\mathcal{F} = \phi\mathcal{R}$$

Also, the magnetomotive force is related to the current i by

$$\mathcal{F} = Ni = 100 \times 2 = 200 \; A\text{-}t$$

which means that the reluctance is

$$\mathcal{R} = \frac{\mathcal{F}}{\phi} = \frac{200}{1 \times 10^{-3}} = 200{,}000 \frac{A\text{-}t\text{-}in^2}{Wb}$$

Therefore, the answer is b.

Additional examples illustrating the concept of magnetic circuits may be found in Chapter 14, Examples 14.2, 14.3, 14.4, 14.6, 14.7, 14.8, 14.11, 14.12, 14.15.

9.   ELECTRICAL MACHINERY

The nonelectrical engineer will be frequently exposed to electrical machinery. This subject is covered in Chapters 15 and 16. The most popular electric machines in engineering applications are the DC motor and the AC induction motor. The former is discussed in the first half of Chapter 15, while the latter is discussed in Chapter 15 (three-phase motors) and in Chapter 16 (single-phase motors).

DRILL EXERCISES

MA.1 A four-pole synchronous motor operating from a 60-Hz supply will have a synchronous speed, in rev/min, of

a. 900   b. 1800   c. 1200   d. 3600   e. 4800

Solution:

Frequency = (no. of poles/2) × (rev/min/60)

Therefore,

60 = (4/2) × (rev/min/60)

or rev/min = 1800. Thus, b is the answer.

MA.2 The armature resistance of a 55 hp, 525 V, DC shunt wound motor is 0.4 Ω. The full-load armature current of this motor is 80 A. If the initial starting current is 175 percent of the full-load value, the resistance of the starting coil in Ω should be nearest to

a. 6.6125   b. 3.75   c. 3.35   d. 4.15   e. 2.75

Solution:

I_{start} (starting current) = 1.75 × 80 = 140 A

Therefore, the total resistance is

$$R = \frac{V}{I_s} = \frac{525}{140} = 3.75 \ \Omega$$

Thus, the resistance of the starter is 3.75 − 0.4 = 3.35 Ω. The correct answer is c.

MA.3 The speed of an AC electric motor

a.   is independent of the frequency.
b.   is directly proportional to the square of the frequency.
c.   varies directly with the number of poles.
d.   varies inversely with the number of poles.
e.   none of the above.

Solution:

The speed varies inversely with the number of poles and directly with the frequency. Thus, a is the correct answer.

In addition to the sample problems given above, the following examples can assist you in your review of electric machines: 15.1, 15.4, 15.5, 15.6, 15.7, 15.9, 15.11, 15.13, 15.14.

Answers to Selected Problems

Chapter 2

2.1   a. $E = 5.184 \times 10^6$ J
      b. $E_{\text{stored}} = 4.32$ MJ

2.5   a. $Q = 864$ C
      b. $E = 1166.58$ J

2.10 $v = -3$ V

2.15   Elements A and B dissipate power and element C supplies power.

2.19 $R_{\text{total}} = 1.2\ \Omega$

2.22 $i = 0.5$ mA, $R = 18$ kΩ, $v = 16$ V, $v_1 = 2$ V

2.26 $V_1 = 30$ V

2.31 $R_{\text{eq}} = 18\ \Omega$

2.36   With terminals c-d open, $R_{\text{eq}} = 400\ \Omega$.
      With terminals c-d shorted, $R_{\text{eq}} = 390\ \Omega$.
      With terminals a-b open, $R_{\text{eq}} = 360\ \Omega$.
      With terminals a-b shorted, $R_{\text{eq}} = 351\ \Omega$.

2.40 $V_{\text{out}} = -4.46$ V

2.44   a. $R_{\text{int}} = 0.061\ \Omega$
      b. $R_{\text{int}} = 84.1\ \Omega$

2.47 $r_m = 899\ \Omega$

2.51 $R_S = 3.98$ kΩ

Chapter 3

3.1 $V_1 = 4.8$ V, $V_2 = 2.4$ V

3.4 $i_1 = 0.5$ A, $i_2 = 0.397$ A, $i_3 = 0.555$ A

3.8 $v_a = 5.36$ V, $v_b = 7.89$ V,
$v_a - v_b = -2.54$ V

3.12 $i_1 = 0.1923$ A, $i_2 = 0.865$ A, $v = 6.727$ V

3.16 $v = 39.29$ V

3.20 $i_1 = -\dfrac{1}{3}$ A, $i_2 = \dfrac{1}{3}$ A, $i_3 = \dfrac{1}{6}$ A

3.25 $V_{OC} = 10$ V, $R_T = 0$ Ω

3.29 $v_{OC} = 2$ V

3.34 $i_N = 1$ A

3.38 $R_T = 2.993$ Ω, $v_{OC} = 12.224$ V

3.42 $R_T = 1001.2$ Ω, $v_{OC} = 10^{-3}V_s$ V

3.48 $V2 = 2.33$ V

3.53 a. $T \approx 0°$
b. $v_{out} \approx 3.5$ V

Chapter 4

4.1   1.155 V, $4.4\dfrac{1}{\sqrt{3}}$ V

4.8   2.87

4.12 0.6 J

4.16 $\dfrac{(10)^{-15}}{36x}$ F

4.24 0.957×10^{-6} J

4.28 $\dfrac{V_1}{(\omega RC)^2 + 1}\sqrt{(\omega RC)^2 + 1} \times \cos(\omega t - \tan^{-1}(\omega RC)$

4.32 $\dfrac{I_m R}{\cos\phi - \omega RC\sin\phi}\cos(\omega t + \tan^{-1}(-\omega RC))$

4.36 $\dfrac{dv_C}{dt} + 66.66v_C = 33.3 \times 60\cos\omega t$

4.40 a. $456.1e^{j74.7°}$
b. $131.8e^{-j99.6°}$

4.44 a. $1 + j5$
b. 510.1 μF
c. 26 Ω

4.48 $j1.67 \times 10^{-5}\ \mathbf{V}_S(\omega)$

4.52 $50.0\sin 100t$ V

4.56 $24.8\angle -23.8°$ V

4.60 $0.2357\cos(2t + 45°)$ A

4.64 2 Ω

4.68 $14.83\angle -8.53°$ V

Chapter 5

5.1   1.6 W

5.4   0.25 W

5.8   0.5 lagging

5.12 a. 0.848 leading
b. 0.9925 lagging
c. 0.08716 leading
d. 0.9487 lagging

5.16 a. $228\angle -40°$ VA
b. 174.66 W

5.20 $L_2 = \dfrac{R_1^2 + w^2 L_1^2}{w^2 L_1}$, $R_2 = \dfrac{R_1^2 + w^2 L_1^2}{R_1}$

5.24 11 μF

5.28 982.4 μF

5.32 a. $6.67\angle 32°$ V
b. 12

5.36 15

5.40 $245.1\angle -179.7°$ V

5.44 $11.448\angle 135.18°$ A

5.48 $6\angle -2.7°$ A, $6\angle -242.7°$, $6\angle -122.7°$

5.52 a. $15.36\angle -50.19°$ A
b. 2832.23 W
c. 0.64
d. 25%

5.57 a. 691.2 W
b. $1.2\cos(377t + 23.13°) - \{1.2\cos(377t + 23.13°) - 2.4\cos(377t + 23.13°) - 2.4\cos(377t_1 - 36.87°)\}e^{-t/2\text{ms}}$ A,
$\{2.4\cos(377t_1 + 63.13°)\}e^{-t/2\text{ms}}$ A,
$-1.2\cos(377t + 23.13°) + \{1.2\cos(377t + 23.13°) + 2.4\cos(377t_1 - 156.87°)\}e^{-t/2\text{ms}}$ A

Chapter 6

6.1 $\left|\dfrac{\mathbf{V}_{out}}{\mathbf{V}_{in}}\right| = \dfrac{1}{\sqrt{1 + 0.01\omega^2}}\phi(\omega)$

$= -\tan^{-1}(0.1\omega)$

6.5 $\left|\dfrac{\mathbf{V}_{out}}{\mathbf{V}_{in}}\right| = \dfrac{0.1\omega}{\sqrt{1 + 0.01\omega^2}}\phi(\omega)$

$= 90° - \tan^{-1}(0.1\omega)$

6.10 $\left|\dfrac{\mathbf{I}_L}{\mathbf{V}_S}\right| = \sqrt{\dfrac{(20)^2 + 0.04\omega^2}{(400)2 + 64\omega^2}}$

$\phi(\omega) = \tan^{-1}\left(\dfrac{0.2\omega}{20}\right) - \tan^{-1}\left(\dfrac{8\omega}{400}\right)$

6.15 a. $\left|\dfrac{\mathbf{V}_o}{\mathbf{V}_i}\right| = \dfrac{0.01\omega}{\sqrt{(1 - 20(10^{-6})\omega^2)^2 + (.01\omega)^2}}$

b. $\omega_{center} = 223.6$ rad/s

d. 3-dB BW= $\omega_2 - \omega_1 = 500$ rad/s

6.20 a. $\dfrac{\mathbf{V}_L}{\mathbf{V}_{LO}} = \dfrac{1}{1 + j\omega(5 \times 10^{-6})}$

b. $\dfrac{\mathbf{V}_L}{\mathbf{V}_{LO}} = 0.02$

c. $\dfrac{\mathbf{V}_L}{\mathbf{V}_{LO}} = 0.8944$

6.24 a. $\dfrac{\mathbf{V}_0(j\omega)}{\mathbf{V}_S(j\omega)} = \dfrac{\frac{j\omega}{2000\pi} + 1}{\frac{0j\omega^3}{(2000\pi)^2} - \frac{2\omega^2}{(2000\pi)^2} + \frac{3j\omega}{2000\pi} + 2}$

6.28 a. $i_L(t) = -4/7e^{-7/3t} + 4/7$

b. $\tau = 3/7$ s

d. $i_L(0) = 0$, $i_L(\tau) = 0.36$ A, $i_L(2\tau) = 0.49$ A, $i_L(5\tau) = 0.57$ A, $i_L(10\tau) = 0.57$ A

6.33 a. $v_c(t) = -8e^{-1/24t} + 8$

b. $\tau = 24$ s

d. $v_c(0) = 0$, $v_c(\tau) = 5.06$ V, $v_c(2\tau) = 6.9$ V, $v_c(5\tau) = 7.95$ V, $v_c(10\tau) = 8$ V

6.38 $\tau = 0.923$ ms $\tau' = 1.333$ ms

6.42 $v_{C1}(t) = 6.67 - 6.67e^{-t/66.7\mu s}$ V $t > 0$

6.47 $V_C(t) = 9.8e^{-100(t-.05)}$ V $t > 0$

6.50 $\dfrac{d^2v_c}{dt^2} + 100\dfrac{dv_c}{dt} + 100{,}000\,v_c = 0$

6.54 $i(t) = 6e^{-t} - e^{-6t}$ $t > 0$

6.57 $v(t) = k_1e^{-t} + k_2e^{(-3)t}$ $t > 0$

Chapter 7

7.1 cut off

7.4 cut off

7.8 $V_{in} > 0$

7.12 a. 0 V

b. 4.76 V

c. 4.545 V

7.17 a. 0.6%

b. $-0.002T° +0.8$

7.21 22 ms

7.27 a. 0.6 V

b. 5 V

c. 0.6 V

d. 0.6 V

7.31 15 mA, 1.143 V

7.38 $-0.75e^{-(t-t1)/\tau}$

7.42 50

7.45 0.5017 V

Chapter 8

8.1 a. active

b. cut off

c. saturation

d. cut off

8.4 a. active

b. cut off

c. saturation

d. active

8.8 a. 6.65 V

b. 5.9 mA

c. 39 mW

8.12 a. 9.8 mA

b. 102 μA

8.16 a. 143

b. 81.5 mA, 6.14 V

8.20 22.45 kΩ

8.24 -2.22

8.28 0.89

8.32 5 kΩ

8.36   −4.15

8.40   3.69 ms

Chapter 9

9.1   a. cut off
          b. saturation
          c. ohmic
          d. saturation
          e. cut off
          f. saturation

9.4   a. triode
          b. triode
          c. saturation
          d. cut off

9.8 $62.5 \times 10^{-6} \dfrac{A}{V^2}$

9.12 1.64 mA

9.16 −4.5

9.20 6 kΩ

9.24 161.3 kΩ

9.28 0.25 mA

9.32 0 V

9.36 $\infty \dfrac{\Delta V_{out}}{\Delta I_C}$

9.41 a. 45.98 W
          b. 50.98 W

Chapter 10

10.1 a. $v_1 = \dfrac{v_g}{4}$

          b. $v_1 = \dfrac{v_g}{2}$

10.5 $i = -3$ A

10.8 $v = 10$ V

10.12 $i = 2\sin 3t$ mA

10.16 $R_S = 1$ kΩ, $R_{Smax} = 1.02$ kΩ, $R_{Smin} = 980$ kΩ

10.20 $V_0 = \dfrac{-R_f}{R_{S1}} V_1 - \dfrac{R_f}{R_{S2}} V_2 = \dfrac{R_f}{R_{S3}} V_3$

10.24 $\dfrac{v_o}{v_{in}} = 0.990$

10.28 a. $\tau = 104$ ms
          b. $\tau = 120$ ms

10.37 $\dfrac{v_{out}}{v_{in}} = -\dfrac{6}{\omega^2 - j5\omega - 6}$

10.42 $CMRR_{min} = 74$ dB

10.51 a. $ACL = 1.9999$
          b. 5.9999

Chapter 11

11.3 $A = 21$

11.6 $R_1 = 133.8$ Ω

11.9 $R = 1$ kΩ and $R_2 = 89$ kΩ

11.12 $Q = \dfrac{1}{2}\sqrt{\dfrac{C_1}{C_2}}$

11.16 $Q = 250\sqrt{2}$, $K \approx 3$

11.36 $C = 0.91$ μF

11.41 $T_+ = 0.9$ ms; $T_- = 0.6$ ms; $R_1 = 43$ kΩ; $R_2 = 87$ kΩ, $C = 10$ μF

11.43 44 kΩ $\le R_B \le 92$ kΩ

11.47 $R_{C2} = 600$ Ω

11.49 2-input NAND gate

11.52 Q_4 together with the resistor are designed to serve as a load resistor for the output transistor Q_3. This yields a faster switching time. D_1 is used to ensure that Q_4 is off when Q_2 and Q_3 are off.

11.54 The two transistors in top are off and the two in bottom are on

11.59 The output of the circuit of Figure 11.41 need to be used as an input for the circuit of Figure P11.7.

Chapter 12

12.1 a. 191, 110010001
          b. 111, 100010001
          c. F, 1111

d. 26, 100110
e. 38, 111000

12.5   a. 11111010
          b. 100010100
          c. 110000100

12.8   a. 01111110
          b. 11111110
          c. 01101100
          d. 11100010

12.14 $F(X,Y,Z) = \bar{X} + Y + \bar{Z}$

12.19 $F = AB + \bar{B}\bar{C}$

12.24 $f(A,B,C) = \bar{A}\bar{B}C + \bar{A}B\bar{C} + A\bar{B}\bar{C} + ABC$

12.30 $f = \bar{A} + \bar{B}$

12.34 $F = \bar{A}C + \bar{A}B + \bar{B}\bar{D}$

12.38 $F = \bar{B} + \bar{A}C + A\bar{C}$

12.42 $F = \bar{A}\bar{C}D + \bar{A}B\bar{C} + ABC$

12.47 $W = 2^8$ and $N = 4$

12.53 This is discussed in the digital counters
          section.

Chapter 13

13.9   1. Digital signal is more accurate.
          2. Digital signal does not drift with time
              and temperature.

13.12 a. $v_a = -3.6$ V
          b. $(v_a)_{max} = -4.5$ V
          c. $\delta v = 0.3$ V
          d. $n = 4$

13.16 $R_f = 133.3\ \Omega$

13.19 a. $v_a = -0.8425$ V
          b. $(v_a)_{max} = -10$ V
          c. $\delta v = 2.44$ mV
          d. $n = 12$

13.23 $n = 9$

13.27 $\delta v = 10$ mV, therefore, $n = 11$

13.32 A 9-bit ADC is required.

13.36 This is discussed in Section 13.5.

13.39 a. 44 69 67 74 61 6C
          b. 43 6F 6D 70 75 74 65 72
          c. 41 73 63 69 69
          d. 41 53 43 49 49

13.44 This is discussed in Section 13.5.

Chapter 14

14.1   a. 0.1458 J, 0.1667 J, 0.667 H
          b. $0.625 + 2.668 \sin(400t + 89.8°)$

14.4   a. 0.6917 H
          b. $\sin t + \cos t\{1 - \dfrac{\sin t}{3} + \dfrac{\sin^2 t}{40}\}$

14.9   5.97 A

14.13 121 H

14.17 1.784 A

14.21 a. 5
          b. $353.6\angle{-0.84°}$ V, $0.066\angle{-50.9°}$ A,
              14.98 W, 80.1%

14.25 a. 7.91
          b. 8.94 V
          c. 70.7 V

14.30 $$\frac{(M_1 - M_2)S}{(L_p S + R_p)\left(2S(L - M) + 1 + 2\frac{R_S}{R_P}\right)}$$

14.34 a. 30 V; 10 A
          b. 1.25 A
          c. 192 Ω
          d. 192 Ω

14.38 $-0.02 \cos 10t$

14.41 5.66 V

Chapter 15

15.1   a. 1552 rev/min
          b. 107.5 N-m

15.4   4.9%

15.8   a. 31.472 W
          b. 70.4 A
          c. 1464 W

15.12   a. 254 A
             b. 415.24 V

15.17   85.8%

15.20   230.1 N-m

15.24 $255.2\angle13.03°$, $13.03°$

15.28   a. $24.3\angle36.87°$
             b. $96.4°$
             c. $-27.9°$

15.32 $134.4\angle66.82°$. 875.1 N-m

15.36   a. 855 rev/min
             b. -45 rev/min
             c. 3 Hz
             d. 0 rev/min

15.41 $11.02\angle-26.2°$, 3426 W, 3121 W, 17.24 N-m

15.46   474.4 A

15.50 $9.39\angle-27.15°$ A, 0.8898

Chapter 16

16.3 $v = Ri + L\dfrac{di}{dt} + K_E\omega$

             $T = (J_m + J_L)\dfrac{d\omega}{dt} + D\omega + T_F + T_L$

16.8   24, 2

16.11   a. integral
             b. fractional
             c. fractional
             d. fractional

16.14   1.97, 2910 rev/min

16.18 $29.9\angle-20.52°$ A, 14.08 N-m

16.22   a. $156.05\angle-19.95°$ V
             b. 1014.3 W
             c. 949.3 W
             d. 64.7%

Index